ENIGMATIC MICROORGANISMS AND LIFE IN EXTREME ENVIRONMENTS

Cellular Origin and Life in Extreme Habitats

Volume 1

Enigmatic Microorganisms and Life in Extreme Environments

Edited by

Joseph Seckbach

KLUWER ACADEMIC PUBLISHERS

DORDRECHT / BOSTON / LONDON

A C.I.P. Catalogue record for this book is available from the Library of Congress.

```
Enigmatic microorganisms and life in extreme environments / edited by
  Joseph Seckbach.
      p.   cm. -- (Cellular origin and life in extreme habitats ; v.
  1)
    Includes index.
    ISBN 0-7923-5492-3 (hbk. : alk. paper)
    1. Extreme environments--Microbiology.   I. Seckbach, J. (Joseph)
  II. Series.
  QR100.9.E54  1998
  579'.17--dc21                                              98-48551
```

ISBN 0-7923-5492-3

Published by Kluwer Academic Publishers,
P.O. Box 17, 3300 AA Dordrecht, The Netherlands.

Sold and distributed in North, Central and South America
by Kluwer Academic Publishers,
101 Philip Drive, Norwell, MA 02061, U.S.A.

In all other countries, sold and distributed
by Kluwer Academic Publishers,
P.O. Box 322, 3300 AH Dordrecht, The Netherlands.

Printed on acid-free paper

Printed in the Netherlands.

This book is dedicated to:
Fern Seckbach, a critique, proofreader, colleague, friend, catalyst and ….wife.

TABLE OF CONTENTS

Halophiles

Living in Enormous pH Ranges

i. Acidophiles

Biodata of **Russell L. Chapman** contributor of the '***Foreword***' for this volume.

Dr. Russ L. Chapman is Professor of Biological Sciences and the Executive Director of the Center for Coastal, Energy, and Environmental Resources (CCEER), and also an Adjunct Professor in the Department of Oceanography and Coastal Sciences at the Louisiana State University (Baton Rouge, Louisiana). He earned his Ph.D. in 1973 at the University of California, Davis (major: Botany [Phycology]). Dr. Chapman's main research interests are: algal ultrastructure and phylogeny, biology of the Trentepohliaceae (Chlorophyta), ribosomal gene sequencing and molecular evolution in algae and bryophytes. He has been affiliated with several national and international professional committees and societies and is a member of various Honor Societies. As a requested speaker and organizer, he has delivered many invited lectures and symposia presentations and has been a member of invited workshops, and consulting and review teams (nationally and internationally).
E-Mail: **btruss@unix1.sncc.lsu.edu**

Foreword

Modern methods and approaches, such as the analysis of molecular sequences to infer evolutionary relationships among organisms, have provided vast new sets of data to further our understanding of living organisms, but there remain enigmas in the biological world that will keep scientists working and thinking for decades. Microorganisms by virtue of their small size and almost unbounded diversity provide ample examples of intriguing mysteries that are being challenged with all of the techniques the modern scientific arsenal can provide. One whole arena of this battle to resolve puzzling mysteries about various microorganisms is the almost unbelievable ability of many micro-organisms to live in extreme environments. Whether the challenge is extreme heat, cold, pressure, hyper-salinity, alkalinity or acidity, some micro-organisms live now where no life might seem possible. This fascinating state of affairs is the context for this present volume edited by Joseph Seckbach.

This Volume is a compilation of many of the especially interesting questions and biological challenges that arise in the consideration of microorganisms in general and the extremophiles in particular. Although the editor has his own special interest in *Cyanidium caldarium* and will never miss the chance to share information about this particular intriguing alga (and its cohorts), he has brought together the work of many specialists and has covered over a dozen special topics ranging from the origins and evolution of the first cells to the varied array of extreme environments in which some microorganisms survive and thrive. Even the special challenges of symbiotic environments and the sometimes controversial topic of exobiology receive attention within the chapters of this book. Whether regarded as a complete review of major topics in our understanding of microorganisms in extreme environments or as a source of detailed coverage of specific topics of particular interest, this book serves as a wonderful source of current information on a wide array of topics on a broad spectrum of organisms. To those who are completely new to the field and to those who seek more information about microorganisms that have already caught their attention, this volume provides a wealth of information as well as a fair measure of mystery and enigma.

Louisiana State University
Baton Rouge, LA
14 May 1998

Russell L. Chapman

Biodata of **Joseph Seckbach,** editor of this volume, and author of "***The Cyanidiaceae: Hot Spring Acidophilic Algae***" and co-author (with M.M Walsh) of " ***The Versatility of Microorganisms.***"

Dr. Joseph Seckbach, served as the first selected occupant (1997/1998) of the John P. Laborde Endowed Chair for Louisiana Sea Grant Research and Technology Transfer, and as visiting Professor in the Department of Life Sciences at Louisiana State University (LSU), Baton Rouge, LA. He earned his Ph.D. from the University of Chicago (Chicago, IL) in 1965, and spent his postdoctoral years in the Division of Biology at Caltech (Pasadena, CA). Then he headed at the University of Californian at Los Angeles (UCLA) a team for searching of extraterrestrial Life. Dr. Seckbach has been appointed to the faculty of the Hebrew University (in Organic Chemistry and later in the School for Over Seas Students) performing algal research and teaching (resp.). He spent his sabbatical periods in Tübingen (Germany), UCLA (Los Angeles), and Harvard University (Cambridge, MA). Among his publications are three books and numerous scientific articles concerning plant ferritin, Cellular evolution, and acido-thermophilic algae. His recent interest is in the field of enigmatic microorganisms and Life in extreme environments.
E-Mail: **seckbach@cc.huji.ac.il**

"O Lord, how manifold are your works! in wisdom have you made them all: the Earth is full of your riches." ***Psalms 104, 24***

PREFACE

The purpose of this treatise is to introduce the researcher, teacher or student as well as the open-minded reader to some of the variability of living microorganisms and life forms, as we do not always know them. In this volume a few dozen authors review their investigations into enigmatic organisms and life in mysterious habitats.

Life is ubiquitous on Earth and some Extremophiles or their products can find their way into biotechnological research and eventually into industry (e.g., extremozymes). Furthermore, the diversity of life under those extreme environments may serve as a model for the possibilities of life on other celestial habitats (e.g., on Mars, or Europa, the moon of Jupiter).

This volume, one link in the new book series being published by Kluwer Academic Publishers (The Netherlands) is entitled *Cellular Origin and Life in Extreme Habitats.* This current book follows and supplements the first treatise on the *Evolutionary Pathways and Enigmatic Algae* (Ed. By J. Seckbach, published by Kluwer, 1994).

The term "enigmatic microorganisms" is a relative concept and refers to life forms that are seen, from the anthropocentric point of view, to be puzzling exceptional creatures. The six sections of this book, containing contributions by over fifty researchers, cover various aspects of these unique mysterious organisms and their environments. The question as whether the *Endosymbiosis* is the only path for *Eukaryogenesis* (the evolvement of the bacterium into the first eukaryotic cell) is being addressed by some of our authors in the book's opening section on *Origin and Evolution of the First Cells*. The chapters on fossilized bacteria and *Kakabekia* provide additional data on earlier/past microorganisms. Other chapters, in sections *Enigmatic Microorganisms* and *Versatile Extension of Life* describe new microbial communities, such as the relatively recently discovered Prochlorophytes, the picoplankton, Cryptomonads, and nanobacteria (also with a relation to man). A similar profile is taken for the extremophilic organisms which colonize bizarre areas of natural uncommon or extreme habitats. A few of our authors present overviews on the general concept of these phenomena and some of them have pointed out that while such environments surely kill the majority of communities, other find it an Elysium.

The Extremophiles can be found at high temperature ranges, for example, the hot springs acidophilic algae (Cyanidiaceae) grow at 45-56^{0}C, while the hyper-thermophilic archaea tolerate a temperature range over boiling point (> 100^{0}C). Other organisms discussed live in cold polar zones, such as the psychrophiles or the Antarctic microbes. We present aspects of microbes living at higher levels of pH, interestingly it has been pointed out here that the insect, rose chafer larvae host microbial community at extreme alkaline pH (>11). While the microorganisms at the hindgut of these insect's larvae live at neutral pH. On the other scale of pH, there are the acidophilic archaea (e.g., *Thermoplasma acidophilum*), or the eukaryotic unicellular alga *Cyanidium caldarium* (Cyanidiaceae) and *Dunaliella* (Chlorophyta) thriving in very acidic habitats (pH ranges 0 - 4.0). Microbial communities have been isolated from hyper saline areas, lakes or salt mines (containing concentrated salt solutions as exit in the Dead Sea in Israel). Such hyperhalophiles may exhibit unusual cell morphologies (and appear as triangular or square archaean cells, see in the section of *Halophiles* in this volume). The investigations with the barophilic cells (tolerating pressure of above 600 atmospheres) are presented in three chapters.

Microorganisms are ubiquitous, they have been observed as airborne cells, (e.g., algae), in guts of higher animals and of insects as well as in very dry habitats. Dry desert environments with their low moisture content and exposure to high levels of UV radiation are considered extreme habitats. Most interesting are the microorganisms associated with metals, growing on petroleum, under various gases such as ammonium or under CO_2.

Several authors covered the field of *Living Together: Symbiosis* where two or more organisms of different species share a single habitat or live as symbionts inside the host. Those partners have developed mutual relationships and have continuous cross "talks" between host-symbionts physiological and behavioral interactions. These relationships are long lasting associations between organisms of different species. For example, termites are known to posses symbiotic protists and bacteria in their intestine. Many cockroach species host ciliates and a few have also flagellates in their hindguts. Symbiotic microbial communities are present in digestive tracts of animals consist of anaerobic bacteria, methanogenic archaea, protists, and anaerobic fungi. Thus the animal (or insects) guts are open ecosystems and these microbes degrade plant polymers that are resistant to host digestive enzymes. Methane emission from animal and insects is due to their methanogenic archaea and bacteria. It has been shown that some of these insects' symbionts can inactivate and detoxificate poisonous diets consume by their hosts. In a bovine rumen there are more than 100 different species of symbiotic bacteria (10^{11} cells per ml rumen fluid) and 50 species of ciliates as well as two-three anaerobic fungi (see the chapter by Hackstein et al. in this volume). Further chapters describe more aspects of the symbiotic phenomena, like the association of cyanobacteria-plants and associations of bacteria-amoeba, algae-protists and invertebrates, bacteria-insects and of fungi-grasses.

This volume's appearance is quite timely as interest in the topic of extremophiles, based on the numbers of people involved in their study, is growing at an ever-increasing rate. Twenty years ago there appeared *Microbial Life in Extreme Environments* (ed. by D.J. Kushner, Academic Press, 1978). More recent works devoted to the topic are *Extremophile Mikroorganismen in Ausgefallenen Lebensräumen* (ed. by K. Hausmann and B.P. Kremer, VCH publisher, 1995) and *Extremophiles: Microbial Life in Extreme Environments* (ed. by K. Horikoshi and W.D. Grant, Willy-Liss Publisher, 1998). Our current book complements them.

Joseph Seckbach

Hebrew University of Jerusalem
Mount Scopus 91905 Israel
<seckbach@cc.huji.ac.il>
August 10, 1998

Acknowledgements

I would like to acknowledge, in compiling this volume, the assistance of my colleagues at Louisiana State University (LSU) at Baton Rouge, LA. (while I served there as the first selected occupant of The John P. Laborde Endowed Chair for Louisiana Sea Grant Research and Technology Transfer, and as visiting Professor in the Department of Life Sciences at LSU). I thank my generous host, **Dr. Jack R. Van Lopik** (Executive Director of Sea Grant at the LSU) for his warm hospitality. The efficient technical contributions of Mrs. **Brenda Henning** (Sea Grant, LS U.) is very much appreciated. Many thanks are due to Dr. **Russell Chapman** (Executive Director of CCEER at the LSU) and to his devoted team in Biological Sciences for providing a suitable atmosphere and for their kind technical services and hospitality during the final organization and compilation of this book. Personal gratitude and thanks are due to **Fern Seckbach**, Dr. **Maud M. Walsh** and to Dr. **Aharon Oren** for their critical reading some of these manuscripts. Last but not least, I acknowledge the warm cooperation of **René Mijs** (Kluwer Publication) as well as many thanks to all the authors for their contributions and patience.

Hebrew University of Jerusalem
Mount Scopus 91905 Israel
<seckbach@cc.huji.ac.il>
August 10, 1998

Joseph Seckbach

Biodata of Donald A. Bryant (coauthor with T.M. Gruber of the chapter on: ***Tracing the Relationships among the Eubacteria***).

Dr. Donald A. Bryant serves as the Ernest C. Pollard Professor of Biotechnology, and Professor of Biochemistry and Molecular Biology at the Pennsylvania State University in University Park, PA. He earned his Ph.D. in Molecular Biology in 1977 from the University of California, Los Angeles. His major research interest include the physiology and molecular genetics of cyanobacteria and green sulfur bacteria. He is the editor of *The Molecular Biology of Cyanobacteria* (published by Kluwer in 1994).
E-Mail: dab14@psu.edu

Biodata of **Tanja Gruber** (author with D.A. Bryant of *Tracing the Relationships among the Eubacteria*).

Dr. Tanja M. Gruber is a 1998 graduate of the Department of Biochemistry and Molecular Biology at The Pennsylvania State University (University Park, PA). She is presently a postdoctoral fellow at the University of California, San Francisco. Her research interests include the characterization of the sigma factors of the cyanobacteria and green bacteria, the use of sigma factors as phylogenetic markers to study eubacteria taxonomy, and structure-function relationships of RNA polymerase.
E-Mail: tgruber@itsa.ucsf.edu

J. Seckbach (ed.), Enigmatic Microorganisms and Life in Extreme Environments, 3–13.

TRACING THE RELATIONSHIPS AMONG THE EUBACTERIA USING σ^{70}-TYPE SIGMA FACTORS

TANJA M. GRUBER AND DONALD A. BRYANT*
Department of Biochemistry and Molecular Biology, and Center for Biomolecular Structure and Function, The Pennsylvania State University, University Park, PA 16802, USA
**Telephone: 814-865-1992, E-mail: dab14@psu.edu*

1. Introduction

Morphological and molecular approaches each have distinct advantages and disadvantages in determining evolutionary relationships. Morphological data can be more readily obtained from ancient fossils and extensive preserved collections (Gauthier *et al.*, 1988). However, sequence information can be more precisely interpreted and is usually more informative of evolutionary relationships than phenotypic information (Woese, 1987). Over the past few decades genotypic information has replaced morphological or other phenotypic patterns as the primary tool to determine evolutionary relationships between microorganisms.

The primary molecule of choice for such molecular systematic studies has been the small rRNA subunit (SS-rRNA). Comparisons of SS-rRNA data have revolutionized knowledge of phylogenetic relationships, especially of microorganisms (see Fox *et al.*, 1980; Pace *et al.*, 1986; Olsen *et al.*, 1994). For example, comparisons of rRNA sequences first established the existence of eubacteria and archaebacteria (now archaea) as two distinct kingdoms among the prokaryotes (Balch *et al.*, 1977). Comparisons of SS-rRNA sequences are useful for molecular systematic studies for several reasons (Woese, 1987; Pace *et al.*, 1986); this molecule is present in all organisms, and SS-rRNA molecules are relatively easily cloned, sequenced and aligned.

There are also a number of potential problems associated with SS-rRNA sequence comparisons. Rothschild et al. (Rothschild *et al.*, 1986) have pointed out several sources of error if SS-rRNA is used as the sole criterion to generate a phylogeny. Some of these include the possibility of gene transfer between species as well as between cellular organelles, intra-organismal heterogeneity of the rRNA genes since they occur in multiple copies in most organisms, and the occurrence of non-random mutations due to the influence of adjacent bases. Misinterpretations of relationships have also been detected due to drastically different G+C contents of species, (Hasegawa and Hashimoto, 1993; Vawter and Brown, 1993). Phylogenetic models established by SS-rRNA sequence comparisons can also be unstable if the organisms studied are very similar (Roux *et al.*, 1997). For these and other reasons, additional molecules are now being used for phylogenetic studies to compare and contrast with results obtained using SS-rRNA. A variety of protein sequences have been used for this purpose, and they include σ factors (Gruber and Bryant, 1997), RecA (Eisen, 1995), (Karlin *et al.*, 1995), citrate synthase (Roux *et al.*, 1997), transketolase and fructosebisphosphate aldolase (van den

Bergh *et al.*, 1996), HSP70 (Gupta and Golding, 1993; Boorstein *et al.*, 1994; Rensing and Maier, 1994; van den Bergh *et al.*, 1996), RNA polymerase (Klenk and Zillig, 1994), elongation factor TU (Delwiche *et al.*, 1995; Ludwig *et al.*, 1994), the β subunit of ATP synthase (Ludwig *et al.*, 1994), photosynthetic reaction centers (Nagashima *et al.*, 1995), and GroEL (Viale *et al.*, 1994). Studies that include sequence data from various enzymes across species have also been carried out to determine the divergence times of the major kingdoms (Doolittle *et al.*, 1996). It seems that many different genes should be investigated from the same organisms to be able to understand their phylogenetic relationships more completely and reliably.

2. Phylogenetic analysis

In our studies group 1 σ^{70}-type sigma factors have been chosen as the marker molecule for eubacterial systematics. The sigma factor protein is a dissociable subunit of the eubacterial RNA polymerase holoenzyme (Burgess *et al.*, 1969); this subunit confers on the RNA polymerase core enzyme promoter specificity for transcription initiation. Sigma factor homologs have not been found in the archaea or in eukaryotes. However, nuclear-encoded sigma factors that are closely related to the corresponding cyanobacterial proteins have been found to function in a number of chloroplasts (Liu and Troxler, 1996), (Tanaka *et al.*, 1996). Recently, a gene that resembles sigma factors has been sequenced in the primitive mitochondrion of the freshwater protozoon *Reclinomonas americana* (Lang *et al.*, 1997).

The group 1 σ^{70}-type sigma factor is the primary, vegetative sigma factor and its members are present in all known eubacteria, in which these proteins are essential for cell viability (for reviews, see (Helmann and Chamberlin, 1988; Lonetto *et al.*, 1992). Groups 2 and 3 include the so-called alternative sigma factors, which are dispensable for cell growth. Group 2 members are rather similar in sequence to primary sigma factors and include proteins such as the stationary phase-specific sigma factor RpoS (Mulvey and Loewen, 1989). Group 3 sigma factors vary more significantly in sequence from the other two groups and include groupings with specific functions such as regulators of heat-shock (Grossman *et al.*, 1984) and sporulation (for review, see Haldenwang, 1995).

Group 1 sigma factors also possess a variety of characteristics that make them excellent candidates as phylogenetic markers. They occur across the eubacterial kingdom and as noted above are essential for cell viability. Functional constraints for these proteins give rise to highly conserved structural features. Even though sigma factor proteins vary in length, four regions of very high sequence conservation have been identified (Helmann and Chamberlin, 1988). These regions span approximately 250 contiguous amino acids and include the subregions that are responsible for recognition of the -10 and -35 promoter elements (Siegele *et al.*, 1989; Waldburger *et al.*, 1990). Alignment of the homologous regions is rather straightforward and appears to be free of significant ambiguities. A further advantage of sigma factors is the diversity and relatively large number of sequences already available. Open reading frames (ORFs) encoding proteins that show high sequence similarity to the functionally characterized *Escherichia coli* σ^{70} proteins (Strickland *et al.*, 1988) have been identified from a rapidly expanding list of species in the last few years. Considering that putative sigma factors from diverse species have been shown to carry out equivalent functions as the sigma factors in *E. coli*, the members of this family are now considered to be homologs (Chang and Doi, 1990; Schyns *et al.*, 1994)). In our analyses a total of 60 group 1 σ^{70}-

type sigma factors have been studied to investigate phylogenetic relationships among eubacteria. Fig. 1 shows the phylogenetic tree obtained using σ^{70} group 1 sigma factors.

Three major groupings become apparent in the sigma factor tree: proteobacteria, gram-positive bacteria, and cyanobacteria. Placement of several other phyla which are not represented by a large number of organisms is well supported in the tree. For example, the chlamydia are positioned on the proteobacterial branch, and *Chloroflexus aurantiacus* and *Chlorobium tepidum* (both photosynthetic bacteria) are placed as relatives of the cyanobacteria. Other less represented phyla are positioned more ambiguously, e.g., *Deinococcus radiodurans* (*Thermus/Deinococcus* group), the spirochetes (*Borrelia* and *Treponema* sp.) and the extreme thermophile *Aquifex aeolicus* (Aquificales group). The Aquificales have been placed as the most deeply branching eubacteria in the SS-rRNA tree (Pace, 1997), and it is highly plausible that the hypothetical root of the sigma factor tree could fall near the position occupied by *A. aeolicus*. Although the bootstrap values supporting the positions of *A. aeolicus* and the spirochetes are quite low, analyses using other regions of the sigma factor protein sequences show that these sequences clearly belong with the proteobacterial lineage to the exclusion of all other groups (T. Gruber, D. Bryant, unpublished results).

2.1 PROTEOBACTERIA

The proteobacteria (formerly called purple bacteria) include most of the traditional gram-negative bacteria (Stackebrandt *et al.*, 1988). The proteobacterial phylum has been divided into five phylogenetically distinct subdivisions (α, β, γ, δ, ε), based on SS-rRNA sequences (Olsen *et al.*, 1994). Using sigma factors as the marker, the arrangement of four of the subgroups (α, β, γ, δ) is essentially identical to that for trees derived from SS-rRNA sequences of the same species composition, but the groups are even more strongly supported statistically (Gruber and Bryant, 1997). The β and γ subgroups are relatively closely related, and the δ group is positioned somewhat apart from the other three subgroups. This pattern has been observed in several other studies (Woese, 1987; Olsen *et al.*, 1994; Eisen, 1995). Another phylogenetic pattern that is consistent is the separation of the *Pseudomonas* from *E. coli*, *Buchnera aphidicola*, *Haemophilus influenzae*, and *Salmonella typhimurium* within the γ subdivision. In this study *Helicobacter pylori,* the only member of the ε subdivision as defined by the SS-rRNA sequences, is placed near *A. aeolicus*. Its position is very weakly supported as can be seen from the very low bootstrap values. More sequences from members of this subgroup will have to be included to draw more statistically sound conclusions.

The chlamydia are grouped closely to the proteobacteria in the sigma factor tree, and this association is very highly supported (bootstrap value of 100%). Based upon both their molecular mass (due to a large insertion between conserved regions 1 and 2) as well as their sequence properties, the sigma factors of chlamydia support their inclusion as a distinctive subgroup of the proteobacteria. Chlamydia are obligate intracellular parasites of eukaryotic cells that cause a number of diseases in mammals and birds (Beatty *et al.*, 1994). Chlamydia possess some unique phenotypic characteristics within the eubacteria, such as a biphasic life cycle and an outer cell membrane that differs from the membranes of other gram-negative bacteria in that it is highly cross-linked by disulfide bridges (Birkelund, 1992). Looking at studies of a number of different molecular markers reveals that the position of the chlamydia within the eubacterial kingdom is not well defined. For example, trees employing elongation

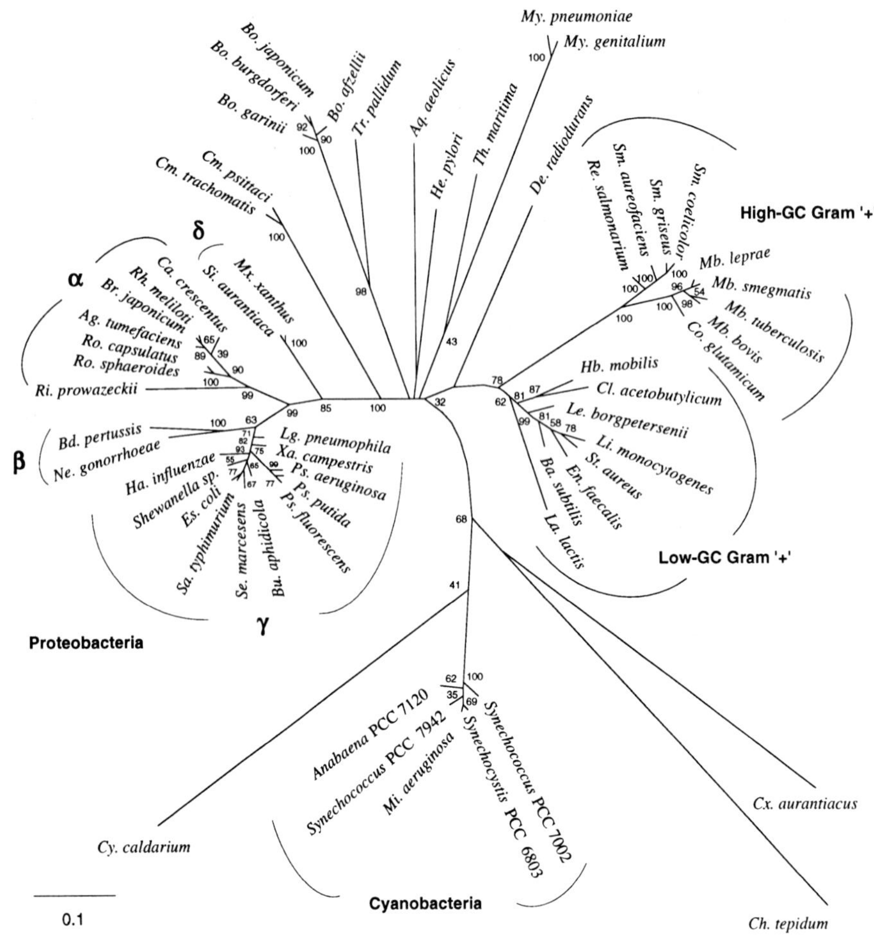

Figure 1. Neighbor-joining tree for Group 1 sigma factors. The tree was generated by the method of Kimura from the multiple sequence alignments encompassing regions 2.1 to 4.2 of group 1 sigma factor proteins (Gruber and Bryant, 1997). The alignment was generated using the *clustalw* multiple sequence alignment program. The distances were calculated using the *protdist* program of the PHYLIP package, using a PAM-matrix based distance correction (Felsenstein, 1989). The scale bar represents 0.1 substitutions per site. Bootstrap values were obtained after 100 replications and are indicated when over 30. Alignments and other information can be requested by the authors. The sequence of SigA of *Aquifex aeolicus* was kindly provided by Dr. Ron Swanson of Diversa Corp. prior to publication.
(*Ag. = Agrobacterium; Aq. = Aquifex; Ba. = Bacillus; Bd. = Bordatella; Bo. = Borrelia; Br. = Bradyrhizobium; Bu. = Buchnera; Ca. = Caulobacter; Ch. = Chlorobium; Cl. = Clostridium; Cm. = Chlamydia; Co. = Corynebacterium; Cx. = Chloroflexus; Cy. = Cyanidium; De. = Deinococcus; En. = Enterococcus; Es. = Escherichia; Ha. = Haemophilus; Hb.= Heliobacillus; He. = Helicobacter; La. = Lactococcus; Le. = Leptospira; Li. = Listeria; Lg. = Legionella; Mb. = Mycobacterium; Mi. = Microcystis; Mx. = Myxococcus; My. = Mycoplasma; Ne. = Neisseria; Ps. = Pseudomonas; Rh. = Rhizobium; Ri. = Rickettsia; Ro. = Rhodobacter; Sa. = Salmonella; Se. = Serratia; Si. = Stigmatella; St. = Staphylococcus; Sm. = Streptomyces; Th. = Thermotoga; Tr. = Treponema; Xa. = Xanthomonas*)

factor Tu as the phylogenetic marker suggest that the chlamydia are most closely related to the spirochetes (Viale *et al.*, 1994; Delwiche *et al.*, 1995). The use of GroEL as the marker positions the chlamydia between the *Bacteroides* sp. and the spirochetes (Viale *et al.*, 1994). Studies with HSP70s place the chlamydia adjacent to the gram-positives and cyanobacteria (Rensing and Maier, 1994; Boorstein *et al.*, 1994), whereas a comparison of RecA sequences places the chlamydia between the *Deinococcus-Thermus* group and the low-GC gram positive cluster (Eisen, 1995). A major problem in positioning the chlamydia within the eubacteria appears to be the small number of organisms isolated and studied from this group. However, it should be noted that of all molecular markers used so far, sigma factor comparisons position the chlamydia with the highest confidence.

2.2 GRAM-POSITIVE BACTERIA

The gram-positive phylum is believed to consist of four subdivisions (Woese, 1987), two of which have been well characterized. These two subgroups, the low-GC and high-GC groups, are well represented in this study, and a member of the third subgroup, namely *Heliobacillus mobilis* of the photosynthetic subdivision, was also included. The fourth subgroup is comprised of the genera *Megasphaera*, *Selenomonas*, and *Sporomusa* (Stackebrandt *et al.*, 1985), but no sigma factor sequences have yet been determined for any of these organisms. There has been much debate over whether the gram-positive bacteria are monophyletic. For example, some studies on rRNA genes (Woese, 1987) and GroEL genes (Viale *et al.*, 1994) suggested that the gram-positive bacteria are of monophyletic origin, whereas other studies on rRNA genes (van der Peer *et al.*, 1994), elongation factor Tu, the β subunit of ATP synthase (Ludwig *et al.*, 1994), and RecA sequences (Eisen, 1995) suggested that these organisms are not monophyletic. With the species represented in studies using sigma factors as the marker, the gram-positive bacteria were found to be of monophyletic origin, and this conclusion is supported by a very high bootstrap value (78%).

The relationships of the organisms within the subgroups are very similar among the different trees (Viale *et al.*, 1994; Eisen, 1995; Gruber and Bryant, 1997). In the high-GC group the *Streptomyces* sp. and the *Mycobacterium* sp. are tightly clustered, and the *Corynebacterium* sp. are positioned between these two clusters. In the low-GC cluster most organisms are also found in equivalent positions in the different trees. The only member of the photosynthetic subdivision, *H. mobilis*, is found within this cluster. Interestingly, in the sigma factor analyses, *Leptospira borgpetersenii* is unambiguously placed within the low-GC group, and this placement is supported by a very high bootstrap value. The *Leptospira* sp. were formerly grouped with the spirochetes in the SS-RNA tree (Olsen *et al.*, 1994). However, in one of the more recent SS-rRNA trees the *Leptospira* sp. form their own group which is placed near the proteobacteria (Pace, 1997).

Mycoplasmas are generally placed within the low-GC gram-positives (Woese, 1987; Weisburg *et al.*, 1989), however, their phylogenetic classification is ambiguous (Razin, 1978). In the tree derived from sigma factor sequences, the two representatives of the mycoplasmas, *Mycoplasma genitalium*, and *Mycoplasma pneumoniae,* are most closely related to thermophile *Thermotoga maritima*. The rapidly evolving nature of the mycoplasmas, as evidenced by the long branch lengths for these organisms, is apparent in Figure 1. The properties of the mycoplasmas are very unusual. They have no cell walls, their genomes are far smaller than those of most other bacteria (Fraser *et al.*,

1995), and they have a number of other cytological and biochemical peculiarities (Razin, 1985). Some microbiologists argue that mycoplasmas constitute a phylogenetically distinctive group that is distantly related to other eubacteria (Razin, 1985), although it is clear that the transcription apparatus of these organisms is of the eubacterial type. It is difficult to draw any definitive conclusions on that issue from the data of this work, since only two mycoplasma-derived sequences are included. However, it is interesting to note that the SigA proteins of both *Mycoplasma* sp. have a unique insertion of five residues which is not found in any other sigma factors (see (Gruber and Bryant, 1997). In a study comparing phylogenetic results for RecA with those for SS-rRNA (Eisen, 1995), it was found that the SS-rRNA data placed the mycoplasmas within the gram-positives, whereas their position in the RecA tree was uncertain. Further sigma factor sequences will have to be included to position the mycoplasmas more accurately and to determine whether phylogenetic analyses of sigma factor protein sequences support the notion that mycoplasmas form a phylogenetically distinctive group within the eubacteria.

2.3 CYANOBACTERIA

Cyanobacteria, including the closely related prochlorophytes (Matthijs *et al.*, 1994), are the only eubacteria that perform oxygenic photosynthesis. They form a coherent clade in the sigma factor tree that contains no non-photosynthetic members. The cyanobacteria are most closely related to several other photosynthetic organisms, namely the green-gliding bacterium *C. aurantiacus* (perhaps better referred to as an anoxygenic flexibacterium or a filamentous, anoxygenic phototrophic bacterium (Pierson and Castenholz, 1995) and the green sulfur bacterium *C. tepidum*. Analyses of a number of other genes have suggested that cyanobacteria and gram-positives might be sister groups, and both protein sequences (Viale *et al.*, 1994; Eisen, 1995) and SS-rRNA sequences (Woese, 1987; van der Peer *et al.*, 1994) have been used to support this interpretation. This pattern, however, is not supported in this tree, in which the cyanobacteria and their relatives form one of the three major groupings of eubacteria. This result appears to be consistent with the fossil record which suggests that cyanobacteria are one of the oldest groups of eubacteria on Earth. The nuclear-encoded, chloroplast-targeted sigma factor of the red alga *Cyanidium caldarium* groups closely with the cyanobacterial proteins, suggesting that the sigma factor genes of *C. caldarium* are of cyanobacterial origin and that these genes were transferred to the nucleus after the primary endosymbiosis occurred (Tanaka *et al.* , 1996; Liu and Troxler, 1996).

2.4 SPIROCHETES

The spirochetes form a coherent phylogenetic group (except *L. borgpetersenii*, see above) whose positioning relative to other groups is not highly supported. According to SS-rRNA data, the spirochetes are most closely related to the *Planctomyces/Chlamydia* group (Olsen *et al.*, 1994). It is interesting to note that in the sigma tree the pathogenic bacteria *Borrelia* sp., the chlamydia, and *H. pylori* are relatively closely related to each other.

2.5 OTHER PHYLOGENETIC GROUPS

A number of phyla are presently represented by only one or two sequences (e.g., the thermotogales, chlamydia, green-gliding bacteria, and green sulfur bacteria) which contributes to the inability to produce a precise placement of some of these organisms within the tree. This is reflected by relatively low bootstrap values for these species. A notable exception is the very strongly supported position of the chlamydia in the sigma factor tree. According to these results, a placement of the chlamydia within or very near the proteobacterial group would seem to be appropriate.

The green sulfur bacterium *C. tepidum* and the green gliding bacterium *C. aurantiacus* are related to each other in this tree and furthermore these organisms are positioned close to the cyanobacteria. Both organisms perform photosynthesis and they resemble one another in having chlorosomes as their light-harvesting antenna system (Gibson *et al.*, 1985); however, their reaction center types are quite different (Pierson and Olson, 1987). According to SS-rRNA comparisons these two groups are not related (Woese, 1987). However, the green sulfur bacteria and the green gliding bacteria are not particularly well-established phylogenetic groups; for example, only a few SS-rRNA sequences have been cataloged (Gibson *et al.*, 1985; Woese, 1987). The branch length for the sigma factor sequence of *C. tepidum* is extremely long, a result that could indicate that this gene is subject to very rapid evolution, that this organism is only distantly related to other eubacteria, or that a sigma factor substitution (or lateral transfer) has occurred. A further indicator of the relatedness of cyanobacteria and *C. aurantiacus* is that they both contain multiple group 2 sigma factors (a trait otherwise found only in certain high-GC gram-positive organisms).

The two thermophilic eubacteria *T. maritima* and *A. aeolicus,* and the radiation-resistant *D. radiodurans* are the lone representatives of their respective phyla and their positions within the tree not highly supported. Based upon SS-rRNA analysis and phenotypic characterizations, it is believed that the Thermotogales and Aquificales are among the most deeply branching of the cultured bacterial lineages (Pace, 1997). Members of the Aquificales are among the most thermophilic members of eubacteria that have been described (Pitulle *et al.*, 1994), for example, *Aquifex pyrophilus* grows at an optimum temperature of 95 °C.

3. Conclusions

Different phylogenetic markers by and large appear to yield very similar phylogenetic groups, but their arrangement relative to one another is not well established, not even in the very extensive sampling of SS-rRNA data (Woese, 1987). A reason for this relative uncertainty for the relationships of the various groups to one another could be because bacterial diversification appears to have occurred mainly from an explosive radiation of the lineages, in contrast to the more sequential divergence of the main groups of the Eukarya (Pace, 1997). Many additional phyla of eubacteria are still being discovered, mainly in geothermal features such as those in Yellowstone National Park. In 1987, 12 phyla were described (Woese, 1987), whereas recently, 25-30 distinct groups were resolved (Pace, 1997).

It becomes apparent that phenotypic characteristics can be quite diverse within a phylogenetic group. For example, the anaerobic *Clostridium* species is found in the same phylum as the aerobic *Bacillus* sp. Photosynthetic members intermingle with non-photosynthetic organisms in a number of phyla (e.g. *Rhodobacter sphaeroides* in the proteobacteria and *H. mobilis* in the gram-positives), lithotrophic eubacteria intermix with organotrophic ones, and thermophilic organisms group together with mesophilic organisms. Genotypic information is a crucial component to determine the evolutionary relationships among such diverse and versatile organisms as the eubacteria. As more sequence data becomes available for diverse bacteria, one can expect that "enigmatic" species will become less so and that a more detailed picture of bacterial evolution will emerge.

4. References

Balch, W.E., Magrum, L.J., Fox, G.E., Wolfe, R.S. and Woese, C.R. (1977) J. Mol. Evol. 9: 305-311.
Beatty, W.L., Morrison, R.P. and Byrne, G.I. (1994) Microbiol. Rev. 58: 686-699.
Birkelund, S. (1992) Danish Medical Bulletin 39: 304-320.
Boorstein, W.R., Ziegelhoffer, T. and Craig, E.A. (1994) J. Mol. Evol. 38: 1-17.
Burgess, R.R., Travers, A.A., Dunn, J.J. and Bautz, E.K.F. (1969) Nature (London) 221: 43-46.
Chang, B.Y. and Doi, R.H. (1990) J. Bacteriol. 172: 3257-3263.
Delwiche, C.F., Kuhsel, M. and Palmer, J.D. (1995) Mol. Phylogenet. Evol. 4: 110-128.
Doolittle, R.F., Feng, D.-F., Tsang, S., Cho, G. and Little, E. (1996) Science 271: 470-477.
Eisen, J.A. (1995) J. Mol. Evol. 41: 1105-1123.
Felsenstein, J. (1989) Cladistics 5: 164-166.
Fox, G.E., Stackebrandt, E., Hespell, R.B., Gibson, J., Maniloff, J., Dyer, T.A., Wolfe, R.S., Balch, W.E., Tanner, R.S., Magrum, L.J., Zablen, L.B., Blakemore, R., Gupta, R., Bonen, L., Lewis, B.J., Stahl, D.A., Leuhrsen, K.H., Chen, K.N. and Woese, C.R. (1980) Science 209: 457-463.
Fraser, C.M., Gocayne, J.D., White, O., Adams, M.D., Clayton, R.A., Fleischmann, R.D., Bult, C.J., Kerlavage, A.R., Sutton, G., Kelley, J.M., Fritchman, J.L., Weidman, J.F., Small, K.V., Sandusky, M., Fuhrmann, J.L., Nguyen, D.T., Utterback, T.R., Saudek, D.M., Phillips, C.A., Merrick, J.M., Tomb, J.-F., Dougherty, B.A., Bott, K.F., Hu, P.-C., Lucier, T.S., Peterson, S.N., Smith, H.O., Hutchison, C.A. and Venter, J.C. (1995) Science 270: 397-403.
Gauthier, J.A., Kluge, A.G. and Rowe, T. (1988) Cladistics 4: 105-205.
Gibson, J., Ludwig, W., Stackebrandt, E. and Woese, C.R. (1985) Syst. Appl. Microbiol. 6: 152-156.
Grossman, A.D., Erickson, J.W. and Gross, C.A. (1984) Cell 38: 383-390.
Gruber, T.M. and Bryant, D.A. (1997) J. Bacteriol. 179: 1734-1747.
Gupta, R.S. and Golding, G.B. (1993) J. Mol. Evol. 37: 573-582.
Haldenwang, W.G. (1995) Microbiol. Rev. 59: 1-30.
Hasegawa, M. and Hashimoto, T. (1993) Nature (London) 361: 23.
Helmann, J.D. and Chamberlin, M.J. (1988) Annu. Rev. Biochem. 57: 839-872.
Karlin, S., Weinstock, G.M. and Brendel, V. (1995) J. Bacteriol. 177: 6881-6893.
Klenk, H.-P. and Zillig, W. (1994) J. Mol. Evol. 38: 420-432.
Lang, B.F., Burger, G., O'Kelly, C.J., Cedergren, R., Golding, G.B., Lemieux, C., Sankoff, D., Turmel, M. and Gray, M.W. (1997) Nature 387: 493-497.
Liu, B. and Troxler, R.F. (1996) Proc. Natl. Acad. Sci. USA 93: 3313-3320.
Lonetto, M., Gribskov, M. and Gross, C.A. (1992) J. Bacteriol. 174: 3843-3849.
Ludwig, W., Neumaier, J., Klugbauer, N., Brockmann, E., Roller, C., Jilg, S., Reetz, K., Schachtner, I., Ludvigsen, A., Bachleitner, M., Fischer, U. and Schleifer, K.H. (1994) Antonie Van Leeuwenhoek 64: 285-305.
Matthijs, H.C.P., van der Staay, G.W.M. and Mur, L.R. (1994) In: D. A. Bryant (ed.) The Molecular Biology of Cyanobacteria, Kluwer Academic Publishers, Dordrecht, pp. 49-64.
Mulvey, M.R. and Loewen, P.C. (1989) Nucleic Acids Res. 17: 9979-9991.
Nagashima, K.V.P., Hanada, S., Hiraishi, A., Shimada, K. and Matsuura, K. (1995) In: P. Mathis (ed.) Photosynthesis: From Light to Biosphere, Kluwer Academic Publishers, Dordrecht, pp. 975-978.
Olsen, G.J., Woese, C.R. and Overbeek, R. (1994) J. Bacteriol. 176: 1-6.
Pace, N.R. (1997) Science 276: 734-740.
Pace, N.R., Olsen, G.J. and Woese, C.R. (1986) Cell 45: 325-326.
Pierson, B.K. and Castenholz, R.W. (1995) In: R. E. Blankenship, M. T. Madigan and C. E. Bauer (eds.) Anoxygenic Photosynthetic Bacteria, Kluwer Academic Publishers, Dordrecht, pp. 32-47.

Pierson, B.K. and Olson, J.M. (1987) In: J. Amesz (ed.) Photosynthesis, Elsevier Science Publishers B. V., pp. 21-42.
Pitulle, C., Yang, Y., Marchiani, M., Moore, E.R.B., Siefert, J.L., Aragno, M., Jurtshuk, P. and Fox, G.E. (1994) Int. J. Syst. Bacteriol. 44: 620-626.
Razin, S. (1978) Microbiol. Rev. 42: 414-470.
Razin, S. (1985) Microbiol. Rev. 49: 419-455.
Rensing, S.A. and Maier, U.G. (1994) J. Mol. Evol. 39: 80-86.
Rothschild, L.J., Ragan, M.A., Coleman, A.W., Heywood, P. and Gerbi, S.A. (1986) Cell 47: 640.
Roux, V., Rydkina, E., Eremeeva, M. and Raoult, D. (1997) Int. J. Syst. Bacteriol. 47: 252-261.
Schyns, G., Sobczyk, A., Tandeau de Marsac, N. and Houmard, J. (1994) Mol. Microbiol. 13: 887-896.
Siegele, D.A., Hu, J.C., Walter, W.A. and Gross, C.A. (1989) J. Mol. Biol. 206: 591-603.
Stackebrandt, E., Murray, R.G.E. and Trueper, H.G. (1988) Int. J. Syst. Bacteriol. 38: 321-325.
Stackebrandt, E., Pohla, H., Kroppenstedt, R., Hippe, H. and Woese, C.R. (1985) Arch. Microbiol. 143: 270-276.
Strickland, M.S., Thompson, N.E. and Burgess, R.R. (1988) Biochemistry 27: 5755-5762.
Tanaka, K., Oikawa, K., Ohta, N., Kuroiwa, H., Kuroiwa, T. and Takahashi, H. (1996) Science 272: 1932-1935.
van den Bergh, E.R.E., Baker, C., Raggers, R.J., Terpstra, P., Woudstra, E.C., Dijkhuizen, L. and Meijer, W.G. (1996) J. Bacteriol. 178: 888-893.
van der Peer, Y., Neefs, J.M., De Rijk, P., De Vos, P. and De Wachter, R. (1994) Syst. Appl. Microbiol. 17: 32-38.
Vawter, L. and Brown, W.M. (1993) Genetics 134: 597-608.
Viale, A.M., Arakaki, A.K., Soncini, F.C. and Ferreyra, R.G. (1994) Int. J. Syst. Bacteriol. 44: 527-533.
Waldburger, C., Gardella, T., Wong, R. and Susskind, M.M. (1990) J. Mol. Biol. 215: 267-276.
Weisburg, W.G., Tully, J.G., Rose, D.L., Petzel, J.P., Oyaizu, H., Yang, D., Mandelco, L., Sechrest, J., Lawrence, T.G., van Etten, J., Maniloff, J. and Woese, C.R. (1989) J. Bacteriol. 171(12): 6455-6467.
Woese, C.R. (1987) Microbiol. Rev. 51: 221-271.

Biodata of **Peter Gogarten** (coauthor with L. Olendzenski of the chapter on: **"Gene transfer in Early Evolution."**)

Dr. Peter Gogarten is a Professor of Molecular and Cell Biology at the University of Connecticut. Currently his main interests are in the role of gene duplications and horizontal gene transfer in evolution. His laboratory studies the horizontal transfer of genes between microorganisms by comparing the molecular evolution of housekeeping genes, in particular proton pumping ATPases and coupling factor ATP synthases. Additionally, research in his lab focuses on the evolution of small gene families in eukaryotes.
E-mail: gogarten@uconnvm.uconn.edu

Biodata of **Lorraine Olendzenski** (contributor, with P. Gogarten of "***Gene Transfer in Early Evolution.***")

Lorraine Olendzenski is doing her graduate research in the laboratory of P. Gogarten at the University of Connecticut. Her research intersts include Evolution and diversity of eukaryotic micro-organisms and the role of horizontal gene transfer in prokaryotic evolution. Currently she is examining the phylogenetic position of deep branching Eukaryotes using proton pumping ATPase sequences as an evolutionary marker and is analyzing Archaeal genes in the Deinococcaceae and Spirochaete lineages.
E-mail: lco95001@uconnvm.uconn.edu.

J. Seckbach (ed.), Enigmatic Microorganisms and Life in Extreme Environments, 15–27.

GENE TRANSFER IN EARLY EVOLUTION

LORRAINE OLENDZENSKI and J. PETER GOGARTEN
University of Connecticut
Dept. of Molecular and Cellular Biology
75 N. Eagleville Rd., U-44
Storrs, CT 06269

1. Introduction

Clues to the early evolutionary history of microorganisms can be found in the macromolecules of extant organisms. Two types of molecular data have proven useful in the study of early evolution: comparative analysis of biochemical pathways (Wächtershäuser, 1992, Morowitz, 1992) and analysis of macromolecular sequence data. Universal phylogenies inferred from nucleic acid or amino acid sequences which compare organisms from distant lineages allow us to infer major events in the early history of life. Burgeoning data available from molecular databases and ongoing genome projects now allow us to examine phylogenies from many different molecular markers. However, phylogenies from different markers do not always yield the same branching pattern. Rather, markers can be grouped into categories that show a concensus relationship, or into those that show conflicting topology.

Conflicting tree topologies complicate the step from molecular phylogeny to species phylogeny and at best, phylogenetic analysis of a single marker can only provide information about the molecule under study. Conflict among molecular phylogenies can arise for three reasons (Gogarten et al., 1996) - insufficient data for resolution, incorrect comparsion of gene sequences thought to be orthologous (i.e., unrecognized paralogy), and horizontal transfer of genes. Conflicting molecular phylogenies between well resolved data sets known to be orthologous provide evidence for the horizontal transfer of genes between distant lineages.

2. Two Prokaryotic Domains?

The best studied molecular marker has undoubtedly been the small subunit ribosomal RNA (Olsen et al., 1994). To date, more than 10,000 different 16S rRNA sequences are available through the Ribosomal Database Project (see http://rdp.life.uiuc.edu, Maidak et al., 1997). Since ribosomes play a central role in cellular information processing and interact specifically with many cellular components, they have coevolved over time with these components in each of the major cell lineages and are unlikely to have been transferred between distantly related organisms. Analysis of small subunit (SSU) rRNA has provided us with a view of life divided into three main groups, or domains (Woese, 1987, Woese et al., 1990), two of which, the Archaea and Bacteria are prokaryotic in organization. Using distance, parsimony and maximum likelihood methods the Archaea (archaebacteria), Bacteria (eubacteria) and Eucarya (eukaryotes) appear as monophyletic groups. However, the relationship of these three groups to each other cannot be determined by RNA analysis alone because of the lack of a suitable outgroup in this data set when considering a universal phylogeny.

Comparisons of ancient gene duplications allow us to root the universal tree of life (Gogarten et al., 1989, Iwabe et al., 1989), and to make inferences about the last common ancestor. By comparing sets of paralogous, or duplicated genes, one set of the paralogues can be used as an outgroup for the other (Schwartz and Dayhoff, 1978). This approach has been used with the catalytic and non-catalytic subunits of H+-pumping ATPases (Gogarten et. al., 1989), elongation factors EF-Tu and EF-G (Iwabe et al., 1989; Cammarano et al., 1992, Baldauf et al., 1996), and aminoacyl-tRNA synthetases (Brown and Doolittle, 1995). All three data sets support the placement of the root between a branch leading to eubacteria, and a branch leading to archaebacteria and eukaryotes (Fig. 1A).

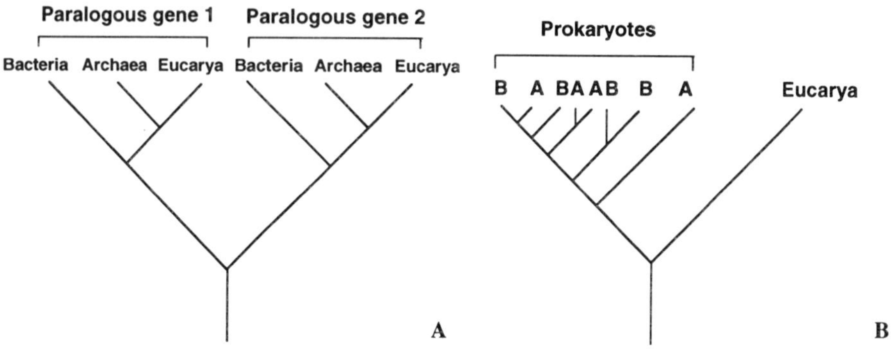

Figure 1 A. Universal tree showing relationship of domains and placement of root based on use of paralogous genes as an outgroup. B. Representative phylogeny showing Bacterial (B) and Archaeal (A) sequences grouping together, separate from Eucaryal homologues.

3. Bacterial Contributions to the Archaea

Since the recognition that the Archaea form a cohesive group (Woese and Fox, 1977), numerous characters have been found which support the idea that the archaebacteria (Archaea) and the eubacteria (Bacteria) are two distinct lineages. These include ribosomal proteins, promotor organization, membrane lipids, cell wall structure, antibiotic sensitivity and flagellins (Zillig et al, 1992). There continues to be debate concerning the possibility that the Archaea might constitute a paraphyletic group, because some or all of them might share a more recent common ancestor with a lineage that gave rise to the eukaryotic nucleocytoplasm (Rivera and Lake, 1992, 1996; Olsen et al., 1996).

At the same time that the unity of the domain Archaea is being accepted, increased availability of molecular sequences is forcing us to take a closer look at the evolution of this group. Many genes in the Archaea have been found to be very similar to their eukaryotic counterparts, especially those involved in genome structure, transcription and translation (Marsh et al., 1994; Langer et al., 1995; Keeling and Doolittle, 1995). These findings support the identification of the Archaea as the group giving rise to the eukaryotic nucleocytoplasmic component, a relationship first indicated through studies of ancient duplicated genes (Gogarten et al., 1989; Iwabe et al., 1989, Fig. 1A).

Evidence is accumulating however, that many other archaeal genes are best described as "prokaryotic" in character. Phylogenetic trees constructed from these markers, many of which are enzymes in core metabolic pathways, do not yield branches which group archaebacteria separately from eubacteria. Rather, they yield trees which show archaebacterial branches interspersed throughout their eubacterial counterparts (Fig. 1B). These include homologues of the heat shock proteins (HSP70) (Gupta and Singh, 1992; Gupta and Golding, 1993), glutamine synthetases (Brown et al., 1994; Kumada et al., 1993; Tiboni et al., 1993), nitrogenases (Souillard et al., 1988; Chien and Zinder, 1994), glutamate dehydrogenases (Benachenhou-Lafha et al., 1993; Hilario and Gogarten, 1993), and an F-ATPase subunit fragment from *Methanosarcina* (Sumi et al., 1992; Hilario and Gogarten, 1993). Distance and parsimony analyses of enolase and phosphoglycerate kinase (glycolytic enzymes), dihydrofolate reductase (pterine biosynthesis pathway) and dihydrolipoamide dehydrogenase also give tree topologies in which the Archaeal sequences group among the Bacteria (Gogarten, unpublished data). In the cases where these phylogenies can be rooted by ancient gene duplications, the root appears to be placed between all of the prokaryotes on one side, and the eukaryotes on the other. In the case of those molecular markers that allow a better resolution among the prokaryotes, the Archaea group as a paraphyletic group within the Bacteria in a clade that is well supported by the respective data (e.g.: Brown et al., 1994; Gupta and Golding, 1993). Molecular phylogenies of these genes suggest a fundamental division between Prokaryotes and Eukaryotes, rather than one between Archaea and Bacteria (Pesole et al., 1995).

Another marker which does not group the archaebacteria as separate from the eubacteria is the internal duplication found in carbamylphosphate synthetase (Schofield, 1993, van den Hoff et al., 1995, Olendzenski et al., in press, Lawson et al. 1996, Lazcano, Puente, and Leguina, unpublished). In this case the first half of the molecule shares homology with the second half and each can be used as an outgroup of the other. One interesting feature of this analysis is that only the molecules from *Methanococcus, Methanosarcina, Methanobacterium* and *Archaeoglobus* group among the eubacteria; the *Sulfolobus* sequence groups with the eukaryotes.

The difference between those molecular markers that group the Archaea within the bacterial domain and those that reflect two distinct prokaryotic domains does not appear to be due to ill resolved branching patterns (Gupta and Golding, 1993; Brown et al., 1994). If the conflicting topology is due to a so far unrecognized ancient gene duplication, the true phylogeny would then be reflected by HSP70 homologues, glutamate dehydrogenases, and glutamine synthetases. Since H+-ATPases, elongation factors, RNA polymerases, aminoacyl-tRNA synthases, and ribosomes have retained comparable functions in all extant organisms it seems unlikely that the last common ancestor would have had duplicate divergent copies for all of these activities. It also seems extremely unlikely that nearly identical sets of these characters were lost several times in the lineages leading to the different bacterial groups, whereas exactly the complementary set would have to have been lost multiple times in the lineages leading to present day archaea and the eukaryotic nucleocytoplasm.

Horizontal gene transfer appears to be the best explanation for the presence of bacterial genes in archaeal genomes (Gogarten et al., 1996). Since a number of genes group the Archaea together with the Bacteria, this suggests that a significant portion of the genome was involved in this transfer. Many of these genes are involved in biosynthetic pathways (e.g. glutamine synthetase, carbamylphosphate synthetase, glutamate dehydogenase). Some other well studied enzymes, for example those involved in nitrogen fixation (Chien and Zinder, 1994; Souillard et al., 1988) also suggest a close association between Gram positives and at least some Archaea. In addition, analysis of the *Methanococcus janaschii* genome shows that 44% of the open reading frames share closest similarity with bacterial rather than eukaryotic genes (Koonin, 1997).

At present, insufficient information is available to pinpoint the phylogenetic positions of the donor and recipient of these horizontal gene transfer events. It is not even possible to decide whether these eubacterial genes were contributed to the archaeal lineage in one massive horizontal transfer (resulting from a fusion of the archaeal ancestor with a bacterium), or if these genes were transferred in a trickle-down fashion (as important innovations evolved in the bacterial domain these were passed horizontally to other organisms that could make use of them). An argument against the formation of a chimera at the root of the archaeal domain is that if there had been a single transfer event of a large number of genes, one would expect the Archaea to form a single group within the Bacteria in phylogenies based on the transferred genes. This is not observed for glutamine synthetase; the sequences from *Sulfolobus* consistently group at the base of the eubacteria, while those from other Archaea group within the eubacteria (Brown et al., 1994). Furthermore, the carbamylphosphate synthetase phylogeny indicates that at least this gene was transferred into the archaeal gene after the split between cren- and euryarcheota had already occurred and that *Sulfolobus* never received the eubacterial gene.

4. Transfer from Archaea to Bacteria: Proton Pumping ATPases

Proton pumping ATPases/ATP synthases are found in all groups of extant organisms. Traditionally these ATPases are divided into three categories (V-, F- and P-type ATPases). F-ATPases occur as the ATP synthase in mitochondria and chloroplasts of eukaryotes and as an ATP synthase or ATP driven proton pump in bacteria. V-ATPases (vacuolar ATPases) are found as proton pumps in the endomembrane system of eukaryotes. Archaea possess H+-ATPases called A-ATPases, which are similar in molecular sequence to V-ATPases but whose function and quarternary structure are more similar to the F-ATPases in that they can synthesize ATP (Gogarten et al., 1992,

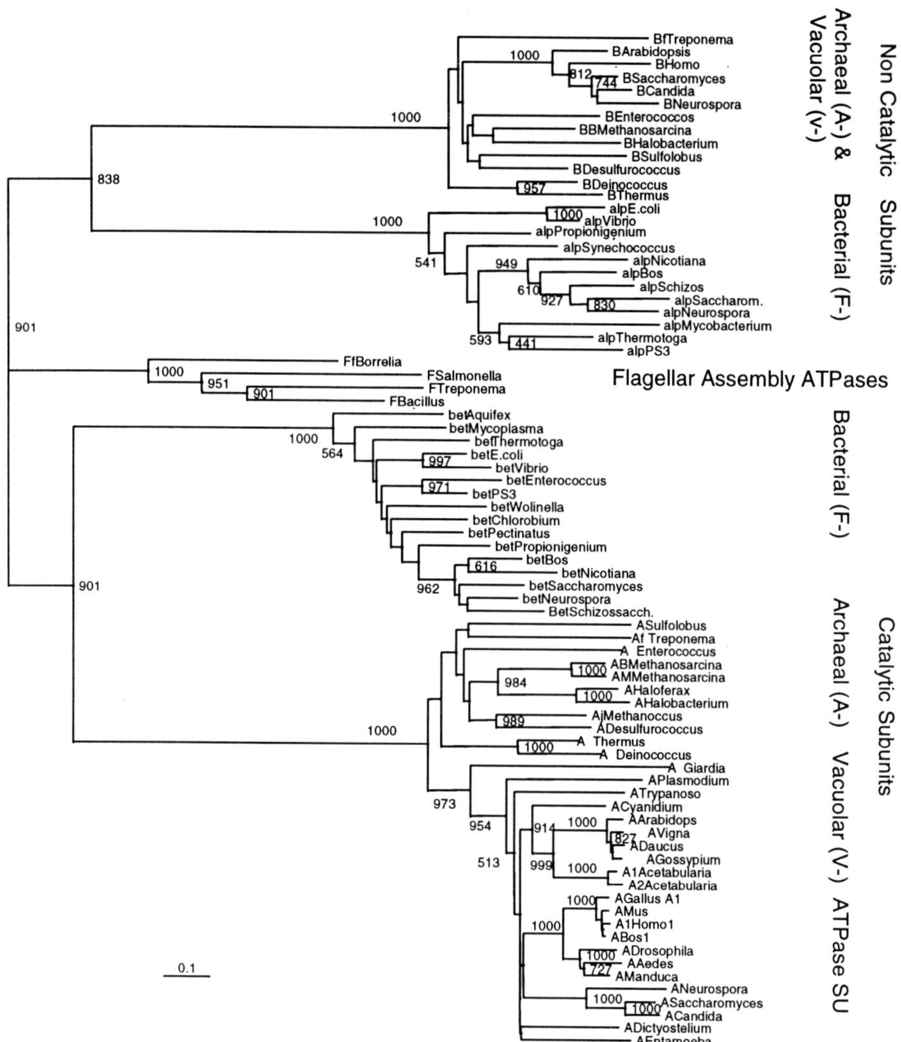

Figure 2. Distance analysis of catalytic (beta and A) and non-catalytic (alpha and B) subunits of F- V- and A-ATPases and the homologous ATPases encoded as part of the bacterial flagella operon (Salf1 and Bacf1). Sequence alignment was calculated using ClustalW1.6 with a BLOSUM series substitution matrix. Gaps were included in the distance analysis. alp=alpha, bet=beta.

Schäfer and Meyering-Vos, 1992, Ihara et al., 1992). P-type ATPases function as cation pumps and are not homologous to the F/V/A type ATPases.

The vacuolar (V-), bacterial (F-), and archaeal (A-) ATPases are homologous enzymes, i.e., they evolved from the same ancestral enzyme (Zimniak, 1988; Gogarten et al., 1992). All are large multisubunit enzymes composed of a water soluble (V/F/A1) portion and a proton conducting membrane spanning complex (V/F/A0) portion. Each hydrophilic sector contains three copies of an ATP hydrolyzing catalytic subunit (β-subunit in F-ATPases, A-subunit in A/V-ATPases) and three copies of a non-catalytic subunit (α-subunit in F-ATPases, B-subunit in A/V-ATPases). The catalytic and the non-catalytic subunits have evolved from the same ancestral gene after an ancient gene duplication and are therefore paralogous subunits. The gene duplication which gave rise to these subunits had already occurred in the last common ancestor of the bacteria, archaea and eukaryotes, allowing comparison of these paralogous subunits to each other to be used to root the tree of life (Gogarten et al., 1989, Gogarten and Taiz, 1992).

Phylogenies calculated from ATPase sequences (Fig. 2) for the most part agree with phylogenies calculated from other markers (e.g., 16S like rRNA, Gogarten et al., 1996, Ludwig et al, 1993) therefore, it appears that the coupling factor ATPases have evolved together with the organisms in which they reside and can be used as reliable phylogenetic markers. The majority of Bacteria have an F-ATPase and the Archaea in which the ATPase has been studied possess the homologous A-ATPase. However, some important exceptions exist: Based on rRNAs and other biochemical characters (cell wall, lipids, antibiotic resistance) *Thermus* is classified as bacterial genus, with closest similarities to the genus *Deinococcus* (Woese, 1987; Hensel et al., 1986). However, *Thermus thermophilus* HB8 does not have an F-ATPase like other eubacteria, but an A/V-ATPase (Yokoyama et al., 1990; Tsutsumi et al., 1991,Gogarten et al., 1992). *Thermus aquaticus* YT-1 and *Meiothermus* (*Thermus*) *chiarophilus* also appear to possess an A/V-ATPase (Radax et al., in press). *Deinococcus* appears to possess the same complement of A/V-ATPase subunit genes, in the same gene order, as *Thermus thermophilus* (The Institute for Genomic Research (TIGR), personal communication).

Other Bacteria also have been found to have A/V-ATPases. In addition to a normal bacterial F-ATPase that groups as expected with other low GC Gram positives, the eubacterium *Enterococcus hirae* has a sodium pumping ATPase that based on its sequence is classified as an archaeal type ATPase (Takase et al., 1994; Hilario and Gogarten, 1993,Fig. 2). Recently, the spirochetes *Borrelia* and *Treponema* have been shown to possess the A and B subunits of an A-ATPase (Fig. 2, TIGR, personal communication). Biochemical, immunological, and N-terminal sequence data suggest that *Clostridium fervidis* (Höner zu Bentrup et al., 1997) possesses a Na+ pumping A/V-ATPase complex.

One possible explanation for the presence of A/V type ATPases in Bacteria is that the last common ancestor already had an A/V- and an F-type ATPase and that in the case of the bacteria, most of the lineages leading to extant organisms lost the A/V-type, whereas the archaeal lineages lost the F-ATPase. The conflicting topology of the tree in Figure 2 then, would be a result of the comparison of unrecognized paralogues (Forterre et al., 1993). It was recently reported that *Thermus scooductus,* based on biochemical, immunological and amino terminal sequence data appears to possess an F-type ATPase (Radax et al., in press) and not an A-ATPase like the other Deinococcaceae (e.g., *Thermus thermophilus, Meiothermus chliarophilus, Deinococcus radiodurans).* This finding might be considered as supporting the simultaneous presence of both an A/V and F-ATPase in the last common ancestor. However, based on 16SrRNA sequences *Thermus scooductus* is a closer relative of *Thermus thermophilus* (Radax et al., in press) than are *Meiothermus chliarophilus* and *Deinococcus radiodurans.* Thus several independent losses of A- and F-ATPases need to be postulated within the Deinococcaceae alone to explain the distribution of ATPases in this family by means of an ancestor that had both ATPase types. Furthermore, both *Aquifex aeolicus* (Ron Swanson, Genomics Recombinant Biocatalysis Inc, San Diego, pers. comm.) and *Thermotoga maritima* (Ludwig et al. 1993, TIGR, personal communication), bacteria whose lineages presumably evolved before that of the Deinococcaceae (Olsen et al., 1994), have typical F-type ATPases, as do the majority of Bacteria so far examined. The complete genomes of *Aquifex aeolicus* (see above), *Escherichia coli* (Blattner et al. 1997), *Synechocystis sp. PCC 6803* (Kaneko et al., 1996), *Mycoplasma pneumoniae* (Himmelreich et al., 1996), *Mycoplasma genitalium* (Fraser et al., 1995), *Haemophilus influenzae* (Fleischmann et al., 1995), and *Bacillus subtilis* (Kunst et al., 1997) reveal the presence of only one complete F-ATPase, with no corresponding A/V-ATPase. On the other hand, complete genome sequences of Archaea available to date show that *Methanococcus jannashii* (Bult et al., 1996), *Methanobacterium thermoautotrophicum* (Smith et al., 1997), and *Archaeoglobus fulgidus* (TIGR, personal communication) possess only an A-ATPase, with no F-ATPase present. Therefore, explaining this distribution of ATPase types by horizontal transfer appears more parsimonious than assuming many independent losses.

5. Other Archaeal Genes in the Deinococcaceae

The presence of an A/V type ATPase in *Thermus* prompted us to look for other archaeal genes that may have been transferred into this organism at the same time. By constructing probes from genomic DNA of *Thermoplasma thermoacidophilum* (an archaeon), we were able to screen a genomic library made from *Meiothermus* (*Thermus*) *ruber*. Plaques which showed increased hybridization against background were picked and sequenced. One of these contained a gene for prolyl tRNA synthetase which showed closer similarity in BLAST search to eukaryotic homologues. We then used this fragment to obtain the corresponding full length gene from *Deincococcus radiodurans* (TIGR, personal commuincation). Distance anaylysis of this gene and its homologues (Fig. 3) shows that the *Deinococcus* prolyl tRNA synthetase groups in a clade with the eukaryotic sequences, not with the majority of the other bacterial sequences. Interestingly, *Borrelia* and *Treponema*, two spirochetes which have A/V type ATPases have different types of prolyl tRNA synthetases. *Borrelia* has a prolyl tRNA synthetase which also groups with the eukaryotic/archaea homologues, while *Treponema* has the bacterial homologue. If we assume that the *Saccharomyces* #2 and

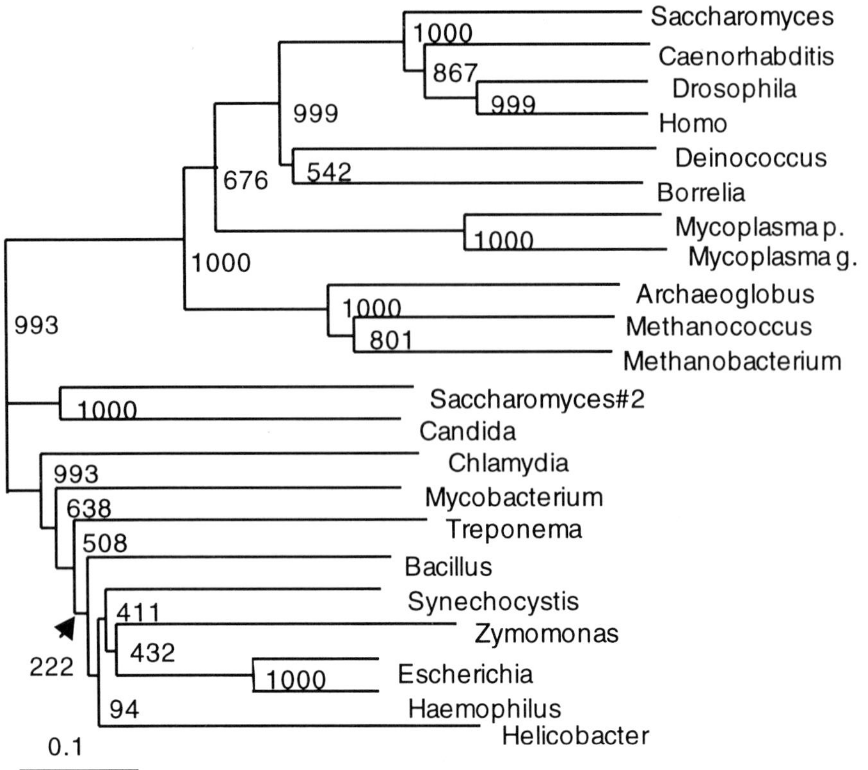

Figure 3. Distance analysis of prolyl tRNA synthetase amino acid sequences. All sequences were obtained from GenBank except for *Deinococcus* and *Archaeoglobus* sequences which were obtained from TIGR using a BLASTX search. Alignments were performed using ClustalW1.6 with a BLOSUM series protein matrix. 1000 bootstrapped data sets were calculated using the phylogenetic trees option of ClustalW. Gaps were excluded from the analysis. Numbers to the right of each node indicate bootstrap replicates out of 1000 supporting that node.

Candida sequences represent eubacterial homologues acquired from mitochondria, then the data suggest that eukaryotic/archaeal homologue found in *Deinococcus*, *Thermus* and *Borrelia* was acquired through horizontal transfer.

The following intriguing observation suggests that the A/V ATPase operon and the prolyl tRNA synthetase gene may have been transferred in a single event: in the archaeon *Sulfolobus solfataricus*, the gene immediately downstream from the A-ATPase operon encodes the prolyl tRNA synthetase (M. Ragan, MRC, Halifax, N.S., personal communication). Although this gene order is not maintained in other Archaea, nor in *Deinococcus*, it suggests that an organism close to the root of the archaeal domain donated a piece of DNA to the ancestor of the Deinococcaceae, and that this piece of DNA encoded both the A-ATPase and the prolyl tRNA synthetase.

The finding of uptake of genes by the Deinoccocacceae is not unprecedented. Both *Thermus thermophilus* and *Deinococcus radiodurans* are naturally competent (Tigari and Moseley, 1980, Hidaka et al., 1994), and laboratory studies have demonstrated the transfer of genes between *Thermus* strains in the lab (Fernandez-Herrero et al., 1995). The *Thermus* malate dehydrogenase sequence is more closely related to eukaryotic homologues than bacterial ones (Iwabe et al., 1989, Gogarten et al., unpublished results.) Biochemical and phylogenetic analyses also indicate that *Thermus thermophilus* contains two cytochrome oxidases, one which is bacterial and one which is more closely related to an archaeal gene (Gogarten et al., 1996, Keightley et al., 1995). Evidence is accumulating that *Thermus* and its relatives within the Deinococcaceae may have acquired a number of genes from Archaea. Further phylogenetic analysis or genes from *Thermus* and *Deinococcus* species, and the availability of the complete *Deinoccoccus* genome (TIGR, in preparation) will allow us to determine the extent to which horizontal gene transfer contributed to the evolution of this early bacterial lineage.

The pivotal role of horizontal gene transfer in shaping microbial communities and microbial genomes is emerging as a new paradigm in the study of microbial evolution. The selfish operon theory (Lawrence and Roth, 1996) which considers horizontal transfer as the main driving force in operon formation was developed based on operons which are under selection only periodically. The data discussed here demonstrate that housekeeping genes like H^+-ATPases/ATPsynthases and aminoacyl tRNA synthases were transferred infrequently between divergent groups of prokaryotes. In addition to the prolyl tRNA synthetases and the A-ATPases discussed above, James Brown (Smith Kline and Beecham, Collegeville, PA, personal communication) has found evidence for the transfer of isoleucine tRNA synthetases across domain boundaries, and indications for the transfer of glutamine/glutamate tRNA synthetases from eukaryotes to prokaryotes were reported by Lamour et al. (1994). In all of these cases the transferred functions can be characterized as housekeeping. One hardly can envision a “modern” organism without pathways to synthesize all 20 aminoacyl-tRNAs, or without an energized plasmamembrane. It appears likely that the receiving organism at least temporarily retained two genes that encoded enzymes with very similar function and that over time one of these copies was lost. This replacement of housekeeping genes with copies obtained from only distantly related organisms has been so infrequent that these transfer events are clearly visible in the comparison of different molecular phylogenies. While these transfer events make it more complicated to reconstruct organismal evolution and to infer properties of ancestral organisms, these transfer events can be utilized to correlate evolution in different parts of the tree of life, and thus they might provide the key to a better understanding of organismal evolution.

Acknowledgments

The authors thank the NASA Exobiology Program for support and Ryan Murphy for work on the screening and sequencing of *Meiothermus ruber* genes. We also thank The Institute for Genomic Research (TIGR) for availablity of sequence data prior to publication.

6. References

Baldauf, S.L., Palmer, J.D., and Doolittle, W.F. (1996) *Proceedings of the National Academy of Sciences. USA* **93**, 7749-7754.

Benachenhou-Lafha, N., Forterre, P. and Labedan, B. (1993) *J. Molecular Evolution* **36**, 335-346.

Blattner F.R, et al. (1997) *Science* **277**, 1453-1474.

Brown, J.R., Masuchi, Y., Robb F. T. and Doolittle W. F. (1994) *J. Molecular Evolution* **38**, 566-576.

Brown, J. R. and Doolittle, W. F. (1995) *Proceedings of the National Academy of Sciences, USA* **92**, 2441-2445.

Bult et al. (1996) *Science* **273**, 1058-1073.

Cammarano, P., Palm, P., Creti, R., Ceccarelli, E., Sanangelantoni, A. M. and Tiboni, O. (1992) *J. Molecular Evolution* **34**, 396-405.

Chien, Y. T. and Zinder, S. H. (1994) *J. Bacteriology* **176**, 6590-6598.

Fleischmann R.D. et al. (1995) *Science* **269**, 496-512.

Forterre, P., Benachenhou-Lafha, N., Confalonieri, F., Duguet, M., Elie, C. and Labedan, B. (1993) *BioSystems* **28**, 15-32.

Fraser C.M., et al. (1995) *Science* **270**, 397-403.

Gogarten, J.P. and Taiz, L. (1992) *Photosynthesis Research* **33**, 137-146.

Gogarten, J.P., Kibak, H., Dittrich, P., Taiz, L., Bowman, E. J., Bowman, B.J., Manolson, M.F., Poole, R.J., Date, T., Oshima, T., Konishi ,J., Denda, K. and Yoshida, M. (1989) *Proceedings of the National Academy of Sciences, USA* **86**, 6661-6665.

Gogarten, J.P. Starke, T., Kibak, H., Fichmann, J. and Taiz, L. (1992) *J. Experimental Biology* **172**, 137-147.

Gogarten, J. P., E. Hilario, and Olendzenski L. (1996), in D.McL. Roberts, P.Sharp, G. Alderson and M. Collins (eds.), *Evolution of Microbial Life, Society for General Microbiology Symposium 54*,. Cambridge University Press, Cambridge, pp. 267-292.

Gupta, R. S. and Golding, G. B. (1993) *J. Molecular Evolution*, **37**, 573-582.

Gupta, R. S. and Singh, B. (1992) *J. Bacteriology*, **174**, 4594-4605.

Hensel, R., Demharter, W., Kandler, O., Kroppenstedt, M. and Stackebrandt, E. (1986) *Int. J. Systematic Bacteriology* **36**, 444-453.

Hensel, R., Zwickl, P., Fabry, S., Lang, J. and Palm, P. (1989) *Canadian J. Microbiology* **35**, 81-85.

Himmelreich R., Hilbert H., Plagens H., Pirkl E., Li B.C., Herrmann R. (1996) *Nucleic Acids Research* **24**:4420-4449.

Hidaka, Y, Hasegawa, M., Nakahara, T., Hoshino, T. (1994) *Bioscience, Biotechnology, Biochemistry* **58**, 1338-1339.

Hilario E.and Gogarten J.P. (1993) *BioSystems* **31**, 111-119.

Höner zu Bentrup, K., Ubbink-Kok, T., Lolkema, J.S., and Konings, W.H. (1997) *J. Bacteriology* **179**, 1274-1279.

Ibba M., Morgan S., Curnow A.W., Pridmore D.R., Vothknecht U.C., Gardner W., Lin W., Woese C.R., Soll D. (1997) *Science* **278**, 1119-1122.

Ihara K., Abe T., Sugimura K.-I., Mukohata Y. (1992) *J. Experimental Biology* **172**, 475-485.

Iwabe, N., Kuma, K-I., Hasegawa, M., Osawa, S. and Miyata, T. (1989) *Proceedings of the National Academy of Sciences, USA* **86**, 9355-9359.

Kaneko T. et al., (1996) *DNA Research* **3**, 185-209.

Keeling, P.J., and Doolittle, W.F. (1995) *Proceedings of the National Academy of Sciences, USA* **92**, 5761-5764.

Keightley, J.A., Zimmermann, B.H., Mather, M.W., Springer, P., Pastuszyn, A., Lawrence, D.M., and Fee, J.A. (1995) *J. Biological Chemistry* **270**, 20345-20358.

Koonin, E.V. (1997) *Molecular Microbiology* **25**, 619-637.

Kumada, Y., Benson, D. R., Hillemann, D., Hosted, T.J., Rochford, D.A., Thompson, C.J., Wohlleben, W. and Tateno, Y. (1993) Evolution of the glutamine synthase gene, one of the oldest existing and functioning genes. *Proceedings of the National Academy of Sciences, USA* **90**, 3009-3013.

Kunst, F. et al (1997) *Nature* **390**, 249-256.

Lamour V., Quevillon S., Diriong S., N'Guyen V.C., Lipinski M., Mirande M. (1994) *Proceedings of the National Academy of Sciences, USA* **91**, 8670-8674

Langer, D., Hain, J., Thuriaux, P., and Zillig, W. (1995) *Proceedings of the National Academy of Sciences, USA* **92**, 5768-5772.

Lawrence J.G., Roth J.R. (1996) *Genetics* **143**, 1843-1860.

Lawson, F.S., Charlebois, R.L., and. Dillon, J.R (1996) *Molecular Biology and Evolution* **13**, 970-977.

Ludwig, W., Neumaier, J., Klugbauer, N., Brockmann, E., Roller, C., Jilg, S., Reetz, K., Schachtner, I., Ludvigsen, A., Bachleitner, M., Fischer, U. Schleifer, K.H. (1993) *Antonie Van Leeuwenhoek J. Microbiology* **64**, 285-305

Maidak, B.L., Olsen, G.J., Larsen, N., Overbeek, R., McCaughey, M.J. and Woese, C.R.(1997) *Nucleic Acids Research* **25**, 109-111.

Marsh, T.L., Reich, C.I., Whitelock, R.B.and Olsen, G.J.(1994) *Proceedings of the National Academy of Sciences, USA*, **91**, 4180-4184.

Morowitz, H. J. (1992) *Beginnings of Cellular Life: Metabolism Recapitulates Biogenesis*, New Haven: Yale University Press.

Olendzenski, L. and Gogarten, J.P., in J. Weigel and M. Adams (eds.), *Thermophiles: The Keys to Molecular Evolutioin and the Origin of Life?*, Taylor and Francis, London in press.

Olsen, G.J., Woese, C.R., and Overbeek, R. (1994) *J. Bacteriology* **176**, 1-6.

Olsen, G.J., Woese, C.R. White, O., and Venter, J.C. (1996) *Science* **274**, 902.

Pesole, G., Gissi, C., Lanave, C. and Saccone, C. (1995) *Molecular Biology and Evolution* **12**, 189-197.

Radax, C., Sigurdsson, O., Hreggvidsson, G., Aichinger, N., Gruber, C., Kristjansson, J.K., Stan-Lotter, H. (1998) *Systematic and Applied Microbiology* in press.

Rivera, M. C., and Lake, J.A. (1992) *Science* **257**, 74-76.

Schäfer G., Meyering-Vos M. (1992) *Annals of the New York Academy of Science* **671**, 293-309.

Schofield, J.P. (1993) Clinical Science **84**, 119-128.

Schwartz, R.M. and Dayhoff, M.O. (1978) *Science* **199**, 395-403.

Smith D.R. et al. (1997) Complete genome sequence *of Methanobacterium thermoautotrophicum* deltaH: functional analysis and comparative genomics. *J. Bacteriology* **179**, 7135-7155

Souillard, N., Magot, M., Possot, O. and Sibold, L. (1988) *J. Molecular Evolution* **27**, 65-76.

Sumi,M.,. Sato, M.H, Denda, K., Date T. and Yoshida M. (1992) *Federation of European Biochemical Societies Letters* **314**, 207-210.

Takase, K., Kakinuma, S., Yamato, I., Konishi, K., Igarashi K. and Kakinuma Y. (1994) *J. Biological Chemistry* **269**, 11037-11044.

Tiboni, O., Cammarano, P. and Sanangelantoni, A. M. (1993) *J. Bacteriology* **175**, 2961-2969.

Tigari, S. and Moseley, B.E.B. (1980) *J. General Microbiology* **119**, 287-296.

Tsutsumi, S., Denda, K., Yokoyama, K., Oshima, T., Date, T. and Yoshida, M. (1991) *Biochimica et Biophysica Acta*, **1098**, 13-20.

van den Hoff, M. J.B., Jonker, A., Beintema, J.J., Lamers, W.H. (1995) *J. Molecular Evolution* **41**, 813-832.

Wächtershäuser, G. (1992) *Progress in Biophysics and Molecular Biology* **58**, 85-201.

Woese, C.R. (1987) *Microbiological Reviews* **51**, 221-271.

Woese, C.R., Kandler, O., Wheelis, M.L. (1990) *Proceedings for the National Academy of Sciences, USA*, 87, 4576-4579.

Yokoyama, K., Oshima, T. and Yoshida, M. (1990) *Journal of Biological Chemistry* **265**, 21946-21950.

Zillig, W., Palm, P. and Klenk, H-P. (1992), in H. Hartman and K. Matsuno (eds.), *The Origin and Evolution of the Cell,* World Scientific, Singapore pp. 163-182.

Zimniak, L., Dittrich, P., Gogarten, J.P., Kibak, H. and Taiz, L. (1988) *J. Biological Chemistry* **263**, 9102-9112.

Biodata of **AWD Larkum** author of " ***The Evolution of Algae.***"

Dr. Tony Larkum is a Professor in the School of Biological Sciences (Plant Sciences) at the University of Sydney (Australia). He received his Ph.D. in 1961 in Oxford (UK) and had since published three books and more than 130 research articles in wide range of areas covering molecular biology, evolution, plant physiology and biochemistry and marine ecology, with a particular interest in photosynthesis and its mechanisms. The areas of his interest are: hazards of ozone depletion in the stratosphere and accompanying increases in UB-B irradiation; evolution and phylogeny of algae; Prochloron and other exotic algae; analysis of plant pigment and coral bleaching.
E-mail: **alark@mail.usyd.edu.au**

J. Seckbach (ed.), Enigmatic Microorganisms and Life in Extreme Environments, 29–48.

THE EVOLUTION OF ALGAE

A.W.D. LARKUM
School of Biological Sciences
University of Sydney
NSW 2006, Australia

1. Introduction

This article deals with the evolution of the algal protists i.e. the eukaryotic algae. It attempts to assess the origins of these algae at the level of the phyla. Necessarily the arguments are based to a great extent on the evolution of prokaryotic organisms and the origin of the protists.

Current hypotheses on the origin and evolution of prokaryotic organisms and protists are grounded on two forms of evidence i) products of expressed genes (including morphological characteristics) and ii) genetic information (mostly DNA) of which the commonest form is the sequence of nucleotides in segments of DNA. Over the last decade the latter evidence i.e. "genetic information" has gained a strong hold in the formulation of evolutionary hypotheses. This is the result of the apparent ease with which the information can be fed into computer programs which will produce "unbiased" phylogenetic trees, when homologous sequences from different organisms are compared.

Algae have at least three sets of DNA - genomic, plastid and mitochondrial DNA and appear therefore to be ideally suited to full phylogenetic characterization by this approach. Unfortunately the results of such phylogenetic tree analyses are neither unbiased nor always informative and therefore the first section here will be devoted to a discussion of this topic. In the following sections various arguments will be discussed in the light of this conclusion. Structural, chemical, biochemical and physiological characteristics will be taken into account as well as genetic attributes and the various viewpoints will first presented clearly and without prejudice. Then evidence for and against the various arguments will be discussed. Finally my own view will be presented as a final summary.

2. Phylogenetic Tree Reconstruction

2.1. THE QUEST FOR A REALISTIC PHYLOGENY OF THE PROKARYOTIC AND EARLY EUKARYOTIC WORLD

A very real fear of current methods of phylogenetic analysis is that they will not be informative on events that took place over 3.5 to 1 BYA. This is because the critical information has been lost by subsequent changes in genomes. A number of these difficulties are discussed below.

2.1.1 *Log Det Method*
An early fault with many analyses was lack of an adequate method to allow for base compositional bias (Lockhart *et al.*, 1992) but this has now been corrected by the formulation of the Log Det method (Lockhart *et al.*, 1994) (available on DNAML and PAUP). However, it is a requirement of this method that only variable sites are used. Many analysts have not adhered to this constraint and so a number of unreliable trees are to be found in the recent literature. This is perhaps understandable in light of the fact that strict adherence to this condition leads to "star pattern" phylogenies with no information (Scarramuzzi *et al.*, 1995; Howe *et al.*, 1995; see Fig. 1). Nevertheless those who do choose to ignore such strictures do so at their own peril. Swofford et al. (1996) discuss implementation of the LogDet method using an "invariable sites model"

2.1.2. *Sites-free-to-vary*
As pointed out above it is necessary to take only variable sites (meaning free to vary i.e. a base/amino acid that is free to change if it can [even if it has not happened yet]) in the correct application of phylogenetic analyses. Lockhart *et al.* (1996b) showed how lack of adherence to this stricture can actually lead to the wrong tree. When they applied the rules to defining the tree which led to the evolution of Chl *a* and Bchl *a* they showed that the priority of either one over the other could not be decided upon, i.e., there was insufficient data. In previous work on the same problem difficulties arose because the sequences compared had variable sites at different positions (probably because the genes compared were of different function). This occurrence led to an extreme tree shape long branches leading to the outgroup (*nifH*), and *bchX* and short branches leading to the other taxa. This is the tree shape that Felsenstein (1978) says will mislead parsimony. Here it also misled maximum likelihood because in addition to the invariable sites at different positions, there were also many invariable sites at the **same** position across all taxa; so maximum likelihood underestimated the extent of sequence difference between the outgroup and *bchX*, on the one hand, and the remaining taxa, on the other. Had the invariable sites been removed across all taxa then maximum likelihood would have identified an extreme tree topology and not have favored one tree over the other. Thus in doing such analyses one should only use sites that are free to vary and the variable sites that are used should be at the same positions in all the compared sequences.

2.1.3. *Rates of Change*
Models of phylogeny depend on equal rates of change at all variable sites. As shown by Feselstein (1988) unequal rates can lead to the wrong tree. In analysis of ancient taxa it is often the case that unequal rates of evolution apply across the widely disparate taxa and specific methods need to be applied to obtain the correct tree. Often this can only be done by careful application of a variety of methods. In this approach the use of SPLITSTREE is very helpful (Lockhart *et al.*, 1995).

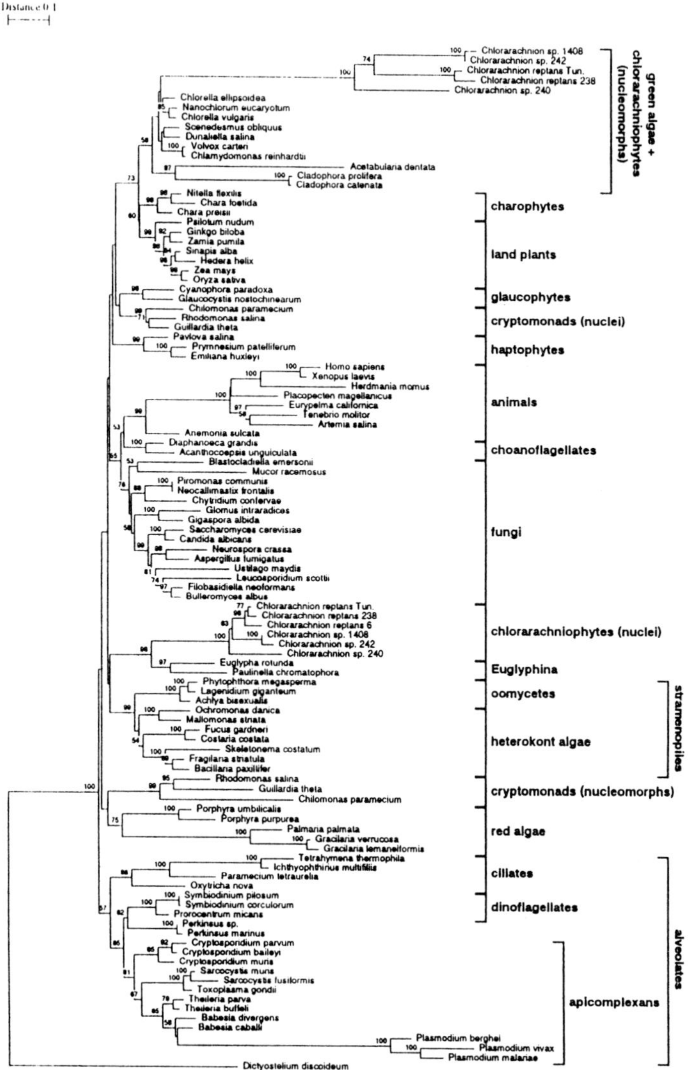

Figure 1. Evolutionary tree constructed by a neighbour-joining method from a distance matrix of 107 eukaryotic SSU rRNA sequences and based on "substitution rate calibration". The slime mold *Dictyostelium discoideum* was used to root the tree. Reproduced from Van de Peer et al. (1996) with permission.

In the analysis of chlorarachniophyte origins, Van de Peer *et al.* (1996) used just such an approach based on a distance method. Their conclusion is worth noting "In conclusion, we want to point out the importance of non molecular data. On the basis of most standard tree inferring methods, the nucleomorph and plastid rRNA phylogenies gave misleading results, sometimes very convincing, regarding the phylogenetic position of the nucleomorph endosymbiont. If there had not been good non-molecular reasons to question these molecular results; rate, calibration and transversion analysis might never have been applied in this case. It is probable that many more such misleading results are present in eukaryotic trees based on SSU -rRNA sequences. Substitutive rate calibration should contribute to the elimination of artifacts caused by fast-evolving sequences". The best tree that these workers produced is reproduced in Fig. 2 but the above comments should be remembered.

2.1.4. *Covariant Analysis*

Central to the idea of differential rates of change at different sites is the covarion model of Fitch and Markowitz (1970). The model shows how these conditions can lead to choosing of the wrong tree (as demonstrated above in the example from Lockhart et al., 1996). In a recent contribution, Waddell *et al.* 1997 have used Hadamard conjugations to model correct evolution where unequal rates apply across sites. They point out that where unequal rates occur across sites, ancient taxa will be particularly prone to misleading results, even for rRNA data. This is a similar problem to that pointed out by Van de Peer *et al.* (1996) (above); that is if sites are changing at different rates (but with the same heterogeneity in the different taxa) then calculations of distances will underestimate the true amount of change and lead to the wrong tree. The same difficulty also misleads parsimony because fast changing sites, where the information is misleading, and slower more reliable sites are not distinguished. Another good example, albeit in a different field, is Phillipe *et al.* (1997).

2.1.5. *Distance Methods and Spectral Analysis*

In certain distance methods the property of producing a dichotomous tree is used (e.g. Split Decomposition). The use of local constructive distance methods such as, for example, Neighbour Joining/UPMGA may be satisfying at first sight because they are fast. However they can be misleading as they tend to impose tree shape artifacts onto data, i.e. they favor the recovery of certain tree shapes. Many dichotomous trees collapse to "star-pattern" trees when other methods such as SPLITSTREE are applied, even with the proper corrections using Log Det and variable sites (Howe *et al.*, 1995, see fig 1). Caution should therefore be used in applying distance methods. As pointed out by Waddell *et al.* (1997) spectral analysis can be used as " one way of evaluating the gap between our hope for the resolving power of the data, and the 'disrespect' which real molecules may show for our simplifying assumptions". Furthermore Van der Peer *et al.* (1996) remind us that non-molecular data should be used just as much as molecular data in informing our decisions.

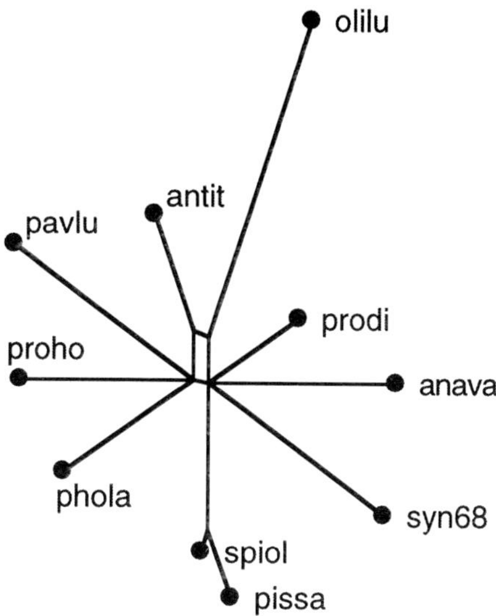

Figure 2. Splitsgraph generated using the Splitstree program showing the relationship of 5 cyanobacteria /prochlorophytes and 5 plastids. Branches are proportional to predicted evolutionary distance. The dataset was constructed to contain a proportion of constant sites that was consistent with the calculated number of invariant sites.

2.1.6. *Finding the Wrong Tree*

There is a subtle allure in tree building even when we arrive at the wrong tree. This is because the trees that are obtained often look very convincing and logical: close relatives fall on neighboring branches and broad patterns seem to be consistent. But no such appeal to commonsense should be allowed to sway our reason. And where clear anomalies exist we should be even more cautious. One example of this is the placement of photosynthetic organisms in the eubacteria SSU-rRNA tree (Woese, 1987). Photosynthetic organisms fell on several disparate branches (and, as noted in Section 3.2, cyanobacteria emerged as a rather recent branch). It is the really deep branches where we should apply the most caution. Clearly several approaches need to be taken to try to resolve specific issues and we should all be cautious of applying what seem like clear-cut results based on a single line of reasoning. And as pointed out by Van der Peer *et al.* (1997) we should not shirk from using non-molecular data to inform us on decisions in the world of molecular inference.

3. Evolution of Photoautotrophic Prokaryotes

3.1. EVOLUTION OF PHOTOSYNTHESIS

All the photoautotrophic prokaryotes occur in the Eubacteria and it must be assumed that photosynthesis arose in this Super Kingdom (the evolution of a light-driven proton pump in *Halobacter* spp. (Archaebacteria) is generally regarded as an independent evolutionary development). Larkum (1991) gave a scheme for the early evolution of photosynthesis. Since that time new evidence has accrued to suggest that the Heliobacteria may be the most primitive photosynthetic group extant in the modern world (Vermaas, 1994). These organisms possess a homodimeric reaction center of the "Quinone" type (i.e. homologous with RCII of alga). At the same time there presumably existed photosynthetic bacteria with the alternate reaction centre the "FeS" type (homologous with RCI of algae), for in any scheme it is likely that some form of anoxygenic photosynthesis preceded oxygenic photosynthesis. The date for oxygenic photosynthesis is at least 3.45 billion years ago (BYA)(Schopf, 1993). This is so long ago and only some 0.4 billion years after life was possible on the Earth (>3.8 BYA; Mojsis *et al.,* 1996) that it is unlikely that evidence exists in any form to chart evolution of photosynthetic organisms prior to this time. Nevertheless around 3.45 BYA oxygenic photosynthesis arose. So it seems inevitable that both types of photosynthetic reaction centre existed and that there arose an amalgamation of the "Quinone" type and the "FeS" type of reaction centers, either by fusion of two different genetic sources or by differentiation of the two systems within the same genetic line (Blankenship, 1992; Larkum, 1992). This engine of photosynthesis appears to have been handed down to the present time with little modification

3.2 EVOLUTION OF CHLOROXYBACTERIA

The evidence on the nature of the first oxygenic cells rests on microfossils and stable isotope ratios of the rocks in which these are found (Schopf, 1993). For this reason it is not possible to assign these cells to Cyanobacteria, which many sources state, because the Cyanobacteria are defined as possessing phycobiliproteins. That all oxygenic prokaryotic phototrophs could not be accommodated within this description was first indicated by the discovery of *Prochloron* in 1974 (see Lewin 1977) which is a prokaryotic oxygenic phototroph possessing Chl *a* + Chl *b* (and no PBP), found as an extracellular symbiont in didemnid ascidians of coral reefs and tropical mangroves. Two other members of the family, *Prochlorothrix hollandica* (Burger-Wiersma *et al.*, 1989 and *Prochlorococcus marinus* (Chisholm *et al.*, 1992) were also discovered. Subsequently, a chlorophyll *c*-like pigment has been detected in *Prochloron* (Larkum *et al.*, 1994) and in *Prochlorococcus* (Goericke and Repeta, 1992) and a gene for phycoerythin has been discovered in certain strains (species?) of *Prochlorococcus* (Hess *et al.*, 1996).

The cyanobacteria are placed on a rather shallow branch of the Eubacteria on phylogenetic tree reconstructions (Woese, 1987), based on SSU-rRNA. Based on the same gene and on rpo it been found that the prochlorophytes fall within the cyanobacterial clade and occur on different branches within the cyanobacterial radiation (Palenik and Haselkorn, 1992; Urbach *et al.*, 1992). These analyses can be criticized in that they do not address a number of problems in phylogenetic analysis such as covarion evolution, sites-free-to-vary, and different rates of change (see section 2.1). More recent attempts to address some of these problems have indicated that most extant cyanobacteria evolved more recently than the ancestors of chloroplasts (Nelissen *et al.*, 1995). Attempts to take full account of these factors results in "star pattern" trees, implying that there is insufficient information in the data sets to predict lineages accurately (Scaramuzzi *et al.*, 1995; Howe *et al.*, 1995), although no detailed attempt has so far been made using SSU rRNA data, because the principles on which this could be done have not been defined.

In a different approach, much attention was given to the question of what type of protein would be involved in binding Chl *b* (and where it occurs, Chl *c*) in prochlorophytes. It has now been shown that Chl *b* (and *c*-like pigment in *Prochloron*) is bound to a protein, which is not homologous with the CAB/CAC family of eukaryotic algae (La Roche *et al.*, 1996). However this is not informative on the question of an ancestral position for prochlorophytes in the lineage to green algae. It is just as possible that Chl *b* from this line eventually became associated with an early CAB/CAC protein (derived form *hliA*) at a prokaryotic stage (and then independently formed green algal chloroplasts) as an origin at the eukaryotic stage of Chl *b*, Chl *c* and CAB/CAC proteins.

The recent discovery of an oxygenic photosynthetic eubacterium with a novel light-harvesting system based on Chl *d* (Miyashita *et al.*, 1996) has further complicated the situation. *Acaryochloris* was discovered as a second symbiont along with *Prochloron didemni* in the didemnid ascidian *Lissoclinum patella*. This is a very significant discovery. It not only demonstrates the existence of Chl *d*, which had previously been doubted but demonstrates further diversity of oxygenic autotrophs at the eubacterial level viz. within the Chloroxybacteria. It will be of great interest to learn where this new organism fits into the eubacterial tree, although, as pointed out in the section above placement may not be a realistic possibility at the present time.

Cyanobacteria (*sensu stricto*), prochlorophytes and *Acaryochloris* need to be placed in a superphylum for organisms with prokaryotic characteristics and oxygenic photosynthesis. The term "Chloroxybacteria" (Lockhart *et al.*, 1996b) is descriptive and will be used throughout this article (the term "Oxychlorobacteria" has also been used: Burger-Wiersma *et al.*, 1989).

4. Evolution of Early Protists

The early evolution of eukaryotes is still obscure. Several hypotheses exist but at the moment reliable evidence is scarce. The "best guess" is that the first eukaryotes arose 1998). Even this hypothesis has been challenged, by the suggestion that lateral gene as anaerobic scavengers from an archaebacterial origin (but see Martin and Muller, transfer occurred between eubacteria and archaebacteria to create a chimaeric origin of this line (Gupta and Golding, 1996). In any event it is envisaged that a scavenging cell became larger and more complex, leading to the development of 1) the endoplasmic reticulum (ER) and 2) the formation of the nucleus, from the ER, and the initiation of chromosome development and of nuclear division.

From this stage it is possible that endosymbiotic events led to the development of mitochondria (from a proteobacterium) and flagella (from spirochaetes) and, at a later stage, plastids (from chloroxybacteria). The endosymbiotic event(s) which led to mitochondria is/are likely to have occurred more that once. At the root of this tree, it should be remembered, none of these events occurred at all, so that even today certain amitochondriate protists occur, where these features are believed to be the primitive condition [diplomonads, microsporidae, but probably not trichomonads (Bui *et al.*, 1996)].

4.1. THE "CROWN TAXA"

Other developments leading along this evolutionary tree are the development of endoplasmic reticulum and centrosomes. Various branches lead to protists while at the crown of the tree are the Plantae, Fungi and Animalae (Metazoans) which possess the most advanced features (see e.g. Palmer & Delwiche, 1996). The interesting question is where the algal protists branch off the trunk of this tree. It is clear that the host cells of algal protists show a diversity of origins, i.e. their origin is polyphyletic. Branching off near the crown of the tree are the green algae and many chromophytes. Red algae on the other hand may be an earlier branch (Perasso *et al.*, 1989; Liaud *et al.* 1995; Stiller and Davison, 1997). However much of this work is based on SSU rRNA and other information such as that based on structural genes needs to be considered.

An indication that the rRNA data lacks heuristic power, comes from information at biochemical and morphological levels, which indicate a very different sequence of events in some cases.

4.2. FLAGELLA

When and how eukaryotic flagella arose is still problematic. The original scavenger seems to have had amoeboid characteristics (implying the development of microtubules). By endosymbiosis it is possible to imagine the initiation of a flagellum (from a spirochaete) although the new structure (9+2 microtubules) which is different from that in spirochaetes (and any other eubacteria) is still a difficulty. The possession of a tripartite flagellum is characteristic of the heterokont algae and these plant protists, and associated animal protists, lie in the middle branches. Red algae with no flagella possibly lie further down the tree. While green algae occupy a position nearer the top, green algae such as Micromonadophytes (Moestrup 1996) would seem to be much more primitive

4.3. THE CELLULOSE CELL WALL

The cell wall is also problematic. As a cellulose covering of polysaccharide, the cell wall occurs in slime moulds, red algae, green algae and many heterokonts but not in alveolates and metazoans. Thus it is a general feature with few exceptions. Its origin is unknown but it is believed to be a derived feature, the primitive feature being a naked cell or one covered in scales as in prasinophytes (Lee, 1980). Cellulose does occur in eubacteria (Koyama *et al.,* 1997). However the eubacterial cell wall is based on muramic acid, an amino derivative of acetyl-glucose. Although the structure of the cell wall in algae is based on cellulose (1,4 β-linked glucose polymer), amorphous material used to infill the cellulose fibers differs from one group to another, being a glucose polymer in the case of green and red algae but a mannose or galactan polymer in chromophytes. Presumably this diversity is a secondary development but the reason for it is not apparent and may suggest lateral gene transfer rather than adaptive evolution.

4.4. PYRENOIDS AND STORAGE GRANULES

Pyrenoids are also a general feature of plant protists. The phylogeny of pyrenoids is bound up with the plastid and along with the storage granules, has been generally ignored by molecular phylogeneticists

4.5. SUMMARY

In summary the plant protists appear to be a diverse, polyphyletic group of protists from at or near the crown of the eukaryotic tree. As a result all contain mitochondria, microtubules, eyespots, nuclei with chromosomes (which divide through a mitotic spindle involving centrioles). All contain flagella except red algae. Cell walls and cell coverings are very variable. Such features and others not mentioned suggest more evolutionary divergence than is indicated by current gene sequence phylogenetic analyses.

5. Plastid Phylogeny

5.1. THE THEORY OF ENDOSYMBIOSIS

The currently accepted hypothesis for the origin of the plastid of plant protists (and the plastids of Plantae) is through endosymbiosis. The sequence of events is suggested to have been i) the ingestion of a chloroxybacterium by a mitochondriate, early protist, ii) the invagination of the "invader" by a host (ER) membrane, iii) the loss of the cell wall and iv) the loss of many genes, either altogether, or to the host nucleus. The loss of genes would have led to reduction in the size of the symbiotic genome; in extant plastids the genome size ranges from 100 to 200 kB containing from 100 to 255 genes (see e.g. Wakasugi *et al.*, 1997)

There are several examples of what can best be construed as organisms of cyanobacterial origin which are in what appear to be early stages of an endosymbiotic event (assigned to Glaucophytes or Glaucocystophytes). These symbionts are called cyanelles; they occur in *Cyanophora paradoxa* and other glaucophytes. The cyanelle genome has a size of 135 kB and contains 182 genes. The evolutionary age of these symbiotic associations is not clear. If they are very young they tell us little about the origin of plastids except by analogy. If they are ancient they might well be informative but a question then arises as to why these associations have remained *in limbo* for so long. These associations form a critical link in some explanatory hypotheses of endosymbiosis (see below).

5.2. A SINGLE EVENT VERSUS MULTIPLE SYMBIOTIC EVENTS: A MONOPHYLETIC VERSUS A POLYPHYLETIC ORIGIN OF PLASTIDS

5.2.1. *Evidence for a Single Event (Monophyletic Hypothesis)*

It is possible to suggest that a single endosymbiotic event gave rise to an ancestral plastid from which all other plastids arose. This "single event" hypothesis embodies the view that the symbiotic event which could lead to plastid formation is so rare that it only occurred once (e.g. Cavalier-Smith, 1993). The timing of this event would have been in the mid-stages of protist evolution – around 1.5-2 BYA (Canfield and Taske, 1992).

It is fair to say that most molecular phylogeneticists support a **single event** at the present time, based on current phylogenetic evidence. This evidence is largely based on chloroplast DNA (cpDNA). Two lines of evidence are used:

a) comparison of a single gene and its sequence, particularly rRNA,
b) gene arrangement, particularly the ATP gene family, and
c).complete chloroplast genome characteristics.

Phylogenetic analyses of Small Sub-Unit (SSU) rRNA indicate that chloroxybacteria are a well-defined group within the Eubacteria, with a fairly "late" origin (that is, they do not branch near the root of the tree). All "Plastid Tree Reconstructions" branch close to the chloroxybacteria, with cyanelles forming an early branch within the plastid clade, and with the plastid clade dividing into two broad families, the green plastid line (green algae, euglenoids, Astasia, and Plantae) and the other groups [red algae and chromophytes [ie. those taxa containing Chl *c* or closely related pigments]) (Fig 2). However it should be remembered that based on RPB1 and other data the red algae branch before the crown taxa of the protist tree (Stiller & Davison, 1997)].

5.2.1.1. *Other phylogenetic approaches*. Gene arrangement studies of the *atp*B/E gene cluster provide strong support for cyanelles as within the group, which gave rise to "the" plastid (Douglas and Murphy, 1994; Kroth Pancic et at., 1995). This because in the position of *atp* genes matches more closely that in plastids than in cyanobacteria. Another recent approach, made possible by the rapid progress in obtaining complete genomes, is to analyze complete chloroplast genomes (Martin *et al.*, 1998). Early conclusions support the monophyletic hypothesis but these need to be closely scrutinised.

5.2.2. *Evidence against a single event*

The evidence against a single event is based on the following arguments:

1) the likelihood of multiple events
2) the difficulty of constructing a single tree

5.2.2.1. *The Likelihood of Multiple Events.* Given the time frame of endosymbiotic event(s) it is unreasonable to suppose that only one event occurred in the last 1.5 to 2 BY. For this event to happen there had to arise a recognition system which allowed the "invader/ingested organism" to be invested in an endoplasmic reticulum-derived Golgi membrane (which developed into an endosome but not a lyzosome). Following this an accommodation would have been necessary with i) the development of specialized transport proteins (which fine-tuned the import/export needs of the host and symbiont) and ii) the integration of the two genomes in order to regulate cell division of host and symbiont. In the case of the plastid, the symbiotic association has clearly broken down to the extent that the host has subjugated the original symbiont "enslaving" it and reducing it to an organelle devoted to photosynthesis.

Each of these events should be considered in any discussion of early symbiotic events;. but this has rarely been done. Thus it is theoretically possible that each step evolved only once: however the sheer diversity of endosymbiosis in prokaryotes and eukaryotes today would suggest that symbiosis has been a common event for a long time and endosymbiosis has arisen, independently, many times. Even if there was just one first event it should be borne in mind that this still does not exclude a multiple event process, since the sharing of that information, to other eubacteria, could have been very widespread. This could have occurred if the original symbiont came out of symbiosis leaving behind genes for endosymbiosis and for other characters (e.g. light-harvesting proteins) in the host. Loss of autonomy by the symbiont to the host seems to be universal, and it seems clear that three steps are crucial: i) the migration of genes to the host genome, ii) the generation of signal peptides for import of essential organellar proteins, and iii) the loss of unwanted proteins, and structures such as the cell wall. The exchange of DNA/protein can of course be a two way process. Genes passed to the nucleus could then be passed back to other symbionts or passed forward to other chimaeras, as symbiont swapping continued. Thus it seems likely that once the initial endosymbiotic accord was initiated various free-living bacteria would have begun the symbiotic route. As bacteria experimented with various associations, sharing of genes would almost certainly have become widespread. Gene exchange in this way could therefore account for some previously puzzling examples, e.g. the occurrence in *Prochlorococcus* with Chl *a*, Chl *b*, a Chl *c*-like pigment and phycoerythrin (Hess *et al.*, 1996).

Thus by the time of the evolution of the present plant protist phyla it is possible that many alternatives were available, even assuming a monophyletic event.

5.2.2.2. *The Difficulty of Constructing a Single Tree.* The most often used phylogenetic characters are the photosynthetic pigments (and their proteins) which have been used to classify algae from the beginning. There are two possible routes in terms of light-harvesting proteins in a monophyletic hypothesis. These are:

i) **The "ark" hypothesis.** Here all the light-harvesting pigments/chromophores and proteins are present in the first plastid (Bryant, 1992) or,

ii) **Later diversification.** Here an ancestor with a single set of light-harvesting genes, say a phycobilisome system as in cyanelles, forms the first plastid which passes these characteristics on to different phyla, which develop alternative light-harvesting systems such as [Chl a + Chl c] (in Chromophytes) or [Chl a + Chl b] in Chlorophytes (see e.g., Melkonian, 1995).

A major difficulty with this latter approach is accommodating the loss, on the one hand, of useful genes such as the phycobiliproteins and the acquisition, on the other, of genes for the synthesis of Chl *b* or Chl *c* (c_1, c_2 and c_3) and their light-harvesting protein (the CAB/CAC family - see Green and Pichersky, 1996). It has been argued that the evolution of Chl *b* and Chl *c* could have easily occurred more than once by separate random mutations. This is probably true. However the "invention" of a protein on which to anchor these chromophores is seen to be more difficult (Larkum, 1991). It would need to have been derived from the cyanobacterial/rhodoplast line. Recently a member of this CAB/CAC protein family has been found in a red algal chloroplast, in association with photosystem 1 (Wolfe *et al.*, 1994). Furthermore a potential ancestor protein has been found in cyanobacteria (Dolganov *et al.*, 1995). Thus it is possible that new systems did evolve to accommodate Chl *b* and Chl *c*, but even so the loss of a phycobilisome system is a major difficulty. It might be argued that the PBS system was lost because of constraints on thylakoid arrangement and the development of appressed thylakoids (Larkum and Barrett, 1983). However, in this regard it should be noted that Cryptophyta accommodate both systems by internalizing the phycobiliproteins in the intra-thylakoid space (Hiller *et al.*, 1993). This strategy yields the most efficient light-harvesting system. As a result it is very difficult to understand why the green algae have lost all trace of a phycobilisome system, if they did evolve along a route where phycobiliproteins were, at one time, present.

Another difficulty concerns the chloroplast envelope. In the chromophytes there are four envelope membranes. Whatley *et al.*, (1979) first proposed that this could be taken as evidence for the Serial Endosymbiosis Theory (SET), with, in this case, a second serial endosymbiosis. This proposal has been given strong support by the demonstration of a reduced nuclear genome between the outer and inner pair of envelope membranes in both the cryptophytes and the chlorarachniophytes (McFadden *et al.*, 1997). Thus for the chromophytes, SET and a monophyletic hypothesis (in either of its forms) proposes a route for the evolution of plastids with Chl *a* and Chl *c* through two or more hosts. Accordingly the chromophytes could have acquired flagella and varied light-harvesting genes from varied hosts and have now lost all trace of earlier nuclei. However, the Chlorophyta would then present a major difficulty, for they have only two chloroplast envelope membranes, have acquired Chl *b*, lost PBP and have acquired flagella. In contrast, the red algae and the cyanobacteria (from which rhodoplasts are derived), show no evidence for the development or acquisition of flagella and do not possess Chl *b* or Chl *c*, or many of the characteristic carotenoids of green algae or chromophytes.

5.2.3. *Evidence in favor of Multiple Events of Plastid Evolution (Polyphyletic Hypothesis).*

Raven (1970) first made the proposal for a polyphyletic origin of plastids. . The proposal was at first rejected as unlikely for lack of evidence (Lee, 1972) but was later given apparent strong support by the discovery of *Prochloron*, *Prochlorothrix hollandica* and *Prochlorococcus marinus* all of which possess Chl *b* in addition to Chl *a*. (see Section 3.2)

Much attention was given to the question of what type of protein would be involved in binding Chl *b*, and where it occurs, Chl *c*, in prochlorophytes. However, as discussed in Section 3.2. this is not informative on the question of an ancestral position for these prochlorophytes and an independent lineage to green chloroplasts.

Based on SSU-rRNA phylogenetic trees (and other gene trees) (see for example Medlin and Bhattacharya, 1995) it has been proposed that the prochlorophytes occur on different branches within the cyanobacterial radiation (Palenik and Haselkorn, 1992; Urbach *et al.*, 1992). These analyses can be criticized in that they do not address a number of problems in phylogenetic analysis such as covarion evolution, sites-free-to-vary, and different rates of change (see section 2.1). Attempts to take such things into account result in "star pattern" trees, implying that there is insufficient information in the data sets to predict lineages accurately (Scarramuzzi *et al.*, 1995; Howe *et al.*, 1995).

This finding provides encouragement for further searches for novel prokaryotic oxygenic autotrophs.

The separate evolution of a green plastid from an autonomous prokaryote is still an attractive proposal. And an ancestor within the prochlorophyte family is the most logical source. Such an ancestor might have Chl *a*, Chl *b* and Chl *c* (or *c*-like pigment) bound to a CAB/CAC protein. It is just possible that it might also carry a phycobiliprotein, such as the phycoerythrin found in *Prochlorococcus* (Hess *et al.*,

1996). Upon entering into symbiosis with a flagellate host the association would have formed an organism which today would be placed in the Prymnesiophyta (Micromonadophyta). This is a vast group of flagellate algae (Sym and Pienaar, 1993) that have plastids which contain Chl *a* and Chl *b* and sometimes DVPP. These algae have primitive features, especially in having scales instead of a cellulose wall. They also lack symmetry. and have often been proposed as coming close to the root of the green algal lineage, according to morphological and fine-structural evidence (Stewart and Mattox, 1978; Moestrup, 1991) and more recently from molecular phylogenetic analysis (Daugbjerg et al., 1995).

It seems clear that evolution within the green algal line has led to the disappearance of DVPP and the close appression of thylakoids (Larkum and Barrett, 1983; Larkum 1991).

It is possible that another line of evolution led from this ancestral stock (possessing Chl *a* + Chl *b* + Chl *c* and PBP) to the chromophytes. Two branches on this line could be proposed. On the one hand there is the possibility for the development of a heterokont (stramenopile) line, presumably by at least two serial endosymbioses (and the loss of phycobiliproteins). On the other hand, evolution of a cryptophyte line could be proposed, again involving at least two serial endosymbioses. However the proposal that the original plastid of cryptophytes is a red algal plastid has the difficulty that the βchain of the phycobiliprotein is unrelated to that in the red algae (Larkum and Howe, 1997) and the thylakoids contain Chl c_2. In fact the evidence merely shows is that the primary host in the cryptophyte line had red algal genomic characters (Douglas *et al.*, 1991).

5.2.3.1. *Difficulties with a Polyphyletic Origin of Plastids*

5.2.3.1.1 *The inverted repeat.* Inverted repeats occur in the plastid genomes of most groups of plant protists, so far studied. It is not present in *Chlorella vulgaris* (Wakasugi *et al.*, 1997), which is in the Chlorophyta and possibly closest to the higher plant plastids (which do mostly possess an inverted repeat). The inverted repeat has been taken by some to imply a single origin of plastids. However, it is possible that the arrangement is analogous and not homologous, resulting from a natural tendency in the stabilizing mechanism of a reduced, circular DNA chromosome. The lack of such a structure in *C. vulgaris* is a clear indication a cautious approach should be taken.

5.2.3.1.2. *The presence of a member of the CAB/CAC family in rhodophytes (Wolff et al., 1994).* This small protein has been found in Photosystem 1 in *Porphyridium* where it plays a minor role, presumably, in light harvesting. It does not bind Chl *b* (only Chl *a*). Thus it is quite possible that it too is an analogous structure independently derived from a cyanobacterial ancestor. It is also possible that it is a result of gene transfer either be some vector or as a result of the exchanging of genes during the early experimental development of symbiosis (Sections 1 and 6). Until more is known of the ancestry of the red algal host cell little can be said on these options.

5.2.4. *Inconsistent characters*

There are few chloroplasts whose parentage is known beyond question. It seems reasonably certain now that the chloroplast of *Chlorarachnion* is descended form a serial endosymbiosis of a "chlorophyte" chloroplast. There are no clear ancestors for the plastids of Chlorophyta, Chromophyta (*sensu lato),* Cryptophyta or Rhodophyta. In the case of the Rhodophyta there are clear links between the plastid and the cyanobacteria. However the presence in rhodoplasts of an extra subunit in the phycoerythrin (γ-subunit) of no clear origin, is clearly indicative of evolutionary distance from the cyanobacteria. The micromonadophyte prasinophytes have a plastid which possesses Chl *a* + Chl *b* and a Chl *c*-like pigment (MgDVP). It could be argued that this pigment complement could have been inherited from a prochlorophyte. However the light-harvesting protein of micromonadophytes is of the CAB/CAC family and does not resemble the LHC of prochlorophytes. There is no direct ancestor of the chromophyte plastid. It has been argued that the cryptophyte chloroplast most nearly resembles a rhodoplast, so that a serial symbiosis from a rhodoplast-bearing host could be considered. However there are considerable differences in the β subunit of the phycobiliproteins between Rhodophyta and Cryptophyta. Moreover the placement of the phycobilins is very different (Larkum and Barrett, 1983) and the possession by cryptophytes of Chl c_2 is a further difference wanting of explanation (see Section 5.2.3).

Thus the hypothesis of polyphyletic origin of plastids by simple endocytosis of free-living photoautotrophs does not easily explain the facts.

It seems therefore that it is necessary to refine our approaches to plastid evolution. The next section puts forwarded a proposal based on the ideas developed in section 5.2.2.1.

6. A Working Scheme for the Evolution of Algal Protists

Previous schemes to assign plastids evolution to either monophyletic or polyphyletic hypotheses suffer from the conflict of evidence as outlined above. This quandary has forced both sides of the debate to relinquish long-held beliefs. For example Stiller and Davison (1997) have recently suggested that the double membrane envelope of plastids of Chlorophyta and Rhodophyta is not a primary character, but secondarily derived. Larkum (1992) put forward a scheme for a polyphyletic origin, which now has to be modified with the discovery of phycobiliproteins in two prochlorophytes and in *Acaryochloris*. The scheme in Fig. 3 attempts to resolve the situation according to the evidence and the ideas expressed earlier in this article. Following Larkum (1990), it is proposed that the earliest functional chlorophylls were Chl *c*-like pigments, followed by Chl *a*. Chl *b* is seen as the next development, followed by phycobiliproteins. However as the oxygenic prokaryotes (Chloroxybacteria) entered into symbiosis with early eukaryotes, a sharing of genetic information occurred which sent back new genetic characters to the Chloroxybacteria. In this way the Chl *b* line may have acquired phycobiliproteins and the cyanobacterial line may have acquired CAB/CAC genes. As the symbioses developed to the stage of complete loss of autonomy, genetic separation became stronger. The set of events described above might be called the **"promiscuous hypothesis"**. At the plant protist level there then developed the possibility for

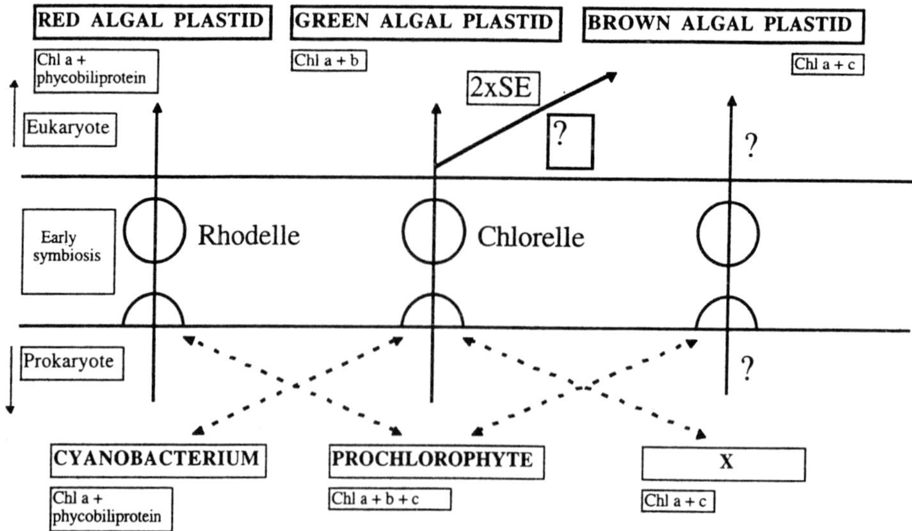

Figure 3. A scheme for the early formation of endosymbiotic events leading to the eventual formation of plastids. It is envisaged that in early experiments in endosymbiosis genes may have been shared by a number of reversible symbiotic events. 2 x SE, refers to 2 serial endosymbioses.

secondary symbioses. Cryptophytes and chlorarachniophytes are certainly examples of this. Many other chromophytes and the Euglenophyta are also probably the result of a similar process. In all these cases there is the possibility of sharing photosynthetic characters between the host nuclei. In the case of Rhodophyta and Chlorophyta the stages in the development of the plastid remains unresolved, but until further evidence is obtained to the contrary it may be assumed that these are the result of a single endosymbiosis. If this is the case then a multiple origin of plastids is still the simplest hypothesis.

7. Acknowledgements

P.J. Lockhart is thanked for stimulating discussions and for making critical comments on a draft of this article. Dr. Yves Van de Peer is thanked for providing Figure 1 and Dr. C.J. Howe for providing Figure 2.

8. References

Bhattacharya, D., and L. Medlin. (1995) *J. Phycol.* **31**, 489-98.
Blankenship, R. E. (1992) Origin and early evolution of photosynthesis. *Photosynth. Res* **33**, 91-111.
Bryant, D. A. (1992). *Current Biology* **2**, 240-242.
Bui, E. T. N., P. J. Bradley, and P. J. Johnson. (1996) *Proc. Nat. Acad. Sci. (US)* **93**, 9651-56.
Burger-Wiersma, T., L. J. Stal, and I Mur. (1989) *Intern. J. Syst. Bact.* **39**, 250-257.
Canfield, D. E., and A Teske. (1996) *Nature* **382**, 127-32.
Cavaler-Smith, T. (1992) *Biosystems* **28**, 91-106.
Chisholm, S. W., S. J. Frenkel, R. Goericke, R. J. Olson, B. Plenik, J. B. Waterbury, L. West-Johnson, and E. R. Zetter. (1992) *Arch. Microbiol.* **157**, 297-300.
Daugbjerg, N., O. Moestrup, and P. Arctander. (1995) *Phycological Research* **43**, 203-13.
Dolganov, N. A., D. Bhaya, and A. R. Grossman. (1995) *Proc. Nat. Acad. Sci. (US)* **92**, 636-40.
Douglas, S. E., and C. A. Murphy. (1994) *J. Phycol.* **30**, 329-40.
Dougls, S. E., C. A. Murphy, D. F. Spencer, and M. W. Gray. (1991) *Nature (London)* **350**, 148-51.
Felsenstein, J. (1988) *Annu. Rev. Genet.* **22**, 521-65.
Fitch, W. M., and E. Markowitch. (1970) *Biochem. Genet.* **4**, 579-93.
Goericke, R., and Repeta. (1992) *Limnol. Oceanogr.* **37**, 425-33.
Green, B. R., and D. G. Durnford. (1996) *Ann. Rev. Pl. Physiol.Mol.Biol.* **47**, 685-714.
Gupta, R. S., and B. G. Golding. (1996) *Trends in Biochemical Science* **21**, 166-71.
Hess, W. R., F. Partensky, G. Van der Staay, J. M. Garcia-Fernandez, T. Booner, and D. Vaulot. (1996) *Proc. Nat. Acad. Sci. (US)* **93**, 11126-30.
Howe, C. J., A. C. Barbrook, and P. J. Laockhart. (1995).In: P. Matthis (ed.) *Photosynthesis: from light to biosphere*, Kluwer Acad. Pub., Dordrecht, Vol. 3, pp. 939-44.
Koyama, M., W. Helbert, T. Imai, J. Sugiyama, and Henrissat. (1997) *Proceedings of the National Academy of Sciences of the United States of America* **94**, 9091-95.
Kroth Pancic, P. G., U. Freier, H. Strotmann, and K V. Kowallik. (1995) *Journal of Phycology* **31**, 962-69.
La Roche, J., G. W. M. Van der Staay, A. Ducret, R Aedersold, R. Li, S. Golden, R. G. Hiller, P. M. Wrench, A. W. D. Larkum, and B. R Green. (1996) *Proc. Nat. Acad. Sci. (US)* **93**, 15244-48.
Larkum, A. W. D. (1991).In: H. Scheer (ed.) *Chlorophylls*, CRC Press, Boca Raton, Vol. 367-83.
Larkum, A. W. D. (1992) In: N. Murata (*ed.)* Kluwer Acad. Pub., Dordrecht, Vol. 3, pp. 475-82.
Larkum, A. W. D., and J Barrett. (1983) *Adv. Bot. Res* **10**, 1-221.
Larkum, A. W. D., C. Scaramuzzi, G. C. Cox, R. G. Hiller, and A. G. Turner. (1994) *Proc. Nat. Acad. Sci. (US)* **91**, 679-83.
Larkum, T., and C. J. Howe. (1997) *Adv. Bot. Res.* **27**, 258-330.
Lee, R. E. (1972) *Nature* **237**, 44-46.

Lewin, R. A. (1977) *Phycologia* **16**, 217.

Liaud, M.-F., U. Brandt, and Cerff. R. (1995) *Plant Molecular Biology* **28**, 313-25.

Liaud, M. F., U. Brandt, M. Scherzinger, and R. Cerff. (1977) *Journal of Molecular Evolution* **44**, S28-S37.

Lockhart, P. J., C. J Howe, D. A. Bryant, T. J. Beanland, and A. W. D. Larkum. (1992) *J. Mol. Evol.* **34**, 153-62.

Lockhart, P. J., A. W. D. Larkum, M. A. Steel, P. J. Waddell, and D. Penny. (1996b) *Proc.Nat. Acad. Sci. (US)* **93**, 1930-1934.

Lockhart, P. J, D. Penny, and A. Meyer. (1995) *J. Mol. Evol.* **41**, 666-74.

Lockhart, P. J., M. A. Steel, M. D. Hendy, and D. Penny. (1996a) *Mol. Biol. Evol.* **11**, 605-12.

Lockhart, P. J., M. A. Steel, M. D. Hendy, and D Penny. (1994). *Mol. Biol. Evol.* **11**, 605-12.

Martin, W. and M. Muller. (1998) Nature **392**, 37-41

Martin, W., B. Stoebe, V. Goremykin, S. Hansmann, M. Hasegawa and K.V. Kowallik (1998) Nature **393**, 162-165.

McFadden, G. I., P. R. Gilson, S. E. Douglas, T Hofmann C. J. B. Cavalier-Smith, and U.-G. Maier. (1997) *Trends in Genetics* **13**, 46-49.

Melkonian, M. (1995) *Progress in Botany* **57**, 281-311.

Miyashita, H., H. Ikemoto, N. Kurano, K Adachi, M. Chihara, and S Miyachi. (1996) *Nature* **383**, 402.

Moestrup, O. (1991) *J. Phycol.* **27**, 119-33.

Mojzsis, S. J., G. Arrhenius, K. D. McKeegean, T. M. Harrison, A. P. Nutman, and C. R. L. Friend. (1996) *Nature* **384**, 55-59.

Nelissen, B, Y. Van De Peer, A. Wilmotte, and R. De Wachter. (1995) *Molecular Biology & Evolution* **12** , 1166-73.

Palenik. B., and R. Haselkorn. (1992) *Nature (London)* **355**, 265-67.

Palmer, J. D., and C. F. Delwiche. (1996) *Proc. Nat. Acad. Sci. (US)* **93**, 7432-35.

Perasso, R., A. Baroin, H. Q. Liang, J. P. Bachellerie, and A. adeloutte. (1989*Nature (London)* **339**, 142-44.

Philippe, H., G. Lecointre, H.L.V Le, and H. Le Guyader. (1997) Mol. Biol Evol. 13, 1174-1186.

Raven, P. H. (1970) *Science* **169**, 641-46.

Scaramuzzi, C, R. G. Hiller, P Wrench, P. J. Lockhart, and A. W. D. Larkum. (1995) In: P. Matthis (ed.) *Photosynthesis: from light to biosphere*, Kluwer Acad. Pub., Dordrecht, Vol. 3, pp. 939-44.

Schopf, J. W. (1993) *Science (Washington D C)* **260**, 640-646.

Stiller, J, and I. R. Davison. (1997) *Proc. Nat. Acad. Sci.(U.S.)* **94**, 4520-4525.

Sym, S D., and R N. Pienaar. (1993) In: F. E. Round and D. J. Chapman (eds.) *Progress in Phycological Research*, Biopress Ltd., Bristol, England, Vol. Vol. 9, pp. 281-376.

Turner, S., T. Burger-Wiersma, S. J. Giovannoni, L. R. Mur, and N. R. Pace. (1989) *Nature (London)* **337**, 380-382.

Urbach, E., D. L. Robertson, and S. W. Chisholm. (1992) *Nature (London)* **355**, 267-70.

Van de Peer, Y., S. A. Rensing, U.-G. Maier, and R. De Wachter. (1996) *Proc. Nat. Acad. Sci. (US)* **93**, 7732-36.

Van der Staay, G. W. M, N. Yurkova, and B. R. Green. (1997) Plant Mol. Biol. **36**, 709-716.

Vermaas, W. F. J. (1994) *Photosyn. Res.* **41**, 285-94.

Waddell, P. J., D. Penny, and T. Moore. (1997) *Mol. Phylogenet. Evol.* **8**, 33-50.

Wakasugi, T., T. Nagai, M. Kapoor, M. Sugita, M Ito, S. Ito, J. Tsudsuki, K. Nakashima, T. Tsudsuki, Y. Susuki, A. Hamada, T. Ohta, A. Inamura, K. Yoshinaga, and M. Sugiura. (1997) *Proc. Nat. Acad. Sci. (US)* **94**, 5967-72.

Whatley, J. M, P. John, and F. R. Whatley. (1979) *Proc. R.Soc.Lond. B.* **204**, 165-87.

Woese, C. R. (1987) *Microbiological Rev.* **51**, 221-71.

Wolfe, G. R., F. X. Cunningham, D. Durnford, B. R. Green, and E. Gantt. (1994) *Nature* **367**, 566-68.

Biodata of **Hakobu Nakamura** author of ***"From Bacteria to Protista"*** and ***"Salt Sensitivity of Cells."***

Dr. Hakobu Nakamura is a Professor in the Biological Institute of Konan University in Kobe, Japan. He is the author of many articles and has written and translated several scientific books (English - Japanese). Dr. Nakamura's main interests (already more than 40 years) in the origin and evolution of the cellular membranes, and in tying up molecular biology with the cellular morphogenesis. He received his Ph.D. in (1961) from the Kyoto University.
Dr. Nakamura is also active in several public institutes and serves as an active member in several Japanese and International Societies. He serves also as an advisor for the Kobe Science Museum.
E-Mail: **<nakamura@konan-u.ac.jp>**

J. Seckbach (ed.), Enigmatic Microorganisms and Life in Extreme Environments, 49–62.

FROM BACTERIA TO PROTISTA

HAKOBU NAKAMURA
Biological Institute, Faculty of Science,
Konan University Higashinada, Kobe 658, Japan

1. Introduction

Life can be defined as a material system that fills the following three conditions simultaneously; (i) it makes and maintains by itself the system that has a specific attribute to be termed so-called "life", (ii) it should replicate and reproduce by itself descendants, and (iii) it can evolve with time. The first such organized system, proto-cell, appeared in primitive sea about four billion years ago. Proto-cells are Prokaryotic in cellular organization and belong to the bacterial world. A remarkable feature of prokaryotic cells is that the nucleoid area containing the genome is not surrounded by a nuclear membrane. In contrast, the genome in eukaryotic cells is located in a nucleus, (mitochondria and chloroplasts possess also DNAs), surrounded by generally two (an inner and outer) nuclear membranes.

The evolution of cells from Prokaryota to Eukaryota is considered to have occurred based on fossil records (Schopf, 1974) about 1.5 billion years ago. The path of cellular evolution that streamed from Prokaryota to Eukaryota namely from bacteria to Protista, is a big river. Protista, the third taxonomic group, that contain Bacteriocota, Eukaryomycota, unicellular algae, Protozoa, and Spongiaria, was first established by Haeckel (1866, 1868). Later, Whittaker (1969) assigned all unicellular Eukaryota to the Protista and all Prokaryota to Monera in his five taxonomic world theory. The taxonomic worlds have been divided into three (Stainer et al. 1970, and Curtis, 1968), four (Copeland, 1938), and five (Margulis, 1974). Here, Whittaker's concept will be discussed.

It is well known that bacterial mRNA has a half-life from 1.8 min whereas eukaryotic mRNA has a half-life of several hours to several days. Therefore, the bacterial turnover of synthetic metabolism is very rapid as compared to the eukaryotic turnover. This suggests that bacteria cells can rapidly respond to environmental changes and are highly adaptable to physiological modifications. Further, bacterial cells can divide very rapidly as compared to eukaryotic cells. In other words, bacterial DNAs can replicate and produce high rates of spontaneous mutations. For example, when a bacterial culture growing exponentially in synthetic medium is shifted to nutrient broth, the growth rates doubles three times per hour. RNA synthesis soon responds and begins to increase rapidly. Syntheses of protein and DNA follow, and finally the rate of cell division increases. On the other hand, when a bacterial culture growing exponentially in nutrient broth at a rate of three doubling per hour is washed by centrifugation with phosphate buffer, and put in synthetic medium, the rate decreases to only 1.2 doublings per hour. With this nutritional shift-down, RNA synthesis initially ceases, then begins gradually to increase with a long lag time. Protein synthesis also responds relatively late to the shift-down, follow by DNA synthesis and cell division. For example, *Escherichia coli* cells have a generation time of 20 min, while *Saccharomyces cerevisiae* have a generation time of 120 min in broth media. The *E. coli* population can reach more than 10^9 cells per ml, while the *S. cerevisiae* population more than 10^8 cells per ml in broth media. Spontaneous mutation occurs independently of environmental conditions, and is considered an error of base pairing during DNA replication. Spontaneous mutations accumulate in the genome with time and eventually produce genetic changes in the phenotype of the organism. According to neo-Darwinism, biological evolution is designed through natural selection under differing environmental conditions. This seems to indicate that bacteria are populationally, metabolically, and genetically more adaptable to changes in the environment than protistan cells. This explains why bacterial species have survived and actively evolved for about four billion years on the earth.

2. Evolution in the Bacterial World

Bacteria are unicellular, and have a much longer history (of about 4 billion years), than eukaryotes (about 1.5 billion years). Protista are also unicellular, though grouped in the eukaryotic world, and have a unique history that begins five hundred million years before the origin of multiple eukaryotic organisms. Therefore, it is apparent that certain fundamental steps exist in the multi-cellularization of protists.

The most remarkable evolution in the prokaryotic world is in metabolisms. Metabolisms are usually classified into anabolism, the synthesis of more complex compounds, directly or indirectly, and catabolism, the breaking-down of compounds into simpler forms. Catabolism generally accompanies the production of energy, coinciding with energy metabolism, while anabolism is an endothermic reaction, and requires a certain level of energy which is supplied by catabolism. Therefore, both are coupled in the cellular metabolisms.

It has been supposed that proto-cells obtained living energy from three sources: (i) organic compounds dissolved in the primitive sea, which were synthesized during the chemical evolution, (ii) sunlight containing a large amount of ultraviolet light, which is harmful to the cells, and (iii) membrane potential produced by the concentration gradient of protons and other ions across the cell membrane, of which phospholipid as its skeleton does not penetrate the ions.

As the primitive atmosphere contained little oxygen gas, the cells had to catabolize organic compounds incorporated from seawater through anaerobic metabolism, that is fermentation. According to Horowitz (1945) and Wittaker (1969), biological evolution is a history of food acquisition. Therefore, an organism whose food source is exhausted must acquire a new organic compound as nutrient. In some cases, a new reaction was formed to catabolize such a compound through mutation and selection. Present-day metabolism evolved over time through selections for many kinds of spontaneous mutants.

Horowitz (1945) proposed that the metabolic sequence of reactions was established in the direction reverse to the present-day flow. Such a step-by-step elongation would result from mutations and selections during genetic and cellular evolution.

2.1 FERMENTATION

The oldest fermentation process in anaerobic bacteria is believed to be the Embden-Myerhof (EM) pathway, which converts glucose to pyruvate (Fig. 1). However, we have proposed that the reactions from glyceraldeyde 3-phosphate to pyruvate, occurring in the latter half of the EM pathway, form the original fermentation metabolism (Nakamura, 1983). The reasoning for this is : (i) all the ATP- productive reactions, the so-called substrate-level phosphorylation, take place in the latter half of the EM pathway, (ii) several fermentative flows, including the EM, Entner-Doudroff, (oxidative) pentose-phosphate, and phosphoketolase pathways, join into the flow from glyceraldehyde to pyruvate, and further (iii) the fermentative metabolisms support the Horowitz hypothesis. Therefore, we have referred to the original reaction as primitive fermentation. Thus, we can assume that the primitive fermentation was established step-wise by elongation of an enzyme arrangement from pyruvate kinase to glyceraldehyde 3-phosphate dehydrogenase. In other words, the pyruvate kinase that catabolizes the conversion of phosphoenolpyruvate to pyruvate accompanied by the generation of one molar of ATP, is the first enzyme in the production of ATP in the fermentation.

2.2. PHOTOSYNTHESIS

The second proto-type metabolism for ATP production is a phosphorylation that couples with proton movement across the plasma membrane. The (phospho)lipid bilayer as the membrane skeleton is not permeable to protons and thus establishes a proton gradient between both sides of the lipid bilayer. On the other hand, ATP synthase, F_0F_1-ATPase or H^+-ATPase, integrated in the plasma membrane can catabolize the reaction ADP + Pi → ATP, in which the F_0 component is the proton channel through the membrane and the F_1 component is the active site of ATP synthesis. When protons pass through the F_0 channel along the gradient, the energy flow activates F_1 to synthesize ATP. Thus, if proto-cells were to contain a proteinous catalyzer of ATP synthesis in the plasma membrane, they could produce ATP. The halophilic bacterium *Halobacterium halobium* has a bacteriorhodopsin in the purple (plasma) membrane, which works as proton pump in bright light. The bacterial ATP synthase can actively synthesize ATP by using bacteriorhodopsin as a generator of the proton gradient. However, *H. halobium* cells can not synthesize sugar because they lack the so-called reduced pentose phosphate cycle.

More evolved photosynthetic bacteria, such as green sulfur (Chlorobiaceae), purple sulfur (Chromatiaceae), and purple non-sulfur (Rhodospirilliaceae) bacteria, have both the basic metabolisms of the photosystem (light reaction) and reduced pentose phosphate cycle (dark reaction) for the photosynthesis that synthesizes sugars from carbon dioxide and reducing agents like hydrogen sulfide in light. Interestingly, the reduced pentose phosphate cycle is a derivative of fermentation, and the cycle reversed, i.e., the oxidative pentose phosphate cycle, is a metabolism to synthesize pentoses, ribose for RNA synthesis

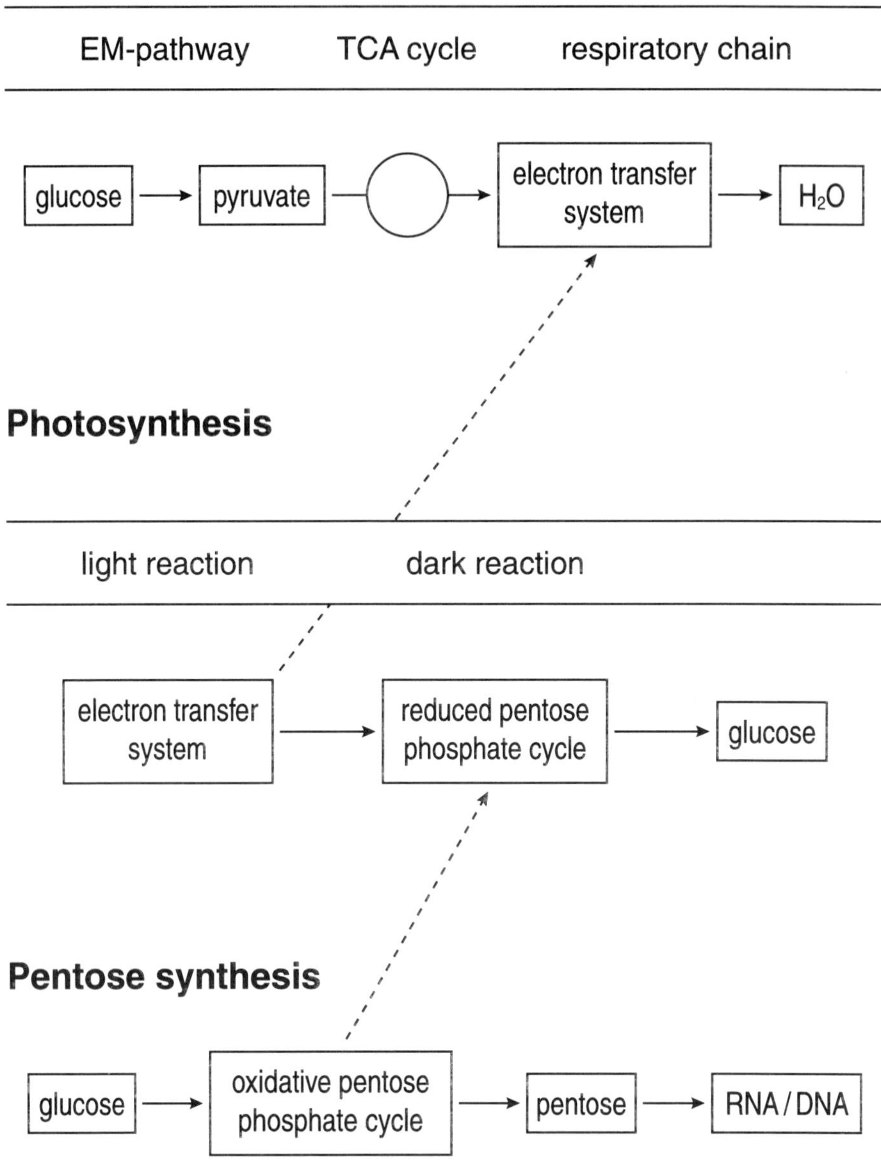

Fig. 1. Assumptive evolution of energy metabolism. Fermentation contains primitive ATP-productive metabolism, referred to as EM-pathway, and respiration consists of EM-pathway - TCA cycle, a fermentative organic acid metabolism, - respiratory chain derived from photosystem in photosynthesis. Reduced pentose phosphate cycle in photosynthesis is a reverse of oxidative pentose phosphate cycle that produces ribose for RNA and deoxyribose for DNA.Chemosynthesis consists of an electron transfer system as chemical reaction (no photosystem) and reduced pentose phosphate cycle.

and deoxyribose for DNA synthesis. Therefore, it is apparent that the dark reaction is a reversal of the oxidative pentose phosphate cycle, one of the most important metabolisms in living cells. The light reaction is an electron transfer system that functions as proton-pumping machinery in the presence of the lipid bilayer (bacterial cell membrane). Thus, this photosystem is considered to have evolved from more primitive machinery, such as bacteriorhodopsin, as a proton pump in halobacteria.

Photosynthetic bacteria have a common photosystem that transfers electrons through ubiquinone (**Q**), cytochrome **b** (b), and cytochrome **c** (c), and uses hydrogen sulfide and others as hydrogen donors for reduction of carbon dioxide. Furthermore, cyanobacteria have a photosystem with the electron transfer system of photosynthetic bacteria, **Q** → **b** → **c**, and use water as a hydrogen donor for reduction of carbon dioxide. Thus, cyanobacteria produce (similar to green plants) oxygen gas as a by-product, while photosynthetic bacteria do not.

It is an important fact that cyanobacterial photosynthesis has been inherited in higher plants, although it was developed extensively in terms of cellular organization, and that the reduced pentose phosphate cycle was inherited from the photosynthetic bacteria to higher plants via the cyanobacteria. We can conclude here that the photosynthesis in cyanobacteria evolved to a point where it could not improve further. In the course of evolution, the cyanobacteria succeeded in employing carbon dioxide, water and sunlight as materials for photosynthesis, and these oxygen-evolutionary (photosynthetic) systems spread explosively. In fact, fossils of such organisms have been discovered all around the earth. In particular, the stromatolites are known to exist in very deep strata in various places.

2.3. RESPIRATION

Respiration consists of three large metabolisms: (i) the EM-pathway, (ii) the TCA cycle, and (iii) the respiratory chain. It is of evolutionary interest that oxygen-respiration involves the fermentative EM-pathway. This is because the end product of fermentation, pyruvate, is used in fermentative organic acid metabolism, the TCA (tricarboxylic acid) cycle. Even in the bacterial world, fermentation is branched into and produces various end products; alcoholic f., lactic acid f., acetic acid f., butyric acid f., and so on. The large metabolism connecting the EM-pathway and TCA cycle is a kind of organic acid fermentation. Therefore, it is reasonable to assume that organic acid fermentation was established before respiration. We have proposed that the TCA cycle is the joining of propionate fermentation and glutamate fermentation, each of which operates independently in some bacterial species. Furthermore, many species of bacteria, protista, and higher plants have an incomplete TCA cycle that is deficient in α-ketoglutarate dehydrogenase, which catalyzes the reaction of α-ketoglutarate to succinyl CoA, referred to as glyoxylate cycle. However, in *Propionibacterium acidipropionici* cells, the metabolisms are joined: that is, in this strictly anaerobic bacterium, the TCA cycle is complete. Therefore, it is apparent that present-day respiration evolved through the connecting of the fermentative metabolisms with an electron transfer system, the respiratory chain.

Metabolism is the flow of reaction sequences, some of which are separated, depending on physiological and genetically conditions. The evolution of respiration had to be dynamically generated by a new environment, such as the oxygenic atmosphere. In principal, metabolism is as dynamic as water flow. In fact, cellular metabolisms, especially in bacteria, are quite changeable physiologically and genetically. According to our experiments, for example, purple non-sulfur bacteria can actively grow by fermentation under anaerobic conditions, by respiration under aerobic conditions, and by photosynthesis under anaerobic conditions in the presence of light.

The respiratory chain is an electron transfer system with the sequence of carriers, **Q** → **b** → **c**. (i) As the oxygenic atmosphere is considered to have been produced by oxygen-evolutionary photosynthesis, and (ii) as the respiratory chain is considered to be inherited from the photosystem, respiration must have been established after fixation of fermentation and photosynthesis. It is an important fact that bacterial cells have accumulated metabolisms in the course of evolution, and thus can actively adapt to various changes in the environment.

3. Prokaryota to Protista

The enzymes of a cell are located in specific structures. Some are integrated in the plasma and intracellular membranes, and others are embedded in the cytoplasm matrix, mitochondrion matrix, chloroplast stroma, and other matrices. However, these locations are functional and, fermentative enzymes are restricted to the matrices, while electronic and proton carriers are embedded in the membranes. Therefore, for example, the EM-pathway in respiration occurs in the cytoplasm matrix while the respiratory chain is in the plasma membrane of bacterial cells and the cristae in mitochondrion. Further, TCA cycle enzymes are located on the protoplasmic surface of the bacterial plasma membrane and in the mitochondria matrices. Location contributes to the smooth flow of intermediates and electrons along the metabolic sequences of reactions. Another significant point is that the membranes compartmentalize the metabolisms and thus the membranous organelles contain specific metabolism(s), although non-membranous organelles such as ribosome and glycosome also contain specific metabolisms. Therefore, the membrane plays a dual role as (i) a border to separate the inside from the outside and (ii) a barrier to ions and electrons. The former in particular developed over the metabolic evolution from prokaryotic to eukaryotic cells.

Prokaryotic and eukaryotic cells are discriminated by the absence and presence of a nuclear membrane around the genome; the naked genome structure being referred to as a nucleoid and membranous genome structure a nucleus. Generally, there are two nuclear membranes, an inner and outer, but there can be one to four depending on phylogeny. Several kinds of intracellular membranes have been observed to develop from the plasma membranes; non-photosynthetic bacteria differentiate mesosomes, photosynthetic bacteria chromatophores, and cyanobacteria thylakoids, all having respiratory and photosynthetic activities. In cyanobacterial cells of some species, thylakoid membranes encircle the nucleoid as shown in Fig. 2. Jensen (1994) emphasizes that the encircling of the nucleoid

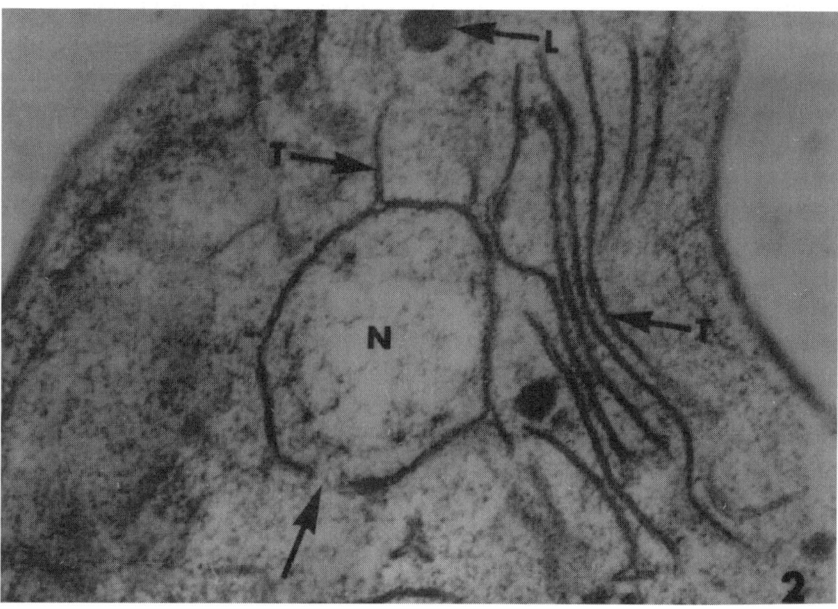

Fig. 2. Encircling of DNA area (nucleoid = N) by thylakoid membrane(T) in cyanobacterial (*Oscillatoria, sp.*) cell. This is considered to be an ancestral nuclear formation (Jensen, 1994). (This author thanks Kluwer publication for permission to use this photo).

is an indication of eukaryogenesis in the cyanobacterial cells. As the nuclear structure in present-day eukaryotic cells is quite stable, except for the pro-metaphase of cell division, the formation of the nuclear membrane must be genetically fixed and expressed dependent on the phase of the cell cycle.

Mitochondrion and chloroplast appear in some phylogenic lines but not others. These DNA-organelles are not specific to the Eukaryota. Furthermore, the metabolisms containing these organelles change according to species. For example, chloroplasts contain various tools for synthetic metabolisms such as purines, pyrimidines, all kinds of fatty acids, the majority of amino acids, and a nitrite reduction system in addition to a photosynthetic system. Alberts et al. (1994) have stated that the importance of these concomitant metabolisms in chloroplasts is beyond photosynthesis. On the other hand, in animal cells, the metabolisms are performed in the cytoplasmic matrix, showing that the compartmentalization of the same metabolism differs between animal and plant cells.

The following hypothesis, the membrane evolution theory, on eukarygenesis was first proposed by us 1975 as an alternative to the symbiotic hypothesis proposed by L. Margulis in 1970. Fig. 3 shows the evolutionary process of membranes. We consider the ancestral Prokaryota in eukarygenesis to be evolved cyanobacterium, which performs metabolic processes for oxygen-evolved photosynthesis with a pathway identical to that of chloroplasts and for oxygen respiration with a pathway identical to that of mitochondrion. The cyanobacterium, generally, has a larger cell and genome size than *Cyanidium*, which belongs to primitive red algal genera (Seckbach, 1994). Such a probable candidate cyanobacterium is *Chlorogloea fritsii*, a species of Cyanophyta, which possesses the respiratory metabolism consisting of the EM-pathway - TCA cycle - respiratory chain, as well as a photosynthetic metabolism consisting of a complete set of photosystems (I and II) and a reduced pentose phosphate cycle.

The cyanobacterial DNA connects with the plasma membrane at several sites. In addition, a specific site on the membrane has been shown to contain the machinery for DNA replication. It is well known that even in the eukaryotic cell, the inner membranes of the nucleus, mitochondrion, and chloroplast connect with their DNAs. The DNA of the ancestral cyanobacterial cell was divided into at least three:(i) a small number of genes responsible for the respiratory metabolism, (ii) a small number of genes responsible for the photosynthetic metabolism, and (iii) almost all the genes of the genome. DNA fragmentation can also be seen in extant prokaryotic cells. Furthermore, the genetic rearrangement that often accompanies cellular evolution is a result of the break-reunion of genomic DNA. For example, episomes and transposones move in and out of the chromosome. These three types of cyanobacterial DNA were wrapped by the membranes and formed the ancestral mitochondrion, chloroplast, and nucleus, and were reorganized as in present-day eukaryotic cells. Seckbach (1994) has proposed that a species, *Cyanidioschyzon merolae*, of the primitive red algae bridges the Prokaryota (cyanobacterium) with Eukaryota.

Animal and fungal cells were derived by the spontaneous elimination of chloroplasts in a certain plant cell. This elimination may have been a kind of mutation in the sense of a natural genomic alteration. This is apparent in *Chlorella* and *Euglena* cells, for example, a newly defined species, *Entosiphon sp.*, usually referred to as color-less

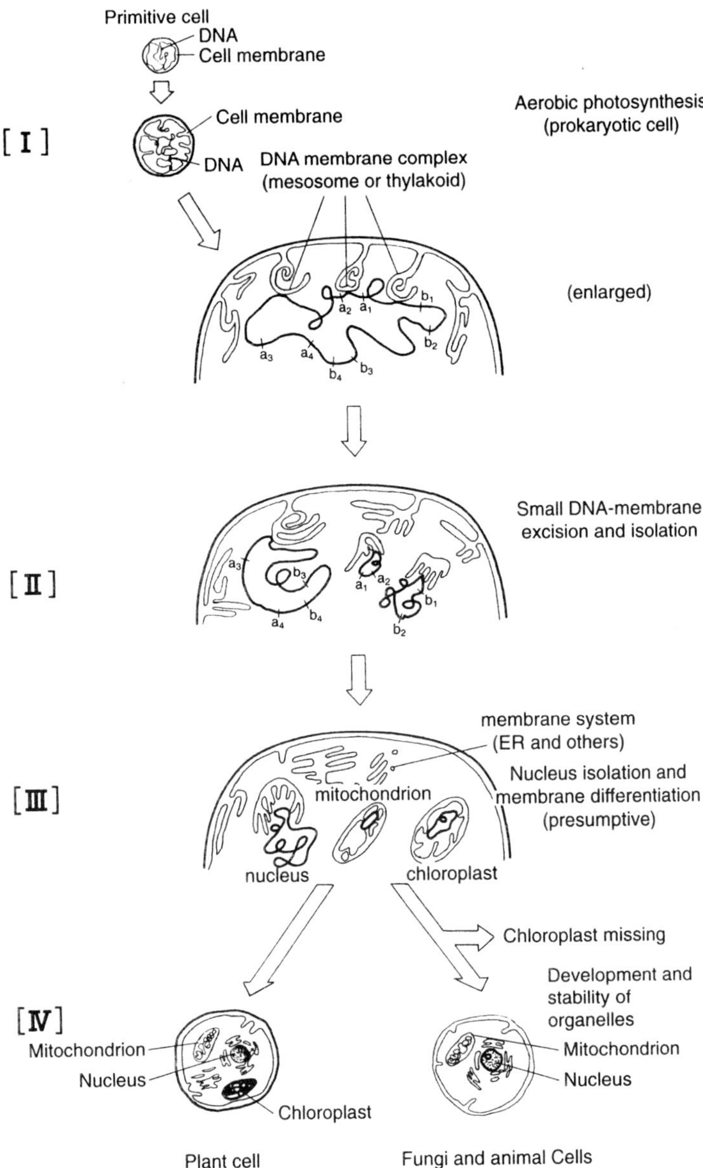

Fig. 3. Membrane evolution theory for eukaryogenesis. Ancestral cell was cyanobacterium and DNA was divided into three parts for mitochondrion, chloroplast, and nucleus with the plasma membrane. Fungi and animal cells are resulted from a spontaneous loss of chloroplast in the plant cell (Nakamura, 1975).

Euglena, seems to be a spontaneously chloroplast-lost mutant of *Euglena*. Another example is color breaking (variegation) plants in which chloroplasts are partially lost by virus infection and other agents.

The metabolic similarities between Prokaryota and Eukaryota were established during the development of photosynthesis and oxygen respiration in the purple non-sulfur bacterial phylogeny. Cyanobacteria evolved to use water as the reducing agents of carbon dioxide in place of hydrogen sulfide in photosynthetic bacteria.

4. Protista World

Protista contains cytological very complex organisms that often confuse taxonomists. The group consists of eukaryotic single cells, large or small, with or without a cell wall. Many cells are motile, some are not, some are photosynthetic (unicellular algae), and others are absorptive (many parasitic forms), or ingestive (engulf large food particles, for example *Amoebae*). The cytomusculature, particularly microfilaments of actin and myosin, provides the motive force for pseudopodial locomotion that is an important characteristic of animals. Furthermore, many protists bear cilia or flagella. Both are constructed of microtubules, having the so-called "9 + 2" arrangement and providing locomotive force by bending. Most protists can reproduce asexually by cell division, but many are capable of sexual reproduction as well. Since the first microscopic observation by A. von Leeuwenhook (1674), more then 30,000 different species have been described, many with radically different and complex cell morphologies, this far outweighs the number of known bacterial species (around 17,000).

Interestingly, the phylum Ciliophora of the protista possesses especially differentiated cells. The cells contain two types of nuclei, a macro nucleus and micro nucleus, which are different morphologically and functionally, and show a typical nuclear dualism. Asexual reproduction occurs by a sideway-binary fission but some species also perform budding or multiple fission. On the other hand, sexual reproduction occurs through conjugation of two types of gametophytes, but some show so-called autogamy or cytogamy. All of the species are heterotrophic, taking in food from a cytostome and excreting it through a cytoproct. Further differentiation of the Ciliophora cells is typical in the genus *Paramecium*. These cells also have curious organelles such as trichocyts, oral grooves, food vacuoles (forming), contractile vacuoles, and even a very primitive nerve system.

Furthermore, among the phytoflagellates, the genus *Euglena* has a light-sensing organelle of a photoreceptor, an "eyespot". The eyespot apparently directs the flagellum, which propels the cells toward light of an intensity appropriate for photosynthesis. However, if *Euglena* is maintained in darkness, it loses its chloroplasts but can still absorb nutrients from its surroundings. The cells contain general eukaryotic organelles.

Ciliophora cells are quite remarkable in respect to eukarygenesis, and seem to have undergone a particular multicellular evolution in spite of being a unicellular organism. Some taxonomists have described Ciliophora as "non-cellular" organisms. Their evolution seems to have taken a blind passage due to the failure of multicellularization.

Some protista eventually achieved multicellularization, evolving into Fungi, Plantae, and Animalia as the rest stages. .

5. Conclusion

The early stages of the evolution of unicellular organisms involved three important progressions; proto-cells: Monera (Prokaryota) - Protista (Eukaryota). Proto-cells were the first life system to develop through the chemical evolution, and are the ancestors of all present-day organisms. In primitive bacteria such as *Mycoplasma*, proto-type constituents and organizations are maintained. We have emphasized that the (phospho)lipid bilayer, a spontaneous assembly of phospholipid molecules, played a most important role as a biomembrane in the origin and evolution of cells. Metabolisms branched into catabolism to harvest energy and anabolism to synthesize cell components with the evolution. On other hand, membranous and non-membranous organelles developed to compartmentalize metabolisms from bacteria to protists, in particular, the membranous evolutions were remarkable. In the protistan world, cellular differentiation erupted radically. However, this was not random but phylogenetic. Therefore, there exists a broad phylogenetic path, which leads to the three multicellular kingdoms: Fungi, Plantae, and Animalia.

Fungi are nutritionally heterotrophic and abortive, Plantae are autotrophic and photosynthesizable, while Animalia are heterotrophic and mostly ingestive. Many species of Protista, referred to as phytoflagelletes, such as *Euglena*, have feature of both animals (a whip and wandering) and plants (photosynthesis). However, it is of interest that such protists did not develop into multi-cellular organisms. A possible explanation for this is that organic compounds as nutrients for heterotrophs and sunlight for photosynthetic autotrophs have been available in abundance throughout history, although mass extinctions occurred several times due to drastic environmental changes like the ice ages. Anyway, single-celled organisms are estimated to account for over half the total weight of living creatures in the world. (This is because when compared to estimation of the population size, the single celled organisms are tremendous in number). Ecologically, they support all forms of heterotrophism.

6. References

Alberts, B., Bray, D., Lewis, J., Raff, M., Roberts, K., and Watson, J. D. (1994) Molecular Biology of the Cell (3rd Edition), Garland Publishing, Inc., New York.

Copelend, H. F. (1956) Classification of the Lower organisms, Pacific Book, Palo Alto, Calif.

Curtis, H. (1968) Biology, Worth, New York.

Horowitz, N. H. (1945) Proc. Natl. Acad. Science., USA 31, 153-157.

Jensen, T. E. (1994) In: J. Seckbach (ed.) Evolutionary Pathways and Enigmatic Algae: *Cyanidium caldarium* (Rhodophyta) and Related Cells, Kluwer Academic Publishers, Dordrecht, pp. 53-66.

Margulis, L. (1970) Origin of Eukaryotic Cell, Yale University Press, New Haven.

Margulis, L. (1974) In: T. Dobzhansky, M. Hecht, and W. C. Steere (eds.) Evolutionary Biology, vol. 7, Plenum Press, New York.

Nakamura, H. Viva Origino 4, 43-61.

Nakamura, H. (1983) Microbial Evolution, Baifukan Publisher, Tokyo.

Nakamura, H. (1994) In: J. Seckbach (ed.) Evolutionary Pathways and Enigmatic Algae: *Cyanidium caldarium* (Rhodophyta) and Related Cells, Kluwer Academic Publishers, Dordrecht, pp. 3-18.

Schopf, J. W. (1974) In: T. Dobzhansky, M. Hecht, and W. C. Steere (eds.) Evolutionary Biology, vol. 7, Plenum Press, New York.

Seckbach, J. (1994) In: J. Seckbach (ed.) Evolutionary Pathways and Enigmatic Algae: *Cyanidium caldarium* (Rhodophyta) and Related Cells, Kluwer Academic Publishers, Dordrecht, pp. 99-112.

Stainer, R. Y., Doudroff, M. and Adelberg, E. (1970) The Microbial World (3rd Ed.), Prentice-Hall, Engle Wood Cliffs (N. J.).

Whittaker, R. H. (1969) Science 163, 150-159.

CV ov **Julian Chela-Flores** Author of "**Eukaryogenesis**"

Julian Chela-Flores was born in Caracas, Venezuela. He studied in University of London (UK). There he obtained B.Sc (Special Hons) in Mathematics (1965), a M.Phil. In Quantum Mechanics (1967) and a Ph.D. also in Quantum Mechanics in 1969. He was a researcher at the Venezuelan Institute for Scientific Research (1971-1978) and Professor at Simon Bolivar University (Caracas) until his retirement in 1990. Dr. Chela-Flores is a Fellow of the Latin American Academy of Sciences, the Third World Academy of Sciences and of the Academy of Creative Endeavors (Moscow).

Dr. Chela-Flores current positions are Staff Associate of The International Center for Theoretical Physics (ICTP) in Trieste, Italy, Professor, International Institute for Advanced Studies (Simon Bolivar University). He is also Research Associate of the School of Theoretical Physics, Dublin Institute for Advanced Studies. He has done research in many fields in the life sciences and physics, but in particular, biophysics and astrobiology. He is organizer of the regular meetings on Chemical Evolution in the ICTP and the editor of several books on **Chemical Evolution and The Origin of Life**. His E-mail: Chelaf@ictp.trieste.it

J. Seckbach (ed.), Enigmatic Microorganisms and Life in Extreme Environments, 63–71.

EUKARYOGENESIS:
The Search for an Evolutionary Transition Towards Intelligence in an Extreme Environmental Habitat of the Outer Solar System.

JULIAN CHELA-FLORES (*)
International Centre for Theoretical Physics, Miramare P.O. Box 586; 34100 Trieste, Italy and Instituto Internacional de Estudios Avanzados (Universidad Simon Bolivar), Apartado 17606 Parque Central, Caracas 1015A, Venezuela.

1. Introduction

The most remarkable period subsequent to the establishment of the genetic code, or more precisely to the establishment of the translation mechanism, concerns the period in which the chromosomes originated at the very beginning of the evolution of cells. It is widely believed that the intron-exon structure of genes was part of the first cell. It has been remarked that this view implies that the genes of the most advanced higher organisms may reflect the primitive genome structure more clearly than the genes of prokaryotes, as during their evolution bacteria and archaebacteria have eliminated their introns (Brenner, 1994). This idea is persuasive enough for looking at the contemporary eukaryotic cell as a potential source of information on the structure and function of the first cell. This viewpoint allows the possibility of facing some of the most difficult questions concerning the primordial cell. Perhaps amongst such difficult questions, the origin and evolution of chromosomes (Maynard-Smith and Szathmary, 1993) stands out. One general guiding line in this important aspect of the origin of the progenote is to explain the past in terms of processes that we may currently be present in cells. It seems likely that changes in a protocell that led to the eukaryotic cell were in some way related to the control of gene expression (Maynard-Smith, 1993). Unfortunately, there is yet no general understanding of this aspect of the eukaryotic cell. On the other hand, sufficient insights have been obtained, as shown in Secs. 2 and 3, to encourage discussions of the origin of the eukaryotic cell in the context of extraterrestrial life, as we shall see in Secs. 4 and subsequent work below.

2. The Origin of the Eukaryotic Cell from the Geologic Record

Several lines of research suggest the absence of current values of O_2 for a major part of the Earth history. There are some arguments, nevertheless, that militate in favour of Archean atmospheres with values of the partial pressure of atmospheric oxygen O_2 about 10^{-12} of present atmospheric level (PAL).Shale is a rock that has played a role in

---

(*) Research Associate at the School of Theoretical Physics, Dublin Institute for Advanced Studies, 10, Burlington Road, Dublin 4, Ireland.

our understanding of biological evolution. Shale is mainly clay that has hardened into rock. The onset of atmospheric oxygen is demonstrated by the presence in the geologicrecord of red shale coloured by ferric oxide. The age of such 'red beds' is estimated to be about 2 thousand million years (Gyr). At that time oxygen levels may have reached 1-2% PAL, sufficient for the development of a moderate ozone protection for the microorganisms of the Proterozoic from ultraviolet radiation. The paleontologic record suggests that the origin of the eukaryotic cell occurred earlier than 1.5 Gyr before the present (BP).

Some algae may even date from 2.1 Gyr BP (Han and Runnegar, 1992), a period comparable to the first onset of red beds. This is still rather late, compared to the earliest available prokaryotic fossils of some 3.5 Gyr BP (Schopf, 1993). However, if we keep in mind certain affinities between eukaryotes and archaebacteria (such as homologous factors in protein synthesis), we may argue that archaebacteria and the stem group of eukaryotes may have diverged at about the same time (Runneger, 1992).

This conjecture, combined with the lightest carbon isotope ratios from organic matter, imply that bacteria capable of metabolically oxidizing methane CH_4 ('methylotrophs') may have been using methane produced by archaebacteria that were able to produce it as a byproduct of the reduction of carbon dioxide ('methanogens'). From the age of such fossils a tentative date of 2.7 Gyr BP, or earlier, may even be assigned to eukaryogenesis.

Metazoans are hypothesized to have arisen as part of a major eukaryotic radiation in the Riphean Period of the of the Late Proterozoic, approximately 0.8-1 Gyr BP (Knoll, 1994), when the level of atmospheric O_2 had reached 4-8% PAL. There is some evidence in the Late Proterozoic during the Vendian Period, for the existence of early diploblastic grades (Ediacaran faunas): These organisms were early metazoans with two germ layers, such as the modern coelenterates (jellyfishes, corals, and sea anemones). Later on, when the level of atmospheric O_2 had reached values in excess of 10% PAL, these grades were overtaken in numbers by triploblastic phyla as the level of atmospheric O_2 had reached 40% PAL. These organisms are called the Cambrian faunas, i.e., Early Paleozoic faunas, some 500 million years before the present (Myr BP), which were mainly metazoans with three germ layers). We may obtain further insights from paleontology: acceleration in the evolutionary tempo is observed after the onset of eukaryogenesis, as it is clearly demonstrated by the microfossils of algae from the Late Proterozoic (Knoll, 1994) and by the macrofossils of the Early Phanerozoic (Cambrian Period of the Paleozoic Era) (Conway-Morris, 1993). Such evolutionary changes within the first billion years of atmospheric oxygen rose the simple prokaryotes to eukaryotes.

3. Biological Aspects of Eukaryogenesis

One way of approaching the intricate problem of eukaryogenesis is to extract from the known phenomena of genetics some general aspects of the relations between the main relevant parameters and to use such knowledge to sketch possible aspects of the earliest chromosomes. A very tentative start in this program has already been published (Chela-Flores, 1987; 1992). The folding problem of the macromolecules of life seems a good starting point in our search for insights of the earliest aspects of the cell. Some insights have been gained in understanding the underlying mechanism that controls the folding of any polypeptide chain into unique three-dimensional protein structures. We may consider the analogous problem with the DNA that makes up the chromosomes, namely, the folding of the 100-Å nucleosome filament. There is a hierarchy of levels of folding beyond the 100-Å nucleosome filament: The next level of complexity is provided by 300-Å filament, which is arranged into a solenoidal configuration with about six nucleosomes per turn (Finch and Klug, 1976). During interphase in the cell cycle it is this solenoidal arrangement that constitutes the most abundant form of chromatin. However, at later stages in the cell cycle this structure serves as the basis for further folding, ending up at the highest degree of folding observed at the metaphase chromosome.

In view of the basic role played by all the stages of the hierarchy, it is of considerable interest to find a formalism by means of which we may anticipate the regular manner of DNA folding. This problem is of considerable difficulty. We have thus restricted our attention to DNA folding in the later stages of chromatin compaction. The biochemical basis for the difference between heterochromatin (the more compact structure of chromatin) and euchromatin (its less compact form), remains unknown. In both cases the 100-Å nucleosome filament contains approximately the same DNA / histone ratio (Watson *et al*, 1987). Models attempt to account for impeding fork movement at a terminus, where such repeated sequences may be observed (Kornbeerg and Baker, 1992). We shall argue that chromatin compaction is a clear hallmark of eukaryogenesis.

4. Eukaryotic Identity

Primitive eukaryotic organisms have been studied in detail and many difficulties are inherent in the eventual design of an assay to identify them in an extraterrestrial environment, a question which begins to be important in view of the forthcoming space missions. In the present paragraph we discuss some properties of organisms that illustrate how difficult is to define a clear test in a robotic search in solar system exploration for the difference between prokaryotes and eukaryotes:

• One such taxon is the family Cyanidiophyceae. These rhodophytes, are known as red algae (Seckbach, 1994). Seckbach has argued that these acido-thermophilic algae may constitute a bridge between cyanobacteria and red algae (Seckbach, 1995). In any case the point we wish to emphasize here is that the morphology of these microorganisms is not radically different from prokaryotes, except for the inner membrane. In particular, *Cyanidium caldarium* may live at temperatures of up to 57°C and shows better rates of growth and photosynthesis when cultured in an 'atmosphere' of pure CO_2 (Seckbach *et al.*, 1970; 1972) .

• Silicification experiments of microorganisms have led to the identification of one artifact, which can cause confusion in the identification of the transition organisms. Frances Westall and co-workers at the University of Bologna have observed an artificial nucleus formed during the process of fossilization (Westall, 1995). If such a phenomenon is preserved in the natural fossil record, then it can lead to a confusion with the eukaryotes.

• Prokaryotic symbionts of the ciliated protozoan Euplotidium itoi has ultrastructure which is more complicated than the majority of prokaryotes (Rosatti *et al.*, 1993). Strictly from the point of view of planning future experiments for teh search of extraterrestrial eukaryotes (SETE), the most striking feature discovered by Giovanna Rosati and co-workers at the University of Pisa is the presence of 'basket tubules', consisting of the protein tubulin, normally present only in eukaryotes.

It is useful to know in which atmospheres extremophiles may survive, as we now know of several Solar-System planets and satellites that have atmospheres. For example, four large planetary satellites in the outer Solar System are known to have atmospheres (Hall *et al.*, 1995). Mars currently has a CO_2 dominated atmosphere. The possibility of extending the biosphere deep into the silicate crust of Mars has an analogy with organisms that have been found to inhabit deep in the silicate crust of the Earth, which may been deposited with the original sediment, and survived over geologic time. McKay (1996) has considered this question from the point of view of geophysics.

Europa is a second candidate for life in the Solar System, after the more prominent candidate, the planet Mars. This follows from the remark that a large proportion of the spectroscopically detectable material on its surface is water (Soffen, 1976; Horneck, 1995). According to the results on the Jupiter system obtained by the Voyager 2 mission, Europa is covered by a layer of ice, under which there may be an ocean of water (Anderson, *et al.*, 1997; Oro *et al.*, 1992), whose temperature is 4°C. From the similarity of the processes that gave rise to the solid bodies of the Solar System, we may expect that hot springs may lie at the bottom of the ocean. Oro's main thesis is that, as Jupiter's primordial nebula must have contained many organic compounds, then possibly, organisms similar to thermophilic archaebacteria could exist at the bottom of Europa's ocean.

However, given the incomplete understanding of the evolution of early life on Earth, at present we should extend the Oro argument in order to include eukaryotes as well.

5. Eukaryogenesis in Solar System Exploration

In the present search for life in the Solar System there are two prominent candidates: The first one is Mars, but the Jovian satellite Europa is a second environment in our own solar system in which we have proposed to test the hypothesis of the ubiquity of eukaryogenesis (Chela-Flores, 1998; 1997a).

The most evident candidate in a program on SETE is Mars. Nevertheless, previous work has suggested that through tidal and radiogenic heating on Europa, an internal ocean has been formed (underneath a relatively thin ice cover) through dehydration of silicates (Reynolds *et al.*, 1983); the heating source due to tidal heating is expected to be of the order of 52 erg $cm^{-2}s^{-1}$with an additional 8 erg $cm^{-2}s^{-1}$, due to radiogenic heating (Reynolds *et al.*, 1987). For the above reasons it was estimated that the temperature underneath the icy crust could be water above the freezing point. From the similarity of the processes that gave rise to the solid bodies of the Solar System, we may expect that hot springs may lie at the bottom of the ocean.

The main thesis of the proponents of the existence of a Europan biota is that, as Jupiter's primordial nebula must have contained many organic compounds, then possibly, organisms similar to thermophilic archaebacteria can evolve at the bottom of Europa's ocean. The previous argument (Oro *et al.*, 1992) correctly pointed out that the most important requirements for the maintenance of life in Europa are the above mentioned conditions (water, an energy source and organic compounds). However, the analogous Earth ecosystems considered by them excluded eukaryotes.

The case of the acidophilic and thermophilic cyanidiophycean algae (Seckbach *et al.*, 1970; 1972) is a warning that we should keep an open mind while discussing the possible degree of evolution of Europan biota. Indeed, plate tectonics has obliterated fossils of early organisms from the crust of the Earth. Additional factors arising from current experience with eukaryotes may contribute to clarify the case for not excluding nucleated cells from the microorganisms to be searched in new Solar System environments; three arguments militate in favour of the ubiquity of eukaryotes in other ecosystems of the Solar System.

We have remarked that the identification of primitive eukaryotes is not straightforward; another difficulty is that the simplest criterion to recognise them, namely, the search for a membrane-bounded set of chromosomes, does not help to identify unambiguously eukaryotes, as there are prokaryotic organisms, that do have a membrane-bounded nucleoid (Fuerst and Webb, 1991).

In addition, work (Rosatti *et al.*, 1993) on prokaryotic symbionts of protozoans having proteins typical of eukaryotes, and silicification experiments of certain microorganisms (Westall *et al*, 1995) suggest that the identification of eukaryotes at the morphological or molecular levels, particularly during the process of fossilization, may lead to some confusion.

Consequently, we have proposed (Chela-Flores, 1997a) that tests for recognising eukaryotes (a SETE programme) should also be envisaged in future space missions. On the other hand, if living microorganisms are present in Europa one may adopt a gene-centered approach (Chela-Flores, 1997b), analogous to the one developed in our earlier work. This may allow us to search for cellular replication in relation to a delay in replication of chromosome segments. Such a phenomenon would confirm that the chromosomes maintain the condensed state during interphase, a eukaryotic hallmark. If frozen or fossilized organisms are encountered, then we need to consider the differences between primitive eukaryotes and members of the other two domains.

6. Discussion and Conclusions

We have reviewed some aspects of the problem of eukaryogenesis in extreme environmental habitats. Our emphasis has been on the possibility of eukaryotic life arising in other Solar System environments, such as Mars, but particularly in Europa, the Outer Solar System Jovian satellite . We have commented on the design of an advanced lander mission that may melt through the ice layer above the Europan ocean, in order to deploy a tethered submersible, which could, in principle, test for signs of evolution of Europan biota, if it exists (Chela-Flores, 1996). Such a submersible seems to be the most appropriate means for implementing a SETE program.

Archaebacteria-like organisms are the most likely biota in the Europa ocean (Delaney *et al.*, 1996). We have argued that evolution should have occurred in Europa and that the experimental test of this conjecture is feasible through the Hydrobot/Cryobot proposal (Trowell *et al.*, 1996). The proposal of a lander space craft capable of penetrating the Europan surface ice-layer by melting through it, does not seem beyond our technological capabilities (Horvath *et al.*, 1997).

Although difficult design issues are involved in the preparation of a package to search for life, we have initiated above the discussion of what would seem to be a reasonable biological experiment.

Some of our thinking has been based on the the fact that we should not consider the possibility of the existence of refuges against evolution (Little *et al.*, 1997). This new insight is based on the remark that Cambrian fauna (lamp shells and primitive molluscs) were maintained during Silurian times by microorganisms that lived in hydrothermal vents. At the present these hot environments have seen the extinction of such fauna. It is then reasonable to believe that the microorganisms themselves would be subject to analogous evolutionary pressures that have led to the first steps in evolution. On the Earth the eukaryotic theme took a little over one billion years to make its first appearance, after the prokaryotic communities (stromatolites) were well established in the Early Archean Eon. It still remains to be confirmed, or rejected, whether the Europan environment may have had liquid water over geologic time that may have sustained a primary archaebacteria community.

To sum up, two ingredients support the present attempt to identify indicators of the degree of evolution of putative Europan biota:

ù the possibility of testing the onset of eukaryogenesis in extraterrestrial environments, and

• the lack of refuges against evolution, and hence the possibility that the eukaryotic transition may be a general consequence of the geological changes coupled to the effect of natural selection (i.e., independent of the fact that microorganisms are part of the Earth biosphere).

7. References

Anderson, J.D., Lau, E.L., Sjogren, W.L., Schubert, G., and Moore, W.B. (1997). Science **276**, 1236-239.

Brenner, S. (1994). Nature, **367**, 228-229.

Chela-Flores, J. (1987). J. Theor. Biol. **126**, 127-136.

Chela-Flores, J.(1992). J. Theor. Biol. **154**, 519-539, and Erratum, J. Theor. Biol. 157, 269.

Chela-Flores, J. (1996). Europa Ocean Conference, Loc. cit. pp. 21-21a, 1996.

Chela-Flores, J. (1997a). In: Astronomical and Biochemical Origins and the Search for Life in the Universe. Eds. C.B. Cosmovici, S. Bowyer and D.Werthimer. Editrice Compositore: Bologna pp.525-532.

Chela-Flores, J. (1997b). Origins Life Evol. Biosphere (in press).

Chela-Flores, J. (1998). Origins Life Evol. Biosphere (in press).

Conway-Morris, S. (1993). Nature **361**, 219-225.

Delaney, J. , Baross, J., Lilley, M. and Kelly, D. (1996). Europa Ocean Conference, Loc. cit. p. 26.

Finch, J.T. and Klug, A. (1976). Proc. Natl. Acad. Sci. USA **73**, 1897-1901.

Fuerst, J.A. and Webb, R.J. (1991). Proc. Natl. Acad. Sci. USA **88**, 8184-8188.

Hall, D.T., Strobel, D.F., Feldman, P.D., McGrath, M.A. and Weaver, H.A. (1995). Nature **373**, 677-679.

Han, T.-M. and Runnegar, B. (1992). Science **257**, 232-235.

Horneck, G. (1995). Planet. Space Sci. **43**, 189-217.

Horvath, J., Carsey, F., Cutts, J., Jones, J., Johnson, E., Landry, B., Lane, L., Lynch, G., Chela-Flores, J., Jeng, T.-W. and Bradley, A. (1997). Proc.Optical Science, Engineering, and Instrumentation **3111** .Instruments, Methods and Missions for Investigation of Extraterrestrial Microorganisms. San Diego, California, USA. pp. 490-500.

Kornberg, A. and Baker, T.A. (1992). DNA Replication, 2nd. ed., W.H. Freeman and Co., NewYork, pp. 508-510.

Knoll, A.H. (1994). Proc. Natl. Acad. Sci. USA **91**, 6743-6750.

Little, C.T.S., Herrington, R.J., Maslennikov, V.V., Morris, N.J. and Zaykov, V.V. (1997). Nature **385**, 146-148.

Maynard Smith, J. (1993). The theory of evolution. Canto Edition, Cambridge University Press, Cambridge (UK). p. 122.

Maynard Smith, J. and Szathmary, E. (1993). J. Theor. Biol. **164**, 437-446.

McKay, C.P. (1996). Oxygen and the rapid evolution of life on Mars. In: Chela-Flores, J. and Raulin, F. (Eds.). Chemical Evolution: Physics of the Origin and Evolution of Life. Kluwer Academic Publishers, Dordrecht, The Netherlands. pp. 177-184.

Oro, J. Squyres, S.W., Reynolds, R.T., and Mills, T.M. (1992). In: Exobiology in Solar System Exploration. G.C. Carle, D.E. Schwartz, and J.L. Huntington, eds. NASA publication SP 512. pp. 103-125.

Reynolds, R.T., Squyres, S.W., Colburn, D.S.and McKay, C.P. (1983), On the habitability of Europa. Icarus, **56**, p. 246-254.

Reynolds, R.T., McKay, C.P. and Kasting, J.F. (1987). Adv. Space Res., 7, p. 125-132.

Rosati, G., Lenzi, P., and Franco, U. (1993).'Micron, **24**, p. 465-471.

Runnegar, B. (1992). In. Schopf, J.W. and Klein, C. (1992). pp. 485.

Schopf, J.W.(1993). Science **260**, 640-646.

Seckbach, J. (1972). Microbios **5**, 133-142.

Seckbach, J. (1994). In: Evolutionary pathways and enigmatic algae: Cyanidium caldarium (Rhodophyta) and related cells. Ed. Seckbach, J. Kluwer Academic Publishers: Dordrecht, The Netherlands. pp. 99-112.

Seckbach, J. (1995). In: Ponnamperuma, C. and Chela-Flores, J. (Eds.). Chemical Evolution: The Structure and Model of the First Cell . Kluwer Academic Publishers, Dordrecht, The Netherlands.pp. 335-345.

Seckbach, J., Baker, F.A. and Shugarman, P.M., (1970) Nature **227**, 744-745.

Soffen, G.A.(1976). Science **194**, 1274-1276.

Trowell , S., Wild, J., Hovrath, J., J. Jones, Johnson, E. and Cutts, J. (1996). Europa Ocean Conference, (Abstracts), 5th. Capistrano Conf. San Juan Capistrano Research Institute [31872 Camino Capistrano, San Juan Capistrano, CA 92675, USA. Fax: (714) 240 0482], 12-14 November, p. 76.

Westall, F., Boni, L.and Guerzoni, E. (1995), Paleontology, **38**, p. 495-528.

Watson, J.D., Hopkins, N.H., Roberts, J.W., Steitz, J.A., and Weiner, A.M. (1987). Molecular Biology of the Gene, 4th. ed., The Benjamin / Cummings Publishing Co., Menlo Park, California, p. 682.

Biodata of **Frances Westall** author of ***"Fossil Bacteria."***

Dr. Frances Westall received her Ph.D. in Marine Geology in 1984 from the University of Cape Town, South Africa. At the Alfred Wagner Institute for Polar and Marine Research in Bremerhaven, Germany (1984-1989) she studied the influence of climatic changes over the last 4 million years on sediments deposited in the Antarctic South Atlantic Ocean. Her subject of study changed to fossil bacteria during research at the University of Nantes, France (1989-1991) and then to the fossilization of bacteria and the search for the oldest fossil bacteria on Earth and in Martian meteorite ALH84001 at the University of Bologna (1991-1997). Currently (1998-1999) she is an NRC Fellow at the Johnson Space Center, Houston, Texas where her research of the oldest forms of Life on Earth as possible analogues for extraterrestrial life continues.

E-Mail: **frances.westall1@jsc.nasa.gov**

J. Seckbach (ed.), Enigmatic Microorganisms and Life in Extreme Environments, 73–88.

FOSSIL BACTERIA

FRANCES WESTALL
University of Bologna
DIPROVAL, via San Giacomo 7, 49125 Bologna, Italy

1. Introduction

The history of bacteria spans at least 3.8 billion years (Ga) and may well extend to 4 billion years or more. The oldest indirect evidence for the presence of life on Earth dates from 3.8 Ga and is provided by $\delta^{13}C$ isotope ratios, indicative of fractionation by bacterial metabolism, measured in sediments from the Isua Supracrustal province in SW Greenland (Mojzsis *et al.*, 1996). If bacteria were already active and interacting with processes on Earth so as to leave traces of their presence already by 3.8 Ga, then their development must have taken place even before that period. Bacteria are, therefore, the earliest forms of life known, to date, on Earth. However, so far no fossils have been found in the Isua sediments and the oldest known fossil bacteria occur in rocks 3.5 Ga old (Knoll and Barghoorn, 1977, Schopf and Walter, 1983; Schopf, 1993;Walsh and Lowe, 1985; Walsh, 1992; Westall, 1997a; Westall *et al.*, 1977a,b; Westall *et al.*, in prep.). Not only are prokaryotes the oldest known form of life, they were the only form of life for more than 70% of the history of life on Earth until the appearance of the eukaryotes sometime around 2.1 Ga.

This chapter on fossil bacteria necessarily covers a vast timescale. I make no attempt to provide an exhaustive review of fossil bacteria, partly for reasons of space and partly because reviews in Schopf (1983), Schopf and Klein (1992), and Knoll (1996) have already dealt amply with occurrences during the Precambrian period. Before describing a representative selection of fossil bacteria from the whole of the history of life on Earth, I will discuss the fossilisation of bacteria: the study of both ancient fossils and modern "potential" fossils, as well as experiments simulating the fossilisation of bacteria, have greatly aided our understanding of the process of fossilisation. I will then sweep through the occurrences of fossil bacteria in chronological order to provide an indicative, but not exhaustive, list of the temporal and spatial distribution of the fossils, concluding with a brief discussion of the possible fossil bacteria described in Martian meteorite ALH84001.

2. The Preservation and Fossilisation of Bacteria

The vast majority of fossil bacteria in the rock record have affinities with modern cyanophytes (Schopf, 1992a,b, 1993) and filamentous, non-cyanophyte bacteria. Very few other types of bacteria have been described, to date. There are a number of reasons for this skewed record. (1) The sheaths of the cyanobacteria are relatively resistant to degradation, thus rendering them more susceptible to "fixation" as fossils in sediments (Golubic and Barghoorn, 1977). (2) The mucopolysaccharides making up the sheath readily chelate metals and minerals which (a) hinder degradation and (b) encrust or nucleate within the sheath, both processes, thus, increasing the preservation potential of the sheaths (Golubic, 1973; Beveridge and Murray, 1976; Pentecost and Riding, 1986; Ferris *et al.*, 1988). (3) Cyanobacteria and filamentous bacteria tend to have relatively large maximum dimensions (generally over 2 μm and often over 10 μm) and are, thus, visible with a high powered light microscope, the most common instrument of study of fossil bacteria. An additional reason for the predominance of cyanophytes in the fossil record has been suggested by Golubic and Barghoorn (1977): at least for the early part of the history of life on Earth cyanobacteria occupied the foot of the trophic pyramid and were, therefore, probably more common than contemporary heterotrophs. In fact, the relatively good preservation potential of these soft, organic life forms is borne out by some superb fossil specimens in the rock record, especially from the Proterozoic (e.g. Knoll, 1985; Butterfield *et al.*, 1994; Fairchild *et al.*, 1996). Older fossils from the Archean, however, are generally less well preserved owing to the effects of diagenesis and metamorphism (e.g. Schopf, 1993; Knoll *et al.*, 1988).

The majority of the described fossils appear to have a carbonaceous composition, although some fossils may be mineralised, i.e.. encrusted, replaced or infilled with minerals such as calcium carbonate (Golubic, 1973; Pentecost and Riding, 1986; Knoll *et al.*, 1993), Fe or Mn (Ghoirse and Hirsch, 1978, 1979), silica (Monty *et al.*, 1991; Westall, 1977a,b; Westall *et al.*, 1977), calcium phosphate (Liebig *et al.*, 1996), siderite (Wuttke, 1983), pyrite (J.H. Oehler, 1977).

Throughout most of the Precambrian period (about 85% of the history of life on Earth), most of the well-preserved fossil bacteria occur in cherts, primary or diagenetic siliceous deposits. The resistant, organic remnants of bacteria, in various stages of degradation, would have been preserved by being permeated with silica and, thus, locked into a silica matrix. Knoll (1985) notes a close association between the organic content of sediments and silicification, speculating that the presence of organic matter might affect the solubility of silica and its precipitation although he also observes that

Figure 1. Experimental silicification of bacteria (Westall *et al.*, 1995; Westall, 1997b). (a) Encrusted bacteria containing silicified collapsed cytoplasm (scale 0.2 μm). (b) Encrusted bacilli (scale 1 μm). (c) Encrusted cast (scale 0.2 μm). (d) Calcified nannobacteria from the Le Zitelle hot springs, Viterbo, Italy (scale 1 μm). Courtesy of R. Folk. (e) Cross section of a compressed colony of thickly encrusted Gram positive *Bacillus laterosporus* (scale 1 μm). (f) Delicately encrusted Gram negative *Pseudomonas vescicularis* (scale 0.2 μm)

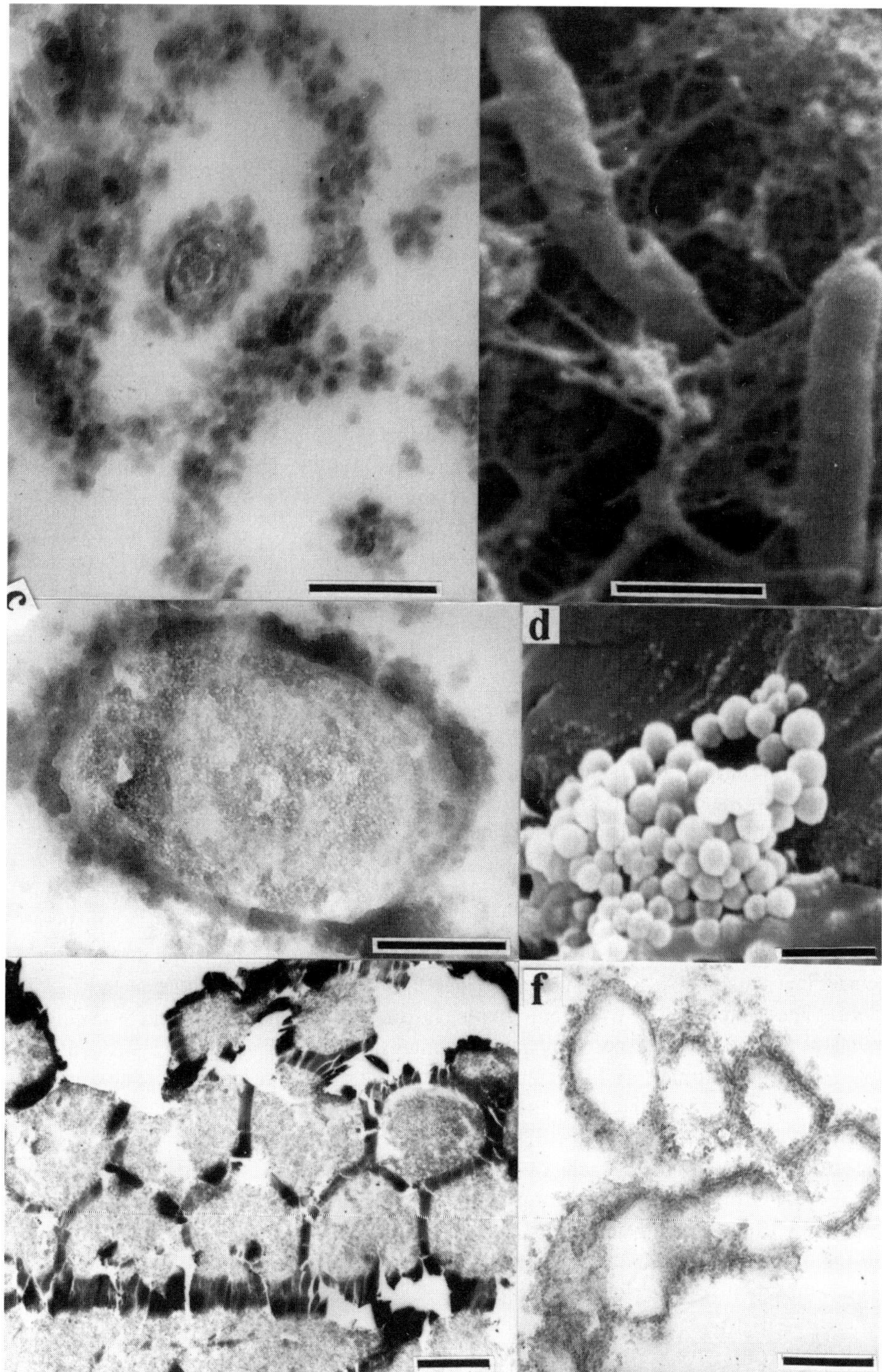
c
d
f

high organic contents are correlated with poor preservation of carbonaceous microfossils (Knoll, 1996). The relationship between organic matter and mineral-replacement of the bacteria degrading the organic matter is documented by the occurrences of phosphatised bacteria covering the soft parts of fossil vertebrates, invertebrates and coprolites in organic-rich Lagerstätte (Wuttke, 1983; Martill and Wilby, 1994; Liebig *et al.*, 1996, Liebig, 1997). Reports of mineral replacements of fossil bacteria, however, are less common, e.g. silicified Archean bacteria from South Africa (Westall, 1997a; Westall *et al.*, 1997), pyritised bacteria from the Proterozoic of Australia (J.H. Oehler, 1977), calcified bacteria from the Holocene of the Ross Sea (Barbieri *et al.*, in press). Bacterial preservation as compressions or impressions in shales is also relatively rare, e.g. the Proterozoic Draken Formation of Spitzbergen, Knoll *et al.* (1991).

Understanding of the process and results of the degradation and fossilisation of bacteria has been aided by experimental simulations of degradation and fossilisation (e.g. Leo and Barghoorn, 1976). Changes in the size and shape of mat-forming cyanobacteria take place during degradation, resulting in structures which were similar to those found in Precambrian cherts (Oehler and Schopf, 1971; Awramik *et al.*, 1972; J.H. Oehler, 1976, Knoll and Golubic, 1979). Other experiments showed that many of the characteristics previously used as parameters for the identification of eukaryotes in thin sections of rocks (e.g. internal spots as the supposed remains of organelles, tetrahedral tetrad arrangements of cells, scars indicative of meiotic cell division, large cell size, and mitotic cell divisions) are actually present in experimentally silicified bacteria (Fig. 1a) (Francis *et al.*, 1978a,b; Westall. *et al.*, 1995). These artifacts complicate the distinction of fossil eukaryotes and prokaryotes, thus, rendering difficult the interpretation of the timing of the first appearance of eukaryotes in the fossil record.

Other experiments documented the mode of silicification of non-cyanophytes, showing that minute crystallites of silica nucleate onto the organic membranes of the microorganisms, resulting in complete replacement of the microorganisms as crusts around the outer cell wall (Fig. 1a, b), casts, encrusted casts (Fig. 1c) or moulds (Westall *et al.*, 1995). Gram positive and Gram negative bacteria exhibit differential responses to fossilisation owing to differences in the composition of the cell wall (Figs. 1e,f), a fact which has consequences for the fossilisation potential of Gram negative bacteria (Westall, 1997b). This will influence their preservation potential in the fossil record. The disappearance of organic matter during the silicification of bacteria (Figs. 1a,c) has also been noted, not only in experiments, but also in modern hydrothermal vents (Cady and Farmer, 1976) and is in sharp contrast to the many descriptions of carbonaceous bacteria in the fossil record (compare also the carbonaceous microfossils in Figs. 2b, c with the clearly non-carbonaceous microfossils in Figs. 2a, 2d and 2e, g). Interestingly, Cady and Farmer (1996) showed how the type of preservation of the cyanophytes (replacement, encrustation, permineralisation) depended on the temperature of the spring water.

It has been proposed that even smaller bacteria ("nannobacteria" 0.05-0.5 μm) may play an important volumetric role in biogenic and sedimentologic processes, as well as providing a substantial presence in the rock record as fossils (Fig. 1d) (Folk, 1993; Folk and Lynch, 1997). There is growing biological evidence for the existence of such small bacteria (Ackerman *et al.*, 1993; A. Pentecost quoted in Folk, 1993) but much work needs to be done to document their distribution, role and importance as fossils in the rock record.

3. Fossil Bacteria through Time

3.1. THE ARCHEAN (4.0-2.6 Ga)

For an exhaustive review of the accepted and refuted fossil bacteria finds in the Archean and the Proterozoic I refer the reader to Schopf (1983), Schopf and Klein (1992), and Knoll (1996). Fossil bacteria from the Archean are rare compared with the later eons, mainly because very few Archean terrains have survived. An additional reason for the paucity of fossils is the difficulty of identifying bona fide fossil structures. Criteria for defining bona fide fossil bacteria include originality (in the sense of being indigenous) and syngenicity to the host rock, and the assured biogenicity (composition, shape, size, association) of the bacteriomorphs (Schopf and Walter, 1983; Schopf, 1992a).

Fossil bacteria from the Archean are known from shallow water and hydrothermal deposits from South Africa, Australia and India. The oldest known fossil bacteria come from two Early Archean terrains of the same age (and possibly originally conjugate, Knoll, 1996): the Kaapvaal Craton (Barberton greenstone belt) in eastern South Africa and the Pilbara Craton in northwestern Australia. Carbon isotope studies provide supporting evidence for the presence of bacteria at this period (Walsh and Lowe, 1985; Schopf, 1993; de Ronde and Ebbesen, 1996; Westall *et al.*, in prep). The poorly diverse assemblages in 3.3-3.5 Ga cherts from the Onverwacht and Fig Tree Groups of South Africa, described from thin section studies, consist of carbonaceous and possibly mineral-encrusted bacteriomorph filaments and some possibly biogenic spheroids (Knoll and Barghoorn, 1977; Schopf and Walter, 1983; Walsh and Lowe, 1985; Walsh, 1992). In addition to the carbonaceous fossils, smaller (0.6-3.8 μm), silicified, apparently non-carbonaceous (TOC of the samples 0.01-0.03%), short rod-shaped bacteriomorphs (Fig. 2a) were found in cherts from a number of different stratigraphic horizons in the Onverwacht Group (Westall, 1977a; Westall *et al.*, 1997; Westall *et al.*, in prep.).

In the Early Archean of Australia, relatively large, carbonaceous bacterial filaments occur in cherts of the Towers Formation (Awramik *et al.*, 1983; Schopf and Walter, 1983) and the overlying Apex Formation (Schopf, 1992a, 1993) in the Warrawoona Group, NW Australia. The taxa identified include those with Oscillatoriacean affinities *Primaevifilum laticellulosum, Archaeoscillatoriopsis grandid*, and *A. maxima* and two non-cyanobacteria taxa (*Archaeotrichion septatum* and *Eoleptonema apex*). Other, non-carbonaceous, mineral-replaced bacteriomorphs (non-cyanophyte and cyanophyte bacteria), similar to those from the Onverwacht Group, have been found in a silicified stromatolite from the type stromatolite locality in the Warrawoona group at North Pole (Westall, 1997a; Westall *et al.*, in prep.).

Late Archean fossil bacteria have been described from Australia (Knoll, 1996), India (Venkatachala *et al.*, 1990), and South Africa (Lanier, 1986; Altermann and Schopf, 1995). The taxa identified include *Oscillatoria spp.*, *Lyngbya spp. Entophysalidacea, Chroococcacea* and possibly the sheathless cyanobacteria *Phormidium.*

Considerations of the atmospheric oxygen levels during the Archean period suggest that, although pockets of relatively oxygen-rich ecosystems may have existed at various times and in various locations, the cverall level of oxygen in the atmosphere was too low to have supported obligate aerobes (Chapman and Schopf, 1983). The majority of Archean microbes were probably (photo) autotrophs (Knoll, 1996).

3.2. THE PROTEROZOIC (2.5-0.57 Ga)

For exhaustive reviews of the Proterozoic fossil bacteria I refer the reader to Hofmann and Schopf (1983), Schopf and Klein (1992) and Knoll (1996). Compared to the Archean biota, the Proterozoic biota is much more diverse and generally very well preserved with fossil exhibiting excellent cellular structure, thus, aiding comparison with extant taxa. Not only are the microorganisms well preserved, but there are many more occurrences, mainly due to more widespread cratonal sedimentation during the Proterozoic. Once again, most of the microfossils are carbonaceous relicts, permineralised in chert or preserved as impressions or compressions in shale (Hofmann and Schopf, 1983; Germs *et al.*, 1986; Butterfield *et al.*, 1994), although some are preserved in pyrite (J.H. Oehler, 1976), goethite and apatite (Butterfield *et al.*, 1994) or carbonate (Klein *et al.*, 1987; Knoll *et al.*, 1991, 1993).

Most Proterozoic fossil bacteria occur in intertidal to shallow water sedimentary facies, many of them with associated stromatolites (largely microbially-mediated deposits), which were either directly permineralised by early diagenetic silica, or subjected to later silicification of carbonate lithologies.

Figure 2. Fossil bacteria through time. (a) Non-carbonaceous, silicified coccobacilli in chert from the Onverwacht Group (3.5-3.3 Ga), South Africa (scale 2 μm). (b) Carbonaceous cyanobacteria (*Synodophycus euthemos*) in chert from the Draken Formation (800-700 Ma), Spitsbergen (scale 10 μm). Courtesy of A.H. Knoll. (c) Carbonaceous filamentous bacteria (*Gunflintia minuta*) from the Deoban Formation (ca. 1000 Ma), India (scale 10 μm). Courtesy of V. Tewari. (d) Silicified colony of filamentous bacteria from the M. Miocene (15-10 Ma) sediments from the deep-sea, S. Atlantic Ocean (Scale 2 μm). (e) (g) Phosphatised, non-carbonaceous coccoid and rod-shaped bacteria from the E. Eocene (49 Ma), Lake Messel, S. Germany (scale 0.2 μm, 1 μm for (e) and (g) respectively). (g) Courtesy of K. Liebig. (f) Coccoid bacteria and fungal spores in a gas bubble in amber from the Baltic (scale 2 μm). Courtesy of R. Kohring.

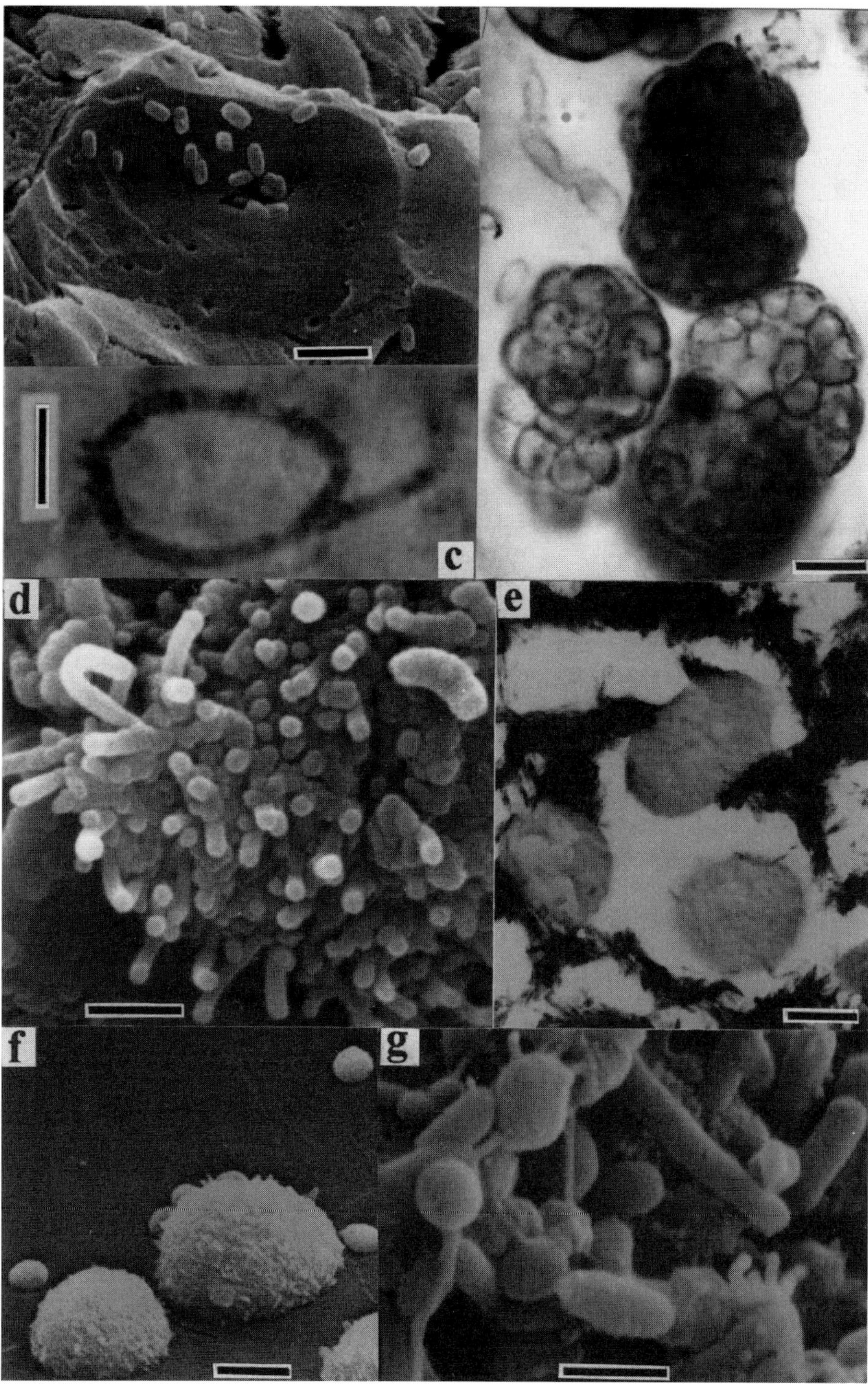
c
d
e
f
g

3.2.1. Early Proterozoic (2.5-1.6 Ga)

The most famous of the early Proterozoic finds is the 2 Ga Gunflint Chert of Ontario, Canada (see reviews in Hofmann and Schopf, 1983; Schopf and Klein, 1992; Knoll, 1996). The Gunflint Formation is characterised by both a stromatolitic biota as well as a rarer, non-stromatolitic biota. The stromatolitic biota includes the non-cyanobacteria, coccoid genus *Eoastrion* and many genera with cyanobacterial affinities: the coccoid genera *Chlamydomonopsis, Corymbococcus, Glenobotrydion, Huroniospora, Megalytrum, Palaeoanacystis*, and *Sphaerophycus*; the septate, filamentous genera *Anabaenidium, Gunflintia, Palaeoscytonema, Palaeospiralis, Palaeospirulina, Primorivularia*, and "*Schizothrix*"; the nonseptate, tubular filament *Entosphaeroides* is of possible cyanobacterial affinity; the branching filament *Archaeorestis* and the bizarre form *Xenothrix* may be bacterial. Of the non stromatolitic biota, only the coccoid *Leptoteichos* has cyanobacterial affinities.

Other Early Proterozoic fossiliferous formations include the Campbell-Rand Subgroup-Kuruman Formation, South Africa (Klein *et al.*, 1987); the Pokegama Quartzite, Minnesota (Cloud and Licari, 1972); the Belcher Supergroup, Canada, from which the oldest known cyanobacterial fossil of diagnostic morphology (*Eoentophysalis belcherensis*) has been described (Hofmann, 1976); and the Dahongyu Formation, Hebei Province, China (Zhang, Y., 1988).

The Early Proterozoic period was important in terms of atmospheric and organic evolution owing to the gradual rise in atmospheric oxygen levels which, for the first time, exceeded the 1-2% PAL (present atmospheric level) considered necessary to support obligate aerobes (Holland and Beukes, 1990). Possible eukaryotes start appearing in rocks 1.9-1.6 Ga in age and the first most likely eukaryotic fossils are leiosphaerid acritarchs from China (Zhang, Z., 1986).

3.2.2. Middle Proterozoic (1.5-1.1 Ga)

The Middle Proterozoic is represented by many occurrences of microfossils (mostly cyanophytes and, increasingly, eukaryotic algae) in Australian formations such as the ca. 2 Ga Duck Creek Dolomite of Western Australia, a non-stromatolitic, subtidal deposit (Knoll and Barghoorn, 1977; Hofmann and Schopf, 1983; Knoll *et al.*, 1988). The fossil assemblage from Duck Creek includes the coccoid bacteria (*Huroniospora sp.* and ? *Leptoteichos golubicii*), the septate filamentous fossils (*Ferrimonilis sp., Gunflintia sp. Rhicnonema sp., Oscillatoriopsis majuscula* [this is one of the largest septate filamentous fossils described from the Proterozoic, Knoll *et al.*, 1988], and *Oscillatoriopsis cuboides*), the tubular, unbranched *Siphonophycus* sp., and the unusual or bizarre microfossils with possible bacterial affinities, *Kakabekia sp.* and *Eoastrion sp.* The younger McArthur Group (1.5 Ga) contains a number of fossiliferous units, e.g. the Amelia Dolomite (Muir, 1976); the Balbirini Dolomite (D.Z. Oehler, 1978); and the H.Y.C. Pyritic Shale Member (J.H. Oehler, 1977). The latter unit is particular in that it consists of a deep-water assemblage of benthic, pyrite-encrusted and -replaced, filamentous bacteria and planktonic carbonaceous cyanobacteria.

Other Middle Proterozoic fossiliferous formations come from China (Zhang, 1985), N. Siberia (Sergeev *et al.*, 1995; Golubic *et al.*, 1995), the Urals (Sergeev, 1994), Arctic Canada (Horodyski and Donaldson, 1980, 1983), and central Brazil (Fairchild *et al.*, 1996).

3.2.3. Late Proterozoic (1.0-0.56 Ga)

Increased diversity of the fossil biota characterises the Late Proterozoic formations and excellent microfossil preservation is shown in a number of formations from the Arctic North Atlantic. The 800-700 Ma Draken Formation of Spitzbergen consists of silicified carbonates and shales deposited in environments ranging from intertidal through lagoonal to quiet shelf (Knoll *et al.*, 1991, 1993). Each ecological zone is characterised by its own assemblage of microfossils. Low diversity, cyanobacteria-dominated mat communities (various species of the genus *Siphophycus* depending on distance from the low tide level) occur in the intertidal zone. Diversity increases towards the lagoon (eukaryote dominated, e.g. *Trachyhystrichosphaera*) while endolithic and epilithic cyanophytes (e.g. *Siphonphycus robustum*) dominate the oolitic sediments of the barriers at the mouth of the lagoon. The fossils are preserved in early diagenetic silicified carbonates (Fig. 2b), as silica moulds in fine-grained carbonates, and as compressions in shales. Well preserved fossil bacteria are found in other formations from Spitsbergen (Butterfield *et al.*, 1994), East Greenland (Green *et al.*, 1988), Australia (Schopf, 1968; Schopf and Blacic, 1971; D.Z. Oehler, 1978; Knoll and Golubic, 1979, Southgate, 1986), China (Wang *et al.*, 1983), India (Fig. 2c) (Kumar and Srivastava, 1995; Shuckla *et al.*, 1986; Kumar and Srivastava, 1992), the Bohemian Platform (Konzalava, 1981), and the Armorican Massif of Brittany and Normandy (Chauval and Mansuy, 1981), but fossil preservation in the Nama Group of Namibia is very poor (Germs *et al.*, 1986).

3.3. THE PHANEROZOIC TO RECENT (0.56 Ga to Present)

Calcified cyanophytes associated with stromatolites and bioherms dominate the Phanerozoic record of fossil bacteria (Riding, 1992; Knoll *et al.*, 1993) but there are also an increasing number of descriptions of non-cyanophyte fossil bacteria from the Cenozoic onwards.

Phanerozoic stromatolitic cyanophytes preserved in dolomite and gypsum (Monty *et al.*, 1987), calcite (Krumbein and Cohen, 1974; Krumbein *et al.*, 1977; Horodyski and Vonder Haar, 1975), and phosphate (Martin-Algarra and Vera, 1994) have been described. Non-cyanophyte bacteria are also preserved in various minerals, e.g. phosphatised (Figs. 2e,g) and sideritised bacteria replacing the soft parts of larger vertebrate and invertebrate fossils from the Eocene volcanic lake Messel in S. Germany (Wuttke, 1983; Liebig *et al.*, 1995; Liebig, 1997) and from the Geiseltal Lagestätte of the same period (Mrugowsky, 1936). Phosphatised cyanobacteria have also been found in Cambrian soil crusts (phoscretes) in Australia (Southgate, 1986).

Cenozoic fossil bacteria have been described from some unusual environments, such as the deep sea or amber. For example, from the deep South Atlantic Ocean (3700 m water depth) exceptionally well-preserved, silica-replaced, non-cyanophyte, rod-shaped and filamentous bacteria (Fig. 2d) were found associated with a manganese nodule horizon in E. Miocene sediments (Monty *et al.*, 1991; Westall, 1994), and there have been a number of reports of bacteria preserved apparently unchanged (not even minerally-replaced) in amber (Fig. 2f) (Nissenbaum, 1991; Kohring, 1995; Poinar, 1992).

Interest in the preservation of bacteria as fossils in hot spring deposits, and in hot spring environments as potential sites where life may have originated on Earth and Mars, has lead to a spate of studies of both ancient and recent examples: the Cambrian Mount Windsor Volcanic Belt, NW Australia (Duhig *et al.*, 1992), the Late Devonian to Late Carboniferous sinters (siliceous hot spring deposits) from Queensland (White *et al.*, 1989); the Devonian Rhynie Chert from Scotland (Knoll, 1985; Trewin, 1994, 1996), the Miocene Ries Impact Crater from S. Germany (Arp, 1995), and recent hot springs on Iceland (Tazaki, 1995; Schultze-Lam *et al.*, 1995; Konhauser and Ferris, 1996), in the Yellowstone Park (Ferris *et al.*, 1986; Cady and Farmer, 1996) and in Italy (Folk, 1993; Guo and Riding, 1994).

4. Bacteria from Mars?

The much disputed Martian meteorite ALH84001 contains tiny globules of carbonate (<200 μm) deposited in impact-produced fractures in the igneous host rock, in which structures apparently similar to "nannobacteria" (20-200 nm) and larger bacteria (up to 1 μm) are described (McKay *et al.*, 1996). The globules are believed to have been formed on the surface of Mars at about 3.6 Ga, which would make the possible fossil microorganisms more or less contemporaneous with the fossil microorganisms discovered in the Early Archean of Australia and South Africa. Further investigations to identify structures of biogenic origin are underway to provide a comprehensive database of the possible microfossils contained in the meteorite and to verify their originality (i.e. to eliminate to possibility of contamination while the meteorite lay on the Antarctic glacier surface).

5. Acknowledgements

I wish to thank Laurita Boni and Roberto Barbieri for critical reading of this chapter; R. Folk, A.H. Knoll; V. Tewari, K. Liebig and R. Kohring for very kindly making available their photographs; and A. Fantazzini for his ever ready photographical help.

6. References

Ackerman, K.K., Juronen, I., and Kajander, I. (1993) *Scanning Electron Microscopy* **15**, **Suppl., 111**, 90-91.

Altermann, W. and Schopf, J.W. (1995) *Precambrian Research* **75**, 65-90.

Arp, G. (1995) *Facies* **33**, 35-90.

Awramik, S.M. (1976) in M. Walter (ed.) *Stromatolites*, Elsevier, Amsterdam, pp. 311-320.

Awramik., S.M., Golubic, S., and Barghoorn, E.S. (1972) *Geological Society of America Abstracts Program* **4**, 438.

Awramik, S.M., Schopf, J.W., and Walter, M.R. (1983) *Precambrian Research* **20**, 357-374.

Barbieri, R., D'Onofrio, S., Melis, R., and Westall, F. (in press) *Palaeogeography, Palaeoclimatology and Palaeoecology*.

Beveridge, T.J. and Murray, R.G.E. (1976) *J. Bacteriology* **127**, 1502-1518.

Buick, R. (1990) *Palaios* **5**, 441-459.

Butterfield, N.J., Knoll, A.H. and Swett, K. (1994) *Fossils and Strata* **34**, 3-83.

Byerly, G.R., Lowe, D.R., and Walsh, M.M. (1986) *Nature* **319**, 489-491.

Cady, S. and Farmer, J (1996) in G.R. Bock and J.A. Goode (eds), *Evolution of hydrothermal ecosystems on Earth (and Mars)*, Wiley, Chichester, pp. 150-173.

Chapman, D.J. and Schopf, J.W. (1983) in J.W. Schopf and C. Klein (eds.) *The Proterozoic Biosphere*, Cambridge University Press, New York, pp. 302-330.

Chauval, J.J. and Mansuy, C. (1981) *Precambrian Research* **15**, 25-42.

Cloud, P.E. Jr. and Hagen, H. (1965) *Proceedings of the National Academy of Sciences USA* **54**, 1-8.

Cloud, P. and Licari, G.R. (1872) *American J. Science* **272**, 138-149.

de Ronde, C.E.J. and Ebbesen, T.W. (1996) *Geology* **24**, 791-793.

Duhig, N.C., Stolz, J., Davidson, G.J., Large, R.R. (1992) *Economic Geology* **87**, 764-784.

Fairchild, T.R., Schopf, J.W., Shen-Miller, J., Guimaraes, E.M., Edwards, M.D., Lagstein, A., Li, X., Pabst, M., and Soares de Melo-Filho, L. (1996) *Precambrian Research* **80**, 125-152.

Ferris, F.G., Beveridge, T.J., and Fyfe, W.S. (1986) *Nature* **320**, 609-611.

Ferris, F.G., Fyfe, W.S., and Beveridge, T.J. (1988) *Geology* **16**, 149-152.

Francis, S., Margulis, L., and Barghoorn, E.S. (1978a) *Precambrian Research* **6**, 65-100.

Francis, S., Barghoorn, E.S. and Margulis, L. (1978b) *Precambrian Research* 7, 377-383.

Folk, R.L. (1993) *J. Sedimentary Petrology* **63**, 990-999.

Folk, R. and Lynch, L (1997b) *J. Sedimentary Research* **67**, 583-589.

Germs, G.J.B., Knoll, A.H., and Vidal, G. (1986) *Precambrian Research* **32**, 45-62.

Ghiorse, W.C. and Hirsch, P. (1978) in W.E. Krumbein (ed.) *Environmental Biochemistry and Geomicrobiology*, Ann Arbor Science, Ann Arbor, Michigan, v3, p. 897-909.

Ghoirse, W.C. and Hirsch, P. (1979) *Archives of Microbiology* **123**, 213-226.

Golubic, S. (1973) in N.G. Carr and B.A. Whitton (eds.) *The Biology of Blue-Green Algae*, Blackwell, Oxford, pp. 434-472.

Golubic, S. and Barghoorn, E.S. (1977) in E. Flügel (ed.) *Fossil Algae*, Springer, Berlin, pp. 1-14.

Golubic, S., Sergeev, V.N., and Knoll, A.H. (1995) *Lethaia* **28**, 285-298.

Green, J.W., Knoll, A.H. and Swett, K. (1988) *J. Paleontology* **62**, 835-852.

Guo, L. and Riding, R. (1994) *Sedimentology* **41**, 499-520.

Hofmann, H. (1976) *J. Paleontology* **50**, 1040-1073.

Hofmann, H. and Schopf, J.W. (1983) in J.W. Schopf and C. Klein (eds.) *The Proterozoic Biosphere*,

Cambridge University Press, New York, pp. 321-360.
Holland, H.D. and Beukes, N.J. (1990) *American J. Science* **290A**, 1-34.
Horodyski, R.J. and Vonder Haar, S.P. (1975) *J. Sedimentary Petrology* **45**, 894-906.
Horodyski R.J. and Donaldson, J.A. (1980) *Precambrian Research* **11** 125-159.
Horodyski R.J. and Donaldson, J.A. (1983) *J. Paleontology* **57**, 271-288.
Klein, C., Beukes, N.J., and Schopf, J.W. (1987) *Precambrian Research* **36**, 81-94.
Kohring, R. (1995) *Neues Jahrbuch für Paläontologie Mh.* **6**, 321-335.
Konhauser, K.O. and Ferris, F.G. (1996) *Geology* **24**, 323-326.
Konzalova, M. (1981) *Precambrian Research* **15**, 43-62.
Knoll, A.H. (1985) *Philosophical Transactions Royal Society, London* **B311**, 11-122.
Knoll, A.H. (1996) in J. Jansonius and D.C. McGregor (eds.) *Palynology: principals and applications*, American Association of Stratigraphic Palynologists Foundation, pp. 51-80.
Knoll, A.H. and Barghoorn, E.S. (1977) *Science* **198**, 396-398.
Knoll, A.H. and Golubic, S. (1979) *Precambrian Research* **10**, 115-151.
Knoll, A.H., Strother, P.K., and Rossi, S. (1988) *Precambrian Research* **38**, 257-279.
Knoll, A.H., Swett, K., and Mark, J. (1991) *J. Paleontology* **65**, 531-570.
Knoll, A.H., Fairchild, I.J., Swett, K. (1993) *Palaios* **8**, 512-525.
Konhauser, K.O. and Ferris, F.G. (1996) *Geology* **24**, 323-326.
Krumbein, W.E. and Cohen, Y. (1974) *Geologische Rundshau* **63**, 1035-1065.
Krumbein, W.E., Cohen, Y., and Shilo, M (1977) *Limnology and Oceanography* **22**, 635-656.
Kumar, S. and Srivastava, P. (1992) *Precambrian Research* **56**, 291-318.
Kumar, S. and Srivastava, P. (1995) *Precambrian Research* **74**, 91-117.
Lanier, W.P. (1986) *Palaios* **1**, 525-542.
Lanier, W.P. (1989) *Precambrian Research* **45**, 291-318.
Leo, R.F. and Barghoorn, E.S. (1976) *Botanical Museum Leaflet Harvard University* **25**, 1-47.
Liebig, K. (1997) *Investigation of ultrastructures from the Eocene Messel Formation (Messel oil shale pit near Darmstadt, Germany) with emphasis on fossil microorganisms*, unpub. PhD dissertation, Univ. Heidelberg.
Liebig, K., Westall, F., and Schmitz, M. (1996) *Neues Jahrbuch für Geologie Paläontologie Mh.* **4**, 218-231.
Martill, D.M. and Wilby, P.R. (1994) in M. Schmitz, K. Ernst (eds.) Microorganisms, Facies Analysis and Fossil Diagenesis - *Kaupia Darmstädter Beiträge z. Naturgeschichte* **4**, 71-77.
Martin-Algarra, A. and Vera, J.A. (1994) in J. Bertrand-Sarfati and C. Monty (eds.) *Phanerozoic Stromatolites II*, Kluwer, Dordrecht, pp. 345-391.
McKay, D.S., Gibson, Jr. E.K., Thomas-Keprta, K.L., Vali, H. Romanek, C.S., Clemett, S.J., Chillier, X.R.D., Maedling, C.R., and Zare, R.N. (1996) *Science* **273**, 924-930.
Mojzsis, S.J., Arrhenius, G., McKeegan, K.D., Harrison, T.M., Nutman, A.P., and Friend, C.R.L. (1996) *Nature* **384**, 55-58.
Monty, C.L.V., Rouchy, J.M., Maurin, A., Bernet-Rollande, M.C., and Perthuisot, J.P. (1987) in T.M. Peryt (ed.) *Evaporitic Basins*, Springer, Berlin, pp. 133-188.
Monty, C.L.V., Westall, F., and van der Gaast, S. (1991) in. P.F. Ciesielski, Y. Kristoffersen *et al.*. (Eds) *Proc. of the Ocean Drilling Program, Scientific Results* **114** , ODP, College Station, pp. 685-710.
Mrugowsky, J. (1936) *Nova Acta Leopoldina, N.F.* **17**, 597-606.
Muir, M.D. (1976) *Alcheringia* **1**, 143-158.
Nissenbaum, A. (1991) in J.C. Berthelin (ed.) *Diversity of Environmental Biogeochemistry (Developments in*

Geochemistry **6**), Elsevier, Amsterdam, pp. 15-21.

Oehler, J.H. (1976) *Geological Society America Bulletin* **87**, 117-129.

Oehler, D.Z. (1976) *J. Paleontology* **50**, 90-106.

Oehler, D.Z. (1978), *Alcheringia* **2**, 269-309.

Oehler, J.H. (1977) *Alcheringia* **1**, 315-349.

Oehler, J.H. and Schopf, J.W. (1971) *Science* **174**, 1229-1231.

Pentecost, A. and Riding, R. (1986) in B.S.C. Leadbeater and R.Riding (eds.) *Biomineralisation in lower plants and animals, Systematics Association Spec. Pub.* **30**, 73-90.

Poinar, G.O. (1992) *Life in Amber*, Stanford University Press, Stanford.

Riding, R. (1992) *J. Geological Society London* **149**, 979-989.

Schopf, J.W. (1968) *J. Paleontology* **42**, 651-688

Schopf, J.W. (1975) *Annual Review of Earth and Planetary Sciences* **3**, 213-249.

Schopf, J.W. (1983) (ed.) *Earth's Earliest Biosphere*, Princeton University Press, New Jersey.

Schopf, J.W. (1990) *J. Paleontology* **44**, 1-6.

Schopf, J.W. (1992a) in J.W. Schopf and C. Klein (eds.) *The Proterozoic Biosphere*, Cambridge University Press, New York, pp. 25-42.

Schopf, J.W. (1992b) in J.W. Schopf and C. Klein (eds.) *The Proterozoic Biosphere*, Cambridge University Press, New York, pp. 195-218.

Schopf, J.W. (1993) *Science* **260**, 640-646.

Schopf, J.W. and Blacic, J.M. (1971) *J. Paleontology* **45**, 925-960.

Schopf, J.W. and Klein, C. (1992) *The Proterozoic Biosphere*, Cambridge University Press, Cambridge.

Schopf, J.W. and Walter, M.R. (1993) in J.W. Schopf (ed.) *Earth's Earliest Biosphere*, Princeton University Press, New Jersey, pp. 214-239.

Schopf, J.W. and Packer, B.M. (1987) *Science* **237**, 70-73.

Schultze-Lam, S., Ferris, F.G., Konhauser, K.O., and Wiese, R.G. (1995) *Canadian J. Earth Science* **32**, 2021-2026.

Sergeev, V.N. (1994) *Precambrian Research* **65**, 231-254.

Sergeev, V.N., Knoll, A.H., Kolosova, S.P., and Kolosov, P.N. (1994) *Stratigraphy and Geological Correlation* **2**, 19-33.

Sergeev, V.N., Knoll, A.H. and Grotzinger, J.P. (1995) *J. Paleontology Mem.* **39**.

Shuckla, M., Tewari, V.C., and Yadav, V.K. (1986) *The Palaeobotanist* **33**, 347-356.

Southgate, P.N. (1986) *J. Sedimentary Petrology* **56**, 429-441.

Tazaki, K. (1995) *J. Geological Society, Japan* **101**, 304-314.

Trewin, N.H. (1994) *Transactions Royal Society Edinburgh* **84**, 433-442.

Trewin, N.H. (1996) in G.R. Bock and J.A. Goode (eds), *Evolution of hydrothermal ecosystems on Earth (and Mars)*, Wiley, Chichester, pp.131-144.

Tyler, H.S.A. and Barghoorn, E.S. (1954) *Science* **119**, 606-608.

Venkatachala, B.S., Shukla, M., Sharma, M., Naqvi, S.M., Srinvasa, R., and Udairaj (1990), *Precambrian Research* **47**, 27-34.

Walsh, M.M. (1992) *Precambrian Research* **54**, 271-293.

Walsh, M.M. and Lowe, D.R. (1985) *Nature* **314**, 530-532.

Wang, F., Zhang, X., and Ruihuan, G. (1983) *Precambrian Research* **23**, 133-175.

Westall, F. (1994) in M. Schmitz, K. Ernst (eds.) Microorganisms, Facies Analysis and Fossil Diagenesis - *Kaupia Darmstädter Beiträge z. Naturgeschichte* **4**, 29-43.

Westall, F. (1997a) The nature of fossil bacteria. *Workshop on Early Mars, Houston, April,* Astract.

Westall, F. (1997b) in C. Cosmovici, S. Bowyer and D. Werthimer (eds.) *Astronomical and Biochemical Origins and the Search for Life in the Universe*, Editori compositrici, Bolgna, pp. 491-504.

Westall, F., Boni, L., and Guerzoni, M.E. (1995) *Palaeontology* **38**, 495-528.

Westall, F. de Wit, M.J. Dann, J. (1997) *28th Lunar and Planetary Society Conference, Houston, March,* Abstract.

Westall, F., de Wit, M.J., Dann, J., van der Gaast, S., and de Ronde, C.E.J. (in prep.) *Fossil bacteria and microbial mats from the Early Archean Onverwacht Group, South Africa.*

White, N.C., Wood, D.G., and Lee, M.C. (1989) *Geology* **17**, 718-722.

Wuttke, M. (1983) *Senckenbergiana lethaia* **65**, 509-527.

Zhang, Z. (1986) *J. Micropalaeontology* **5**, 9-16.

Zhang, Y. (1985) *Precambrian Research* **30**, 277-302.

Zhang, Y. (1988) *Precambrian Research* **38**, 105-175.

Biodata of **Barbara Z. Siegel** contributor of "***Growth of Organisms in Ammonia: Kakabekia' A microbial Enigma.***"

Dr. Barbara Z. Seigel received her Ph.D. from Yale University in Biology (Physiology and Biochemistry). For 28 years she was on the faculty of the University of Hawaii where she served as faculty, Dean, and Director of Research. In addition to her teaching at the University of Hawaii, she has tough in Finland, Germany, the former Yugoslavia, Israel, Italy, China and the Federated States of Micronesia. She is the author or coauthor of more than 150 peer-reviewed scientific papers and sits on the editorial board of two international Journals. Her major research and professional interest include environmental public health with an emphasis on natural and man-made disasters. She has also worked on the sources, sinks and mitigation of environmental mercury, lead and other pollutants. As part of her studies on environmental stress physiology (heat, salt, radiation, heavy metals, pollutants, etc.) she has used the enzyme peroxidase as an indicator that an organism is under stress. She has been publishing on the physiology and evolution of peroxidases for more than 30 years. Dr. Siegel has also contributed to elucidating peace processes through improving the general health and the quality of life in developing countries.

E-Mail: **bsiegel@zianet.com**

J. Seckbach (ed.), Enigmatic Microorganisms and Life in Extreme Environments, 89–96.

GROWTH OF ORGANISMS IN AMMONIA:
Kakabekia, A Microbial Enigma

B.Z. SIEGEL
Ecotrophics
4215 Superstition Drive
Las Cruces, NM 88011

1. Introduction

This paper focuses on a few of the organisms, including a relict form, that utilize ammonia as their sources of nitrogen or can live under conditions with a minimum of 50% ammonia -50% air at pHs above 9. Though many bacteria can grow in ammonia and some are even ammonia requiring, only a few heterotrophic flagellate protozoans have been reported to survive with ammonia as their only nitrogen source. Bacteria in ruminants also utilize ammonia and other comparatively simple organic compounds such as urea, which can act as a source of nitrogen for synthesizing amino and nucleic acids within the host. A number of higher plants, especially *Allium sp.* are capable of germination in high ammonia environments (Siegel and Daly, 1996). A critical element in any community metabolic system is nitrogen, as it constitutes a major component of amino and nucleic acids.

The reservoir of inorganic nitrogen is the gaseous N_2 molecule which comprises roughly 75% of the atmosphere. However N_2 has very little biological activity as it enters and leaves cells. Most of the unstable nitrogen for the earth's ecosystems is formed through the process of nitrogen fixation. Many nitrogen-fixing bacteria live in close symbiotic relationships with root nodules of higher plants. Legumes are the best known for possessing nitrogen fixing nodules (*Rhizobium*). Other non-leguminous higher plants which can fix nitrogen include *Hippoplae, Shepherdia, Myrica,* Scots pine and *Ceanothus* (Keeton, 1967).

Some nitrogen-fixing organisms are capable of ammonia production as well, along with a number of decomposers which also produce ammonia as part of their catabolic process. The availability of ammonia-rich microhabitats should be anticipated and not be allowed to act as a barrier to research and the finding of organisms such as *Kakabekia.* It should be remembered that all of today's organisms are equidistant from the event(s) that occurred at the origin of life, and that both *Kakabekia* and their niches may often still be found today. One of the most striking discoveries was the 1958 appearance of the Coelancanth, a primitive fish that was thought to have become extinct 150 million years ago.

2. *Kakabekia* – a Living Fossil

"Living fossils" appear to extend to the most ancient forms of life (Day, 1979). The

Precambrian microfossil, *Kakabekia umbellata* first described by Barghoorn and Taylor (1965) was found in the two billion-year-old Gunflint chert at the northern shore of Lake Superior in Ontario, Canada. The cherts contained the microfossil, *Kakabekia*, along with other microfossils and they have been assigned to the mid-Precambrian, about 2 x 10^9 years of age. *Kakabekia* is considered as an organism with uncertain taxonomic affinities. It is an unusual form which consists of an umbrella-like cap with a slender stalk and basal bulb. In the fossil-and extant-form the width of the organism is approximately of 5-10 microns across the cap and the stalk of stipe is approximately the same length.

K. barghoorniana is a rare living member of the genus and was first discovered in soils from Harlech., Wales in 1964 (Siegel and Guimarro, 1965, 1966). The similarities between the fossil and extant form are striking but not identical. *K. umbellata* has its cap highly incised, whereas the living form, *K. barghoorniana*, usually consists of scalloped or entire circles of protoplasm. The bulbous end of the stalk is extremely rare in *K. barghooniana* but has been seen though smaller than described in the fossil form. The living organism has a flat disk-like form with a thick peripheral ring of about 2 microns thick. These forms have a central prominence, possible a "stipe primordium."

The two known species of *Kakabekia* should most likely be placed in the Kingdom Monera and in the division Schizomycetes of uncertain phylogenetic affinities, and they are often grouped with other stalked bacteria. In the Bitter Spring formation of Australia (reported older than the Gunflint chert) Schopf (1979) found a number of microfossils including three species of bacteria, 20 blue-green algae (cyanobacteria) and some possible fungal arrays, but none were similar to the earlier reported *Kakabekia* (Siegel, 1977).

The established location of the *Kakabekia* fossils from within southern Ontario chert and a living species originally from the topsoil of Wales argued for a wider distribution of the living forms; their restrictions to ammoniacal micro habitats notwithstanding.

Between the fossil *K. umbellata* and the living *K. barghooniana* stands some 2 x 10^9 years and some four thousand miles. It is reasonable to assume that some of the genera have had ample opportunity for a wider distribution on earth. The fact that it has not been reported in soil cultures previously may merely reflect an unusual requirement for ammonia (and/or high pH). It may also reflect the failure of *Kakabekia* to survive competition with the large numbers of rapidly growing soil microorganisms under the routine culture conditions used in soil microbiology.

Almost since its discovery, *K. barghoorniana* has routinely been grown in 15 M aqueous ammonia with 0.5% glucose and at 7^0C. A number of culture media have been tested but the greatest yields of *Kakabekia* have been found in protocols using 0.5 grams of soil in 10 ml of the ammonia/glucose media. Initially inoculated media were incubated at 20^0C in filled, stoppered vials, excluding most free air space. Improved numbers of organisms were seen for longer periods of time when the cultures were maintained at 7^0C rather than at warmer temperatures (Siegel, B.Z. and Siegel, 1971b). When initially sampled, no *Kakabekia*-like forms were apparent in the growth medium, but a peak concentration of *K. Barghoorniana* appeared about day 14. The concentration of organisms slowly disappear and on days 25-30 a secondary peak, though reduced from the original one, was noted. Sometimes a third peak was seen at about 40 days but much reduced in numbers compared to the earlier ones. Under these growth conditions an increase in the total number of organisms was not seen compared to the original flush after about 14 days in laboratory culture. Prolonged culture led to a multiradiate mantle 5-15 microns in diameter, but degeneration of the entire structure followed. In every case so far a fresh inoculum has been required for *K. barghoorniana* to appear after inoculations after about 6 weeks.

Three separate soils samples taken a year apart from the Harlech Castle area in Wales during 1967 confirmed the original reported data. Observations suggest that the umbellate form is a mature stage. Barghoorn reported possible similar ontogenetic changes in the fossil with the contents of the stipe apparently moving toward the cap. The precursor in dry soils and in culture up to 5 days was not evident (Siegel and Siegel, 1967).

Table 1.
Inocula from various sites that yielded the growth of *Kakabekia* (after Siegel B.Z. and Siegel, S.M. 1979)

Location	Presence of *Kakabekia* after 7 days culture in	
	15 M aqueous NH_3	0.001 M NaOH
Lake Hakone Area near Mitochondrion. Fuji	-	-
Low hills near Taipei	-	-
	Elevation above 2000 meters	
Himalayan foothills near Katmandu, Nepal	+	-
Outskirts of Mexico City	+	
Bogota, Columbia	+	-
Quito, Equador	+	
Mauna Kea, Hawaii (4000 m)	+	+
Haleakala, Maui (2800 m)	+	
	Latitude above 45^{0}N	
Sapporo, Japan	+	
Showa Shinzan, Japan	+	
Point Barrow, Alaska	+	+
Mendenhall Glacier, Alaska	+	+
Iceland-Reykjanes peninsula	+	+

In water mounts, the extant *Kakabekia*–like forms stained readily with eosin, crystal violet, and other dyes, but staining did not significantly increase structural detail. Extensive soil surveys over much of the globe have been made to detect *Kakabekia*. In all but one Iceland sample, no living specimens have been found. In most of the cases, the typical *Kakabekia* form is not seen until the soils have been cultured for about one week in an ammonia–glucose solution.

In cultures grown from a wide varieties of environmental conditions, two new conclusions could be drawn. One was that for some forms of *Kakabekia* the pH was more critical than the ammonia *per se*, and second that cool temperatures were probably critical for the maintenance of *Kakabekia* in nature. Table 1 lists some of the locations from where *Kakabekia* has been cultured. either the living organism came from a site above 2000 meters or is found at latitudes above 40^0N (Table 1). Samples from the southern hemisphere, New Zealand and McMurdo (Antarctica) yielded no identifiable specimens of *Kakabekia*.

All of the living forms found up to this point tend to be about 80% entire and 20% incised like most of the fossil forms. These morphological differences might have resulted from the fossilization process and/or different growth conditions. The microorganisms (companion forms) found in association with both the fossil form and the cultured organism include similar cyanophytes (cyanobacteria), diatom-like algae plus some fungus-like structures (Siegel, SM., 1965).

During several decades a number of attempts have been made to find a medium which would allow *K. barghoorniana* to be isolated in pure culture and generally to increase the overall yield. Inorganic nutrients, amino acids and vitamins, separately and in combination; Eagle-Hanks medium; and complex organic fluids including blood, milk, coconut milk and urine were all tested. Liquid culture media, both shaken and stagnant were compared with the semi-solid media. Heat and radiation (Co^{60}) prevented the appearance of significant *Kakabekia*, and rather than enriching the organism under study, these conditions allowed for the overgrowth of blue-green algae (cyanobacteria) and eubacteria.

The extant *Kakabekia* was stained for it ability to carry out a number of enzymatic reactions, as well as for Feulgen-positive material. Peroxidase, catalase and cytochrome-oxidase were positive in about 20% all of the *K. barghoorniana* forms. They were 100% negative for starch and lipids in their cytoplasm. Feulgen-positive (DNA?) material was seen as "clots" throughout the cytoplasm and no distinct membranes were seen internally. These findings would place the organism into the prokaryotes with an ability to survive, not only the ammonia, but also weakly oxidizing and cool environments. Efforts at electron microscopy were not successful, since the thick ring shattered during efforts at thin-sectioning. Selective incineration (700^0C) of the material gave some indication that the ring found internal to the cap of *K. barghoorniana* might be siliceous.

Soil analysis (Siegel, S.M. et al., 1967) was conducted by spectrography (calculated as oxides) yielded in mg/gm dry weight: silica as 766-779, FeO as 11-71, Al_2O_3 as 3-5, K_2O as 4.8-9.6, etc. In neither the principal inorganic nor trace inorganic chemical compounds was any unexpected data forthcoming. The general profile of soil constituents are similar to the criteria used for identifying loams (Bear).

In summary, the similarities between the fossil form, *Kakabekia umbellata*, and the extant variety, *Kakabekia barghooniana*, include size, basic morphology, and the presence of a alga-like bodies, morphologically identical to modern blue-green algae (cyanobacteria such as Oscillatoriaceae) both as microfossils in the microflora of the Gunflint chert and as part of the living samples form Harlech and other soil samples.

3. Life in Ammonia

As part of the early space program, culturing organisms in high ammonia environments was part of a program to elucidate the types of organisms that could possibly grow in these harsh environments. Ammonia was of special interest since it is known to be high on Jupiter, as well as being a likely component in the earth's early evolution. This work led to the discovery of *Kakabekia* but also revealed unusual capabilities of certain species of higher plants. Whether these tolerances are residual capabilities for the survival in more extensive ammonia niches in past history, are attributes bestowing a current evolutionary advantage or incidentally bred into the commercial varieties, remains unanswered.

In animals the toxicity of ammonia is almost equivalent to carbon monoxide. When seeds were incubated under NH_3 (50%) and air (50% by volume), it was found that germination was totally and irreversibly inhibited in various members of the Gramineae (wheat, rice, rye), Compositae (lettuce, sunflower), Leguminosae (bean, pea), Cruciferae (turnip, cabbage, radish), Labiateae (mint, sage, coleus) and representatives of other Angiosperm families.

In spite of the strong phytotoxic record, NH_3 failed to inhibit completely seeds of *Allium sp.* A resistant fraction, under 15% of various populations, constitutes a physiologically distinct group. Twenty-five cultivars of commercial onion (*Allium cepa*) and 4 of leek (*A. porrum*) as well as wild onion (*A. cernuum*) [were] collected in Virginia. Of this group about 80% of the onion species, all of the leek and none of the wild seeds showed germination. The responses were largely imbibitional swelling but leek in 100% ammonia showed about 5% dividing cells compared to 14% in water/air, and onion (Evergreen cultivar) had 4% dividing cells compared to water/air controls. The chromosomes from root cells from the seeds germinated under ammonia were abnormal having the appearance of highly attenuated elongated spirals, and metaphases were poorly defined (Siegel, 1965). Though these experiments with *Allium* species did not distinguish between ammonia-tolerance and the tolerance to high pHs, it did indicate that even everyday organism have tolerances to exceedingly harsh environments.

Once the search for life on other planets is seriously begun, these experiments will support the concept that life will be found whether as relicts or highly adaptable species now extant.

4. Summary

Ammonia may have been part of the earth's primitive atmosphere as well as a constituent of current Jupiter's gases. The 2 x 10^9 year old the microfossil from the Gunflint cherts of Ontario, Canada and its possible affinity to a living form with ammonia/air requirements are of importance in the study of life in the universe. *K. umbellata*, the fossil form, is similar in morphology, size, and associated microorganism very similar to that found with the living forms. The extant form, *K. barghoorniana* is a cryophile with a temperature optimum of 7^0C and a requirement for high ammonia levels (minimum of about 50% in water by volume). Other organisms which grow and/or survive in ammonia are also discussed, in an effort to elucidate environmental niches where unusual and unstudied organism may exist. The unique conditions may provide habitats for microorganisms with more direct lineal ties to ancient forms than have hitherto been believed impossible.

5. References

Barghoorn, E. and Tyler, S (1965) Science 147, 563-577.
Day, W. (1979) Genesis on Planet Earth, The House of Talos Publishers, East Lansing, Mich.
Keeton, W.T. (1967) Biological Science. Northon and Company. New York.
Schopf, J. W. (1979) In: C. Ponnampeuma (ed.) Exobiology. North Holland Publishing Company.
Siegel, B.Z. (1977) In: Chemical Evolution and the Early Precambrian. Academic Press, Inc. New York.
Siegel, B.Z. and Siegel S.M. (1971a) Nat. Aeronaut. Space Admin. Tech. Memo X-62,009.Section 8.11. pp 117-121.
Siegel, B.Z and Siegel, S.M. (1971b) Abstract, Cryobiology 8:388 (3rd Int. Congr. Cryobiology, Washington, September 1971).
Siegel, B.Z. and Siegel S.M. (1979) National Geographic Society Research reports 12, 639-646.
Siegel, S.M. (1965) National Aeronautics and Space Administration Semi-Annual Report. Contract No. NAS-w-767.
Siegel, S. M. and Giumarro, C. (1965) Icarus 4, 37-40.
Siegel, S.M. and Giumarro, C. (1965) Proc. Nat. Acad. Sci. (USA) 55, 349-353.
Siegel, S.M., Roberts, K., Nathan, H. and Daly, O. (1967) Science 156, 1231-1234.
Siegel, S.M. and Siegel, B.Z. (1968a) Proc. Nat. Acad. Sci (USA) 67, 1005-1010.
Siegel, S.M. and Siegel, B.Z. (1968b) Amer. J. Bot. 55, 684-687.

II. ENIGMATIC MICROORGANISMS

Biodata of **David J. Chapman**, contributor of ***Enigmatic Unicellular Protista.***

Dr. **David J. Chapman** is a Professor in the Department of Ecology, Evolution and Marine Biology at the University of California, Santa Barbara. He earned his PhD (1965) at the Scripps Institute of Oceanography (San Diego, CA) in Marine Biology. His research interests cover a broad spectrum of algal and protistan biology. Particular interests include algal (and animal) pigments, the role of algal natural products in the marine environment, applied uses of algae. The evolution of biochemical systems in protists is his other major interest. The emphasis is the interdisciplinary approach, tied to the evolution/changes that occurred in the atmosphere, hydrosphere during the Proterozoic and early Palaeozoic eras.
E-Mail: dchapman@descartes.ucsb.edu

J. Seckbach (ed.), Enigmatic Microorganisms and Life in Extreme Environments, 99–111.

ENIGMATIC UNICELLULAR PROTISTA: ARE THEY REALLY ENIGMATIC? THE ALGAE CASE.

DAVID J. CHAPMAN
Department of Ecology, Evolution and Marine Biology
University of California
Santa Barbara, CA 93106 U.S.A

1. Introduction

The Oxford Dictionary describes "enigmatic" as "puzzling". This contribution could perhaps then be retitled . Are there any truly puzzling algae? Such algae would be those that are distinguished from their presumptive relatives by differences in their physiological, biochemical, structural/ultrastructural and other cellular and molecular features such that their phylogeny and evolution does not seem to fit into any obvious pattern. Frequently phycologists are tempted to place an algal oddity in a pigeon-hole (= taxon) before all the data are assembled or sufficient knowledge is available to make an informed decision. These algae often do not seem to fit into the existing delineations of the higher taxa, particularly those of Class and Division. This lack of a comfortable fit or pigeon hole is often a result of our own lack of flexibility, our own slowness in assembling the necessary data needed to make the informed decision, our reluctance to embrace new techniques of analysing the questions and to challenge established dogma. Much of this is steeped in historical precedent. Current algal classification has its origins in the system proposed by Pascher (1931) some 65 years ago. The tools available at that time were those of light microscopy and very rudimentary and basic chemical analyses. Later , improved methods of chemical analyses provided a measure of increased detail and refinement. These provided the data bases upon which the system was built and subjected to minor modifications. The next level of improved sophistication, applied in the early 1960's was that of electron microscopy and greatly improved chemical analyses brought about by the physical techniques of nmr and mass spectrometry. These were followed by the next level, namely that of protein and gene sequencing. These increasing levels of sophistication and refinement have allowed us to base our systematics on criteria that more closely approach the genome. This is perhaps the real goal of a sound systematics and taxonomy, the importance of which was emphasised by Ragan & Chapman (1978). This experimental trend has sometimes been paralleled by the concerted and focussed efforts of researchers dedicated to expanding the data base and asking what particular information should improve the chances of resolving the unanswered questions.

2. Case Studies Of Enigmatic Algae

2.1. THE BASIC THESIS

The thesis of this contribution is that "enigmatic " algae are that simply because of our own ignorance and a temptation to "assign" all algae to one of the existing classes originally proposed by Pascher, or a "spin-off" class thereof. The problems generated by these are further compounded by the fact that the first representatives to be analysed and described in depth are often assumed to be the norm or typical representative of that class or divisional taxon or at least assigned that status by default. To state it quite simply, the more one learns about these "enigmatic" algae, the less enigmatic they become. In support of this proposition I will consider three examples: The **Cyanidium, Galdieria, Cyanidioschyzon** complex, the Glaucocystophyceae (**Cyanophora, Glaucocystis, Gloeochaete**) and four classes (or divisions), the photosynthetic representatives of which are currently considered to have derived as a result of secondary symbioses, i.e. Euglenophyceae, Cryptophyceae, Chlororachniophyceae and the Dinophyceae. I am deliberately restricting the discussion to eukaryotes/protists and avoiding any discussion of the Prochlorophyceae.

I am following my own admonition (Chapman, 1992) that the algae are not a natural taxonomic group, and that the term is one of convenience to simply describe those protists that are photosynthetic or their obviously very closely related colourless counterparts. A similar view was expressed by Gibbs (1993, page 107) who commented that "The algae are not a natural assemblage of organisms. Rather they are a diverse group of protists and fungi which have acquired chloroplasts in various ways." We all know, or pretend to know, what or who are the algae even if it is impossible and fruitless to attempt to provide a satisfactory definition. This article is not intended to be a literature review of all the pertinent data. In the last few years there have been a number of review articles or edited volumes dedicated specifically to the genera or groups that are considered below. The reader is referred to these sources for detailed information

2.2. CYANIDIUM, GALDIERIA, CYANIDIOSCHYZON

A very detailed review of all aspects of these three genera has recently appeared (Seckbach, 1994a) and serves as an excellent data-base for our current knowledge of these genera. That compendium also includes a complete literature compilation so there is no need to provide or reference the earlier literature. Within this volume four reviews (Ott & Seckbach, 1994a,b; Seckbach, 1994b; Seckbach & Ott, 1994) trace the systematic and phylogenetic proposals and arguments for these genera. An earlier review (Seckbach, 1992) had discussed specific aspects of the phylogeny of **Cyanidium** with particular emphasis on proposals and ideas dealing with the possible symbiotic nature of this alga. **Cyanidium** has undergone revison, going from a single genus to a split into **Cyanidium, Galdieria** (Merola et al., 1981) and **Cyanidioschyzon** (De Luca et al., 1978) and then a subsequent coalescing of **Cyanidium** and **Galdieria** (Ott & Seckbach,1994b) into **Cyanidium** again with **Cyanidioschyzon** aligned with the

Porphyridiales (Ott & Seckbach, 1994a). There is no evidence to suggest that these three genera are not very closely related. Molecular evidence (Ohta et al., 1997) indicates a close relationship between the two latter genera. However, the development and changing patterns of all these ideas, from the systematics of the genera to the assignment to particular orders and classes, serve as a quintessential example of how the increasing resolution provided by techniques that have allowed us to use criteria closer and closer to the genome have gradually resolved the systematic and phylogenetic conundrums to the point that the description enigmatic is no longer appropriate.

The genus **Cyanidium** was originally established by Geitler (1933) for an alga that had been first described in 1889. By any comparison with traditional algae (eukaryotic protists) it occupied an unusual ecological habitat, being acidophilic and thermophilic and typically found in hot springs. In its early history the genus did not seem to belong comfortably in any existing taxon. By any of the taxonomic criteria then in use, **Cyanidium** always stood out as an exception or very different from other genera regardless of what higher taxa to which it was assigned. In retrospect we can now see that much of that was simply a reflection of our ignorance of the diversity of classes and divisions and perhaps our own desires to keep classes and divisions within tight and easily defined boundaries. **Cyanidium**, after its formal establishment by Geitler, was shunted around among the Chlorophyceae (e.g. Hirose, 1950; Allen, 1959) the Cryptophyceae (Lewin, 1961), and Rhodophyceae (Hirose, 1958). Admittedly all these authors had some reservations about their assignment. While there had been numerous earlier publications suggesting either chlorophytan or cyanobacterial affinities the work of Hirose and then Allen must be considered the seminal starting points for the ultimate resolution of the systematics: Hirose for his emphasis on the presence of a plastid (i.e. the eukaryotic nature) and Allen for her placing the organism in culture and the subsequent preliminary physiological/biochemical studies that identified many of its unusual attributes. These early systematic assignments were based largely upon gross morphological characteristics, then ultrastructure, and more refined chemical analyses especially of pigments, cell walls and storage reserves all of which were basic chemical criteria for algal classifications at that time. However, even then, our data bases were limited particularly, also, at the level of identified chemical structures so that even the approaches later suggested by Birch (1973), such as the use of biosynthetic pathways, for the handling of chemical data with discrimination were really precluded. Despite these problems there was a trend to a rhodophycean classification. Chapman (1965) proposed an order Cyanidiales within the Rhodophyceae which was later amended to a family Cyanidiaceae within the Porphyridiales (Chapman, 1974). Despite a side excursion for a possible placement with the cyanelle-containing genera, **Glaucocystis, Cyanophora** and **Gloeochaete** (and hence not Rhodophyceae), proposed by Kremer et al. (1978) and Kremer (1982), **Cyanidium** (and latterly the more recently described and related genera **Galdieria** and **Cyanidioschyzon**) has been accepted as a red alga for the last 25-30 years. A number of authorities have accepted this but with greater or lesser degrees of assurance and acceptance. Part of the hesitancy is obviously due to the unusual environment (by traditional rhodophycean standards), and physiology and biochemistry (e.g. acidophilic, thermophilic, heterotrophic capability). In these regards the genera are certainly distinct from the traditional single-celled red algal genera such as **Porphyridium**. But these attributes, per se, are not grounds for exclusion from the red algae. The Rhodophyceae are not defined on the basis of habitat or such physiological features as heterotrophic capability.

It was to be expected that molecular data, i.e protein and gene sequences, would help clarify the status of these genera. Certainly there has been no disappointment, starting with the ferredoxin sequences of two strains (Hase et al., 1985; Nagashima et al, 1989) of **Cyanidium**. Sequences subsequently determined for the psbA gene and plastid 16S rRNA gene (Maid & Zetsche, 1990, 1991), and the genes for both large and small subunits of Rubisco (Valentin & Zetsche, 1990) demonstrate that the plastid of **Cyanidium** is clearly that of a Rhodophyceae. Concatenated amino acid sequences of the mitochondrial proteins COX1-COX-COB place the **Cyanidium** mitochondrion with those of the red algae (one other genus sequenced) (LeBlanc et al. 1997), although it must be pointed out that the data base in this regard is very small. The available gene and protein phylogenies confirm the evolutionary relationship of these three genera to the red algae. When taken in the totality of data for other algal groups it is clear that these genera are appropriately classified within the red algae. The recent proposal, therefore, of Ott and Seckbach (1994b) for a classification within the Rhodophyceae, as an order Cyanidiales, would appear justified. These algae are no longer 'enigmatic": instead we are now compelled to regard the Rhodophyceae as more diverse than previously accepted.

The lesson from this example is simple. The more one learned about these algae and the closer one got to the genome, the less enigmatic they became. Indeed they are no longer such. Questions as to whether they are bridge red algae, primitive or simple must await the accumulation of larger data-bases. By comparison with other red algae they are unusual; but this designation derives largely from the fact that our own prejudices about red algae have always associated them as almost entirely multicellular marine forms, i.e. members of a narrowly defined ecological environment. In fact only about 3% of the genera and species of Rhodophyceae are not marine and there are few truly single-celled genera and species. It is reasonable to expect that a representative from an acid hot spring would be different with regard to some aspects of its physiology and biochemistry.

2.3. CYANOPHORA, GLAUCOCYSTIS, GLOEOCHAETE, AND PAULINELLA

These four genera represent another example where a lack of information and perhaps a reluctance to move outside of the Pascherian or traditional class systematics of the algae resulted in their being considered as puzzling (i.e. enigmatic) or at the very least most unusual. The result was a checkered systematic history that appears now to have been resolved in favour of establishing their own separate class as proposed by some early researchers. Before beginning the discussion the reader is referrred to three key reviews that summarise early knowledge of physiology, biochemistry and ultrastructure (Trench, 1982), the systematic gyrations (Kies & Kremer, 1990) and the current status of the molecular biology of the cyanelle (Löffelhardt & Bohnert, 1994). The enigmatic status of these genera can be traced to the dispute or uncertainty over the nature of the blue-green inclusions within the cell. At the time of their original discovery there was disagreement as to whether these inclusions were cyanobacteria or plastids. The principal researchers during the first half of the century, Geitler and Pascher (see Kies &

Kremer, 1990) regarded the inclusions as a modified cyanobacterial endosymbiont for which the names cyanelles or syncyanosen were coined (see also Geitler, 1959 and Pringsheim, 1958). When electron microscopy was first added to the investigative arsenal the cyanelle continued to be regarded as cyanobacterial in nature and were in fact assigned their own systematic designation. The cyanelle of **Cyanophora** became **Cyanocyta** (Hall & Claus, 1963), while that of **Glaucocystis** became **Skujapelta** (Hall & Claus, 1967). Were these in fact cyanobacterial entities then it was to be expected that the "host" should have a taxonomic designation and indeed this had occurred. The **Cyanophora** "host" was most generally assigned to the Cryptophyceae while that of **Glaucocystis** had been delegated variously to the Chlorophyceae or Rhodophyceae. **Gloeochaete** had been considered a chlorophyte while **Paulinella** was easily regarded as a rhizopod. It was clear from ultrastructural and biochemical observations that these cyanellae would be atypical if they were either cyanobacterial or plastidic in nature. Not unexpectedly these observations fueled speculation that these were extant "in progress" representatives of the endosymbiotic acquisition of chloroplasts.. The disproportionate influence of the cyanelles, regrettable in hindsight, on the endosymbiotic theory was specifically addressed by Chapman (1992). As researchers sought answers and offered different explanations the idea that these algae were quite different from existing classes and deserving of their own higher taxon gained credence. This represented a reversion back to suggestions earlier promulgated by Skuja (1954), but which had subsequently been discarded, for a separate division, Glaucophyta. Kies (1979) suggested that these genera be assigned to their own class Glaucocystophyceae which was later formally typified (Kies & Kremer, 1986). Although a number of individuals accepted this proposal, the definitive evidence in support of this solution was lacking. The genus **Glaucosphaera**, occasionally regarded in the past as a cyanelle-containing entity has since been removed to the Rhodophyceae. **Paulinella** is unquestionably a member of the Rhizopoda, a fact that raises interesting questions about the evolution of "cyanelles" (Bhattacharya et al., 1995b).

As has been the case with the **Cyanidium** complex, molecular evidence has been the key to resolving the question. Beginning with the work on the cell biology by Burnap and Trench (1989a,b,c) and ferredoxin sequence data (Stevanovic et al., 1989) and work emanating from a number of other laboratories (summarised by Löffelhardt & Bohnert, 1994) there is now persuasive evidence to regard the cyanelles as an offshoot branch of the lineage from the semiautonomous endosymbiont to the primitive rhodophycean plastid (Löffelhardt & Bohnert, 1994). This line to the cyanellae branches off before the primitive rhodophyceaean plastid. This position has been supported most recently by Helmchen et al. (1995) who showed, using plastid 16S rRNA sequences, that the cyanelles of the three genera are a distinct evolutionary lineage with a probable monophyletic origin. These results provide further support for the idea that cyanelles represent the first diverging group after the endosynbiotic event. Because cyanelles possess a peptidoglycan wall the next event was presumably the loss of such wall. It is worth noting that the plastid of **Glaucosphaera** positions with those from red algae. If there is a caution it is that cyanelles should not be considered as modified plastids. Comparable studies (Bhattacharya et al., 1995a) using sequences of the nuclear SSU rRNA gene demonstrate that the genera are a phylogenetically distinct group. The preponderance of evidence now supports the position that these genera are in fact a separate lineage and taxon within the protists whose designation as the Glaucocystophyceae/Glaucophyta is justified. Even more recent work by Bhattacharya & Weber (1997) using the actin gene in **Cyanophora** provides further support for this position. These same investigations support the removal (originally proposed by McCracken et al., 1980) of **Glaucosphaera** to the Rhodophyceae. The status of the cyanelle from **Paulinella** still awaits reolution, but the outcome will surely be of interest since it now appears from nuclear SSU rRNA that the nuclear phylogeny indicates a relationship to the Euglyphidae (which includes **Paulinella** and Chlorarachniophyceae), (Bhattacharya et al., 1995b), quite distinct from the Glaucocystophyceae.

This particular example, that of the cyanelle-bearing algae, reinforces the thesis that algae are enigmatic only so long as we lack the appropriate information and comparative data bases and are unwilling to break away from preconceived notions.

2.4. CRYPTOPHYTA, EUGLENOPHYTA, CHLORARACHNIOPHYTA; THE SECONDARY SYMBIOSES

Although these three algal classes/divisions have never been considered enigmatic, the determination of their evolutionary position and relationships has in the past been a problem. Evolutionary relationships based upon non-plastid characters such as flagellar structure and other biochemical features provided one answer, and a logical one at the time. However the evolutionary relationships that one would infer from chloroplastidic data were in conflict,. or vice versa. However these three divisions illustrate, in a very cogent fashion, how a seemingly intractable puzzle of evolutionary relationships is readily solved if one is again willing to ask the right questions, to be bold or daring and depart from the traditional norms. In these cases everything appears to have fallen into place with the acceptance of the idea of “secondary symbioses”. In a succinct summary it can be said that the chloroplast characters of euglenoids, in essence those of a “green-algal-type”, posed problems because the non plastid characters were clearly of a non-green algal type. The “red-algal-type” chloroplast of the cryptomonads was puzzling because the remainder of the cellular features clearly precluded any close relationship to the Rhodophyta. The Chlorarachniophyta, with “green-algal-type” chloroplasts was entirely anomalous for rhizopod-type cells. It was easy to place these classes in isolated phylogenetic positions, but this begged the question of explaining the inconsistencies or anomalies of the plastid versus non-plastid features or characters. In the light of the primary symbiosis theory for the origin of the chloroplast from cyanobacteria, the biochemistry of this organelle was the anomalous feature which precluded a tidy fit with existing ideas on algal phylogeny. Ragan & Chapman (1978) had discussed the evolutionary possibilities that had been considered up to that time, which was the threshold of the molecular era. The combination of biochemical and non-biochemical data, all considered reliable, or at least the best, evolutionary characters at that time, provided unsatisfactory solutions. The seminal research that provided the impetus to think afresh came from some work and ideas on some dinoflagellates that were anomalous in containing pigments that were typically non-dinoflagellate and possessing two nuclei (see Ragan & Chapman, 1978 for summary). The work of Tomas & Cox

(1973) and Tomas et al. (1973), followed by Jeffrey et al. (1975), provided a simple and elegant explanation; that these dinoflagellates were derived from a colorless heterotrophic dinoflagellate that had ingested a chrysophyte and thus established a new "evolutionary type" by retaining permanently part of the ingested chrysophyte. Schnepf (1993) has distinguished between "kleptochloroplasts" (i.e. those of temporary retention and use, such as in **Amphidinium poecilochroum**,) and those of an obligate association, such as the two **Peridinium** species. It has now been demonstrated very recently, Chesnick et al., (1997) that the chloroplasts of **Peridinium balticum** and **Peridinium foliaceum** are derived from those of a diatom.

This proposal of secondary symbioses, or the acquisition of organelles such as chloroplasts by ingestion of an already exisiting chloroplast-containing eukaryote, was obviously a major departure from the hitherto accepted notion that all chloroplasts were originally derived from cyanobacteria (or in some special cases some other possible photosynthetic prokaryotye) by endosymbiosis, the traditional thesis of a chloroplast evolving from an ingested photosynthetic prokaryote (e.g. cyanobacterium). This paradigm had driven evolutionary thinking and for a long time researchers seemed reluctant to challenge this dogma, such that chloroplasts, from an evolutionary point of view, had to be "fitted" to a cyanobacterial origin, and perhaps by implication a "single" event.

Following on these explanations for the anomalous dinoflagellates came a similar suggestion to explain the conundrums of the euglenoids. It was initiated by results of ultrastructure observations (Gibbs 1978). Again the problems of the triple membrane and chloroplast biochemistry were the stumbling blocks to a nice tidy evolutionary explanation for the euglenoids. It is enlighteneing to review the history of suggestions for euglenoid phylogeny and evolution (Walne & Kivic, 1990). The nature of the eukaryote symbiont that was the ancestral source for the euglenoid plastid has still to be resolved. While the early empirical data (e.g. ultrastructural, chemical) would have suggested a green algal origin for the plastid (e.g. chlorophycean or prasinophycean), the molecular data have not been able to resolve this (Morden et al., 1992). Depending upon the application of protein or SSU rRNA sequence data the euglenoid plastid clusters with either green algae or chromophyte algae.

Shortly after the presentation of the idea of a secondary symbiosis for the **Euglena** plastid a similar idea was promulgated to explain the evolutionary position of the cryptomonads. In this case the driving force was not the plastid but the nucleomorph and the ultrastructural considerations thereof, e.g. Gillott & Gibbs (1980), Morrall & Greenwood (1982), Ludwig & Gibbs (1985), which prompted suggestions for a red algal involvement in the secondary symbiosis. These observations were followed by a series of cell and molecular investigations that supported this idea and which culminated in the demonstration (Douglas et al., 1991) that the nucleomorph was probably derived from the red algae, since the SSU rRNA showed the nucleomorph clustering with the red algal nuclear SSU rRNA.. These authors proposed that the cryptomonads were "chimaeras" of two phylogenetically distinct eukaryotes. Just prior to this proposal were those observations from plastid sequences (rbcL) that indicated a relationship to the red algal plastid for that of the cryptomonads (Douglas et al., 1990). Numerous other observations have all supported this notion of a secondary symbiosis of a red alga resulting in the evolutionary chimaera of the Cryptophyceae. Cavalier-Smith et al., (1994) have recently demonstrated that the nucleomorph of the aplastidic cryptomonad **Chilomonas paramecium** is closely related to the nuclear branch of the red algae. Cavalier-Smith et al. (1996) have now demonstrated a close relationship of the nucleomorph for a number of cryptomonads to the red algal nucleus and that the cryptomonad nucleomorphs are monophyletic, For a full review the reader is referred to McFadden (1993). This author has pointed out that the key to the secondary endosymbiosis scenario has been the nucleomorph that is closely associated with the chloroplast within the cell which in turn is surrounded by four membranes, in contrast to the double membrane of chloroplasts usually considered as derivatives of the primary (or cyanobacterial) endosymbiosis. As this concept has come into general acceptance the question of the evolutionary explanation for the photosynthetic chlorarachniophytes has been quickly resolved. These amoeboid rhizopods were identified in 1984 (Hibberd & Norris, 1984) as possessing a green plastid (presumptively of green algal origin) and a nucleomorph. There is now persuasive evidence (McFadden et al., 1994 , McFadden & Gilson, 1995, McFadden et al., 1995, McFadden et al., 1995, Van De Peer et al., 1996) that the chlorarachniophytes are but another example of secondary endosymbioses in which the ingested eukaryote was a chlorophyte alga. Bhattacharya et al (1995) have taken this scenario one step further. Their data from nuclear SSU rRNA show that within the amoeboid Rhizopoda there is a distinct lineage containing **Euglypha, Paulinella and Chlorarachnion.** While the nuclear genome phylogeny provides for a discrete lineage the plastid component probably is not. **Paulinella** contains a cyanelle while **Chlorarachnion** contains a chlorophytan type plastid. These authors have postulated as one possible scenario the ingestion of a glaucocystophyte as the origin of the **Paulinella** cyanelle.

Twenty years ago plastid origins were a tidy example of cyanobacterial endosymbiosis with a few anomalies, or perhaps enigmas present in the system. All attempts to accommodate these enigmas were built upon this scenario, but with very obvious degrees of success. The acceptance of the secondary endosymbiosis phenomenon has resolved those particular problems but created successive re-evaluations of the protistan phylogeny (e.g. Cavalier-Smith, 1982, 1986,1989 1993a,b) and numerous postulates and discussions on the question of single or multiple origins of chloroplasts (e.g. Martin et al, 1992, Valentin et al., 1993, Douglas, 1994, Loiseaux de Goër, 1994, Bhattacharya & Medlin, 1995). Perhaps the message from all of this work is that the evolution of photosynthetic protists is far more involved and diverse than we were willing to give credit for and that in considering the evolution of the organism we must look at not only the nuclear or "host" genome, but also the organellar or "symbiont" genome.

3. Summary

While we are dealing with the evolution and phylogeny of extant protists we must remember that they have been present for about 2 billion years (Knoll, 1992, Han & Runnegar, 1992) and the earliest multicellular form is a bangiophyte red alga dating perhaps as far back as 1.26 billion years. It is perfectly reasonable, therefore for us to expect and to entertain exceptional diversity and a multiplicity of evolutionary mechanisms such as described above. We need but the courage of our convictions and the data to support them.

At the beginning of this essay I posed the question" Are there any truly puzzling algae?" I would now suggest an answer. There have been, but they existed as puzzles only because of our lack of knowledge and/or our reluctance to entertain new ideas. The examples that have been discussed show how puzzling genera or groups of algae have advanced from the status of enigma to an accepted resolved status that does not depend upon exceptions or fanciful scenarios. In retrospect the logic of the explanations is such that we might well ask why was not this thought of earlier. When the full arsenal of techniques, ultrastructural, biochemical, cellular and molecular are applied to carefully posed questions or hypotheses the answers quickly fall into place and yesterday's enigma becomes today's example of a new facet of protistan evolution.

4. Acknowledgements

I express my great appreciation to Drs. Mark Ragan and Robert Trench for their reading of this contribution and their thoughtful comments and suggestions for improvement.

5. References

Allen, M.B. (1959) *Arch. Mikrobiol.* **32,** 270-277.
Bhattacharya, D., Helmchen, T. and Melkonian, M. (1995a) *J. Euk. Microbiol.* **42,** 65-69.
Bhattacharya, D., Helmchen, T., Bibeau, C and Melkonian, M, (1995b) *Mol. Biol. Evol.* **12,** 415-420.
Bhattacharya, D. and Medlin, L, (1995) *J. Phycol.* **31,**489-498.
Bhattacharya, D. and Weber, K. (1997) *Curr.Genet.* **31,**439-446.
Birch, A.J. (1973) *Pure Appl. Chem.* **33,** 17-38.
Burnap, R.L. and Trench, R.K. (1989a) *Proc. Roy. Soc. Lond.* **238B,** 53-72.
Burnap, R.L. and Trench, R.K. (1989b) *Proc. Roy. Soc. Lond.* **238B,** 73-87.
Burnap, R.L. and Trench, R.K. (1989c) *Proc. Roy. Soc. Lond.* **238B,** 89-102.
Cavalier-Smith, T. (1982) The origins of plastids, *Biol. J. Linn. Soc.* **17,** 289-306.
Cavalier-Smith, T. (1992) *Biosystems* **28,** 91-106.
Cavalier-Smith, T. (1993a) in R.A. Lewin (ed.), *Origins of Plastids.* Chapman & Hall, New York, pp 291-348.
Cavalier-Smith, T. (1993b) *Microbiol. Revs,* **57,** 953-994.
Cavalier-Smith, T., Allsopp, M.T.E.P., and Chao, E.E. (1994) *Proc. Natl. Acad. Sci. USA.* **91,** 11368-11372.
Cavalier-Smith, T., Couch, J.A., Thorsteinson, K.E., Gilson,. Deane, J.A., Hill, D.R.A. & McFadden, G.I. (1996) *Eur. J. Phycol.* **31,** 315-328.
Chapman, D.J. (1965) PhD Dissertation. University of California San Diego. pp XI + 169
Chapman, D.J. (1974) *Nova Hedw.* **25,** 341-346.
Chapman, D.J. (1992) in J.W.Schopf & C. Klein (eds.), *The Proterozoic Biosphere.* Cambridge Univ. Press.,London. pp.477-484.
Chapman, D.J. and Ragan, M.A. (1984) in T.V. Desikachary & V.N. Raja Rao (eds.), *Taxonomy of algae.* ICAR. India, pp.65-87.
Chesnick, J.M., Kooistra, W H.C.F., Wellbrock, U. and Medlin, L.K. (1997) *J. Euk. Microbiol* **44,** 314-320.
De Luca, P., Taddei,R, and Varano, L. (1978) *Webbia* **33,** 37-44.
Douglas, S.E. (1992) *Biosystems* **28,** 57-68.
Douglas, S.E. (1994) in D.A. Bryant (ed.) *Molecular Biology of Cyanobacteria.* Kluwer Academic Publishers, Dordrecht. pp 91-108.

Douglas, S.E., Murphy, C.A., Spencer, D.F. and Gray M.W. (1991) *Nature* **350,** 148-151.
Geitler, L. (1933). *Arch. Hydrobiol. Suppl.* **Bd XII**. 622-634.
Geitler, L. (1959) in W. Ruhland (ed.) *Handbuch der Pflanzenphysiologie* Vol. **11:** . Springer Verlag, Berlin. pp 530-545.
Gibbs, S.P. (1978) *Canad. J. Bot.* **58,** 2883-2889.
Gibbs, S.P. (1993) in R.A. Lewin (ed.), *Origins of Plastids.* Chapman & Hall. New York. pp. 107-121.
Gillott, M.A. aand Gibbs, S.P. (1980) *J. Phycol.,* **16,** 558-568.
Hall, W.T. and Claus, G. (1963) *J. Cell Biol.* **19,** 551-563.
Hall, W.T. and Claus, G. (1967) *J. Phycol.* **3,** 37-51.
Han, T-M, and Runnegar, B. (1992) *Science* **257,** 232-235.
Hase, T., Wakabayashi, S., Wada, K., Matsubara, H., Juettner, F., Rao,K., Fry, I. and Hall, D.O. (1985) *FEBS Letters* **96,** 41-44.
Helmchen, T.A., Bhattacharya, D. and Melkonian, M (1995) *J. Mol. Evol.* **41,** 203-210.
Hibberd, D.J. and Norris R.E. (1984) *J Phycol.* **20,** 310-330.
Hirose, H. (1950) *Bot. Mag. (Tokyo)* **63,** 107-111.
Hirose, H. (1958) *Bot. Mag (Tokyo)* **71,** 347-352.
Jeffrey, S.W. Sielicki, N. and Haxo, F.T. (1975) *J. Phycol.* **11,** 374-384.
Kies, L. (1979) *Ber.dtsch.bot.Gesellsch,* **92,** 445-454.
Kies, L and Kremer, B.P. (1986) *Taxon* **35,** 128-135.
Kies, L and Kremer, B.P. (1990) in L. Margulis, J.O. Corliss, M. Melkonian and D.J. Chapman (eds.), *Handbook of Protoctista.* Jones & Bartlett, Boston. pp.152-166.
Knoll, A.H. (1992) *Science* **256,** 622-627.
Kremer, B.P. (1982) *Brit. Phycol. J.* **17,** 51-61.
Kremer, B.P., Feige, G.B. and Schneider, H.A.W. (1978) *Naturwiss.* **65,** 17.
LeBlanc, C., Richard, O., Kloareg, B., Viehmann, S., Zetsche, K. and Boyen, C. (1997) *Curr. Genetics* **31,** 193-207.
Lewin, R.A. (1961) *Phycol. News Bull.* **14:** 6-7.
Löffelhardt, W. and Bohnert, H.J. (1994) in D.A. Bryant (ed.), *The Molecular Biology of Cyanobacteria.* Kluwer Academic Publishers, Dordrecht. pp.65-89.
Loiseaux-de-Goër, S. (1994) *Progr. Phycol. Res.* **10,** 137-177.
Ludwig, M, and Gibbs, S.P. (1985) *Protoplasma* **127,** 9-20.
Maid, U. Valentin, K. and Zetsche,K. (1990) *Curr. Genetics* **17,** 255-259.
Maid, U. and Zetsche, K. (1991) Plant Mol. Biol., **16,** 537-546.
Martin, W., Somerville, C.C. and Loiseaux-de Goër, S. (1992) *J. Mol. Evol.* **35,** 385-404.
McCracken, D.A., Nadakavukaren, M.J. and Cain, J.R. (1980) *New Phytol.* **86,** 39-44.
McFadden, G.I. (1993) *Adv. Bot. Res.* **19,** 190-230.
McFadden, G.I., Gilson, P.R., and Douglas, S.E. (1994) *J. Cell Sci.* **107,** 649-657.
McFadden, G.I., Gilson. P.R., Hofmann., C.J.B., Adcock, G.J., and Maier, U-G. (1994) *Proc. Natl. Acad. Sci. USA,* **91,** 3690-3694.
McFadden, G. and Gilson, P. (1995 *TREE. Trends Ecol. Evol.* **10,** 12-17.
McFadden, G.I., Gilson, P.R. & Waller, R.F. (1995) *Arch Protistenk.* **145,** 231-239.
Merola, A., Castaldo, R., De Luca, P., Gambardella, R., Musacchio, A. (1981) *Giorn. Bot. Ital.* **115,** 189-195.
Morden, C.W., Delwiche, C.F., Kuhsel, M. and Palmer, J.D. (1992) *Biosystems* **28,** 75-90.
Morrall, S. and Greenwood, A.D. (1982) *J. Cell Sci.* **54,** 311-328.
Nagashima, H., Sakai, K., Kamo, M. and Tsugita A. (1989) *Protein Sequ. Data Anal.* **2,** 457-460.
Ohta, N., Sato, N., Ueda, K. and Kuroiwa, T. (1997) *J. Plant Res.* **110,** 235-245.
Ott, F.D. and Seckbach, J (1994a) in J. Seckbach (ed.), *Evolutionary pathways and enigmatic algae: Cyanidium caldarium (Rhodophyta) and related cells.* Kluwer Academic Publishers, Dordrecht pp.113-132.
Ott, F.D. and Seckbach,J. (1994b) in J. Seckbach (ed.), *Evolutionary pathways and enigmatic algae: Cyanidium caldarium (Rhodophyta) and related cells.* Kluwer Academic Publishers, Dordrecht. pp. 145-152.
Päscher, A. (1931) *Beih. Bot. centr.* **48,** 317-332.
Pringsheim, E.G. (1958) in *Studies in Plant Physiology.* Praha pp. 165-184.
Ragan, M.A. and Chapman, D.J. (1978) *A Biochemical Phylogeny of the Protists.* Academic Press, New York pp. X + 317.
Schnepf, E. (1993) in R.A. Lewin (ed.), *Origins of Plastids,* Chapman & Hall, New York, pp.53-76 .
Seckbach, J. (1992) in W. Reisser (ed.), *Algae and Symbioses: Plants, Animals, Fungi, Viruses, Interactions Explored,* Biopress Ltd., Bristol, UK. pp. 400-426.

Seckbach, J. (1994a) Edit. *Evolutionary Pathways and Enigmatic Algae*: *Cyanidium caldarium (Rhodophyta) and Related Cells.* Kluwer Academic Publishers, Dordrecht XII + 349.
Seckbach, J. (1994b) in J. Seckbach (ed.)..*Evolutionary Pathways and Enigmatic Algae: Cyanidium caldarium (Rhodophyta) and Related Cells. Kluwer Academic Publishers, Dordrecht.* pp.99-112.
Seckbach, J. and Ott, F.D. (1994) in J. Seckbach (ed.), *Evolutionary Pathways and Enigmatic Algae: Cyanidium caldarium (Rhodophyta) and related Cells.* Kluwer Academic Publishers, Dordrecht, pp.133-143.
Skuja, H. (1954) in H. Melchior & E. Werdermann (eds.), *Syllabus der Pflanzenfamilien. Vol.1.* Gebruder Borntraeger, Berlin. pp 56-67.
Stevanovic, S., Bayer, M.G., Troger, W. and Schenk, H.E.A. (1989) *Endocytobiosis & Cell Res.* **6,** 219-226
Tomas, R.N. and Cox, E.R. (1973) *J. Phycol.* **9,** 304-323.
Tomas, R.N., Cox, E.R. and Steidinger, K.A. (1993) *J. Phycol.* **9,** 91-98.
Trench, R.K. (1982) *Progr. Phycol. Res.* **1,** 257-288.
Valentin, K.J. and Zetsche, K (1990) *Mol. Gen. Genet.* **222:** 425-430.
Valentin. K., Cattolico, R.A. and Zetsche, K (1993) in R.A. Lewin (ed.), *Origins of Plastids.* Chapman & Hall, New York, pp.193-221.
Van De Peer, Y., Rensing, S.A., Maier, U-G. & De Wachter, R. (1996) *Proc. Natl. Acad. Sci. USA,* **93,** 7732-7736.
Walne, P.A. and Kivic, P.A. (1989) in L.Margulis, J.O. Corliss, M. Melkonian and D.J. Chapman (eds.), *Handbook of Protoctista.* Jones & Bartlett, Boston. pp. 270-287.

Biodata of **Anton Post** contributor of "***The prochlorophytes - An Algal Enigma.***"

Dr. Anton Post is a senior lecturer at the Hebrew University, stationed at the H. Steinitz Marine Biology Laboratory in Eilat (Israel). He earned his Ph.D. in Microbiology (1987) from the University of Amsterdam (The Netherlands). His current research focuses on the biological oceanography, phytoplankton ecology, photosynthesis, molecular regulation of nutrient acquisition of marine cyanobacteria and prochlorophytes and assessment of the nutrient(s) limiting primary production in the sea.

E-Mail: **anton@vms.huji.ac.il**

J. Seckbach (ed.), Enigmatic Microorganisms and Life in Extreme Environments, 113–125.

THE PROCHLOROPHYTES - AN ALGAL ENIGMA
Biology of Chlorophyll* a/b *Containing Photosynthetic Prokaryotes

A.F. POST
Moshe Shilo Minerva Center for Marine Biogeochemistry,
H. Steinitz Marine Biology Laboratory,
The Interuniversity Institute for Marine Science,
Coral Beach, P.O. Box 469, 88103 Eilat, ISRAEL

1. Introduction

Enigmas like the ancestry of the chloroplast, the existence of a prokaryotic light harvesting antenna employing chlorophyll *a/b* and the origin of divinyl chlorophyll *a* in the ocean have kept sections of the scientific community puzzled for years and sometimes decades. Partial answers have been obtained with the discovery of each of the Prochlorophyte species. In turn, features of the Prochlorophytes have provided some new enigmas, waiting to be addressed. Prochlorophytes were defined as photosynthetic prokaryotes containing a chlorophyll *a/b* pigment complement instead of the phycobiliproteins known from the cyanobacteria. The three species known to date are in historical order: the marine tunicate symbiont *Prochloron didemnii*, the filamentous fresh water *Prochlorothrix hollandica* and the unicellular marine *Prochlorococcus marinus*. Interestingly, the discovery of each new Prochlorophyte species has pushed interest in the previous one to the background. I present in this chapter the history of the discovery of Prochlorophytes and discuss the shift in focus that each new species has brought along, together with the contribution which each has made to widely different areas such as molecular phylogeny, photosynthesis and biological oceanography. The different focal points are discussed against the background of the ecological importance and characteristics of Prochlorophytes that distinguish them from other photosynthetic organisms.

2. Prochlorophytes in a Historical Perspective.

During the early seventies the marine botanist Ralph Lewin (and coworkers) noticed the yellow to pale-green color associated with some didemnid tunicate specimens. Upon closer examination this color appeared to be associated with a unicellular, prokaryotic endosymbiont which originally was named *Synechocystis* (Lewin, 1975; Lewin and Whithers, 1975; Shulz - Baldes and Lewin, 1976), but later renamed

Prochloron (Lewin, 1976). All attempts to culture this organism have proven unsuccessful thereby hampering many physiological, biochemical and molecular studies and rendering others inconclusive. To the surprise of many this prokaryotic endosymbiont appeared to be lacking phycobiliproteins and instead contained both chlorophyll *a* and *b* (Lewin and Whithers, 1975; Thorne et al., 1977). This pigment complement was thought to be an exclusive trait of eukaryotic algae and higher plants. It was established that chlorophyll *b* is involved in excitation energy transfer to photosystem II leading up to oxygen evolution and carbon fixation (Whithers et al., 1978; Whithers et al., 1978; Critchley and Andrews, 1984; Alberte et al., 1986), much like in green chloroplasts. Based on these observations the hypothesis emerged that *Prochloron* and the green chloroplast may share common ancestry and the term Prochlorophyte was rapidly accepted as a popular characterization of this prokaryote.

The discovery of *Prochlorothrix* showed some resemblance to that of *Prochloron*. Despite the green appearance of cultured isolates of a photosynthetic filamentous organism from a shallow Dutch lake it was at first considered a cyanobacterium. Only after confirming both its prokaryotic nature and the presence of chlorophyll *a* and *b* (Burger-Wiersma et al., 1986) was this filamentous, free-living freshwater prokaryote identified as a Prochlorophyte and named *Prochlorothrix hollandica* (Burger-Wiersma et al., 1989). The wish to provide a taxonomic umbrella for the widely different marine, unicellular, obligate endosymbiont *Prochloron* and the freshwater, filamentous, autotrophic *Prochlorothrix* even triggered the short-lived declaration of Prochlorophyta as a new phylum (Burger-Wiersma et al., 1989).

Prochlorococcus was observed but not recognized in the early electron micrographs of natural marine plankton communities showing photosynthetic prokaryotes, chroococcoid cyanobacteria, that appeared to be lacking in phycobilisomes (Johnson and Sieburth, 1979). During the same period HPLC pigment analyses of marine phytoplankton samples showed an unknown chlorophyll *a* derivative which

Table 1. Pigment composition and chlorophyll ratios for cultured isolates of three known Prochlorophytes and the recently discovered *Acaryochloris*. Chlorophylls a_1,b_1 - monovinyl chlorophylls *a* and *b*; chlorophylls a_2, b_2 - divinyl chlorophylls *a* and *b*. PE - phycoerythrin; APC - allophycocyanin; PC - phycocyanin. Data after (Lewin and Whithers, 1975; Whithers et al., 1978; Burger-Wiersma and Post, 1989; Larkum et al., 1994; Miyashita et al., 1996; Marquardt et al., 1997; Partensky et al., 1997).

Organism	chlorophylls	chlorophyll ratio (*a/b* or *a/d*)	phycobiliproteins	carotenoids
Prochloron	a_1,b_1,c	4-7	-	β-carotene zeaxanthin cryptoxanthin echinenone mutatochrome
Prochlorothrix	a_1,b_1	7-18	-	β-carotene zeaxanthin
Prochlorococcus	a_2, b_2,b_1,c	1-15	PE	α-carotene zeaxanthin
Acaryochloris	a_1,c,d	0.03-0.09	PC, APC	α-carotene zeaxanthin

could not be attributed to any known algal species (Gieskes and Kraay, 1983; Gieskes and Kraay, 1986; Gieskes et al., 1988). Only with sensitive flow cytometry analyses of seawater samples was a small sized, dimly fluorescent picoplankton component recognized (Chisholm et al., 1988). The subsequent cultivation of this picoplankter in clonal cultures (Chisholm et al., 1992) revealed that the third prochlorophyte *Prochlorococcus marinus* carried the unique pigment complement of divinyl chlorophylls *a* and *b* (Goericke and Repeta, 1992).

Recently, the first cracks appeared in the once solid definition of Prochlorophytes as chlorophyll *a/b* containing photosynthetic prokaryotes (Table I). Not only does *Prochloron* carry chlorophyll *a* and *b*, it also contains a chlorophyll *c*-like pigment (Larkum et al., 1994). *Prochlorococcus* possesses a chlorophyll *c*-like pigment next to the divinyl chlorophyll *a/b* complement (Goericke and Repeta, 1992). The discovery of *Acaryochloris marina*, a prokaryote rich in chlorophyll *d* (Miyashita et al., 1996), added to the notion that plasticity in chlorophyll composition is a common feature of photosynthetic prokaryotes. In addition it was shown that a *Prochlorococcus* strain contains the cyanobacterial pigment phycoerythrin (Hess et al., 1996). Flow cytometric analysis of deep *Prochlorococcus* populations may show fluorescence emission typical of phycoerythrin (Hess et al., 1996). Also *Acaryochloris* was recently reported to carry phycobiliproteins (Marquardt et al., 1997). It may well be that the near future will witness a drive for unification of Prochlorophytes and Cyanobacteria into a more general description like Oxyphotobacteria or Oxychlorobacteria, which can accommodate all different chlorophyll and phycobilin pigment types.

3. *Prochloron*, Erstwhile Ancestor of the Chloroplast.

As mentioned above the original discovery of *Prochloron* led to the hypothesis that this endosymbiont and the green chloroplast of algae and higher plants share common ancestry. The logic behind this hypothesis was intuitively obvious, but both biochemical and molecular studies provided evidence for its rejection rather than for its support. Although parallels between *Prochloron* and green chloroplasts exist in thylakoid structure/function relationships, the *Prochloron* antenna protein is different in size from all known chlorophyll *a/b* antenna proteins of green chloroplasts (Bullerjahn et al., 1990). Moreover, polyclonal antibodies raised against such proteins failed to cross-react with the *Prochloron* chlorophyll *a/b* antenna protein. Later studies (see below) showed that Prochlorophyte species carry a different, prokaryotic, chlorophyll *a/b* light harvesting complex to that known in higher plants and algae with the apoproteins being encoded by a gene family distinct from the eukaryotic *cab* genes (LaRoche et al., 1996).

On another front, molecular phylogeny studies based on 5S rRNA sequence analysis indicated that *Prochloron* was not part of the chloroplast lineage (McKay et al., 1982). It rather lined up with known cyanobacterial sequences. Studies using much larger databases of 16S rRNA confirmed this conclusion (Seewaldt and Stackebrandt, 1982). The discovery of *Prochlorothrix* triggered renewed discussion concerning the ancestry of the green chloroplast. 16S rRNA sequence analysis showed no close relationship between *Prochlorothrix* and the green chloroplast (Turner et al., 1988; Turner, 1992). However, *Prochlorothrix psbA* sequence predicted a truncated C-terminal stretch of the D1-protein as commonly encountered in chloroplasts but not in the cyanobacteria (Morden and Golden, 1989; Morden and Golden, 1989). Other genes encoding proteins of the photosynthetic machinery like *psbBH* (photosystem II), *petBD* (cytochrome b_6/f complex), *atpBE* (ATPase), *petE* (plastocyanin) and *rbcLS* (RUBISCO) all seemed to confirm a position for *Prochlorothrix* among the cyanobacteria (Greer and Golden, 1991; Morden and Golden, 1991; Greer and Golden, 1992; Lockhart et al., 1992; Arudchandran et al., 1994). The matter came to rest with the discovery of *Prochlorococcus*. Comparison of 16S rRNA and DNA-dependent RNA polymerase sequences showed conclusively that Prochlorophytes do not form a monophyletic group, but rather a non-related group scattered within the cyanobacterial radiation (Palenik and Haselkorn, 1992; Urbach et al., 1992). Today most phylogenetic trees picture *Prochloron* with the other Prochlorophytes as a diverse group with no special affinity to plastid sequences. As such the questions remain as to whether we will able to identify the ancestor of the green chloroplast and whether chlorophyll *a/b* based light harvesting complexes have emerged more than once among the cyanobacteria.

4. *Prochlorothrix*, Identification of the Prokaryotic Chlorophyll *a/b* Antenna.

Photosynthesis in *Prochlorothrix* shows distinct eukaryotic and prokaryotic traits. Light-shade adaptation and characteristics of state I to state II transitions closely resemble those observed in higher plants and eukaryotic algae (Burger-Wiersma and Post, 1989). However, anoxygenic photosynthetic activities and sulfide tolerance of photosynthesis are clearly prokaryotic trademarks (Post and Arieli, 1997). The sulfide tolerance correlated with the presence of a hydrophobic manganese stabilizing protein embedded deeply in the thylakoid membrane (Mor et al., 1992; Mor et al., 1993). Most importantly, studies of structure-function relationships in photosynthetic membranes of *Prochlorothrix* lead to the identification and characterization of the prokaryotic chlorophyll *a/b* antenna. Thylakoid fractionation studies of *Prochloron* revealed that the antenna chlorophylls *a* and *b* were bound in a 2.2-2.4 ratio to an abundant 30-34 kD membrane protein (Schuster et al., 1984; Hiller and Larkum, 1985). As in *Prochloron* the chlorophyll *a/b* antenna of *Prochlorothrix* consists of hydrophobic membrane proteins of 30-35 kD apparent molecular weight that bind chlorophyll *a/b* in a 2.4-2.5 ratio (Post and Bullerjahn, 1994). These findings contrasted with observations on chloroplasts which carry more hydrophilic proteins of 22-27 kD molecular weight and bind chlorophyll *a* and *b* in a ratio of <1.5. Antibodies raised against eukaryotic antenna proteins did not recognize any of the *Prochlorothrix* antenna components (Bullerjahn et al., 1987). Interestingly, an antibody raised against the 30 kD antenna apoprotein of *Prochlorothrix* cross-reacted with all three antenna proteins in this organism as well as with the 34 kD apoprotein of *Prochloron* (Bullerjahn et al., 1990) and faintly, but positively with 30-38 kD thylakoid apoproteins in two *Prochlorococcus* strains, which were also identified as antenna polypeptides (Lichtle et al., 1995; Partensky et al., 1997). The antibody does not cross-react with any antenna polypeptide from chloroplastic origin. The antibody cross-reaction patterns in all Prochlorophytes led to the conclusion that the prokaryotic chlorophyll *a/b* antenna apoproteins may form a homologous group of proteins distinct from their eukaryotic counterparts (Bullerjahn and Post, 1993; Post et al., 1993). This conclusion was later confirmed from sequence analyses of the various Prochlorophyte *pcb* genes encoding the antenna polypeptides (LaRoche et al., 1996). Interestingly, the *pcb* gene family shows a high degree of homology to the *psbB* gene and the iron stress regulated *isiA* gene of both cyanobacteria and Prochlorophytes.

Strangely enough, a major fraction of the *Prochlorothrix* chlorophyll *a/b* antenna complexes containing the 35 and 30 kD proteins was found to copurify with photosystem I on non-denaturing gels, whereas a smaller fraction containing the 33 kD apoprotein was found in association with photosystem II (Post et al., 1992). Such an organization would imply that the bulk of excitation energy would be funneled to photosystem I, in apparent contradiction with the low quantum efficiencies for oxygen evolution reported for *Prochlorothrix* (Burger-Wiersma and Post, 1989). These considerations prompted the hypothesis that photosystems I and II in *Prochlorothrix* (and possibly in *Prochloron* and *Prochlorococcus*) share a common antenna and so form a supercomplex (Post et al., 1993). Such a configuration would yield a random distribution of antennae within the thylakoid membranes in contrast with the lateral heterogeneity encountered in eukaryotic chloroplasts. Immunogold labeling of thylakoids with the antibody against the 30 kD antenna apoprotein indeed showed a random distribution of this antenna type (Bullerjahn et al., 1990). The hypothesis further predicts that energy transfer from chlorophyll *b* to photosystem I is substantial, that action spectra of photosystem I and II are (nearly) identical and thus that Emerson-enhancement cannot occur in such systems. Energy coupling between chlorophyll *b* and photosystem I is much stronger in *Prochlorothrix* and in *Prochlorococcus* than in green chloroplasts (Post et al., 1993; Partensky et al., 1997). Emerson-enhancement could not be observed in *Prochlorothrix* from modulated fluorescence (Post and Canaani, unpubl. res.) and oxygen evolution measurements (Gavrjushina and Bullerjahn, unpubl. res.). Ultrastructural studies have documented some degree of lateral heterogeneity of reaction centers, but the generally low levels of membrane stacking suggests that such heterogeneity is limited in all Prochlorophytes (Giddings et al., 1980; Miller et al., 1988; Van Der Staay and Staehelin, 1994). The organization and features of the photosynthetic apparatus in *Prochloron*, *Prochlorothrix* and *Prochlorococcus* differ in every detail from that found in chloroplast systems, but allows for all known functions and adaptations in photosynthesis to be performed in full. Whether the recently discovered *Acaryochloris marina* which carries a chlorophyll *a/d* pigment complement (Miyashita et al., 1996; Miyashita et al., 1997) exhibits these features awaits future investigation. It was recently reported that this organism contains a different type of reaction center I which employs chlorophyll *d* instead of P_{700} chlorophylls (Itoh et al., 1997).

Observations on isolated cells of *Prochloron* showed the occurrence of light independent phosphorylation of the 34 kD antenna protein (Schuster et al., 1984). Early studies on *Prochlorothrix* indicated the presence of a redox-, but not light-controlled thylakoid protein kinase (Van Der Staay et al., 1989). It was subsequently shown that this kinase activity is under both light and redox control and that the 35 kD antenna protein forms its major target (Post et al., 1992). It was further shown from modulated fluorescence and from fluorescence induction studies that state I to state II transitions occur in *Prochlorothrix* (Burger-Wiersma and Post, 1989; Post et al., 1993). The state transitions correlated with the phosphorylated state of the 35 kD antenna apoprotein in a fashion similar to that known from green chloroplasts (Post et al., 1993; Van Der Staay and Staehelin, 1994). Light-controlled reversible phosphorylation forms the structural basis for state transitions in the latter, during which phosphorylated antennae are thought to move away laterally from photosystem II and reassociate with photosystem I. A functionally similar model has been proposed for *Prochlorothrix* based on its different thylakoid organization. Photosystem II together with the antenna fraction carried by the 33 kD apoprotein are thought to become disconnected from the bulk antenna during state I to state II transitions (Post et al., 1993; Van Der Staay and Staehelin, 1994), leaving the latter in a tight association with photosystem I only. The antenna's association/dissociation is then mediated through the light-dependent phosphorylation of the 35 kD antenna component (Post et al., 1993). As such it is clear that light dependent phosphorylation as a regulatory mechanism in photosynthesis had evolved long before the symbiotic relationship between the eukaryotic plant cell and its chloroplast.

5. *Prochlorococcus*, Global Importance and Unique Pigmentation.

The Prochlorophytes have gained an ecological significance since 1988, when *Prochlorococcus* was established as a numerically dominant component of phytoplankton mainly in the deep chlorophyll maximum of stratified (sub)tropical waterbodies. It was subsequently shown that this Prochlorophyte constitutes a significant fraction of phytoplanktonic communities in marine waters between latitudes of roughly 40 °S and 40 °N. e.g. *Prochlorococcus* forms a significant component of the photic layer picoplankton of stratified waters in the Western and Central Pacific, Tropical and Northern Atlantic and the Mediterranean Sea, (Chisholm et al., 1988; Olson et al., 1990; Veldhuis and Kraay, 1990; Li et al., 1992; Campbell and Vaulot, 1993; Li et al., 1993; Veldhuis et al., 1993; Li, 1995; Binder et al., 1997). Based on cell numbers prochlorophytes may contribute up to 90% of total phytoplankton (Campbell and Vaulot, 1993; Veldhuis et al., 1993). Their highest cell densities were reported for the Red Sea with numbers exceeding 250,000 cells.ml^{-1} (Veldhuis and Kraay, 1993; Lindell and Post, 1995). Other parameters also indicate that this group forms an important component of the primary producers. Prochlorophytes contribute 11-13 % of phytoplankton carbon biomass (Li et al., 1992;

Li, 1995); 30-50 % of chlorophyll *a* biomass (Goericke and Repeta, 1992; Goericke and Welschmeyer, 1993; Letelier et al., 1993; Lindell and Post, 1995; Suzuki et al., 1995) and 25 % of yearly productivity (Goericke and Repeta, 1992). Mortality rates of *Prochlorococcus* are roughly equal to its intrinsic growth rates (DiTullio et al., 1993; Landry et al., 1995; Liu et al., 1995). The numerical dominance of *Prochlorococcus* in (sub)tropical waters so far remains a reported but unexplained phenomenon. Other questions which now are being addressed, relate to the absence of *Prochlorococcus* from deeply mixed water bodies and the understanding of the dominant presence of *Prochlorococcus* in the surface layers of stratified waters down to depths of 150-200 m, far below the 0.1% light depth.

Little attention has so far been given to the analysis of populations of the dimly fluorescent *Prochlorococcus* from mixing water columns. *Prochlorococcus* population dynamics in mixing water bodies and its physiological responses to mixing conditions are little understood. Observations in the Mediterranean show that *Prochlorococcus* dominance among picophytoplankton persists in water bodies which are mixed to a depth of at least 150 m (Vaulot et al., 1990; Vaulot and Partensky, 1992). *Prochlorococcus* abundance in the Sargasso Sea decreases significantly following periods of mixing down to 200-300 m (Olson et al., 1990; Goericke and Welschmeyer, 1993) Mann and Chisholm, pers. comm.]. Not even a small summer population of *Prochlorococcus* is found in the shortly stratified water bodies of the Atlantic Ocean above 42 °N that undergo frequent mixing events throughout the year (Veldhuis and Kraay, 1993). The Gulf of Aqaba in the northern part of the Red Sea forms an excellent model system for studies of mixing effects on *Prochlorococcus* populations. This deep basin is stably stratified in summer, but undergoes deep convective mixing in winter. These mixing events are nearly always deeper than 350 m and may reach to depths of >650 m in unusually cold winters like that of 1992/1993 (Genin et al., 1995). Analysis of ultraphytoplankton communities (Lindell and Post, 1995) during the 1992/1993 seasonal cycle yielded a unique succession phenomenon in which *Prochlorococcus* populations varied between virtual absence during winter mixing to numerically dominant in summer (Fig. 1). Especially interesting is that upon stratification of the Gulf of Aqaba a *Synechococcus* bloom is observed whereas *Prochlorococcus* appears long after stratification has been established. At the height of summer *Prochlorococcus* contributed about 50% of the chlorophyll *a* standing stock (Lindell and Post, 1995). How the mixing event affects *Prochlorococcus* populations is as yet not known, but it appears that, following the onset of stratification, arrival of *Prochlorococcus* populations depends to a large extent on advection of surface waters from the permanently stratified Red Sea into the Gulf of Aqaba (Post et al., unpublished results).

Prochlorococcus dominance throughout the photic zone has prompted the hypothesis that *Prochlorococcus* has high ecological plasticity with respect to its light requirements (Vaulot et al., 1990; Veldhuis and Kraay, 1993). Cellular contents of divinyl chlorophyll *a* and *b* are inversely related to depth and thus to ambient light (Veldhuis and Kraay, 1990). Due to its small size and relatively high contents of divinyl chlorophyll *b* *Prochlorococcus* is a highly efficient light absorber and well suited for light harvesting near the bottom of the photic zone (Morel et al., 1993). The increasing dominance of *Prochlorococcus* in deeper layers was attributed to its optical properties (Morel et al., 1993), the spectral properties of divinyl chlorophyll *b*

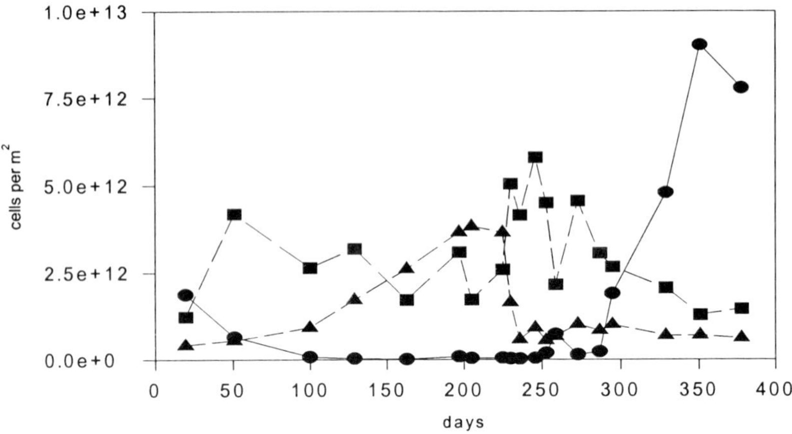

Figure 1. Change in total integrated cell numbers of unicellular *Prochlorococcus* (●), *Synechococcus* (■) and eukaryotic algae (▲) during 1992/1993 seasonal cycle in the Gulf of Aqaba, Red Sea. Note the delayed arrival of *Prochlorococcus* in the newly stratified water column. Further note the sharp drop in eukaryotic algal number and marked increase in *Synechococcus* number immediate after onset of stratification (after Lindell and Post, 1995).

which absorbs best those wavelengths that penetrate deepest in ocean waters (Goericke and Repeta, 1992) and the high cellular contents of divinyl chlorophyll *b* (Chisholm et al., 1988). Later culture studies confirmed the ability of *Prochlorococcus* to adapt to changes in light intensity by adjusting cellular pigment contents and ratios (Partensky et al., 1993; Moore et al., 1995). These observations imply that the variability of physiological characteristics found among natural *Prochlorococcus* populations in a given water parcel, could be accounted for by one *Prochlorococcus* type. This however does not to seem to be the case. Campbell and Vaulot (1993) reported the co-occurrence of dual *Prochlorococcus* populations near the deep chlorophyll maximum in the subtropical North-Pacific. The divinyl-chlorophyll *a/b* ratios observed in natural samples (Goericke and Welschmeyer, 1993) spanned a much wider range than those of individual strains (Partensky et al., 1993; Moore, 1997). It was shown from restriction fragment length polymorphism that deep water *Prochlorococcus* isolates are genetically different from their shallow counterparts (Scanlan et al., 1996). Less pigmented *Prochlorococcus* isolates in general originated from shallow waters, whereas well pigmented strains were mostly isolated from deeper parts of the photic zone (Moore, 1997). Coexistence of such genetically and physiologically distinct *Prochlorococcus* types in the same waterbody has recently been reported (Moore, 1997; Urbach et al., 1997). The co-occurrence of multiple *Prochlorococcus* types with different physiological ranges of light acclimation could account for observed variation in pigment contents and ratios of natural *Prochlorococcus* populations. Moreover, the recent discovery of the gene encoding phycoerythrin in *Prochlorococcus* and the presence of phycoerythrin in cells grown at low light or sampled from great depth (Hess et al., 1996) opens the prospect that phycobiliproteins play a light harvesting role in Prochlorophytes after all.

6. Concluding Remarks

Studies of the three known Prochlorophytes has enabled the elucidation of their phylogenetic affinity to cyanobacteria, the identification of a common prokaryotic light harvesting chlorophyll *a/b* antenna and establishment of *Prochlorococcus* as an important contributor to primary productivity in the sea. However, the relationship of Prochlorophytes to the chloroplast is still an open question. Structural aspects of the prokaryotic chlorophyll *a/b* antenna and its apparent strong connection to photosystem I remain unclear, as do energy coupling of phycoerythrin and nutrient acquisition in *Prochlorococcus*. Moreover, in depth studies of Prochlorophytes and the discovery of a prokaryote containing mainly chlorophyll *d* necessitate a reassessment of their original grouping. I propose that a broader description is adopted, like Oxyphotobacteria or Oxychlorobacteria, which may accomodate all pigment types found among the oxygenic photosynthetic prokarytes.

7. Acknowledgements

I am grateful to G.S. Bullerjahn, A.W.D. Larkum and D. Lindell for helpful comments. A.F. Post received financial support for the study of the *Prochlorothrix hollandica* light harvesting antenna from grant #88-00341 by the US-Israel Binational Science Foundation and for research on *Prochlorococcus marinus* ecology in the Red Sea from grant #GR1307 by the German Ministry of Science and Technology.

8. Literature

Alberte, R. S., Cheng, L. and Lewin, R. A. (1986). Mar. Biol. **90**: 575-587.

Arudchandran, A., Seeburg, D., Burkhart, W. and Bullerjahn, G. S. (1994). Biochim. Biophys. Acta **1188**: 447-449.

Binder, B., Chisholm, S. W., Olson, R. J., Frankel, S. L. and Worden, A. Z. (1997). Deep-Sea Res. **43**: 907-931.

Bullerjahn, G. S., Jensen, T. C., Sherman, D. M. and Sherman, L. A. (1990). FEMS Microbiol. Lett. **67**: 99-106.

Bullerjahn, G. S., Matthijs, H. C. P., Mur, L. R. and Sherman, L. A. (1987). Eur. J. Biochem. **168**: 295-300.

Burger-Wiersma, T. and Post, A. F. (1989). Plant Physiol. **91**: 770-774.

Burger-Wiersma, T., Stal, L. J. and Mur, L. R. (1989). Internat. J. for Syst. Bacteriol. **39**: 250-257.

Burger-Wiersma, T., Veenhuis, M., Korthals, H. J., Van Der Wiel, C. C. M. and Mur, L. R. (1986). Nature **320**: 262-264.

Campbell, L. and Vaulot, D. (1993). Deep-Sea Res. **40**: 2043-2060.

Chisholm, S. W., Frankel, S. L., Goericke, R., Olson, R. J., Palenik, B., Waterbury, J. B., West-Johnsrud, L. and Zettler, E. R. (1992). Arch. Microbiol. **157**: 297-300.

Chisholm, S. W., Olson, R. J., Zettler, E. R., Goericke, R., Waterbury, J. B. and Welschmeyer, N. A. (1988). Nature **334**: 340-343.

Critchley, C. and Andrews, T. J. (1984). Arch. Microbiol. **138**: 247-250.

DiTullio, G. R., Hutchins, D. A. and Bruland, K. W. (1993). Limnol. Oceanogr. **38**: 495-508.

Genin, A., Lazar, B. and Brenner, S. (1995). Nature **377**: 507-510.

Giddings, T. H., Whithers, N. W. and Staehelin, L. A. (1980). Proc. Natl. Acad. Sci. USA **77**: 352-356.

Gieskes, W. W. C. and Kraay, G. W. (1983). Limnol. Oceanogr. **28**: 757-765.

Gieskes, W. W. C. and Kraay, G. W. (1986). Mar. Biol. **91**: 567-576.

Gieskes, W. W. C., Kraay, G. W., Nontji, A., Setiapermana, D. and Sutomo (1988). Netherlands J. Sea Res. **22**: 123-137.

Goericke, R. and Repeta, D. J. (1992). Limnol. Oceanogr. **37**: 425-433.

Goericke, R. and Welschmeyer, N. A. (1993). Deep-Sea Res. **40**: 2283-2294.

Greer, K. L. and Golden, S. S. (1991). Plant Mol. Biol. **17**: 915-917.

Greer, K. L. and Golden, S. S. (1992). Plant Mol. Biol. **19**: 355-365.

Hess, W. R., Partensky, F., Van Der Staay, G. W. M., Garcia-Fernandez, J. M., Boerner, T. and Vaulot, D. (1996). Proc. Natl. Acad. Sci. USA **93**: 11126-11130.

Hiller, R. G. and Larkum, A. W. D. (1985). Biochim. Biophys. Acta **806**: 107-115.

Itoh, S., Iwaki, M., Hu, Q. and Miyashita, S. (1997). IX Internat. Symp. on Phototrophic Prokaryotes., Vienna.

Johnson, P. W. and Sieburth, J. M. (1979). Limnol. Oceanogr. 24: 928-935.

Landry, M. R., Kirshtein, J. and Constantinou, J. (1995). Mar. Ecol. Prog. Ser. **120**: 53-63.

Larkum, A. W. D., Scaramuzzi, C., Cox, G. C., Hiller, R. G. and Turner, A. G. (1994). Proc. Natl. Acad. Sci. USA **91**: 679-683.

LaRoche, J., et al. (1996). Proc. Natl. Acad. Sci. USA **93**: 15244-15248.

Letelier, R. M., Bidigare, R. R., Hebel, D. V., Ondrusek, M., Winn, C. D. and Karl, D. M. (1993). Limnol. Oceanogr. **38**: 1420-1437.

Lewin, R. A. (1975). Phycol. **14**: 153-160.

Lewin, R. A. (1976). Nature **261**: 697-698.

Lewin, R. A. and Whithers, N. W. (1975). Nature **256**: 735-737.

Li, W. K. W. (1995). Mar. Ecol. Prog. Ser. **122**: 1-8.

Li, W. K. W., Dickie, P. M., Irwin, B. D. and Wood, A. M. (1992). Deep-Sea Res. **39**: 501-519.

Li, W. K. W., Zohary, T., Yacobi, Y. Z. and Wood, A. M. (1993). Mar. Ecol. Prog. Ser. **102**: 79-87.

Lichtle, C., Thomas, J. C. and Spilar, A. (1995). J. Phycol. **31**: 934-941.

Lindell, D. and Post, A. F. (1995). Limnol. Oceanogr. **40**: 1130-1141.

Liu, H., Campbell, L. and Landry, M. R. (1995). Mar. Ecol. Prog. Ser. **116**: 277-287.

Lockhart, P. J., Beanland, T. J., Howe, C. J. and Larkum, A. W. D. (1992). Proc. Natl. Acad. Sci. USA **89**: 2742-2746.

Marquardt, J., Senger, H., Miyashita, H., Miyachi, S. and Morschel, E. (1997). FEBS Lett. **410**: 428-432.

McKay, R. M., Salgado, D., Bonen, I. and Stackebrandt, E. (1982). Nucleic Acid Res. **10**: 2963-2970.

Miller, K. R., Jacob, J. S., Burger-Wiersma, T. and Matthijs, H. C. P. (1988). J.Cell Sci. **91**: 577-586.

Miyashita, H., Adachi, K., Kurano, N., Ikemoto, H., Chihara, M. and Miyachi, S. (1997). Plant Cell Physiol. **38**: 274-281.

Miyashita, H., Ikemoto, H., Kurano, N., Adachi, K., Chihara, M. and Miyachi, S. (1996). Nature **383**: 402.

Moore, L. R. (1997). PhD thesis, Dept. of Civil end Environmental Engineering, Massachusetts Institute of Technology, Cambridge: 183 p.

Moore, L. R., Goericke, R. and Chisholm, S. W. (1995). Mar. Ecol. Prog. Ser. **116**: 259-275.

Mor, T. S., Post, A. F. and Ohad, I. (1992). In: Regulation of Chloroplast Biogenesis. (J. A. Argyroudi, ed.). New York, Plenum Press: pp. 433-437.
Mor, T. S., Post, A. F. and Ohad, I. (1993). Biochim. Biophys. Acta **1141**: 206-212.
Morden, C. W. and Golden, S. S. (1989). Nature **339**: 400.
Morden, C. W. and Golden, S. S. (1989). Nature **337**: 382-385.
Morden, C. W. and Golden, S. S. (1991). J. Mol. Evol. **32**: 379-395.
Morel, A., Ahn, Y.-H., Partensky, F., Vaulot, D. and Claustre, H. (1993). J. Mar. Res. **51**: 617-649.
Olson, R. J., Chisholm, S. W., Zettler, E. R., Altabet, M. A. and Dusenberry, J. A. (1990). Deep-Sea Res. **37**: 1033-1051.
Palenik, B. and Haselkorn, R. (1992). Nature **355**: 265-267.
Partensky, F., Hoepffner, N., Li, W. K. W., Ulloa, O. and Vaulot, D. (1993). Plant Physiol. **101**: 285-296.
Partensky, F., LaRoche, J., Wyman, K. and Falkowski, P. G. (1997). Photosynthesis Res. **51**: 209-222.
Post, A. F. and Arieli, B. (1997). Appl. Environm. Microbiol. **63**: 3507-3511.
Post, A. F. and Bullerjahn, G. S. (1994). FEMS Microbiol. Rev. **13**: 393-414.
Post, A. F., Gal, A., Ohad, I., Milbauer, K. M. and Bullerjahn, G. S. (1992). Biochim. Biophys. Acta **1100**: 75-82.
Post, A. F., Ohad, I., Warner, K. M. and Bullerjahn, G. S. (1993). Biochim. Biophys. Acta **1144**: 374-384.
Scanlan, D., Hess, W. R., Partensky, F., Scanlan, J. and Vaulot, D. (1996). Eur. J. Phycol. **31**: 1-9.
Schuster, G., Owens, G. C., Cohen, Y. and Ohad, I. (1984). Biochim. Biophys. Acta **767**: 596-605.
Seewaldt, E. and Stackebrandt, E. (1982). Nature **295**: 618-620.
Shulz-Baldes, M. and Lewin, R. A. (1976). Phycol. **15**: 1-6.
Suzuki, K., Handa, N., Kiyosawa, H. and Ishizaka, J. (1995). Limnol. Oceanogr. **40**: 983-989.
Thorne, S. W., Newcomb, E. H. and Osmond, C. B. (1977). Proc. Natl. Acad. Sci. USA **74**: 575-578.
Turner, S. (1992). Cyanones **8**: 5-6.
Turner, S., Burger-Wiersma, T., Giovannoni, S. J., Mur, L. R. and Pace, N. R. (1988). Nature **337**: 380-385.
Urbach, E., Robertson, D. L. and Chisholm, S. W. (1992). Nature **355**: 267-270.
Urbach, E., Scanlan, D. J., Distel, D. L., Waterbury, J. B. and Chisholm, S. W. (1997). J. Mol. Evol. in press.
Van Der Staay, G. W. M., Matthijs, H. C. P. and Mur, L. R. (1989). Biochim. Biophys. Acta **975**: 317-324.
Van Der Staay, G. W. M. and Staehelin, L. A. (1994). J. Biol. Chem. **269**: 24834-24844.
Vaulot, D. and Partensky, F. (1992). Deep-Sea Res. **39**: 727-742.
Vaulot, D., Partensky, F., Neveux, J., Mantoura, R. and Llewellyn, C. (1990). Limnol. Oceanogr. **35**: 1156-1164.
Veldhuis, M. J. W. and Kraay, G. W. (1990). Mar. Ecol. Prog. Ser. **68**: 121-127.
Veldhuis, M. J. W. and Kraay, G. W. (1993). Netherlands J. Sea Res. **31**: 135-145.
Veldhuis, M. J. W., Kraay, G. W. and Gieskes, W. W. C. (1993). Deep-Sea Res. **40**: 609-626.
Whithers, N. W., Alberte, R. S., Lewin, R. A., Thornber, J. P., Britton, G. and Goodwin, T. W. (1978). Proc. Natl. Acad. Sci. USA **75**: 2301-2305.
Whithers, N. W., Vidaver, W. and Lewin, R. A. (1978). Phycol. **17**: 167-171.

Biodata of **Paul Kugrens** contributor of "*Cryptomonad Systematics - An Algal Enigma?.*"

Dr. Paul Kugrens is a Professor of Biology and the Chair of Faculty Council at Colorado State University (at Fort Collins, Colorado). He gained his Ph.D. in 1971 from the University of California, Berkeley. Initially he studies the ultrastructure of red algal parasites, then began studies on Cryptomonads. Dr. Kugrens is the Vice President/President-Elect of the Phycological Society of America.
E-mail: **pkugrens@lamar.colostate.edu**

J. Seckbach (ed.), Enigmatic Microorganisms and Life in Extreme Environments, 127–138.

CRYPTOMONAD SYSTEMATICS - AN ALGAL ENIGMA?

PAUL KUGRENS
Department of Biology
Colorado State University
Fort Collins, Colorado 80523, USA

1. Introduction

The name assigned to the cryptomonads infers that this is an obscure group of organisms (crypto = hidden). However, a considerable amount of information has accumulated, particularly with regard to cellular structure. Unfortunately, phycologists have not utilized this information in a consistent manner to establish a coherent classification scheme, thus the systematics of this group remain enigmatic, although considerable information exists regarding their phylogeny. Consequently, it is the intent of this short chapter to summarize the known information about cryptomonad genera and to present those aspects which need more research to make this group less enigmatic.

2. Ecology

The information regarding cryptomonad ecology usually is general and anecdotal, with few definitive studies having been conducted. Cryptomonads are important primary producers in freshwater and marine habitats (Gillott 1990, Klaveness 1988), they are cosmopolitan in their distribution and they often assume dominant phytoplankton status in temperate lakes and reservoirs where they may dominate the sub-ice, early spring and late fall populations. They are common in phytoplankton samples but their cells are extremely delicate and rupture when fixatives are added or when temperatures are elevated. As a consequence, their numbers generally are low in preserved samples, and therefore they are assumed to be a small and obscure taxonomic group of protists.

One cryptomonad, *Cyanomonas*, has been reported to cause massive catfish deaths in Texas ponds (Pfiester & Holt 1978). Neither the organism nor the toxin were isolated, therefore the identity of *Cyanomonas* remains in doubt.

3. Evolution of the Cryptomonad Cell

Except for *Goniomonas*, cryptomonads are one of the most complex cells known, having originated from a eukaryote host cell and a eukaryotic endosymbiont (Douglas, Murphy, Spencer and Gray 1991, Gillott 1990, Ludwig and Gibbs 1985a, 1989, McKerracher and Gibbs 1982). As a result of this secondary endosymbiosis, there are two prokaryotic endosymbionts (the mitochondrion & chloroplast) and a eukaryotic endosymbiont, presumably derived from a red algal ancestor.

Based on the phylogenetic information available for cryptomonads, the host nuclear encoded rRNA appears to be related to the protozoan *Acanthamoeba* and the green algal lineage, whereas the nucleomorph encoded rRNA appears to be related to red algae (Douglas, Murphy, Spencer and Gray 1991). Again, this information is rather tenuous because too few protists have been examined to resolve the issue.

4. Unique Cryptomonad Characteristics

4.1. CELL ARCHITECTURE

The external architecture of the cell is determined primarily by a furrow gullet/complex which imparts an asymmetrical shape to the cells. Cell shapes may be oval, compressed, lunulate, caudate, acute, elongate, sigmoid, reniform or contorted. All cells possess a vestibulum from which the flagella originate on the right side. In addition, there may only be a tube-like gullet, some type of furrow, or a combination of a furrow-gullet present. A furrow is a ventral groove, of variable length, that begins in the vestibular region of the cell and extends posteriorly, terminating somewhere in the anterior half of the cell (Hill and Wetherbee 1986, 1989, Kugrens, Lee and Andersen 1986, Munawar and Bistricki 1979). A tubular invagination called the gullet may extend posteriorly from the vestibulum or the end of the furrow (Hill and Wetherbee 1986, 1989). Ejectisomes flank the furrow and/or gullet. Several types of furrows have been described in cryptomonads (Kugrens, Lee and Andersen 1986) and these variations have become, and should remain, major features in cryptomonad systematics (Hill 1990, 1991b, Hill and Wetherbee 1986, 1989). The gullet probably evolved from the fusion of a furrow (Kugrens and Lee 1991). Consequently, the combination of a furrow and gullet may represent intermediate stages in the evolution of a gullet (Kugrens & Lee 1991).

4.2. PERIPLAST STRUCTURE

Periplasts are unique cell coverings found only in cryptomonads. Periplasts consist of inner and surface components (Hill 1990, 1991, Hill and Wetherbee 1986, 1989, Kugrens and Lee 1987, 1991, Wetherbee, Hill and McFadden 1986) and both can be variable in their composition. The inner periplast component (IPC) consists either of multiple plates having various shapes (Hill 1990, 1991, Hill and Wetherbee 1986, 1989, Kugrens and Lee 1987) or it may consist of a single sheet. Multiple plated periplasts may have hexagonal, square, oval/round, rectangular or irregular-shaped plates. The plates of the inner periplast component are connected to the cell membrane by intramembrane particles or proteins (Kugrens and Lee 1987, Wetherbee, Hill and McFadden 1986). The surface periplast component (SPC) may consist of plates, heptagonal scales, mucilage or a combination of any of the preceding. Variations in the shapes of these plates and the type of surface component are major features in establishing genera.

4.3. FLAGELLA & FLAGELLAR APPARATUS

With the exception of *Goniomonas*, cells have two subapically inserted, subequal flagella with bipartite tubular hairs on at least one of the flagella (Kugrens, Lee & Andersen 1987). There appear to be at least five variations in the arrangement of tubular and non-tubular hairs on the flagella (Kugrens, Lee and Andersen 1987), but the most common arrangement of hairs is one in which the longer or dorsal flagellum bears two laterally opposed rows of tubular hairs and the shorter or ventral flagellum bears a single row of hairs. In addition flagella in some species may bear heptagonal scales (Lee and Kugrens 1986). Instead of tubular hairs, *Goniomonas* has a unilateral row of curved spikes on one of its flagella and fine, non-tubular hairs on both flagella (Kugrens and Lee 1987, 1991). The flagellar transition region is unique and consists of a doublet system of septa in all cryptomonads (Grain, Mignot and Puyturac 1988, Kugrens and Lee 1991). A rhizostyle and a striated rootlet are integral components of the flagellar apparatus in most cryptomonads. The rhizostyle may consist of winged microtubules that originate near the basal bodies and then extend posteriorly into the cell, or a second type of rhizostyle which lacks wings on the microtubules. Usually, this latter type of rhizostyle terminates anterior to the nucleus.

4.4. EJECTISOMES

Ejectisomes are the extrusive organelles of all cryptomonads, and they appear to be identical in all genera that have been investigated. Large ejectisomes are associated with the furrow/gullet and small ejectisomes are found in other regions of the cell. Ejectisomes consist of two different-sized tightly coiled and tapered ribbons, which are joined, and both are surrounded by a membrane. Upon discharge the ribbons roll up to form a tube. A crystalline substructure has been reported (Kugrens, Lee and Corliss 1994). Ejectisome function remains speculative but it has been suggested that their discharge may be an evasive mechanism from predators or a stinging mechanism to repel predators.

4.5. MITOCHONDRIA

A single reticulate mitochondrion with flattened cristae (Gillott 1990, Kugrens and Lee 1991) apparently occurs in all cryptomonad cells.

4.6. CHLOROPLASTS

Chloroplasts may be olive green, brownish, blue-green or red depending on the accessory pigments present. Pigments consist of chlorophylls a and c_2, alpha and beta carotene, alloxanthin, diadinoxanthin and either of the phycobiliproteins Cr-phycocyanin or Cr-phycoerythrin. Chloroplasts are surrounded by a double membraned endoplasmic reticulum called the periplastidial envelope, periplastidial compartment, periplastidal complex, starch compartment or chloroplast endoplasmic reticulum, which originates from an evagination of the outer membrane of the nuclear envelope and surrounds the chloroplast, starch grains and a reduced nucleus known as the nucleomorph (Ludwig and Gibbs 1985). Thylakoids in the chloroplasts usually are arranged in pairs (Gantt 1979) or sometimes in groups of three (Hill 1991), or in stacks of variable number (Hill 1991). Either Cr-phycoerythrin or Cr-phycocyanin is located in the intrathylakoidal lumen (Hill & Rowan 1989, Gantt 1979). *Chilomonas* has a reduced chloroplast which lacks pigments and is called a leucoplast. *Goniomonas* lacks plastids and a nucleomorph, consequently it also lacks the periplastidial compartment.

4.7. NUCLEOMORPHS

Nucleomorphs are located in the periplastidial compartment (McKerracher and Gibbs 1982, Ludwig and Gibbs 1985). A nucleomorph represents a vestigial nucleus, which remains from a red algal endosymbiont (Douglas, Murphy, Spencer, Gray 1991, Cavalier-Smith, Couch, Thorsteinsen, Gilson, Deane, Hill and McFadden 1996).

5. Nutrition

With the exception of *Goniomonas* and *Chilomonas*, which are heterotrophs, all cryptomonads are obligate phototrophs. *Goniomonas* is phagocytic, whereas *Chilomonas* is strictly osmotrophic because it does not ingest particulate materials as was previously assumed. One species of *Chroomonas* is mixotrophic (Kugrens and Lee 1990), but there is no evidence that any other cryptomonads also have this mode of nutrition.

6. Reproduction & Life Cycles

Reproduction usually occurs by mitotic divisions, although sexual cycles have been documented for *Proteomonas* (Hill and Wetherbee 1986) and *Chroomonas* (Kugrens and Lee 1988). In addition, *Proteomonas* may have a diplobiontic life history. Cysts are also produced in some genera to withstand adverse conditions.

During mitotic divisions microtubules proliferate near the flagellar bases at the onset of mitosis. The nuclear envelope disaggregates as the microtubules move away from the basal bodies to form a spindle. At metaphase the chromosomes appear as a solid mass, in which the chromosomal microtubules terminate (McKerracher and Gibbs 1982), but kinetochores have not been observed. At anaphase the tightly appressed chromosomal mass splits and the two masses move to the poles. A cytokinetic ring (contractile ring?) is formed at metaphase which cleaves the cells into two daughter cells.

7. Systematics of Cryptomonads (Table 1 and Figure 1)

By using the variations inherent in the furrow/gullet complex and the periplast 15 genera of cryptomonads are currently recognized. For convenience and brevity, the genera are grouped according to coloration. Diagrams of the most common genera are depicted in Figure 1 and their major characteristics are provided in Table 1.

7.1. COLORLESS GENERA (*Chilomonas*, *Goniomonas*)

There are only two colorless genera. *Goniomonas* lacks any vestige of a plastid, however, *Chilomonas* contains two leucoplasts, indicating that its colorless nature and nutrition are secondarily derived. Therefore *Goniomonas* appears to be an ancestral form to the pigmented forms.

7.2. BLUE-GREEN GENERA (*Chroomonas*, *Falcomonas*, *Komma*)

Chroomonas and *Komma* lack a furrow and are separated on the basis of periplast plate differences. *Chroomonas* has rectangular inner and outer plates whereas *Komma* has rectangular plates. *Falcomonas* also has hexagonal plates, but the presence of a furrow distinguishes it from *Komma.*

7.3. BROWN COLORED GENERA (*Campylomonas*, *Cryptomonas*, *Geminigera*, *Guillardia*, *Teleaulax*)

The chloroplasts in these genera are several shades of brown. *Cryptomonas* is unique within this group because it is the only one that has multiple periplast plates and a complex furrow. The others have an inner periplast sheet and simple furrows.

TABLE 1. Cryptomonad genera and the the most significant cellular characteristics used in systematics.

GENUS	Furrow/Gullet	Inner Periplast Plate Component	Furrow Stoma	Furrow Type/	Rhizostyle Location	Nucleomorph Color	Chloroplast
Falcomonas	Furrow only	Hexagonal	Fibrous	Simple/yes	None	Free	Blue-green
Goniomonas	Furrow only	Rectangular	????	Simple/yes	Non-keeled	None	No Plastid
***Proteomonas* (n)**	Furrow only	Hexagonal	????	Simple/??	Non-keeled	Free	Red
***Proteomonas* (2n)**	Furrow only	Sheet	Scalariform	Simple/no	Keeled	Free	Red
Teleaulax	Furrow only	Sheet	Scalariform	Simple/??	Keeled	Free	Brown
Rhodomonas	1/2 Furrow	Square	Fibrous	Simple/no	Keeled	In pyrenoid	Red
Cryptomonas	2/3 Furrow	Oval	Fibrous	Complex/yes	Non-keeled	Free	Brown
Campylomonas	1/3 Furrow	Sheet	Scalariform	Simple/no	Keeled	Free	Brown
Chilomonas	1/5 Furrow	Sheet	Scalariform	Simple/no	Keeled	Free	Colorless
Geminigera	3/4 Furrow	Sheet	??????	Simple/no	Keeled	In nucleus	Brown
Plagioselmis	3/4 Furrow	Hexagonal	??????	Simple/yes	None	Free	Red/Brown
Chroomonas	Gullet only	Rectangular	None	None	None	Free	Blue-green
Guillardia	Gullet only	Sheet	Scalariform	None	Keeled	Free	Brown
Komma	Gullet only	Hexagonal	Fibrous	None	None	Free	Blue-green
Hemiselmis	Gullet only	Hexagonal	Fibrous	None	None	Free	Red
Rhinomonas	Gullet only	Hexagonal	Fibrous	None	None	In pyrenoid	Red
Storeatula	Gullet only	Sheet	Scalariform	None	Keeled	In pyrenoid	Red

The remaining genera are separated on the basis of their furrow/gullet architecture. For instance, *Teleaulax* lacks a gullet, whereas the others have different length furrows in relation to the gullet. In addition, specific cytological characters aid in separating these genera, and some may not warrant generic status.

7.4 RED COLORED CRYPTOMONADS (*Hemiselmis, Plagioselmis; Proteomonas, Rhinomonas, Rhodomonas, Storeatula*)

This is the most variable aggregation of genera and makes them difficult to identify with the light microscope. *Hemiselmis* cells are reniform in shape with the vestibulum located 1/3 the cell length from the anterior (Wetherbee, Hill and McFadden 1986). *Rhodomonas* has a deep gullet, a short furrow and square inner periplast plates with beveled corners. *Storeatula* and the diploid phase of *Proteomonas* both have a single inner periplast sheet, but are separated by the presence of a gullet only in *Storeatula* and a furrow only in *Proteomonas*. The haploid phase of *Proteomonas* has hexagonal plates and a furrow only. *Rhinomonas* has inner hexagonal periplast plates and a gullet only, separating it from the haploid phase of *Proteomonas*.

8. Future Research

As mentioned earlier, electron microscopic studies have revealed many unique features of cryptomonads and due to their unicellularity and small size, current and future classification schemes for cryptomonads must continue to rely upon ultrastructural features. Based on the limited knowledge regarding cryptomonad systematics & phylogeny, the focus of future research should include the following: 1) Several described genera need to be reinvestigated ultrastructurally. *Plagioselmis, Hillea, Leucocryptos/Katablepharis, Cryptochrysis and Cyanomonas* are genera that are still recognized (Bourelly 1970, Butcher 1968) but their validity as distinct genera is in doubt. For instance, there is some doubt as to whether *Hillea, Cyanomonas* and *Cryptochrysis* actually exist since they have not been isolated and cultured. The status of *Plagioselmis* also remains unresolved, even though there have been some recent observations on this genus (Novarino, Lucas, Morrall 1994); 2) Since the diversity of cryptomonads is underestimated (Andersen 1992), taxonomic studies need to be intensified to determine the range of genera & species in this group; 3) Cryptomonad systematics is unsettled and numerous cryptomonads placed in existing genera actually belong in different genera, therefore heterogeneous complexes need to be separated; 4) The range of genera that occurs among the freshwater cryptomonads is unknown. Will any of the strictly marine genera also be found in freshwater? Many exceptions probably will be found, when more critical examinations of freshwater cryptomonads are conducted; 5) Are the colorless flagellate *Katablepharis/Leucocryptos* and its relatives cryptomonads? These differ from

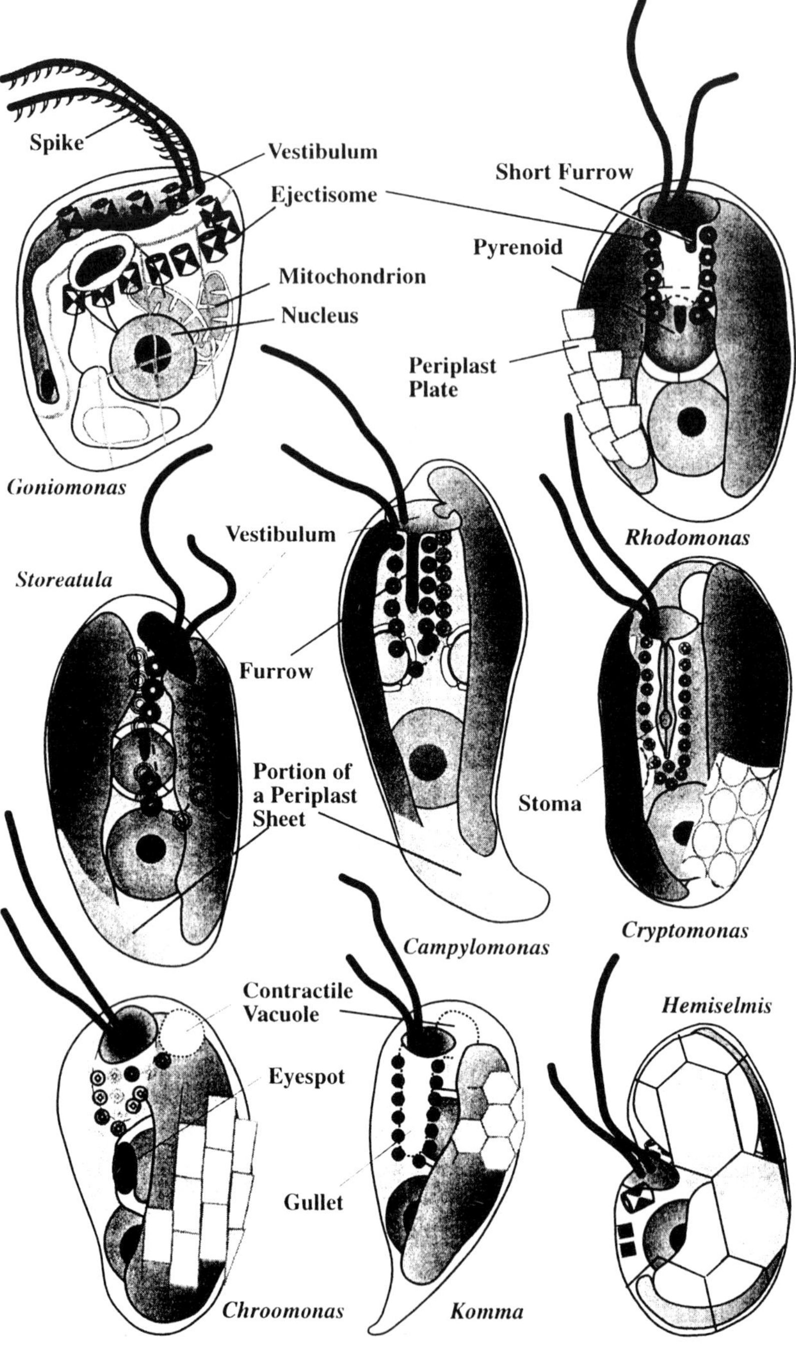

Figure 1. Diagrams of the most common cryptomonads and their characteristics.

cryptomonads in several ultrastructural aspects, including the presence of tubular mitochondrial cristae and the purported absence of a subsidiary scroll in the ejectisomes. A re-evaluation of *Katablepharis* is warranted since the feeding apparatus resembles an apical complex (Lee, Kugrens and Mylnikov 1991); 6) The true phylogenetic position of the cryptomonads with respect to other eukaryotes needs to be assessed. The most problematic polychotomy remaining in the eukaryotic tree is represented by a large number of protist groups whose precise relationships have yet to be worked out. Among this diverse group are the cryptomonads for which only four ssrRNA sequences are known. Ultrastructurally they appear related to the 'stramenopiles' sensu Patterson (Heterokonta sensu Cavalier-Smith) by virtue of possessing tubular hairs and chloroplasts not free in the cytosol, but are delimited by two extra membranes. They do, however, have flat cristae thereby demonstrating their affinities with other platycristate eukaryotes. Maximum parsimony (MP) and Maximum likelihood (ML) trees have produced topologies that allude to this relationship (Cavalier-Smith, Allsopp, and Chao 1994) but the bootstrap value for the MP tree is extremely low and thus the branch is considerably weak. Neighbor joining (NJ) trees have yielded either indeterminate topologies (Cavalier-Smith, Allsopp and Chao 1994, Cavalier-Smith, Couch, Thorsteinsen, Gilson, Deane, Hill and McFadden 1996) or have hinted of a sister relationship with the Rhizopoda, specifically the genus *Acanthamoeba*. However, the NJ tree reported by McFadden et. al. (1994) shows only weak support for this branching pattern (bootstrap value of 52) and two other published NJ trees (Eschbach, Wolters, and Sitte 1991; Douglas, Murphy, Spencer, and Gray 1991) do not include bootstrap values. Therefore these do not provide any real significant insights into the previously mentioned polychotomy. The lack of "robustness" regarding these results may reflect the paucity of sequence data available and should be resolved when more genera are analyzed.

9. References

Andersen, R.A. (1992) *Biodiversity Conserv.* 1:267-292.

Cavalier-Smith, T., Allsopp, M.T.E.P., and Chao, E.E. (1994) *Proc. Natl. Acad. Sci.* USA. 91,11368-11372.

Cavalier-Smith, T., Couch, J.A., Thorsteinsen, K.E., Gilson. P., Deane, J.A., Hill, D.R.A. and McFadden G.I. (1996) *Eur. J. Phycol.* 31,315-328.

Douglas, S.E., Murphy, C.A., Spencer, D.F. and Gray, M.W. (1991) *Nature* 350,148-151.

Eschbach, S., Wolters, J., and Sitte, P. (1991) *J. Mol. Evol.* 32,247-252.

Gantt, E. (1979) In: Levandowsky, M. and Hunter, S.H. (ed.), Biochemistry and Physiology of Protozoa Vol. 1. Academic Press, New York and London, pp. 121-138.

Gillott, M. (1990) In: Margulis, L., Corliss, J.O., Melkonian, M., & Chapman, D.J. (ed.), *Handbook of Protoctista*, Jones & Bartlett Publishers, Boston, pp. 139-151.

Grain, J., Mignot, J.P., and Puytorac, P. (1988) *Biol. Cell* 63,219-237.

Hill, D.R.A. (1990) *J. of Phycol.* 26,133-145.

Hill, D.R.A. (1991) *Phycologia* 30,170-188.

Hill, D.R.A. and Wetherbee, R. (1986) *Phycologia* 27,521-543.

Hill, D.R.A. and R. Wetherbee. (1989) *Phycologia* 28,143-158.

Klaveness, D. (1988) In: Sandgren, C.D. (ed.). *Growth and Reproductive Strategies of Freshwater Phytoplankton*, Cambridge Univ. Press, New York, pp. 105-133.

Kugrens P. and Lee, R.E. (1987) *J. Phycol.* 23,365-76.

Kugrens, P. and Lee, R.E. (1988). *J. Phycol.* 24,510-518.
Kugrens, P. and Lee, R.E. (1990) *J. Protozool.* 37,263-267.
Kugrens, P. and Lee, R.E. (1991) In: Patterson, D.J. and Larsen, J. (ed.) *The Biology of Free-Living Heterotrophic Flagellates*, The Systematics Association Special Volume 45, pp. 219-233.
Kugrens, P., Lee, R.E. and Andersen, R.E. (1986) *J. Phycol.* 22,512-522.
Kugrens, P., Lee, R.E. & Corliss, J.O. (1994) *Protoplasma.* 181,164-190.
Lee, R.E. and Kugrens, P. (1986) *J. Phycol.* 22,549-52.
Lee, R.E., Kugrens, P., and Mylnikov, A.P. (1991) *J. Phycol.* 27,725-733.
Ludwig, M. and Gibbs, S.P. (1985) *Protoplasma* 127,9-20.
Ludwig, M. and Gibbs, S.P. (1989) *J. Cell Biol.* 108,875-884.
McFadden, G.I., Gilson, P.R., and Hill, D.R.A. (1994) *Eur. J. Phycol.* 29,29-32.
McKerracher, L. and Gibbs, S.P. (1982) *Can. J. Bot.* 60,2440-02452.
Munawar, M. and Bistricki T. (1979) *Scanning Electr. Microscopy* 3,247-252.
Novarino, G., Lucas, I.A.N. & Morrall, S. (1994) *Cryptogamie, Algologie* 15,87-96.
Pfiester, L.A. and Holt, J.R. (1978) *Southwestern Naturalist* 23,103-110.
Wetherbee, R., Hill, D.R.A., and McFadden, G.I. (1986) *Protoplasma* 131,11-22.

The Biodata of **Niji Ohta** contributor of the chapter on:
"Molecular Living Fossil..."

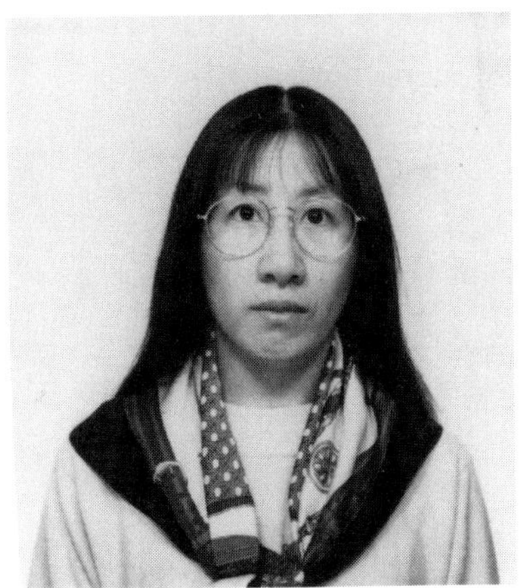

Dr. Niji Ohta is a research fellow of the Japan Society of the Promotion of Science at the School of Human Science, Waseda University (Saitama, Japan). She received her D.Sc. degree in 1993 from the University of Tokyo (Hongo, Tokyo, Japan). Dr. Ohta's research interest is in the origin and evolution of organelles and has studied the plastid and mitochondrial genomes of the ultra small in very primitive rhodophytes (*Cyanidium caldarium* and *Cyanidioschyzon merolae*). She is the first molecular biologist to complete the analyses of the whole plastid and mitochondrial genomes sequences of *C. merolae*. Her publications deal with the molecular sequences of those organisms.
E-mail: **<niji@human.waseda.ac.jp>**

J. Seckbach (ed.), Enigmatic Microorganisms and Life in Extreme Environments, 139–149.

THE ORGANELLAR GENOMES OF *Cyanidioschyzon merolae*

NIJI OHTA
School of Human Science
Waseda University
2-579-15 Mikajima, Saitama, 359 Japan

NAOKI SATO
Department of Biochemistry and Molecular Biology
Faculty of Science
Saitama University
255 Shimo-Ohkubo, Urawa, Saitama Prefecture, 338 Japan

TSUNEYOSHI KUROIWA
Department of Biological Sciences
Graduate School of Sciences
University of Tokyo
7-3-1 Hongo, Tokyo, 113 Japan

1. Introduction

In a eukaryotic cell, the genetic materials are located in the cell nucleus, the mitochondrion and the plastid, and each of the three has their own transcriptional and translational apparatus. Origin of the organellar genomes, or the organelles themselves, has long been an important subject of biological research. Two major hypotheses have been considered. One is the autogenous origin hypothesis: the two organelles arose through a process of intracellular compartmentalization and functional specialization within a single hypothetical eukaryotic progenitor cell (Raff and Mahler, 1972). The other is the endosymbiotic hypothesis: the two organelles had originated from a free-living organisms that were subsequently engulfed by a eukaryotic progenitor cell (Margulis, 1981; Gray and Doolittle, 1982). In the light of the present knowledge of molecular phylogeny, endosymbiotic hypothesis is thought to be more acceptable than the other.

One approach to the study of the origin and the evolution of organelles is a search for most "primitive" eukaryotes. The analysis of the organellar genomes of these organisms might shed light on the link between the prokaryotic progenitors and the organelles. One promising candidate of such algae whose organelles can be a link between prokaryotes and organelles of higher organisms includes a group of red algae that belong to Cyanidiaceae (Seckbach, 1992). Cyanidiaceae contain very primitive red algal species, namely, *Cyanidioschyzon merolae*, *Cyanidium caldarium* and *Galdieria sulphuraria*. Their major habitats are located in acidic hot springs. *C. caldarium* has been extensively studied. It can thrive under anaerobic conditions (Seckbach *et al.*, 1970). Molecular phylogenetic analysis of its plastid genome was reported (Ohta *et al.*, 1994a, b; Maid *et al.*, 1992).

2. Cytological Evidence for the Primitiveness of Cyanidiaceae

Cyanidioschyzon merolae, which was first isolated from acidic hot springs (De Luca *et al.*, 1978), has been described as a unicellular rhodophyte that belongs to Cyanidiaceae (Seckbach, 1992). This alga is closely related to *Cyanidium caldarium* that has long been thought as one of the most primitive plants. *C. merolae* and *C. caldarium* have both a very simple intracellular organization, but cellular traits suggest that *C. merolae* is more primitive than *C. caldarium*. The size of the cell-nuclear genome (Takahashi *et al.*, 1993), The mode of cell division (Merola *et al.*, 1981) and so forth.

We previously showed that the plastid genome of *C. caldarium* strain RK-1 contains the genes that are not in the plastid genome of chlorophytes (i.e., land plants and green algae) such as *trxM* (Ohta *et al.*, 1994a) and *trpA* (Ohta *et al.*, 1994b). These as well as other plastid genes are thought to be transferred from the organelle to the cell nucleus during the evolution of plants and algae. Transfer of mitochondrial genes to the cell nucleus is also well documented (Baldauf and Palmer, 1990). The primitive cellular traits of *C. merolae* suggest that the organelle genomes of this alga also contain several genes that are not present in the organelle genomes of chlorophytes. Complete nucleotide sequences of the mitochondrial and plastid genomes of *C. merolae* now shed light on the origin and the evolution of this red alga.

3. The Plastid Genome

3.1. GENERAL VIEW OF THE PLASTID GENOME

In the plastid genomes of chlorophytes, approximately 150 genes including the genes for tRNA and rRNA are described (Sugiura, 1992). The numbers and kinds of genes differ only slightly, and most of the open reading frames are conserved. The difference of the kind of the genes in the plastid genome is thought to have been the results of the evolutionary gene transfer. Several genes are known to have been transferred from the plastid to the cell nucleus step by step during the evolutionary process. In rhodophytes, such gene transfer might have occurred, but the detail remains unknown.

The plastid genomes of *Porphyra purpurea* and *Cryptomonas* ϕ have been found to contain the genes that are not found in those of chlorophytes (Wang and Liu, 1991; Reith and Munholland, 1993). In Cyanidiophyceae, the plastid genome of *C. caldarium* contained *trpA*, which was found actively transcribed in the cell (Ohta *et al.*, 1994a). The *trxM* gene was also detected in the plastid genome of *C. caldarium* (Ohta *et al.*, 1994b).

3.2. PHYLOGENETIC RELATIONSHIP OF THE PLASTID OF *C. merolae* TO THE PLASTID GENOME OF OTHER ALGAE AND PLANTS

A phylogenetic tree inferred from deduced amino acid sequences of the *rbcL* genes has been constructed (Ohta *et al.*, 1997a). The amino acid sequence of *rbcL* of *C. caldarium* RK-1 has been reported by Langsdorf *et al.* (P37395 of SwissProt database), but was not identical to that of the strain of the same name (RK-1) reported by Ohta *et al.* (1997a), and therefore, both of the sequences were used to construct the tree as two different kinds of sequence. Three main groups were recognized: a group of rhodophytes and chromophytes, a group of b-proteobacteria, and a group of chlorophytes, cyanobacteria and glaucophytes. An anomaly was the position of *Cyanophora paradoxa*, a glaucophyte containing chlorophyll *a* and phycobilins, which is placed within the group of chlorophytes. The sequence of *C. merolae* and the two sequences of *C. caldarium* strain (see above) made up a single lineage.

Although *C. merolae* and *C. caldarium* belong to Rhodophyta, they did not form a single group with other species of rhodophytes such as *Porphyra purpurea*. The results suggest that *C. merolae* and the two strains of *C. caldarium* are closely related and that they form a sister group distinct from other rhodophytes.

3.3. TRANSLOCATION OF *str* CLUSTER IN THE PLASTID GENOMES OF RHODOPHYTES-CHROMOPHYTES LINEAGE

We then used the gene cluster of the ribosomal proteins in the plastid genome in the phylogenetic analysis, because the order of the genes and their nucleotide sequence of the gene cluster is conserved in various organisms. The gene order was the same as those of *P. purpurea* and *Odontella sinensis* (chromophytes). The gene cluster made of *rps12*, *rps7* and *tufA* that we call *str* cluster is separated from the groups of other ribosomal protein gene clusters of the plastid genomes of most plants and algae analyzed except for rhodophytes and chromophytes. Interestingly, the *str* cluster is located at the end of the ribosomal cluster in the lineage of rhodophytes and chromophytes. The results suggest that the translocation of the *str* cluster occurred early in the lineage of rhodophytes and chromophytes after these groups had been separated from chlorophytes and glaucophytes. We were able to construct a phylogenetic tree that explains all the deletions and translocations of ribosomal genes in the cluster (Ohta *et al.*, 1997b). Plastids made up a single group, and the cyanobacterium Synechocystis was a sister group of the plastid group. This result suggests that an ancestral cyanobacterium might be the origin of plastid as proposed in previous studies (Bergsland and Haselkorn, 1991; Morden *et al.*, 1992; Helmchen *et al.*, 1995). In the plastid group, *C. merolae* is located within the group of rhodophytes. This analysis was totally based on amino acid sequences, and the result was consistent with the result obtained from the analysis of gene order as described above. This seems exceptional, since gene order is usually highly variable and is not considered to be a good marker for phylogenetic analysis.

3.4. THE STRUCTURE OF THE PLASTID GENOME OF *C. merolae*

We then succeeded in determining the nucleotide sequence of the plastid genome of *C. merolae*. This information is useful in assessing the phylogenetic status of this organism.

3.4.1. *Size and Overall Structure of the Plastid Genome of C. merolae*

The plastid genome of *C. merolae* is a circular molecule composed of 149,647 bp. This size is within the range of the sizes of plastid genomes (from 80 to 200 kbp).

The plastid genome of *C. merolae* does not have an inverted repeat sequence (IR) containing rRNAs. In rhodophytes, *Griffithsia pacifica* does not have an IR (Li and Cattolico, 1987), while that of *Porphyra purpurea* has a direct repeat (Reith and Munholland, 1995). The plastid genome of *C. caldarium* has a typical IR (Ohta *et al.*, 1994). The IR region in the plastid genomes, if it is present in the original endosymbiont as in the case of cyanobacteria, might have been preferentially lost during the process of evolution in the lineage of rhodophytes. The IR of *C. caldarium* escaped from loss for some unknown reason. Again, this is an evidence that the plastid genome of *C. merolae* was subject to substantial modification during the evolution from the primary endosymbiont.

3.4.2. *Gene Content*

The results of nucleotide sequence analysis indicated that the plastid genome of *C. merolae* contains 212 protein-coding genes, 3 genes for rRNA and 31 genes for tRNA. None of the genes contain an intron. Spacer region between genes are generally quite short and several genes even overlap each other. The plastid genome contained the genes not only for photosynthesis and gene expression but also for various kinds of biosynthesis that take place within the plastid. It also contains several genes that have not been found in other plastid genomes analyzed so far (Table 1). We will summarize some interesting features of the genes in more detail.

Genes for RNA Polymerases. The plastid genome of *C. merolae* contains *rpoA*, *rpoB* and *rpoC* as genes for the RNA polymerase. The genome of *Synechocystis* and the plastid genome of various plants and algae contain *rpoA*, *rpoB*, *rpoC1* and *C2*. It is interesting to note that the beta' subunit is encoded by *rpoC* in *C. merolae*, as in the case of most eubacteria except cyanobacteria. In cyanobacteria and most plastids, the β' subunit is split into two subunits, β' and β", which form a complex to function as a single β' subunit. There are two explanations for the presence of a single *rpoC* in *C. merolae*: either it has been generated by gene fusion of *rpoC1* and *rpoC2*, which had been present in the ancestral endosymbiont, or it came from another bacteria by horizontal transfer. This point should be studied in more detail to elucidate the origin and evolution of the split *rpoC* gene.

Genes for Enzymes Involved in Various Biosynthetic Pathways. Plastid plays important roles in the metabolism of nitrogen and sulfur including the biosynthesis of amino acids, as well as the biosynthesis of fatty acids and lipids, and the biosynthesis of vitamins and biologically active molecules. Most of the genes for the proteins involved in these processes are encoded by the cell nucleus. This is especially true in chlorophytes. The plastid genome of rhodophytes, chromophytes and glaucophytes, however, contain some of the genes for enzymes involved in these biosynthetic pathways. The plastid genome of *C. merolae* contains 7 genes for the biosynthesis of amino acids (*hisH*, *trpA*, *trpG*, *ilvB*, *ilvH*, *argB*, *gltB*), 5 genes for fatty acid metabolism (*accA*, *accB*, *accD*, *acpP*, *desA*), two genes for chlorophyll biosynthesis (*chlI*, *cobA*), a carotenoid biosynthesis gene (*crtE*), a lipoate biosynthesis gene (*lipB*), 5 genes for the biosynthesis of vitamin K (*menA*, *menB*, *menD*, *menE*, *menF*), a gene for molybdopterin synthesis (*moeB*), a gene for thiamin synthesis (*thiG*), a gene for thioredoxin (*trxM*), two genes for the secretion of proteins (*secA*, *secY*), 3 genes for lipopolysaccharide biosynthesis (*glmS*, *lpxA*, *lpxC*), a gene for carbamoyl-phosphate synthase (*carA*), two genes for pyruvate dehydrogenase (*odpA*, *odpB*).

It should be noted that many genes in this category are also contained in the plastid genome of *P. purpurea*, but several are unique to *C. merolae* (see below).

Genes for DNA-binding Proteins. The plastid genome of *C. merolae* contains several genes that encode putative DNA-binding proteins. This point is especially important in the autonomy of the plastid genome. The HU-like protein gene (*hupA*) has been found in *Cryptomonas* ϕ, but no other plastid genome has been known to contain genes for DNA-binding proteins, if we exclude RNA polymerase from this category. The plastid genome of *C. merolae* contains *hupA* and a gene for putative DNA-binding protein that is homologous with the global regulator for nitrogen assimilation NtcA of cyanobacteria. Since we still have many unknown ORFs, the number of genes that encode putative DNA-binding proteins will be increased. Since HU-like protein has been found in the plastids of higher plants (Yurina *et al.* 1995), the *hupA* gene has been transferred from the plastid genome to the cell nucleus during the evolution of plants and algae. The presence of the *ntcA* homologue in the plastid genome of *C. merolae* is interesting, because there are a number of genes that are involved in amino acid and other nitrogenous compounds in the plastid genome of this organism.

Genes not Known in Other Plastid Genomes. The plastid genome of *C. merolae* contains 13 genes which have not been found in other plastid genomes: *menA*, *menB*, *menD*, *menE*, *menF*, *desA*, *cobA*, *yhbG*, *lipB*, *thdF*, *glmS*, *lpxA* and *lpxC*.

The presence of the *lpxA* and *lpxC* genes in the plastid genome of *C. merolae* is an enigma, because these genes are not expected to have a function within the plastid itself. The products of these genes in eubacteria are known to be involved in the biosynthesis of lipid A. Lipid A molecules constitute the outer monolayer of the outer membrane of most of Gram-negative bacteria. *Synechocystis* sp. PCC6803, whose genome has been completely sequenced, also contains the full set of genes involved in the biosynthesis of lipid A including *lpxA* and *lpxC*. If a cyanobacterial ancestor had been the origin of endosymbiont that became plastid later, the *lpxA* and *lpxC* genes in *C. merolae* can be considered to be remnants of the genome of cyanobacterial symbiont. Therefore, we might be able to call these genes "molecular living fossil".

4. Mitochondrial Genome

4.1. GENERAL VIEW OF THE MITOCHONDRIAL GENOME

The mitochondrial genome of each of the three kingdoms, Animalia, Eukaryomycota and Plantae, have its own characteristics. Animal mitochondrial genomes are in a range 16-19 kbp, do not have an intron, and the number of genes is generally limited to the minimum. The size of fungal mitochondrial genomes ranges 17-176 kbp. The reason for the variable size is not explained by the coding capacities of the genes but by the presence of introns and spacer regions that are variable in length. The fungal mitochondrial genomes contain almost as many genes as those of animals do. In clear contrast to these mitochondrial genomes, the size of plant mitochondrial genomes is much more variable. It ranges 16-2400 kbp (Gillham, 1994). The variety of the gene content, molecular structure and the variation of the length of the spacer regions and the introns are largest among the three kingdoms. The mitochondrial genome of angiosperms is especially large and sometimes include plastid DNA sequences that had been transferred during the evolution.

Until now, complete nucleotide sequence of mitochondrial genome has been determined in four species of plants and algae, namely, *Chlamydomonas reinhardtii* (Boer and Gray, 1991), *Marchantia polymorpha* (Oda *et al.*, 1992), *Prototheca wickerhamii* (Wolff *et al.*, 1994) and *Chondrus crispus* (Leblanc *et al.*, 1995).

Complete nucleotide sequence has been determined in two species of protozoa, *Acanthamoeba castellanii* (Burger *et al.*, 1995) and *Reclinomonas americana* (Lang *et al.*, 1997), and in two species of fungi, *Hansenulla wingei* (Sekito *et al.*, 1995) and *Saccharomyces cerevisiae* (de Zamaroczy and Bernardi). Ohta *et al.* (1997c) also determined complete nucleotide sequence of the mitochondrial genome of *C. merolae*. Important features of this genome are summarized below.

4.2. MITOCHONDRIAL GENOME OF *C. merolae*

4.2.1. *Size and Structure*

The mitochondrial genome of *C. merolae* is a circular molecule composed of 32,211 bp. In this genome, we detected 34 protein-coding genes, three genes for rRNA and 25 genes for tRNA. None of these genes contain an intron, though intron(s) have been found in some of the mitochondrial genes in *C. crispus*, *P. wickerhamii*, *M. polymorpha*, *A. castellanii* and *H. wingei*.

Spacer regions between genes in the mitochondrial genome of *C. merolae* are generally very short. Intergenic region was only 4.5 % of the entire mitochondrial genome sequence, showing a high coding capacity of the genome. Absence of intron and high coding capacity of the mitochondrial genome are the features shared by the mitochondrial genomes of animals.

4.2.2. *Gene Content*

Mitochondrial genome of *C. merolae* contains the standard set of mitochondrial protein-coding genes, namely, genes for cytochrome oxidase, cytochrome b, subunits of ATPase and NADH dehydrogenase complexes. The genome also contains additional protein genes: genes related to c type cytochrome biogenesis (*ccmA*, *ccmC*, *ccmF*), succinate dehydrogenase subunits (*sdhB*, *sdhC*), genes for ribosomal proteins, some other ORFs homologous to other mitochondrial genes and unique to *C. merolae*. Among these, *ccmA* (cytochrome c type biogenesis ATP-binding protein) was first identified as a gene encoded in the mitochondrial genome of plant.

In the mitochondrial genome of *C. merolae*, several genes that work together make up single clusters. The two genes for the subunit of succinate dehydrogenase, *sdhB* and *sdhC*, are located in tandem. The genes of apocytochrome b (*cytb*) and cytochrome oxidase subunits (*cox1, 2, 3*) are present on the same strand and located clustered. Among the seven genes for NADH dehydrogenase subunits, *nad1, 2, 3, 4, 4L, 5, 6*, that were identified, five genes, *nad1, 2, 3, 4, 5*, are grouped together and have the same transcriptional orientation, though *nad4L* and *nad6* are encoded by the opposite strand and located apart from the nad gene cluster. The genes for seven ribosomal proteins out of eleven make up a ribosomal gene cluster.

5. Concluding remarks

C. merolae is considered to be one of the most primitive algae from the viewpoint of cytological traits. Analysis of the mitochondrial and plastid genomes showed us interesting features that suggest the origin and the evolution of the two organelles in this alga.

The results of phylogenetic analysis using *rbcL* and the ribosomal protein gene cluster clearly suggest that *C. merolae* is considered to be classified in the rhodophyte-chromophyte lineage. The plastid genome of *C. merolae* presents both ancestral and non-ancestral characteristics. The ancestral characteristics are the followings: (1) It contains many genes that are not known in other plastid genomes such as *lpxA*, *lpxC*, *desA* and *hupA* which might be able to call these genes "molecular living fossil". (2) The presence of genes that encode DNA-binding proteins (*hupA* and *ntcA*) is a sign of limited autonomy of the plastid genome, which might be remaining after most of the genes that regulate the expression of plastid genomes have been transferred to the cell nucleus. (3) The plastid genome contain many unidentified ORFs that are shared only by cyanobacteria. These genes are considered to be in the cell nucleus in other species.

The plastid genome of *C. merolae* is, however, thought to be subjected to various modification after the formation of ancestral rhodophyte. (1) It does not contain *chlB*, *chlL* and *chlN* which are contained in the plastid genome of *P. purpurea*. These genes that are involved in the light-independent chlorophyll biosynthesis must have been lost after *C. merolae* separated from other algae. It should be examined if the cell of *C. merolae* becomes etiolated in the dark. (2) Intergenic regions of this plastid genome are usually short, and there are many overlapping regions. Absence of inverted repeat which is typical to the plastid genomes also increases the coding capacity of the plastid genome of *C. merolae*. The economy of coding capacity must have been elaborated during the evolution of this alga, under the selection pressure that favored small size of cell. (3) An interesting example of modification of the genome is the *rpoC* gene, which seemed to have been produced as a result of fusion of *rpoC1* and *rpoC2*.

The mitochondrial genome of *C. merolae* is a circular molecule of 32 kbp, and the genome contains standard set of the genes as the mitochondrial genome as well as the genes for several ribosomal proteins, succinate dehydrogenase subunits and cytochrome c biogenesis. It also contains several ORFs which are in common to the genes of the mitochondrial genome of higher plants suggesting that the two mitochondria originate from the same ancestor. The gene content resembles that of plant mitochondrial genome, whereas the genome organization resembles those of animals: it does not contain an intron, and the intergenic regions are small. The results suggest that the mitochondrial genome of *C. merolae* is standard as the plant mitochondrial genome, but the size is shortened. Therefore, the coding capacity of the mitochondrial genome of *C. merolae* is extremely high as is the plastid counterpart.

Further detailed analysis of the entire sequence of plastid and mitochondrial genomes in *C. merolae* will shed a new light on the origin and evolution of mitochondrion and plastid. In addition, analysis of the cell-nuclear genome will tell us more about the evolution of this organism.

Table 1. Genes identified in the plastid genome of *C. merolae*.

Gene	Gene product		
accA	acetyl-CoA carboxylase carboxyltransferase α	*apcE*	phycobilisome LCM core-membrane linker
		apcF	allophycocyanin B18 subunit
accB	biotin carboxyl carrier protein of acetyl-CoA carboxylase	*argB*	N-acetylglutamate kinase
		atpA	ATP synthase alpha subunit
accD	acetyl-CoA carboxylase carboxyltransferase β	*atpB*	ATP synthase beta subunit
acpP	acyl carrier protein	*atpD*	ATP synthase delta subunit
apcA	allophycocyanin α chain	*atpE*	ATP synthase epsilon subunit
apcB	allophycocyanin β chain		
apcD	allophycocyanin α -B chain	*atpF*	ATP synthase subunit b

atpG ATP synthase subunit b'
atpH ATP synthase subunit c
atpI ATP synthase subunit a
carA carbamoyl-phosphate synthetase small chain
cfxQ CfxQ protein homolog
chlI magnesium-chelatase subunit ChlI
clpC Clp protease regulatory subunit
cobA uroporphyrin-III C-methyltransferase
cpcA C-phycocyanin alpha subunit
cpcB C-hycocyanin beta subunit
cpcG phycobilisome rod-core linker polypeptide
crtE geranylgeranyl pyrophosphate synthase
cysT sulfate transport system permease protein
cysW sulfate transport system permease protein
desA fatty acid desaturase
dnaB replicative DNA helicase
dnaK DnaK protein
ftrC ferredoxin reductase catalytic chain
ftsH cell division protein FtsH homolog
glmS L-glutamine:D-fructose-6-P amidotransferase
gltB ferredoxin-dependent glutamate synthase
groEL 60kD chaperonin
hisH N-acetyl-glutamate-gamma-semialdehyde dehydrogenase
hupA DNA binding protein HU
ilvB acetolactate synthase large subunit
ilvH acetolactate synthase small subunit
infB initiation factor IF-2
infC initiation factor IF-3
lipB lipoate-protein ligase B
lpxA acyl-[acyl-carrier-protein]--UDP- N-acetylglucosamine acyltransferase
lpxC UDP-3-0-acyl N-acetylglcosamine deacetylase
menA menaquinone biosynthesis protein
menB naphthoate synthase
menD 2-succinyl-6-hydroxy-2,4- cyclohexadiene-1-carboxylate synthase
menE O-succinylbenzoic acid-CoA ligase
menF isochorismate synthase
moeB molybdopterin biosynthesis MoeB protein
ntcA global nitrogen regulator
odpA pyruvate dehydrogenase E1 alpha subunit
odpB pyruvate dehydrogenase E1 beta subunit
petA apocytochrome f
petB cytochrome b6
petD cytochrome b6-f complex subunit 4
petF ferredoxin
petG PetG subunit of the cytochrome b6/f complex
petJ cytochrome c553
psaA P700 apoprotein subunit Ia
psaB P700 apoprotein subunit Ib
psaC photosystem I subunit VII
psaD photosystem I subunit II
psaE photosystem I subunit IV
psaF photosystem I subunit III
psaI photosystem I subunit VIII
psaJ photosystem I subunit IX
psaK photosystem I subunit X
psaL photosystem I subunit XI
psaM photosystem I reaction centre subunit XII
psbA photosystem II D1 protein
psbB photosystem II P680 chlorophyll A apoprotein
psbC photosystem II 44kD reaction center protein
psbD photosystem II D2 protein
psbE cytochrome b559 a subunit
psbF cytochrome b559 b subunit
psbH photosystem II PsbH protein
psbI photosystem II PsbI protein
psbJ photosystem II PsbJ protein
psbK photosystem II PsbK protein
psbL photosystem II PsbL protein
psbN photosystem II PsbN protein
psbT photosystem II PsbT protein
psbV cytochrome c550
psbW photosystem II PsbW protein
psbX photosystem II PsbX protein
rpl1, 2, 3, 4, 5, 6, 11, 12, 13, 14, 16, 18, 19, 20, 21, 22, 23, 24, 27, 28, 29, 31, 32, 33, 34, 35, 36 50S ribosomal protein
rps2, 3, 4, 5, 6, 7, 8, 9, 10, 11, 12, 13, 14, 16, 17, 18, 19, 20 30S ribosomal protein
rpoA RNA polymerase alpha chain
rpoB RNA polymerase beta chain
rpoC RNA polymerase beta dash chain
rbcL ribulose bisphosphate carboxylase large chain
rbcS ribulose bisphosphate carboxylase small chain
secA preprotein translocase SecA subunit
secY preprotein translocase SecY subunit
thdF thiophene and furan oxidation protein
thiG ThiG protein
trpA tryptophan synthase alpha chain
trpG anthranilate synthase component II
trxM thioredoxin type m
tsf elongation factor TS
tufA elongation factor Tu
yhbG ABC transporter

6. References

Baldauf, S. L. and Palmer, J. D. (1990) *Nature* **344**: 262-265.

Bergsland K. J. and Haselkorn, K (1991) *J. Bacteriol* **173**: 3446-3455.

Boer, P. H. and Gray, M. W. (1991) *Curr. Genet.* **19**: 309-312.

Burger, G., Plante, I., Lonergan, K. M., Gray, M. W (1995) *J. Mol. Biol.* **245**: 522-537.

De Luca, P., Taddei, R. and Varano, L. (1978) *Webbia* **33**: 37-44.

Delwiche, C. F. and Palmer, J. D. (1996) *Mol. Biol. Evol.* **13**: 873-882.

De Zamaroczy, M. and Bernardi, G. (1986) *Gene* **47**: 155-177.

Gillham, M. W. (1994) *Organelle genes and genomes.* Oxford University Press. New York.

Gray, M. W. and Doolittle, W. F. (1982) *Microbiol. Rev.* **46**: 1-42.

Helmchen, T. A, Bhattacharya, D. and Melkonian, M. (1995) *J. Mol. Evol.* **41**: 203-210.

Kowallik, K. V., Stoebe, B., Schaffran, S. I., Kroth-Pancic, P. and Freier, U. (1995) *Plant Mol. Biol. Rep.* **13**: 336-342.

Lang, B. F., Burger, G., O'Kelly, C. J., Cedergren, R., Golding, G., B., Lemieux, C., Sankoff, D., Turmel, M. and Gray, M. W. *Nature* **387**: 493-497 (1997).

Leblanc, C., Boyen, C. Richard, O., Bonnard, G., Grienenberger, J.-M., Kloareg, B. (1995) *J. Mol. Biol.* **250**: 484-495.

Maid, U. and Zetsche, K. (1992) *Plant Mol. Biol.* **8**: 327-336.

Margulis, L. (1981) *Symbiosis in cell evolution: Life and its environment on the Early Earth.* W. H. Freeman, San Francisco.

Merola, A., Castaldo, R., De Luca, P., Gambardella, R., Musacchio, A. and Taddeei, R. (1981) *G. Bot. Ital.* **115**: 189-195.

Morden, C. W., Delwiche, C. F., Kuhsel, M. and Palmer, J. D. (1992) *BioSystems* **28**: 75-90.

Oda, K., Yamato, K., Ohta, E., Nakamura, Y., Takemura, M., Nozato, N., Akashi, K., Kanegae, T., Ogura, Y., Kohchi, T., Ohyama, K. (1992) *J. Mol. Biol.* **223**: 1-7.

Ohta, N., Kawano, S. and Kuroiwa, T. (1994a) *Curr. Genet.* **26**: 136-138.

Ohta, N., Sato, N., Kawano, S. and Kuroiwa, T. (1994b) *Curr. Genet.* **25**: 357-361.

Ohta, N., Sato, N., Ueda, K. and Kuroiwa, T. (1997a) *J. Plant Res.* **110**: 235-245.

Ohta, N., Sato, N., Nozaki, H. and Kuroiwa, T. (1997b) *J. Mol. Evol.* 45: 688-695.

Ohta, N., Sato, N. and Kuroiwa, T. (1997c) *Plant Physiol.* **114**: 172.

Raff, R. A. and Mahler, H. R. (1972) *Science* **177**: 575-582.

Reith, M. and Munholland, J. (1993) *Plant Cell* **5**: 465-475.

Reith, M. and Munholland, J. (1995) *Plant Mol. Biol. Rep.* **13**: 333-335.

Seckbach, J., Baker, F. A. and Shugarman, P. M. (1970) *Nature* **227**: 744-745.

Seckbach, J. (1992) In: Reiser W (ed), *Algae and symbioses: plants, animals, fungi, viruses, interactions explored.* Biopress Limited, Bristol, England. pp. 400-426.

Sekito, T., Okamoto, K., Kitano, H. Yoshida, K. (1995) *Curr. Genet.* **28**: 39-53.

Stirewalt, V. L., Michalowski, C. B., Loffelhardt, W., Bohnert, H. J. and Bryant, D. A. (1995) *Plant. Mol. Biol. Rep.* **13**: 327-332.

Sugiura, M. (1992) *Plant Mol. Biol.* **19**: 149-168.

Takahashi, H., Suzuki, K., Ohta, N., Suzuki, T., Takano, H., Kawano, S. and Kuroiwa, T. (1993) *Cytologia* **58**: 477-482.

Wang, S. and Liu, X -Q. (1991) *Proc. Natl. Acad. Sci. USA* **88**: 10783-10787.

Wolff, G., Plante, I., Lang, B. F., Kuck, U. and Burger, G. (1994) *J. Mol. Biol.* **237**: 75-86.

Yurina, A. N., Belkina, G. G., Karapetyan, N. V. and Odintsova, M. S. (1995) *Biochem. Mol. Biol. Int.* **36**: 145-154.

III. THE VERSATILE EXTENSION OF LIFE

Biodata of Dr. **Maud Walsh** author (with co-author Dr. **Joseph Seckbach**) of the chapter entitled "***The Versatility of Microorganisms: A mini Survey.***"

Dr. Maud M. Walsh is an Assistant Professor, Research, in the Institute for Environmental Studies at Louisiana State University (Baton Rouge). She earned a B. A. in English from Bryn Mawr College in 1975, a M.L.S. in Library Science from Louisiana State University in 1976, and a Ph.D. in Geology and Geophysics from Louisiana State University in 1989. Her research interests span across the field of geomicrobiology, from bioremediation of hazardous waste using microorganisms to the study of early life on Earth. She has studied some of the oldest organisms on Earth as preserved in the rock record of the Swaziland Supergroup, South Africa. Dr. Walsh is currently president of the Louisiana State University chapter of ***Sigma Xi,*** the Scientific Research Society, and is publicity officer of the Baton Rouge chapter of the Association for Women in Science.
E-mail: **evwals@unix1.sncc.lsu.edu**

J. Seckbach (ed.), Enigmatic Microorganisms and Life in Extreme Environments, 153–162.

THE VERSATILITY OF MICROORGANISMS

MAUD M. WALSH
Institute for Environmental Studies
Louisiana State University
Baton Rouge, LA 7080. U.S.A.

JOSEPH SECKBACH[1]
Hebrew University of Jerusalem
Jerusalem 91904. Israel

1. Introduction

Living organisms are ubiquitous; they are observed in almost every ecological niche, from the air to various habitats on land and deep in the oceans. The abiding presence of microorganisms has also a temporal dimension. Living microorganisms have been found from 100 year-old beer bottles to 40 million-year-old amber. Fossil evidence suggests that the Earth's earliest organisms were themselves "extremophiles." Extremophiles can be defined as those organisms observed in uncommon habitats (from our anthropocentric viewpoint). Kristjánsson and Hreggvidsson (1995) define an extremophile as one whose optimal growth conditions are found outside of "normal" environments, with normal being those that have a temperature between 4 and 40^{0}C, pH between 5 and 8.5, and with a salinity between that of freshwater and that of seawater. Extremophiles include not only bacteria and archaea but also some eukaryotic organisms, see chapters by Roberts, Elster, Chretiennol-Dinet (q.v.).

Microorganisms have the ability to adapt to many extreme environmental conditions and even to the presence of toxic compounds. Prokaryotes as a group have the ability to utilize almost any redox couple for deriving energy and are composed of a relatively simple structure that can impart a greater resistance to extreme conditions of temperature, pH and salinity (Nealson, 1997). This ability, and evidence that prokaryotes can survive for long periods without water or nutrients, make the study of the extremophiles central to our search for extraterrestrial life.

2. Prokaryotes at Extreme Temperatures

2.1. THERMOPHILES (SOME LOVE IT HOT)

Thermophilic microorganisms thrive in high temperature environments, growing fastest above 40°C, and are found in hot waters, sun-headed soils and geothermal areas Stetter, 1998a). Hyperthermophilic bacteria and archaea grow fastest between 80 and 100°C

[1] Visiting **Professor** & John P. Laborde **Chair** of Louisiana Sea Grant, LSU, Baton Rouge,LA. 70803 U.S.A.

and are generally unable to grow at temperatures below 60°C (Stetter, 1998b; Rachel, this volume). In the hot springs of Yellowstone National Park (USA) several species, including the bacterium *Thermus aquaticus*, grow at greater than 70°C (Brock, 1978a, b). In the deep ocean and the deep Earth, high pressure allows the existence of liquid water at temperatures well in excess of 100°C. Archaea and bacteria have been detected thriving at 115°C in the vicinity of submarine hydrothermal vents (Corliss et al., 1979). One hyperthermophile that lives in deep-sea chimneys is the methane-producing Archaea *Methanopyrus*. This species of Archaea, like many of the methanogens and thermophiles, lies near the root in the tree of life (Pace, 1997; Madigan and Marrs, 1997).

Thermophilic sulfate-reducing bacteria have been isolated from 75°C oil field waters in the North Sea that originated from formations 2,000 to 4,000 m below the sea floor (Rosnes et al., 1991; Beeder et al., 1994). Hyperthermophilic archaea and bacteria have been cultured from production fluids with temperatures up to 110°C from oil reservoirs 3,000 meters below the bed of the North Sea and below the permafrost surface of the North Slope of Alaska (Stetter et al., 1993).

2.2. PSYCHROPHILES (SOME LOVE IT COLD)

The psychrophiles, or the cold loving organisms, are on the other side of the temperature scale from the thermophiles. Psychrophiles are organisms that have optimum growth at temperatures of approximately 15°C, a maximum growth temperature of about 20°C and a minimum of 0° or lower (Morita, 1975). They live in cold soils and waters and even in sea ice (Vishniac, q.v.). An abundance of psychrophilic microorganisms thrive in Antarctica (Friedmann, 1993; Onofri, q.v.), but the largest reservoirs of psychrophiles are deep ocean waters and sediments, as ocean water masses are fairly uniformly cold below the thermocline (Herbert, 1986).

3. Microbial Cells at Extreme pH Ranges

3.1. ALKALIPHILES (LIFE AT ELEVATED pH LEVELS)

Alkaliphiles, bacteria that exhibit optimum growth at pH values above 9 and grow either slowly or not at all at neutral pH values (or at levels below pH 6.5 values) are widely distributed throughout nature and have been isolated from soils, feces, and deep-sea sediments (Horikoshi, 1998; see also Kamekura, q.v.). Soda lakes and deserts, such as those in East Africa, are the most promising sources of alkaliphiles (Grant and Tindall, 1980). One modern alkaliphilic eukaryote has been presented as an analog to a fossil described from rocks approximately 2 billion years old (Siegel, q.v.). Enzymes isolated from alkaliphiles have been used extensively in the detergent industry (Horikoshi, 1998).

3.2. ACIDOPHILES (SOME LIKE IT ACIDIC)

Among the best-studied prokaryotic acidophiles are the iron- and sulfur-oxidizing cells of *Thiobacillus ferroxidans* because of their role in the production of acid rock drainage in mine dumps, where low pH is produced by the weathering of sulfide minerals (Brierley and Brierley, 1997). Although the environmental damage caused by the activity of these bacteria has spurred research on their control, the same bacteria have also been utilized for "bioleaching," which releases metals into solution for eventual recovery. *Thiobacillus ferroxidans* is a mesophile, growing best at moderate temperatures. Thermophilic and hyperthermophilic acidophilic bacteria and archaea thrive in areas of acid hot springs, associated with subsurface magmatic activity (Yellowstone Park) or surface activity (mid-ocean ridges). *Sulfolobus*, which grows in the temperature range of 50 to 80°C under acidic conditions, has been reported from hot, acid springs from around the world (Brock, 1978b). Also found thriving in acid hot springs are the Cyanidiaceae, primitive eukaryotes (*Cyanidium caldarium* and its cohorts) that have been proposed as the "bridge" between prokaryotic cyanobacteria and eukaryotic red algae (Seckbach, 1994, Seckbach et.al. 1983; and see q.v. Chela-Flores, Gross, Ohta et al.; Itoh and Kuroiwa, Seckbach [1998]). It is interesting the *Cyanidium caldarium*-like alga has been discovered in caves which grows in ambient pH conditions (Hoffman, 1994). Furthermore, there some other algae like the chlorophyte, *Dunaliella acidophila* known to be a very acidophilic green alga (Pick, q.v.).

4. Microbes In Other Extreme Conditions

4.1. BAROPHILES (THE PRESSURE LOVERS)

The best-studied environments of barophiles (or piezophiles, as they are termed by some investigators--see Yayanos, 1998, Bartlett and Bidle, q.v.) are the deep oceans, in which hydrostatic pressure increases 1 atmosphere every 10 m in depth; ocean depths average about 380 m but range up to over 10,000 m (Morita, 1986). In the deep oceans, bacteria and archaea range from barotolerant (able to grow at high pressures, but grow more quickly at ambient pressure of 1 atm) and true obligate barophiles (grow more quickly or only at high pressure). In the marine environment, barophilic organisms are also likely to live in extreme temperatures, with the barophilic psychrophiles inhabiting most of the deep oceans (see Bartlett and Bidle, q.v.), and the barophilic thermophiles in the vicinity of hydrothermal vents (see Lee, q.v.). In the past 20 years, sophisticated sampling and growth equipment have allowed the elucidation of changes in membranes as a response to increased pressure (Bartlett et al., 1989). Another area in which thermophilic barophiles are found are in the deep subsurface. Bacteria and archaea have been collected from kilometers deep under the earth's surface, where pressures may reach 200-500 atmospheres (Rosnes et al., 1991; Stetter et al., 1993; see also Onstott et. al. q.v.).

4.2. HALOPHILES (THE SALT LOVERS)

Halophilic microorganisms thrive in habitats characterized by extreme hypersalinity (greater than 7%) and commonly the precipitation of evaporite minerals (Javor, 1989). The most common habitats of halophilic bacteria and archaea are hypersaline lakes, such as the Dead Sea (see q.v. chapters by Ventosa, Ben Amotz, and by Oren) and the Great Salt Lake; hypersaline lagoons and sabkhas, such as the Gavish Sabkha;

hypersaline, alkaline lakes, such as Lake Magadi; and man-made salterns or salinas. One other environment in which halophilic organisms are found are deep sea brines, such as are present in the Red Sea and in parts of the Gulf of Mexico. The salt in deep sea brines derives from the dissolution of ancient evaporite deposits below the sea floor (Javor, 1989). There is evidence that bacteria may survive in ancient salt deposits (Vreeland and Rosenzweig, 1998, q.v.). The halophilic archaea that "virtually define the hypersaline niche" (Javor, 1989) are the halobacteria, which are morphologically distinctive in that they commonly have triangular or square cellular shapes rather than the regular spherical shapes (Oren, q.v.).

4.3. GROWTH IN NUTRIENT- OR WATER-POOR ENVIRONMENTS

Although water and nutrients are essential for all life forms, some prokaryotes are able to survive with a minimum of one or the other or both. A variety of bacteria and cyanobacteria, particularly those found in hot or cold desert conditions, are able to survive desiccation for long periods of time, either in soil or within rocks (see Seckbach, 1997 and see q.v. chapters by Flechtner, by Van Tielen and Garbary, and by Garty). Bacteria are also found in low-nutrient water and soil environments. The organisms adapt to the low-nutrient conditions in a variety of ways, including miniaturization, surface attachment, and storage polymers (Morgan and Dow, 1986).

Some microbes, bacteria and archaea live under various gases such as pure CO_2, or NH_3, (Siegel, q.v.) some may grow chemoautotrophically, obtaining energy from oxidizing chemicals, e.g. S, Fe in reducing CO_2. They may be found in anoxic areas such as superficial and aquatic mud layers and municipal waste. Furthermore, Seckbach et al. 1970, see also Seckbach, 1994) reported on the thriving of thermo-acidophilic algae under pure carbon di-oxide.

Microbial ecosystems that are independent of photosynthetic primary production has been reported from deep basalt aquifers (up to 1400m) in the northwestern United States by Stevens and McKinley (1995). Evidence for microbial activity was also found in a granitic aquifer in Sweden at depths of approximately 207m below sea level (Pedersen et al., 1997).

One adaptation for growth in conditions less than favorable is symbiosis, in which organisms of different species have a close association with beneficial results for at least one of the species. Termites possess symbiotic protists and bacteria in their intestine (Margulis, 1992) and many cockroach species host cilitates and even flagellates in their hindguts. Symbiotic anaerobic microbial communities are present in the digestive tracts of higher organisms (Hackstein, q.v.).Symbiotic microorganisms in the insects and the cattle degrade plant polymers that are resistant to host digestive enzymes. It has been shown that some of these insects' symbionts can inactivate and detoxify poisons in the hosts' diets. Crucial symbiotic relationships are also found in the ocean depths. In the vicinities of deep-sea hydrocarbon seeps, chemosynthetic bacteria form the basis of communities of mussels, clams and tube worms (Brooks et al., 1987).

4.4. MICROORGANISMS RESISTING TOXINS

The search for approaches for remediating man-made contamination of soil and water has expanded our knowledge of the ability of microorganisms to survive and even use as carbon or energy sources many pollutants. Naturally-occurring microbial degradation of hydrocarbons compounds has been widely documented in areas subject to frequent man-made spills or natural hydrocarbon seeps, such as in the Gulf of Mexico, and in other extreme environments (Foght and McFarlane, q.v.). Manipulation of environmental conditions has been used to initiate or accelerate cleanup of hydrocarbon spills (Leahy and Colwell, 1990). Chlorinated hydrocarbons can also be degraded by aerobic and anaerobic bacteria or archaea (Vogel et al., 1987). Microorganisms from a variety of environmental settings have demonstrated abilities to degrade manmade compounds such as chlorobiphenyls or chlorobenzenes (van der Meer at al., 1992). Studies indicate that microorganisms have genetic mechanisms which allow evolutionary changes in existing metabolic pathways, allowing adaptation to newly-introduced contaminants (van der Meer at al., 1992).

Many microorganisms have mechanisms, such as biomethylation, volatilization, and precipitation, that allow them to detoxify heavy metals which are toxic to most living organisms (Collins and Stotsky, 1989). Indeed, microorganisms that can tolerate extreme conditions of metal exposure and/or pH have been used in a variety of ways in the mining industry (Brierley and Brierley, 1997).

5. Discussion

The study of microorganisms in extreme environments has greatly expanded our understanding of physiological processes in all living organisms and provided us with tools used in scientific inquiry and for our applications in everyday life. Exploration of extreme environments has also contributed to our understanding of both early life on Earth and the possibilities of life on other planetary bodies. It has been hypothesized that life originated at submarine hot springs like those found at modern mid-ocean ridges (Corliss et al., 1981). Indeed, the oldest fossils on Earth, remnants of coccoidal and filamentous prokaryotes, are found in settings that suggest that these early organisms inhabited areas of volcanic hydrothermal activity and periodic meteorite impact (Walsh, 1992; Schopf, 1993; see also Westall, q.v.). Recent reports of possible remnants of life on Mars (McKay et al., 1995; Chela Flores, q.v.) and indicators that water, essential for life as we know it, may be present on Europa, one of Jupiter's moons (Svitil, 1997; Holden, 1998, Seckbach et al. 1998), have stimulated interest in the types of organisms that could have survived conditions considered "extreme." It has, however, been suggested (Gorbushina and Krumbein,q.v.) that it might be more appropriate to look for analogs for extraterrestrial life forms among the "poikilotrophs," organisms that can endure extreme conditions but will flourish when "normal" conditions exist. Hot vent environments and rock pore or crevice environments, either of which might have harbored life on Mars, are typically "poikilotrophic." In either case, the examination of organisms in extreme environments will play an essential role in elucidating biological processes on the early Earth and on other planets.

6. Acknowledgements

The coauthor (J. Seckbach) thanks several people who assisted and encouraged him during the compiling of this volume while serving as a Visiting Professor in the Biological Department at LSU (Baton Rouge, LOUISIANA). He appreciates and thanks Dr. John P. Laborde for establishing the Endowed Chair for Sea Grant Research and Technology Transfer (Where J. Seckbach has been the first elected distinguished Professor for this position). He appreciates the kindness of LSU people, Dr. Jack Van Lopik (Executive Director of Sea Grant) for warm friendship, warm hospitality and full cooperation. Dr. Russell L. Chapman (Exc. Dir. of CCEER in LSU) for providing a productive and stimulating atmosphere to proceed with this volume. Many thanks are due to Mrs. Brenda Henning (Sea Grant) for general faithful service, to Mr. Ronald Bouchard & Mrs. Cindy Henk (microscopic section of Biol. Science.) for providing general information and assisting in the material for presentation, and to the efficient team of the Biol. Sciences' Office for everything they help.

7. References

Bartlett, D., Wright, M., Yayanos, A. A. and Silverman, M. 1989. Nature 342: 572-574.

Bartlett, D. and Bidle, K.A. (1998) In: J. Seckbach (ed.), Enigmatic Microorganisms and Life in Extreme Environments, Kluwer Academic Publishers. The Netherlands (this volume).

Beeder, J., Nilsen, R. K., Rosnes, J. T., Torsvik, T. and Lien, T. (1994) Applied and Environmental Microbiology 60(4): 1227-1231.

Ben Amotz, A. (1998) In: J. Seckbach (ed.), Enigmatic Microorganisms and Life in Extreme Environments, Kluwer Academic Publishers. The Netherlands (this volume).

Brierley, C. L. and Brierley, J. A. (1997) In: C. J. Hurst (ed.) Manual of Environmental Microbiology. American Society for Microbiology, Washington, D. C., pp. 830-841.

Brock, T. D. 1978a. Science 158: 1012-1019.

Brock, T. D. 1978b. Thermophilic microorganisms and life at high temperatures. New York, Springer Verlag.

Brooks, J. M., Kennicutt, M. C. II, Fisher, C. R., Macko, S. A, Cole, K., Childress, J. J., Bidigar, R. R. and Vetter, R. D. (1987) Science 238: 1138-1142.

Chela-Flores, J. (1998) In: J. Seckbach (ed.) Enigmatic Microorganisms and Life in Extreme Environments, Kluwer Academic Publishers. The Netherlands. (this volume).

Chretiennol-Dinet, J. In: J. Seckbach, (ed.) Enigmatic Microorganisms and Life in Extreme Environments, Kluwer Academic Publishers. The Netherlands. (this volume).

Collins, Y. E. and Stotsky, G. (1989) In: Beveridge, T. J. and Doyle, R. J. (eds.), Metal Ions and Bacteria. Wiley, New York, pp. 31-90.

Corliss, J. B., Baross, J. A. and Hoffman, S. E. 1981. Oceanologica Acta: Proceedings of the 26th International Geological Congress, Geology of Oceans Symposium, Paris, July 7-17, 1980, 56-69.

Corliss, J. B., Dymond, J., Gordon, L. I., Edmond, J. M., von Herzen, R. P., Ballard, R. D., Green, K., Williams, D., Bainbridge, A., Crane, K. and van Andel, T. H. (1979) Science 203: 1073-1083.

Elster, J. 91998) In: J. Seckbach, (ed.): Enigmatic Microorganisms and Life in Extreme Environments, Kluwer Academic Publishers. The Netherlands. (this volume).

Flechtner, V. (1998) In: J. Seckbach (ed.), Enigmatic Microorganisms and Life in Extreme Environments. Kluwer, The Netherlands (this volume).

Foght, J. and McFarlane, D.M. (1998) In: J. Seckbach, (ed.) Enigmatic Microorganisms and Life in Extreme Environments. Kluwer Academic Publishers. The Netherlands. (this volume).

Friedmann, E. I. (1993) Antarctic Microbiology. New York: Wiley-Liss. 634 p.

Garty, J. (1998) In: J. Seckbach (ed.), Enigmatic Microorganisms and Life in Extreme Environments. Kluwer, The Netherlands (this volume).

Gorbushina, A. A. and Krumbein, W. E. (1998) In: J. Seckbach, (ed.) Enigmatic Microorganisms and Life in Extreme Environments. Kluwer Academic Publishers. The Netherlands. (this volume).

Grant, W. D. and Tindall, B. J. (1980) In: G. W. Gould and J. E. L. Corry (eds.) Microbial Growth and Survival in Extremes of Environment, Academic Press, New York, pp. 27-38.

Gross, W. (1998) In: J. Seckbach (ed.) Enigmatic Microorganisms and Life in Extreme Environments, Kluwer Academic Publishers. The Netherlands. (this volume).

Hackstein, J. (1998) In J. Seckbach, (ed.): Enigmatic Microorganisms and Life in Extreme Environments, Kluwer Academic Publishers. The Netherlands. (this volume).
Herbert, R. A. (1986) In: Microbes in Extreme Environments, R. A. Herbert and G. A. Codd (eds.). Academic Press, New York, pp. 1-23.
Hoffmann L. (1994) In: J. Seckbach (ed.) Evolutionary Pathways and Enigmatic Algae: Cyanidium caldarium (Rhodophyta) and Related Cells. Kluwer Academic Publishers. PP. 175-182.
Holden, C. (1998) Science 270: 1639.
Horikoshi, K. (1998) In: J. Horikoshi and W. D. Grant (eds.), Extremophiles: Microbial Life in Extreme Environments, Wiley, New York, pp. 155-179.
Itoh, R. and Kuroiwa, T. (1998) In: J. Seckbach, (ed.) Enigmatic Microorganisms and Life in Extreme Environments, Kluwer Academic Publishers. The Netherlands. (this volume).
Javor, B. (1989) Hypersaline Environments: Microbiology and Biogeochemistry. Springer-Verlag, New York, 328 p.
Kamekura, M. (1998) In: J. Seckbach, (ed.) Enigmatic Microorganisms and Life in Extreme Environments, Kluwer Academic Publishers. The Netherlands. (this volume).
Kristjänsson, J. K. and Hreggvidsson, G. O. (1995) World Journal of Microbiol. & Biotechnology 11: 17-25.
Leahy, J. G. and Colwell, R. R. (1990) Microbiological Reviews 54: 305-315.
Lee, R. (1998) In: J. Seckbach (ed.) Enigmatic Microorganisms and Life in Extreme Environments, Kluwer Academic Publishers. The Netherlands (this volume).
Madigan, M. T. and Marrs, B. L. (1997) Scientific American 276 (4) 66-71.
Margulis, L. (1992) In: L. Margulis and L. Olendzenski (eds.) Environmental Evolution,. Cambridge, Mass.: The MIT Press, pp. 149-172.
McKay, D. S., Gibson Jr., E. K., Thomas-Keprta, K. L., Vali, H., Romanek, C. S., Clemett, S. J., Chillier, X. D. F., Maechilng, C R. and Zare, R. N. (1996) Science 273: 924-930.
Morgan, P. and Dow, C. S. 1986. In: R. A. Herbert and G. A. Codd (eds.), Microbes in Extreme Environments, Academic Press, New York, pp. 187-214.
Morita, R. Y. (1975) Bacteriological Reviews 39: 146-167.
Morita, R. Y. (1986) In: R. A. Herbert and G. A. Codd (eds.), Microbes in Extreme Environments, Academic Press, New York, pp. 171-185.
Nealson, K. H. (1997) Journal of Geophysical Research 102 (E10): 23,675-23,686.
Ohta, N., Sato, N. and Kuroiwa, T. (1998) In: J. Seckbach (ed.) Enigmatic Microorganisms and Life in Extreme Environments, Kluwer Academic Publishers, The Netherlands (this volume).
Onofri, S. (1998) In: J. Seckbach (ed.) Enigmatic Microorganisms and Life in Extreme Environments, Kluwer Academic Publishers, The Netherlands (this volume).
Onstott, T. C. (1998) In: J. Seckbach (ed.) Enigmatic Microorganisms and Life in Extreme Environments, Kluwer Academic Publishers, The Netherlands (this volume).
Oren, A. (1998) In: J. Seckbach. (ed.), Enigmatic Microorganisms and Life in Extreme Environments. Kluwer, The Netherlands (this volume).
Pace, N. R. (1997) Science 276: 734-740.
Pedersen, K., Ekendahl, S., Tullborg, E-L., Furnes, H., Thorseth, I. and Tumyr, O. (1997) Geology 25: 827-830.
Pick, U. (1998) In: J. Seckbach (ed.) Enigmatic Microorganisms and Life in Extreme Environments Kluwer Academic Publishers. The Netherlands. (this volume).
Rachel, R. (1998) In: J. Seckbach. (ed.) Enigmatic Microorganisms and Life in Extreme Environments. Kluwer, The Netherlands (this volume).
Roberts, D. In: J. Seckbach (ed.) Enigmatic Microorganisms and Life in Extreme Environments, Kluwer Academic Publishers. The Netherlands. (this volume).
Rosnes, J. T., Torsvik, T. and Lien, T. (1991) Applied and Environmental Microbiology 57: 2302-2307.
Schopf, J. W. (1993) Science 260: 640-646.
Seckbach, J. (1994) editor. Evolutionary Pathways and Eigmatic Algae: Cyanidium caldarium (Rhodophyta) and related Cells. Kluer Academic Publishers. The Netherlands.
Seckbach, J. (1997) In: C.B. Cosmovici, S. Bowyer and D. Werthimer (eds.) Astronomical and Biochemical Origins and the Search for Life in the Univrse. Poceeding of the 5th International Confrerence on Bioastronomy IAU Colloquium No. 161 (Capri, July 1-5, 1996). Editrice Compsitori, Italy. pp 511-523.
Seckbach, J. (1998) In: J. Seckbach (ed.) Enigmatic Microorganisms and Life in Extreme Environments Kluwer Academic Publishers. The Netherlands. (this volume).
Seckbach, J., Baker, F. A. and Shugarman, P.M. (1970) Nature 227: 744-745.

Seckbach, J., Fredrick, J.F. and Garbary, D.J. (1983) In: H.E.H. Schenk and W. Schwemmler (eds.) Endocytobiology II. Intracellular Space as Oligogenetic Spaces. Proceeding of the second Colloquium on Endocytobiology. Tubingen, Germany. April 10-15, 1983.Walter de Gruyter, Berlin. pp. 947-962.

Seckbach, J. Jensen, T.E., Matsuno, K., Nakamura, H., Walsh, M.M. and Chela-Flores, J. (1998) In: J. Chela-Flores and F. Raulin (ed). V Trieste Conference on Chemical Evolution. Exobiology: Matter, Energy, and Information in the Origin and Evolution of Life in the Universe. Kluwer Academic Publishers. The Netherlands.

Siegel, B. 1998. In: J. Seckbach (ed.), Enigmatic Microorganisms and Life in Extreme Environments, Kluwer Academic Publishers. The Netherlands (this volume)

Stetter, K. O. (1998a) In: E. Wagner et al. (eds) Proceedings of Endocytobiology VII, April 1998, Freiburg, Germany.

Stetter, K. O. (1998b). In: J. Horikoshi and W. D. Grant (eds.), Extremophiles: Microbial Life in Extreme Environments, Wiley, New York, pp. 1-24.

Stetter, K. O., Huber, R., Block, E., Kurr, M., Eden, R. D., Fielder, M., Cash, H. and Vance, I. (1993). Nature 365: 743-745.

Stevens, T. O. and McKinley, J. P. (1995) Science 270: 450-454.

Svitil, K. A. (1997) Discover 18: 86-88.

van der Meer, J. R., de Vos, W. M., Harayama, S. and Zehnder, A. J. B. (1992) Microbiological Reviews 56: 677-694.

Van Tielen, N. and Garbary, D. J. (1998) In: J. Seckbach. (ed.) Enigmatic Microorganisms and Life in Extreme Environments. Kluwer, The Netherlands (this volume).

Ventosa A. (1998) In: J. Seckbach (ed.), Enigmatic Microorganisms and Life in Extreme Environments, Kluwer Academic Publishers. The Netherlands (this volume).

Vishniac, H. S. (1998) In: J. Seckbach, (ed.), Enigmatic Microorganisms and Life in Extreme Environments, Kluwer Academic Publishers, The Netherlands (this volume).

Vogel, T. M., Criddle, C. S. and McCarty, P. L (1987) Environmental Science and Technology 21: 722-736.

Vreeland, R. and Rosenzweig, W. D. (1998) In: J. Seckbach (ed.), Enigmatic Microorganisms and Life in Extreme Environments, Kluwer Academic Publishers. The Netherlands (this volume).

Walsh, M. (1992) Precambrian Research 54: 271-293.

Westall, F. (1998) In: J. Seckbach, (ed.), Enigmatic Microorganisms and Life in Extreme Environments, Kluwer Academic Publishers, The Netherlands (this volume).

Yayanos, A. A. (1998) In: J. Horikoshi and W. D. Grant (eds.), Extremophiles: Microbial Life in Extreme Environments, Wiley, New York, pp. 47-92.

Biodata of **Dave Roberts** contributor of ***"Eukaryotic Cells Under Extreme Condition."***

Dr. Dave Roberts is head of the division of Protista and Mathematics at the **Natural History Museum in London.** He gained his Ph.D. from the University of Surrey in 1982. His work revolves around the identification of coherent evolutionary groups amongst the unicellular eukaryotes. To this end he is currently actively studying high-temperature environments, both terrestrial volcanic and deep-sea hydrothermal. Both of the environments are normally oxygen poor and because the earliest-diverging eukaryotic branches of the SSU rDNA tree of life are presumably amitochondrial, this forms a natural interest in anaerobic eukaryotes. Dr. Roberts other active research interests are in waste-water treatment and bioinformatics.
E-Mail: **d.roberts@nhm.ac.uk**

J. Seckbach (ed.), Enigmatic Microorganisms and Life in Extreme Environments, 163–173.

EUKARYOTIC CELLS UNDER EXTREME CONDITIONS

D.M[c]L. ROBERTS
The Natural History Museum
Cromwell Road, London SW7 5BD, United Kingdom.

1. Introduction

Modern biology has taken a few laboratory strains to its heart and has sought to extrapolate from these some general rules for life. But, the laboratory workhorses of *Escherichia coli* and *Saccharomyces cerivisae* cannot be taken as typical or representative of a diversity that has evolved over thousands of millions of years. This may be self evident, yet how often do you see sweeping statements of the form "eukaryotes are ..." or "eukaryotes have ..." based on observations from yeast, mouse and man? *Saccharomyces* is an unusual eukaryote in that it is an osmotrophic heterotroph with a very fast growth cycle, for instance.

The modern reductionist fashion in biology, bringing all life down to a discussion of genes and their products, can obscure the variations that eukaryotic life displays: this in turn is an indication of the possibilities that have arisen over evolutionary time. In this review, issues of mere survival in inimical environments will not be addressed. The question to be explored is the extent to which eukaryotic physiology can be stretched and modified; in other words, what are the environmental limits to normal cell growth? What can these limits tell us about the organisation of the eukaryotic cell? Any given population has physiological limits and optimal growth conditions, and in microbial studies it is well known that these optima can be changed by application of sufficiently strong selective pressure, within a chemostat for instance. It is also well known that quantum leaps in physiological capacity can be made by some individuals in a population by mechanisms such as lateral transfer or symbiosis, for example plastids in eukaryotic phototrophs. Many eukaryotes carry symbionts which extend their capabilities, although such interactions are beyond the scope of this review.

Organisms growing in conditions commonly viewed as extreme, within the microbial literature, will be reviewed: anaerobes, thermophiles, psychrophiles, acidophiles, alkalophiles, halophiles, barophiles and xerophiles (Lincoln *et al.*, 1982). This review is based on the World Wide Web page which may be found at HTTP://WWW.NHM.AC.UK/ZOOLOGY/EXTREME.HTML, where an updated list of references is maintained.

The following review will illustrate the flexibility and evolutionary adaptability of the eukaryotes in colonising various extreme habitats. It should be remembered that phagocytosis is a character shared by all extant eukaryotes, except certain obligate phototrophs (algae) or osmotrophs (fungi) where the capacity has been lost. A problem facing a phagotroph is physical access to the environment. Eukaryotes are generally larger than prokaryotes and can therefore be excluded from some environments by purely physical means, for instance in fine sediments, and it should be borne in mind that such physical exclusion can occur even though the chemical conditions are suitable.

2. Anaerobes

Eukaryotic cells carry mitochondria and therefore normally use oxidative phosphorylation as a primary means of ATP generation. Some eukaryotes, collectively known as the Archezoa (Cavalier-Smith, 1983; Cavalier-Smith, 1987), were believed to have diverged before the endosymbiotic event which gave rise to the mitochondrion (Margulis, 1970) and thus have never possessed the capacity for aerobic growth. But these lineages have recently been shown to carry genes in their nuclei characteristic of mitochondria (Bui *et al.*, 1996; Germot *et al.*, 1996; Horner *et al.*, 1996; Roger *et al.*, 1996; Germot *et al.*, 1997; Giezen *et al.*, 1997; Hirt *et al.*, 1997) and this is strong evidence that all extant eukaryotic lineages required oxygen in their evolutionary history (Vogel, 1997). True anaerobic growth is present in a number of lineages and must have evolved independently (Embley *et al.*, 1992; Esteban *et al.*, 1993; Brul and Stumm, 1994; Embley *et al.*, 1995). There are no known eukaryotic anaerobic phototrophs (Fenchel, 1996). See Fenchel (1996) for a review of the physiological implications for eukaryotes.

3. Thermophiles

The best studied high-temperature eukaryote is probably the acidophilic phototroph *Cyanidium caldarium*; for a review see Seckbach (1994). Its optimal growth temperature is 45°C and the maximum temperature at which growth can occur is 57°C (Brock, 1978); earlier reports of growth at much higher temperatures, such as 75 - 80°C (Copeland, 1936) have been attributed either to measurement of temperature away from the organisms themselves or the organisms were observed in the process of dying. *Cyanidium* can be found in all hot, acid soils and waters around the world (except Hawaii, where such ecosystems seem to be devoid of life) (Brock, 1978) and isolates from various sites have the same growth optimum and maximum, implying a strict physiological limit. *Cyanidium* does not have any known competition, for the niche it occupies, *i.e.* hot and very acid, which suggests that its growth is on the limit of what is possible.

Brock (Brock, 1978; Tansey and Brock, 1978) stressed the difficulties of estimating growth temperatures from ecological observations. There are many thermophilic fungi which have been isolated from compost and similar environments where temperatures can exceed 80°C, but until laboratory cultures demonstrate growth at these temperatures, the observations should be treated with caution (Tansey and Brock, 1978). The upper limit for thermophilic fungi seems to be in the region of 60°C (Tansey and Brock, 1978).

Unlike the situation in *Cyanidium*, three species of flagellates have been raised to grow at 70°C by incremental increases in cultivation temperature (Dallinger, 1887). This study reported more than 100 years ago has not been repeated, but Dallinger noted that increasing the temperature in too large steps, or before the cells had fully adapted to the current temperature, killed the cultures. This phenomenon of incremental adaptation to growth temperature has also been noted for psychrophiles (Lee and Fenchel, 1972), (see below).

There is a vigorous debate about whether it is possible for eukaryotic architecture to evolve an hyperthermophile (Forterre *et al.*, 1995). The crux of this revolves around the ability of the central biochemical machinery, nucleic acid transcription and translation, to operate at these high temperatures. Clearly the cell's membrane composition must change to retain the required degree of fluidity for proper function (Sprott *et al.*, 1991). Also, all known hyperthermophiles contain reverse gyrase, which induces positive super-coiling of DNA enhancing its thermal stability (Bouthier de la Tour *et al.*, 1991; Forterre *et al.*, 1995). No eukaryote has been shown to contain a reverse gyrase, although a possible candidate has been identified in *Saccharomyces cerivisae* (Gangloff *et al.*, 1994). Clearly we do not yet understand all the protective mechanisms which allow cells like the archaean *Pyrococcus* to grow above 100°C, and indeed what the actual upper limit for life might be (Stetter *et al.*, 1990). The half-life of unprotected RNA falls very rapidly with increasing temperature (Forterre, 1995) and as a consequence eukaryotes face two major problems. First, following transcription many eukaryotic genes are subjected to post-transcriptional modification, a process which take a certain amount of time. However, a number of hyperthermophilic archaeans post-transcriptionally modify the product of their tRNA genes in a manner which is reminiscent of the eukaryotes (Edmonds *et al.*, 1991). Furthermore, the essential organisation of the transcriptional apparatus predates the divergence of the Archaea and the Eucarya (Ciaramella *et al.*, 1995). Second, the mRNA has to make its way out of the nuclear membrane in order to be translated by ribosomes. Any hyperthermophile would have to possess a mechanism to protect the mRNA from hydrolysis. From an evolutionary perspective, if the Archaea and the Eucarya are sister taxa then the potential to colonise high temperature environments presumably existed in their last common ancestor, and there was therefore the potential in the eukaryotic lineage to do so too.

4. Psychrophiles

Life at low temperatures seems to present less of a challenge since there is substantial abundance and diversity of life at close to freezing point. Water is the solvent for life and must be present for growth to occur. When water solidifies, normal cellular chemistry (*i.e.* growth) must be suspended: freezing point is the *de facto* lower limit for growth. None the less, snow can be coloured by *Chlamydomonas nivalis*, *Chloromonas* (*Scotiella*), *Ankistrodesmus*, *Raphionema*, *Mycanthococcus* and certain dinoflagellates (Prescott, 1978).

One adaptation to life in the cold has been studied in *Heteromita globosa* which is an heterotrophic flagellate growing in Antarctic fellfields whose physical environment is characterised by highly variable moisture and temperature regimens, including short-term freeze-thaw cycles and diurnal temperature fluctuations of up to 20°C (occasionally up to 40°C). At least 24 species of protists, and some fruticose lichens and mosses grow in these conditions, in addition to bacteria (Smith, 1984; Cowling and Smith, 1986). *Heteromita* is a very common soil microflagellate with a world-wide distribution and an optimum temperature for growth around 23°C. Under Antarctic conditions it demonstrates adaptations which permit survival in freeze-thaw cycles, by rapid and temperature-sensitive encystment and excystment, allowing utilisation of resources during short periods of activity and allows this temperate species to grow actively at mean temperatures close to zero (Hughes and Smith, 1989).

Adaptation to growth at low temperature has been most extensively studied in the Antarctic sea-ice, where a wide taxonomic range of flagellates has been observed. Here, in contrast to *Heteromita*, efforts to cultivate these strains, which normally grow at around -2°C, failed completely if the temperature was raised above 2°C, even when the cultures were established at lower temperatures and raised in small steps (pers. comm; Leakey, Brit. Antarctic Servey). Lee and Fenchel (1972) observed that Antarctic sea-ice contained some new ciliate species and others which were morphologically identical to well-known marine ciliates. In studies with the ciliate *Euplotes*, they compared 3 species isolated from Antarctic sea-ice, temperate waters and tropical waters. The temperature range for the three species overlapped: the Antarctic species *E. antarcticus* was unable to survive above 17°C, although there was no clear cut-off for growth, because the range could be enhanced by slow adaptation. The ciliate *Holosticha* sp. was reported to be unable to divide above -2°C (Lee and Fenchel, 1972).

Garrison and Buck (Garrison and Buck, 1989b) reported that in the Antarctic Weddell Sea (south of the S. Orkney Islands), at an unrecorded temperature but close to the edge of the sea-ice, the heterotrophic biomass was dominated by flagellates and ciliates, other protists and micrometazoa making up only a small fraction. The abundance was greatest in a well-developed ice-edge bloom in the spring. They also reported (Garrison and Buck, 1989a) that the sea-ice itself contains a rich and varied population of microbes, which included phototrophs (diatoms and flagellates) and phagotrophic heterotrophs (flagellates, ciliates and micrometazoa). The presence of the phagotrophs indicates an active food web.

Modifications to low temperature growth must, as in high temperature systems (see above), involve substantial modification of the cell's lipid or fatty acid composition in order to retain membrane fluidity. Presumably, in the stable, low temperature of the sea, the flagellates have lost the ability to synthesise some components, perhaps the lipids or fatty acids, rendering them unable to grow at higher temperatures. Some low-temperature environments are also characterised by low water-availability, such as hypersalinity associated with sea-ice (Garrison *et al.*, 1986) or the arid conditions associated with fellfields (Smith, 1984; Cowling and Smith, 1986).

5. Acidophiles

Until recently only 4 organisms, all eukaryotes, were known to grow near pH 0; *Cyanidium caldarium* (see above under thermophiles), and three fungi, *Acontium cylatium*, *Cephalosporium* sp., and *Trichosporon cerebriae* (Schleper *et al.*, 1995). It has been observed that *Cyanidium* maintains its internal milieu at close to neutral pH (Brock, 1978; Beardall and Entwisle, 1984) which is, presumably, a primary adaptation to this niche. Recently two prokaryotic members were added to this exclusive set, *Picrophilus oshimae* and *P. torridus*, thermophilic archaeans from Japanese solfataras (Schleper *et al.*, 1995). It remains an important physiological question to determine how this is achieved, either with a strong proton pump or a low proton membrane permeability?

At more moderate pH values there is a great profusion of protistan and fungal organisms, as found for instance in the rumen. The phototroph *Dunaliella acidophila* is capable of growth below pH 3, for example, see the chapter by Pick (this volume).

6. Alkalophiles

Two African soda lakes, with a pH of about 10, have been studied for their microbial populations, Lake Nakuru and Lake Simbi (Curds *et al.*, 1986; Finlay *et al.*, 1987). The former supported a very high population of flamingos, counted in millions, feeding largely on cyanobacteria (Brown, 1975) (dominated by *Spirulina*), growing in the lake. The flamingos have been calculated to return about 15.6 tonnes dry weight of faecal and urinary matter to the lake each day, which resulted in a standing crop of non-cyanobacterial prokaryotes in the region of $3x10^8$ cells ml^{-1}. In this rich environment, there were at least 20 different heterotrophic species of protist, and 3 species of rotifers. Lake Simbi, on the other hand, was stratified with an extensive hypolimnion and there were far fewer flamingos, less than 20 individuals during the study period. The diversity and abundance of eukaryotic species was much lower than at Lake Nakuru. Samples of these populations (from Lake Nakuru) grew readily in the laboratory in a medium designed to mimic the ionic strength and pH of the lake, where several species not included in the above surveys were observed after enrichment (Roberts, unpublished observation). These figures for abundance and diversity are in the same range as similar studies on non-soda lakes in the same region of Africa (Curds *et al.*, 1986). Alkalophiles, including many eukaryotes, have been reviewed by Kroll and Edwards (1990).

Extrapolating from the observation that *Cyanidium* maintains its internal milieu at close to neutral pH (see above), one might speculate that cells growing at pH 10 are able to resist the influence of the external medium on their internal chemistry.

7. Halophiles

Hypersaline conditions support a wide variety of eukaryotic microbes, including diatoms (Rothschild *et al.*, 1994). There are many salt-adapted flagellates (Ruinen, 1938), including numerous heterotrophs. The African soda-lakes are also hypersaline (Lake Nakuru was reported being 8% Na; see above under alkalophiles), and as discussed above, can support a very diverse microbial population if there is an ample energy input (Finlay *et al.*, 1987). Halophilic and halotolerant algae have been reviewed (Gilmour and Edwards, 1990), but Grant (1991) took the view that there were no eukaryotes among the extreme halophiles, although this view must depend on an exact definition of "extreme" since *Dunaliella salina* grows up to saline saturation (see below). A broad range of eukaryotic taxonomic groups is found in hypersaline environments although few have been studied in culture (Grant, 1991).

The major problem facing halophiles is control of their osmotic pressure, without which they might lose water to the surrounding environment. *Dunaliella salina* synthesises high concentrations of intracellular glycerol to balance the external osmotic pressure (Avron and Ben-Amotz, 1978). The control of water relations has been reviewed elsewhere (Smith, 1978; Brown, 1990).

8. Barophiles

High pressure is used as a means of sterilising drugs and foodstuffs (Ludwig *et al.*, 1996). The mechanisms by which pressure affects cellular physiology are many and varied (Marquis and Matsumura, 1978; Markley *et al.*, 1996). Ciliates can certainly withstand rapid, repeated cycling (up and down in unit minutes) from 0.1 to 0.3 MPa without ill-effects (Roberts, unpublished data). The fungus *Magnaporthe grisea* is capable of producing an internal pressure of 8 MPa during the process of mechanical penetration of its host plant (de Jong *et al.*, 1997) by the synthesis of glycerol (*cf Dunaliella* above). There are abundant eukaryotic communities on the continental shelf (up to 20 MPa). Indeed, there are metazoa present in the deepest oceans (Bruun, 1977), so there seems to be no fundamental reason why heterotrophic eukaryotic microbes should not be present at great depths given sufficient food. Holothurians (Sea Cucumbers; members of the phylum Echinodermata) are reported to be abundant in the world's deepest location, the Challenger Deep (in the West Pacific, 400 Km SW of Guam; approximately 110 MPa) (Kato, 1997).

Surface organisms cannot necessarily survive great pressures; amoebae, for instance, become progressively less able to form pseudopodia with increasing pressure, becoming spherical and motionless at about 40 MPa (Marsland, 1958; Kitching, 1970). The mechanism for this loss of mobility is unknown. Increasing pressure causes acidification of the central vacuole in the yeast *Saccharomyces cerevisiae* at about 40 MPa (Abe and Horikoshi, 1995).

9. Xerophiles

The driest regions can support eukaryotic life, for instance in the lichens which grow on stones, or even in the Negev Desert (Palmer and Friedmann, 1990) or the deserts of the western USA (Fletcher, this volume). However, the ability to grow in conditions out of liquid water seems to be restricted in the microbes to the fungi. It is an everyday observation in food spoilage that the first colonisers are normally fungi, especially of foods with reduced water activity as a means of preservation (jams, marmalades and similar conserves). The fungi are also able to utilise the filamentous habit to grow through regions unsuitable for growth including bridging air spaces in the search of suitable habitats. The capacity to distribute nutrients through the filamentous colony is clearly an important adaptation for this niche as well as the capacity to withstand desiccation whilst growing out of water.

10. Conclusions

A central problem in the study of all microbial eukaryotes is the lack of reliable cultivation methods. Only a tiny fraction of the organisms that can be observed in environmental samples can be cultured in the laboratory, even from mesophilic environments. Extreme environments are generally more difficult to replicate in the laboratory and more difficult to keep stable. The ability to bring these fascinating creatures living into the laboratory is currently the biggest stumbling block to advancing knowledge.

Eukaryotic microbial life may be found actively growing in almost any extreme condition where there is a sufficient energy source to sustain it, with the exception of high temperature (>70°C). For most eukaryotes, therefore, a central requirement for growth in an habitat is sufficient energy flowing through the biosphere to support a second trophic level, as illustrated by the soda lakes Simi and Nakuru discussed above. If it were not for the high productivity of Lake Nakuru, the low diversity in Lake Simi could easily have been attributed to its extreme pH.

Colonisation of extreme habitats is not normally restricted to a single taxonomic group, with the exception of xeropholous habitats which are only tolerated by the fungi. Eukaryotic cells are exceedingly adaptable and not notably less adaptable than the prokaryotes, although most habitats have not been sufficiently well explored for sound generalisations.

11. References

Abe, F. and Horikoshi, K. (1995) *FEMS Microbiol. Lett.*, **130**, 307-312.

Avron, M. and Ben-Amotz, A. (1978) In: M. Shilo (ed.) *Strategies of Microbial Life in Extreme Environments*, Berlin: Dahlem Konferenzen. pp. 83-91

Beardall, J. and Entwisle, L. (1984) *Phycologia*, **23**, 397-399.

Bouthier de la Tour, C., Portemer, C., Huber, R., Forterre, P. and Duguet, M. (1991) *J. Bact.*, **173**, 3921-3923.

Brock, T. D. (1978) *Thermophilic microorganisms and life at high temperatures*, pp. 465. Springer-Verlag: New York.

Brown, A. D. (1990). *Microbial water stress physiology. Principles and perspectives.* John Wiley & Sons: Chichester.

Brown, L. H. (1975) In: J. Kear and N. Duplaix-Hall (eds.) *Flamingos.* T and A D Poyser: Berkhamsted.

Brul, S. and Stumm, C. K. (1994) *Trends Ecol. Evol.*, **9**, 319-324.

Bruun, A. F. (1977) In: A. Gibor (ed.) *Conditions for Life: readings from Scientific American.* San Fransisco: W.H. Freeman and Company. pp. 208-215.

Bui, E. T. N., Bradley, P. J. and Johnson, P. J. (1996) *Proc. Natn. Acad. Sci. U.S.A.*, **93**, 9651-9656.

Cavalier-Smith, T. (1983) The Kingdom Protozoa - Classification and Phylogeny. *Protistologica,* **19**, 472-472.

Cavalier-Smith, T. (1987) *Ann. N. Y. Acad. Sci.*, **503**, 17-54.

Ciaramella, M., Cannio, R., Moracci, M., Pisani, F. M. and Rossi, M. (1995) *World Journal of Microbiology and Biotechnology,* **11**, 71-84.

Copeland, J. J. (1936) *Ann. N.Y. Acad. Sci.*, **36**, 1-229.

Cowling, A. J. and Smith, H. G. (1986) In: *Colloque sur les Ecosystèmes Terrestres Subantarctiques*, Paimpont: Comité National Français des Recherches Antarctiques. pp. 205-213.

Curds, C. R., Bamforth, S. S. and Finlay, B. J. (1986) *Insect Sci. Appl.*, **7**, 447-449.

Dallinger, W. H. (1887) The president's address. *J. R. Microsc. Soc.*,, 185-199 + plate (facing p185).

deJong, J. C., McCormack, B. J., Smirnoff, N. and Talbot, N. J. (1997) *Nature, Lond.*, **389**, 244-245.

Edmonds, C. G., Crain, P. F., Gupta, R., Hashizume, T., Hocart, C. H., Kowalak, J. A., Pomerantz, S. C., Stetter, K. O. and McCloskey, J. A. (1991) *J. Bact.*, **173**, 3138-3148.

Embley, T. M., Finlay, B. J., Dyal, P. L., Hirt, R. P., Wilkinson, M. and Williams, A. G. (1995) *Proc. R. Soc. Ser. B*, **262**, 87-93.

Embley, T. M., Finlay, B. J., Thomas, R. H. and Dyal, P. L. (1992) *J. Gen. Microbiol.*, **138**, 1479-1487.

Esteban, G., Finlay, B. J. and Embley, T. M. (1993) *FEMS Microbiol. Lett.*, **109**, 93-100.

Fenchel, T. (1996) In: D.M. Roberts, P. Sharp, G. Alderson and M. Collins (ed.) *Evolution of microbial life*. Cambridge: Cambridge University Press, pp. 185-203..

Finlay, B. J., Curds, C. R., Bamforth, S. S. and Bafort, J. M. (1987) *Arch. Protistenk.*, **133**, 81-91.

Forterre, P. (1995) *C. R. Acad. Sci. Paris, Sciences de la vie/Life sciences*, **318**, 415-422.

Forterre, P., Confalonieri, F., Charbonnier, F. and Duguet, M. (1995) *Origins Of Life and Evolution Of the Biosphere*, **25**, 235-249.

Gangloff, S., McDonald, J. P., Bendixen, C., Arthur, L. and Rothstein, R. (1994) *Molec. Cell. Biol.*, **14**, 8391-8398.

Garrison, D. L. and Buck, K. R. (1989a) *Polar Biology*, **10**, 211-219.

Garrison, D. L. and Buck, K. R. (1989b) *Polar Biology*, **9**, 341-351.

Garrison, D. L., Sullivan, C. W. and Ackley, S. F. (1986) *BioScience*, **36**, 243-250.

Germot, A., Philippe, H. and Leguyader, H. (1996) *Proc. natn. Acad. Sci. U.S.A.*, **93**, 14614-14617.

Germot, A., Philippe, H. and LeGuyader, H. (1997) *Molec. Biochem. Parasitol.*, **87**, 159-168.

Giezen, M. v. d., Rechinger, K. B., Svendsen, I., Durand, R., Hirt, R. P., Fevre, M., Embley, T. M. and Prins, R. A. (1997) *Molec. Microbiol.*, **23**, 11-21.

Gilmour, D. (1990) In: C. Edwards. Milton Keynes, U. K. (eds.) *Microbiology of Extreme Environments*, Open University Press, pp. 147-177.

Grant, W. D. (1991). In: K. Horikoshi and W.D. Grnat (eds.) *Superbugs: Microorganisms in Extreme Environments*, Tokyo: Japan Scientific Societies Press, pp. 15-37.

Hirt, R. P., Healy, B., Vossbrinck, C. R., Canning, E. U. and Embley, T. M. (1997) *Current Biology*, **7**, 995-998.

Horner, D. S., Hirt, R. P., Kilvington, S., Lloyd, D. and Embley, T. M. (1996) *Poc. R. Soc. Ser. B*, **263**, 1053-1059.

Hughes, J. and Smith, H. G. (1989) In: R. B. Heywood (ed.) *University Research in Antarctica. Proceedings of British Antarctic Survey Antarctic Special Topic Award Scheme Symposium, 9-10 November 1988*, British Antarctic Survey, Natural Environment Research Council, Cambridge. pp. 117-122.

Kato, C. (1997) *Umi-ushi Letter*, **14**, 12 (in Japanese).

Kitching, J. A. (1970) In: A.M. Zimmerman (ed.) *High pressure effects on cellular processes*, New York: Academic Press, pp. 155-177.

Kroll, R. G. (1990) In: C. Ddwards (ed.) *Microbiology of Extreme Environments*, Milton Keynes, U. K.: Open University Press. pp. 55-92.

Leakey, R. (1995). British Antarctic Survey.

Lee, C. C. and Fenchel, T. (1972) *Arch. Protistenk.*, **114**, 237-244.

Lincoln, R. J., Boxhall, G. A. and Clark, P. F. (1982). *A Dictionary of ecology, evolution and systematics*, Cambridge University Press: Cambridge. pp. 298.

Ludwig, H., Scigalla, W. and Sojka, B. (1996) In: J.L. Markley, D.B. Northrop and C.A. Royer (eds.) *High-pressure effects in molecular biophysics and enzymology*, New York, Oxford University Press, pp. 346-363

Margulis, L. (1970). *Origin of Eukaryotic Cells*. Yale University Press: New Haven.

Markley, J. L., Northrop, D. B. and Royer, C. A. (1996). High-pressure effects in molecular biophysics and enzymology, New York: Oxford University Press, pp. 381.

Marquis, R. E. and Matsumura, P. (1978) In: D.J. Kushner (ed.) *Microbial life in extreme environments*,.London: Academic Press, pp. 105-158.

Marsland, D. (1958) *Scientific American*, **199**, 36-43.

Palmer, R. J. and Friedmann, E. I. (1990) *Microbial Ecol.*, **19**, 111-118.

Prescott, G. W. (1978). *How to know the freshwater algae*, 3rd edn. Wm. C. Brown: Dubuque, Iowa.

Roger, A. J., Clark, C. G. and Doolittle, W. F. (1996) *Proc. natn. Acad. Sci. U.S.A.*, **93**, 14618-14622.

Rothschild, L. J., Giver, L. J., White, M. R. and Mancinelli, R. L. (1994) *J. Phycol.*, **30**, 431-438.

Ruinen, J. (1938) *Arch. Protistenk.*, **90**, 210-258.

Schleper, C., Pühler, G., Kühlmorgen, B. and Zillig, W. (1995) *Nature, Lond.*, **375**, 741-742.

Seckbach, J. (1994) editor, Evolutionary pathways and enigmatic algae: *Cyanidium caldarium* (Rhodophyta) and related cells. In *Developments in Hydrobiology*, Dordrecht Kluwer Academic Publishers, pp. 349.

Smith, D. W. (1978) In: D.J. Kushner (ed.) *Microbial Life in Extreme Environments*, London: Academic Press, pp. 369-377.

Smith, H. G. (1984) *British Antarctic Survey Bulletin*, **64**, 55-61.

Sprott, G. D., Meloche, M. and Richards, J. C. (1991) *J. Bact.*, **173**, 3907-3910.

Stetter, K. O., Fiala, G., Huber, G., Huber, R. and Segerer, A. (1990) *FEMS Microbiology Reviews*, **75**, 117-124.

Tansey, M. R. and Brock, T. D. (1978) In: D. J. Kushner (ed.) *Microbial life in extreme environments*,. London: Academic Press, pp. 159-194.

Vogel, G. (1997) *Science*, **277**, 1604.

Dr. Anna A. Gorbushina and Prof. Wolfgang Krumbein are the authors of: **The Poikilotroph Micro-organism and Environment: Microbial Strategies of Establishment, Growth and Survival.**

Anna A. Gorbushina studied Biology, Botany and Mycology in the Faculty of Biology and Soil Science of St. Petersburg State University in Russia. She received her Ph.D. from St. Petersburg in 1997. Her research interests include mycological, ecophysiological and molecular ecological studies of microbial communities involved in biodeterioration and biofilm formation of the mineral surfaces (with special reference to microscopic fungi). She is currently at the University of Odenburg in Germany and joined a chapter with W. Krumbein.

Wolfgang E. Krumbein coauthor with Anna Gorbushina of: **The Poikilotroph Microorganisms and Environment: Microbial Strategies of Establishment, Growth and Survival**

Wolfgang E. Krumbein is a Professor of Geomicrobiology at the university of Oldenburg (Germany). His Ph.D. was received from the University of Wuerzburg (1966). His major interests are microbiology of rocks and sediments and geophysiology. Dr. Krumbein has studied the following organisms: cyanobacteria, eubacteria, fungi and actinomycetes. He has published hundreds of articles and edited 11 books including 'Microbial Geochemistry'.

E-mail: wek@africa.geomic.uni-oldenburg.de

J. Seckbach (ed.), Enigmatic Microorganisms and Life in Extreme Environments, 175–185.

THE POIKILOTROPHIC MICRO-ORGANISM AND ITS ENVIRONMENT

Microbial Strategies Of Establishment, Growth And Survival

A. A. Gorbushina and W. E. Krumbein
Geomicrobiology, ICBM,
Carl von Ossietzky University Oldenburg
P.O. Box. 2503, D-26111 Oldenburg, Germany

1. The poikilo-environment as the true environment of extremes

The planet Earth has an evolutionary history of about 4,700 million years and perhaps a history of 3,700 million years for life on Earth. During these periods of time the physico-chemical surface conditions must have changed considerably and in some cases this is documented through rock records or by other means of extrapolation. Whether or not abiogenesis or other means for the establishment of life occurred is not yet clear. It is, however, believed, that abiogenesis, i. e. the beginning, formation, and evolution of life does not take place on Earth as it is at present. Nor is it clear whether it ever occurred (and whether it generates itself through the elimination of other abiogenetical experiments). There seems to be agreement that life on Earth is monophyletic with strong tendencies of symbiotic and other recombinations of one and the same principle or basic form of life. However, we shall not be continuing this line of argument further.

In the present article we do want to review the physico-chemical conditions of life. What we can assume, however, is that humankind probably would not have been able to survive the environmental conditions of life on Earth 3,000 million years ago. The explanation is simple: the absence of free oxygen in the atmosphere at that time would represent an almost fatally extreme deviation from the environmental requirements of the human species.

The endothermic (37°C) and special catabolic (triple environmental heat production) behaviour of humankind as one representative of the homoiotherm group means that humankind created the first environment which could be called extreme in terms of "normal" conditions for life on Earth. This "living-room" environment is a very well defined and very well protected one which differs strongly from any other environment on Earth. Micro-organisms hitherto regarded as "extremophiles" such as the thermophilic, psychrophilic, halophilic or extremely oligotrophic and endemic inhabitants of very peculiar environments (e.g. South-American Tepuis or the hind-gut of a termite) can be regarded as organisms which are highly adapted to very narrow limits of environmental conditions. Thus these organisms should be called "normaphilic" because they can only live in a very restricted environmental space or condition which sets extremely narrow environmental limits.

The truly extremotolerant organism will be in turn the real expert in survival and metabolic capacity in very wide limits of environmental conditions. It will not set norms or narrow standards. On the contrary, it will try to meet all requirements for survival and metabolic activity under "any" condition. Therefore, these organisms will be the "true" extremophiles. The environments to look for such organisms will not be permanently extremely salty, highly acidic or alkaline, hot or cold, dry or wet. They will not be free of oxygen or saturated with hydrogen sulphide, loaded with methane, exhibiting a lack of carbon dioxide, nor will they be almost free of inorganic and organic nutrients or saturated with the latter. The environment for this "extremophile" will be either very "normal" or exhibit major or even extreme shifts between the highest and the lowest at regular, irregular or episodic time intervals.

In 1862 a temperature of 55°C was expected to be too extreme to allow the survival of any micro- or macro-organism (Cohn, 1862), in 1978 it was 91°C (Brock, 1978). Today, we may talk about nearly 120°C (Karl, 1995) and assumptions go as far as 150°C as an upper temperature limit for metabolic ability and activity. Similarly, our concept of life in salty or bitter water has advanced from biblical time to today. Formerly it was thought that salt or salt water (pickling brine) sterilised food. Now we know that bacteria and fungi can metabolise even inside large crystals of halite.

Hausmann and Kremer (1994) approached the topic of "extreme environments" in a way which was already partially adopted by Brock (1978). They realised that the assignment "extreme" to any relatively stable environment and "extremophile micro-organism" to any microbe living in such an environment is extremely anthropocentric and even inadequate. They further concluded that humankind, in its self-centred views does not realise that all environments on Earth are "normal" for living beings and will be inhabited by organisms fit for their proper environment in more or less narrow limits. The micro-organism will metabolise quite normally and successfully in any appropriate niche.

The environment of the "true extremophile" is actually not determined by space, concentration or any other physical/chemical parameter. The environment of this organism will be an environment determined by time lapse. The organism will have to exhibit characteristics which enable it to overcome unfavourable times and to continue growth when favourable conditions occur: patience, the ability to wait, the equipment for surviving under irregular changes of environmental conditions. We will give some examples of such microbiota, their habitats and conditions under which they live. It could be more appropriate to use instead of the terms "true extremophile" or "extremotolerant" a new term, which we derived from the nutrient demands of micro-organisms, namely the term "poikilotrophic" (Krumbein, 1988; Gorbushina et al., 1996). This term can, however, be applied to any of the environmental demands mentioned.

What then is a "poikiloenvironment"? The prefix "poikilo-" is well known from the poikilothermic behaviour of certain groups of animals (snakes, lizards) or the poikilohydric behaviour in lichens (Nash, 1996). "Poikilo-" means variegated or various. Strong variations and deviations from a midpoint are designated to it as well as changes in behaviour which occur only when the environmental conditions demand it. Thus, in scientific terms, poikiloenvironment may more appropriately describe environmental conditions deviating considerably (or drastically) from the narrow range of normal micro-organismic or human requirements. Jaag (1945) gave an excellent example and provided superb data for the changing temperature and humidity conditions on bare rocks in the

Swiss Alps. Most Middle Europeans and Americans usually define any temperature above 30°C and below -10°C as extremely hot and extremely cold respectively. Many a micro-organism will find a temperature below 30°C or above 50°C (with a total variation of 20 degrees tolerated) as inbearable (e.g. *Escherichia coli* which is well adapted to the temperature range of human intestines). Thus, the temperature extremes of *E. coli* are nothing compared to those of a human being. As we all know, humankind has learned to protect itself so efficiently that it can now stand the temperatures of space and those of boiling water when sufficiently equipped and protected. The temperature ranges reached in this latter case are in excess of 200 Celsius degrees. This temperature range, however, can also be tolerated by the well-equipped and well-protected spores of some fungi.

Let us now turn to water. In the long run, no life on Earth is possible without water and the certain level of water activity needs to be guaranteed for an active metabolism. A second question is the time span of survival for a cell, a spore, a propagule or a cell cluster without a supply of water in order to enable this unit to resume active metabolism after a pause or hiatus forced upon it. One brilliant example was given by the botanist Michael Evenari who studied the survival potential of desert plant seeds and the duration of humid period needed to cause germination. A desert seed will not be tricked by an accidental rain even if it were a strong one. It senses the amount of humidity transmitted to the soil and germinates only when the conditions are really suitable. Freeze-dried bacterial cells or spores can survive decades, centuries and even millennia without being touched by the level of water activity insufficient to cause metabolic activity. They will, however, ultimately and invariably die when heated to a temperature above 140°C. The same limitations hold true for toxic substances, pH values, insufficient nutrients and energy sources and other parameters available at certain times in a given place and then lacking for extended periods of time.

From these and other considerations we have defined the limits of environments as they may occur on Earth. A normal environment is defined as any environment which keeps the same physical and chemical conditions stable or within time-dependent variations suitable and helpful for the metabolism and multiplication of any given normal species. A normative environment can also be designated when any changes of the environmental parameters tolerated are low on an arbitrary logarithmic scale of conditions suitable for growth.

An extreme environment is an obsolete term in our view and the logical opposite of a normal (normative, standardised) environment would be poikiloenvironment. The poikiloenvironment is characterised by sudden, very delayed or extremely rarely occurring conditions for the optimal metabolism or propagation of a given species. The environmental conditions in such environments need to be described as unpredictable or sporadic. This is the environment of unexpected and drastic changes of the conditions for life. Many enigmatic and hitherto uncultivated organisms in such environments are waiting to be isolated and characterised.

Where then do such poikiloenvironments exist on Earth? Any environment which is stable within certain limits of growth and nutrition will be called a normal environment. This holds true for extreme heat, coldness, dampness, aridity, acidity, alkalinity, areas rich in oxygen, areas poor in oxygen and other gasses etc., where the organisms living in it prefer to do so or metabolise readily under such strictly defined conditions. The environment of the "true extremophile" or the "poikilotrophic" micro-organism will be a changing environment. Such environments will be characterised as follows: (1) Water potential and water activity are extremely low for extended periods of time. This will happen in deserts, on rock surfaces exposed to the sun e.g. buildings and monuments, and on high mountain ranges. It can also be realised in arctic and antarctic cold regions, where active (liquid) water occurs only occasionally. (2) Due to rain and tide events the water available may be extremely salty or concentrated in osmotically active substances and turn suddenly or gradually into fresh-water or even close to distilled water. (3) Due to the harsh environmental conditions any organic energy source may only occur episodically and will be practically equal to disturbance (desert with migrating animals). However, such environments can also be envisaged when water is always available. Hot vent environments are typical poikilotrophic environments, for the hot vent may occur in dead spots of the ocean bottom after a long interval of cool and nutrient free conditions. Characteristically, the rock pore or rock crevice environment can also be called poikilotrophic because it may take thousands of years until a slow migration of energy sources and/or electron acceptors may suddenly reach a spot where germs were merely biding their time, waiting for a period of activity.

Hypothetical poikiloenvironments may be imagined for the large plains of Mars or other planets. In such places, it may be that sunlight suitable for photosynthesis occurs at considerably different intervals or that temperatures for metabolic activity occur only once in 25 years. Since Earth is a well-balanced and biologically amply buffered normal environment, we may imagine that on Mars and other planets poikilotrophic environments and poikilotrophy are the normal case of life as compared to life on Earth, where the normative environment and normative organisms are established and mutually keep each other within their chosen and narrow limits for life. Again, the "living-room" is the appropriate term since many micro-organisms shape their environment. It may be a principle of life to create or shape environments in such a way that they enable metabolic activity within certain narrow limits and at ample supplies of nutrients and energy sources. Thus, it is even more astonishing that within these elegantly balanced environmental settings poikilotrophic environments still exist and the organisms which are fit to settle within them still await future investigation.

2. The poikilotrophic principle and typical poikilotrophic micro-organisms

The analysis of more than 300 monuments and rock outcrops in Central Europe, Africa, the Americas, parts of Asia and Australia has shown that micro-organisms inhabiting rock can be classified into three groups of survival strategies. The normal soil flora, which is frequently observed on rock surfaces at the initial steps of exposure or at ample water and nutrient supplies is characterised by two classic micro-organism types or strategies namely *oligotrophic* (i.e. growing and metabolising slowly but steadily at low concentrations of energy sources, nutrients and water) and *copio-* or *zymotrophic* (i. e. growing and metabolising quickly at high energy source, nutrient and water levels but being forced into survival and escape strategies like spores and conidia when the conditions on and in the rock change). With the continued air-exposure environmental conditions on the rock surface are characterised by a general lack of nutrients, energy sources and extreme dryness. These are, however, rarely or sporadically interrupted by short periods of wetness and supply and deposition of organics from the surrounding atmosphere. In this hostile, changing and unpredictable poikiloenvironment the newly described poikilotrophic type or strategy will be more and more established on and in the rocks exposed to Mediterranean and desert environmental conditions.

The photosynthetic representatives (e.g. *Chroococcidiopsis*) can grow at low light and water activity (Dor et al., 1991); the chemoorganotrophic representatives (e.g. *Geodermatophilus* or *Coniosporium* and *Phaeococcomyces)* grow slowly, usually by budding or similar division methods and exhibit typical small and leathery tough microcolonies, which can develop into intertwined mycelial strands not unlike the slime-embedded bundles of another representative of the phototrophic poikilotrophic type (*Microcoleus sp* from desert soil). The maintenance of each individual cell formed on one place for prolonged periods of time reaching up to 50 years or more is a typical feature (Eppard et.al., 1996).

The poikilotrophic micro-organisms are characterised by generally slow growth, dense and usually irregular node-like packs of cells of irregular shape. They rarely or never form specialised propagative or survival cells or structures. By contrast, each individual cell and cell cluster is equipped and structured to survive as a whole. New cells produced by active growth (e.g. in budding rock inhabiting fungus *Coniosporium* or in *Chroococcidiopsis*) quickly gain a thick pigmented cell wall enabling them to resort to dormancy if conditions abruptly get more hostile and for extended periods of time.

Some of them have advantages by abundant excretion of extra-cellular polysaccharides. It brings about some immediate advantages: high water storage potential through thick cell walls and extra-cellular polymeric substances (EPS) nutrient adsorption to EPS layers, resistance to desiccation etc. This slimy matrix, however, also gives a possibility for other organisms to attach to the substrate in the immediate proximity of the founder organism. This way polysaccharide substances are serving as a prerequisite to microbial interaction in an interface environment (Potts, 1994; Little et al., 1997).

Another classic feature is the pleomorph behaviour which is observed in all groups of organisms inhabiting rock. This enables them to change both the mode of their vegetative growth and sporulation in accordance with environmental conditions. Pleomorphism has recently been subject to much attention in connection with poikilotrophic cyanobacteria (Dor et al., 1991; Palinska, 1996). Rock-dwelling fungi adopt a strategy of producing highly polymorphic protected units (sclerotia, chlamydospores, cell clusters etc.) which, after the alleviation of hostile conditions, are transformed into growing mycelium. This life strategy corresponds with the so-called latent invasion mechanism described by Cooke and Rayner (1984) for phytopathogenic fungi. Constant stress on the rock surface brings about a considerably prolonged use of the units originally designed for short existence. Morphological characteristics of vegetative poikilotrophic cells are coming close to those of dormant structures, but keep an option of reversing from dormancy to growth without considerable changes in their morphology.

One could mention, as a feature which unifies rock-dwelling micro-organisms, the formation of micro-colonies (Staley et al., 1982; Gorbushina et al., 1993; Wollenzien et al., 1995; Sterflinger and Krumbein, 1995) resulting from meristematic growth. Micro-colonial growth may be considered to impart the following advantages:

- Nutrients aggregate at the interface between the air and the solid interface. Consequently, any organism attached to the substratum which has remained localised to the proximity of

the interface will have selective advantages (Krumbein, 1966; Krumbein and Gorbushina, 1996; Little et al., 1997).
- By the production of dense and compact multi-cellular units and the functional differentiation of some cells in the complex, the survival potential of the colonial morphotype extend beyond that expressed by a single cell of the same species.

The formation of tough and very small colonies by rock-inhabiting micro-organisms can also be interpreted as a spreading mechanism for the vegetative structure with the same genotype. For instance, the formation of satellite colonies (Sterflinger, 1995) or appendix-like irregularities of colony surface strongly resembles the formation of vegetative propagation structures (isidia) by lichens.

The microbial community of the rock surface and interior develops with the course of time. Stones freshly excavated from the Earth are still protected from the most harmful species by the presence of an abundant, diverse and active soil micro-flora (e.g. the Caesarea marbles in Israel have only been exposed to the atmosphere in the past 25 years. Therefore, the poikilotrophic flora has not had sufficient time to establish itself). Yet prolonged air-exposure and established poikiloenvironment bring about the spread of a more damaging flora of poikilotrophic escapers and survivors, whose impact on the surface structure can be regarded as highly biodeteriorative.

3. Activities of poikilotrophic microbiota on rock surfaces

One of the most important features of the poikilotrophic micro-organism is the capacity of scavenging nutrients and even humidity from the nearby environment. It has been known since Humboldt (1793) and Crace-Calvert (1871) that, astonishingly, phototrophic micro-organisms live in deep mines where they are hit only occasionally by the light of a miner's lamp and that chemotrophic micro-organisms not only survive but thrive in distilled water. Fry (1990) has given a brilliant review of the strategies of growth of oligotrophic micro-organisms. Even more enigmatic is the approach to micro-organisms which actually have the potential for growth in nutrient-rich as well as in nutrient-poor environments. These, however, will never metabolise as fast as the classic eutrophic micro-organism, but are capable of utilising a wide variety of substrates.

The most important feature of rock-dwelling poikilotrophic micro-organisms, however, is their potential for matching a harsh physical environment and its sudden changes with a high capacity for physical responses and work. The rock-dwelling poikilotrophic is an aggressive organism. It is capable of mechanically boring and pushing itself into the rock or pushing the rock itself outward. It will undermine the rock texture and structure both physically and chemically, in order to create a niche for survival (Krumbein and Jens, 1981). Furthermore, all of them seem to have the tendency and/or capacity to create or mechanically penetrate into fissures, cavities and biopits in their habitat, be it rock (Sterflinger and Krumbein, 1997), glass (Krumbein et al., 1996) or metal. Thus, the typical poikilotrophic micro-organism is extremely destructive in terms of the durability of materials and represents a true danger e.g. for monuments. In its natural environment, it contributes majorly to natural biogeomorphogenesis and to the erosion of rock in desert environments. In fact, all physical factors of weathering and rock decay in the desert and in high mountain ranges are considerably accelerated by the activities of poikilotrophic rock flora. It has been shown that poikilotrophic micromycetes can penetrate into marble at an unexpected speed of 1mm per annum (Sterflinger and Krumbein, 1997). Thus in 100 years a rock surface can be mechanically (or chemically) pitted to a depth of a maximum of 10 cm. Other poikilotrophs have a tendency to establish themselves inside the material they colonise in such a way and at such places that optimal conditions for water activity and light intensity can be maintained over prolonged periods of time. This depends on the penetration depth of water and humidity as well as the retention potential of the material for liquid or condensed water (Friedmann, 1971, 1993; Krumbein, 1966, 1969; Potts, 1994). In general, this leads to the phenomenon of growth parallel to the surface at a certain depth. This in turn creates chipping and exfoliation. Thus, depending on the environmental conditions, one and the same rock type under one and the same climatic conditions may exhibit different communities of poikliotrophic micro-organisms and hereby different bio-erosional patterns. One of the most conspicuous environments to demonstrate this principle is the Negev desert in Israel. Limestones are exposed by geotectonic forces in valleys (or wadis) which run North-South or East-West (Krumbein, 1969). At a certain level and inclination one individual rock type may be exposed and within a few square kilometres may exhibit totally different biota and thus biodeterioration patterns (Danin and Garty, 1983). Many of the micro-organisms dwelling in such environments and environmental conditions are yet to be cultivated and identified. Furthermore the study of whether these micro-organisms create their own environment (i.e. the biopit, the biopatina, or the endolithic biofilm) by mechanical and/or chemical means is only in ist early days (Krumbein and Jens, 1981; Dornieden et al., 1997).

4. Planetary and Biotechnological aspects

We wish to touch briefly on the debated topic of extra-terrestrial life and the chances and methods to search for it or to detect it. Doubtlessly, the conditions for life on Mars, some of the moons of Jupiter etc., exhibit physico-chemical and general climatic conditions differing considerably from those on the geophysiologically-buffered and controlled planet Earth. Two possible means of creating a scenario for the search of life are imaginable:

1) Looking for "extreme environments" and micro-organisms adapted to them. This has been done by researchers in desert environments, highly alkaline and acid environments as well as in hypersaline environments and those which deviate considerably from the present atmospheric conditions.

2) Looking for micro-organisms and the metabolic potential of micro-organisms which have an extremely high survival potential under constantly, periodically or sporadically changing conditions for life. This is typical for poikilotrophic micro-organisms. The search for poikilotrophs on Earth has only just begun and bears a great potential for the future. Furthermore the study of poikilotrophs in their natural environment hints to the fact that any remote sensing device capable of detecting metabolic activities such as respiration and fermentation and the detection of respiration/fermentation products should also account for the fact that, in a given place, a specific micro-organism may wait for the right time and conditions to resume metabolic activity for many years. This in turn brings us to the problem of the Indian scout sitting behind a bush and overlooking a landscape for hours and hours without being capable of recognising that he is surrounded by enemies who take

care not to move and not to breathe. This somewhat awkward comparison should make it clear that we have to refine our techniques of life detection much further before we can be sure that life does not exist on Mars, for example.

In summary: poikilotrophic micro-organisms are specialists in survival in extremely changable environmental conditions and specialists in colonising environments which appear to be too hostile for settlement. Thus, they are in turn good candidates for testing the outer limits of life here on Earth and elsewhere. Additionally, they are good candidates for testing fouling and anti-fouling materials and techniques because they are persistent and self-sufficient enough to overcome the most difficult environmental conditions imaginable on Earth.

5. Acknowledgements

We acknowledge the financial support of the EC for the analysis of microbiota on mural paintings as well as the DFG for supporting research on micro-organisms on glass and marble. We further acknowledge the discussion of poikilotrophy with Prof. Gromov and other colleagues.

6. References

Brock, T. D. (1978) Thermophilic microorganisms and life at high temperatures. Springer, NY., 465p.

Crace-Calvert, F. (1871) Trans. Br. Assoc. Sci. (Edinburgh), 123-124.

Cohn, F. (1862) *Abhandlungen der Schlesischen Gesellschaft fuer vaterlaendische Kultur*, Abt.Naturwiss. Med., II, 35-55.

Cooke, R.C. and Rayner, A.D.M. (1984) Ecology of Saprotrophic Fungi. Longman: London.

Danin, A. and Garty, J. (1983) *Z. Geomorphol. N.F.* 27 (4), 423-444.

Dor, I., Carl, N. and Baldinger, I. (1991) *Algological Studies*, 64, 411-421.

Dornieden, Th., Gorbushina, A. and Krumbein, W. E. (1997) *Int. Journal for Restoration of Buildings*, 3, 441-456.

Eppard, M, Krumbein, WE, Koch, C, Rhiel, E, Staley, J, Stackebrandt, E. (1996) *Archives of Microbiology*, 166, 12-22.

Friedmann, E. I. (1971) *Phycologia*, 10 (4), 411-428.

Friedmann, E. I. (ed.) (1993) *Antarctic Microbiology*, Wiley NY, 634p.

Fry, J. C. (1990) In. E. Edwards (ed.), Microbiology of Extreme Environments. Open University Press, Milton, Keynes, UK, p. 93-116.

Gorbushina, A. A., Krumbein, W.E., Hamann, C. H., Panina, L., Soukharjevski, S. and Wollenzien, U. (1993) *Geomicrobiology Journal* 11, 205-221.

Gorbushina, A. A., Krumbein, W. E., Vlasov, D.Y. (1996) In: R. Pancella (ed.), Preservation and restoration of cultural heritage. Proc. of the 1995 LPC Congress. EPFL, Lausanne, 319-332.

Gorbushina, A. A. (1997) Ph.D. thesis, St. Petersburg (in Russian), 148p.

Hausmann, K. and Kremer, B.P. (eds.) (1994) Extremophile. VCH, Weinheim, 419p.

Humboldt, A. v. (1793) Rottmann Berlin, 189p.

Jaag, O. (1945) *Beitr. Kryptogamenfl. Schweiz*, 9 (3), Buchler& Co., Bern, 560p.

Karl, D.M. (ed.) (1995) The microbiology of deep-sea hydrothermal vents. CRC Press, Boca Raton, 299p.

Krumbein, W. E. (1966) Ph.D. Thesis, Wuerzburg 130p.+ appendices.

Krumbein, W. E. (1969) *Geol. Rdsch.*, 58, 333-363.

Krumbein, W. E. (1988) *Durability of Building Materials*, 5, 359-382.

Krumbein, W. E. and Gorbushina, A. A. (1996) In: R. Pancella (ed.) Preservation and restoration of cultural heritage. Proc. of the 1995 LPC Congress. EPFL, Lausanne, 277-284.

Krumbein, W.E. and Jens, K. (1981) *Oecologia*, 50, pp. 25-38.

Krumbein, W.E., Gorbushina, A.A., Palinska, K.A. (1996) 2. Statuskolloquium des Deutsch-Franzoesischen Forschungsprogramms fuer die Erhaltung von Baudenkmaelern, Filtz J.-F. (ed.), 39-45.

Little, B. J., Wagner, P. A., Lewandowski, Z. (1997) *Reviews in Mineralogy*, 35, 123-159.

Nash, Th.H. III (1996) Lichen biology. Cambridge University Press, 303p.

Palinska, K. A. (1996) Ph.D. thesis, Oldenburg, 122p.

Potts M. (1994) *Microbiol. Reviews*, 58, 755-805.

Staley, J.T., Palmer, F. and Adams, J.B. (1982) *Science*, 215, 1093-1095.

Sterflinger, K. (1995) Ph.D. Thesis, Oldenburg, 137p.

Sterflinger, K., and Krumbein, W.E. (1995) *Botanica Acta*, 108, 490-496.

Sterflinger, K. and Krumbein, W.E. (1997) *Geomicrobiology Journal*, 14, 219-230.

Wollenzien, U., de Hoog, G.S., Krumbein, W. E. and Urzi, C. (1995) *Sci. of the Total Environ.* 167, 287-294.

Biodata of **Thomas E. Jensen** (contributor, with W.A. Corpe of "***The Study of Enigmatic Microbial Communities***").

Dr. Thomas E. Jensen is a Professor in the Department of Biological Sciences, Lehman College, The City University of New York (Bronx, NY). He earned hid Ph.D. from Iowa State University. His long-standing interest is in the fine structure of the Cyanobacteria. He is also interested in heavy metal binding by bacterial cells especially by polyphosphate bodies. A recent interest has also been on the autogenous origin of the eukaryotic cell.
E-Mail: **dovlc@cunyvm.cuny.edu**

Biodata of **William A. Corpe** (co-author with T.E. Jensen of "***The Study of Enigmatic Microbial Communities***)."

Dr. William A. Corpe is a Professor Emeritus, Biological Sciences, Columbia University (N.Y.C.). He obtained his Ph.D. from Pennsylvania State University. His research has been concerned with aspects of general physiological and microbial ecology.

J. Seckbach (ed.), Enigmatic Microorganisms and Life in Extreme Environments, 187–194.

THE STUDY OF ENIGMATIC MICROBIAL COMMUNITIES

T.E. JENSEN
Department of Biological Sciences
Lehman College, The City University of New York
250 Bedford Park Boulevard West, Bronx, New York 10468

W.A. CORPE
Department of Biological Sciences
Columbia University, New York, New York 10027

1. Introduction

The study of microbial communities in natural aquatic environments has always been a major undertaking in as much as a very substantial portion of the microbes counted by direct light microscopic means could not be matched by plate count or other procedures meant to estimate numbers of viable cells. The direct-count method using fluorescent dye and the epifluorescent microscope has been modified many times since Daley and Hobbie's (1975) work. They have been useful for estimation of phototrophic and non-phototrophic cells in both the picoplankton and nano-plankton size fractions of the same samples (Sicko-Goad and Stoermer, 1984), but little effort given to the direct study of species diversity.

The greater proportions of microbes are just too small and featureless at the limit of resolution of the light microscope. With the newly developed interference video system, workers (Allen *et al.*, 1985) claim to be able to see a single microbtubule, which would mean resolution of about 23 nm. Even with this kind of resolving power most bacteria will still appear featureless.

The recognition of *bacterial sized plankton* as an important part of the biomass in oceanic waters was made by Johnson and Sieburth (1979) and Waterbury *et al.* (1979). Johnson and Sieburth (1982) used electron microscopic methods to explore the picoplankton fraction recovered from large volumes of seawater by centrifugation after screening and filtration to remove cells larger than 2 µm. The value of the approach was that one could study in situ, the relative abundance of morphological and ultrastructural types of picoplanktonic organisms in nature by direct means. Ultimately we would hope this approach will contribute to physiological and molecular (Leadbetter, 1997; Amann *et al.*, 1995) recognition of diverse species in aquatic environments.

2. Electron Microscopic Methods

Most microbiologists and perhaps many molecular taxonomists studying aquatic environments would be aided immeasureably by "seeing" the objects of interest. We became interested in learning if it would be possible to explore various aquatic microorganisms directly with the electron microscope taking advantage of images produced by high magnification and great resolution, that is to say images that are morphologically and ultrastructurally distinctive. Today transmission electron microscopes (TEM) generally guarantee resolution of about 3.0Å. In order to "see" a small bacterium of 0.5 μm in diameter with the TEM the sample must be very thin, in the range of 60 nm and a contrasting agent must be present in the sample. *Contrast* is provided using osmium which is deposited in the cell mainly at double bonds in lipid membranes (Wischnitzer, 1970). *Thin sectioning* of cells is accomplished with a proper embedding medium which could withstand the vacuum and heat produced in the TEM. Excellent preservation of bacteria is obtained when cells are prepared according to the so-called Kellenberger procedure developed by Pankranz and Bowen (1963). Luft's (1961) introduction of Epon followed by other Epoxys' overcame the problem. With the excellent preservation of cell structure afforded by aldehyde fixation especially glutaraldehyde, microbiologists were finally able to observe in detail the structure of internal and external surfaces of microbes, which was especially useful for the study of bacteria. Cells in most cases can be considered randomly sectioned so counts of the different sizes of cells can be made. Sicko-Goad and Stoermer (1984) recommended general use of the terms nanoplankton for organisms 2 to 20 μm and picoplankton for organisms 0.2 to 2 μm used earlier by Johnson and Sieburth (1982). We routinely prepare the samples in three ways (1) fixed and embedded (2) negatively stained and (3) air dried onto formvar coated grids. The ability to produce quality electron micrographs requires experience and several pieces of complicated equipment, the ultramicrotomes and the TEM itself, which must be properly operated before quality can be accomplished. A range of excellent images of microorganisms particularly bacteria have become available for study. These are usually prepared from pure cultures. A collection of these images with accompanying references appears in The Prokaryotes (Balows *et al.*, 1992), a valuable resource for all microbiologists.

3. Element Analysis Using an Energy Dispersive X-ray Spectrometer (EDX)

As negatives of the cells are being taken elemental analysis of items of interest can also be carried out using the associated EDX system. To accomplish this the unit is operated in the scanning transmission mode (STEM) and analysis carried out with the spot setting. The different elements present and their amounts can then be determined and from this, the item can often be identified.

The *air dried sample* of a fixed cell suspension can be observed readily in the TEM mode of the system and quite surprising details are observed in this unconventional method. The same kind of structures, which were analyzed by EDX in thin section, can be analyzed from the air-dried samples. Internal items such as polyphosphate bodies

(PPB's) and magnetite inclusions can be analyzed with the STEM-EDX system as can cell surface accumulated iron particles. We have found little difference in the elemental composition of PPB's in fixed and embedded cells or in air dried cells (Jensen and Corpe, 1994). The EDX system can also be utilized to determine the mass of items one observes whether cells or particles (Heldal, 1985; Goldberg *et al.*, 1996). The *negatively stained preparations* can also be examined and surface structural detail of the cells can be observed. These can be correlated with cell shape and correlations with the thin-sectioned cells can sometimes be made.

4. Evaluation of Picoplankton Biomass

The majority of the picoplanktonic community of microorganisms found in a small lake (Corpe and Jensen, 1992) were unable to form visible colonies on common agar media. Some 70% of the cells examined by electron microscopy had an average diameter of 0.5 μm. The maximum diameters ranged from 0.56 to 1.8 μm. Using the details of ultrastructure cells of two cyanobacterial morphotypes were recognized by the presence of thylakoids and polyhedral bodies (Fig. 1a). Eukaryotic picoplankton were also observed (nucleus, chloroplast, mitochondria, thick cell walls (Fig. 1b). The arrays of intracytoplasmic membranes seen in about 20% of the sectioned cells were suggestive of methanotrophs and chemoautotrophs (Fig. 1c). Methanotrophs (Whittenbury *et al.*, 1970) and nitrifying bacteria (Gehardt *et al.*, 1994) were detected using enrichment culture methods (unpublished work). *Prosthecomicrobium* sp. were tentatively recognized with characteristic prosthecae (Corpe and Jensen, 1996) and both *Hyphomicrobium* and *Caulobacter spp.* were occasionally seen with partial prosthecae in sections (Fig. 2a). Their presence were confirmed by enrichment culture methods as described in Gerhardt *et al.* (1994). Among the smallest cells were those that contained 4-6 magnetosomes per cell (Fig. 2b) which were confirmed by EDX spectroscopy of thin sections. Iron was also detected in the envelopes of some cells.

The smallest cell type, shown in Fig. 2c, contained a single polyphosphate body and a dense cytoplasm at the protoplast membrane inner surface. The cell appeared as a protoplast surrounded by a convoluted outer membrane which suggests that the cell we see is the product of starvation of a larger cell. Attempts to reverse such a starvation process have been unsuccessful to date.

5. Molecular Systematics and Study of Diversity

According to Stahl (1996) the biopolymers of cells include DNA, RNA and protein or polypeptides. They are used in various ways to evaluate genomic relatedness. DNA-DNA hybridization reflects total similarity and can be applied to evaluate environmental diversity within a community of organisms. Comparative sequencing of rRNAs has been most widely applied in providing a taxonomic basis and a measure of environmental diversity. One of the interesting or at least seeming successes of molecular probing is the discovery of organisms not previously described in a particular habitat.

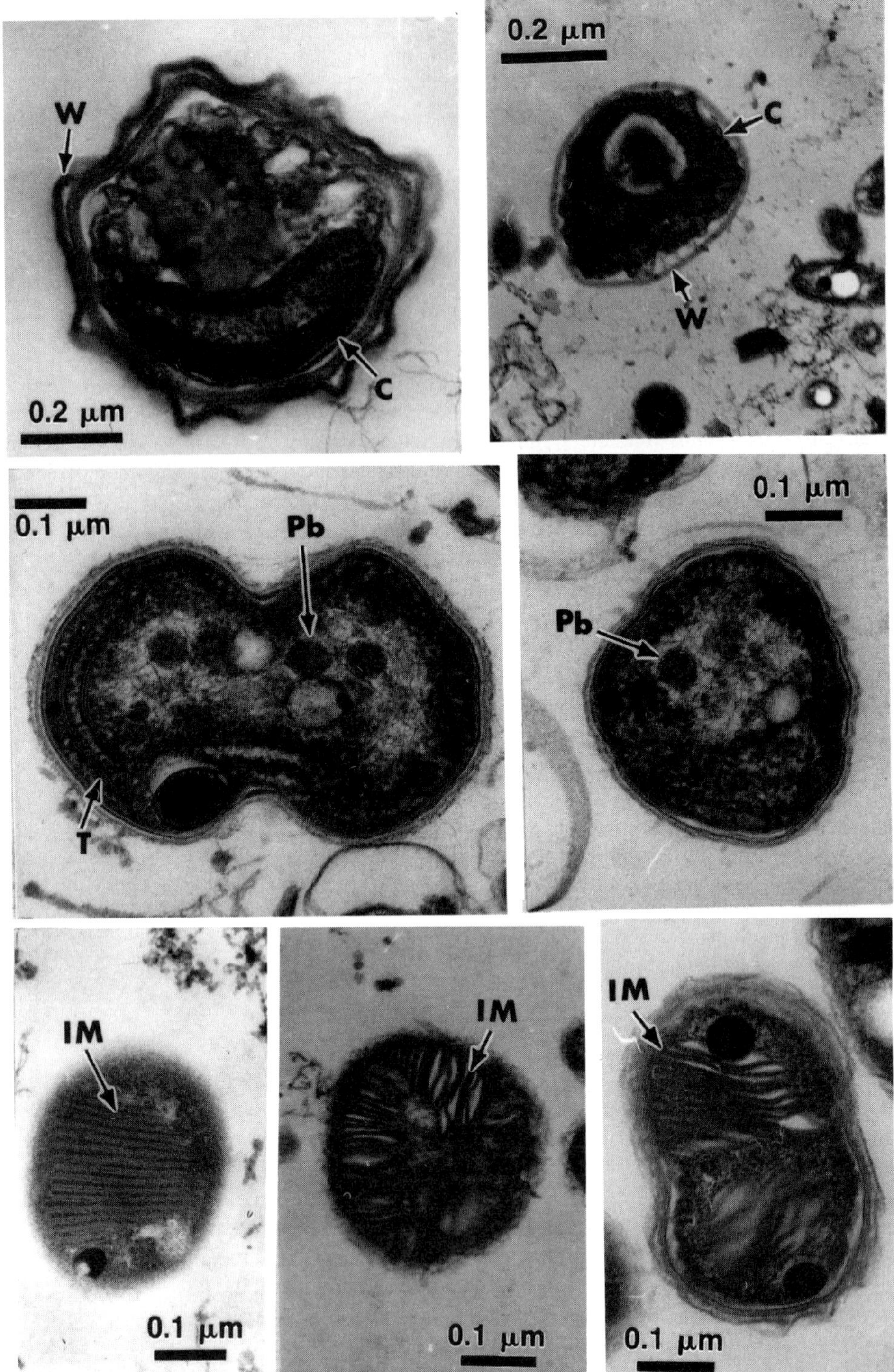

Figure 1. (a) Eukaryotic phototrophs that show chloroplast (C) and thick cell wall (W). (b) Cyanobacteria photoautotrophs showing thylakoid membranes (T) and polyhedral bodies (Pb). (c) Bacterial cells with arrays of intracytoplasmic membranes (Im) probably methanotrophs and chemoautotrophs.

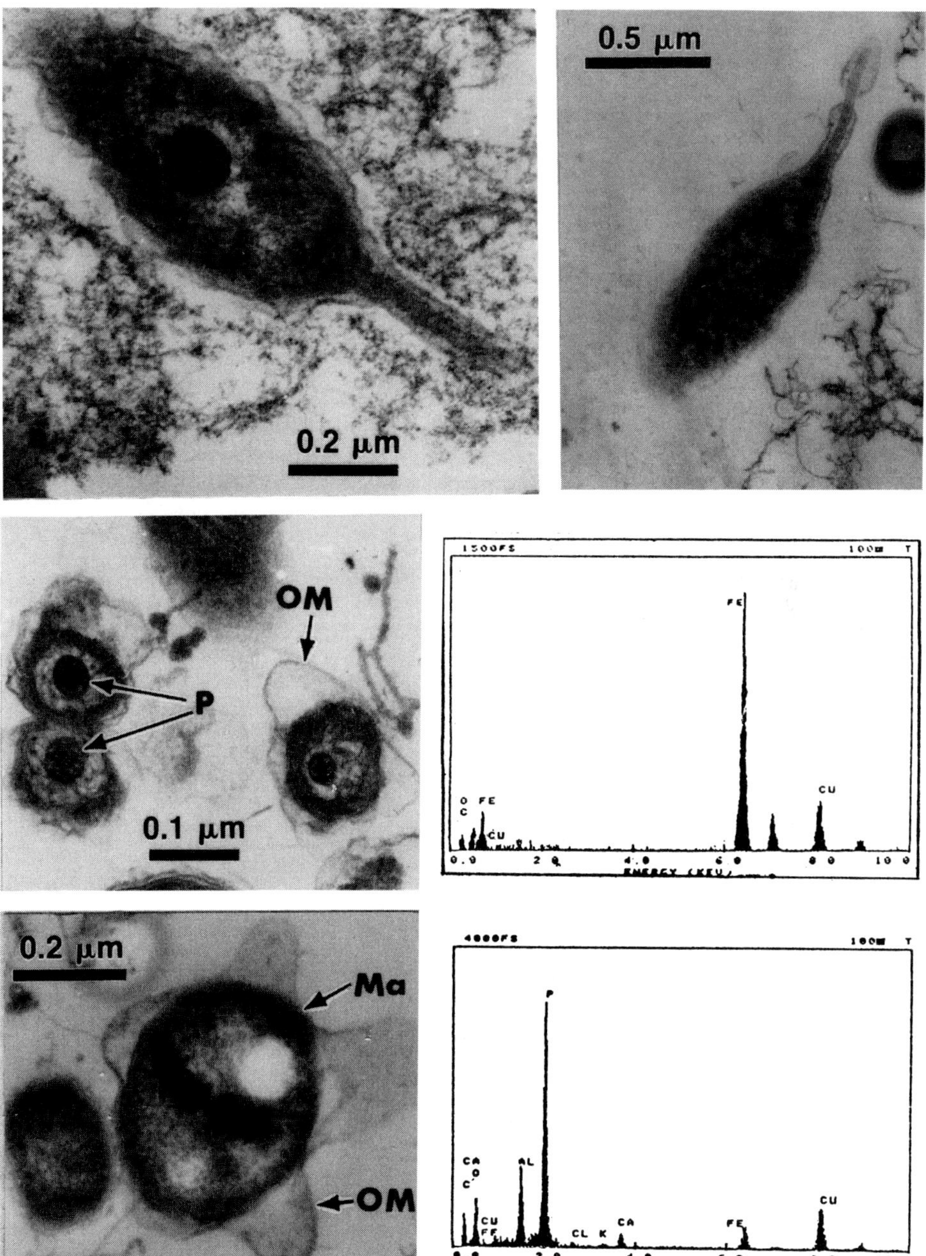

Figure 2. (a) Prosthecate bacteria, *Caulobacter* cells in thin sections with partial stalks. (b) Magnetotactic bacterium. A protoplast with magnitite (Ma) granules surrounded by a large convoluted outer membrane (OM). EDX spectrum of magnitite granules (on the right). (c). An abundant bacterial cell type showing polyphosphate bodies (P). The small cells are enveloped by an outer membrane (OM). EDX Spectrum of polyphosphate body (on the right).

We have mentioned the problem of identifying and characterizing the very minute bacterial cells many of which may in fact not be new organisms but rather because of "starvation" have been "reduced" from their former condition. They may still be identified by molecular techniques (Amann *et al.*, 1995).

6. Acknowledgements

This work was supported by the Black Rock Forest Consortium to T.E. Jensen and W.A. Corpe, Faculty Award Program Grant 6-63161 to T.E. Jensen and NSF Grants BCS-91-17165 and BCS-9313317 to W.A. Corpe and T.E. Jensen.

7. References

Allen, R.D., Weiss, D.G., Hayden, J.H., Brown, D.T., Fujiwake, H., and Simpson, M. (1985) *J. Cell Biol.* **100**, 1736-1752.

Amann, R.I., Ludwig, W. and Schleifer, K.H. (1995) *Microbiol. Rev.* **59**, 143-169.

Balows, A., Truper, H.G., Dworkin, M., Harder, W. and Schleifer, K.H. (ed) *The Prokaryotes, 2nd ed.Handbook On the Biology of Bacteria.* Vol. 1-4. Springer-Verlag, New York.

Corpe, W.A., and Jensen, T.E. (1992) *Microb. Ecol.* **24**, 181-197.

Corpe, W.A. and Jensen, T.E. (1996) *Appl. Microbiol. Biotechnol.* **46**, 622-630.

Daley, R.J. and Hobbie, J.E. (1975) *Limnol. Oceanogr.* **20**, 875-882.

Gerhardt, P. (ed.) (1994) *Methods for General and Molecular Bacteriology*, Amer.Soc. Microbiol., Washington, D.C.

Goldberg, J, Gonzalez, H., Jensen, T.E., and Corpe, W.A. (1996) Quantitative elemental analysis of bacterial polyphosphate bodies using scanning transmission electron microscopy and energy dispersive x-ray spectroscopy. Proc. 54th Ann. Meeting Microscopy Soc. Amer. San Francisco Press, CA. 802-803.

Heldal, M., Norland, S., and Tumyr, O. (1985) *Microbiology* **50**, 1251-1257.

Jensen, T.E. and Corpe, W.A. (1994) *Water Res.* **28**, 901-907.

Johnson, P.W. and Sieburth, J.McN. (1979) *Limnol. Oceanogr.* **24**, 928-935.

Johnson, P.W. and Sieburth, J.McN. (1982) *J. Phycol.* **18**, 318-327.

Leadbetter, E.R. (1997) Prokaryotic diversity: form, ecophysiology and habitat. In: C.J. Hurst (ed.-in-chief) *Manual of Environmental Microbiology.* ASM Press, Washington, D.C.

Luft, J.H. (1961) *J. Biophys.Biochem. Cytol.* **9**, 409- 414.

Pankrantz, H.S. and Bowen, C.C. (1963) *Am. J. Bot.* **50**, 387-389.

Sicko-Goad, L., and Stoermer, E.F. (1984) *J. Great Lakes Res.* **10**, 90-93.

Stahl, D.A. (1996) In: C.J. Hurst (ed.) *Manual of Environmental Microbiology.* Part II, pp. 102-114, ASM Press, Washington, D.C.

Waterbury, J.B., Watson, S.W., Guillard, R.R.L., and Brand, L.E. (1979) *Nature* **277**, 293-294.

Whittenbury, R., Philllips, K.C. and Wilkinson, J.F. (1970). *J. Gen. Microbiol.* **61**, 205-218.

Wischnitzer, S. (1970) *Introduction to Electron Microscopy,* Academic Press.

Biodata of **E. Olavi Kajander,** contributor of "Nanobacteria and Man" with Neva Ciftcioglu and Mikael Björklund.

E. Olavi Kajander (MD, Ph.D) is an assistant Professor in medical biochemistry, University of Kuopio, Kuopio, Finland. His scientific work concentrates on methionine and selenium metabolism, methylation reactions, antibody tecnology, pathogenic fungi and discovery of nanobacteria. Recently his group has started clinical analysis of nanobacteria in human patients.
E-Mail: <**kajander@messi.uku.fi**>

Biodata of **Neva Ciftcioglu**, (coauthor with E.O. Kajander and M. Björklund on the chapter of "Nanobacteria and Man".

Dr. Neva Ciftcioglu is a microbiologist and researcher in the University of Kuopio (Finland). She earned her Ph.D. from the University of Ankara (Turkey) in 1991 and worked six years in nanobacteria project, scientific work on microbiology and allergy and has discovered social behavior and extraordinary growth phases of nanobacteria under serum-free culture conditions.

Biodata of **Mikael Björklund,** coauthor with E.O. Kajander and N. Ciftcioglu on the chapter of "Nanobacteria and Man".

Mikael Björklund is a biotechnology student (making his master thesis in nanogroup.

J. Seckbach (ed.), Enigmatic Microorganisms and Life in Extreme Environments, 195–204.

NANOBACTERIA AND MAN

E. OLAVI KAJANDER, MIKAEL BJÖRKLUND AND NEVA ÇIFTÇIOGLU
Department of Biochemistry and Biotechnology, University of Kuopio, Savilahdentie 9F, 70210 Kuopio, Finland

1. Introduction

Nanobacteria are very small mineral forming bacteria recently discovered in mammalian blood and tissues. They are the first mineral forming bacteria found in blood and the first heat and gamma irradiation resistant bacteria detected in man. They seem to have a genetic organization, which permits them to compensate for their small size and slow metabolism, and endure in their environmental niche. These properties together with their very slow growth rate meant that they were not detected until about 10 years ago. Their extraordinary nature also hindered attempts to publish this data. These autonomously replicating microorganisms were approaching the theoretical limit of self-replicating life with a size of only one hundredth of that of usual bacteria (Kajander, 1992; Åkerman et al, 1993, Kajander et al, 1994). Morphologically similar organisms had been discovered by Folk (1993) in sedimentary rocks and in hot-springs in travertine, and these he had given the name Nanobacteria. Such bacteria seem to contribute to the formation of many kinds of minerals. These bacteria may secrete molecules acting as nucleation centers for depositing biogenic minerals, e.g., apatite, around them in aquatic environments (Mojzsis et al, 1996). In a similar way, nanobacteria from blood can surround themselves with mineral deposits.

2. Nanobacteria

Nanobacteria were named as *Nanobacterium sanquineum* referring to their surprisingly small size and apparently vivid movements, and their existence in blood. The organism was reposited in the German Collection of Microorganisms (DSM No: 5819-5821). These organisms were isolated from 'sterile' commercial human and bovine sera. Fetal bovine serum was found to be ubiquitously contaminated with nanobacteria so that about 80% of serum batches contained culturable organisms. Nanobacteria were found in about 5% of human serum samples (Çiftçioglu et al, 1997a). Thus nanobacteria are the most common cell culture contaminant and a major cause for bacteria. Yet they have remained undiscovered, most likely because of their unique properties.

2.1. DISCOVERY

E. Olavi Kajander, helped by several enthusiastic students, discovered these carbonate/hydroxyapatite mineral-forming organisms in blood and sera for cell culture. Bacteria-like particles were initially observed in long-term mammalian cell cultures. It was only with some batches of serum that the particles could be seen. The cultured mammalian cells became vacuolated (Figure 1A) and died within four weeks. The harmful agent could be passaged and infected other cells with the same outcome. Intracellular bacteria-like particles were observed (Figure 1B) that were similar to those cultured without mammalian cells (Figure 1C). These cultures had a white macroscopic layer of bacteria at the bottom but the medium pH was not greatly changed. The novel bacteria replicated whether mammalian cells were present or not.

Microbiological tests including mycoplasma assays were negative. Bacteriological staining methods initially failed because of difficulties in their fixation with classical flame and alcohol techniques. Nanobacteria could not be cultured in standard microbiological media, but grew in all commercial mammalian cell culture media at 37-45^{0}C under 5-10% CO_2 - 90-95% air. Anaerobic cultures produced no growth. Serum concentrations of 10-50% supported growth optimally. Very poor growth was obtained on solid media. Their multiplication rate followed logarithmic growth, after a lag period, with a doubling time between 1 to 5 days. They produced slime as shown in Figure 1D. This slime explains their observed adherence to glass, plastic and other materials. Thus, nanobacteria can form a biofilm, which is associated with mineral formation.

A typical time-course of a nanobacterial culture is: Near the bottom of the culture vessel, a few very tiny particles, either alone or in small groups are seen with light microscopy after about a one week culture period. After two weeks, they become more numerous, bigger and form groups visible even to the untrained microscopist. After one month, many are in clumps and start to attach to the bottom and by two months most exist in a white-colored biofilm visible to the naked eye. The biofilm becomes bone-like after 3 months but sometimes it becomes detached after half a year. Pellicles can be seen in old cultures.

Specific monoclonal antibodies were produced against a bovine nanobacterial isolate. These were shown to recognize all cultured human and bovine nanobacteria thus indicating the presence of common surface antigens. Before then, nanobacteria could be well distinguished with electron microscopy (Figure 1E). However, antibodies let to the development of a convenient detection method (Çiftçioglu and Kajander, 1997). Contaminated cells show the presence of star-like particles in immunofluorescence staining whereas healthy cells do not (Figure 1G and F).

Decreased serum concentration in nanobacterial culture resulted in their social growth as communities. The total removal of serum revealed a novel growth phase as seen in Figure 1H and 2A. Nanobacteria produced thick common apatite walls around small communities. These 'castles' were large, sometimes even bigger than mammalian cells. Nanobacteria were truly culturable in the new growth phase: they have been passaged monthly for over five years. These special forms of nanobacteria could be released from their apatite 'castles' by adding serum to the culture medium. These forms had very bizarre morphology (Figure 1I), were found to react with the monoclonal antibodies and to release small nanobacteria later into the culture (Çiftçioglu et al, 1997b).

2.2. PHYLOGENETIC POSITION

Based upon 1406 base long 16S rRNA gene sequences, deposited now in a gene bank (EMBL Entry X98418 and X98419), nanobacteria belong to the alpha-2 subgroup of

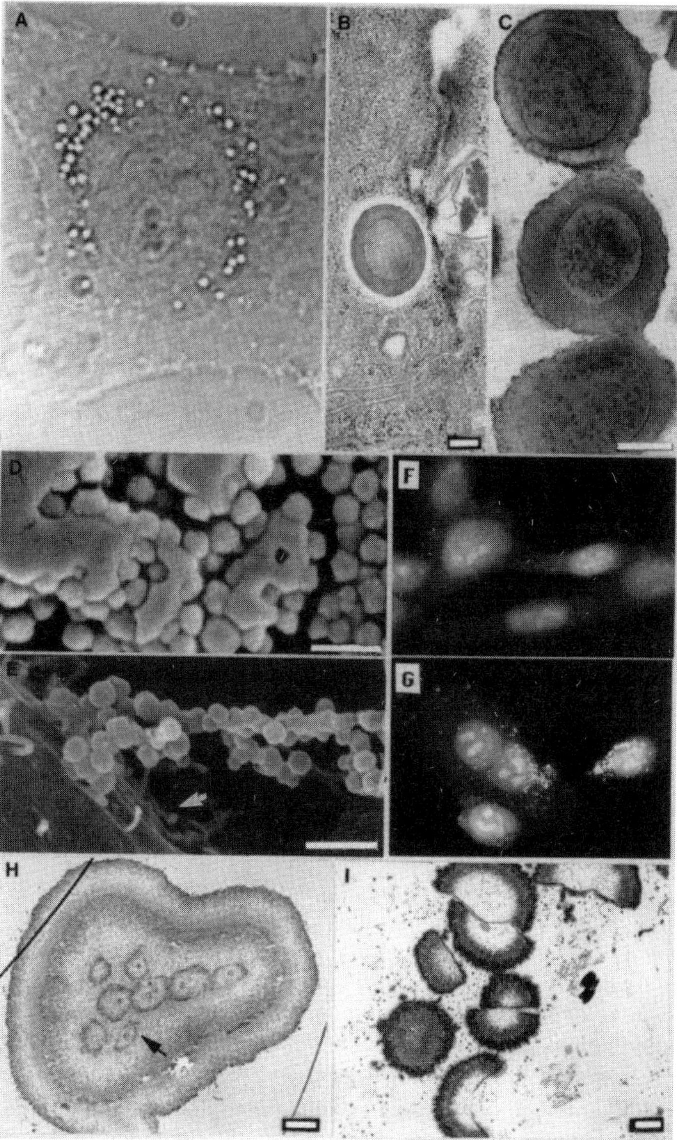

Figure 1. Ultrastructure of nanobacteria and their interaction with fibroblasts. (A) Perinuclear vacuolization of an infected 3T6 cell under phase-contrast microscope; (B) TEM image of a nanobacterium in a BHK cell; (C) cultured nanobacteria (bars 200 nm). (D) SEM image of nanobacteria attached to culture vessel; (E) nanobacteria on a fibroblast surface (arrow: the fibroblast; bars 1 μm). (F) Indirect immunofluorescence staining of healthy 3T6 cells with a monoclonal antibody against nanobacteria; (G) 3T6 cells inoculated with nanobacteria; (H) TEM of a nanobacterial population in a serum-free culture (arrow shows a D-shaped nanobacterium); (I) D-shaped nanobacteria after culture in serum-containing medium (bars 1 μm). Reproduced from Vaccines 97, Cold Spring Harbor Laboratory Press.

Proteobacteria distinct from their nearest relatives which are Phyllobacterium, Thiobacillus, Brucella, Bartonella, Rhizobium and Agrobacterium genera (Kajander et al, 1997). The phylogenetic location is further supported by other findings: nanobacteria are highly resistant to gamma irradiation and to heat, suggesting a recent habitat in hot-springs. Thiobacillus sp., one of their nearest relatives, is known to form iron and sulphur mineral deposits in hot springs (Wood and Kelly, 1985). Furthermore, antibodies against nanobacteria cross-reacted with Bartonella henselae and quintana and antibodies against B. quintana cross-reacted with nanobacteria in immunoassay. Thus, known human pathogens Bartonella and Brucella are relatives of nanobacteria. All share certain common properties: they are very fastidious, impermeable to stains and invade mammalian tissues, with even a preference for fetal tissues. Brucella is known to pass into urine and the Kuopio research group has recently proven that this also happens with nanobacteria (Åkerman et al, 1997). Brucella and Bartonella invade animals and human cells and tissues, other members of the group exist in soil and invade plant cells, except for some Thiobacilli which inhabit hot-springs. Interestingly, mitochondria have been speculated to be derived from an ancestor of this group.

3. Unique Properties

Nanobacteria are one of the most distinctive organisms (Kajander et al, 1997), as summarized in Table 1. Their poor culturability and extremely long doubling time can be compared only to some Mycobacteria, such as *M. leprae*. The average diameter of nanobacteria measured with electron microscopic techniques, 0.2-0.3 µm, is smaller than that of any known cell-walled organism. Several viruses, e.g. Vaccinia, are larger than nanobacteria that can show sizes as small as 0.1 µm. Ultrafiltration methods produce even smaller estimates of size, since nanobacteria pass readily through 0.1 µm pore size. Apparently, nanobacteria can pass through pores smaller than their own size and they must therefore have flexible cell walls. We have also shown that during filtration they lose part of their cell wall (apatite mineral layer). The theoretical minimum diameter of a cell, based on size of those macromolecular components now considered to be necessary and sufficient for a living cell, is about 0.14 µm (Mushegian and Koonin, 1996). Nanobacteria have a size approaching the lower limit of the theoretical size for a living organism. The very slow growth rate and requirement of a very rich medium may be their way of adapting to make their small size compatible with life.

Nanobacteria are highly heat and γ-irradiation resistant and thus the first example of such organisms isolated from mammalian tissues. Bacteria with these properties are found in environments where these traits are essential. Nanobacteria must have emerged from a special environmental source, perhaps only recently in evolutionary terms.

High doses of aminoglycosides, 1 mg/ml, effectively blocked their replication indicating the presence of bacterial type of protein synthesis. Their resistance to these antibiotics may be due to the impermeable cell wall, which also makes their microbiological staining very difficult. The impermeability of the cell wall is further increased by mineral deposits on the cell wall. These deposits are composed of biogenic carbonate or hydroxyl apatite. Figure 2A shows the several micrometer thick apatite formation made and inhabited by a small community of nanobacteria. The cave in the apatite formed indicates its biogenic origin. Under the protection of such 'castles' nanobacteria are almost indestructible. It is noteworthy that nanobacteria cannot be detected by standard sterility testing methods using DNA-stains. We have developed improved methods and started to unravel the nature of nanobacterial nucleic acids, e.g., they show a UV-absorbance shift from 260 to 270 nm (Kajander, 1992; Kajander et al, 1997).

Nanobacteria have a distinctive way of invading mammalian fibroblasts. They trigger cells that are not normally phagocytic to engulf them. After receptor-mediated adherence, followed by internalization, apoptotic cell death occurred in all tested fibroblasts in a dose-dependent way. The cytotoxic potential depended on culture passage number, primary isolates being the most cytotoxic. Adherence and cytotoxicity are rare among cell culture contaminants (Çiftçioglu and Kajander, 1997).

TABLE 1. Extraordinary characteristics of nanobacteria

Doubling time	» 3 days
Culturability	mammalian cell culture medium
Size and shape	0.08-0.50 μm, coccoid
Cell wall	very thick, flexible
Filterability	0.1 μm in high yield
Heat resistance	90°C 1 h
γ-irradiation resistance	1.5 Megarads
Antibiotic sensitivity	resistant to aminoglycosides at 100 μg/ml
Mineral formation	biogenic apatite formations
Resistance to lysis	most lysis methods fail
Nucleic acids	need special methods
Interaction with mammalian cells	adherence, internalization, apoptosis

4. Do Nanobacteria Have A Role in Human Diseases?

Nanobacteria are found in human and cow blood and blood products. This means that bacteria were present but seems to have caused no manifest acute illness. Bacteria may last for a long time, since we have followed a human case for several years with positive nanobacterial tests from blood. However, bacteria would be expected to cause disease(s) at least in some of the infected individuals, since there are no known examples of long-lasting bacteria, which do not cause harmful effects. The apatite coat of nanobacteria may protect them from the immune system since apatite is a normal bodily constituent and also binds serum proteins further hiding the bacteria. This might explain the long-lasting bacteria. On the other hand, it might indicate special pathogenic role in calcifications. In cell culture, nanobacteria attach onto fibroblasts and are internalized so that intracellular calcifications are also observed. Nanobacteria can induce apoptotic cell death. Are fibroblasts attacked by nanobacteria also *in vivo?*

4.1. TISSUE DISTRIBUTION IN VIVO

In rabbits, nanobacteria labeled with ^{99m}Tc and injected intravenously showed a tissue-specific distribution with kidneys being the major target. Surprisingly, there were no significant amounts of nanobacteria in muscle, skin, bone and other tissues with fibroblastoid cells. Nanobacteria were found to pass through glomerular and tubular cells, so that nanobacteria could be isolated from urine (Åkerman et al, 1997). This implies that nanobacteria may cause diseases in the urinary tract. Is there any evidence for this proposal?

4.2. NANOBACTERIA IN HUMAN STONES

Nanobacteria produce apatite minerals, which are very common in kidney stones. In fact, apatite may play a key role in the formation of all kidney stones. The crystalline components of urinary tract stones are: calcium oxalate, calcium phosphate, bacterial related, purines, or cystine. The majority of urinary stones are admixtures of two or more components, with the primary admixture being calcium oxalate with apatite (Mandel, 1996). Furthermore, fermentor model studies have shown that calcium phosphate nidi are always initially formed, and may subsequently be coated by calcium oxalate (Leusmann and Sabinski, 1996) or other components. There is a remarkable similarity in the size and morphology of nanobacterial apatite and apatite found in human kidney stones (Figure 2A and B, respectively). Nanobacteria were found commonly in the 30 human kidney stones that we have now screened. Immunologically similar bacteria could be found also in samples of dental pulp stone (our unpublished data). If a bacterial cause for stone formation could be proven, the disease may lend itself to treatment with antibiotics.

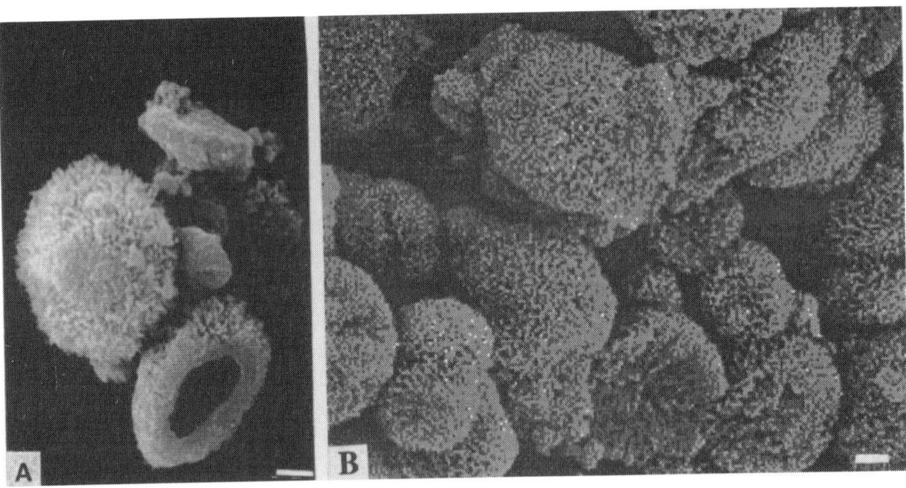

Figure 2. Scanning electron micrographs of nanobacterial dwellings (serum free culture) after scraping from the bottom of the culture vessel (A) and calcium phosphate human kidney stone (B). Both pictures are presented in the same scale, bar 1 μm. (B) is reproduced from http://www.herringlab.com/sems/sems2.html. Energy dispersive X-ray microanalysis (Kajander et al, 1997) revealed that material in (A) was also composed of calcium phosphate.

4.3. TISSUE CALCIFICATIONS AND AUTOIMMUNE DISEASES

Malacoplakia is a rare chronic inflammatory disease of unknown cause, but a bacterial factor has been strongly implicated. The disease typically occurs as tumoral growth in urogenital system characterized by an intensive infiltrate of histiocytes containing intra and extracellular calcospherules, so called Michaelis-Gutmann bodies which are composed of apatite (Ho, 1989). The structure of these spherules closely resembles our calcified nanobacteria. We propose that malacoplakia may be caused by nanobacteria.

Tissue calcifications are found in several diseases such as ovarian serous tumor, papillary adenocarcinoma of the endometrium, breast carcinoma, papillary carcinoma of the thyroid, duodenal carcinoid tumor, and craniopharyngioma (see Ho, 1989). Many malignant cells have receptors for nanobacterial adherence (Çiftçioglu and Kajander, 1997). They could introduce nanobacteria into the tumor with subsequent calcification. Furthermore, some dividing cells under inflammatory stimuli may have receptors for adherence, e.g., in atherosclerotic plaques known to have calcium phosphate accumulation.

Alzheimer plaques have been found to be labeled with anti-nanobacterial polyclonal antibodies (Kajander et al, 1993). These polyclonal antibodies contain some autoantibodies and we have also obtained some monoclonal autoantibodies in nanobacterial immunizations. Slow bacterial infection has been suggested to play a role in autoimmune diseases (Rook and Stanford, 1992). Nanobacteria are a new example of slowly growing organisms infecting man for long periods of time. The apatite structure and anomalous nucleic acids may contribute to abnormalities in immune response to this infection.

5. Conclusions

Nanobacteria are novel emerging pathogens and may be related to small mineral forming bacteria found in sedimentary rocks, linking medicine to geology. They produce biogenic apatite in vitro and also seem to do so *in vivo*. Since apatite is considered to be the main nidus initiating the formation of most kidney stones, nanobacteria seem to be excellent candidates for triggering this process. Nanobacteria injected to blood circulation were shown to penetrate through kidney cells and pass into urine. In urine, apatite formation by nanobacteria is further increased. Other minerals may thereafter bind onto this nidus. The focus of future work should be to clarify their role also in intracellular calcifications and tissue calcinosis, processes involved in many major diseases. Since nanobacteria are cytotoxic, their possible role in kidney damage also warrants further work. Our data indicate that the pathophysiology of stone formation in human diseases has now to be re-evaluated to include nanobacteria in the etiology and this may lead to new treatment strategies and therapies for stones and calcifications.

6. References

Çiftçioglu, N., and Kajander, E.O. (1997) Pathophysiology, in press.

Çiftçioglu, N., Kuronen, I., Åkerman, K., Hiltunen, E., Laukkanen, J., and Kajander, E.O. (1997a) In F. Brown, D. Burton, P. Doherty, J. Mekalanos and E. Norrby (eds.), Vaccines 97, Cold Spring Harbor Laboratory Press, Cold Spring Harbor, pp. 99-103.

Çiftçioglu, N., Pelttari, A., and Kajander, E.O. (1997b) SPIE Proceedings 3111, 429-435.

Folk, R.L. (1993) J. Sedim. Petrol. 63, 990-999.

Ho, K.-L. (1989) Arch. Patol. Lab. Med. 113, 874-879.

Kajander, E,O. (1992) Culture and detection method for sterile-filterable autonomously replicating biological particles. US patent No: 5,135,851, pp. 1-16.

Kajander, E.O., Liesi, P., and Çiftçioglu, N. (1993) Viruses and Virus-Like Agents in Disease, A Karger Symposium, Basel, March 7-9,. Abstract No: M 10, pp. 41.

Kajander, E.O., Tahvanainen, E., Kuronen, I., and Çiftçioglu, N. (1994) Zentralbl. Bakteriol. 26 (Suppl.), 147-149.

Kajander, E.O., Kuronen, I., kerman, K., Pelttari, A. and Çiftçioglu, N. (1997) SPIE Proceedings 3111, 420-428.

Leusmann, D.B., and Sabinski, F. (1996) Urol. Res. 24, 73-78.

Mandel, N. (1996) Semin. Nephrol. 16, 364-374.

Mojzsis, S.J., Arrhenius, G., McKeegan, T.M., Nutman, A.P., and Friend, C.R.L. (1996) Nature 384, 55-59.

Mushegian, A.R., and Koonin, E.V. (1996) Proc. Natl. Acad. Sci. U.S.A. 93, 10268-10273.

Rook, G.A.W., and Stanford, J.L. (1992) Immunol. Today 13, 160-164.

Wood, A.P., and Kelly, D.P. (1985) Int. J. Syst. Bacteriol. 35, 434-437.

Åkerman, K.K., Kuronen, I., and Kajander, E.O. (1993) Scanning 15, (Suppl. 3) 90-91.

Åkerman, K.K., Kuikka, J.T., Çiftçioglu, N., Parkkinen, J., Bergstr"m, K.A. Kuronen, I., and Kajander, E.O. (1997) SPIE Proceedings 3111, 436-442.

Biodata of M. -J. Chretiennot–Dinet author of *"An Enigma in Marine Nanoplankton."*

Dr. Marie-Josèphe Chrétiennot-Dinet is an algologist employed as a CNRS Research Scientist at the Observatoire Océanologique de Banyules, Laboratoire Arago (France). She works in the Department of Oceanography (Laboratoire d'Océanographie Biologie), run by the University of Paris 6 and by the CNRS/INSU (Centre National de la Recherché Scientifique/Institut National des Sciences de l'Univers).
Dr. Chrétiennot-Dinet has been working on nanoplankton for 20 years and more recently on picoplankton, discovering the smallest free-living eukaryote in the Thau Lagoon. She is the author of the third volume of the *Atlas du Phytoplancton Marin*, treating of ten algal classes and is particular expertise in the ultrastructure of small marine organisms. Her involvement includes also the survey of potentially toxic algae along the French coasts, particularly for *Phaeocystis* or other harmful Prymesiophyceae.
E-mail: **mjdinet@arago.obs-banyuls.fr**

J. Seckbach (ed.), Enigmatic Microorganisms and Life in Extreme Environments, 205–213.

AN ENIGMA IN MARINE NANOPLANKTON

The role of star-like structures produced by Phaeocystis

M.-J. CHRETIENNOT-DINET
Laboratoire d'Océanographie Biologique de Banyuls,
UPMC/INSU/CNRS -Laboratoire Arago,
O.O.B., B.P. 44 - 66651 Banyuls-sur-Mer Cedex, France

1. Introduction

Phaeocystis has long been known to be a nuisance for fish, man and environment (Moestrup & Larsen, 1992). It is one of the very few colony-forming members of the Prymnesiophyceae Hibberd, 1976 (= Haptophyceae Christensen, 1962). These are golden-brown algae mostly found in marine waters. Because of their small size (around 5 µm), the motile cells are considered as part of the nanoplankton as defined by Sieburth *et al.* (1978), i.e., comprising cells 2-20 µm in size. The ultrastructure of motile cells was described by Parke *et al.* (1971). They are covered by minute organic scales and possess two flagella and a short haptonema, the latter organelle characteristic of the class (Christensen 1962, Chrétiennot-Dinet, 1990), but in addition, they have the unique characteristic of producing star-like filaments (Figure 1). The exact number of species of *Phaeocystis* is still under debate (Sournia, 1988, Baumann *et al.*, 1994, Medlin *et al.*, 1994, Vaulot *et al.*, 1994), but only two types of star-like structure have been found so far. A five-rayed star, as illustrated in Figure 1, is found in all species except *P. scrobiculata* Moestrup (Moestrup, 1979) which produces a nine-rayed star by addition of a second filament at the branching of each star-tip except one. Nine-rayed star-like structures were observed by the author in Mediterranean waters (Bay of Banyuls), suggesting a large distribution of *P. scrobiculata*, though this species was initially described from New Zealand (Moestrup, 1979). Davidson and Marchant (1992) stated that : 'These stars remain something of an enigma and few reports shed light on their mode and rate of synthesis or their release of the cells, let alone their function'.

2. Life Cycle and Colony Formation in *Phaeocystis*

2.1. LIFE CYCLE OF *PHAEOCYSTIS*

The life-cycle of *Phaeocystis* is complex (Figure 2) and involves an alternation of motile flagellate stages and non-motile colony-forming cells (Rousseau *et al.*, 1994, Lancelot and Rousseau, 1994). Several strains are now maintained in culture, and it appears that some of them are unable to form colonies whilst in others the colonial stage

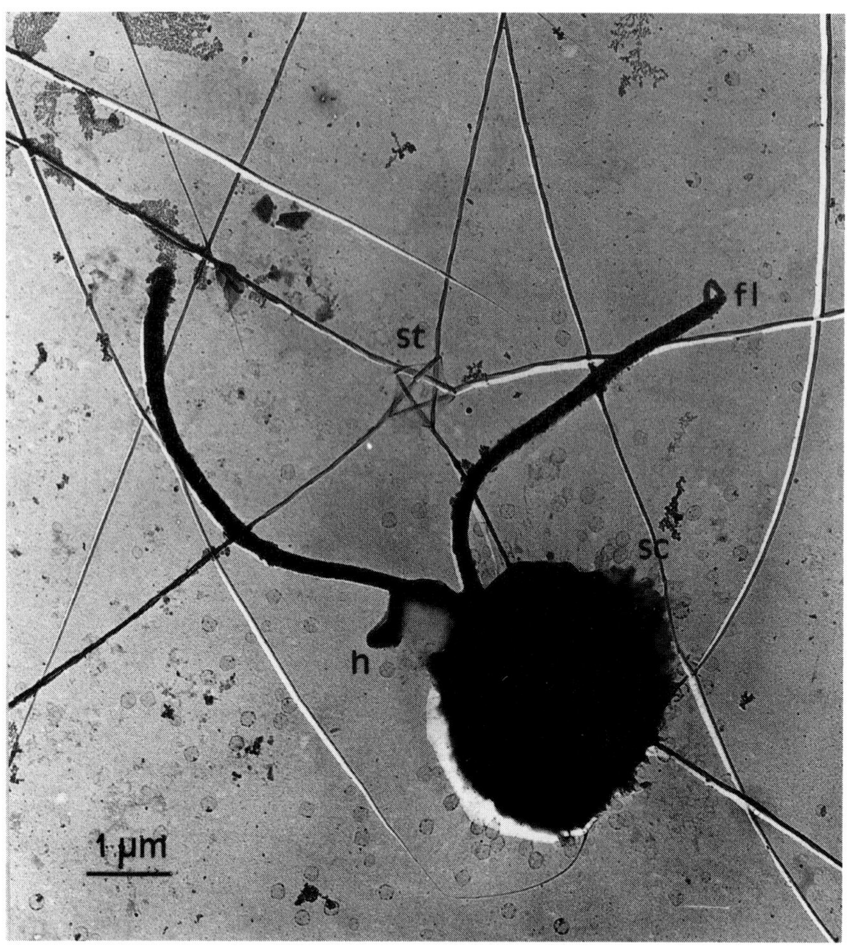

Figure 1. Shadowcast whole mount (transmission electron microscopy) , illustrating the main features of a *Phaeocystis* motile cell :
sc = organic scales, fl = flagella, h = haptonema, st = star-like structure

is dominant (Vaulot *et al.*, 1994). The conditions for switching back and forth between flagellate and colonial forms are unknown, both in culture and in nature. For example, *Phaeocystis scrobiculata* is only known in its motile stage up to now. Life cycles in Prymnesiophyceae may show alternation of morphologically distinct generations (Billard, 1994), but the position of meiosis is difficult to observe and a haplo-diploid cycle is considered to be an adaptation to different environmental conditions. From flow cytometric analyses of the DNA content of the different cell types of several strains of *Phaeocystis*, the motile stage usually appears to be haploid, and such cells are considered as microzoospores, while colonial cells are diploid (Vaulot *et al.*, 1994). However, some flagellate cells also appear to be diploid and are considered as macrozoospores. The presence or absence of filaments and stars in the different stages was verified using cultured material and almost all motile cells were found to produce filaments. In contrast, cells in colonies were devoid of vesicles containing filaments (author's unpublished results). Such results suggest a possible reason for the presence of these structures in the motile cells.

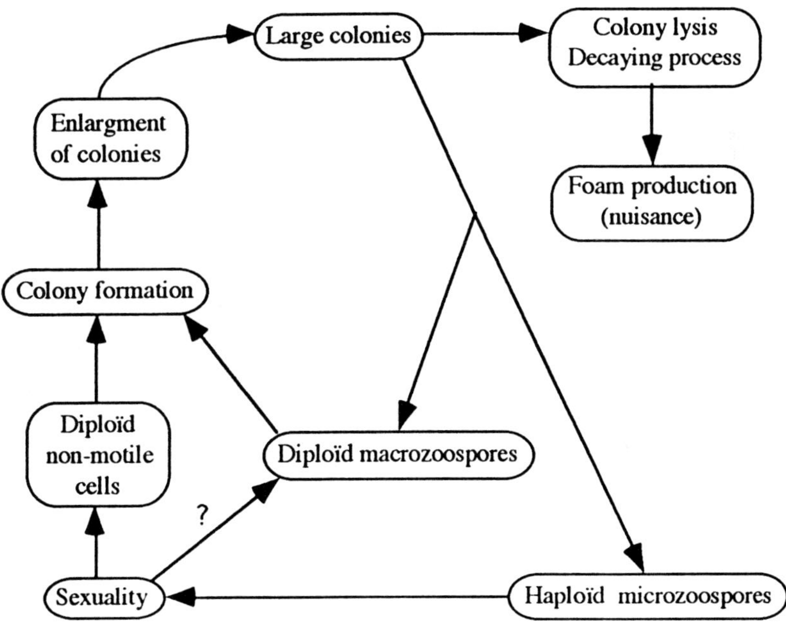

Figure 2. Simplified Life Cycle of *Phaeocystis* (from Rousseau *et al.*, 1994).

2.2. IMPORTANCE OF COLONY FORMATION IN MARINE ENVIRONMENTS

The gelatinous colonies of *Phaeocystis* may form dense, almost monospecific blooms, dominating a marine ecosystem (Lancelot *et al.*, 1992) and the subsequent collapse of such blooms often leads to environmental problems arising from copious foam formation (Lancelot *et al.*, 1987) and the release of volatile sulphur compounds as the colonies disintegrate (Malin *et al.*, 1994). The environmental conditions required for colony formation have been intensively investigated and appear to be related mainly to nutrient enrichment and meteorological conditions (Lancelot *et al.*, 1991; for a review on colony formation, see Rousseau *et al.*., 1994 and Cariou *et al.*, 1994). However, although numerous studies have been devoted to the initiation of *Phaeocystis* blooms, little is known about the attachment of motile cells to a substrate (Davies *et al.*, 1992, Cariou *et al.*, 1994). It is known that *Phaeocystis* blooms frequently follow the spring diatom bloom and young colonies (Figure 3) are often found attached to *Chaetoceros* setae (Lancelot and Rousseau, 1994). A biochemical stimulus was suggested by Boalch (Boalch, 1987) to explain this specific association with a diatom, but these long expanded siliceous appendages probably represent a suitable substrate for the mechanical attachment of the motile cells by means of the star-like structures and adjacent filaments.

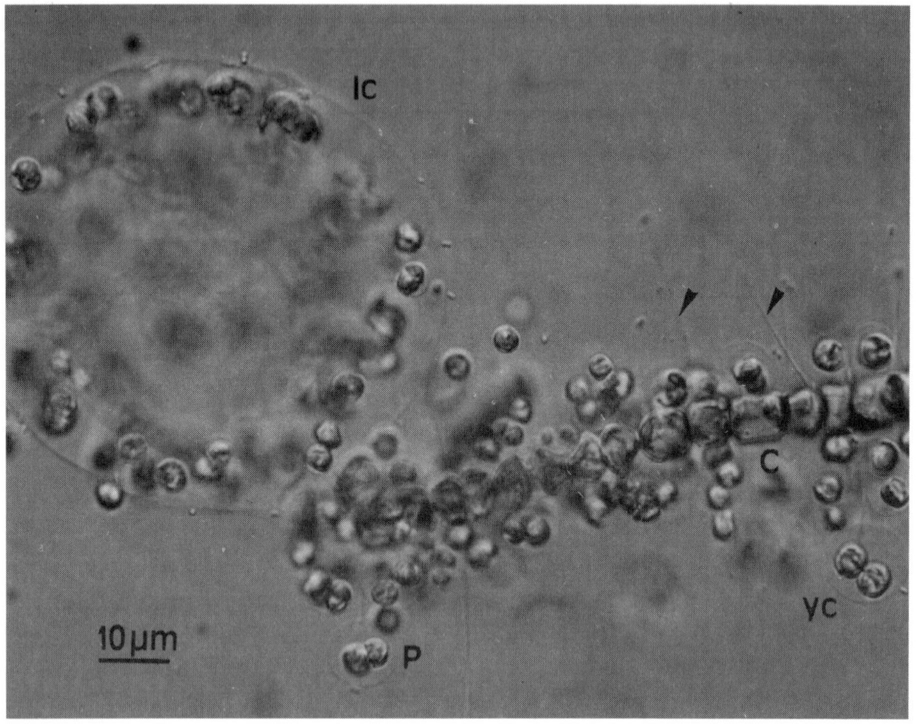

Figure 3. Different stages of colony formation around *Chaetoceros* setae (arrows).
P = *Phaeocystis*, yc = young colony, lc = large colony, C = *Chaetoceros*.
(Light Micrograph, by courtesy of C. Billard)

3. Trichocyst-like production and the nature of the filaments extruded by *Phaeocystis*

3.1. TRICHOCYST PRODUCTION IN MARINE ALGAE.

Some marine unicellular algae, e.g., those belonging to the Dinophyceae, Chrysophyceae or Cryptophyceae (in a few cases to the Prasinophyceae), are known to eject superficial structures, called trichocysts (Hovasse and Mignot, 1975). Such organelles are usually extruded under adverse conditions (observation under a coverslip in light microscopy or the addition of a fixative) and are therefore considered as defensive structures. In the Cryptophyceae, the ejectosomes are situated in the periplast and are ejected outside the cell as long ribbons, illustrated here in Figure 4. The case of *Phaeocystis* is somewhat unique in that it possesses trichocyst-like structures, located in superficial protruding vesicles, but completely different from classical trichocysts. As described above, they appear in most cases as five-rayed filaments producing an almost perfect star in their central part after release from the cell (Fig.1). The nature and function of these filaments remained unknown for a long time but their composition has been recently elucidated (Chrétiennot-Dinet *et al.*, 1997) and suggests an explanation for their function.

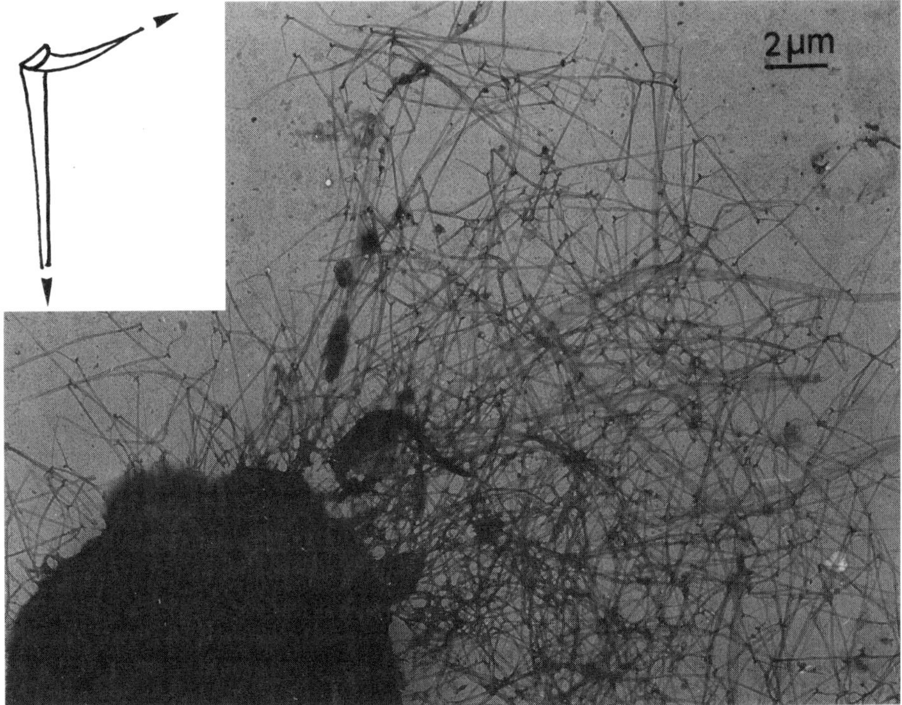

Figure 4. Numerous trichocysts ejected from a Cryptophycean cell. Insert : schematic representation of a single trichocyst (from Hovasse and Mignot, 1975).

3.2. THE COMPOSITION OF THE FILAMENTS

A striking feature about *Phaeocystis* filaments is that they are very resistant. They can be visualized in light microscopy by Alcian blue staining on living material (Vaulot *et al.*, 1994) and they resist dessication. They are found in coastal and open sea samples examined by scanning or transmission electron microscopy and they appear to be present in the water column down to 1000 m depth (author's unpublished data). They are easily collected by simple filtration of sea-water on Nuclepore or Nylon filters and do not need a specific treatment for preservation. Thus, as motile cells together with their flagella and haptonema are destroyed if critical point drying is not used on the sample, the filaments provide a reliable means of detecting the presence of *Phaeocystis* in a body of water devoid of colonies. Italian workers suggested that these filaments were siliceous in nature (Ramani *et al.*, 1994), but a more careful study revealed that they were actually composed of α-chitin, a unique example among unicellular algae (Chrétiennot-Dinet *et al.*, 1997). The chitinous nature of the filaments gives them rigidity and resistance while the central star makes the whole structure a floating anchor with five arms that can grip on to a substrate.

3.3. HOW ARE THEY PRODUCED ?

Although the number of stellate structures and the frequency of their extrusion by the motile cells are not known exactly, it has been shown that several vesicles containing filaments could be found on a single cell (Pienaar, 1991) and up to three vesicles were seen on sections in transmission electron microscopy by the author. A model for their storage and ejection has been recently proposed (Chrétiennot-Dinet *et al.*, 1997). Each vesicle contains five coiled filaments, each of them being wound on itself approximately 5 to 6 times. The first winding corresponds to the initial attachment of the filaments in such a way that uncoiling produces the typical star in the center and the filaments behave as coiled springs once the vesicle is broken.

3.4. HYPOTHESIS ABOUT THEIR ROLE

Up to now, no suggestion has been put forward to give an explanation for the presence of such structures and very little was known on their nature and production. It is assumed here that the filaments ejected by *Phaeocystis* act as floating anchors for attachment and have, therefore, a functional role in colony initiation. The length of each filament is up to 25 μm and the whole structure covers then a circular area of about 50 μm in diameter. The chitinous nature of these filaments, as recently demonstrated, gives an explanation to their strength and explains why they are almost never disrupted when observed under the light or electron microscope. As several of these star-like structures can be ejected simultaneously, the cell becomes surrounded by floating elements likely to become attached to solid structures represented by, for example, the diatom *Chaetoceros* and its long setae. Such a device probably gives the settling cell the stability required for colony initiation.

4. References

Baumann, M.E.M., Lancelot, C., Brandini, F.P., Sakshaug, E., and John, D.M. (1994) *J. Mar. Syst.* **5**, 5-22.

Billard, C. (1994) in J.C. Green and B.S.C Leadbeater (eds.), *The Haptophyte Algae*, The Systematics Association, Clarendon Press, Oxford, pp. 194-

Boalch, G.T. (1987) *Rapports et Procès-verbaux des Réunions. Conseil International pour l'Exploration de la Mer* **187**, 94-97.

Cariou, V., Casotti, R., Birrien, J.L., Vaulot, D. (1994) *J. Plankt. Res.* **16**, 457-470.

Chrétiennot-Dinet, M.J. (1990) *Atlas du Phytoplancton Marin* **3**, Editions du CNRS, Paris.

Chrétiennot-Dinet, M.J., Giraud-Guille, M.M., Vaulot, D., Putaux, J.L., Saito, Y., and Chanzy, H. (1997) *J. Phycol.* **33**, 666-672.

Christensen, T. (1962) in T.W. Böcher, M. Lange and T. Sorensen (eds.), *Botanik (Systematik Botanik)*, Copenhagen, Bd 2, pp 1-178.

Davidson, A.T. and Marchant, H.J. (1992) in F.E. Round and D.J. Chapman (eds.), *Progress in Phycological Research*, Biopress Ltd., Bristol, **8**, pp. 1-40.

Davies, A.G., De Madariaga, I., Bautista, B., Fernandez, E., Harbour, D.S., Serret, P. and Tranter P.R.G. (1992) *J. mar. Biol. Ass. U.K.* **72**, 691-708.

Hibberd, D.J. (1976) *Bot. J. Linn. Soc.* **72**, 55-80.

Hovasse, R. and Mignot, J.P. (1975) Ann. Biol. **14**, 397-422.

Lancelot, C., Billen, G. and Barth, H. (1991) *Water Pollution Research Report* **23**.

Lancelot, C., Billen, G., Sournia, A., Weisse, T., Colijn, F., Veldhuis, M.J.W., Davies, A. and Wassmann, P. (1987) *Ambio* **16**, 38-46.

Lancelot, C. and Rousseau, V. (1994) in J.C. Green and B.S.C Leadbeater (eds.), *The Haptophyte Algae*, The Systematics Association, Clarendon Press, Oxford, pp. 229-245.

Lancelot, C., Wassmann, P. and Barth, H. (1992) *Mar. Pol. Bull.* **24**, 56-57.

Malin, G., Liss, P.S. and Turner, S.M. (1994) in J.C. Green and B.S.C Leadbeater (eds.), *The Haptophyte Algae*, The Systematics Association, Clarendon Press, Oxford, pp. 303-320.

Medlin, L.K., Barker, G.L.A., Baumann, M., Hayes, P.K. and Lange, M. (1994) in J.C. Green and B.S.C Leadbeater (eds.), *The Haptophyte Algae*, The Systematics Association, Clarendon Press, Oxford, pp. 393-411.

Moestrup, Ø. (1979) *New Zeal. J. Bot.* **17**, 61-95.

Moestrup, Ø. and Larsen, J. (1992) ICES Identification Leaflets for Plankton **179**: *Intern. Council Explor. Sea, Copenhagen*, 11pp.

Parke, M., Green, J.C. and Manton I. (1971) *J. Mar. Biol. Ass. U.K.* **51**, 927-941.

Pienaar, R.N. (1991) *Electr. Micr. Soc. S. Africa* **21**, 135-136.

Ramani, P., Cardinali, S., Cabrini, M. and Del Negro, P. (1994) *Giorn. Bot. Ital.* **128**, 1065-1066.

Rousseau, V., Vaulot, D., Casotti, R., Cariou, V., Lenz, J., Gunkel, J. and Baumann, M. (1994) *J. Mar. Syst.* **5**, 23-39.

Sieburth, J.Mc N., Stemacek, V. and Lenz, J. (1978) *Limnol. Oceanogr.* **23**, 1256-1263.

Sournia, A. (1988) *Nova Hedwigia* **47**, 211-217.

Vaulot, D., Birrien, J.L., Marie, D., Casotti, R., Veldhuis, M.J.W., Kraay, G.W. and Chrétiennot-Dinet, M.J. (1994) *J. Phycol.* **30**, 1022-1035.

Biodata of **Josef Elster** Author of: **Algal Versatility in Various extreme Environments.**

Dr. Josef Elster is a member of the scientific Staff of the Institute of Botany, Academy of Sciences of the Czech Republic. He is also a lecturer at the University of South Bohemia, Faculty of Biological Sciences, České Budějovice, Czech Republic. His interest is in ecology and ecophycology of algae in extreme, natural and man-influenced habitats. Dr. Elster has been studying the mechanisms of mass occurrence and development of algae in several extreme habitats created by intensive crop farming, animal husbandry and fish farming in ponds. During the last few years he has participated in international research teams (Svalbard-Norway, Ellesmere Island - High Canadian Arctic, King George Island, South Shetlands - Polish Research Station) which were entrusted with the special task to study the ecological role of algae in polar environments. E-Mail: jelster@butn.cas.cz

Key Word for the above art

Algae,
Desiccation
Extreme Environment,
Light,
Salt,
Temperature,
Toxicity.
Water Deficiency,

J. Seckbach (ed.), Enigmatic Microorganisms and Life in Extreme Environments, 215–227.

ALGAL VERSATILITY IN VARIOUS EXTREME ENVIRONMENTS

JOSEF ELSTER
Academy of Sciences of the Czech Republic. Institute of Botany CZ - 379 82 Třeboň. University of South Bohemia. Faculty of Biological Sciences. CZ-378 01 České Budějovice. Czech Republic

1. Introduction

Cyanobacteria (Cyanoprokaryota, Cyanophyta) and algae (to avoid duplication of terms, "algae" and "algal" in the text also includes Cyanobacteria, unless further specified), remain at the beginning of plant evolution. Due to their evolutionary antiquity they are widely adapted to all extremes related with changes in geological time. Although clearly defined studies are not abundant, because much of the research so far has concentrated on only a few species, prokaryotes are much better adapted to all environmental extremes than eukaryotes. In comparison with other ancient microorganisms, algae possess oxyphototrophic type of photosynthesis. The physiology of photosynthesis and its versatility under various extreme conditions is one of the most important features of their adaptation and algae thus form a considerable part of the primary production of the world ecosystem.

In this chapter, an overview of the ecology and ecophysiology of those algae which occur in extreme terrestrial and aquatic habitats is made. No phycology textbook has yet cleared up what we understand by the term extreme environment. Here, it is proposed to establish in phycology research two categories of extreme environments:

A. *Objectively "extreme" but stable environments*

Habitats where algae live near the limit of their physiological potential. Their life processes provide an example of adaptation to those extremes in which algae exist for long periods of time, and successive development and selection of genotypes have occurred. These special habitats are each inhabited by their own well-adapted algae (e. g. psychrophiles - ice algae, thermophiles - algae of hot springs, acidophiles - metaphyton of peat bogs, halophiles - phytoplankton of hypersaline lakes, etc.). In objectively extreme but stable environments, algal existence may be prolonged and stable, but in fluctuating and unstable extreme environments, metabolic cost of survival may be high and probability of death and extinction considerable (Friedmann 1993b).

B. *"Marginal" unstable types of environment.*

The seasonal and diurnal variation of the environment can impose severe conditions, and thus algae have to overcome a series of environmental patterns to occupy such terrestrial and some aquatic habitats. In marginal habitats, the seasonal and diurnal environmental changes are dramatical. Examples are the life conditions experienced by the periphyton of polar rivers and streams, cryptoendolithic communities of polar and hot deserts, aerophytic corticulous algal community, etc. These environmental patterns may differ in periodicity, amplitude, synchrony and regularity, and may initiate a number of different adaptations in algae. For two of the dominant ecological parameters, light and temperature, daily fluctuations are irregular, while their seasonal fluctuation is more regular and therefore predictable.

However, in terrestrial and aquatic habitats, both natural and those influenced by man, a series of transient ecological situations also exist. The ecological literature commonly uses the term adaptation, but the present phycological literature has not yet given its precise definition. Three terms are commonly cited: adaptation - acclimatization - acclimation. Yet their meanings are not always unambiguous. According to Lawlor and Keys (1993), the terms adaptation and acclimation, as applied to plants' response to the environment, are synonymous. On the other hand, in zoology, several types of organism responses to the environment are distinguished. For example, concerning temperature, Clarke (1991) has defined adaptation, acclimatization and acclimation. Our understanding of algal responses to environmental extremes for both objective extremes as well as marginal situations, is the following: The term adaptation is used for a genetically-fixed (phylogenetic) algal response to environmental conditions (e. g., ice diatoms species of the genera *Corethron, Synedra, Nitzschia* adapted to high salinity, low temperature, lack of light and nutrients, etc.). Acclimatization is understood as a response of algae (ontogenetic), (e. g., of the genera *Klebsormidium, Stichococcus* etc.), to sporadic seasonal environmental extremes such as low temperature or dryness. Finally, acclimation we consider to be a response of algae to laboratory conditions (e. g. in algal collection).

From many, clearly-defined studies in algal ecology, it seems that species diversity decreases with increasing harshness of certain environmental factors (Elster and Komárek 1993, Pokorný *et al.* 1994, Elster *et al.* 1994, Elster and Svoboda 1996, Elster *et al.* 1997, etc.). Algae, either as organisms or their groupings (community or population), react to stress; those algae which can maintain the highest assimilation and have the most successful metabolism are able to improve their own physiology to a new level of homeostasis and survive the new conditions. Some algal species are included among microorganisms which are the most sensitive to any environmental stress. On the other hand, other algae have evolve to such a high ecological plasticity that enables them to prosper under almost all environmental conditions worldwide. Their adaptation - acclimatization response to changes in the environment is generally fast and persistent both at the genotypic and phenotypić level. Previously published monographs which have paid attention to the role algae play in various extreme environments by Castenholz (1973), Friedmann and Galun (1974), Alexander (1976), Borowitzka (1981), Gerdes *et al.* (1985), Sullivan *et al.* (1985), Broady (1986), Brock (1986), Vincent (1988), Wynn-Williams (1990), Reed (1990), Galinski and Tindall (1992), Friedmann (1993), Hausmann and Kremer (1995), etc. Other authors have applied the problem of extreme environments to exobiology, i. e. existence of life outside of the Earth: Friedmann and Ocampo-Friedmann (1984), Friedmann (1986), Friedmann *et al.* (1987), Friedmann and Koriem (1989), Friedmann (1993a), Friedmann *et al.* (1993), McKay *et al.* (1992), McKay (1993), etc.

Although the latest ecological, physiological and biochemical studies of algal adaptation - acclimatization have brought a lot of new information, it would not be correct to say a particular line of research is now complete. There is still much that can be done in the study of the processes and mechanisms of adaptation - acclimatization - acclimation. However, the data on marine algae are more detailed than the available data on algae from terrestrial and freshwater habitats.

An overview is given here of the algal responses to the following extreme environmental factors: high and low temperature, light effects (UV irradiation), relationship to water (and dryness) and mineral salts (toxicity), with particular respect to cyanobacterial and algal ecology and their ecophysiology in terrestrial and aquatic habitats. Other extreme environmental factors such as nutritional deficiencies or pH ranges, gases, etc., are not included because of the limited scope of this chapter. Although each environmental factor is treated under a separate heading, for convenience, it should be remembered that all factors interact with one another to a greater or lesser extent.

2. Extremes of Temperature

As mentioned before, temperature, light and moisture are the major determinants of algal growth. Each of these factors can be limiting under specific circumstances. Frequently, however, especially in marginal habitats, their impacts are coincident. Individual algal species can survive only within relatively narrow temperature ranges. Algae have a similar temperature span of 30 - 40 °C as other microorganisms and higher organisms. However, the temperature limits for growth (minimum, optimum and maximum) range from almost sub-zero to boiling point temperatures. The optimum for growth of some snow and ice algae is 1 °C (Hohan 1975) while for some thermal Cyanobacteria it is 74 - 75 °C (Brock 1967). However, the lower and upper survival limits can often greatly exceed these values.

From the above, it is evident that algae as a group of organisms possess a wide ecological and physiological adaptation - acclimatization which enables them to survive or prosper at both lower and upper temperature extremes.

Although algae occupy such a wide temperature range, classification of individual species according to their temperature requirements is not yet clear. For use in phycology, according to the temperature range for growth, the following three terms are proposed: psychrophilous, mesophilous and thermophilous. A similar classification is widely used in microbial ecology (Russell 1990). However, for low temperatures, terms such as psychrophiles and psychrotrophs, or eurythermal and stenothermal, are also common. The definitions of psychrophile and psychrotroph that have received most widespread acceptance are those by Morita (1975). He proposed that psychrophiles should include organisms having optimum growth temperatures < 15 °C and maximum tolerated temperatures < 20 °C, whereas psychrotrophs (which are still capable of growing at or close to 0°) have optimum growth temperatures > 15 °C and upper temperature limits as high as 40 °C in a few cases. Sometimes, the terms obligate and facultative psychrophiles are also used. A similar classification can be used for the algae living in high upper temperature environments; thermophiles and thermotrophs. Compared with psychrotrophs and psychrophiles, thermotrophs and thermophiles tend to occur and grow within narrower temperature ranges. In our opinion, other terms such as euthermal, mesothermal, oligothermal, metathermal, should not be used because of problems with their definition.

When living at the lower and upper temperature extremes, algae have to reduce the intensity of their metabolic processes. At the low extremes, algae occurring in all cold aquatic and terrestrial habitats (polar regions as well as seasonally cold habitats of temperate zones) are successful in competition mainly thanks to their survival under light-limited conditions at low temperatures. This survival ability is mainly due to their very low respiration rates and, consequently, positive carbon budget. Changes in the lipid composition of membranes is another feature of adaptation/acclimatization to low temperature. In the case of membranes, the degree of unsaturation and length of fatty acid chains, the phospholipid composition, and the presence of sterols, can be used as indicators of cold acclimatization or adaptation (Kirst and Wiencke 1995).

At the upper end of the temperature range, the main problem concerns the protection of proteins against denaturation (Marre 1962). In general, thermophiles have enzymes which are much more resistant to high temperatures than individual proteins. However, as at low temperatures, thermophily is also positively related to the stability of larger membrane structures that are not broken down by high temperatures (Brock 1967).

Much more detailed information, however, concerning the effects of low and high temperatures in microbial ecology and physiology can be found in the reviews by Brock (1967), Castenholz (1969), Russell (1990), Kirst and Wiencke (1995), etc.

3. Light Effects (Uv Irradiation)

Light, in comparison with other environmental limiting factors, is one of the most extreme in terms of variation in time, both in intensity and spectral quality. This variation has led, in the course of evolution, to the acquisition of the light-harvesting pigment systems together with their biochemical reactions. The evolution of such a photosynthetic apparatus is also of major significance for understanding the phylogeny and taxonomy of different algal groups.

The liquid aquatic medium is the commonest environment for the majority of both stable, and unstable, extreme algal habitats. In water, algae have to compete for solar radiation, not only with one another, but also with all the other light-absorbing and scattering components. However, competition for irradiance within the underwater environment, as well as within various terrestrial environments (such as soil, endolithic communities, epilithic communities), has resulted in vertical distribution arrangements and in the evolution of particular light-harvesting pigment systems.

Chromatic adaptation theory (Engelmann 1883) is probably one of the first demonstrations of ecological adaptation, describing the changing spectral distribution of light with water depth as determining the vertical arrangement of algae. Later, Harder (1923) found that chromatic adaptation also involved light intensity adaptation, also involving decrease in respiration rate with decreasing irradiance.

In all waters, irradiance in the red wavebands rapidly diminishes, and underwater light becomes first blue-green and then blue. In deep water with green light, brown algae, with the presence of fucoxanthin, or some green algae containing siphonoxanthin, prevail. At the deepest levels, where the spectral range becomes quite narrow, red algae and cyanobacteria with biliprotein pigments become the most successful algal groups.

Ontogenetic adaptation was also described at the beginning of this century by Gaidukow (1902). He observed that Oscillatoria rubescens was red in color when grown in green light, while blue-green when grown in orange light. Boresh (1921) showed that the color changes were associated with shifts in the types of synthesized biliproteins. The red cells contained predominantly phycoerythrin, the blue-green cells do mainly phycocyanin. Similar pigment changes in the phycoerythrin to chlorophyll ration are also known in some eukaryotic algae (Rhodophyta, Cryptophyta). Green algae react to changes in spectral quality of light by changing their chlorophyll a : b ratio. Changes in the total amount of pigments per cell or in the ratios between different pigments, or both, are common among algae. However, it appears in most algae, that the increase in chlorophyll content during their adaptation to shading is due to an increase in the number of photosynthetic units and in the proportion of photosystems I and II. The photosystem II/photosystem I ratio can shift from near unity in the low-light cells to greater than two in high-light cells.

The radiant intensity of full sunlight can cause photoinhibition, or can even result in photooxidation of cells. Nevertheless, evidence exists that carotenoids, especially xanthophylls not involved in light harvesting apparatus, protect the cells against damage from excessive light.

In broad overall terms, the significance of phylogenic adaptation, ontogenic acclimatization and the protective consequences of light-harvesting umbrellas is that algal species are able to exploit a wide range of habitats, including all sorts of extreme situations. As mentioned above, the processes of chromatic adaptation, acclimatization and acclimation have been widely studied. In aquatic ecosystems, the present state of knowledge has been summarized mainly by Richardson *et al.* (1983), Kirk (1983), Falkowski and LaRoche (1991), Kirk (1994), etc. Barber and Andersson (1994) have summarized our present knowledge of photosynthesis at the physiological and molecular biological level and state that the photosynthetic apparatus is the first of the cell's organelles that has been described in detail at the atomic level (its structure, function and regulation).

The continuing degradation of the Earth's ozone layer by atmospheric pollutants has brought about an increase in the amount of solar ultraviolet radiation received at the Earth's surface. The most dangerous part of ultraviolet radiation, UV-B, can potentially cause a wide range of effects on both aquatic and terrestrial ecosystems. However, solar UV radiation has been significantly variable throughout evolutionary time, and algae are able to develop a variety of protective strategies against possible damage. Algal cells possess four lines of defense against the effects of UV-B. Some species are able to avoid UV exposure by having a migration strategy, while others produce a sunscreening pigment that filters UV wavelengths such as mycosporine-like amino acids and carotenoids. As mentioned before, similar mechanisms are commonly protective against high solar irradiance as well. In addition, most algal cells have at least some ability to identify and repair UV-caused damage to DNA and other biomolecules.

During recent times, the problem of UV-B radiation effects on algal communities has been reviewed by several authors (Karentz *et al.* 1991, Garcia-Pichel *et al.* 1993, Post and Larkum 1993, Vincent and Roy, 1993, Vincent and Quasada, 1994, Wynn-William, 1994, Lesser *et al.* 1994, Xiong *et al.* 1995), most of whom have worked mainly in polar regions. Cold regions of both hemispheres are under the strong influence of UV-B irradiation.

4. Water Deficiency (Water Stress)

During their phylogeny, algae have been exposed to a variety of stress, two kinds of which are closely interrelated: water deficiency or water stress, and extremes of salt concentration or salt stress. Adaptation/acclimatization to water stress has a long history in geological time. It dates from the period when the first oxyphototrophic organisms started to expand to the terrestrial environment.

Life originated in water and all biochemical processes can only take part there. For the algae occurring in water (phytoplankton and the majority of phytobenthos), desiccation is potentially much more dangerous than for terrestrial algae. The Cyanobacteria are traditionally the best adapted to drought and desiccation. In plant ecology, the term "ecological opportunist" is often used, however in algal ecology the use of this term is convenient only for terrestrial species. Desiccation-tolerant algal cells include zygotes, akinets, cysts and spores. In particular, Cyanobacteria may have unmodified vegetative cells that are resistant to desiccation. Cellular changes recorded as a result of de- and rehydration are the possession of a thick cell wall and production of mucilage. The classical studies of Fritsch (Fritsch 1922, Fritsch and Haines 1923) and of Stokes (1940) still remain the most direct ecological examinations of drought-tolerant organisms.

Physiological responses to water and salt stress are traditionally considered separately from osmotic and specific ion effects. The specific ion effects have been subdivided into toxicity and nutritional inbalance (Epstein 1985, Lauchli and Epstein 1990). Microbial cell membranes are selectively permeable to extra- and intracellular solutes, but freely permeable to water. Generally, K^+, Na^+ and Cl^- often occur at relatively high concentrations both inside and outside the cell and contribute to the total flux of ions (Dainty 1976). The cells exposed to desiccation have to cope with matrix forces, and because of evaporation of water, also with an increase in the concentration of dissolved solutes, which in turn affects the osmotic and electro-chemical potential (Reed 1984). Two distinct ways of metabolic adjustments, ionic regulation and synthesis of osmotically-active compounds, are considered. These adjustments eventually result in a state in which the enzymatic machinery can function optimally under the new conditions.

Firstly, ionic regulation is considered as a result of a decrease in external water potential, a water flux directed outwards occurs, and results in cell shrinkage and a decrease in cell turgor (Reed and Walsby 1985). The changes in cell volume caused by this flux are mitigated by a solute transfer, in particular Na^+, across the plasmalemma. In normally growing cells, the trend is towards a higher intracellular K^+ and a lower Na^+ concentration than in the growing medium. Any subsequent potential increase in Na+ concentration can be counterbalanced by a simultaneous uptake of K^+. These processes are active and occur against the electrochemical H^+ gradient produced across the membrane (Allison and Walsby 1981, Reed *et al.* 1984a). The proton gradient is generated by ATP-ase which depends upon photosynthetic and respiratory energy (Blumwald *et al.* 1984) or by an inducible, enhanced level of cytochrome oxidase in the cell membrane (Peschek *et al.* 1989). However, it has been shown that the signal for K+ accumulation is caused by the loss of turgor and reduction of the cytoplasmatic volume (Reed and Stewart 1985a).

The production of osmotically-active compounds, rather the accumulation of inorganic ions, plays a major role in the maintenance of a positive turgor (Reed *et al.* 1984b). These compounds, which accumulate to high concentrations in osmotically-stressed cells do not inhibit enzyme activity and metabolic function and have been termed compatible solutes. Three groups of these compounds, according environment, have been identified: freshwater, marine and hypersaline. They can also be classified according to their habitat (Reed *et al.* 1984a): (A) the least halotolerant (polyhaline-euhaline), accumulate sucrose and trehalose (Reed *et al.* 1984b, Reed and Stewart 1985b); (B) intermediate (metahaline) accumulate glucosylglycerol (Reed *et al.* 1984b, Reed and Stewart 1985b); and (C) (hypersaline) accumulate glycine-betaine or glutamate-betaine as their major intracellular organic solute (Reed *et al.* 1984a, Reed *et al.* 1984b). Osmotic adjustment is not limited to the accumulation of a single carbohydrate but generally the disaccharides, sucrose or trehalose, are synthesized more rapidly than heteroside. The maintenance of cellular organization at an ultrastructure level may explain how cells recover from a state of desiccation. The stability of membranes is connected with the presence of sugars, particularly trehalose, and an increase in fluidity is due to a special type of fatty acid. Also especially-stressed hydrophobic proteins could maintain a reduced environment within cells. This may prevent protein aggregation or denaturation. However, the retention of water within the immediate cellular environment is of critical suggestion.

For the majority of terrestrial algae, the strongly oscillating de- and rehydration forces are also of critical significance for growth. Although water is an absolute discriminatory factor for growth, present knowledge of resistance to desiccation suggests that distinct groups of organisms have different physiological strategies towards the changing availability of water. This chapter has tried to simplify these various aspects of the problem to one basic simple model. Several authors (Brown 1976, Bewley 1979, Borowitzka *et al.* 1980, Bewley and Krochko 1982, Drews and Weckesser 1982, Reed and Stewart 1988, De Winder 1990, etc.) have reviewed the problem of how algae protect themselves against desiccation. However, it is a very complicated story and very intense research is still going on.

5. Salt (Toxicity)

As mentioned above, water stress and salt stress are closely interrelated. While water stress is most common in terrestrial algae, salinity stress is typically a local, rather than a global factor in coastal regions, especially within the intertidal zone, estuarine rock pools, salinas or mineral and hot springs, and in a variety of similar salt-rich, shallow habitats. Here, special microbial mats develop and algae are often the main component of these communities. The concentrations of sodium chloride, other salts, heavy metals and nutrients often reach unusually high levels (Dexter-Dyer *et al.* 1984). In addition, temperature, light and pH can also fluctuate erratically within a wide range. Apart from these natural habitats, there is also a variety of polluted localities influenced by the emissions from mining and smelting industries, or chemical factories, and by the widespread use of metal contaminants, etc.

This section is closely linked with the previous one on water deficiency (water stress). While the previous section summarized the present information on the ecology and physiology of stress due to dryness or salinity, salt toxicity will be discussed here. A wide variety of review articles emphasizes various aspects of the ecology and physiology of salt stress (Gesner and Schramm 1971, Raven 1975, Kauss 1977, Zimmermann and Steudle 1980, Wyn Jones and Gorham 1983, Munns *et al.* 1983, Kirst 1990, etc.).

Algae occurring in chronically-polluted environments (with salts, heavy metals and other potential toxicants) have a tendency to accumulate these chemicals within their cells, sometimes to a dangerous degree. The pollutant contents in algal cells can even be used as an indicator of water pollution. Several reviews, by Whitton 1970, Leland *et al.* 1979, Cain *et al.* 1980, Rai *et al.* 1981, Peterson *et al.* 1984, Peterson 1993, etc., have been published on this subject.

As mineral ions, including those of heavy metals, are needed by living cells for various metabolic processes, extremely varying strategies are adopted by algae to avoid, restrict or alleviate toxic conditions, i. e., high concentrations of the ions. No general principles governing the restriction of toxic conditions have yet been described. Restrictive processes will be directed by the species, and by the chemistry of the ions being accumulated. The mechanisms involved in plant tolerance range from chelation with organic ligands and/or biotransformation of inorganic ions to organic substances, to changes in the rates of physiological and biochemical processes, such as differences in enzyme specificity, ion transport, etc. However, in the latest research on metal phytotoxicity it has been proposed that metal toxicity leads to the oxidative breakdown of the lipid fraction of biomembranes (Van Addche *et al.* 1990).

In addition to the above, several ecological factors affect and determine the toxicity concentration for poisoning by various ions. For example, the toxicity of heavy metals at low pH is greatly increased. On the other hand, water hardness, the concentration of calcium, magnesium and phosphorus, can alleviate the toxicity of metals (Peterson *et al.* 1984).

6. Conclusions

The aim of this chapter was to briefly summarize our present knowledge concerning algal adaptation, and acclimatization processes to both stable and unstable environmental extremes such as temperature, light (UV irradiation), lack of water (and dryness) and mineral salts (and toxicity). During the last decade, these processes were widely studied because of climatic and antropogenic changes of the world ecosystem and also because of our curiosity about the “Martian ecosystem”. Algae are often used as model organisms for such type of studies because of their experimental simplicity. Although, studies of the light adaptation - acclimatization processes have almost one hundred year history, our knowledge of these processes are still very limited. The lack of information is due to the extremely high diversity of algal species and life forms or strategies. However, it is possible to summarize the problem of algal adaptation and acclimatization under the following points:

A. Changes in species diversity in time (reduction of species diversity, various migration strategies, etc.).

B. Changes in the rate of physiological and biochemical processes (reduction of photosynthesis, respiration, etc.).

C. Changes in the biomembrane structure/composition and consequently changes in their fluidity (extremes of low and high temperature, desiccation, salt and metal toxicity, etc.).

D. Production of special protective substances (sunscreening pigments and amino acids against excessive light, enzymes protecting against protein denaturation, carbohydrates or hydrophobic proteins as osmotic adjustments, and chelation or biotransformation of toxicants into organic substances, etc.)

However, it should be noted that light extremes have special protective models, which is due to the special evolution history of photosynthetic apparatus.

7. Acknowledgement

This work was funded by the Grant Agency of the Czech Republic (Grant 204/93/1177). The author wishes to thank Dr. J. Květ and Dr. L. Kováčik (Institute of Botany, Academy of Sciences of the Czech Republic, Trebon) for critical comments. The recommendation of Dr. O. Lhotsk² (editor of Algological Studies) to write these chapter is also appreciated.

8. References

Alexander, M (1976) Natural selection and the ecology of microbial adaptation in a biosphere, in M.R. Heinrich (ed.), *Extreme Environments. Mechanisms of Microbial Adaptation,* Academic Press, New York, San Francisco, London, pp. 3-56.

Allison, E. M. and Walsby, A. E. (1985) J. exp. Bot. 32, 241-249.

Barber, J. and Andersson, B. (1994) Nature, 370, 31-34.

Bewley, J. D. (1979) Ann. Rev. Plant Physiol. 30, 195-238.

Bewley, J. D. and Krochko, J. E. (1982) in: O. L. Lange, P. S. Nobel, C. B. Osmond and H. Ziegler (eds.), *Physiological plant ecology, II Encyclopedia Plant Physiology*, Vol. 12B, Springer Berlin, pp. 325-378.

Blumwald, E., Wolosin, J. M. and Packer, L. (1984) Biochem. Biophys. Res. Comm. 122, 452-459.

Boresh, K. (1921) Arch. Protistenkde, 44, 1-70.(1981)

Borowitzka, L. J (1981) in L. G. Paleg, D. Aspinall (eds.), *Physiology and Biochemistry of Drought Resistance in Plants*, Academic Press, Melbourne, pp. 97-130.

Borowitzka, L. J., Demmerle, S., Mackay, M. A. and Norton, R. S. (1980) Science 210, 650-651.

Broady, P. A. (1986) in: J. Pickard (ed.), *Antarctic Oasis. Terrestrial Environments and History of the Vestfold Hills*, Academic Press, New York, Sydney, pp. 165-202..

Brock, T. D. (1967) Science 158, 1012-1019.

Brock, T. D. (1986) *Thermophiles: General, Molecular and Applied Microbiology*, Wiley & Sons, New York.

Brown, A. D. (1976) Bact. Rev. 40, 803-846.

Cain, J. R., Paschal, D. C. and Hayden, C. M. (1980) Arch. Environm. Contam. and Toxicity 9, 9-16.

Castenholz, R. W. (1969) Bacteriological Reviewes, Vol. 33, No. 4, 476-504.

Castenholz, R. W. (1973) in: N.G. Carr and B.A. Whitton (eds.), *The Biology of Blue-green Algae*, Blackwell, Oxford, pp. 379-414 .

Clarke, A. (1991) Cold adaptation. J. Zool., Lond.on, 225, 691-699.

Dainty, L. N. (1976) in: U. Luttge and M. G. Pitman (eds.), *Encyclopedia of plant physiology*, NS Vol. 2A, Springer, Berlin, pp. 168-201.

De Winder, B. (1990) *Ecophysiological strategies of drought tolerant phototrophic micro-organisms in dune soils*, Academisch Profschrift, University of Amsterdam, pp. 95.

Dexter-Dyer, B., Kretzschmar, M. and Krumbein, W. E. (1984) J. Geol. Soc. London 141, 251-262.

Drews, G. and Weckesser, J. (1982) in: N.G. Carr and B.A. Whitton (eds.), *The biology of cyanobacteria.* Blackwell Scient. Publ., Oxford, pp. 333-359.

Elster, J. and Komárek, J. (1993) Arch. Hydrobiol./ Suppl. Algolog. Stud. 68, 107-122.

Elster, J., Komárek, J. and Svoboda, J. (1994) Scripta Fac. Sci. Nat. Univ. Masaryk. Brun., 24, (Geography), 13-24.

Elster, J. and Svoboda, J. (1996) Natl Inst. Polar Res., Spec. Issue, 51, 99-118.

Elster, J., Svoboda, J., Komárek, J. and Marvan, P. (1997) Arch. Hydrobiol./ Suppl. Algolog. Stud. 85, 57-93. Engelmann, T. W. (1883) Bot. Z. 41, 1-13.

Epstein, E. (1985) Plant Soil 89, 1-12.

Falkowski, P. G. and LaRoche, J. (1991) J. Phycol. 27, 8-14.

Friedmann, E. I. (1986) Adv. Space Res. 6(12), 265-268.

Friedmann, E. I. (1993) in: I.E. Friedman (ed.) *Antarctic Microbiology*, Wiley-Liss, New York, Chichester, Brisbane, Toronto, Singapore.

Friedmann, E. I. (1993a) Giornale Botanico Italiano, 127(3), 369-376.

Friedmann, E. I. (1993b) in: R. Guerrero and C. Pedr≤s (eds.), *Trends in Microbial Ecology*, Spanish Society for Microbiology, pp. 9-12.

Friedmann, E. I. and Galun, M. (1974) in: G.W. Brown (ed.), Desert Biology, Vol. 2, Academic Press, New York, pp. 165-212.

Friedmann, E. I., Hua, M. and Ocampo-Friedmann, R. (1993) J. Brit. Implement. Soc., 46, 291-292.

Friedmann, E. I. and Koriem, A. M. (1989) Adv. Space Res. 9(6), 167-172.

Friedmann, E. I., McKay, C. P. and Nienow J. A. (1987) Polar Biology 7, 273-287.

Friedmann, E. I. and Ocampo-Friedmann, R. (1984) Origin of Life 14, 771-776.

Fritsch, F. E. (1922) Ann. Bot. 36, 1-20.

Fritsch, F. E. and Haines, F. M. (1923) Ann. Bot. 37, 683-728.

Gaidukov, N. (1902) Abh. Preuss. Akad. Wiss. Berlin, No. 5, Quoted by Bogorad (1975).

Galinski, E. A. and Tindall B. J. (1992) in: R.H. Herbert and R. J. Sharp (eds.), *Molecular Biology and Biotechnology of Extremophiles*, Blackie & Son, Glasgow, pp. 76-114.

Garcia-Pichel, F. and Castenholz, R. W. (1993) Appl. Environ. Microbiol. 59, 163-169.

Gessner, F. and Schramm, W. (1971) in: O. Kinne (ed.), *Marine Ecology*, Vol. 1 (2) Environmental Factors, London, Wiley Interscience, pp. 705-1083.

Gerdes, G., Krumbein, W. E. and Holtkamp, E. (1985) in G. M. Friedman and W. E. Krumbein (eds.), *Hypersaline Ecosystem. The Gavish Sabkha*, Springer-Verlag, Berlin, Heidelberg, New York, Tokyo, pp. 219-238.

Harder, R. (1923) Z. Bot. 15, 305-361.

Harder, R. (1923) Z. Bot. 15, 305-360.

Hoham, R. W. (1975) Arctic Alpine Res. 7, 13-24.

Karentz, D., Cleaver, J. E. and Mitchell, D. L. (1991) J. Phycol., 27, 326-341.

Kauss, H. (1977) in: D. H. Northcote (ed.), *International Review of Biochemistry, Plant Biochemistry II*, Baltimore, University Park Press, pp. 119-140.

Kirk, J. T. O. (1983) in: *Light and Photosynthesis Aquatic Ecosystems.* Cambridge University Press.

Kirk, J. T. O. (1994) *Light and Photosynthesis in Aquatic Ecosystem.* Cambridge University Press.

Kirst, G. O. (1990) Annu. Rev. Plant Physiol. Plant Mol. Biol. 41, 21-53.

Kirst, G. O. and Wiencke, Ch. (1995) J. Phycol. 31, 181-199.

Lauchli, A. and Epstein, E. (1990) in: K. K. Tanji (ed.), *Agricultural Salinity Assessment and Management*, ASCE Manuals and Reports on Engineering No. 71, New York, Society of Civil Enginners, pp. 113-137.

Lawlor, D.W. and Keys, A.J. (1993) in: L. Fowden, T. Mansfield and J. Stoddart (eds.), *Plant Adaptation to Environmental Stress*, Chapman & Hall, London, Glasgow, New York, Tokyo, Melbourne, Madras, pp. 85-106.

Leland, H. V., Luoma, S. N. and Fielden, J. M. (1979) J. Water Pollution Control Feder. 51, 1952-1616.

Lesser, M. P., Cullen, J. J. and Neale, P. J. (1994) J. Phycol. 30, 183-192.

Marre, E. (1962) Temperature, in R. A. Lewin (ed.), *Physiology and Biochemistry of the Alage*, Academic Press, New York, pp. 541-550.

McKay, C. P. (1993) in: Friedmann, I. E. (ed.), Antarctic Microbiology, Wiley-Liss, New York, pp. 593-601.

McKay, C. P., Friedmann, E. I., Wharton, R. A. and Davis, W. L. (1992) Adv. Space Res. 12(4), 231-238.

Morita, R. Y. (1975) Bacteriol. Rev. 29, 144-167.

Munns, R., Greenway, H. and Kirst, G. O. (1983) in: O. Lange, P. S. Nobel, C. B. Osmond and H. Ziegler, *Encyclopedia of Plant Physiology* (NS), Vol. 12C, Physiological Plant Ecology III, Springer, Berlin, Heidelberg, New York, pp. 59-135.

Peschek, G. A, Wastyn, M., Molior, V., Kraushaar, H., Obinger, C., and Matthijs, H. C. P. (1989) in: Kotyk *et al.* (eds.), *Highlights in modern biochemistry*, Vol. I, VSP Int. Sc. Publ. Zeist, The Netherlands, pp. 893-902.

Peterson, P. J. (1993) in: L. Fowden, T. Mansfield and J. Stoddart (eds.), *Plant Adaptation to Environmental Stress*, Chapman & Hall, London, Glasgow, New York, Tokyo, Melbourne, Madras, pp. 155-171.

Peterson, G. H., Healey, F. P. and Wagemann, R. (1984) Can. J. Fish. Aquat. Sci. 41, 974-979.

Pokorn², J., Elster, J. and Hammer, L. (1994) Arch. Hydrobiol., Suppl. Algolog. Stud. 73, 99-110.

Post, A. and Larkum, A. W. D. (1993) Aquatic Botany 45, 231-243.

Rai, L. C., Gaur, J. P. and Kumar, H. D. (1981) Biol. Rev. 56, 99-151.

Raven, J. A. (1975) Algal cells, in: D. A. Baker and J. A. Hall (eds.), *Transport in Plant Cells and Tissues*, Amsterdam, North-Holland, pp. 125-60.

Reed, R. H. (1984) Cell Env. 7, 165-170.

Reed, R. H. (1990) in: K.M. Cole and R.G. Sheath (eds.), *Biology of the Red Algae*, Cambridge Univ. Press, Cambridge, pp. 147-170.

Reed, R. H., Chudek, J. A., Foster, R. and Stewart, W. D. P. (1984a) Arch. Microbiol. 183, 333-337.

Reed, R. H., Richardson, D. L., Warr, S. R. C. and Stewart, W. R. C. (1984b) J. gen. Microbiol. 130, 1-4.

Reed, R. H. and Stewart, W. D. P. (1985b) Mar. Biol. 88, 1-9.

Reed, R. H. and Stewart, W. D. P. (1988) in: L. J. Rogers and J. R. Gallon (eds.), *Biochemistry of Algae and Cyanobacteria*. Proc. Phytochem. Soc. Europe 28, Claredon Press, Oxford, pp. 217-231.

Reed, R. H. and Walsby, A. E. (1985) Arch. Microbiol, 143, 290-296.

Richardson, K., Beardall, J. and Raven, J. A. (1983) New. Phytol., 93, 157-191.

Russell, N. J. (1990) in: R. M. Laws and F. Franks (eds.), *Life at Low Temperatures*, London, The Royal Society, pp. 79-92.

Stokes, J. L. (1940) Soil Science 49, 171-184.

Sullivan, C. W., Palmisano, A. C., Kottmeier, S., McGroath Grossi, S. and Moe, R. (1985) in: W.R. Siegfried, P.R. Condy and R.M. Laws (eds.), *Antarctic Nutrient Cycles and Food Webs*, Springer-Verlag, Berlin, Heidelberg, New York, pp. 78-83.

Van Assche, F., Vangronsveld, J., and Clijsters, H. (1990) in: J. Barcelo (ed.), *Proc. 4th Internat. Conf. on Environmental Contamination*, Edinburgh, CEP Consultants, pp. 246-250.

Vincent, W. F. (1988) *Microbial ecosystem of Antarctica*, Cambridge University Press, London.

Vincent, W. F. and Roy, S. (1993) Environ. Rev. Vol. 1, 1-12.

Vincent, W. F. and Quesada, A. (1994) Antarctic Research Series, Vol. 62, 111-124.

Whitton, B. A. (1970) Phykos 9, 116-125.

Wyn Jones, R. G. and Gorham, J. (1983) Plant Cell Environ. 9, 35-58.

Wynn-Williams, D. D. (1990) in: K. C. Marshall (ed.), *Advances in Microbial Ecology*, Plenum, New York, pp. 71-146.

Wynn-Williams, D. D. (1994) Antarctic Research Series, Vol. 62, 243-257.

Xiong, F., Lederer, F., Lukavsk², J. and Nedbal, L. (1996) J. Plant Physiol. Vol. 148, 42-48.

Zimmermann, U. and Steudle, E. (1980) in: R. M. Spanswick, W. J. Lucas and J. Dainty (eds.), *Plant Membrane Transport: Current Conceptual Issues*, Elsevier, North-Holland Biomedical Press, Amsterdam, New York, Oxford, pp. 113-127.

IV. MICROORGANISMS IN EXTREME ENVIRONMENTS

Biodata of **Dr. Valerie Flechtner**, author of "Enigmatic Desert Soil Algae".

Dr. Valerie Flechtner, Professor at the Department of Biology at John Carroll University, University Heights, Ohio. Her research focuses on the algal communities present in crusted and non-crusted soils from arid and semiarid regions of North America. She is interested in determining the level of diversity in crust communities, whether common taxa exist in all crust communities, how similar crust communities from different hot deserts are to one another, how similar the crust communities of hot deserts are to the communities in cold deserts, and whether it is possible to correlate the presence or absence of particular algal taxa with specific physical or chemical soil characteristics. Dr. Flechtner's results suggest that while there are some taxa which appear in most of the locales she has examined (three cold desert sites and five hot desert sites), each locale does have some distinctive aspects to its algal flora, and new taxa have been isolated from several of the sites. The newest thrust of her research is to use nucleotide sequence data to determine the taxonomic relatedness of two groups of cyanobacteria (the Scytonemataceae and members of the genus *Nostoc*) both within and between sites.

E-Mail: <**flechtner@jcvaxa.jcu.edu**>

J. Seckbach (ed.), Enigmatic Microorganisms and Life in Extreme Environments, 231–241.

ENIGMATIC DESERT SOIL ALGAE

Soil algal flora of the western U.S.A. and Baja California, Mexico

V. R. FLECHTNER
Department of Biology
John Carroll University
University Heights, Ohio, 44118

1. Introduction

Vast expanses of arid and semiarid land characterize many areas of the world including the western United States of America (U.S.A.) and Mexico. Subject to high temperatures and limited water availability, these areas would appear to offer inhospitable habitats for living organisms, and indeed vascular plant growth is sparse compared to more temperate regions (Skujins 1995). But microbial communities are present in even the harshest environments, often forming water-stable surface soil aggregates known as cryptogamic or microbiotic crusts. Composed of bacteria, microalgae, fungi, protozoans, and lichens, crust communities contribute to the environment in many ways including: (1) soil particle formation (2) production of organic products through photosynthesis (3) nitrogen fixation and (4) nutrient cycling (Reviewed by Johansen 1993). It may well be that the most important contribution made by cryptogamic crusts is to stablizie the soil by preventing erosion.

Interest in understanding the microbial ecology of arid soils with the possible practical application of manipulating microbial communities to increase soil arability is growing. Because both prokaryotic and eukaryotic microalgae play important roles in crust formation and physiological activity (Johansen 1993), our laboratory has pursued floristics of algae present in crust communities. My purpose in this paper is to present a review of the literature regarding edaphic (soil surface) algae in North American deserts and then to discuss our recent findings including characterizations of several new and enigmatic algal taxa.

2. Review of the Literature

Prior to 1980, the majority of the work on characterization of the microalgae present in arid and semiarid soils in the western U.S.A. and Mexico was done primarily by three main research groups: Roy Cameron and co-workers, Harold Bold and co-workers, and several researchers associated with Brigham Young University, most notably Sam Rushforth and Jeff Johansen.

2.1. ROY CAMERON

Cameron (1964, 1969) investigated the abundance and nature of terrestrial algae in a number of deserts world wide. A floristic study of the terrestrial algae in southern Arizona includes both edaphic (growing on the soil surface) and endolithic (growing in rocks) algae and records over 60 different species, including 50 cyanophyte taxa, 8 chlorophyte taxa and 8 diatom taxa (Cameron 1964). His quantitative studies of deserts in the western U.S.A. and Mexico (Cameron 1969) show algal concentrations ranging from 10^2 to 10^6 algae/g soil; minimal floristic data are presented.

2.2. HAROLD C. BOLD

One of the most active North American laboratories devoted to the study of edaphic algae was that of Harold C. Bold of the University of Texas. In an extensive research effort which spanned 20 years, Bold focused on the identification of new algal taxa and the careful delineation of a variety of non-diatom algal groups. During the 1960's and 1970's, Bold and his students published a series of monographs; several of these focused on one or several genera of soil algae (Archibald and Bold 1970, Bischoff & Bold 1963, Brown & Bold 1964, Chantanachat & Bold 1962, Deason & Bold 1960, Groover & Bold 1969, Kantz & Bold 1969, Lee & Bold 1974). This work significantly extended the published description of terrestrial algae and provided an important resource for investigators working in North America.

2.3. SAM RUSHFORTH AND JEFF JOHANSEN

Work initiated by Sam Rushforth at Brigham Young University and extended by Jeff Johansen and co-workers at John Carroll University focused primarily on algae present in cryptobiotic crusts with the intent of determining the ecological importance of these communities. These investigators studied semiarid deserts and shrub-steppe communities of the Great Basin of Utah (Anderson & Rushforth 1976, Johansen & Rushforth 1985, Johansen et al. 1993) and the Lower Columbia Basin of Washington State (Johansen et al. 1993). The identification of cyanobacteria and diatoms dominate the initial studies. Rushforth and Merkley (1988) published a comprehensive list of the algae of Utah which record nearly 1,900 taxa at the species and varietal level. Of these taxa, nearly 1000 are diatoms and another 550 are green algae. However, few of the studies cited in the review considered non-aquatic habitats and of the 1900 taxa identified, only 160 taxa are identified from terrestrial environments. These include: 43 cyanophyte taxa, 2 euglenophyte taxa, 12 chlorophyte taxa, 1 xanthophyte taxon and 102 diatom taxa.

In the most thorough characterizations of a single terrestrial site published to date, Johansen, examining a cryptogamic crust from a sagebrush steppe community in the Lower Columbia Basin near Richland, Washington, identifies 90 different algal taxa including 47 chlorophyte, 26 cyanophyte, 9 xanthophyte and 8 diatom taxa (Johansen et al. 1993). Unlike previously studied crusts, this algal community is dominated by chlorophytes rather than cyanophytes.

3. Recent Work of Valerie Flechtner and Jeff Johansen

During the last five years, a research team headed by Jeff Johansen and myself has examined the algal communities in cryptogamic crusts and non-crusted soils collected from seven distinct desert locales in North America. Three of these areas are semiarid shrublands located in the state of Utah and include: (1) Dugway Proving Ground (DPG), an army facility in the Great Basin Desert, (2) Buckhorn Wash (BHW), a canyon in the San Rafael Swell (SRS) which is part of the Colorado Plateau and (3) Wedge Overlook (WO) in the SRS, located about 25 miles west of BHW.

Three of the other four locales are mainland hot, arid deserts where rainfall is less than 10 cm annually. Two are located in the Sonoran desert; one site is the Cataviña region of the Central Desert in Baja California, Mexico (BAJA) and the other the Yuma Proving Ground army facility in Yuma, Arizona (AZ). The third arid, hot desert site is in the Chihuahuan desert at Fort Bliss, an army facility in New Mexico (NM). The last site is San Nicolas Island (SN) off the coast of Los Angeles, California.

3.1. METHODS

Isolation of algae involved suspending soil samples in buffer and applying aliquots of suitable dilutions to agar solidified Z-8 Medium (Carmichael 1986) for cyanophytes or Bold's Basal Medium (BBM, Bold & Wynne 1978) for eukaryotic algae. Prokaryotic algae were typically identified by direct microscopic observation of clonal isolates from the initial platings. Identification of eukaryotic algae required subculturing so that the life cycle of the alga could be observed.

3.2. QUANTITATIVE RESULTS

Soil algal densities in our samples ranged from 3.3×10^2 in the NM site to 1.1×10^6 algae/g at DPG. These densities are similar to those reported by other investigators (Cameron 1969). As might be expected, the algal densities were lowest in the hot dry deserts found in the southwestern U.S.A. and highest in the semiarid desert sites in Utah (Table 1).

3.3. ALGAL DIVERSITY

We observed surprisingly diverse algal communities in all desert sites. The greatest difference between the hot, arid deserts and the semiarid sites is in the diversity of the prokaryotic algae. The mainland hot desert sites (AZ and NM) had fewer than half the cyanobacterial taxa observed in the Utah sites (Table 2).

Several algal genera were widespread, appearing in every site when cultures were plated on Z-8 and BBM. The cyanophyte *Schizothrix calcicola* was found in every site which we examined and was often the most abundant cyanobacterial taxon. *Microcoleus vaginatus* and members of the genera *Nostoc* and *Scytonema* also were recovered from most of the sites.

Table 1. .Algal density from seven North American desert locales. (Flechtner and Johansen, unpublished data)

Site	Total algae/gm soil
SN[a]	$3.1 X 10^3 - 3.0 X 10^5$
BAJA	$8.3 X 10^3 - 6.4 X 10^4$
AZ	$1.3 X 10^3 - 6.9 X 10^3$
NM	3.5 X 102 - $1.6 X 10^5$
DPG	$2.9 X 10^4 - 1.1 X 10^6$
BHW	$3.3 X 10^2 X 2.3 X 10^5$
WO	$5.6 X 10^3 X 2.6 X 10^5$

[a]Site abbreviations: SN, San Nicolas Island, CA; BAJA California, MX; AZ, Arizona; NM, New Mexico; DPG, Dugway Proving Ground, UT; BHW, Buckhorn Wash, SRS, UT; WO, Wedge Overlook, SRS, UT.

Among the non-diatom eukaryotic algae recovered, members of the genera *Myrmecia, Chlorosarcinopsis, Bracteacoccus, Chlorella, Stichococcus* and *Diplosphaera* were found in most sites. Members of the genera *Diplosphaera* and *Myrmecia* were the most frequently recovered algae in this study. The high frequency of *Myrmecia* isolates is quite different from the flora reported by other investigators who cite *Bracteacoccus, Chlorococcum, Chlorosarcinopsis,* and *Spongiochloris* as the most abundant taxa.

The data presented in Table 3 demonstrate that each of the sites examined exhibited a distinctive profile of algal flora. The BHW and DPG sites in Utah shared a flora in which the genera *Tolypothrix, Schizothrix* and *Xanthonema* were abundant. They werealso characterized by the greatest species richness and are the only sites in which members.

Table 2. Number of species by algal division observed in soils from seven North American desert locales. (Flechtner and Johansen, unpublished data.)

	Arid				Semiarid		
Division	BAJA[1]	AZ	SN	NM	DPG	BHW	WO
Chlorophyta	35	23	23	14	37	34	14
Cyanophyta	18	19	23	10	42	49	47
Eustigmatophyta	1	0	1	0	0	0	0
Xanthophyta	2	1	1	0	8	8	0

[1]See Table 1 for site abbreviations

of the genus *Xanthonema* were isolated. Dominant chlorophytes included *Chlorella vulgaris* and members of the genera *Stichococcus, Diplosphaera* and *Bracteacoccus.* The

WO site yielded many *Nostoc* isolates; we observed high cyanophyte diversity and low chlorophyte diversity in this site.

The Baja and San Nicolas Island sites were the only sites in which eustigmatophytes were isolated. Eustigmatophytes are characterized by the presence in vegetative cells of a polygonal pyrenoid projecting from the face of the chloroplast by a stalk and the production of flask-shaped, uniflagellate zoospores with prominent anterior eye spot. The Arizona site was unusual in the high frequency of *Coccomyxa confluens*. Also noteworthy is the absence from this site of the genera *Diplosphaera, Myrmecia,* and *Stichococcus* which predominate several of the other sites. The Baja site yielded several enigmatic algal taxa (see below), some of which were found in several plots within a one square kilometer area. The New Mexico site had the lowest algal density and had the fewest number of species. It was the only site from which no isolates of *Bracteacoccus* were recovered and in which *Chlorococcum* was the predominant genus.

3.4. ENIGMATIC ALGAE.

We have isolated many taxa during the course of our studies which do not fit recognized taxonomic designations (unpublished data). Descriptive notes on several of these taxa are presented below; manuscripts containing formal descriptions of these organisms are in review or in preparation.

3.4.1. *Isolate 1* (Figs. 1 and 2).

Regularly spiraled, slightly tapering filaments, 16-20 μm at their widest point and about 75-80% of maximum diameter at tapered end; distance between regular spirals 8-12(-16) μm. Sheath thin in young cultures, thickening in older cultures; clear. Single and double false branching in conjunction with and independent of heterocyst formation. Trichomes slightly.
constricted to not constricted at the cross walls. Filaments (cells plus sheath) 8-13 μm wide, trichomes (cells alone) 6.5-9 μm wide and 1/3-1X as long as wide. Heterocyst formation intercalary (in the internal part of the trichome) and basal (at the end of the trichome), mostly single, occasionally double, rarely triple. Site: San Rafael Swell Wedge Overlook. This taxon was isolated from a highly crusted, undisturbed site in the WO. We believe it is a new genus in the family Scytonemataceae sensu Castenholtz.

3.4.2. *Isolate 2* (Fig. 3)

Cells rectangular, 4-5 μm wide by 6-20 μm long; cells solitary, pairs or short chains (eight or less). Chloroplast parietal, lining rim of cell; no pyrenoid. Some cells irregular or slightly bent. Site: DPG, UT. This taxon, is similar to *Xanthonema hormoides*; it differs from the nominate variety in cell length (our isolate has longer cells) and the presence of irregularly shaped cells.

3.4.3. *Isolate 3.* (Fig. 4)

Cells oval, 5-6.5 μm in diameter and 6-14 μm long with tapering ends; appears triangular, especially during cell division. Elongate, parietal chloroplast with indistinct pyrenoid. Cells embedded in amorphous mucilage matrix. Reproduction through autospores. Site:

Table 3. Relative abundance of algal taxa in seven North American desert sites. (Flechtner and Johansen, unpublished data)

Algal taxon	DPG[1]	BHW	WO	SN	NM	AZ	BAJA
Cyanophyte	NO[2]	P	F	F	NO	NO	NO
Microcoleus vaginatus	F	F	F	F	F	F	F
Nostoc	F	F	F	F	R	R	R
Schizothrix calcicola	F	F	F	F	F	F	F
Scytonema	NO	P	P	P	F	R	R
Tolypothrix	F	P	P	F	NO	R	NO
Chlorophyta							
Bracteacoccus	P	F	P	P	NO	R	F
Chlorococcum	NO	NO	R	NO	F	P	NO
Chlorella vulgaris	F	F	P	P	F	R	R
Chlorosarcinopsis	P	P	NO	R	F	F	F
Coccomyxa confluens	P	NO	NO	NO	NO	F	NO
Diplosphaera	F	F	F	F	P	NO	R
Myrmecia	P	P	F	F	F	No	F
Spongiochloris	NO	NO	NO	NO	F	P	R
Stichococcus	F	P	P	P	NO	NO	R
Eustigmatophytes	NO	NO	NO	F	NO	NO	F
Xanthophytes							
Xanthonema	F	P	NO	NO	NO	NO	NO

[1]See Table 1 for site abbreviations
[2]NO = not observed; F = frequent; P = present; R = rare

Baja California, Mexico. This isolate may be a member of the genus *Elakatothrix*; of the 15 species in this genus, only two species are terrestrial (Komárek and Fott 1995). Our species differs from both of these reported species by having shorter cells which are not as clearly tapered as is typical for the genus; it is most similar to *E. gelatinosa*.

3.4.4. *Isolate 4.* (Fig. 5)

Cells 3.5-6.5 µm wide and 6-8.5 µm long. Cells arranged in masses, cubical packets or tetrads with tightly appressed walls in young cultures; becoming nearly spherical as cultures age. Extracellular mucilage matrix present. Cell wall thin in young cultures, becoming thickened to 4 µm in senescent cultures. Parietal chloroplast with one or more eccentric pyrenoids associated with large starch granules. Granular cytoplasm. Biflagellate, walled, ellipsoidal zoospores with median to anterior stigma, parietal chloroplast and a posterior nucleus. Site: Baja California, Mexico. This isolate is probably a new species in the genus *Fasciculochloris*. The taxon is most similar to *F. boldii*, differing in the shape of the zoospore, the position of the stigma, and the absence of large oil droplets in vegetative cells.

3.4.5. Isolate 5. (Fig. 6)
Cells solitary, cylindrical with rounded ends, 10-15 μm wide, 14-56 μm long. Cell wall thin. Nucleus central. Chloroplast elongated with longitudinal, lobed ridges, constricted in the center to form two halves, each with a pyrenoid. Zoospores not observed. Site: Baja California, Mexico. Isolate 5 appears to be a new variety of *Cylindrocystis brebisonii* differing from the nominate variety *C. brebisonii* Menegh. in cell size and chloroplast morphology. This is the first report on any *Cylindrocystis* species from North American desert soils.

3.4.6. *Isolate 6.* (Fig. 7)
Single cells spherical to subspherical, 5-8μ diameter. Parietal chloroplast with an indistinct pyrenoid. Cells appear primarily in flat clusters composed of tetrads; short chains, pairs and single cells also visible. Reproduction by autospores only. This isolate was very widespread and often is the predominate taxon at a site (e.g. DPG). It is similar to *Diplosphaera chodatii* in that it possesses an indistinct pyrenoid and vegetative cell division resulting in the formation of packets and short chains. But, our isolate differs from the published drawings of *D. chodatii* (Ettl & Gartner 1995 Fig.175b) in that it does not form ellipsoidal cells. It also shares many similarities with vegetative members of the genus *Desmococcus* (Ettl & Gartner 1995, Fig. 176d); however, production of warty sporangia, a diacritical characteristic of this genus, has never been observed. Finally, cells of our isolate are morphologically similar to the drawings of three undescribed species of *Protococcus* which have been observed in lichen symbionts (Ettl & Gartner 1995, Fig. 177c). Since no written descriptions accompany the drawings, further comparisons are not possible.

4. Discussion.

Our experience in working with soils from a variety of arid and semiarid sites in the western USA and Mexico is consistent with the observations of Bold (Deason and Bold 1959): organisms that we identify as new taxa because they do not fit closely previously published descriptions can be isolated from most sites. Identification of these taxa is important for those investigators who are interested in arid and semiarid regions of North America. Few comprehensive studies of the algal flora of North American desert soils exist. Our work will significantly increase the number of North American taxa identified and facilitate comparative studies.

It will be interesting to determine the degree of similarity of the algal flora of sites with similar soil chemistry and climatic conditions. Our studies suggest that some algal taxa are indeed widespread, being isolated from all or most of the sites in this study. But it also appears that each site has some distinctive character to its algal flora. As more sites are studied in depth, it may be possible to determine whether specific environmental and soil parameters are correlated with the presence or absence of specific algal taxa.

Taxonomic diagnosis of isolate six has been particularly challenging. Prior to 1980, many coccoid aerophytic green algae are identified as members of the Protococcales and the most frequently encountered organism is identified as *Protococcus viridis* (e.g., Johansen and Rushforth 1981). More recently, the recognition of *Protococcus* as a legitimate genus has been called into question. In their discussion of Protococcales, Komarék and Fott state that the use of this taxon is illogical and is inadmissible according to the International Code of Botanical Nomenclature (Komarék & Fott 1993 p. 21). They equate the species *P. viridis* to *Desmococcus vulgaris* but provide no further discussion. In their key differentiating the genera of lichenized algae, Ettl and Gärtner (1995) treat "*Protcoccus*" as a "Sammelgattung" or collection of genera including *Appatococcus* and *Desmococcus*. They place some previously described members of the genus (e.g. *P. vulgaris*) in the genus *Desmococcus* and others (e.g. *P. chodatii*) in the genus *Diplosphaera*. Additionally, illustrations of several poorly described *Protococcus* species are provided. Given the frequency with which we have recovered isolate 6 from both cold and hot deserts, it is clear that this alga is a common inhabitant of the soil and it is probable that earlier investigators would have encountered this organism, most likely reporting it as "*Protococcus*". Because little descriptive information is included with earlier citations of the organism(s) identified as "*Protococcus*", it is unlikely that discerning which of the isolates represent members of the genus *Desmococcus* and which represent members of the genus *Diplosphaera* will be possible.

Identification of algal species is often an inexact science. Cameron cautioned that species lists compiled by authors working in different parts of the world is problematic because "...the criteria for naming and identifying species depends too much on the bias of the investigator" (Cameron 1964). One problem for phycologists in the U.S.A. is that many of the authoritative taxonomic sources for both prokaryotic and eukaryotic algae were written by Europeans, frequently in German; these sources have not yet been translated into English. Since historically, many of the phycologists interested in taxonomy were European, a majority of the taxa described in these sources were isolated in Europe (Geitler 1930-2, Komárek & Fott 1983, Ettl & Gärtner 1995). It is not unusual to find that alga taxa isolated from North American soils do not match the published descriptions of algae from other parts of the world. Identification is particularly challenging for those algae which do not form zoospores as part of their life cycle. In some cases it is difficult to even determine to which class these algae belong. Determination of whether or not new isolates are, in fact, new taxa or whether they represent genetic variants of previously described taxa may require molecular analysis.

5. References

Anderson, D.C. and Rushforth, S. R.. (1976): *Nova Hedwigia* 28:691-729.
Archibald, P.A. and Bold, H. C. (1970). *Univ. of Texas Public.* 7015.
Bischoff, H.W. and Bold, H. C. (1963). *Univ. of Texas Public.* 6318.
Brown, R.M. Jr. and Bold, H. C. (1962) *Univ. of Texas Public.* 6417.
Cameron, R.E. (1960) *J. Ariz. Acad. Sci.* 1:85-8.
Cameron, R.E. (1969) Jet Propulsion Lab. Tech. Rep. 2-1378.
Carmichael, W.W. (1986) *Fundamental Research in Homogenous Catalysis*, Gordon and Breach, NY.
Chantanachat, S. and Bold, H.C.. (1962) *Univ. of Texas Public.* 6218.
Deason, T.R. and Bold, H. C. (1959) *Univ. of Texas Public.* 6022.
Ettl, H. and Gärtner, G. (1995) *Syllabus der Boden-, Luft- und Flechtenalgen*, Gustav Fischer Verlag,

Stuttgart..
Geitler, L. (1930-2) Koeltz Scientific Books, Koenigstein, Germany. 1196 pp.
Groover, R.D. and Bold, H. C. (1969) *Univ. of Texas. Public.* 6907.
Johansen, J.R. (1993) *J. Phycology* 29:140-7.
Johansen, J.R., Ashley, J. and Rayburn, W. R. (1993) *Great Basin. Natur.* 53: 73-88.
Johansen, J.R. and Rushforth, S. R. (1981) *Nova Hedwigia* 34:333-90.
Kantz, T. and Bold, H. C. (1969) *Univ. of Texas Public.* 6924.
Komarék, J. and Fott, B. (1983) *Das Phytoplankton des Süßwassers, 7, Schweizerbart, Stuttgart.*
Lee, K.W. and Bold, H. C. (1974) *Univ. of Texas Public.* 7043.
Skujiņš, J., Ed. (1991) *Semiarid Lands and Deserts.* Marcel Dekker, Inc.

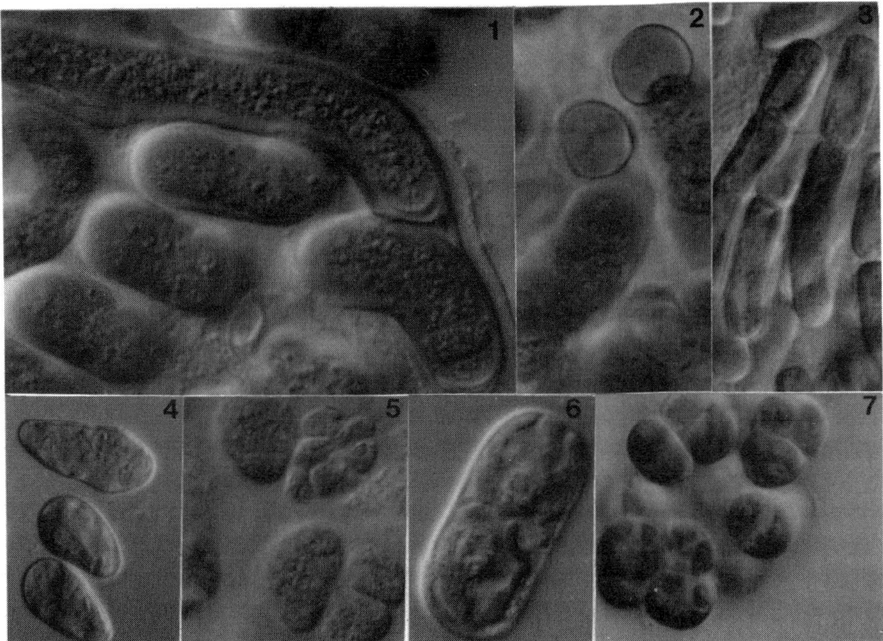

Figure legends. All figures taken at 1000X magnification.. Fig. 1, 2, Isolate 1. Fig. 1. Note spiraling filaments and false branching. Fig. 2. Basal heterocysts. Fig. 3. Isolate 2. Fig. 4, Isolate 3. Note trinagualr shape. Fig.5, Isolate 4, Zoospores, upper right; vegetative cells, lower right. Fig. 6, Isolate 5. Note bipartate, lobed chloroplast. Fig.7. Isolate 6. Iodine stained to reveal faint pyrenoid.

Biodata of **Nocha Van Thielen**, contributing (with David J. Garbary) the chapter on: ***Life in the Rock: Endolithic Algae.***

Dr. Noch Van Thielen is at the Department of Botany, University of Toronto (Canada). and previously investigated the ecology and systematics of endolithic algae. Her current interest lies in the molecular genetic analysis of hormone signal transduction pathways of higher plants and on endophytic algae. She discovered a kelp gametophyte that grow in the cell wall of filamentous red algae.

Biodata of **David J. Garbary,** coauthor with Thielen of the chapter on: ***Life in the Rock: endolithic Algae.***

Dr. David J. Garbary is a Professor of Biology in St. Francis Xavier University at Antigonish, Nova Scotia, Canada. Dr. Garbary's background is primarily in the systematics, ecology and evolution of marine macroalgae. Current research interests include fractal modeling of algal development, cell biology of red algal morphogenesis, endolithic algae and the evolution of land plants. E-mail: dgarbary@juliet.stfx.ca

J. Seckbach (ed.), Enigmatic Microorganisms and Life in Extreme Environments, 243–253.

LIFE IN THE ROCKS - ENDOLITHIC ALGAE

N. VAN THIELEN[1] and D. J. GARBARY[2]

[1]*Department of Botany, University of Toronto, 25 Willcocks Street, Toronto, Ontario, Canada, M5S 3B2*

[2]*Department of Biology, St. Francis Xavier University, Antigonish, Nova Scotia, Canada, B2G 2W5*

1. Introduction

With their extreme exposure to light and extreme temperature fluctuations, rock surfaces would not appear to be a benign environment for algae (or any other organisms). One strategy used by some organisms to overcome these environmental constraints is to become endolithic, i.e., to survive within the rock matrix. Recent studies have shown that bedrock and any deep earth habitats that are sufficiently cool (upper limits of about 150°C have been suggested) are likely the home of extensive bacterial[1] populations (Gold 1992, Ghiorse 1997, see Gould 1996 for popular review). The extent of these bacterial populations was estimated at 2 x 10^{14} tons, equivalent to more than all surface flora and fauna combined (Gold 1992). Although much of this production is associated with deep aquifers (Szewzyk *et al.*, 1994, Stevens and McKinley 1995), >500 m deep in ocean sediments (Parkes *et al.*, 1994) or in deep oil reservoirs (L'Haridon *et al.*, 1995), some of this is associated with the interstices of actual rock, i.e., the endolithos (Gold 1992). The extreme tip of the endolithic 'iceberg' are the endolithic algae (see earlier reviews by Friedmann and Ocampo-Friedmann 1984, Bell 1993).

Photosynthetic algae, unlike most bacteria, have a particular environmental constraint, a requirement for light, which limits their colonization of endolithic substrata. Hence, if algae are to survive in rock, must live near the surface where light is sufficient to maintain photosynthesis. In this review we discuss the interior of rocks as a habitat for algae. The emphasis is on the biological diversity, ecology and physiological constraints of surviving in this poorly understood habitat. We limit the discussion to endolithic algae in rocks associated with terrestrial environments, although analogous situations exist in marine ecosystems (e.g., *Hyella*, Lukas and Golubic 1983, *Ostreobium*, Shashar *et al.*, 1997).

Within the endolithos there is a wide range of occurrence and degree of conspicuousness of algal communities. The most developed of these form a distinct green layer within the rock. These typically begin one to several millimeters below the rock surface, and are one to several millimeters thick. Such well developed communities have been described from a wide range of geographic and climatic areas (e.g., Negev Desert, Friedmann 1971; Antarctica, Friedmann 1982, southwestern United States, Bell *et al.*, 1986; Canada, Gerrath *et al.*, 1995, Garbary *et al.*, 1996). The biology of these algal communities will be the primary focus of this review.

[1] for the purposes of this paper, bacteria will refer to noncyanobacterial prokaryotes

2. The Endolith Environment and its Biota

2.1. CLASSIFICATION OF LITHOBIONTS

Organisms intimately associated with rock substrata are referred to as lithobiontic. Lithobiontic organisms are associated with the surface or the interior of rocks; here we focus on the latter forms which comprise the endolithic biota. Three types of relationship between the individual organism and its substratum were defined by Golubic *et al.* (1981) and Bell (1993). Chasmoendoliths live in cracks and fissures in the rock. Cryptoendoliths live in the pores of the rock, completely enclosed. Euendoliths create their habitat by actively burrowing into the rock. These habitats and their biotas are not mutually exclusive; for example, a euendolith may also be cryptoendolithic or chasmoendolithic. A euendolith or chasmoendolith in limestone may later become cryptoendolithic as the surrounding matrix is solubilized and then recrystalizes.

2.2. THE ROCK ENVIRONMENT

The environments in which endoliths have typically been found were previously considered unsuitable for algae. The pioneering work on these organisms was carried out in the Negev Desert (Friedmann *et al.*, 1967) and Antarctica (Friedmann and Ocampo 1976, Friedmann *et al.*, 1988, Broady 1981b). These surface environments are extremely inhospitable with extremes of temperature (> 50° C range) and little available water. However, by growing within the rock, the rigor of the external environment is avoided (Friedmann 1982). The ability to 'retreat' into the rock enables a major extension of habitat for algal growth. For example, in the Vestfolds of Princessland, Antarctica, only 9.1% of the sites had epilithic growth, whereas 75% of the sites had endolithic growth (Broady 1981a, b). On the Colorado Plateau >90% of sampled sites had endoliths (Bell *et al.*, 1986, 1988)

There has been a basic assumption that rocks suitable for endolithic algae are relatively soft or porous. Thus the bulk of the observations come from sandstone, limestone and gypsum. The fact that gypsum is both soft and porous may explain the high species diversity (42 species) found from only three sites in Nova Scotia (Garbary *et al.*, 1996). The recent characterization of an endolithic community in granite (Bell *et al.*, 1996) suggests that even extremely hard rocks with low porosity can allow the development of endoliths. In addition to hardness, colour plays an important factor; light rocks allow about four times the light penetration to the algal layer (5-66 versus <22 mol photons m^{-2} s^{-1} in light and dark sandstone respectively, Bell 1993). Consequently, light rocks are able to support greater algal diversity and apparent biomass (Bell *et al.*, 1986, Ferris and Lowson 1997). The most extreme example of this comes from two sites in Nova Scotia with light and dark coloured gypsum, the former with 30 species and an obvious algal layer, the latter with two species and no algal layer (Van Thielen *et al.*, in preparation). Rock colour may also be important in moderating temperature since dark rocks will absorb more radiant energy. For example, sandstones of the southwestern United States may have interior temperatures up to 12° C higher than the rock surfaces (Bell 1993).

2.3. BIOTA OF ENDOLITHIC COMMUNITIES

About 100 species from 70 algal genera have been described as endoliths from around the world. This is a highly conservative estimate of species diversity since in a recent study 13 of the 42 taxa identified were newly described as endoliths (Van Thielen *et al.*, in preparation). As more temperate zone communities are studied from nonarid sites, this number will likely expand considerably. The bulk of endolithic taxa are cyanobacteria and green algae, although several chrysophytes and diatoms also are reported (Table 1). Many taxa are free-living, although some form the phycobionts of cryptoendolithic lichens. Bell (1993) suggested that, in less extreme environmental conditions, green algae predominate over cyanobacteria. The endolithic gypsum flora in Nova Scotia (a moderate climate compared to that of the Negev Desert, Antarctica and Colorado Plateau) runs counter to this trend with five times as many species of cyanobacteria as green algae (Table 1).

The most widely reported endolithic genus is the cyanobacterium, *Chroococcidiopsis*, which is a component of all major endolithic communities described to date (Table 1). Other genera commonly present include *Gleothece*, *Myxosarcina* and *Plectonema* (cyanobacteria), and *Chlorella* and *Stichococcus* (green algae). The lichen-phycobiont green algal genus *Trebouxia* is a common endolith in both warm and cold deserts (Negev Desert and Antarctica, respectively, Friedmann *et al.*, 1967, Friedmann 1982)

The identification of many taxa is equivocal, and many cyanobacteria are reported simply at generic rank. It is difficult therefore to conclude if the same species is being reported from different sites. Of those endolithic taxa identified from culture, it is often difficult to identify species from taxonomic manuals largely developed from field samples. Thus, it is unclear whether many endoliths represent new species, or if the forms in culture are phenotypic variants of nonendolithic taxa. Critical systematic work remains to be done on most of these organisms. The lack of consistency in the application of generic concepts for cyanobacteria is a major problem. General use of the generic concepts by Anagnostidis and Komárek (1985, 1988) and Komárek and Anagnostidis (1986, 1989) provide an important contribution to resolving this issue.

Table 1. Numbers of species of algae and their taxonomic affiliation in endolithic communities

Location	Cyanobacteria	Green algae	other algae	Total Number of Species	Reference
Antarctica 1	17	2	0	19	Friedmann *et al.*, 1988
Antarctica 2	6	10	1	17	Broady 1981b
Greece	14	?	17	>31	Anagnostidis *et al.*, 1983
Southwestern United States	6	14	0	20	Bell *et al.*, 1986
South Africa	13	4	0	17	Wessels and Büdel 1995
Ontario, Canada	7	6	1	14	Gerrath *et al.*, 1995
Nova Scotia, Canada	37	6	0	43	Garbary *et al.*, 1996

Until further systematic work is carried out, and equivalent modern treatments are available at the species level, the biogeographic interpretation of apparent distributions and specific identifications will remain largely problematic.

The characterization of species composition based on culture studies may be necessary; however, it may lead to misinterpretation of community composition *in situ*. Neither the occurrence nor the abundance of a species in culture may be a true measure of actual contribution to the community. Where only a few species are present this may not be critical. When many species can be isolated from the rock it would be of interest to know which species are the major contributors to the community as well as their relative spatial distribution.

In addition to the algae, other life forms are present as part of the endolithos. Heterotrophic bacteria are observed (Friedmann 1982, Ferris and Lowson 1997), although these are not routinely identified. Filamentous fungi are present in many habitats; however, these have only been well characterized in communities where they form lichens (e.g., Antarctica, Friedmann 1982). Characterization of these organisms and their interactions with the algae are an important part of understanding the overall functioning of the endolithos.

3. Physiology and Ecology

3.1. PHYSIOLOGICAL ECOLOGY AND PRODUCTIVITY

Even in well developed endolithic communities, algal biomass is low. Measures of biomass range from 1.7-27.1 mg chl *a* m^{-2} (Friedmann *et al.*, 1980) and 66-130 mg chl *a* m^{-2} (Kappen and Friedmann 1983) in the Antarctic to 11 mg chl *a* m^{-2} in southern Canada (Matthes-Sears *et al.*, 1997) to 59-106 mg chl *a* m^{-2} in the southwestern United States (Bell and Sommerfeld 1987). In addition to low biomass, all measurements of productivity show the endolithic environment to be one of the least productive photosynthetic communities (Bell and Sommerfeld 1987, Nienow *et al.*, 1988, Johnston and Vestal 1991, Friedmann *et al.*, 1993, Matthes-Sears *et al.*, 1997). However, even low levels of productivity may represent a significant contribution when compared to the above ground productivity of a bare rock face (Bell and Sommerfeld 1987, Matthes-Sears *et al.*, 1997).

Productivity of an endolithic community will vary depending on the community composition, i.e., whether the community is composed mainly of primary producers or consumers, on nutrient availability, and on parameters such as light, temperature and water. It has been shown that nutrients (e.g., nitrogen, phosphate, manganese) are not limiting (Johnston and Vestal 1991, Friedmann and Kibler 1980). Nutrient availability may be enhanced by fungal activity. For example, production of oxalic acid increases solubilization of several essential elements (Johnston and Vestal 1993). Carbon in gaseous form or dissolved as bicarbonate appears to be the only carbon source for the photoautotrophs (Vestal 1988, Ferris and Lowson 1997).

Light attenuates as it passes through the rock matrix (Nienow *et al.*, 1988. In Antarctic communities, the algal layer is found ca. five millimeters below the surface. A mathematical model predicts that 0.05 to 10 μmol photons m^{-2} s^{-1} reach this algal layer (Nienow *et al.*, 1988). Matthes-Sears *et al.* (1997) suggest that a maximum of

0.1 μmol photons m^{-2} s^{-1} of full sunlight (1600 μmol photons m^{-2} s^{-1}) penetrate to the organisms within the limestone of the Niagara Escarpment. These levels are below the light compensation point for most algae, and are limiting for endolithic communities (Nienow *et al.*, 1988, Vestal 1988). Actual photosynthetic rates range from 0.15 mg CO_2 mg chl a^{-1} h^{-1} for the Niagara Escarpment (Matthes-Sears *et al.*, 1997) to 0.48 mg CO_2 chl a^{-1} h^{-1} for the southwestern United States (Bell and Sommerfeld 1987).

In extreme environments water may also be limiting. Palmer and Friedmann (1990) suggest that some endolithic organisms subsist on limited available water. Endolithic lichens of Antarctica, on the one hand, photoassimilate CO_2 using water vapor as the source of moisture (Palmer and Friedmann 1990). Endolithic cyanobacteria, on the other hand, require the water potential of liquid water to photoassimilate CO_2 (Potts and Friedmann 1981, Vestal 1988, Palmer and Friedmann 1990). Despite low water requirements, they are not always met (Friedmann *et al.*, 1993). Many cyanobacteria have extraordinary abilities to withstand desiccation, and then rapidly absorb water when it becomes available (Potts *et al.*, 1983, Scherer *et al.*, 1984). Some cyanobacterial cells can withstand up to six years of desiccation without apparent damage to internal structures (Potts *et al.*, 1983). After two years of desiccation, respiration, photosynthesis and nitrogenase activity are quickly resumed upon addition of water, indicating that vegetative cells as opposed to endospores are surviving (Scherer *et al.*, 1984). A cyanobacterial population in gypsum reached maximum photosynthesis within one minute of water addition (Van Thielen *et al.*, unpublished).

Temperature is also an important physical parameter in constraining some endolithic communities. In Antarctica the endolithic communities experience cycles of freezing and thawing, and the temperature rarely rises above 0° C (McKay and Friedmann 1984). Lichens, cyanobacteria and green algae from these communities are tolerant of repeated freezing and thawing (Meyer *et al.*, 1988, Ocampo-Friedmann *et al.*, 1988). The rock acts as a buffer and moderates the temperature (Friedmann 1977, McKay and Friedmann 1984). Water within the rock is able to supercool, remaining liquid at -3.8° C (Meyer *et al.*, 1988), and allowing photosynthesis to occur at sub zero temperatures (Kappen and Friedmann 1983, Vestal *et al.*, 1984, Vestal 1988). However, it has also been suggested that the optimum temperature for photosynthesis may not be reached in the natural environment, making temperature a limiting factor for photosynthesis (Kappen and Friedmann 1983, Vestal *et al.*, 1984). Despite low ambient temperatures, only the green algae tend to be psychrophilic, with cyanobacteria tending to be mesophilic (i.e., with growth optima between 20-30° C, Ocampo-Friedmann *et al.*, 1988).

When various parameters were considered, Matthes-Sears *et al.* (1997) estimated the number of optimum photosynthetically active hours for the endolithic community in the Niagara Escarpment as 1095 h per year; in contrast to the 1600 hours of bright sunlight estimated for an "above ground" ecosystem at a similar latitude when the temperature is above 0° C (Phillips 1990). A further consideration relating to net productivity is dark respiration, which can greatly reduce net productivity. The carbon loss from dark respiration in the Niagara Escarpment community as a whole can be five times the carbon gain from photosynthesis, resulting in the community being a net consumer Matthes-Sears *et al.* 1997). This may be expected as the community is comprised mainly of heterotrophic fungi (Matthes-Sears *et al.*, 1997). The origin of the fixed carbon that maintains this system is unclear. In Antarctica, 420-1042 h per year are available for metabolic activity (Friedmann *et al.*, 1993). The community consists mainly of primary producers and contributes a net 3 mg C m^{-2} y^{-1}. This community is also in a colder climate, where dark respiration does not occur to the same extent as in warmer climates (Kappen and Friedmann 1983, Friedmann *et al.*, 1993). Thus, regardless of climatic conditions, overall productivity of endolithic communities is extremely low.

3.2. ECOLOGY

Aside from the biotic diversity represented in endolithic algal communities, this habitat has a limited but intriguing ecological significance. A primary ecological impact of this community is the increased rock erosion resulting from algal and fungal growth (Friedmann 1982, Friedmann and Weed 1987). In temperate and polar climates the penetration of water around pores occupied by algal cells and their subsequent freezing increases the breakdown of rock surfaces (Friedmann 1982, Avinoam 1983, Danin and Garty 1983, Danin 1986). The loosened rock particles are part of the soil creation process (Marthe and Chaudhari 1975). Similar phenomena have been associated with man-made structures, and endolithic algae may be contributing to the erosion of the Parthenon and other historic and more recent constructions (Anagnostidis *et al.*, 1983).

Much of the initial research on endolithic algae was carried out in the Negev Desert (hot desert) and in Antarctica (cold desert), both extreme environments. This work was later extended to a variety of arid regions in the southwestern United States (e.g., Bell 1993) and in South Africa (Critchley *et al.*, 1987, Wessels and Büdel 1995). More recent studies in cool temperate regions show that endolithic communities are more widespread (e.g., Ontario and Nova Scotia in Canada, Gerrath *et al.*, 1995, Garbary *et al.*, 1996; Germany, Büdel *et al.*, 1991). These more recent studies suggest that endolithic communities can develop wherever bare rock is exposed which does not become colonized with epilithic organisms that attenuate the light (e.g., lichens, bryophytes) (Matthes-Sears *et al.*, 1997). The prime sites are likely to be the exposed faces of limestone, sandstone and gypsum cliffs, and the rubble at the cliff bases. We have also noted the presence of endolithic communities in large sandstone and limestone boulders (glacial eratics) at or near the seashore in Nova Scotia.

The only experimental analysis of colonization of the endolithic habitat is by Bell (1993). He denuded sandstone surfaces and found colonization rates of up to five centimeters per year from the margins. Thus vegetative growth from the margins is more successful than aerial colonization. If so, these communities may be extremely long lived and very stable. Additional evidence for this comes from turnover times for carbon in phospholipids which is estimated at ca. 20,000 years (Johnston and Vestal 1991, Prince 1992). These observations support the hypothesis that species in endolithic biotas are evolutionarily distinct from nonendolithic populations of the same species.

If algae are the primary producers in this community, are there also consumers that make this a more complex system? Heterotrophic bacteria are certainly present, as are fungi (Friedmann 1982, Johnston and Vestal 1993, Ferris and Lowson 1997). The latter are particularly important in many Antarctic sites. In the limestone cliffs in

Ontario fungi may be the most common organisms (Matthes-Sears *et al.*, 1997). Fungi were not observed in Nova Scotia gypsum (Van Thielen *et al.*, in preparation). Herbivores are typically not apparent within these communities, suggesting that food webs and carbon cycling are based on decomposition or fungal parasitism. An exception is the Negev Desert where 'rock-eating' snails are consumers of endolithic organisms (Shachak *et al.*, 1987, Jones and Shachak 1990).

In Antarctica some endolithic algae become associated with endolithic fungi and form cryptoendolithic lichens (Friedmann 1982). Both of the components are endolithic, and in many entities this is the permanent condition. Cyptoendolithic lichens include species of *Buellia*, *Lecidea* and *Acarospora*. These symbioses form layered thalli in the rock similar to nonendolithic lichens, although both the algal and fungal components can exist without apparent connection. In exceptionally 'mild' environments these lichens can become epilithic, appearing as crustose lichens. Cryptoendolithic lichens are also known from a desert in Israel (Jones and Shachak 1990).

An important human impact on endolithic algal communities might be associated with the decrease in ozone levels in the upper atmosphere, and the subsequent increase in ultraviolet (UV) radiation impacting the earths surface. Wherever large endolithic communities are present, they are typically one to several millimeters subsurface. This may result from photooxidation or extremes of temperature; however, an alternative explanation may be excessive UV radiation. Endozoic algae (i.e., growing inside animals) in marine habitats do not produce UV protection substances; this is done by the host animal (e.g., Shashar *et al.*, 1997). Aquatic cyanobacteria are very sensitive to UV-B radiation (Häder and Worrest 1997). Endolithic cyanobacteria (and other endolithic algae) may have a similarly low tolerance for UV radiation. Increased levels of UV radiation will clearly penetrate deeper into the rock surfaces. Increases in UV may preclude survival of algal cells at a level in the rock necessary for cells to reach photosynthetic compensation points.

4. Paleobiology, Evolutionary Significance and Exobiology

The interior of rocks may not be an important habitat today for photosynthetic organisms except for those in deserts and polar regions. This was not always the case. Cyanobacteria have a long evolutionary history, and in aquatic systems their fossil record in stromatolites makes them among the oldest life forms (Stewart and Rothwell 1993). Putative ancient cyanobacteria occur in stromatolites to ca. 3,500 m.y. ago, and a diverse fossil flora of endoliths was recorded from the late Proterozoic (Knoll *et al.*, 1986). Nonmarine stromatolites are recorded from the late Proterozoic (ca. 2,800 m.y. ago) (Stewart and Rothwell 1993).

Plants in terrestrial habitats have a fossil record to no earlier than the Ordovician (beginning ca. 500 m.y. ago). These fossils consist of single spores and spore tetrads (Gray and Shear 1992). Since endolithic algal communities almost always include cyanobacteria, it is reasonable to hypothesize that endolithic cyanobacteria were present for a period lasting at least 2,300 m.y. (i.e., the time between the origin of nonmarine stromatolites and the development of terrestrial plants). Hence for a period lasting roughly two thirds of the entire history of life, cyanobacteria could have formed the dominant, terrestrial, photosynthetic communities. With the origin of unicellular green algae in the late Precambrian (Schopf 1970), these algae may have been added to endolithic communities. Fossil lichens are known from the Early Devonian (Taylor *et al.*, 1995). It would of interest to know when filamentous fungi joined the endolithic community. The presence of filamentous fungi in rocks and their associations with algae provide interesting analogs for the evolution of lichens and fungal symbioses with terrestrial plants.

Although productivity rates are extremely low, and little biomass is actually present, prior to the evolution of terrestrial plant communities on rock and soil surfaces, endoliths would have been the dominant terrestrial photosynthetic life. As terrestrial plants diversified, they covered the rock surfaces and left only pockets of substratum for endoliths to colonize. In addition to their importance in early colonization of the Earth's terrestrial habitats, Friedmann (1982, 1993) argued that the endolithos provides a model system for the study of potential life on Mars. The recent report of chemical residues of possible biological derivation in a meteorite of Martian origin (McKay *et al.*, 1996) gives credence to Friedmann's speculations.

5. Acknowledgments

We thank Joe Gerrath and Doug Larson for introducing us to these fascinating organisms, and Andrea Gilpin, Kris Bentvelsen, Dianna Padilla, John Coleman and Johan Hellebust for their comments on various drafts of the manuscript. This work was completed while D.J.G. was a visiting researcher at Friday Harbor Laboratories, University of Washington. The Natural Sciences and Engineering Research Council of Canada provided research support to D.J.G.

6. References

Anagnostidis, K., Economou-Amilli, A., and Roussomoustakaki, M. (1983) *Nova Hedwigia* **38**, 227-289.
Anagnostidis, K. and Komárek, J. (1985) *Arch. Hydrobiol./ Supp. 71, Algol. Stud.* **38/39**, 105-114.
Anagnostidis, K. and Komárek, J. (1988) *Arch. Hydrobiol./ Supp. 80, Algol. Stud.* **50-53**, 327-472.
Avinoam, D. (1983) *Z. Geomorph. N.F.* **27**, 413-421.
Bell, R.A. (1993) *J. Phycol.* **29**, 133-139.
Bell, R.A., Wilkins, S., and Ruttman, S. (1996) *J. Phycol.* **32** (3, suppl.), 7.
Bell, R.A., Athey, P.V., and Sommerfeld, M.R. (1986) *J. Phycol.* **22**, 429-435.
Bell, R.A., Athey, P.V., and Sommerfeld, M.R. (1988) *Southwest. Nat.* **33**, 315-322.
Bell, R.A. and Sommerfeld, M.R. (1987) *Amer. J. Bot.* **74**, 294-297.
Broady, P.A. (1981a) *Br. Phycol. J.* **16**, 257-266.
Broady, P.A. (1981b) *Phycologia* **20**, 259-272.
Büdel, B., Mollenhauer, D., and Mollenhauer, R. (1991) *Algol. Stud.* **64**, 357-360.
Critchley, A., Wood, J., Horiguchi, T., and Bruton, A. (1987) *Proc. Electr. Micros. Soc. S. Afr.* **17**, 101-102.
Danin, A. (1986) *Proc. R. Soc. Edinb.* **89B**, 243-253.
Danin, A. and Garty, J. (1983) *Z. Geomorph. N. F.* **27**, 423-444.
Ferris, F.G. and Lowson, E.A. (1997) *Can. J. Microbiol.* **43**, 211-219.
Friedmann, E.I. (1971) *Phycologia* **10**, 411-428.
Friedmann, E.I. (1977) *Antarct. J. U.S.* **12**, 26-29.
Friedmann, E.I. (1982) *Science* **215**, 1045-1053.
Friedmann, E.I. (1993) *Giorn. Bot. Ital.* **127**, 369-376.
Friedmann, E.I., Hua, M., and Ocampo-Friedmann, R. (1988) *Polarforschung* **58**, 251-259.
Friedmann, E.I., Kappen, L., Meyer, M.A., and Nienow, J.A. (1993) *Microbial Ecol.* **25**, 51-69.
Friedmann, E.I. and Kibler, A. P. (1980) *Microbial Ecol.* **6**, 95-108.

Friedmann, E.I., Lipkin, Y., and Ocampo, R. (1967) *Phycologia* **6**, 185-200.
Friedmann, E.I. and Ocampo, R. (1976) *Science* **193**, 1247-1249.
Friedmann, E.I. and Ocampo-Friedmann, R. (1984) Endolithic microorganisms in extreme dry environments: analysis of lithobiontic microbial habitat, in M.J. Klug and L.A. Reddy (eds.), *Current Perspectives in Microbial Ecology*, Amer. Soc. Microbiol., Washington, pp. 177-185.
Friedmann, E.I. and Weed, R. (1987) *Science* **236**, 703-705.
Garbary, D.J., Van Thielen, N., and Miller, A.G. (1996) *J. Phycol.* **32** (3, suppl.), 17.
Gerrath, J.F., Gerrath, J.A., and Larson, D.W. (1995) *Can. J. Bot.* **73**, 788-793.
Ghiorse, W.C. (1997) *Science* **275**, 789.
Gold, T. (1992) *Proc. Natl. Acad. Sci. USA* **89**, 6045-6049.
Golubic, S., Friedmann, E.I., and Schneider, J. (1981) *J. Sedimentary Petrology* **52**, 475-478.
Gould, S.J. (1996) *Full House*, Three Rivers Press, New York.
Gray, J. and Shear, W. (1992) *Amer. Scient.* **80**, 443-456.
Häder, D.-P. and Worrest, R.C. (1997) Consequences of the effects of increased solar ultraviolet radiation on aquatic ecosystems, in D.-P. Häder (ed.), *The Effects of Ozone Depletion on Aquatic Ecosystems*, Academic Press and R.G. Landes Co., Austin and San Diego, pp. 11-30.
Johnston, C.G. and Vestal, J.R. (1991) *Appl. Environm. Microbiol.* **57**, 2308-2311.
Johnston, C.G. and Vestal, J.R. (1993) *Microbial Ecol.* **25**, 305-319.
Jones, C.G. and Shachak, M. (1990) *Nature* **346**, 839-841.
Kappen, I. and Friedmann, E.I. (1983) *Pol. Biol.* **1**, 227-232.
Komárek, J. and Anagnostidis, K. (1986) *Arch. Hydrobiol./ Supp. 73, Algol. Stud.* **43**, 157-226.
Komárek, J. and Anagnostidis, K. (1989) *Arch. Hydrobiol./ Supp. 82, Algol. Stud.* **56**, 247-345.
Knoll, A.H., Golubic, S., Green, J., and Swett, K. (1986) *Nature* **321**, 856-857.
L'Haridon, S., Reysenbach, A.-L., Glénat, P., Prieur, D., and Jeanthon, C. (1995) *Nature* **377**, 223-224.
Lukas, K.J. and Golubic, S. (1983) *J. Phycol.* **19**, 129-136.
Marthe, K.V. and Chaudhari, P.R. (1975) *J. Ecol.* **63**, 65-69.
Matthes-Sears, U., Gerrath, J.A,. and Larson, S.W. (1997) *Int. J. Pl. Sci.* **158**, 451-460.
McKay, C.P. and Friedmann E.I. (1984) *Antarc. J. U.S.* **19**, 170-172.
McKay, D.S., Gibson, E.K., Thomas-Keprta, K.L., Vali, H., Romanek, C.S., Clemett, S.J., Chillier, X.D.F., Maechling, C.R., and Zare, R.N. (1996) *Science* **273**, 924-930.
Meyer, M.A. Huang, G.H., Morris, G.J., and Friedmann, E.I. (1988) *Polarforschung* **58**, 113 -119.
Nienow, J.A., McKay, C.P., and Friedmann, E.I. (1988) *Microbial Ecol.* **16**, 271-289.
Ocampo-Friedmann, R. Meyer, M.A., Chen, M., and Friedmann, E.I. (1988) Polarforschung **58**, 121-124.
Palmer, R.J. Jr. and Friedmann E.I. (1990) *Microbial Ecol.* **19**, 111-118.
Parkes, R.J., Cragg, B.A., Bale, S.J., Getliff, J.M., Goodmann, K., Rochelle, P.A., Fry, J.C., Weightman, A.J., and Harvey, S.M. (1994). *Nature* **371**, 410-413.
Phillips, D. (1990) *The Climate of Canada*, Canadian Government Publishing Centre, Ottawa.
Potts, M. and Friedmann, E.I. (1981) *Arch Microbiol.* **130**, 267-271.
Potts, M., Ocampo-Friedmann, R., Bowman, M. A., and Tözun, B. (1983) *Arch. Microbiol.* **135**, 81-90.
Prince, R.C. (1992) *TREE* **7**, 211.
Scherer, S., Ernst, A., Chen, T., and Boger, P. (1984) *Oecologia* **62**, 418-423.
Schopf, J. W. (1970) *Biol. Rev.* **45**, 319-352.
Shachak, M., Jones, C.G., and Granot, Y. (1987) *Science* **236**, 1098-1099.
Shashar, N., Banaszak, A.T., Lesser, M. P., and Amrami, D. (1997) *Pacific Sci.* **51**, 167-173.
Stevens, T.O. and McKinley, J.P. (1995) *Science* **270**, 450-454.
Stewart, W. N. and Rothwell, G. W. (1993) *Paleobotany* and *the Evolution of plants, 2nd ed.*, Cambridge University Press, Cambridge.
Szewzyk, U., Szewzyk, R. and Stenström, T.-A. (1994) *Proc. Natl. Acad. Sci. USA* **91**, 1810-1813.
Taylor, T.N., Hass, H., Remy, W., and Kerp, H. (1995) *Nature* **378**, 244.
Vestal, R. (1988) *Appl. Environm. Microbiol.* **54**, 961-965.
Vestal, J.R., Federle, T.W., and Friedmann, E.I. (1984) *Antarc. J. U.S.* **19**, 173-174.
Wessels, D.C.J. and Büdel, B. (1995) *Botanica Acta* **108**, 220-226.

Biodata of **Jacob Garty** author of ***"Lithobionts in the Eastern Mediterranean."***

Dr. J. Garty is an associate Professor in the Department of Plant Sciences at the Tel Aviv University. He earned his Ph.D. in 1971 in the Tel Aviv University. Dr. Garty's main interest is in the fields of lichenology, botany and plant ecophysiology.
E-mail: garty@post.tau.ac.il

J. Seckbach (ed.), Enigmatic Microorganisms and Life in Extreme Environments, 255–276.

LITHOBIONTS IN THE EASTERN MEDITERRANEAN

J. GARTY
Department of Plant Sciences and Institute for Nature Conservation Research, The George S. Wise Faculty of Life Sciences, Tel Aviv University, Tel Aviv 69978, Israel

1. Introduction

Floristic publications on the eastern Mediterranean listed hundreds of lithobiont species, whereas only a few representatives were dealt with in environmental studies.

The objective of the present chapter is to focus attention on a few lithobiontic micro-organisms and to characterize their interaction with the environment.

2. Epilithic Lichens, Green Algae and Cyanobacteria

Of the 250 lichen species contained in the checklist for Israel and Sinai (Galun and Mukhtar, 1996), about 70% are lithophytic and for the greater part epilithic; 75% of the lichen species listed for Jordan (El-Oqlah, 1992) are lithophytic, namely saxicolous.

In the framework of ecophysiological studies in the Negev Desert, Israel, attention centered on the epilithic lichen *Ramalina maciformis*. In the last three decades, since the publication of Lange and Bertsch (1965) on photosynthesis and water conditions of *R. maciformis*, a great effort was made to investigate the water relations, photosynthesis, respiration, productivity and growth of this lichen (*e.g.* Lange, 1969, 1980; Lange *et al.*, 1968, 1970a,b, 1975; Kappen *et al.*, 1979, 1980; Lange and Tenhunen, 1981; Palmer and Friedmann, 1990; Büdel and Lange, 1991; Cowan *et al.*, 1992; Green *et al.*, 1994; Green and Lange, 1994; Scheidegger *et al.*, 1995).

Some attention was paid to the dependence or *R. maciformis* on its substrate. This lichen, together with *Buellia canescens, B. sorediosa* and *Caloplaca ehrenbergii* forms an association which is found in great abundance on flintstones located on all sides of hills and mountains in northern and central regions of the Sinai and Negev Deserts, at heights of approx. 250 m and above (Galun, 1970; Galun and Garty, 1972). *R. maciformis* grows also on hard limestone facing north or west (Garty and Galun, 1974) and on desert rock varnish. It was also found growing with its associates, on non-siliceous stones. These stones have a low capacity for the retention of moisture. The available amount of water on the substrate depends on the moisture retention capacity of the substrate in addition to the kinetics of the evaporation process. The harder, less permeable stones retain a thicker and more durable water film on their surface. This may account for the difference in lichen establishment. The thickness and durability of the water film play an important role in the early stage of thallus initiation, in addition to its indispensability for the existence and development of mature colonies of *R. maciformis, B. canescens, B. sorediosa* and *C. ehrenbergii*. The first three of these lichens rarely produce fruiting bodies in the Negev Desert and are defined as sorediose species. Their mode of thallus initiation is presumably by soredia whereas *C. ehrenbergii* is reproduced by germinating spores or thallus fragments.

The amounts of mineral elements in the thallus of *R. maciformis* (Garty, 1985) and the spatial and temporal changes in the concentration of Fe, Zn, Mn, Cu and Pb (Garty *et al.*, 1995) and of K, Na, Mg and Ca (Garty *et al.*, 1996) in epilithic and decom-posing detached thalli, were studied to investigate the potential role of this lichen in the cycling of elements in the Negev Desert. We have found that mineral elements differ in their accumulation rate in *in situ* thalli and in their release from decomposing detached thalli, and that the input/output of mineral elements to/from lichen thalli is influenced by extreme climatic conditions such as prevail in the Negev Desert. A signi-ficant decrease in the concentration of Na and Mg in *in situ* thalli of *R. maciformis* was found after two exceptionally wet and cold winters in the central Negev Highlands. A leakage of 1% of K per month was detected in decomposing, detached thalli placed in litterbags on western aspects of hills. Thalli which were detached from flintstones, kept in litterbags and fixed on southern, hotter and drier aspects of hills, lost 1.6% K per month.

To examine the role played by *R. maciformis* in the cycling of mineral elements in the Negev Desert Highlands, we calculated the element budget of this lichen using data obtained upon estimation of the standing phytomass and the concentrations of K, Na, Mg and Ca in *in situ* and detached thalli on west-facing slopes. The accumulation of K from aerial sources by *in situ* thalli of *R. maciformis* is 0.59 mg m^{-2} $month^{-1}$ or 70.8 g ha^{-1} $year^{-1}$. These values demonstrate the importance of *in situ* epilithic thalli of *R. maciformis* as traps for airborne K. A similar calculation led us to assume that the K loss from decomposing detached thalli located on western aspects may reach 0.48 mg m^{-2} $month^{-1}$ or 57.6 g ha^{-1} $year^{-1}$. Thus, the role of detached and decomposing thalli of *R. maciformis* is demonstrated as a means for the gradual release of K on the soil sur-face, especially at times of extreme temperatures in summer and in cold and rainy winters.

High concentrations of heavy metals such as Fe, Zn, Mn, Cu and Pb deriving from geochemical sources like dust storms, in addition to anthropogenic sources like traffic, accumulated in epilithic and detached thalli of *R. maciformis,* deserve our attention, since the black dwarf goat grown by the Bedouins is observed to graze on this lichen in the Negev and Northern Sinai Deserts.

An additional epilithic lichen which attracted our interest is a crustose lichen, *Caloplaca aurantia* var. *aurantia* which colonizes carbonatic rocks all over Israel (Galun, 1970) and Northern and Central Sinai (Galun and Garty, 1972). The alga-fungus association in three ecotypes of this lichen was studied by Ben-Shaul and co-workers (1969) by means of electron microscopy. It was found that the degree of association between the phycobiont (*Trebouxia* sp.) and mycobiont differed in the three ecotypes. Under moderate climatic conditions prevailing in the Mediterranean territory of Israel, the contact between the two symbionts was limited to a close attachment of their cell walls, whereas differing degrees of haustorial penetration into the algal cells, in specimens growing under xeric conditions, were detected in desert locations in the Negev and the Sinai Peninsula, belonging to the Irano-Turanian and Saharo-Arabian territories. Similar results were obtained by an examination of the epilithic lichen *Lecanora radiosa,* found in the Sinai Peninsula and in the Upper Galilee, Israel, belonging to the Mediterranean territory (Galun *et al.*, 1970).

The growth rate of epilithic crustose lichens in an eastern Mediterranean desert is demonstrated by two studies. Growth measurements of *C. aurantia* were performed by Lange and Evenari (1971) and Lange (1990). Over a period of 23 years, the growth rate of *C. aurantia* in the Negev Desert was 0.371 mm yr^{-1}. Thus, a *C. aurantia* thallus of 10 cm in diameter has presumably an age of 135 years and the largest thallus of this species found near Avdat must have an age of about 500 years (Lange, 1990).

Unlike the saxicolous lichen *R. maciformis* which colonizes for the greater part natural substrates, *C. aurantia* is the most common lichen not only on rocks but also on man-made substrates, especially roof-tiles, in the eastern Mediterranean. This lichen, although it expands on roof tiles of houses in rural and urban sites in the Mediterranean territory of Israel where 800-900 mm of annual rain-fall are common, seems to face rather harsh conditions, as roof-tiles, either of clay or of concrete, keep their moisture following rain episodes for a minimal time. In addition to drought con-ditions prevailing especially on south- and east-facing roof-slopes, the sun radiation creates pseudo-desert conditions for epiliths inhabiting this biotope. The lichen *C. aurantia* was proved to be well adapted to such conditions as it covers roof-tiles sometimes at a rate of 100%. It was proved also to resist air pollution as it grows in the most polluted parts of Israeli cities like Tel Aviv. A comparative study of the heavy metal content in thalli of *C. aurantia* growing on roof-tiles in urban, suburban and rural regions in Israel (Garty *et al.*, 1977) indicated that their content of Mn, Zn, Fe, Pb, Ni, Cu, Cr and Cd was higher in an urban environment over and above rural areas. Among these metals Ni and Zn were found most suitable for a distinction between "town" and "village". Whereas the uptake of heavy metals correlated with the ambient metal pollution, the substratum contained much lower concentrations (Garty *et al.*, 1986). Thus, the metals Fe, Mn, Zn, Cu, Cr, Ni and Cd were absorbed for the greater part directly from the atmosphere whereas the uptake of these metals from the roof-tiles was almost negligible. According to this study, the maximum uptake of Pb by *C. aurantia* from the tile glaze was 11% of the total amount. Metal-containing particles were found to be entrapped in thalli of *C. aurantia* growing on roof-tiles, a fact reflected by high coefficients of variation (SD/mean concentration) especially in polluted areas (Garty *et al.*, 1979).

The formation of new lichen thalli of *C. aurantia* and of additional epilithic lichens growing on roof-tiles was observed by SEM which detected initial stages of thallus establishment (Garty and Delarea, 1987). Bare debris from the upper surface of sampled tiles present the marks of pits and ruts of which many are colonized by a unicellular green alga, presumably *Trebouxia*, one of the most common photobionts in lichens. In many cases, *C. aurantia* ascospores were found to germinate close to the algal cells. In many samples, such ascospores as well as ascospores of the lichens *Lecanora dispersa, Protoblastenia immersa* and *Candelariella aurella* occurred on the substrate in groups of eight (Garty ad Delarea, 1988). These findings testify that the release of ascospores of epilithic lichens under natural conditions occurs in clusters. In many cases, germinating ascospores in clusters were found in close contact with unicellular green algae, probably *Trebouxia*.

The rate of colonization of concrete tiles by *C. aurantia* was assessed by photographing a 50×50 cm area on the west-facing slope of roofs of one-storey buildings in a rural site in Israel (Garty, 1988). The percentage of colonization by *C. aurantia* on roof tiles aged 30, 45 and 60 years was found to be 2.46±0.73%, 22.9±7.3% and 48.5±6.7%, respectively.

The alga-fungus associations reflected by the electron-microscopic studies cited above (Ben-Shaul *et al.*, 1969; Galun *et al.*, 1970), in addition to other patterns of contact of symbionts in different climatic zones, may be compared with another aspect of evolution in lichens. Recently we have found (Nevo *et al.*, 1997) that natural selection causes microscale allozyme diversity in *C. aurantia* at the 'Evolution Canyon' in the lower Nahal Oren microsite, Mount Carmel, Israel. The allozymic diversity was studied for proteins encoded by 13 loci in 78 different thalli of the lichen. The samples were collected from three stations: two on the south-facing slope and one on the north-facing slope. Intense solar radiation on the SF- relative to the NF-slope makes it warmer, drier, spatiotemporally more heterogeneous and climatically more fluctuating and stressful. Significant inter- and intraslope allozymic differentiation was found in *C. aurantia* with generally higher polymorphism, heterozygosity, allele and gene diversity on the more variable and stressful upper station of the SF-slope, as expected by the niche-width variation hypothesis, and the environmental theory of genetic diver-sity. Differences in solar radiation, temperature and aridity-stress caused interslope and intraslope differences on the SF-slope in genotypes and phenotypes of *C. aurantia* as was the case for wild barley (*Hordeum spontaneum*), beetles, diplopods and earthworms. Diversifying natural microclimatic selection appears to be the major evolutionary driving force, causing interslope and SF-intraslope adaptive genetic divergence.

Cryptogamic lithobionts in the eastern Mediterranean are often endangered and annihilated by wildfires raging annually over large areas of forest in the hot and dry summer. Very little is known of the re-establishment of pioneer lithobiontic microorganisms in the Mediterranean. The postfire recovery of rock-inhabiting algae, microfungi and lichens on the Carmel mountain, Israel, was studied for the first time after a severe forest fire which raged for 3 days in November 1983 and destroyed 3.3 km^2 of a natural pine forest (*Pinus halepensis*). About 4 years postfire, we observed that the first lichen thalli on burnt rocks, visible to the naked eye, belong to *Xanthoria parietina* (which grows usually as an epiphytic lichen in the Middle East) together with *C. aurantia* (Garty and Binyamini, 1990).

The two forest fires that occurred in 1983 and in 1989 in the Beth-Oren area, provided us with a unique opportunity to prove that *Trebouxia*, the most common lichen photobiont, belongs to the free-living algal community (Mukhtar *et al.*, 1994). By employing highly sensitive immunological methods, we were able to detect among the many different algae that developed in surface crevices of rocks, still devoid of lichens, three years after the fire, *Trebouxia* cells with surface molecules similar to those of intact photobionts grown in culture: The *Trebouxia* isolations were performed from *Xanthoria parietina* and from *Buellia* sp. growing close to each other, on a rock exposed to the fire in 1983. Free-living *Trebouxia* were also isolated from a rock exposed to fire in 1989. The antiserum against one isolate from *Xanthoria parietina* (XpT_1) recognized only two isolates from this lichen (*i.e.* its homologue XpT_1 and XpT_2), but did not react with the three additional isolates (XpT_3, XpT_4 or XpT_5), thus suggesting that *X. parietina* may contain two (or more) different photobionts. This suggestion was strongly supported by the similar cross-reactivity with the antiserum αBT_1. The antiserum also recognized the two isolated XpT_1 and XpT_2, but produced no reaction with the three other *X. parietina* isolates (XpT_3-XpT_5). These results led to the conclusion that the *Buellia* sp. contains the same photobiont as the *X. parietina* specimen. As the existence of *Trebouxia* as a free-living organism in nature, has long since been a subject of debate, we believe that the presence of *Trebouxia* cells on a completely bare rock sterilized by forest fires before the establis-hment of the host lichens e.g. *X. parietina* and *Buellia* sp., provides unequivocal evidence to conclude this long-lasting dispute.

Walls of buildings in arid and semi arid environments seem to provide rather harsh conditions for epilithic microorganisms. Many walls of one- to four-storied houses in Israel are protected by a whitish mortar which consists of lime, white cement and calcium carbonate powder derived from calcareous rocks ground in local quarries. The first visible colonization of this mortar is by cyanobacteria which form blackened triangles expanding downwards. These biofilms consist initially of both unicellular cyanobacteria and multicellular colonies of the cyanobacterium *Gloecapsa biformis* st. *dermochrous.* Blackened surfaces on mortar walls, colonized for the greater-part by these cyanobacteria, were found to have temperatures up to 8°C higher than non-colonized areas, thus suggesting that these microorganisms may affect the temperature regime in built-up areas (Garty, 1990).

3. Euendoliths

Euendolithic microorganisms, which are very common in the eastern Mediterranean, were defined by Golubic and coworkers (1981) as "those which penetrate actively into the interior of rocks forming tunnels that conform with the shapes of their bodies". Euendolithic green algae and chroococcoid cyanobacteria were detected in the Negev Desert (Friedmann *et al.*, 1967; Friedmann and Galun, 1974). Euendolithic coccoid cyanobacteria and cyanophilous lichens were found to create pits under desert conditions with a precipitation of 20-100 mm in the Negev (Danin *et al.*, 1982; Danin and Garty, 1983). The same microorganisms were found in cliffs and walls in the Mediterranean territory of Israel with a precipitation over 300 mm (Danin, 1986). According to Danin, these microorganisms develop mainly in depressions and make for a continuous weathering of the local substratum. In Jerusalem, euendolithic coccoid cyanobacteria and cyanophilous lichens occupy the major part of west- and southwest-facing walls aged 100 years and more (Danin and Caneva, 1990). It is indicated that southeast-, east- and northeast-facing walls are brown, yellowish-brown or yellow being protected from rain-bearing winds and covered by a crust composed of iron oxide and clay. The rate of rock-weathering by euendolithic coccoid cyanobacteria in Jerusalem is 5 μm/year (Danin, 1983).

Euendolithic lichens are able to withstand severe environments and abound under extreme ecological conditions (Kushnir *et al.*, 1978). Some of the euendolithic lichens which grow in the Negev Desert colonize small stones up to 10 cm^3. These stones bear many pinhead holes filled with fruiting bodies (Danin and Garty, 1983; Danin, 1985, 1989).

The effect produced by the colonization of limestone and dolomite rock-faces by another group of euendolithic lichens is an intricate pattern which is reminiscent of a jigsaw-puzzle (Danin *et al.*, 1982). When, for any reason, these rocks are covered with soil, the lichens die within a few years but the characteristic jigsaw-puzzle pattern persists (Danin *et al.*, 1983).

The spatial position and the volume of the carbonatic substrate are dominant factors in the determination of the kind of lithophyte about to colonize a certain niche. In the Negev Desert Highlands, euendolithic lichens forming a jigsaw-puzzle often colonize medium to large-sized stones (10 cm^3 to 30,000 cm^3, Danin and Garty, 1983). The most common species reported in this habitat is *Caloplaca alociza* sometimes in asso-ciation with five other euendolithic species (Kushnir, 1979). It was found that the volume of the substrate is a determinant of the daily temperature regime. The maximum day-time temperature of the above-mentioned "rolling stones" is higher than that of the large rocks facing south, whereas the minimum night-time temperature is lower (Danin and Garty, 1983). It appears that euendolithic lichens growing on small stones on south-facing slopes profit, in cool nights, from a greater amount of moisture (dew) than that available for euendolithic and chasmoendolithic cyanobacteria inhabiting massive rocks.

Studies of the fungus-alga relationship in euendolithic lichens by means of transmission electron microscopy reveal that the mode of contact of the symbionts is in most cases the same as for epigenous lichens (Kushnir and Galun, 1977): *Caloplaca alociza* and *Caloplaca* sp., for example, were detected in a relative moderate environment of the Mediterranean territory whereas other specimens of *Caloplaca alociza* were found in the Negev and Sinai deserts. The fungus of the latter had pegs which penetrated into 10-13% of the algal cells whereas in the two other endolithic *Caloplaca* specimens the components were only juxtaposed as seen in other *Caloplaca* species growing under moderate environmental conditions (Ben-Shaul *et al.*, 1969).

Euendolithic lichens play an important role in the terrestrial food chain in the Negev Desert. Different species of snails (*Euchondrus* spp.) eat euendolithic lichens growing under the surface of limestone rocks. According to Shachak and his co-workers (1987), this type of herbivory has the capacity to weather the surface at a rate of 0.7 to 1.1 metric tons per hectare per year. Jones and Shachak (1990) found that snails feeding on endolithic lichens ingest both rock and lichens and deposit their faeces on the soil under the rocks. Snails transfer between 22 and 27 mg nitrogen per m^2 per year to the soil, an amount which constitutes about 11% of the total soil nitrogen input, at least 18% of net soil inputs, and a minimum of 27% of the nitrogen annually accumulated by endolithic lichens from dust.

The palaeoclimatic significance of the weathering patterns of euendolithic lichens in limestone and dolomite is demonstrated by a few studies conducted in Israel. Remnants of the jigsaw-puzzle-like weathering characteristic of Mediterranean condi-tions were observed in semiarid and arid zones indicative of relative humid conditions prevailing in the past (Danin *et al.*, 1982). According to Danin (1985, 1986) several Neolithic sites in the Jordan Valley have remnants of euendolithic lichens on stones and boulders buried 9,000-10,000 years ago. As the present climate at these sites permits the development of a more drought-resistant lithobiont community, it is concluded that the climate at the time of Neolithic settlement was more humid than the present. In the same manner, fossil deep pitting and channeling observed in the Mediterranean zone on rock surfaces now protected from weathering are assumed to be indicative of an arid climate in previous times (Danin *et al.*, 1982).

The microrelief shaped by euendolithic microorganisms colonizing carbonatic rocks is related to the recolonization of pioneer lithobiontic microorganisms in the same habitats. Burnt rocks observed in the Carmel Mountain following forest fires were found to exhibit ten kinds of weathering elements of which nine were crypto-gamic imprints corroded by saxicolous unicellular green algae, free-living microfungi and fruiting bodies, or microgrooves of euendolithic lichens in prefire periods (Garty, 1992). The calculated volume of a few weathering elements whose remnants could be observed on burnt rocks, and the consecutive amount of removed rock material is not negligible. An area of 1 ha covered by 100% chalk rocks corroded by one kind of endolithic lichens bearing apothecia measuring 0.8 mm in diameter should yield up to 1,740 kg of rock material. Amounts of this size present an idea of the possible contribution of endoliths to the pedogenesis in the Mount Carmel area as in other Mediterranean ecosystems.

4. Chasmoendolithic Consortia

Chasmoendoliths were defined as microorganisms colonizing fissures and cracks with-in rocks (Golubic *et al.*, 1981). An example of the colonization of fissures and cracks in a very short time, is provided by studies of the postfire recovery of lithobiontic microorganisms in Israel (Garty and Binyamini, 1990; Garty 1990, 1992, 1995).

In forest fires in the Carmel Mountains, chalk rocks cracked because of the heat which reached approx. 700°C. Ten years later it was possible to observe upon removal of rock fragments deposited on the massive rock, blue-green rings at the open edges of the fragments. These rings consisted of coccoid cyanobacteria, fungal hyphae and fungal spores. Soil accumulated between the rock fragments and the surface of the massive rock. The presence of non-lichenic chasmoendolithic fungal-algal associations in the above-mentioned habitat may be compared with other lithic consortia, *e.g.* those described by Turian (1977). In our case, it is unclear whether the fungal hyphae are able to use assimilates produced by adjacent cyanobacteria. It is possible that the organic matter deriving from forest litter and trapped between the rock fragments and the massive rock reaches the fungal hyphae.

Widespread consortia, consisting of chasmoendolithic microorganisms, cause or accelerate the biodeterioration of lithic substrates of historical monuments and archaeological sites. Pieces of detached marble were observed in the temple of Apollo in Side, Southern Turkey. The microorganisms detected underneath the marble fragments were unicellular green algae, coccoid cyanobacteria and other bacteria. Soil particles, *e.g.* clay, which accumulate between the fragmented and the massive marble, absorb humidity, thus intensifying the process of exfoliation. A similar pattern of weathering, associated for the greater part with chasmoendolithic cyanobacteria, was observed in the temple of Apollo in Didim, ca. 200 km south of Izmir, Turkey, by Danin (1992): the stone covering the microorganisms is easily removed as a result of the exfoliation of the marble. The result of this process of biodeterioration in sites of historical importance is the gradual disappearance of the beautiful marble and the creation of flat depressions in walls, columns and statues all over the site.

The bioexfoliation of carbonatic substrates is very common in limestones in the Negev Desert. Chasmoendolithic cyanobacteria make for exfoliated faces (Friedmann and Galun, 1974; Danin and Garty, 1983). Exfoliation is found frequently in south-facing outcrops of massive rocks in desert areas with a precipitation below 200 mm (Danin, 1986). The bioexfoliation of hard limestones appeared to be brought about by chasmoendolithic unicellular green algae and coccoid cyanobacteria in samples collected in the Negev Desert Highlands, in Wadi Arava, Southern Israel and in Northern and Central Sinai. Near the archaeological site at Petra, ca. 260 km south of Amman, Jordan, the detachment of pieces of sandstone was also found to be associated with the presence of chasmoendolithic unicellular green algae, cyanobacteria and other bacteria.

Chasmoendolithic microorganisms combine into associations of different components. Chasmoendolithic biofilms consist of bivalent or trivalent consortia involving sometimes two kinds of C-autotrophs and sometimes C-autotrophs together with C-heterotrophs. The source of carbon and energy for the heterotrophic endoliths seems to be linked with primary procuders, *i.e.* green algae and cyanobacteria.

Our findings indicating the occurrence of unicellular green algae as chasmoendolithic microorganisms in eastern Mediterranean deserts do not agree with the statement made by Friedmann and Ocampo-Friedmann (1985) that in hot deserts, the euendolithic environment supports only prokaryotes and the only photosynthetic organisms present are blue-green algae.

5. Hypolithic Green Algae and Cyanobacteria

Hypolithic green algae and cyanobacteria are common in the Negev and grow on the lower surface of flintstones and opaque limestones, underneath pieces of gypsum lying on the soil and in shells of snails which died recently and are lying on the soil (Friedmann *et al.*, 1967). According to Berner and Evenari (1978), a significant difference was found in the distribution of algae underneath dark transparent and light opaque pebbles: 46.8% of the dark flints bear algae up to a thickness of 40 mm whereas the respective figure for the light flints was 20.9% up to a thickness of 30 mm. The authors assumed that the dark flint has a greater amount of light penetrating into a deeper thickness. The difference in light penetration should explain the widespread distribution under dark flint and the lesser distribution under light flint.

Hypolithic cyanobacteria such as *Chroococcidiopsis* collected in the Negev Desert exhibit a remarkable capability to survive in extremely dry environments. Desiccation experiments show that their thick cell-sheaths are useful in the prevention of water loss (Grilli Caiola *et al.*, 1993, 1996).

6. Cryptoendolithic Microorganisms

Cryptoendoliths are those microorganisms which colonize structural cavities within porous rocks, including spaces produced and vacated by euendoliths (Golubic *et al.*, 1981). Sandstones collected by Friedmann in Sinai and in the Negev near Timna, Israel, contained cryptoendolihic cyanobacteria growing in-between quartz crystals and small plate-like matrix crystals. Such rocks have a distinct porous structure but near the surface the air spaces between the rock particles are filled with inorganic debris that forms a solid crust (Friedmann, 1980). A similar dense mineral substance which forms a dense outer crust was observed on Nubian sandstones sampled in Petra, Jordan, containing cryptoendolithic cyanobacteria (Garty, 1995, unpublished; Danin, 1997, personal communication). Based on a preliminary comparison with findings of Friedmann and Ocampo-Friedmann (1984, 1985), it seems that cryptoendolithic cyanobacteria growing in sandstones near Petra are species of the world-wide genus *Chroococcidiopsis*.

Cryptoendolithic cells of *Chroococcidiopsis* sp. collected from Nubian sandstones in National Park Timna, Israel, were able to resist long periods of nitrogen limitation and starvation. According to Billi and Grilli Caiola (1996), cells grown in medium with only 62 μg N ml^{-1} or without nitrogen altogether, persisted as isolated viable forms after three and one months of growth, respectively. These isolated cells had a multilayered envelope similar to that described for aged, desiccated cultures of *Chroococcidiopsis* (Grilli Caiola *et al.*, 1993) and may represent survival forms (spore-like cells or akinetes) appearing in stressful conditions.

7. Rock Varnish in Eastern Mediterranean Deserts

Desert varnish appears as a brownish-black film found on rocks in arid and semiarid environments. According to Krinsley and co-workers (1990) this <10 to >500 μm thick layer is composed principally of clay minerals, iron and manganese oxides apart from various minor and trace constituents. Manganese-rich desert varnishes were reported to occur in many of the sites studied in the Negev and Sinai deserts (Krumbein and Jens, 1981; Hungate *et al.*, 1987; Grote and Krumbein, 1992a,b). Seventy-four of 79 bacterial isolates obtained from rock varnish samples of the Negev Desert were found to oxidize manganese (MnII→MnIV) under laboratory conditions (Hungate *et al.*, 1987).

According to this study, the predominant genera involved in the formation of desert varnish were *Bacillus, Geodermatophillus, Arthrobacter* and *Micrococcus.*

The precipitation of manganese by fungi and chemoorganotrophic bacteria isolated from desert rocks was shown to depend on increasing NaCl concentrations *in vitro* (Grote and Krumbein, 1992a). These findings support the hypothesis on the microbial genesis and stabilization of rock varnish and the patination of ancient petroglyphs, especially by manganese-precipitating fungi, *e.g.,* species belonging to *Alternaria, Cladosporium, Coniothyrium* and *Pleospora.*

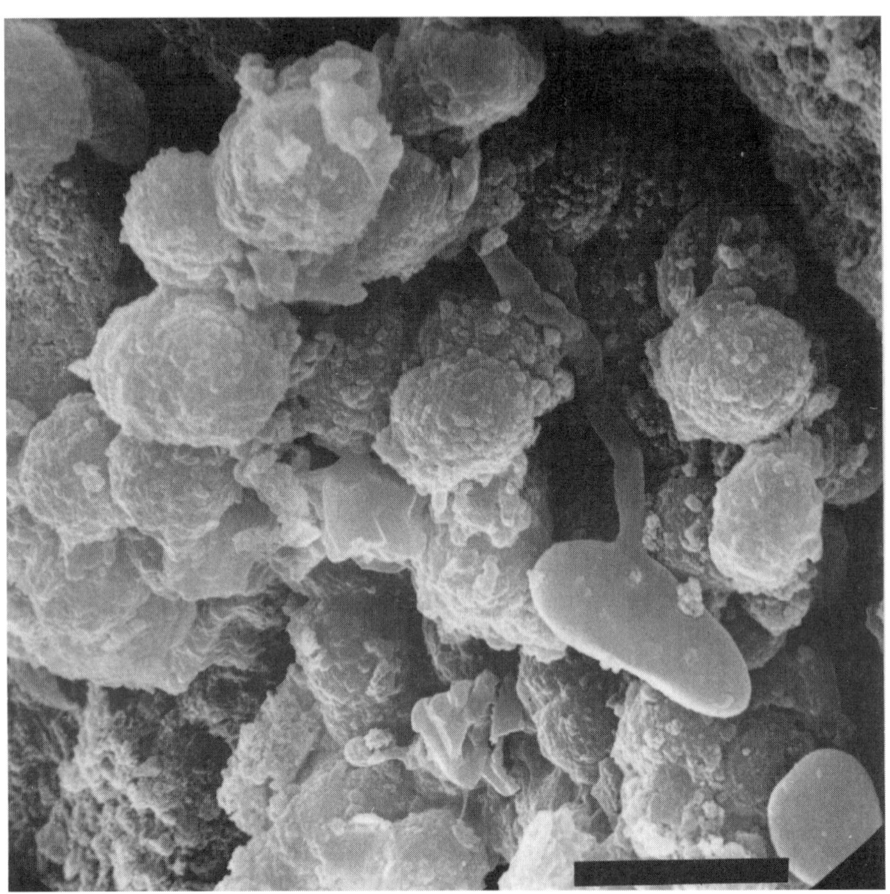

Figure 1. Germinating ascospore, produced and dispersed by an epilithic lichen, and unicellular green algae on concrete roof tile, sampled in Ganei Am, Israeı. Bar = 10 μM.

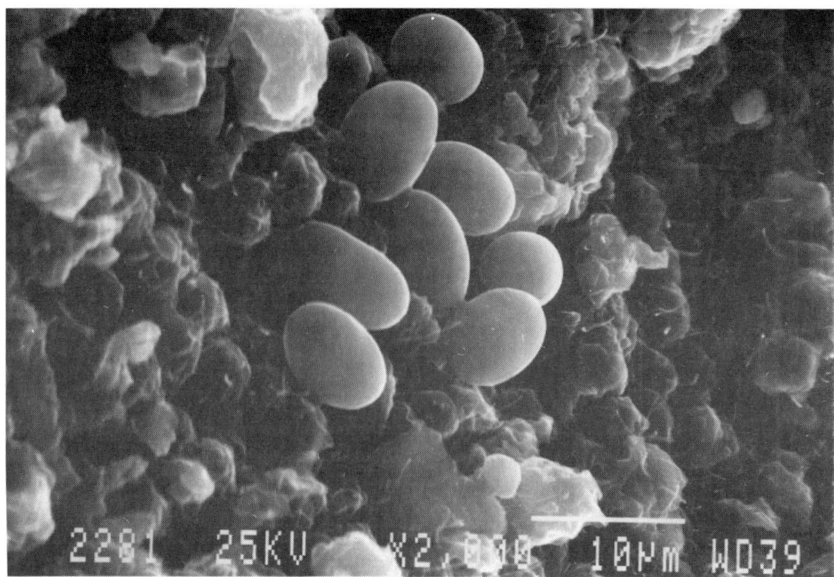

Figure 2. Cluster of lichen ascopores produced and dispersed by *Caloplaca* sp., observed on burnt chalk rock surface ten years after fire of 1983, Carmel Mountain, Israel. Bar = 10 µm.

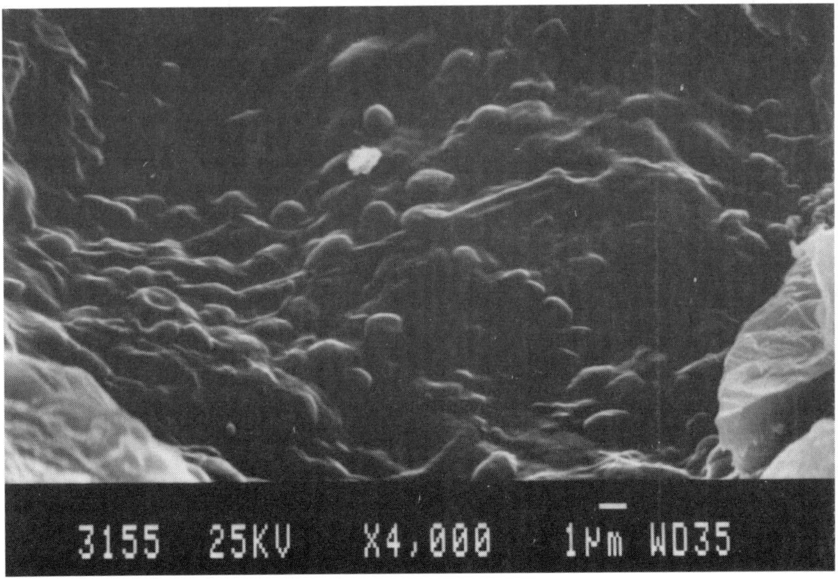

Figure 3. Euendolithic cyanobacteria, limestone, Sede Boker, Negev Desert, Israel. Bar = 1 µm.

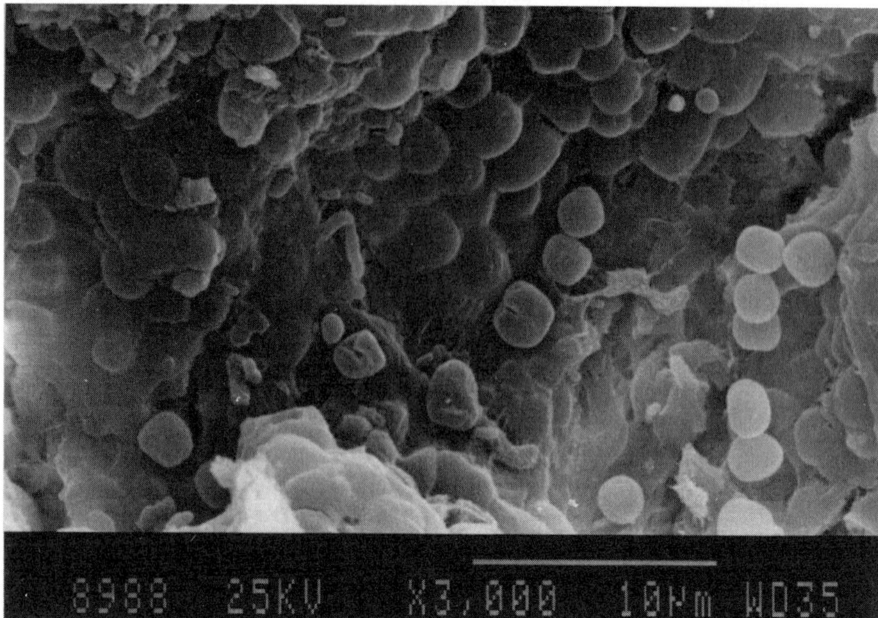

Figure 4. Chasmoendolithic coccoid cyanobacteria in cracks open to the surface of chalk rocks burned in the Carmel Mountain, Israel, November 1983. Sample collected in November 1995. Bar = 10 μm.

Figure 5. Chasmoendolithic green algae, marble, Apollo Temple, Side, Turkey. Bar = 10 μm.

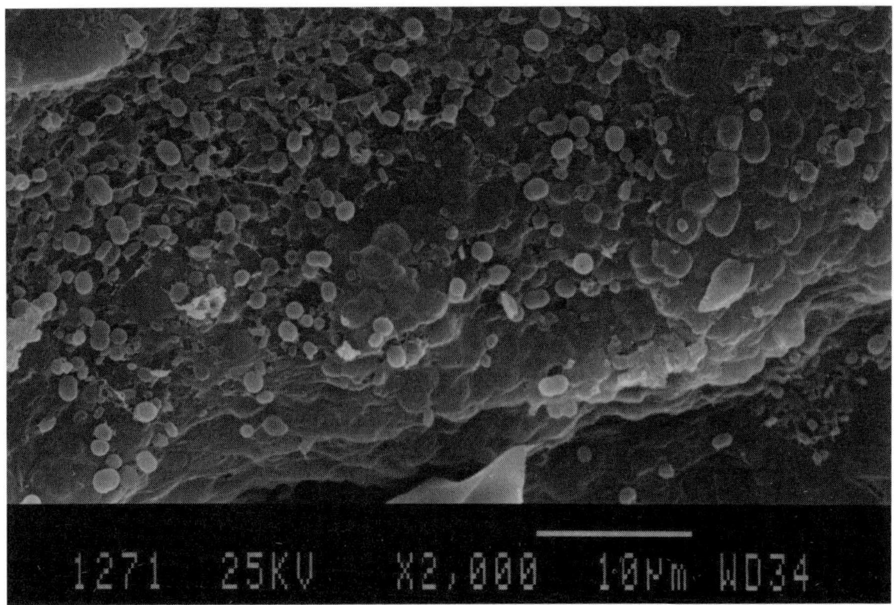

Figure 6. Chasmoendolithic cyanobacteria, marble, Apollo Temple, Side, Turkey. Bar = 10 µm.

Figure 7. Chasmoendolithic cyanobacteria, limestone, Wadi Arava, Israel. Bar = 10 µm.

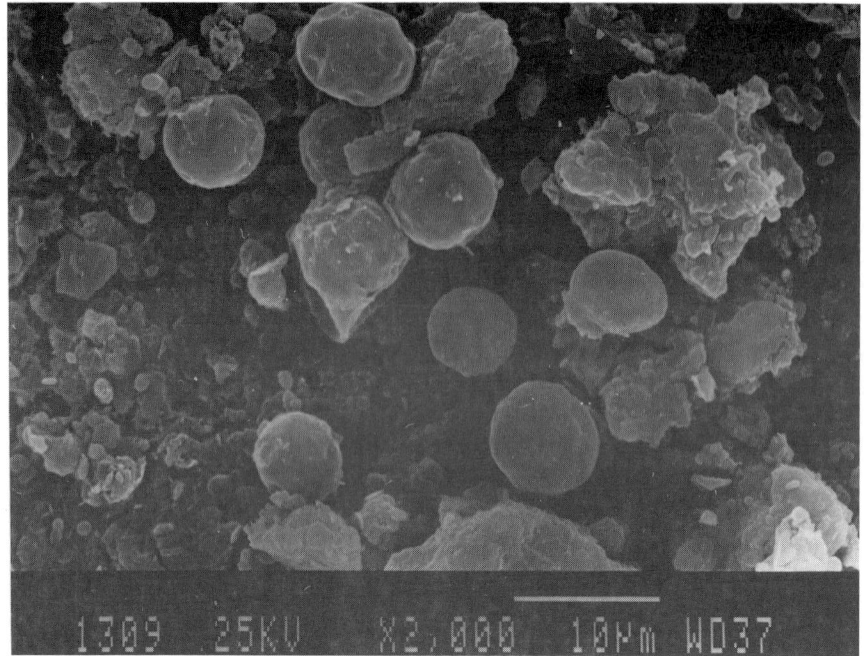

Figure 8. Chasmoendolithic green algae, Nubian sandstone, Petra, Jordan. Bar = 10 μm.

Figure 9. Chasmoendolithic bacteria and a few cyanobacteria, Nubian sandstone, Petra, Jordan. Bar = 10 μm.

Figure 10. Euendolithic cyanobacteria, limestone, Sede Boker, Negev Desert, Israel. Bar is in millimeters. Photo: A. Shoob

Figure 11. Exfoliation of marble. Apollo Temple, Side, Turkey.

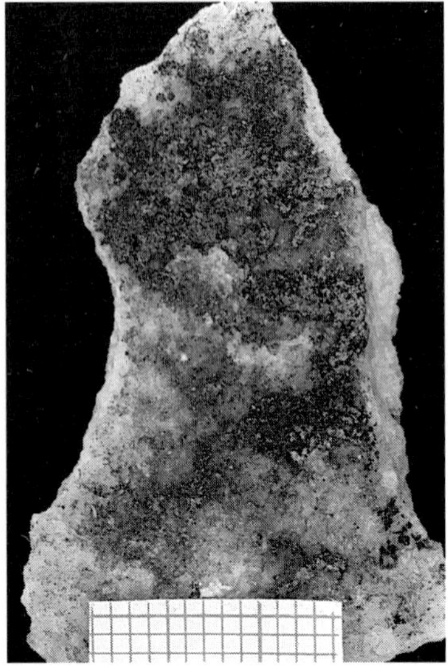

Figure 12. Chasmoendolithic biofilm. Marble, Apollo Temple, Side, Turkey. Bar is in millimeters. Photo: A. Shoob

Figure 13. Chasmoendolithic cyanobacteria, limestone, Wadi Arava, Israel. Bar is in millimeters. Photo: A. Shoob

Figure 14. Chasmoendolithic biofilm. Nubian sandstone, Petra, Jordan. Bar is in millimeters. Photo: A. Shoob

8. Acknowledgment –
I thank Rachel Garty-Spitz for her valued contribution to the written manuscript.

9. References

Ben-Shaul, Y., Paran,. N. and Galun, M. (1969) *J. Microscopie* **8**, 415-422.

Berner, T. and Evenari, M. (1978) *Oecologia* **33**, 255-260.

Billi, D. and Grilli Caiola, M. (1996) *New Phytol.***133**, 563-571.

Büdel, B. and Lange, O.L. (1991) *Bot. Acta* **104**, 361-366.

Cowan, I.R., Lange, O.L. and Green, T.G.A. (1992) *Planta* **87**, 282-294.

Danin, A. (1983) *Isr. Land Nature* **8**, 94-98.

Danin, A. (1985) *Bull. Amer. School Orient. Res.* **259**, 33-43.

Danin, A. (1986) *Proc. Roy. Soc. Edinburgh Biol. Ser.* **89B**, 243-253.

Danin, A. (1989) In E. Spanier (ed.). *Environmental Quality and Ecosystem Stability*, Proceedings of the Fourth International Conference of the Israel Society for Ecology and Environmental Quality Sciences, Jerusalem, Vol. IV/B, pp. 53-64.

Danin, A. (1992) In A. Gasith, A. Adin and Y, Steinberger (eds.). *Environmental Quality and Ecosystem Stability*, Proceedings of the Fifth International Conference of the Israel Society for Ecology and Environmental Quality Sciences, Jerusalem, Vol. V/B, pp. 675-681.

Danin, A. and Garty, J. (1983) *Z. Geomorph.* N.F. **27**, 423-444.

Danin, A. and Caneva, G. (1990) *Inter. Biodeterioration* **26**, 397-417.

Danin, A. Gerson, R., Marton, K. and Garty, J. (1982) *Palaegeogr., Palaeoclimatol., Palaeoecol.* **37**, 221-233.

Danin, A., Gerson, R. and Garty, J. (1983) *Soil Sci.* **136**, 213-217.

El-Oqlah, A.A. (1992) *Dirasat* **19B**, 109-121.

Friedmann, E.I. (1980) *Origin of Life* **10**, 223-235.

Friedmann, E.I. and Galun, M. (1974) In G.W. Brown (ed.). *Desert Biology*. Academic Press, San Francisco, Vol. II,, pp. 165-212.

Friedmann, E.I. and Ocampo-Friedmann, R. (1984) In M.J. Klug and C.A. Reddy (eds.). *Current Perspectives in Microbial Ecology*, Proceedings of the Third International Symposium on Microbial Ecology, American Society for Microbiology, Washington D.C., pp. 177-185.

Friedmann, E.I. and Ocampo-Friedmann, R. (1985) *Arch. Hydrobiol. Suppl.* **71**, 1/2 (algological studies 38/39) 349-350.

Friedmann, I., Lipkin, Y. and Ocampo-Paus, R. (1967) *Phycologia* **6**, 185-200.

Galun, M. (1970) *The Lichens of Israel.* The Israel Academy of Sciences and Humanities, Jerusalem.

Galun, M. and Garty, J. (1972) *Isr. J. Bot.* **21**, 243-254.

Galun, M. and Mukhtar, A. (1996) *Isr. J. Bot.* **44**, 195-218.

Galun, M., Paran, N. and Ben-Shaul, Y. (1970) *J. Microscopie* **9**, 801-806.

Garty, J. (1985) *Environ. Pollut.* Series B. 287-300.

Garty, J. (1988) *Stud. Geobot.*. **8**, 13-21.

Garty, J. (1990) *Can. J. Bot.* **68**, 1349-1353.

Garty, J. (1992) *Can. J. Bot.* **70**, 301-312.

Garty, J. (1995) In F.J.A. Daniels, M. Schulz and J. Peine (eds.). *Flechten Follmann, Contribution to Lichenology in Honour of Gerhard Follmann*, The Geobotanical and Phytotaxonomical Study Group, Botanical Institute, University of Cologne, Cologne, pp. 111-121.

Garty, J. and Galun, M. (1974) *Flora* **163**, 530-534.

Garty, J. and Delarea, J. (1987) *Symbiosis* **3**, 49-56.

Garty, J. and Delarea, J. (1988) *Can. J. Bot.* **66**, 2171-2177.

Garty, J. and Binyamini, N. (1990) *Environ. Exp. Bot.* **30**, 127-139.

Garty, J., Fuchs, C., Zisapel, N. and Galun, M. (1977) *Water, Air, Soil Pollut.* **8**, 171-188.

Garty, J., Galun, M. and Kessel, M. (1979) *New Phytol.* **82**, 159-168.

Garty, J., Galun, M. and Hochberg, Y. (1986) *Lichenologist* **18**, 257-263.

Garty, J., Harel, Y. and Steinberger, Y. (1995) *Arch. Env. Contam. Toxicol.* **29**, 247-253.

Garty, J., Steinberger, Y. and Harel, Y. (1996) *Environ. Exp. Bot.* **36**, 83-97.

Golubic, S., Friedmann, E.I. and Schneider, J. (1981) *J. Sed. Petrol.* **51**, 476-478.

Green, T.G.A. and Lange, O.L. (1994) In E.D. Schulze and M.C. Caldwell (eds.). *Ecophysiology of Photosynthesis*, Springer-Verlag, Berlin-Heidelberg-New York, pp. 319-341.

Green, T.G.A., Lange, O.L. and Cowan, I.R. (1994) *Crypt. Bot.* **4**, 166-178.

Grilli Caiola, M., Ocampo-Friedmann, R. and Friedmann, E.I. (1993) *Phycologia* **32**, 315-322.

Grilli Caiola, M., Billi, D. and Friedmann, E.I. (1996) *Eur. J. Phycol.* **31**, 97-105.

Grote, G. and Krumbein, W.E. (1992a) *Geomicrobiol. J.* **10**, 49-57.

Grote, G. and Krumbein, W.E. (1992b) *Restauro* **99**, 111-114.

Hungate, B., Danin, A., Pellerin, N.B., Stemmler, J., Kjellander, P., Adams, J.B. and Staley, J.T. (1987) *Can. J. Microbiol.* **33**, 939-943.

Jones, C.G. and Shachak, M. (1990) *Nature* **346**, 839-841.

Kappen, L., Lange, O.L., Schulze, E.D., Evenari, M. and Buschbom, U. (1975) In J. Cooper (ed.). *Photosynthesis and Productivity in Different Environments*, International Biological programme, Cambridge University Press, Vol. 3, pp. 133-143.

Kappen, L., Lange, O.L., Schulze, E.D., Evenari, M. and Buschbom, U. (1979) *Flora* **168**, 85-108.

Kappen, L., Lange, O.L., Schulze, E.D., Buschbom, U. and Evenari, M. (1980) *Flora* **169**, 216-229.

Krinsley, D.H., Dorn, R.I. and Anderson, S.W. (1990) *Phys. Geogr.* **11**, 97-119.

Krumbein, W.E. and Jens, K. (1981) *Oecologia* **50**, 25-38.

Kushnir, E. (1979) Studies in endolithic lichens of Israel. Ph.D. thesis (in Hebrew). Tel Aviv University, Tel Aviv.

Kushnir, E. and Galun, M. (1977) *Lichenologist* **9**, 123-130.

Kushnir, E., Tietz, A. and Galun, M. (1978) *Protoplasma* **97**, 47-60.

Lange, O.L. (1969) *Flora* **158**, 324-359.

Lange, O.L. (1980) *Oecologia* **45**, 82-87.

Lange, O.L. (1990) *Isr. J. Bot.* **39**, 383-394.

Lange, O.L. and Bertsch, A. (1965) *Naturwissenschaften* **9**, 215-216.

Lange, O.L. and Evenari, M. (1971) *Flora* **160**, 100-104.

Lange, O.L., Schulze, E.D. and Koch, W. (1968) *Naturwissenschaften* **12**, 658-659.

Lange, O.L. and Tenhunen, J.D. (1981) *Oecologia* **51**, 426-429.

Lange, O.L., Schulze, E.D. and Koch, W. (1970a) *Flora* **159**, 38-62.

Lange, O.L., Schulze, E.D. and Koch, W. (1970b) *Flora* **159**, 525-528.

Lange, O.L., Schulze, E.D., Kappen, L., Buschbom, U. and Evenari, M. (1975) In N.F. Hadley (ed.). *Environmental Physiology of Desert Organisms*, Dowden, Hutchinson and Boss, Stroudsburg, Pennsylvania, pp. 20-37.

Lange, O.L., Geiger, I.L. and Schulze, E.D. (1977) *Oecologia* **28**, 247-259.

Mukhtar, A., Garty, J. and Galun, M. (1994) *Symbiosis* **17**, 247-253.

Nevo, E., Apelbaum-Elkaher, I., Garty, J. and Beiles, A. (1997) *Heredity* **78**, 373-382.

Palmer, R.J. and Friedmann, E.I. (1990) *New Phytol.* **116**, 597-603.

Shachak, M., Jones, C.G. and Granot, Y. (1987) *Science* **236**, 1098-1099.
Scheidegger, C., Schroeter, B. and Frey, B. (1995) *Planta* **197**, 399-409.
Turian, G. (1977) *Ber. schweiz. bot. Ges.* **87**, 25-33.

Biodata of **Reinhard Rachel** author of "***Fine Structure of hyperthermophilic Prokaryotes.***"

Dr. Reinhard Rachel is an assistant Professor at the University of Regensburg (Germany). He earned his Ph.D. degree at the Technical University of Munich (1987) following his investigation at the Max-Planck-Institute (MPI) for Biochemistry, Martinsried (near Munich). Then he had a postdoctoral grant from the Boehringer-Ingelheim-Fonds, working at the MPI (Martinsried) and an EMBO longterm fellowship for the Medical Research Council, Laboratory of Molecular Biology, Cambridge, England. The main interests of Dr. Rachel are the ultrastructure of the extracellular "Cannulae" - network of *Pyrodictium*, a hyperthermophilic archaeum; ultrastructure of hyperthermophilic microorganisms; and structure determination of biological macromolecules by electron microscopy.
E-mail: **reinhard.rachel@biologie.uni-regensburg.de**

J. Seckbach (ed.), Enigmatic Microorganisms and Life in Extreme Environments, 277–289.

FINE STRUCTURE OF HYPERTHERMOPHILIC PROKARYOTES

R. RACHEL
Lehrstuhl für Mikrobiologie and Archaeenzentrum,
Universität Regensburg, Universitätsstr. 31,
D-93053 Regensburg, Germany

1. Introduction

Detailed comparison of the sequences of the 16S rRNA (Burggraf et al., 1997) and the 23S rRNA (Ludwig and Schleifer, 1994) shows that all prokaryotic microorganisms belong to two superkingdoms, the bacteria and the archaea (Woese and Fox, 1977; Woese et al., 1990). The hyperthermophiles, with an optimum growth temperature of $T_{opt} \geq 80°C$ (Stetter, 1996a), belong to phylogenetically distant groups. Most of them have an optimal growth temperature at 80°C to 105°C, but are unable to grow below 60°C. Their upper temperature border for growth is 110°C (*Pyrodictium*; *Methanopyrus*) and 113°C (*Pyrolobus*). In both the archaeal and the bacterial domain, the hyperthermophiles occupy the shortest and deepest branches of the phylogenetic tree, suggesting that the common ancestor of life was a hyperthermophile (Stetter, 1996a).

In a number of original descriptions and reviews, many aspects of the biology of hyperthermophilic microorganisms, like the environments where the original samples were taken, the cultivation of the organisms under laboratory conditions, their phylogeny, taxonomy, and physiology, have been described in detail (see e.g. Stetter, 1996a, b). In this chapter, the ultrastructural characterization of ten hyperthermophilic prokaryotes, belonging to different genera, is summarized.

2. Strains and Preparation Methods

The species presented here are representatives of all major lineages of hyperthermophilic prokaryotes (Stetter, 1996a). All the strains were taken from the culture collection of the Archaeenzentrum at the University of Regensburg. Their taxonomic position is based on the complete 16S rRNA sequence (Stetter, 1996a,b; Burggraf et al., 1997).

Microorganisms were processed for ultrathin sections as follows: *Thermotoga maritima*, *Aquifex pyrophilus*, and *Metallosphaera sedula* were fixed and embedded at room temperature, essentially as outlined (Huber et al., 1992; Fuchs et al., 1995).

Cells of *Pyrobaculum* and *Pyrolobus* were processed by freeze-substitution as described (Blöchl et al., 1997). *Pyrodictium* cells were grown in cellulose capillary tubes, high-pressure frozen, and freeze-substituted (Rieger et al., 1997). Cells representing the hyperthermophilic genera of the Euryarchaeota (*Thermococcus*; *Methanococcus*; *Archaeoglobus*; *Methanopyrus*) were processed as described recently for *Archaeoglobus veneficus* (Huber et al., 1997).

Domain BACTERIA:

Order *Thermotogales*: *Thermotoga maritima* MSB8 (Huber et al., 1986) DSM 3109
Order *Aquificales*: *Aquifex pyrophilus* Kol5a (Huber et al., 1992) DSM 6858

Domain ARCHAEA:

I. Kingdom Crenarchaeota:
Order *Sulfolobales*: *Metallosphaera prunae* Ron12/II (Fuchs et al., 1995) DSM 10039
Order *Desulfurococcales*: *Pyrodictium abyssi* TAG11 (Rieger et al., 1995) DSM ---
Pyrolobus fumarii 1A (Blöchl et al., 1997) DSM 11204
Order *Thermoproteales*: *Pyrobaculum aerophilum* IM2 (Völkl et al., 1993) DSM 7523
II. Kingdom Euryarchaeota
Order *Thermococcales*: *Thermococcus chitonophagus* GC74 (Huber et al., 1995) DSM 10152
Order *Methanococcales*: *Methanococcus igneus* Kol5 (Burggraf et al., 1990b) DSM 5666
Order *Archaeoglobales*: *Archaeoglobus veneficus* SNP6 (Huber et al., 1997) DSM 11195
Order *Methanopyrales*: *Methanopyrus kandleri* AV19 (Kurr et al., 1991) DSM 6324

3. Ultrastucture of Hyperthermophilic Bacteria

Among the bacteria, there are two major orders, the heterotrophic *Thermotogales* and the autotrophic *Aquificales*, which include hyperthermophilic genera.

All species of the five genera of the *Thermotogales* are rod-shaped cells, most of them with a single flagellum. Detailed studies have been carried out on the cell envelope architecture of *Thermotoga maritima*. The cytoplasmic membrane contains unusual, novel lipids (Manca et al., 1992). The murein is sensitive to lysozyme and not synthesized in the presence of antibiotics like Penicillin G or Ampicillin. It cannot be isolated as a rigid sacculus, but only in small patches of peptidoglycan, suggesting a low degree of cross-linking between the chains (Huber et al., 1986). The periplasmic space (Graham et al., 1991b) is spanned by a unique rod-like α-helical protein, Ompα. It has a length of approx. 40 nm (Engel et al., 1992) and is assembled in the form of a parallel four-stranded coiled-coil (Lupas et al., 1995). A characteristic feature of *Thermotoga* is the sheath-like outer membrane (Huber et al., 1986), adhering to the cylindrical part of the cell and overballooning at the cell poles (Fig. 1a). In the case of the close relative *Fervidobacterium*, the outer membrane surrounds several cells in a chain. The flagellum is inserted into the cell wall in the area where the outer membrane starts to balloon out from the cell body. The major constituent of the outer membrane is a porin-like protein, Ompβ (Engel et al., 1993; Schulz, 1996). Like other porins, it has a high content of β-sheets in its secondary structure. In vivo, it is tightly packed in patches of two-dimensional crystalline arrays of trimeric porin complexes (Rachel et al., 1990). The *Thermotoga* porin has, when incorporated into lipid bilayers, a high selectivity for cations (Engel et al., 1993).

Aquifex pyrophilus (Huber et al., 1992) and *A. aeolicus* (Deckert et al., 1998) are marine hyperthermophiles of the order *Aquificales*. The cells are rod-shaped (length: 2 to 6 μm, 0.5 μm wide), with up to 8 flagella. The flagella are structurally similar to those from other bacterial species (19 nm in diameter), and built up from a single flagellin polypeptide of 54 kDa (Behammer et al., 1995). The cell envelope (Fig. 1b) consists of a cytoplasmic membrane with glycerol alkyl ethers, a 20 nm wide periplasmic space, containing peptidoglycan, and an outer membrane, possibly also with porin-like proteins (see the full genome sequence from *A. aeolicus*: Deckert et al. 1995). The cell surface is completely covered by an S-layer, a well-ordered two-dimensional array of surface-layer proteins, arranged on a hexagonal lattice, with a centre-to-centre spacing of 18 nm (Huber et al., 1992).

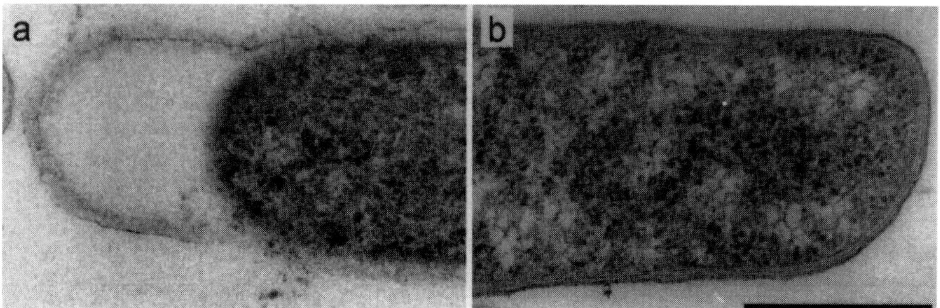

Figure 1: Ultrastructure of hyperthermophilic bacteria: ultrathin sections of *Thermotoga maritima* (a) and *Aquifex pyrophilus* (b), both at the same scale. Bar: 0.5 μm.

4. Ultrastructure of Hyperthermophilic Archaea

Most of the hyperthermophilic archaea exhibit a rather densely stained cytoplasm in ultrathin sections after freeze-substitution. Ribosome-free areas, where the DNA might be located, have been observed occasionally in *Pyrodictium* (Rieger et al., 1995). DNA and histone-like proteins were shown to be unevenly distributed in *Sulfolobus* and *Thermoplasma* (Bohrmann et al., 1994; Bohrmann and Kellenberger, 1994). In archaea, DNA is packed by histones, homologous to the histones found in eukaryotes, into archaeal nucleosomes (Reeve et al., 1997). Reverse gyrases introduce positive super-helical turns into the DNA helix (Bouthier de la Tour et al., 1990).

Many (not only the hyperthermophilic) archaea possess flagella which are structurally and biochemically different to those from bacteria (Typke et al., 1988; Jarrell et al., 1996). A new structure (15 to 20 nm thin) which connects individual cells have been found in two hyperthermophilic archaea, *Thermoproteus* and *Pyrobaculum*, in samples not subjected to centrifugation (Rieger et al., 1997).

The cell envelope of most hyperthermophilic archaea is composed of (a) the cyto-plasmic membrane, (b) many identical copies of a (glyco-)protein, arranged on a two-dimensional lattice, the surface layer or S-layer (Baumeister and Lembcke, 1992), and (c) spacer molecules, anchoring the S-layer into the cytoplasmic membrane. This arrangement results in a space between the cytoplasmic membrane and the S-layer sheet. The compartment was named the '(quasi-)periplasmic space' of the archaeal cell (Sumper, 1987; Baumeister et al., 1989; Baumeister and Lembcke, 1992) and is here referred to as the periplasm.

The cytoplasmic membrane of hyperthermophilic archaea is approx 6 to 10 nm thick. This value is, in general, difficult to determine only by electron microscopy. Archaeal membranes contain lipids of unique chemical composition, namely alkyl glycerol di- or tetraethers. The alkyl chains are isopranoid branched C_{20} or C_{40} molecules (Gambacorta et al., 1994). It is not known, whether this results in a different, possibly tighter packing of the lipid leaflets, in comparison with bacterial membranes. It is noteworthy that bacteriorhodopsin, a molecule spanning the cytoplasmic membrane of the mesophilic archaeum *Halobacterium salinarum*, has a thickness of 5.8 nm (Kimura et al., 1997).

4.1 THE CRENARCHAEOTA

All genera of the Crenarchaeota are thermophiles or hyperthermophiles. This archaeal kingdom comprises three orders, the *Sulfolobales*, the *Thermoproteales*, and the *Desulfurococcales* (Burggraf et al., 1997).

Among the *Sulfolobales*, four genera are known today, *Sulfolobus*, *Metallosphaera*, *Acidianus*, and *Stygiolobus* (Fuchs et al., 1996). The cells are lobed cocci. Cells of *Sulfolobus* and *Metallosphaera* have been shown to possess flagella (Grogan, 1989; Fuchs et al., 1995). In all genera, a cytoplasmic membrane is surrounded by a periplasm, approx. 20 to 25 nm wide, and an S-layer, completely covering the cell surface (Fig. 2a). In cells of *Sulfolobus*, *Metallosphaera*, and *Acidianus*, the S-layer complexes are arranged on a p3 lattice with a centre-to-centre distance of 20 to 21 nm (Baumeister and Lembcke, 1992; Fuchs et al., 1995). The architecture of the S-layer of the fourth genus, *Stygiolobus*, has not been investigated so far.

The order of the *Thermoproteales* (Zillig et al., 1981) contains three genera, *Thermoproteus*, *Pyrobaculum*, and *Thermofilum* (Burggraf et al., 1997). Cells are rod-shaped, up to 10 to 20 μm long or even longer, with nearly rectangular ends, and 0.4 to 0.5 μm (*Thermoproteus*, *Pyrobaculum*) and 0.15 to 0.35 μm (*Thermofilum*) in diameter. *Thermoproteus* and at least some species of *Pyrobaculum* exhibit up to 2 to 3 flagella at the cell poles. The cytoplasmic membrane, with tetraether and diether lipids, is surrounded by a periplasm and an S-layer (Fig. 2b). The thickness of the periplasm is approx. 25 nm. The centre-to-centre distance between the protein complexes of the S-layer is approx. 30 nm (28 - 32 nm, depending on the species; Wildhaber and Baumeister, 1987; Phipps et al., 1990). The three-dimensional architecture of the S-layers from the two genera, *Thermoproteus* and *Pyrobaculum*, is indistinguishable by electron crystallography, but different to all other S-layers described so far. It can be regarded as an ultrastructural feature aiding in the identification of these genera. This S-layer can readily be isolated due to its resistance to denaturing conditions like 2% SDS at T=100°C and is, therefore, most likely a cross-linked protein sacculus. The isolated sheets retain the shape of the cell and are considered to play a role in shape maintenance of the cells. While this S-layer is the outermost boundary of *Thermoproteus* cells, various *Pyrobaculum* species have been shown to be covered by additional layers, either by a second S-layer with same symmetry (p6) but different lattice parameters (Phipps et al., 1991), or by a fibrous or densely stained surface coat (Phipps et al., 1990; Fig. 2b).

Within the *Desulfurococcales*, two families are described so far, the *Desulfurococcaceae* and the *Pyrodictiaceae* (Burggraf et al., 1997). Of the first family, detailed ultrastructural descriptions are available for the species *Desulfurococcus* and *Staphylothermus*. Ultrathin sections, freeze-etching, and electron crystallography have shown that the cell wall of *Desulfurococcus* consists of an S-layer with four-fold symmetry and a centre-to-centre distance of 18 nm (Wildhaber et al., 1987), built up of protein complexes in the shape of a cross, which enclose a periplasm with a width of approx. 30 nm. The cell envelope of *Staphylothermus* is unusual in having an almost 60 nm wide periplasm (Peters et al., 1995). It is spanned by long stalks of the S-layer protein, the so-called tetrabrachion, a protein complex with four-fold symmetry. The stalks are made up of four copies of an α-helical protein, to which a thermostable protease is tightly bound (Mayr et al., 1996). Ultrathin section of other species of the *Desulfurococcaceae* (*Aeropyrum*: Sako et al., 1996; *Sulfophobococcus*: Hensel et al., 1997; *Stetteria*: Jochimsen et al., 1997) reveal that they have the typical cell envelope

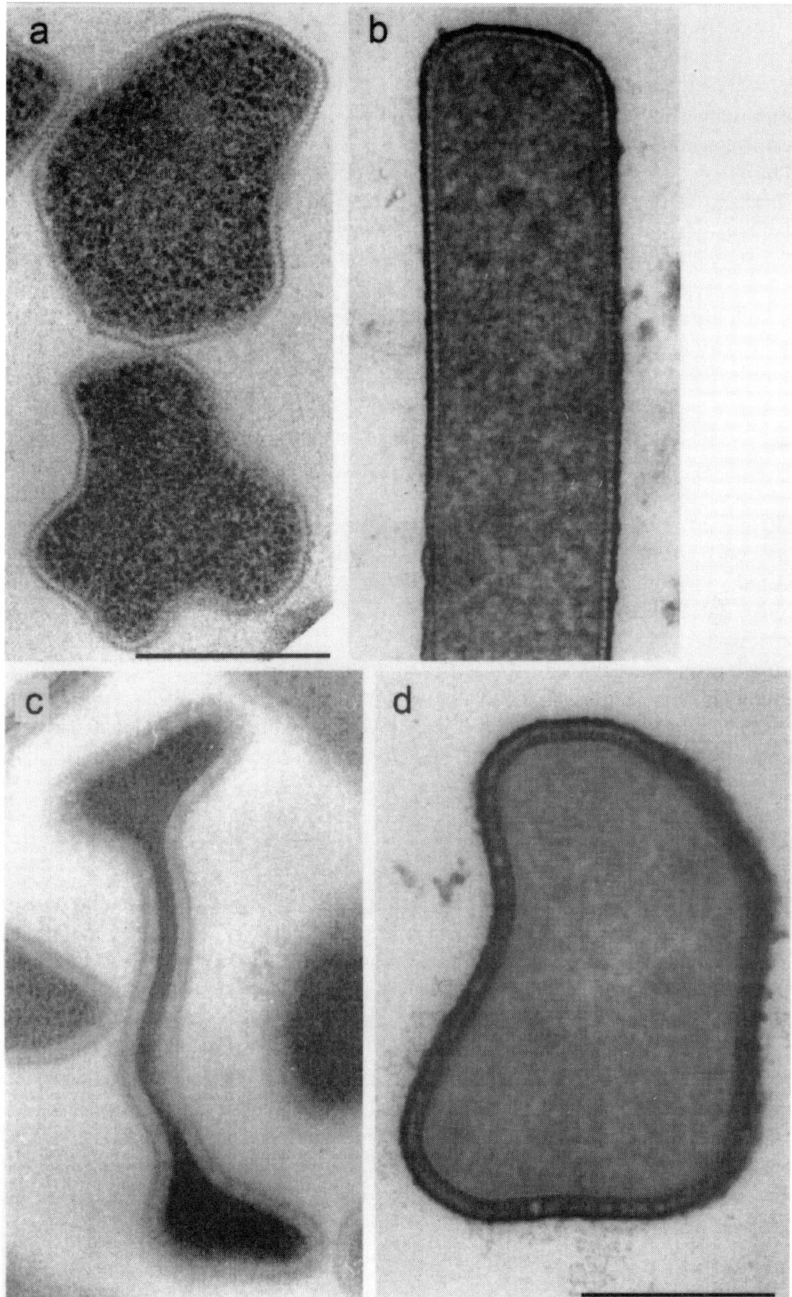

Fig. 2: Ultrastructure of hyperthermophilic Crenarchaeota: ultrathin sections of *Metallosphaera prunae* (a), *Pyrobaculum aerophilum* (b), *Pyrodictium abyssi* (c), and *Pyrolobus fumarii* (d), all at the same scale. Bar: 0.5 μm.

ultrastructure. The cytoplasmic membrane is surrounded by a periplasm with a width of approx. 30 to 40 nm. The outer boundary is a thin layer tentatively named an S-layer, in analogy with the structure seen in sections of other Crenarchaeota. Electron microscopy of freeze-etched cells or of isolated S-layer sheets, or biochemical studies of cell wall components, have not yet been performed on these microorganisms.

The ultrastructure of the species of the *Pyrodictiaceae* has been studied in some detail. The three genera described today are all lobed cocci, with cell diameters of 0.3 µm to 2.5 µm. The periplasm has a width of 30 to 35 nm (*Pyrodictium*: Fig. 2c; *Hyperthermus*: Zillig et al., 1990) or 40 nm (*Pyrolobus*: Fig. 2d), and cells are covered by an S-layer. The structural analysis reveals characteristic differences between these genera. *Pyrodictium* grows in liquid cultures in the form of flakes, 1 to 10 mm in diameter, which can easily be seen by the naked eye. Electron microscopy has shown that it is an organism consisting of highly lobed cells, which are interconnected by a network of hollow tubes (cannulae), 25 nm in diameter (Stetter, 1982; König et al., 1988; Rieger et al., 1995). The network is best visualized by scanning electron microscopy. The cell wall is covered by an S-layer with sixfold symmetry and a centre-to-centre distance of 21 nm (Dürr et al., 1991; Rieger et al., 1995). Ultrathin sections of high-pressure-frozen, freeze-substituted cells (Fig. 2a) show that the cells are indeed extremely lobed (Fig. 2c), thereby increasing the surface-to-volume ratio of the individual cell. *Hyperthermus butylicus* (Zillig et al., 1990) and *Pyrolobus fumarii* (Blöchl et al., 1997) are both closely related organisms, regarding the sequence of their 16S rRNA, but lack the extracellular network of hollow cannulae. *Hyperthermus* cells are also covered by an S-layer with hexagonal symmetry, but in contrast to *Pyrodictium*, with a larger centre-to-centre distance of the complexes (25 vs. 21 nm) and with a different three-dimensional structure (Baumeister et al, 1990; vs. Dürr et al., 1991). *Pyrolobus* cells are covered by an S-layer with different symmetry (p4), similar centre-to-centre distance (19 nm), and markedly different surface relief (Blöchl et al., 1997).

4.2 THE EURYARCHAEOTA

Thermococcaceae is a large family of thermophilic Euryarchaeota, with two genera, *Thermococcus* and *Pyrococcus*. All species known so far are cocci, approx. 0.8 to 1.5 µm in diameter, sometimes with single or few flagella, others with tufts of up to 30 flagella. The ultrastructure of *Thermococcus chitonophagus* was investigated by freeze-etching and by ultrathin sections after freeze-substitution (Huber et al., 1995). The cell membrane and a 10 nm wide periplasm are surrounded by an S-layer with sixfold symmetry, with a centre-to-centre distance between the protein complexes of 18 nm (Fig. 3a). Similar data have been described for the isolated S-layer of *Thermococcus celer* (Baumeister et al., 1990). The other species of the *Thermococci* (see, e.g. Erauso et al., 1993; Fiala et al., 1986; Gonzales et al., 1995; Miroshnichenko, 1989; Keller et al., 1995), as well as the *Pyrococcus* species described so far, have basically the same ultrastructure. In some *Thermococci*, additional cell wall components in the form of densely stained layers have been observed. One example is a thick layer of unknown composition covering the S-layer of *Thermococcus peptonophilus* (Gonzales et al., 1995). A non-proteinaceous cell wall polymer, an exopolysaccharide built up of mannose subunits, was found in *Thermococcus litoralis* (Rinker and Kelly, 1996).

The two hyperthermophilic members of the *Methanococcaceae* are *Methanococcus jannaschii* (Jones et al., 1983; Bult et al., 1996) and *Mc igneus* (Burggraf et al.,

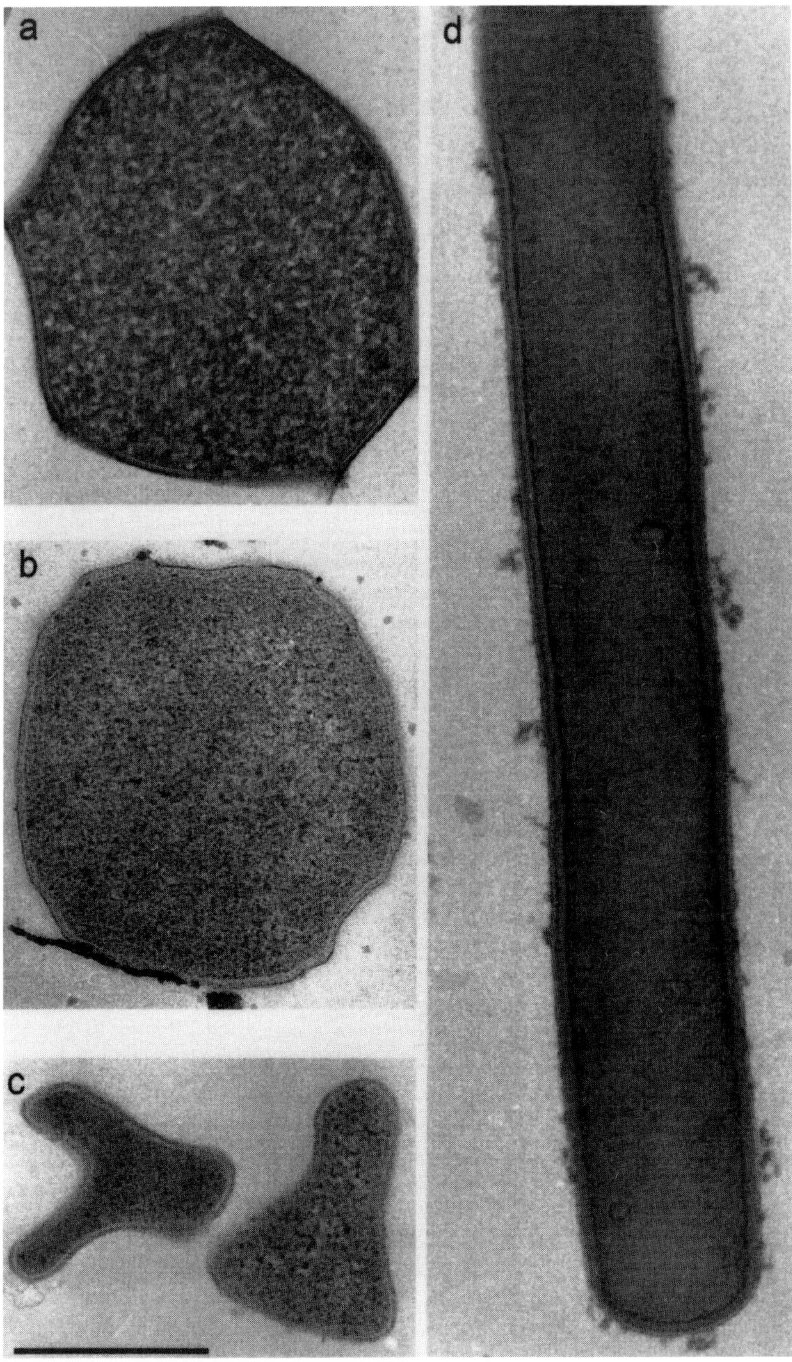

Fig. 3: Ultrastructure of hyperthermophilic Euryarchaeota: ultrathin sections of *Thermococcus chitonophagus* (a), *Methanococcus igneus* (b), *Archaeoglobus veneficus* (c), and *Methanopyrus kandleri* (d), all at the same scale. Bar: 0.5 µm.

1990b). Both are flagellated cocci. *Mc. igneus* has a single flagellum, while *Mc. jannaschii* has two bundles of at least 20 flagella. Their cell wall ultrastructure, as seen in ultrathin sections (Fig. 3b), is basically similar to that of the *Thermococcaceae*. The periplasm, approx. 10 nm wide, is surrounded by an S-layer with sixfold symmetry; the centre-to-centre distance of the protein complexes has not been determined yet.

Regarding ultrastructure, *Archaeoglobus veneficus* (Fig. 3c; Huber et al., 1997) is a representative of the other two species, *A. fulgidus* (Stetter, 1988; Klenk et al., 1997) and *A. profundus* (Burggraf et al., 1990a). The cells are lobed cocci, with one or two flagella. Their cell wall consists of a periplasm, 10 nm wide, and a surface layer protein with hexagonal symmetry and a centre-to-centre distance between the protein complexes of 18 nm. The surface relief of the S-layer is dome-shaped. The connectivities between the protein complexes are barely visible in the reconstructions (Kessel et al., 1990). *Ferroglobus placidus*, a recently described genus of the order *Archaeoglobales*, has basically the same ultrastructure, but cells are covered by an S-layer with fourfold symmetry (Hafenbradl et al., 1996).

Methanopyrus kandleri has the highest growth temperature of all methanogens known today (110°C; Huber et al. 1989; Kurr et al., 1991). The cells are rodshaped, with an average length of 8 to 10 µm and a diameter of approx. 0.5 µm. The composition and architecture of the cell envelope is different and more complex compared to the species described so far. The lipid content of the cell membrane is unusual, as only glycerol diethers were found, including a novel unsatürated lipid (Hafenbradl et al., 1993). In ultrathin sections (Fig. 3d), the outer leaflet of the cytoplasmic membrane is more densely stained than the inner one. The periplasm with the pseudomurein (König et al., 1994) appears as a 15 nm thick, homogeneously stained band. The surface coat, with a total width of approx. 10 to 15 nm, is composed of two layers. The inner layer is weakly stained, the outer one more densely. On the surface of many cells, a 'fuzzy' coat, whose composition is unknown, can be seen. Similar in its ultrastructure, but different in the sequence of its 16S rRNA is *Methanothermus fervidus*, the other hyperthermophilic rod-shaped methanogen (Stetter et al., 1981). In ultrathin sections, the cell wall profile is hardly distinguishable from that of *Methanopyrus* cells. From *Methanothermus*, a surface protein has been isolated and sequenced (Bröckl et al., 1991). Freeze-etching studies of *Methanothermus* and *Methanopyrus* gave no evidence for that the protein on the cell surface is regularly arranged on a two-dimensional lattice and, therefore, could bona fide be named 'S-layer' (Rachel, in preparation).

5. Concluding remarks

The ultrastructure of hyperthermophilic bacteria and archaea is described, based on electron microscopical investigations using metal shadowing, ultrathin sectioning, and freeze-etching, and on data obtained by biochemical methods. Two known groups of the superkingdom bacteria are included (*Thermotoga* and *Aquifex*), as well as all groups of the crenarchaeota and four groups of the euryarchaeota.

The cell wall of hyperthermophilic bacteria is similar to the cell walls of other members of the superkingdom bacteria. They possess a periplasm, an outer membrane, and, in the case of *Aquifex*, an S-layer as the outermost layer.

The cell envelope of the hyperthermophilic Crenarchaeota is composed of three parts, the cytoplasmic membrane, the periplasm (20 to 35 nm wide, in one case up to 60 nm), and an S-layer, with either fourfold symmetry (many species of the

Desulfurococcales), threefold symmetry (most, if not all, species of the *Sulfolobales*), or sixfold symmetry (some species of the *Desulfurococcales*; all species of the *Thermoproteales*). On cells of some genera, additional fibrous or densely stained surface coats and cell-to-cell connections can be found.

The cell wall of most hyperthermophilic Euryarchaeota has a similar architecture, with subtle but noticeable differences to the Crenarchaeota. The periplasm is 5 to 10 nm thin, and the S-layer (if present) usually has sixfold symmetry (fourfold symmetry found only in one species, so far). The rod-shaped hyperthermophilic methanogens have a different, complex cell wall, including a pseudomurein, but lacking an S-layer sheet.

It will be interesting to see to which extent our current understanding of the structure of hyperthermophilic microorganisms will be modified when cryopreparation techniques (Humbel and Müller, 1986; Hohenberg et al., 1994) will be employed routinely, as done in ultrastructural studies of bacterial cells (see e.g., Hobot et al., 1984; Graham et al. 1991a; Hoiczyk and Baumeister, 1995) and of novel (hyper-)thermophilic prokaryotes (Huber et al., 1995; Huber et al., 1996; Blöchl et al., 1997; Huber et al., 1997).

6. Acknowledgments

The author wish to thank Prof. Dr. K.O. Stetter for continous support, Drs. R. Huber and H. Huber for fruitful collaboration and providing cells, P. Hummel for excellent technical assistance, Dr. M. Müller and Dr. R. Hermann (ETH Zürich) for the opportunity to use their equipment, H. Faust, B. Kattenbeck, G. Rieger, W. Eder, K. Teiner, S. Weber, I. Wyschkony, S. Lechner for processing the various archaeal and bacterial cells, and Dr. Ch. Fredriksson for carefully reading the manuscript.
This work was supported by grants of the Deutsche Forschungsgemeinschaft (SFB 43 and Ra 751/1-1) and by the Fonds der Chemischen Industrie.

7. References

Baumeister, W., I. Wildhaber, and B. M. Phipps (1989) Can. J. Microbiol. 35:215-227.

Baumeister, W., U. Santarius, S. Volker, R. Dürr, G. Lembcke, and H. Engelhardt (1990) System. Appl. Microbiol. 13:105-111.

Baumeister, W., and G. Lembcke (1992) J. Bioenerg. Biomembr. 24:567-575.

Behammer, W., Z. Shao, W. Mages, R. Rachel, K.O. Stetter, and R. Schmitt (1995) J. Bacteriol. 177:6630-6637.

Blöchl, E., R. Rachel, S. Burggraf, D. Hafenbradl, H. W. Jannasch, and K.O. Stetter (1997) Extremophiles 1:14-21.

Bohrmann, B., E. Kellenberger, B. Arnold-Schulz-Gahmen, K. Sreenivas, T. Suryanarayana, D. Stroup, and J. N. Reeve (1994) J. Struct. Biol. 112:70-78.

Bohrmann, B., and E. Kellenberger (1994) J. Histochem. Cytochem. 42:635-643.

Bouthier de la Tour, C., C. Portemer, M. Nadal, K.O. Stetter, P. Forterre, and M. Duguet (1990) J. Bacteriol. 172:6803-6808.

Bröckl, G., M. Behr, S. Fabry, R. Hensel, H. Kaudewitz, E. Biendl, and H. König (1991) Eur. J. Biochem. 199:147-152.

Bult, C.J., W. White, G.J. Olsen, L. Zhou, R.D. Fleischmann, G.G. Sutton, J.A. Blake, L.M. FitzGerald, R.A. Clayton, J.D. Gocayne, A.R. Kerlavage, B.A. Dougherty, J.-F. Tomb, M.D. Adams, C.I. Reich, R. Overbeek, E.F. Kirkness, K.G. Weinstock, J.M. Merrick, A. Glodek, J.L. Scott, N.S.M. Geoghagen, J.F. Weidman, J.L. Fuhrmann, D. Nguyen, T.R. Utterback, J.M. Kelley, J.D. Peterson, P.W. Sadow, M.C. Hanna, M.D. Cotton, K.M. Roberts, M.A. Hurst, B.P. Kaine, M. Borodovsky, H.-P. Klenk, C.M. Fraser, H.O. Smith, C.R. Woese, and J.C. Venter (1996) Science 273:1058-1073.

Burggraf, S., H.W. Jannasch, B. Nicolaus, and K.O. Stetter (1990a) System. Appl. Microbiol. 13:24-28

Burggraf, S., H. Fricke, A. Neuner, J. Kristjansson, P. Rouvier, L. Mandelco, C.R. Woese, and K.O. Stetter (1990b) System. Appl. Microbiol. 13:263-269.

Burggraf, S., H. Huber, and K.O. Stetter (1997) J. System. Bacteriol. 47:657-660.

Deckert, G., P.V. Warren, T. Gaasterland, W.G. Young, A.L. Lenox, D.E. Graham, R. Overbeek, M.A. Snead, M. Keller, M. Aujay, R. Huber, R.A. Feldman, J.M. Short, G.J. Olsen, and R.V. Swanson (1998) Nature 392:353-358.

Dürr, R., R. Hegerl, S. Volker, U. Santarius, and W. Baumeister (1991) J. Struct. Biol. 106:181-190.

Engel, A.M., Z. Cejka, A. Lupas, F. Lottspeich, and W. Baumeister (1992) EMBO J. 11:4369-4378.

Engel, A.M., M. Brunen, and W. Baumeister (1993) FEMS Microbiol. Lett. 109:231-236.

Erauso, G., A. L. Reysenbach, A. Godfroy, J. R. Meunier, B. Crump, F. Partensky, J. A. Baross, V. Marteinsson, G. Barbier, N. R. Pace, and D. Prieur (1993) Arch. Microbiol. 160:338-349.

Fiala, G. and K.O. Stetter (1986) Arch. Microbiol. 145:56-61.

Fuchs, T., H. Huber, K. Teiner, S. Burggraf, and K.O. Stetter (1995) System. Appl. Microbiol. 18:560-566.

Fuchs,T., H. Huber, S. Burggraf, and K.O. Stetter (1996) Syst. Appl. Microbiol. 19:56-60.

Gambacorta, A., A. Trincone, B. Nicolaus, L. Lama, and M. de Rosa (1994) *In* F. Pfeifer, P. Palm, K.H. Schleifer (eds.), Molecular biology of archaea. Fischer Verlag, Stuttgart. pp 18-27.

Gonzales, J. M., C. Kato, and K. Horikoshi (1995) Arch. Microbiol 164:159-164.

Graham, L. L., R. Harris, W. Villiger, and T. J. Beveridge (1991a) J. Bacteriol. 173:1623-1633

Graham, L. L., T.J. Beveridge, and N. Nanninga (1991b) TIBS 16:328-329.

Grogan, D.W (1989) J. Bacteriol. 171:6710-6719.

Hafenbradl, D., M. Keller, R. Thiericke, and K.O. Stetter (1993) System. Appl. Microbiol. 16:165-169.

Hafenbradl, D., M. Keller, R. Dirmeier, R. Rachel, P. Rossnagel, S. Burggraf, H. Huber, and K.O. Stetter (1996) Arch. Microbiol. 199:308-314.

Hensel, R., K. Matussek, K. Michalke, L. Tacke, B. J. Tindall, M. Kohlhoff, B. Siebers, and J. Dielenschneider (1997) System. Appl. Microbiol. 20:102-110.

Hobot, J., E. Carlemalm, W. Villiger, and E. Kellenberger (1984) J. Bacteriol. 160:143-152

Hohenberg, H., K. Mannweiler, and M. Müller (1994) J. Microsc. 175, 34-43

Hoiczyk, E., and W. Baumeister (1995) J. Bacteriol. 177:2387-2395.

Huber, R., T.A. Langworthy, H. König, M. Thomm, C.R. Woese, U.B. Sleytr (1986) Arch. Microbiol. 144:324-333.

Huber, R., M. Kurr, H.W. Jannasch, and K.O. Stetter (1989) Nature 342:833-834.

Huber, R., T. Wilharm, D. Huber, A. Trincone, S. Burggraf, H. König, R. Rachel, I. Rockinger, H. Fricke, and K.O. Stetter (1992) Syst. Appl. Microbiol. 15:340-351.

Huber, R., J. Stöhr, S. Hohenhaus, R. Rachel, S. Burggraf, H.W. Jannasch, and K.O. Stetter (1995) Arch. Microbiol. 164:255-264.

Huber, R., P. Rossnagel, C. R. Woese, R. Rachel, T. A. Langworthy, and K. O. Stetter (1996) System. Appl. Microbiol. 19:40-49.

Huber, H., H. Jannasch, R. Rachel, T. Fuchs, and K. O. Stetter (1997) System. Appl. Microbiol. 20:374-380.

Humbel, B. M., and M. Müller (1986) *In* M. Müller, R.P. Becker, A. Boyde, and J.J. Wolosewick (eds.), The science of biological specimen preparation 1985. SEM, AMF O'Hare, Ill., USA. p. 175-183.

Jarrell, K. F., D.P. Bayley, and A.S. Kostyukova (1996) J. Bacteriol. 178:5057-5064.

Jochimsen, B., S. Peinemann-Simon, H. Völker, D. Stüben, R. Botz, P. Stoffers, P.R. Dando, and M. Thomm (1997) Extremophiles 1:67-73.

Jones, W.J., J.A. Leigh, F. Mayer, C.R. Woese, and R.S. Wolfe (1983) Arch. Microbiol. 136:254-261.

Keller, M., F.-J. Braun, R. Dirmeier, D. Hafenbradl, S. Burggraf, R. Rachel, and K.O. Stetter (1995) Arch. Microbiol. 164:390-395.

Kessel, M., S. Volker, U. Santarius, R. Huber, and W. Baumeister (1990) System. Appl. Microbiol. 13:207-213.

Kimura, Y., D. G. Vassylyev, A. Miyazawa, A. Kidera, M. Matsushima, K. Mitsuoka, K. Murata, T. Hirai, and Y. Fujiyoshi (1997) Nature 389:206-211.

Klenk, H. P., R.A. Clayton, J.-F. Tomb, O. White, K.E. Nelson, K.A. Ketchum, R.J. Dodson, M. Gwinn, E.K. Hickey, J.D. Peterson, D.L. Richardson, A.R. Kerlavage, D.E. Graham, N.C. Kyrpides, R.D. Fleischmann, J. Quackenbush, N.H. Lee, G.G. Sutton, S. Gill, E.F. Kirkness, B.A. Dougherty, K. McKenney, M.D. Adams, B. Loftus, S. Peterson, C.I. Reich, L.K. McNeil, J.H. Badger, A. Glodek, L. Zhou, R. Oberbeek, J.D. Gocayne, J.F. Weidman, L. McDonald, T. Utterback, M.D. Cotton, T. Spriggs, P. Artiach, B.P. Kaine, S.M. Sykes, P.W. Sadow, K.P. D'Andrea, Ch. Bowman, C. Fujii, S.A. Garland, T.M. Mason, G.J. Olsen, C.M. Fraser, H.O. Smith, C.R. Woese, and J.C. Venter (1997) Nature 390:364-370.

König, H., P. Messner, and K.O. Stetter (1988) FEMS Microbiol. Lett. 49:207-212.

König, H., E. Hartmann, and U. Kärcher (1994) In: Molecular Biology of Archaea. Ed. F. Pfeifer, P. Palm, and K.-H. Schleifer. Fischer, Stuttgart. p 10-17.

Kurr, M., R. Huber, H. König, H. W. Jannasch, H. Fricke, A. Trincone, J. K. Kristjansson, and K. O. Stetter (1991) Arch. Microbiol. 156:239-247.

Lam, J. S., L. L. Graham, J. Lightfoot, T. Dasgupta, and T. J. Beveridge (1992) J. Bacteriol. 174:7159-7167.

Ludwig, W., and K.-H. Schleifer (1994) FEMS Microbiol. Rev. 15:155-173.

Lupas, A., S. Müller, K. Goldie, A.M. Engel, A. Engel, and W. Baumeister (1995) J. Mol. Biol. 248:180-189.

Manca, M.C., B. Nicolaus, V. Lanzotti, A. Trincone, A. Gambacorta, J. Peter-Katalinic, H. Egge, R. Huber, and K.O. Stetter (1992) Biochim. Biophys. Acta 1124:249-252.

Mayr, J., A. Lupas, J. Kellermann, Ch. Eckerskorn, W. Baumeister, and J. Peters (1996) Current Biology 6:739-749.

Miroshnichenko, M. L., E. A. Bonch-Osmolovskaya, A. Neuner, N. A. Kostrikina, N. A. Chernych, and V. A. Alekseev (1989) System. Appl. Microbiol. 12:257-262.

Peters, J., M. Nitsch, B. Kühlmorgen, R. Golbik, A. Lupas, J. Kellermann, H. Engelhardt, J. P. Pfander, S. Müller, K. Goldie, A. Engel, K. O. Stetter, and W. Baumeister (1995) J. Mol. Biol. 245:385-401.

Phipps, B. M., H. Engelhardt, R. Huber, and W. Baumeister (1990) J. Struct. Biol. 103:152-163.

Phipps, B. M., R. Huber, and W. Baumeister (1991) Mol. Microbiol. 5:253-265.

Rachel, R., A.M. Engel, R. Huber, K.O. Stetter, and W. Baumeister (1990) FEBS Lett. 262:64-68.

Reeve, J.N., K. Sandman, and C.J. Daniels (1997) Cell 89:999-1002.

Rieger G., R. Rachel, R. Hermann, and K.O. Stetter (1995) J. Struct. Biol. 115:78-87.

Rieger G., K. Müller, R. Hermann, K.O. Stetter, and R. Rachel (1997) Arch. Microbiol. 168:373-379

Rinker, K. D., and R. N. Kelly (1996) Appl. Environm. Microbiol. 62:4478-4485.

Sako, Y., N. Nomura, A. Uchida, Y. Ishida, H. Morii, Y. Koga, T. Hoaki, and T. Maruyama (1996) Internat. J. System. Bacteriol. 46:1070-1077.

Schulz, G. E (1996) Current Opinion in Structural Biology 6:485-490.

Stetter, K.O., M. Thomm, J. Winter, G. Wildgruber, H. Huber, W. Zillig, D. Janecovic, H. König, P. Palm, and S. Wunderl (1981) Zbl. Bakt. Hyg., I. Abt. Orig. C2, 166-178.

Stetter, K.O (1982) Nature 300:258-260.

Stetter, K.O (1988) System. Appl. Microbiol. 10:172-173.

Stetter, K.O (1996a) FEMS Microl. Rev. 18:149-158.

Stetter, K.O (1996b) In: Evolution of hydrothermal ecosystems on Earth (and Mars?). Wiley, Chichester (CIBA Foundation Symposium 202) p 1-18.

Sumper, M (1987) Biochim. Biophys. Acta 906:69-79.

Typke, D., M. Nitsch, A. Möhrle, R. Hegerl, M. Alam, D. Grogan, and J. Trent (1988) Inst. Phys. Conf. Ser. No. 93: Vol 3, Ch. 14, pp 379-380

Völkl P., R. Huber, E. Drobner, R. Rachel, S. Burggraf, A. Trincone, and K.O. Stetter (1993) Appl. Environ. Microbiol. 59:2918-2926.

Wildhaber I., and W. Baumeister (1987) EMBO J. 6:1475-1480.

Wildhaber, I., U. Santarius, and W. Baumeister (1987) J. Bacteriol. 169:5563-5568.

Woese, C.R., and G.E. Fox (1977) Proc. Natl. Acad. Sci. USA 74:5088-5090.

Woese, C.R., O. Kandler, and M.L. Wheelis (1990) Proc. Natl. Acad. Sci USA 87:4576-4579.

Zillig, W., K. O. Stetter, W. Schäfer, D. Janekovic, S. Wunderl, I. Holz, and P. Palm (1981) Zentralbl. Bakteriol. Mikrobiol. Hyg. 1. Abt. Orig. C 2:205-227.

Zillig W., I. Holz, D. Janekovic, H.-P. Klenk, E. Imsel, J. Trent, S. Wunderl, V. H. Forjaz, R. Coutinho, and T. Ferreira (1990) J. Bacteriol. 172:3959-3965

Biodata of Debashish Bhattacharya
author of: ***The Phylogeny of thermophiles and Hyperthermophiles and the Three Domains of Life.***

Dr. D. Bhattacharya is an Assistant Professor in the Department of Biological Sciences at the University of Iowa (Iowa City). He received his Ph.D. from Simon Fraser University in Burnaby, British Columbia (Canada). Subsequently, he joined the lab of Mitch Sogin (Woods Hole, MA) for three years as a postdoc in the field of Molecular Evolution. Then he spent six years conducting research in Germany, first at the University of Cologne and then at the Max Plank Institute where he also taught at the University of Göttingen.
e-mail: dbhattac@blue.weeg.uiowa.edu

J. Seckbach (ed.), Enigmatic Microorganisms and Life in Extreme Environments, 291–304.

THE PHYLOGENY OF THERMOPHILES AND HYPERTHERMOPHILES AND THE THREE DOMAINS OF LIFE

The Phylogeny of Thermophiles

DEBASHISH BHATTACHARYA
University of Iowa
Department of Biological Sciences
Biology Building, Iowa City, Iowa 52242-1324
United States

THOMAS FRIEDL
Department of Biology, General Botany
University of Kaiserslautern
P.O. Box 3049, D-67653 Kaiserslautern, Germany

HEIKO SCHMIDT
Deutsches Krebsforschungszentrum
Theoretische Bioinformatik
Im Neuenheimer Feld 280 , D-69120 Heidelberg, Germany

1. Introduction

The nature of the first cells and the environment in which they lived are two of the most interesting problems in evolutionary biology. All living things are descendents of these primordial cells and are divided into three fundamental lineages or domains, Archaea (formerly known as Archaebacteria), Bacteria (formerly known as Eubacteria), and the Eucarya (formerly known as Eukaryotes, Woese et al. 1990). The Archaea and Bacteria are prokaryotic domains whereas the Eucarya includes all other living things that have a nucleus (i.e., the genetic material is separated from the cytoplasm by a nuclear envelope). The observation of the three primary domains, first made on the basis of small subunit (i.e., 16S, 18S) ribosomal DNA (rDNA) sequence comparisons (Woese 1987), has created a framework with which the nature of the last common ancestor (LCA) can be addressed. In this review we present phylogenies of the prokaryotic domains to understand the origin and distribution of the thermophiles (organisms able to grow in temperatures > 45°C) and the hyperthermophiles (organisms able to grow in temperatures > 80°C). Hyperthermophiles are limited to the Archaea and Bacteria. In addition, we inspect the distribution of extremophiles within the cyanobacteria. The cyanobacteria are unique in being able to tolerate rapidly fluctuating environmental conditions. This capacity has presumably allowed some cyanobacteria to survive in niches unexplored by other prokaryotes and eukaryotes which have a more restricted environmental tolerance. We do not deal with eukaryotic extremophiles in this chapter because no multicellular plant or animal yet found can survive in temperatures of about 50°C and no protists can tolerate long-term exposure to temperatures in excess of 60°C (Madigan and Marrs 1997).

The phylogenetic data regarding thermophiles and hyperthermophiles are used to gain insights into the evolutionary relationship of extant cells to the LCA and the environment in which the LCA may have lived. These attempts at "retelling" evolution are purely hypothetical exercises since there is no way to "glimpse" into the distant past and the fossil record offers no clear answers. The analysis of living taxa to infer the characteristics of long-dead ancestors is however a remarkably powerful way to recreate evolutionary history.

2. The Origin of the Three Domains of Life

Fossilized cyanobacterial mats (stromatolites) have been found in strata dating to 3.5 billion years before the present (BYP, Schopf 1993) suggesting that photosynthetic prokaryotes were already abundant on the Earth. This hypothesis is supported by the presence of iron oxide in banded formations that date from the same time (Holland and Beukes 1990). The oxidization is believed to have arisen from the biological production of oxygen via photosynthesis in the marine ecosystem. Since modern cyanobacteria presently form only one branch of the bacterial lineage it would follow that the stromatolites are the remnants of a well-evolved bacterial world (Schopf 1994). The common ancestor of these bacteria must therefore already have existed for millions of years. Alternatively, the cyanobacteria in the stromatolites may be a long-dead branch of photosynthetic life which has left no survivors among the modern-day bacteria. This would be one explanation for the protein sequence trees of Doolittle et al. (1996) who have suggested that the LCA existed only about 2 BYP. The Doolittle et al. hypothesis is however controversial for a number of reasons including the questionable use of protein sequence divergence rates to place dates on deep branches in the universal phylogeny (see Golding 1996, Martin 1996 for comments). In addition, the strong similarity between living cyanobacteria and the fossilized stromatolite cells, and the existence of living stromatolites with dominant cyanobacterial communities (e.g., Shark Bay, Australia; Schopf and Walter 1982, Whitton and Potts 1982) argue for an unbroken evolutionary line connecting fossilized cyanobacteria with those existing today. The current consensus opinion remains that life has probably existed for at least 3.5 billion years and that the earliest recognizable life-form resembles modern-day cyanobacteria. The eukaryotes, which first appear about a billion years later in the fossil record (Knoll 1994), apparently evolved from a prokaryotic ancestor though likely not from a highly evolved cyanobacterium but rather from a heterotrophic common ancestor, be it a Bacteria or an Archaea.

The recognition that rDNA coding regions are universally distributed, highly conserved, and encode significant evolutionary signal has led to their usage to create a molecular phylogeny of living things. This work, set in motion by Woese and his colleagues have, for example, shown that the prokaryotes are comprised of two distinct lineages (i.e., Archaea, Bacteria) that are as distinct from each other as either is from eukaryotes (Woese 1987). The evolutionary distinctness of the three domains has since then been confirmed with comparisons of a number of protein sequences (e.g., ATPase subunits, elongation factors [Iwabe et al. 1989], isoleucyl-tRNA synthetases [Brown and Doolittle 1995]). All of these protein coding regions were presumably also found in the LCA and, due to their fundamental importance in the cell, have maintained their original function. In spite of the exemplary characteristics of rDNA genes, one issue was however left unresolved with these phylogenetic analyses, the position of the root of the universal tree. This problem arises because there are no outgroups of rDNA genes since all taxa contain the same genes and there is no unbiased way to decide which domain came first in evolution. The rooting problem has since been solved using protein coding regions that underwent duplications in the LCA prior to the radiation of the three domains. One set of gene duplication products can then be used as a "natural" outgroup of the other (e.g., EF-Tu/G, ATPase V/F [see Gupta and Golding 1996 for review]). With this approach, the root of the universal tree has been placed in the branch leading to the Bacteria, thereby supporting a monophyletic origin of the Archaea and Eucarya (see Iwabe et al.1989, Gogarten et al. 1989). This view has since then been widely accepted and was used as the basis for the formal recognition of the three domains of life (Woese et al. 1990) and for the rooting of subsequent rDNA trees (see Fig. 1A). Like the fossil record, the molecular systematic scheme then also supports a prokaryotic LCA. In many molecular trees there is additional evidence that the Archaea diverged from the LCA and subsequently gave rise to the eukaryotes.

The hypothesis that the Eucarya diverge from within the Archaea is however not universally accepted. The most convincing evidence against this scheme comes from comparisons of the 70 kDa heat schock proteins (hsp70/dnaK) that suggest a sister group relationship between eukaryotes and gram-negative bacteria. The hsp70, as well as the glutamine synthetase and the glutamate dehydrogenase, trees show clearly that gram-negative bacteria gave rise to prokaryotes whereas gram-positive bacteria are directly related to the Archaea. In global phylogenies using comparisons of a number of conserved protein sequences, Gupta and Golding (1996) have also shown that about one-half of the trees support the Eucarya/Gram-Negative clade whereas about the same number support a Eucarya/Archaea clade. None of the phylogenies are consistent with an Eucarya/Gram-Positive clade. Since two universal trees of life cannot be correct, Gupta and Singh (1994) and Gupta and Golding (1996) have suggested a chimeric model for the origin of the eukaryotic cell. In this model, a gram-negative bacterium engulfed a "eocyte"-type (see Lake and Rivera 1994 for details) thermoacidophilic archaebacterium (see Fig. 1B). Fusion of the chromosomes of these taxa led to the retention/loss of certain genes from each partner. The usage of the surviving genes in sequence analyses lead therefore to different positions of the eukaryotes (i.e., either with the gram-negative bacteria or with the Archaea) in the phylogenies. The high similarity in both primary sequence and tertiary structure of the proteins and protein complexes involved in DNA transcription (Langer et al. 1995, Rowlands et al. 1994, Kosa et al. 1997) found between Eucarya and Archaea could be explained if the DNA transcription machinery traced its ancestry to the archaeal partner in the original cell fusion. The fusion hypothesis would also explain why eukaryotes do not appear until relatively late in the fossil record. The eukaryotes would be recognizable as such only after the fusion event that had created them and the ensuing intracellular differentiation leading to the evolution of a nucleus and a endomembrane system (Golding 1996). In conclusion, the molecular data do not yet clarify the role of

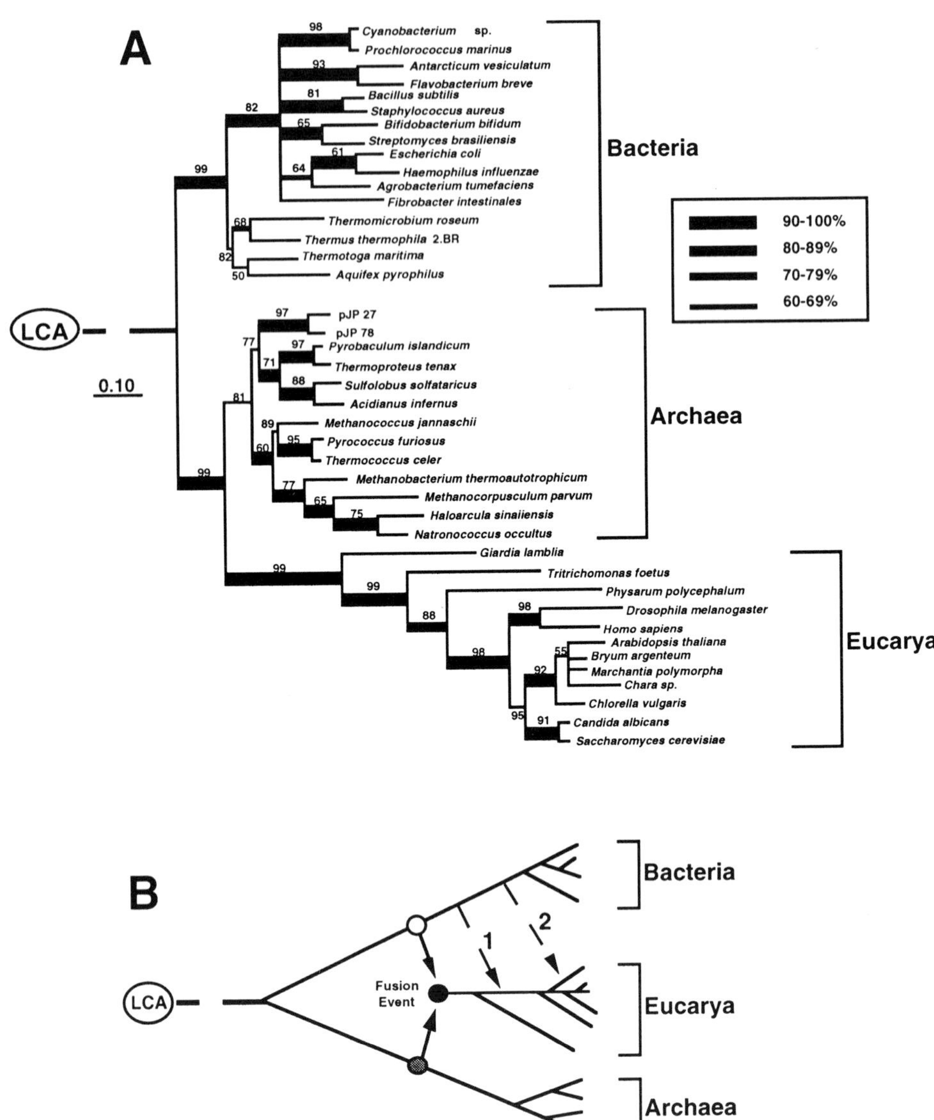

Figure 1. A. Universal tree of life inferred from small subunit rDNA sequence comparisons. This tree was inferred using quartet puzzling and rooted within the branch leading to the Bacteria. Puzzle values are shown above the branches. Results of a distance bootstrap analysis (1000 replications) are shown as branch lengths of differing thicknesses (see legend). B. Origin of the eukaryotes via a fusion involving an Archaea and a gram-negative bacterial cell (based on Gupta and Golding 1996). The numbers 1 and 2 are the putative origins of the mitochondrion and plastid, respectively, in eukaryotes via endosymbiosis. This figure supports a highly chimeric origin of the eukaryotic cell. LCA refers to the last common ancestor of the three domains.

direct filiation from an archaeal ancestor and/or the fusion of archaeal and gram-negative (or some other) bacterial cells in the origin of the Eucarya. It seems relatively certain however that the first cells were prokaryotic in nature (i.e., lacked a nucleus). In the following sections, we present phylogenies of the Archaea and Bacteria, and discuss the kind of environment that may have existed when the prokaryotic LCA existed. These sections will concentrate on the evolution of thermophilic and hyperthermophilic prokaryotes.

3. Phylogeny of the Archaea

3.1 INTRODUCTION

Archaea seem to occupy almost all harsh habitats. They are found in extremely hot niches which until recently were considered to be sterile. Some hyperthermophiles grow between 60°C and 115°C and have an optimal growth temperature near the boiling point of water. Other Archaea are found in marine environments with temperatures around the freezing point. Other extreme environments like those strictly anaerobic, extremely halophilic (3-20% salt) sometimes related to the presence of heavy metals (e.g., at the bottom of the Dead Sea), acidophilic at pH<2 or alkaliphilic at pH>9, or "barophilic" (Pennisi 1997) at high pressures of 1100 atmospheres are also home to Archaea. Often two or more of those environmental factors occur together where Archaea live (Lowe et al. 1993, Stetter 1996).

The Archaea resemble the Bacteria in cellular structure and organisms in both these domains contain single circular genomes with very few introns. Archaeal genes are arranged in polycistronic operons and transcription and translation is coupled like in the Bacteria. The Archaea have however distinguishing characteristics such as cell walls made of protein, glycoprotein, carbohydrate, or pseudomurein, but not peptidoglycan, and isopranyl glycerol ether lipids rather than fatty acid glycerol ester lipids. The ether-linked 40-carbon long lipids form lipid monolayers rather than bilayers (Driessen et al. 1996). The Archaea also share a number of biochemical and genetic features with the Eucarya which are not found in Bacteria (Keeling and Doolitle 1995, Zillig 1991). The best studied of these characters is DNA transcription, the components of which share high sequence and overall tertiary structure similarity among the two domains (see Rowlands et al. 1994, Langer et al. 1995, Kosa et al. 1997). In addition, although archaeal ribosomes are of the 70S bacterial type, translation is initiated by an unformylated methionine like in the Eucarya. So, Archaea share characteristics with both Eucarya and Bacteria which together seem to indicate an intermediate evolutionary position in the universal tree (for reviews, see Zillig 1991, Woese 1987, Lowe et al. 1993).

3.2 PHYLOGENETIC METHODS

We used aligned small subunit rRNA sequences from the rRNA WWW server in Antwerp (Van de Peer et al. 1997) for the phylogenetic analyses of the Archaea. Trees were constructed with the pfastDNAml 2.1.4 (a new parallel version of fastDNAml published by Olsen et al. in 1994) for maximum likelihood analyses (ML), puzzle 3.1 for ML analyses using quartet puzzling (Strimmer and von Haeseler 1996), and CLUSTAL W 1.7 (Thompson et al. 1994) for the neighbor-joining analyses (NJ). Bootstrap analyses were done using 100 pseudosamples with the pfastDNAml method and using 1000 pseudosamples with the neighbor-joining method. Gaps within the alignment columns were excluded from all analyses.

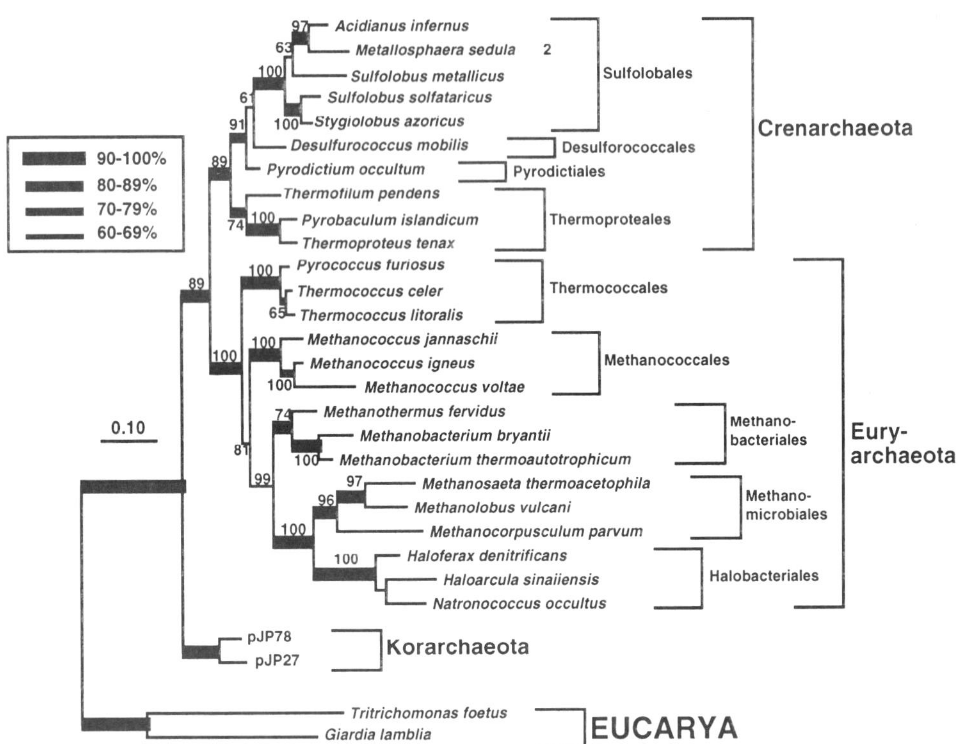

Figure 2. Phylogeny of Archaea inferred from small subunit rDNA sequences. This tree has been inferred with the pfastDNAml program and rooted with the eukaryotic sequences. Bootstrap values (100 replications) from the ML analysis are shown above branches whereas results of distance analyses using 1000 bootstrap pseudosamples are shown as branch lengths of differing thicknesses (see legend).

3.3 PHYLOGENY

The rDNA sequence analyses suggest that the Archaea is divided into two major kingdoms, the Euryarchaeota and the Crenarchaeota (see Fig. 2). The Euryarchaeota include several methanogenic, halophilic, and some sulfur-metabolizing hyperthermophiles. Since all known Crenarchaeota are sulfur-dependent hyperthermophiles, the members of this kingdom are thought to not be widespread but restricted to hyperthermophilic habitats (Woese et al. 1990). There have been hyperthermophilic Archaea also found in cold water conditions, suggesting that hyperthermophilic organisms might be able to exist in the cold, although they cannot grow under such conditions (Stetter 1996).

In the past, Archaea from environmental samples were cultured prior to PCR and sequence analysis of the rDNA coding regions from single isolates. It has been shown however that about 99% of the organisms from an environmental sample are lost during culturing (Fuhrman et al. 1992). Since the development of methods for isolating rDNA directly from environmental samples by PCR amplification (without culturing), a large number of new sequences have been identified and used for phylogenetic analyses. This method led to the identification of mesophilic Archaea (e.g., soil Archaea [Bintrim et al. 1997]). These taxa group with the Crenarchaeotes in the phylogenetic analyses thereby showing that Crenarchaeota are not exclusively hyperthermophilic. In addition, analyses using environmental DNA clones have shown that Archaea are highly abundant in the marine picoplancton, constituting up to 34% of the prokaryotic biomass in coastal arctic surface waters (DeLong et al. 1994). Interestingly, a pair of putative Archaea were also found by cloning environmental rDNA (Barns et al. 1996) that branch before the dichotomy of euryarchaeotes and crenarchaeotes. This group is provisionally called Korarchaeota, and is apparently another kingdom or Archaea (Fig. 2).

4. Phylogeny of the Extremophilic Cyanobacteria

4.1 INTRODUCTION

The broad environmental tolerance range of most extremophile cyanobacteria is in contrast to most other prokaryotic and eukaryotic organisms from extreme environments which occur only within a small window of almost constant conditions, exemplified by hyperthermophilic or acidophilic Archaea. From an evolutionary point of view it appears that the prokaryotic cyanobacteria (also called the "blue-green algae" by some authors) are among the oldest organisms on earth (see above). Many extant cyanobacteria may have been forced by the "modern" eukaryotes to withdraw themselves into habitats of extreme and fluctuating conditions (Whitton 1992).

4.2 PHYLOGENETIC METHODS

We compared small subunit rDNA sequences from the cyanobacteria, with emphasis on extremophiles/thermophiles, to reconstruct the phylogeny of this lineage (see also Wilmotte 1994, Turner 1997). Most phylogenetic relationships were, however, resolved with only poor support from bootstrap analyses. This may be due to the fact that the majority of available cyanobacterial sequences are incomplete. In this study, representatives of almost all groups that were resolved by rDNA phylogenies presented by Wilmotte (1994) and Turner (1997) were selected, with preference for those taxa that exist in extreme habitats. The monophyletic groups resolved in our trees are consistent with those shown in the more comprehensive analysis of Turner (1997).

The alignment of cyanobacterial small subunit rDNA was taken from the rRNA WWW servers in Antwerp (Van de Peer et al. 1997) and from the Ribosomal Database Project (Maidak et al. 1997). The alignment was slightly modified after the addition of some strains of *Synechococcus* using conserved small subunit rRNA secondary structure models. Trees were reconstructed with a weighted (rescaled consistency index over an interval of 1-1000) maximum parsimony method and with the neighbor-joining method with matrices inferred with the LogDet (Lockhart et al. 1994) or "HKY85" (Hasegawa et al. 1985) corrections using a test version of PAUP (version 4.0d59,written by D.L. Swofford and with permission of the author). Phylogenies inferred with maximum likelihood methods (as used in Figs. 1, 2) were almost identical with the phylogeny shown in Fig. 3. A total of 988 unequivocally aligned sequence positions were included in the phylogenetic analyses from which the constant sites were removed resulting in 530 variable sites (403 parsimony-informative characters). The stability of internal nodes was tested with the bootstrap method (500 replications) using both neighbor-joining and parsimony methods.

4.3 PHYLOGENY

Oscillatoria limnetica may serve as a model for cyanobacteria that are adapted to rapidly fluctuating environments and thus have multipotential metabolic patterns. This species is, for example, capable of shifts from oxygenic to anoxygenic photosynthesis, having multiple dark energy-generating systems, and fixing nitrogen anaerobically (Shilo 1980). In Fig. 3, *O. limnetica* clusters with other filamentous non-heterocystous cyanobacteria from hypersaline habitats, *Arthrospira platensis*, *Lyngbya aestuarii* and *Microcoleus chthonoplastes* which form the OSCILLATORIA group (Turner 1997, Fig. 3). These cyanobacteria are primary producers in saline marshes and mangrove communities in which evaporation-dilution cycles result in broad salinity fluctuations (Javor 1989). Although heterocysts are lacking in these cyanobacteria, it is their nitrogen-fixing capacity (Villbrandt et al. 1990) that allows them to colonize sandy estuaries that are extremely poor in nutrients and thus they play an important role in the formation and stabilization of intertidal sediments. The various ways in whoch nitrogen fixation can be performed and is regulated in the abscence of heterocysts in cyanobacteria has been reviewed by Mitsui et al. (1986), and Villbrandt et al. (1990). That *Microcoleus chthonoplastes* is dominant in intertidal microbial communities ("microbial mats") is consistent with its capacity to tolerate enormous changes in salinity and forming various ecotypes within the species (Karsten 1996) which, however, are genetically rather uniform (Garcia-Pichel et al. 1996). *M. chthonoplastes* strains may be phylogenetically removed from other species of that genus as is indicated by a chemosystematic study of carotenoids and mycosporine-like amino acid compounds in various strains of *Microcoleus* (Karsten and Garcia-Pichel 1996). The paraphyletic nature of the genus *Microcoleus* is shown by the rDNA analyses (Fig. 3), *M. chthonoplastes* appears to be distant from *Microcoleus* sp. strain 10mfx. Another organism from hypersaline evironments is the unicellular *Synechocystis* sp. strain PCC 6906 which, however, is within another cluster (SY/PLEU/MY, Fig. 3). Its close relationship with *Microcystis* sp. strain PCC 7806 from a non-extreme environment is well supported in the rDNA-tree.

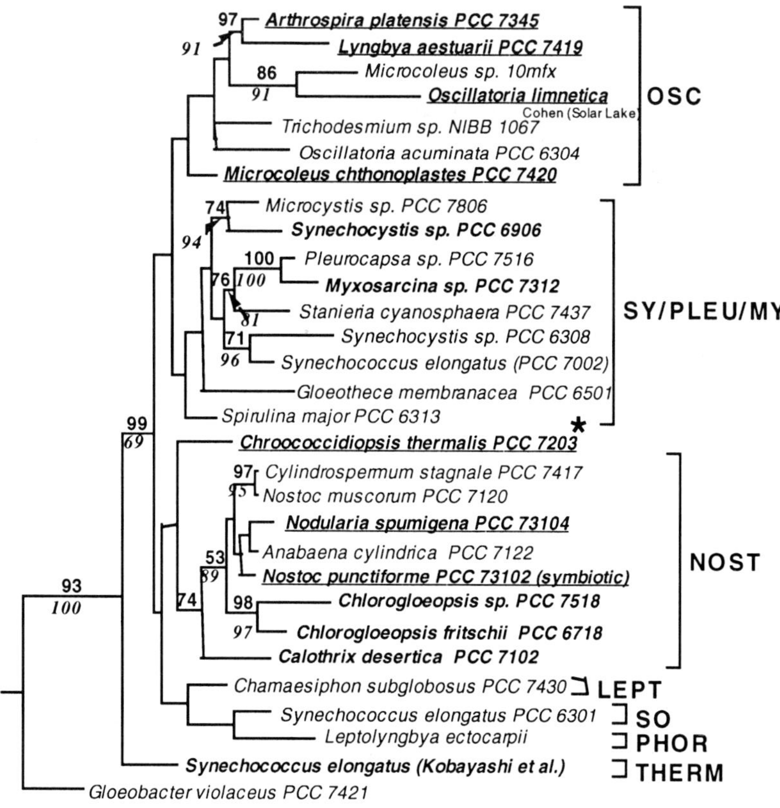

Figure 3. Phylogeny of extremophile/thermophile cyanobacteria inferred from small subunit rDNA sequences. The phylogeny shown is from a weighted maximum parsimony analysis and has a consistency index of 0.663. The values above the internal nodes are from a neighbor-joining bootstrap analysis (500 replications) whereas values shown in italics below the internal nodes were inferred from a weighted maximum parsimony bootstrap analysis (100 replications). Bootstrap values are shown only for those groupings that are resolved by both the neighbor-joining and parsimony methods in more than 60% of replications. Taxa from extreme habitats are underlined, thermophiles are shown in bold face. This phylogeny has been rooted within the branch leading to the green sulfur bacteria (*Chlorolobium limicola* and *C. vibrioforme*), and other bacteria (*Agrobacterium tumifaciens*, *Alcaligines faecalis*, *Escherichia coli*, and *Pseudomonas andropogonis*; not shown). Abbreviations for groupings (bold face and upper case letters) follow Turner (1997): OSC, OSCILLATORIA group (filamentous non-heterocystous cyanobacteria); SY/PLEU/MY, SYNECHOCYSTIS/PLEUROCAPSA/MICROCYSTIS group (unicellular and filamentous cyanobacteria); NOST, NOSTOC group (filamentous heterocystous cyanobacteria); SO, SYNECHOCOCCUS group; LEPT, LEPTOLYNGBYA group; PHOR, PHORMIDIUM group; THERM, THERMOPHILIC unicellular group. An asterisk indicates the uncertain position of *Chroococcidiopsis*.

Cyanobacteria are pioneer organisms in nitrogen-depleted sandy desert soils where they are exposed to high temperatures and dryness. Some filamentous cyanobacteria, such as *Calothrix desertica* and species of *Microcoleus*, may play an important role in desert ecosystems for stabilizing soil and carbon input (Garcia-Pichel & Belnap 1996). *Nodularia spumigena* strain PCC 73104, another filamentous cyanobacterium that has been isolated from alkaline soils, is phylogenetically removed from *Calothrix desertica* and the two species of *Microcoleus* (Fig. 3). Cyanobacteria also frequently occur beneath the surface of light-colored porous rock in a discrete band that follows the surface contours of the rock. Upon microscopic examination, the cyanobacteria are seen as free-living colonies attached to rock crystals or in loose lichen associations (Büdel & Wessels 1991). Extremes of insolation and temperature, scouring winds and moisture stress which make the rock surface uninhabitable are thought to be the likely driving force behind colonization of the endolithic habitat. In hot deserts, the combination of temperature increase in a hydrated state and the concomitant loss of water imposes a severe environmental stress to cryptoendolithic cyanobacteria that eucaryotic organisms do not seem to be able to tolerate (Friedmann & Ocampo-Friedmann 1984).

Thermophilic cyanobacteria from hot springs are distributed over two independent lineages in Fig. 3. A thermophilic strain of the polyphyletic *Synechococcus* (Turner 1997) groups together with *Gloeobacter violaceus* strain PCC7421 at the base of the phylogram in Fig. 3 and appears to be phylogenetically distant from *S. elongatus* PCC 6301 which originates from a freshwater mud sample. Further thermophilic cyanobacteria shown in Fig. 3 are the filamentous heterocystous species of *Chlorogloeopsis* (that includes the former genus *Mastigocladus*); these cluster within the NOSTOC group.

5. A Hot Origin of Life?

The discovery of organisms that can live and grow at temperatures in excess of 100°C and branch deeply in rDNA trees has fueled the idea that these taxa are the relatives of the LCA which was, as well, thermophilic or hyperthermophilic. Eukaryotes cannot live in temperatures greater than about 60°C and in any case appear to have originated relatively "late" in evolution. The real thermophiles are within the prokaryotic domains, in particular the Archaea. The idea of a "hot" origin of life is however controversial. It has been pointed out, for example, that a hyperthermophile world for the LCA is not consistent with the RNA-world that is believed to have preceded it (Forterre 1996). RNA is highly unstable at the boiling point of water (however, see Riera et al. 1997). In addition, there is data from the distribution and origin of reverse gyrase, a topoisomerase that stabilizes topologically-closed DNA and is found in all prokaryotes living at temperatures above 80°C, that suggests that hyperthermophily is a derived and not an ancestral trait (see Forterre 1995, 1996 for details). This result is supported by other studies which show that protection of molecules against heat damage in hyperthermophiles can be achieved by relatively simple base modifications (tRNAs, Kowalak et al. 1994) or increases in the number of ion pairs in protein tertiary structures (e.g., *Taq* polymerase, Korolev et al. 1995). This suggests that thermophily may have evolved from a mesophilic ancestor via thermoadaptation.

In all our phylogenies however, the thermophilic organisms show up as early divergences. This was also shown by other analyses (Stetter 1996). The (known) korarchaeotes also appear to be thermophilic with respect to the hot environment from which they were extracted (Barns et al. 1996) and the culturing and characterization of pJP27 as a thermophilic Archaea by Burggraf et al. (1997). These observations imply that the common ancestor of the Archaea might have been thermophilic. Although many small subunit rRNA sequences of putative mesophilic crenarchaeotes have been extracted directly from the environment, most of those species have not yet been cultured and further characterized. Therefore, we do not yet know which of these are really mesophilic or perhaps are thermophiles existing in a mesophilic environment. The lower branches of the bacterial domain are also occupied by hyperthermophilic bacteria such as *Aquifex* and *Thermotoga* (see Fig. 1A). This branching order, also shown by Barns et al (1996) and Stetter (1996), is consistent with a hyperthermophilic ancestry of the Bacteria. Given the thermophilic life-style of the (known) korarchaeotes, it is plausible therefore that the common ancestor of the Archaea and the Bacteria (therefore also the LCA) may have been a thermophile or hyperthermophile. Within the cyanobacteria, we also find thermophiles as early diverging (*Synechococcus elongatus*) and as more derived (*Chlorogloeopsis* sp.) taxa suggesting that thermophily was a primitive trait in these derived Bacteria.

In spite of these results the origin of life need not have happened in a thermophilic environment. Another explanation for our results is that all (currently known) organisms may have had a common ancestor that was adapted to the hyperthermophile environment (Forterre 1996). In this scenario, life evolved in a mesophile world that later became hot and led to the evolution of the hyperthermophilic LCA from which all extant cells are derived. To gain a better understanding of the origin of thermophiles in the phylogeny of life and in the Archaea in particular, we need analyses of sequences drawn from a larger diversity of environments. The results published in the last few years promise that we will have the possibility to reconstruct a much more precise tree of life in the near future.

6. Acknowledgements

This research was supported in part from a grant from the University of Iowa to DB. We thank the German National Research Center for Information Technology (GMD, Sankt Augustin), NEC Europe Ltd. (C&C Research Laboratories, Sankt Augustin) and the German Cancer Resarch Center (DKFZ, Heidelberg) for access to computing facilities. TF thanks Burkhard Büdel (Kaiserslautern) for valuable discussions.

7. References

Barns, S.M., Delwiche, C.F., Palmer, J.D., and Pace, N.R. (1996) *Proc. Natl. Acad. Sci. USA* **93**, 9188-9193.
Bintrim, S.B., Donohue, T.J., Handelsman, J., Roberts, G.P., and Goodman, R.M. (1997) *Proc. Natl. Acad. Sci. USA* **94**, 277-282.
Brown, J.R. and Doolittle. W.F. (1995) *Proc. Natl. Acad. Sci. USA* **92**, 2441-2445.
Büdel, B., Wessels, D.C.J., and Mollenhauer, D. (1993) *Arch. Protistenkd.* **143**, 229-235.
Burggraf, S., Huber, H., and Stetter, K.O. (1997) *Int. J. Syst. Bacteriol.* **47**, 657-660.
DeLong, E.F., Wu, K.Y., Prézelin, B.B., and Jovine, R.V.M. (1994) *Nature* **371**, 695-697
Doolittle, R.F., Feng, D.-F., Tsang, S., Cho, G., and Little, A. (1996) *Science* **271**, 470-477.
Driessen, A.J.M., van de Vossenberg, J.L.C.M, and Konings, W.N. (1996) *FEMS Microbiol. Rev.* **18**, 139-148
Forterre, P. (1996) *Cell* **85**, 789-792.
Forterre, P. (1995) *C R Acad Sci III* **318**, 415-422.
Friedmann, E.I. and Ocampo-Friedmann, R. (1984) in: M.J. Klug, C.A. Reddy (eds.) *Current perspectives in microbial ecology*, Amer. Soc. Microbiol., Washington D.C., pp.177-185.
Fuhrman, J.A., McCallum, K., and Davis, A.A. (1992) *Nature* **356**, 148-149.
Garcia-Pichel, F. and Belnap, J. (1996) *J. Phycol.* **32**, 774-782.
Garcia-Pichel, F., Prufert-Bebout, L., and Muyzer, G. (1996) *Appl. Environ. Microbiol.* **62**, 3284-3291.
Gogarten, J.P., Kibaj, H., Dittrich, O., Taiz, I., Bowman, E.J., Bowman, B.J., Manolson, M.F., Poole, R.J., Date, T., Oshima, T., et al. (1989) *Proc. Natl. Acad. Sci. USA* **86**, 6661-6665.

Golding, B. (1996) *Curr. Biol.* **6**, 679-682.
Gupta, R.S. and Golding, G.B. (1996) *Trends Biochem. Sci.* **21**, 166-171.
Gupta, R.S. and Singh, B. (1994) *Curr. Biol.* **4**, 1104-1114.
Hasegawa, M., Kishino, H., and Yano, T. (1985) *J. Mol. Evol.* **22**, 32-38.
Holland, H. and Beukes,, N. (1990) *Am. J. Sci.* **290A**, 1-34.
Iwabe, N., Kuma, K., Hasegawa, M., Osawa, S., and Miyata, T. (1989) *Proc. Natl. Acad. Sci. USA* **86**, 9355-9359.
Javor, B. (1989) Springer-Verlag, Berlin, pp. 134-146.
Karsten, U. (1996) *J. Phycol.* **32**, 501-506.
Karsten, U. and Garcia-Pichel, F. (1996) *System. Appl. Micobiol.* **19**, 285-294.
Keeling, P.J. and Doolittle, W.F. (1995) *Proc. Natl. Acad. Sci. USA* **92**, 5761-5764.
Knoll, A.H. (1994) *Proc. Natl. Acad. Sci. USA* **91**, 6743-6750.
Korolev, S., Nayal, M., Barnes, W.M., Di Cera, E., and Waksman, G. (1995) *Proc. Natl. Acad. Sci. USA* **92**, 9264-9268.
Kosa, P.F., Ghosh, G., DeDecker, B.S., and Sigler, P.B. (1997) *Proc. Natl. Acad. Sci. USA* **94**, 6042-6047.
Kowalak, J.A., Dalluge, J.J., McCloskey, J.A., and Stetter, K.O. (1994) *Biochemistry* **33**,7869-7876.
Lake, J.A., and Rivera, M.C. (1994) *Proc. Natl. Acad. Sci. USA* **91**, 2880-2881.
Langer, D., Hain, J., Thuriaux, P., and Zillig, W. (1995) *Proc. Natl. Acad. Sci. USA* **92**, 5768-5772.
Lockhart, P.J., Steel, M.A., Hendy, M.D., and Penny, D. (1994) *Mol. Biol. Evol.* **11**, 605-612.
Lowe, S.E., Jain, M.K., and Zeikus, G. (1993) *Microbiol. Rev.* **57**, 451-509.
Madigan, M.T. and Marrs, B.L. (1997) *Sci. Amer.* **276**, 82-87.
Maidak, B.L., Olsen, G.J., Larsen, N., Overbeek, R., McCaughey, M.J., and Woese CR (1997) *Nucleic Acids Res.* **25**, 109-111.
Martin, W.F. (1996) *BioEssays* **18**, 523-527.
Mitsui, A., Kumazawa, S., Takahashi, A., Ikemoto, H., Cao, S., and Arai, T. (1986) *Nature* **323**, 720-722.
Olsen, G.J., Matsuda, H., Hagstrom, R., and Overbeek, R. (1994) CABIOS **10**:41-48.
Pennisi, E. (1997) *Science* **276**, 705-706
Riera, J., Robb, F.T., Weiss, R., and Fontecave, M. (1997) *Proc. Natl. Acad. Sci. USA* **94**, 475-478.
Rowlands, T., Baumann, P., and Jackson, S.P. (1994) *Science* **264**, 1326-1329.
Scherer, S., Chen, T.W., and Böger, P. (1988) *Plant Physiol.* **88**, 1055-1057.
Scherer, S. and Zhong, Z.P. (1991) *Microb. Ecol.* **22**, 271-283.
Schopf, J.W. (1993) *Science* **260**, 640-646.
Schopf, J.W. (1994) *Proc. Natl. Acad. Sci. USA* **91**, 6735-6742.
Schopf, J.W. and Walter, M.R. (1982) in N.G. Carr and B.A. Whitton (eds.) *The biology of cyanobacteria.* Blackwell, Oxford, pp. 543-564.
Shilo, M. (1980) *Naturwissensch.* **67**, 384-389.
Stetter, K.O. (1996) *FEMS Microbiol. Rev.* **18**, 149-158.
Strimmer, K. and von Haeseler, A. (1996) Mol. Biol. Evol. **13**:964-969.
Thompson, J.D., Higgins, D.G., and Gibson, T.J. (1994) *Nucleic Acids Res.* **22**, 4673-4680.
Turner, S. (1997) in D. Bhattacharya (ed.), *Origins of Algae and Their Plastids*, Springer-Verlag, Wien, pp. 13-52.
Van de Peer, Y., De Rijk, P., and De Wachter, R. (1997) SSU rRNA database. Department of Biochemistry, University of Antwerp http://rrna.uia.ac.be/rrna/ ssuform.html#Eukarya.
Villbrandt, M., Stal, L.J., and Krumbein, W.E. (1990) *FEMS Microbiol. Ecol.* **74**, 59-72.
Whitton, B.A. (1992) in N.H. Mann and N.G. Carr (eds.), *Photosynthetic Prokaryotes*, Plenum Press, New York, pp. 1-51.
Wilmotte, A. (1994) in A. Bryant (ed.) *The Molecular Biology of Cyanobacteria*, Kluwer Academic Publishers, Dordrecht, pp. 1-25.
Whitton, B.A. and Potts, M. (1982) in N.G. Carr, and B.A. Whitton (eds.) *The biology of cyanobacteria.* Blackwell, Oxford, pp. 515-542.
Woese, C.R (1987) *Microbiol. Rev.* **51**, 221-271.
Woese, C.R., Kandler, O., and Wheelis, M.L. (1990) *Proc. Natl. Acad. Sci. USA* **87**, 4576-4579.
Zillig, W. (1991) *Curr. Opin. Genet. Dev.* **1**, 544-551.

BioData of **Weiss Bizzoco R.L.**
Contribution: ***Life in the Extreme: New Prokaryotes Living in High Temperature, Low pH Environments***.

Richard L. Weiss Bizzoco is a Professor of Biology in the San Diego State University. He earned his PhD from Indiana University (1972) and spent his postdoctoral years in Berkeley, Harvard and in Irvine Medical School. His current interest is characterization of new hyperthermal acidophiles growing at 70° to 90°C at pH 2-3.
E-Mail: <rbizzoco@sunstroke.sdsu.edu>

J. Seckbach (ed.), Enigmatic Microorganisms and Life in Extreme Environments, 305–314.

LIFE IN THE EXTREME: NEW PROKARYOTES LIVING IN HIGH TEMPERATURE LOW pH ENVIRONMENTS

R.L. WEISS BIZZOCO
Department of Biology
San Diego State University
San Diego, CA 92182

1. Extreme Environments

1.1. NEW ORGANISMS

There are a number of new organisms that live in acid hot springs that we now know about. These organisms live at about the temperature optimum of *Sulfolobus* (~70°C) and significantly above that optimum. These organisms grow well below the 105°C temperature optimum of *Pyrolobus fumarii*, the most extreme thermophile known (Stetter, 1982). The new prokaryotes may belong both to Archaea, like *Sulfolobus* and to the Bacteria. They were discovered in the acid hot springs of Yellowstone National Park, WY, U.S.A. In hot spring outflow channels with sulfur crystals the presence of *Sulfolobus* was rapidly confirmed in most samples, and at temperatures below 80°C rod shaped organisms were also present. In mixing pools rods were present at temperatures up to about 89°C. My analysis of these new organisms appears in the sections below.

2. New Archaea-like organisms

2.1. A ROD SHAPED ORGANISM WITH A SURFACE FIBRILLAR LAYER

We examined hot spring samples by electron microscopy of negatively stained cells and by thin section electron microscopy. On one rod shaped organism prepared by negative staining of whole mount samples (*Fig. 1A*) we found short hair-like protrusions surrounding the organism. Such structures on other organisms are formed of protein, serve as attachment structures, and are known as pili. Although difficult to visualize (*Fig. 1B, arrows*), they are more evident at the pole of the cell. In some areas the *Acidobacillus* wall structure is particularly obvious. Here, the semicrystalline cell wall is prominent, as are individual wall subunits at the cell margin. There are three new features associated with *Acidobacillus fibrocaldarius:* 1) it is the first rod shaped organism to be reported from acid hot springs with a cell wall array that is reminiscent of the *Sulfolobus* wall (Brock *et al.*, 1972); 2) it is the only Archaea-like organism with peritrichous pili; 3) a cell wall constriction noted near one end of the cell could be related to an unusual mode of division. Although this latter property is the most uncertain of the three, we interpret the constriction and the presence of a smaller cell as cell division.

Although this organism has an Archaea-like cell wall, another unique feature is the fibrillar layer of pili that appears in the background extending away from the cell.

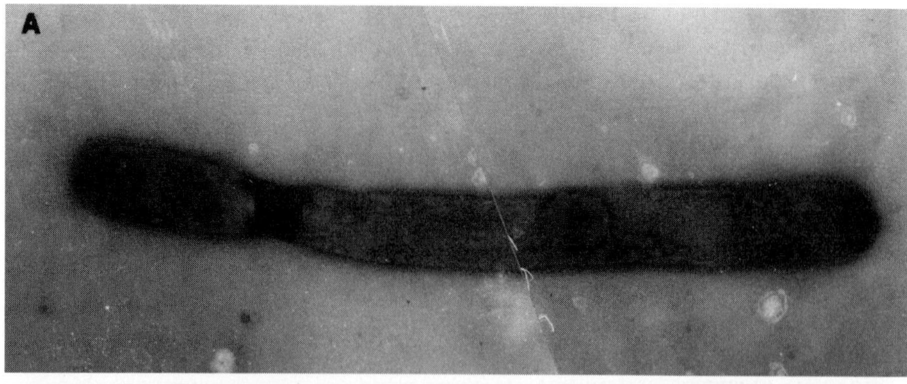

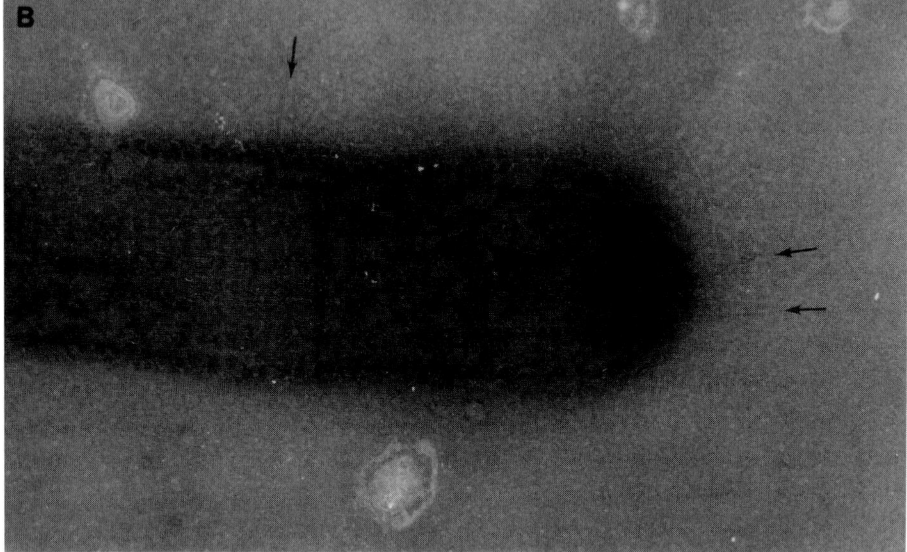

Figure 1. Acidobacillus fibrocaldarius. A. Cell division. *B.* Enlargement showing subunit cell wall and pili (arrows) surrounding the cell. Locomotive Spring, YNP: 87.5°C; pH 2.0. *A.* X 17, 000. *B.* X 55, 000.

These pili are evident at the edge of the cell where they are seen to vary only slightly in length, and all have a similar diameter of about 5.67 $\pm$ 0.49 nm. Their appearance indicates that they cover the entire cell surface. Another similar organism, *Thermoproteus*, (Stetter and Zillig, 1985) has a few lateral pili; *Acidobacillus* pili are shorter, less distinct, and much more numerous than those of *Thermoproteus*. Their abundance immediately suggests that the nature of the cell surface is controlled by these structures. It is presently uncertain whether they might function in attachment to sulfur.

Comparison of the surface layer of pili with *Sulfolobus* pili seems appropriate, since the two organisms live in the same habitat. *Sulfolobus* has large numbers of pili on cells attached to sulfur in flowing streams or outflow channels (Weiss, 1973). These pili emerge in the same general direction which might be expected if they were functioning to hold the organism onto the sulfur crystal. Their diameter of 4.94 $\pm$ 0.27 nm is similar to that of *Acidobacillus* pili, but, they are substantially longer. The layer of pili surrounding the *Acidobacillus* cell is distinctive and could serve as an attachment layer. It may be that this is an Archaea-like organism with pili that were broken off during ongoing mixing in the hot spring, as this organism was sampled in a mixing spring at a high temperature and low pH (87.5°C; pH 2.0).

The proposed name, *Acidobacillus fibrocaldarius,* is derived from Latin. *Acido* is Latin for acid; *bacillus* Latin for rod, *fibro* is Latin for fiber, and *caldarius* is Latin for hot. This organism is a bacillus from acid environments that has fibers and lives in hot springs.

2.2. A ROD SHAPED ORGANISM WITH TERMINAL LOBES

By means of direct electron microscopic examination of thin sections we encountered a second new organism growing in hot acid habitats. This organism, *Lobobacillus acidocaldarius,* was filamentous, but the ends of the cell were lobe-like and exhibited

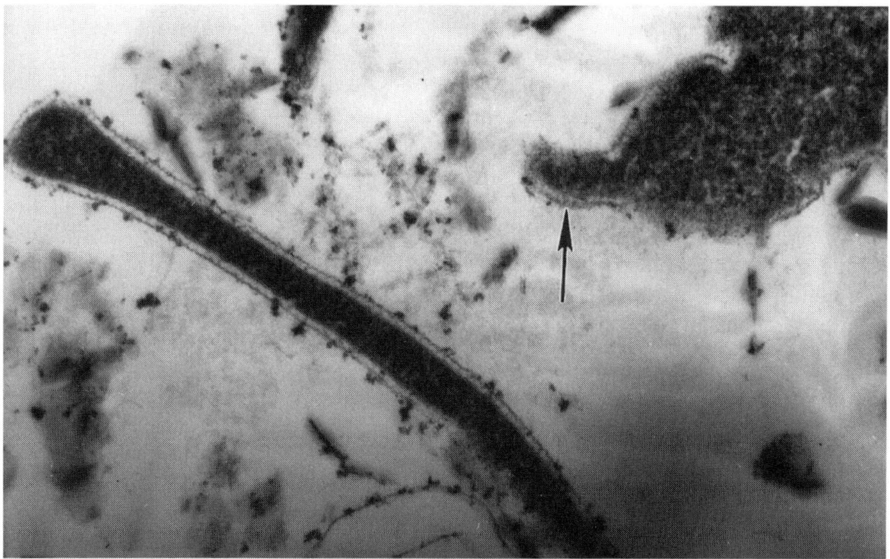

Figure 2. Lobobacillus acidocaldarius, long thin cell with lobed end. Compare wall to *Sulfolobus* type wall (arrow). Amphitheater Spring 2, YNP: 78°C; pH 2.0. X 51, 000.

their own distinctive shape, depending on the habitat in which they appeared. In Sulfur Cauldron we observed a short rod (not shown). One end of the cell had a lobe that was flattened terminally. This new cell type was also seen by directly sampling the spring bottom at Amphitheater Springs, YNP and then examining thin sectioned cells in the electron microscope. The cells appear for a long distance in section (*Fig. 2*). Alongside a typical *Sulfolobus* type wall (arrow) shown on a nearby organism, the wall of the *Lobobacillus* is obviously similar, but this new Archaea-like organism is thin and long and has a very different shape when compared to *Sulfolobus.* The cell shape tends to resemble the microorganisms living at 90°C and pH 9 in Boulder Spring,YNP. Brock *et al.* (1971) described thin filaments with terminally flattened lobes extending from both ends of the cell. In our thin sections we did not always observe obvious flattening at the ends of the long filaments, but if it occurred, flattening varied with the habitat examined, i.e. it was sampling site specific.

The proposed name, *Lobobacillus acidocaldarius,* is derived from Latin. *Lobo* is Latin for lobe; *bacillus* Latin for rod, *acido* is Latin for acid, and *caldarius* is Latin for hot. This organism is a lobed bacillus that grows in acid thermal habitats.

3. New bacteria

3.1. A DIMPLED FILAMENT

There may be one or more features on each of these new bacteria that bear some similarity to those of organisms already reported, but the distinct properties of the organisms themselves are either new or unique or both. The first organism we encountered was a Gram negative rod growing at Amphitheater Springs, YNP. This bacterium was able to grow at 75°C and pH 2.0. These conditions are much more restrictive than *Thiobacillus* has been reportedly able to sustain in acid thermal habitats (Fliermans and Brock, 1972).

In this habitat Gram negative bacteria we observed grew as long thin filaments. They were distinguished from other Gram negative microorganisms only by the presence of dimples located on alternate sides of the sectioned filament (*Fig. 3*). The presumed role of these sites is cell division, suggested by the presence of membranous material (arrowhead) lying just below the depressions. In all other ways these Gram negative filaments were unremarkable structurally. They appeared to have a normal morphology with respect to the multilayered cell envelope (*Fig. 3*, inset) and a compact cytoplasm which contained ribosomes and other cytoplasmic features of normal morphology. No virus particles, inclusions or other unusual features were seen.

3.2. A POLAR COLLAR

Samples from Norris Annex, YNP were fixed, embedded and thin sections were examined by electron microscopy. As shown in *Figure 4A* a wide Gram negative rod with an unusual polar collar (arrowheads) was discovered. This structure is located at one end of the cell directly adjacent to the plasma membrane; thin striations (small arrowheads, *Fig. 4A)* appear in one area near the base. The structure resembles a paraboloid and is truncated; as seen in *Figure 4B* in oblique section, it has an open collar-like structure. Here too, the collar is closely associated with the plasma membrane. *Figure 4C* (arrow) illustrates that the polar collar may also associate with membrane that lies in the nearby cytoplasm. This membranous material is seen near the truncated portion of the polar collar and appears to originate near the plasma membrane; it assumes a layered arrangement as it is synthesized.

Several possibilities can be considered to explain the function of the polar collar. First, it might be directly involved in sulfur oxidation. Second, The polar location of this collar might somehow modify the outer surface for attachment. The outermost wall layer itself is unusual in having a thickened coat. Third, the polar location could indicate that the collar has a role in cell division such as membrane assembly or cell separation. Finally, the polar collar could represent accumulated cell membrane or wall precursors. This type of accumulation could be expected to occur when environmental conditions limit the available nutrients, growth or metabolism.

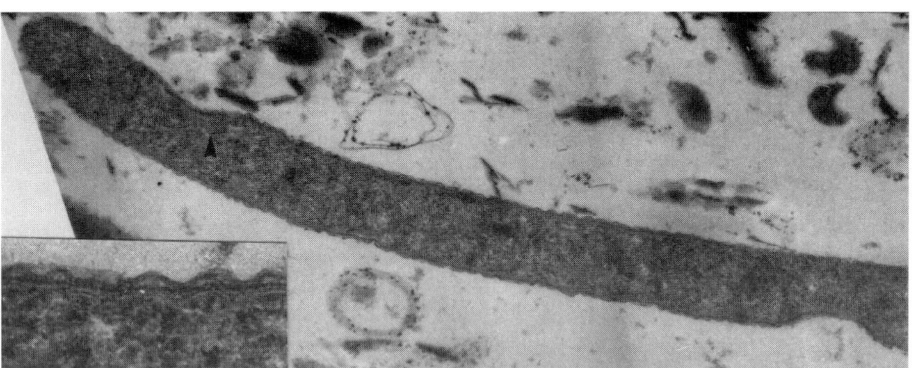

Figure 3. Gram negative filament with depressions and membranes (arrowhead) lying below. Inset. Cell envelope. Amphitheater Spring 1, YNP: 78°C; pH 2.0. X 28, 500. Inset. X 118, 000

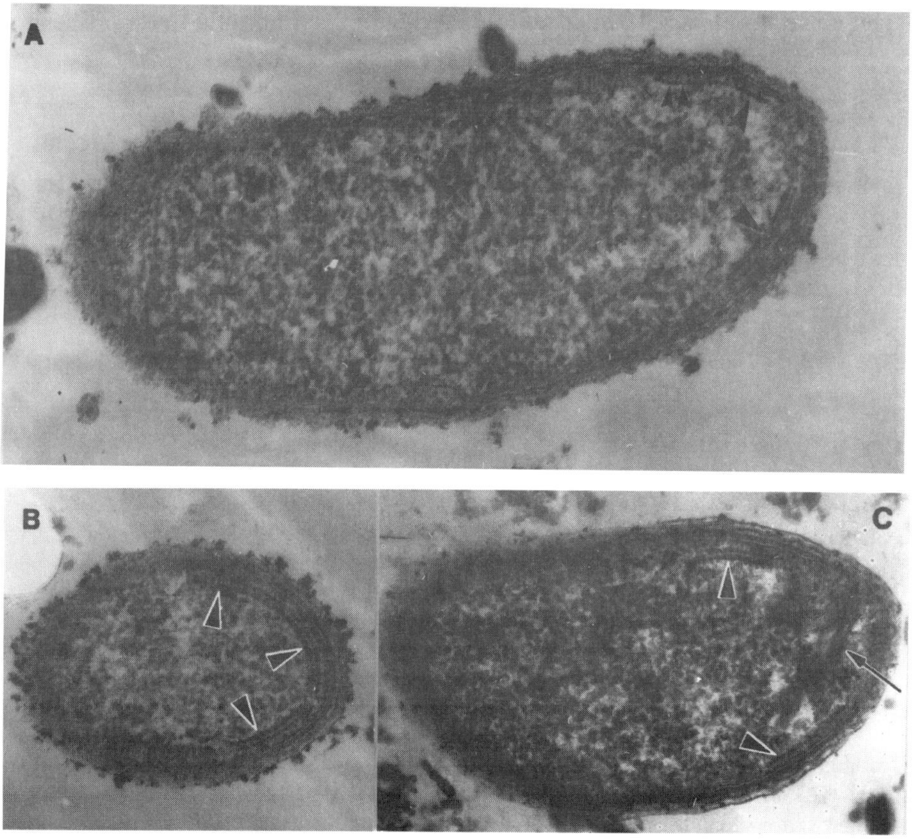

Figure 4. Polar collar in a Gram negative bacterium. *A.* Polar collar appears as a truncated paraboloid (arrowheads) with thin striations (small arrowheads). *B.* Polar collar (arrowheads) in oblique section. *C.* Polar collar (arrowheads) has nearby layered membranes (arrow). Norris Annex, YNP: 78°C; pH 2.3. A-C X 60,000.

The only other comparable structure identified in acidophilic thermophilic organisms is present in *Thermoproteus* Hvv5. Small filaments slightly longer than 1 μm appear in thin sections (Stetter and Zillig 1985). Although they extend from the pole of the cell they are unlike the structure reported here in that they are filamentous, travel through the cytoplasm and are not directly adjacent to the cell membrane.

3.3. A WAVY CELL WALL

Figure 5 illustrates another new organism found in the Norris Geyser Basin, YNP at 89°C and pH 2.0. This new organism is rod shaped with a wavy outer cell wall. The cell envelope can perhaps be likened to a Gram negative cell envelope in that it has a dense straight thin layer lying below the outer wavy layer. The fine structural details, however, are not fully resolved so that it cannot be determined if this straight layer is the equivalent of the peptidoglycan layer of Gram negative cell walls or represents the outer surface of the plasma membrane of a *Sulfolobus* type cell. *Sulfolobus* appears in section above this organism. Comparison to the *Sulfolobus* wall (S) emphasizes their individual differences. Although not resolved here as individual subunits the *Sulfolobus* wall

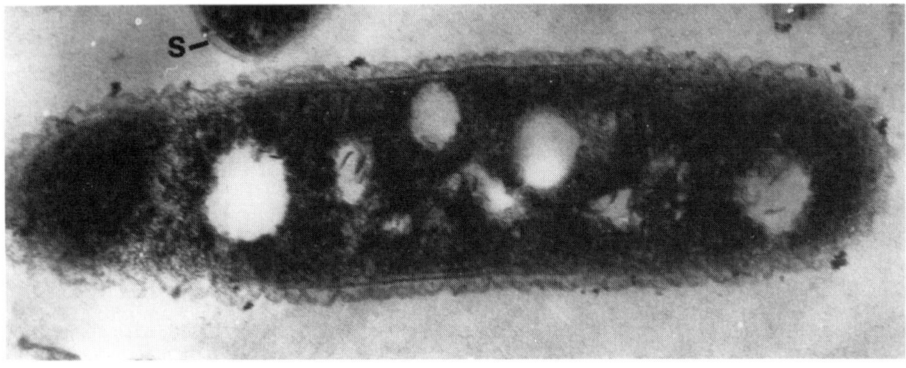

Figure 5. Bacteria-like cell with a wavy outer wall. *Sulfolobus* wall (S) is marked. Growler Spring, YNP: 89°C; pH 2.0. X 128, 000.

forms a single layer directly above the plasma membrane. Even a brief comparison to the more characteristic layered Gram negative cell wall shown in *Figure 3* (inset) suggests that obvious differences exist here as well. Although the wavy wall seen here may have the outer membrane of the Gram negative cell wall as a counterpart, this wavy layer is reminiscent of the subunits that form the *Sulfolobus* wall.

The habitat in which this microbe was discovered is more extreme than most habitats in which both bacteria and the Archaean, *Sulfolobus,* exist. It thus remains to be seen if this organism is a bacterium, merely surviving the extremes or is a member of the Archaea, optimally adapted to the extreme temperature and pH of this habitat.

When we first began this work, we were able to easily obtain enrichment cultures of several rod shaped organisms simply by adding 0.1% yeast extract to the hot spring water or by incubation of a hot spring sample in basal medium containing 0.1% yeast extract (Brock *et al.*, 1972). Over the years the rod shape organisms originally cultured were lost. Table 1 identifies the properties for each new organism originally isolated or observed at a specific sampling site along with the sample temperature and pH.

TABLE 1. New Organisms in acid thermal habitats of Yellowstone National Park, WY, U.S.A.

Location	(°C)	pH	Properties
Amphitheater Spring 1	78	2.0	Gram negative wall thin filaments alternating cell wall depressions
Amphitheater Spring 2	78	2.0	*Lobobacillus acidocaldarius* terminal lobes subunit wall
Norris Geyser Basin			
Growler Spring	89	2.0	Wavy cell wall
Locomotive Spring	87.5	2.0	*Acidobacillus fibrocaldarius* wall array surface fringe terminal cell
Norris Annex	78	2.3	Gram negative wall polar collar
Sulfur Cauldron	76	1.5	*Lobobacillus acidocaldarius* terminal, flat lobes subunit wall

4. Summary

I have identified 5 new organisms in acid hot springs. The Archaea-like organism, *Acidobacillus fibrocaldarius*, has peritrichous pili and a distinct cell wall array. Also Archaea-like was *Lobobacillus acidocaldarius,* a thin rod shaped microbe with lobes and a *Sulfolobus* type wall. The next organism we observed was a filamentous bacterium that exhibited a typical Gram negative cell envelope. It was unremarkable structurally except for depressions on the cell surface. The fourth organism we found was also a Gram negative bacterium. It was rod shaped with a polar collar. In favorable sections membranous material appeared in the collar area at the end of the cell. The structure itself sometimes appeared to be membranous. The significance of the membrane or of the collar is unclear, but it is unlike any other structure so far identified in any microorganism, regardless of habitat. The last organism we found was Bacteria-like; it was rod shaped with a wavy cell wall. The details of the wall structure could not be clearly resolved. Therefore, we are unsure whether it fits in the Bacteria or Archaea. The cell envelope appeared to resemble that of Gram negative bacteria, but it was unclear whether it had a layered cell envelope or a *Sulfolobus* type wall. If the cell wall is formed of subunits, it has a distinctive structure in thin sections; this might be expected since this organism comes from the most extreme conditions yet at pH 2.0 and 89°C.

In our early studies we found these new organisms were readily isolated and easily maintained in culture. The only organism so far reported at temperatures above 70°C and at low pH is *Sulfolobus*. Taken together this collection of new organisms from high temperature, low pH environments is truly unique and warrants further investigation.

5. Acknowledgments

I wish to acknowledge T.D. Brock who provided opportunity, support and lab facilities. The National Park Service is gratefully acknowledged for allowing access to the hot springs at Yellowstone National Park, WY, U.S.A. T.W. Carroll at Montana State University provided electron microscope facilities. RLWB was supported by NIH and TDB's lab received NSF support.

6. References

Brock, T.D., Brock, K.M., Belly, R.T., and Weiss, R.L. (1972) *Sulfolobus*: a new genus of sulfur-oxidizing bacteria living at low pH and high temperature, Arch. Mikrobiol. **84**, 54-68.

Brock, T.D., Brock, M.L., Bott, T.L., and Edwards, M.R. (1971) Microbial life at 90C: The sulfur bacteria of Boulder Spring, J. Bacteriol. **107**, 303-314.

Fliermans, C.B. and Brock, T.D. (1972) Ecology of sulfur-oxidizing bacteria in hot acid soils, J. Bacteriol. **111**, 343-350.

Stetter, K.O. and Zillig, W. (1985) *Thermoplasma* and the thermophilic sulfur dependent Archaebacteria, The Bacteria **8**, 85-170.

Stetter, K.O. (1982) Ultrathin mycelia-forming organisms from submarine volcanic areas having an optimum growth temperature of 105°C, Nature **300**, 258-260.

Weiss, R.L. (1973) Attachment of bacteria to sulphur in extreme environments, J. Gen. Microbiol. **77**, 501-507.

7. Glossary

Pyrolobus The most extreme thermophile known. Grows optimally at 105°C with an upper temperature limit of 113-115°C. Isolated from deep sea hydrothermal vents. Archaea.

Sulfolobus Grows in extremely hot acid springs as lobed spheres about 1-2µm in diameter. Growth at temperatures from 50-80°C with an upper limit of 88-92°C at pH 2-4. It is a facultative autotroph that oxidizes sulfur and can live on yeast extract or other carbon sources. Archaea.

Thermoproteus. Grows anaerobically from 80-92°C at pH 2.5-6 in deep waters and mud of acid hot springs. The temperature optimum is 88°C and some species grow from pH 5.5-7.5. Rods have a diameter of 0.5µm and grow from 1-80 µm in length. Branching possible. Archaea.

Thiobacillus- Gram negative nonphotosynthetic rod shaped cells that obtain energy by oxidation of sulfur and reduced sulfur compounds. Cells are 0.5 X 1-2 µm in size. The upper temperature limit for growth is about 55°C. Bacteria.

Acidobacillus A new rod shaped organism with an Archaea-like cell wall. It is 0.5 X 1.55 µm in size. It grows in the presence of elemental sulfur in mixing hot springs and has been observed at pH 2 and 87.5°C.

Lobobacillus. A new Archaea-like organism. It forms slightly curved rods with terminal lobes and grows in shallow flowing springs in nature at about 78°C; pH 2.3. It is 0.4 X 1-2µm in size and has a typical Archaea-like cell wall.

Biodata of **Helen Vishniac** contributor of the chapter on ***"Psychrophilic Yeasts."***

Dr. Helen Vishniac is a Professor (Emerita) at the Department of Microbiology and Molecular genetics, Oklahoma State University, Stillwater OK (USA). She received her Ph.D. in 1950 from Columbia University. Her primary interest is the systematics and ecology - they go hand in hand- of soil yeasts. Dr. Vishniac became interested in psychrophiles working with the soil samples returned from Antarctica after her husband's (Wolf Vishniac) premature death there.
E-mail: **hsvishi@osuunx.ucc.okstate.edu**

J. Seckbach (ed.), Enigmatic Microorganisms and Life in Extreme Environments, 315–321.

PSYCHROPHILIC YEASTS

H. S. VISHNIAC
Department of Microbiology and Molecular Genetics
Oklahoma State University
Stillwater OK 74078 USA

1. Introduction

Psychrophilic yeasts, as the easiest of single-celled eukaryotes to grow and manipulate, should have been in the forefront of research on cold physiology and genetics. Instead, their role has been primarily the confirmation of concepts developed using other organisms. A major part of the reason may lie in the absence of known sexual reproduction, or, when teleomorphs were discovered, in the tediously lengthy procedure required (Fell et al., 1969). These considerations need not deter the "postmodern" geneticist. It is to be hoped that they will not!

2. Distribution (and taxonomic considerations)

Psychrophilic yeasts are listed in table 1, as are, for comparisons, some psychrotolerant species which meet van Uden's (1984) definition of psychrophily in yeasts. Van Uden (1984) gave 25C as the limiting temperature for growth of psychrophilic yeasts. While such yeasts are more widely considered psychrotolerant, such usage has as much ecological justification as referring to *Hansenula polymorpha* (T_{max} 49-50C) as thermophilic (as mycologists do), when it is widely known that procaryote thermophiles can grow at more than twice this T_{max} . The more generally accepted definition is that of Morita (1975). Morita specified a maximum temperature for growth (T_{max}) $\leq$ 20C, an optimum temperature for growth $\leq$ 15C and a minimum temperature for growth $\leq$ 0C. It is the T_{max} specified which appears to be most generally available. No definition is without problems; temperature characteristics are continuous in nature and strains belonging to a single yeast species typically vary in T_{max} by 5 to 6 centigrade degrees (Vidal-Leira et al., 1979). Greater variations have been successfully predicted to indicate heterogeneity in a species.

Five species appear to include only psychrophilic strains: *Candida psychrophila* [T_{max} 17C (Goto et al., 1969)], *Leucosporidium antarcticum* [T_{max} 17C (Fell et al., 1969)], *Mrakia frigida* [T_{max} 19-20C (Fell et al., 1969), T_{min} < -7 (Larkin and Stokes, 1968)], *Sporobolomyces dracophyllus* [T_{max}17-18C (Hamamoto and Nakase, 1995)], and *Tausonia pamirica* [T_{max} 18-19C1 (Babjeva, 1997)] (It should be noted that the last two, being recently described, are little known.) The others listed include strains with T_{max} slightly to significantly higher. Some strains of *Candida austromarina* grew weakly at 21C (Fell and Hunter, 1974). The psychrophilic strain of *Cryptococcus antarcticus* (T_{max} >15C and <20C) exhibited 73% nDNA similarity to the type strain

Table 1. Psychrophilic yeasts

Species	Discovered in:
Maximum temperatures for growth ≤ 20C	
Candida psychrophila (Goto et al.) S. A. Meyer & Yarrow	Antarctica (penguin dung)
Candida austromarina * (Fell & Hunter) S. A. Meyer & Yarrow	Antarctic ocean
Cryptococcus antarcticus * Vishniac & Kurtaman	Antarctic (dry valley soil)
Cryptococcus vishniacii * Vishniac & Hempfling	Antarctic (dry valley soil)
Leucosporidium antarcticum Fell et al.	Antarctic ocean
Leucosporidium scottii * Fell et al.	Australia (soil, chilled beef)
Mrakia frigida Yamada & Komagata	Antarctic (snow, soil)
Sporobolomyces dracophyllus Hamamoto & Nakase	New Zealand (dead leaves)
Tausonia pamirica Babjeva	Tadjikistan (taiga)
Maximum temperature for growth >20C, ≤ 25C	
Bullera megalospora * Nakase & Suzuki	Japan (dead leaves)
B. punicea (Komagata & Nakase) Nakase & Suzuki	Japan (frozen salmon)
Cryptococcus aquaticus (Jones & Slooff) Rodrigues de Miranda & Weijman	England (lake scum)
C. curiosus (Komagata & Nakase) Rodrigues de Miranda & Weijman	Japan (frozen plant material)
Cystofilobasidium capitatum (Fell et al.) Oberwinkler and Bandoni	Far southern oceans
C. infirmo-miniatum * (Fell et al.) Hamamoto et al.	Tokyo (air)
Nadsonia commutata Golubev	Falkland Is. (soil)
Rhodosporidium malvinellum * Fell & Hunter	Far southern oceans
Rhodotorula lignophila * (Dill et al.) Roeijmans et al.	Chile (rotting wood)
Sporobolomyces coprosmicola Hamamoto & Nakase	New Zealand (dead leaves)
S. dimennae Hamamoto & Nakase	New Zealand (dead leaves)
*S.gracilis** Derx	Netherlands and France (leaves)
S. novazealandicus Hamamoto & Nakase	New Zealand (dead leaves)
S. yuccicola Nakase & Suzuki	Vancouver, Canada (dead leaf)
*Trichosporon pullulans** (Lindner) Diddens & Lodder	Germany (air, beer casks)

* Some strains of these species have T_{max} within the 5 centigrade degrees above that specified. Strains of *Leucosporidium scottii* reportedly have T_{max} 15-30C.

(T_{max} >20C and <25C) (Vishniac and Kurtzman, 1992), so is undoubtedly conspecific. The nDNA of *Cryptococcus vishniacii* Vishniac and Hempfling (1979) (named in honor of W. V. Vishniac) is, within experimental limits, identical (unpublished data of H. S. Vishniac) to that of species segregated (Baharaeen and Vishniac, 1982; Vishniac and Baharaeen, 1982) from it -- *Cryptococcus lupi*, *C. asgardensis*, *C. baldrensis*, *C. hempflingii*, *C. tyrolensis*, and *C. wrightensis* -- as well as to that of *Cryptococcus consortionis* and *C. socialis* Vishniac (1985). The only psychrophilic strain appears to be the type of the species.

Leucosporidium scottii presents a problem: Fell et al. (1969) list T_{max} of 15-30C for this species. The evidence for conspecificity of isolates other than the types is the extensive mating experiments of Fell's laboratory, experiments which could not include the homothallic type of the anamorph, *Candida scottii*. However, *C. scottii* , described as having a T_{max} of "ca.15C" (di Menna, 1960), possessed close similarity in partial 18S and 26S rRNA sequences (Yamada and Nakagawa, 1990) with the type of *L. scottii* itself, which grew well at 25C and mated successfully at "room temperature" (Fell et al., 1969).

The absence of some species from this list may surprise readers. *Mrakia frigida* appears, but not *M. gelida* , *M. nivalis* or *M. stokesii*. Although the hybridizations which have served to define yeast species have not been performed, these species are probably synomyous with *M. frigida*. They significantly share both rRNA characteristics (Fell, 1993) and whole-cell protein electrophoretic patterns (Vancanneyt et al., 1992). *Cryptococcus albidus* (Saito) Skinner, "psychrophilic strains" of which have been repeatedly isolated from Antarctic sites (see Vishniac, 1996), is an extremely unlikely colonizer in the Antarctic. The type strain has a T_{max} near 35C (Vishniac and Kurtzman, 1992). Nuclear DNA hybridizations have shown that several isolates attributed to *C. albidus* were not conspecific with it (Martini, 1991). It has therefore been omitted from table 1.

Truly psychrophilic yeast species appear therefore to be relatively few. The majority of them were discovered in or around Antarctica (table 1), as were the species including psychrophilic strains. Some of these species are known only from Antarctic regions. *Candida psychrophila* is known both from Cape Royds (Ross Island) penguin dung and Arena Valley (Ross Desert) soil (Atlas et al., 1978) *Candida austromarina* similarly is known only from the beautifully detailed floristic studies of yeast distribution in southern oceans (Fell, 1974). *Cryptococcus antarcticus* is known only from University Valley (Ross Desert), though some of the many Antarctic collections of "*Cryptococcus albidus*" may well have been strains of this species. *Cryptococcus vishniacii* occurs widely in the Ross Desert ("dry valleys"), but is unknown outside of Antarctica. The highlands of these valleys are arid and, except for glacial melt streams, productive only where cryptoendolithic life exists. It has been suggested (Vishniac and Kurtzman, 1992) that *Cr. antarcticus* and *Cr. vishniacii* have been selected for varying degrees of oligotrophy. Their insensitivity to aridity is presumeably a property of all encapsulated yeasts. *Cryptococcus albidus* has served as a model in experiments demonstrating the ability of such yeasts to compete successfully with bacteria in soil of low water content (Vishniac, 1995). The unique multiple pressures of their habitat may well be the factor confining these species to the Antarctic.

The *Leucosporidium* and *Mrakia* species are best known from better watered, consequently more productive, Antarctic areas as well as from sites outside Antarctica. *Leucosporidium scottii*, credited to Australia because the type of "*Candida scottii*" came from soil in the vicinity of a chilled beef plant (and from the chilled beef), was also collected from Antarctic soil and water by di Menna (1966), Fell et al. (1969, see also Fell, 1974), and Goto et al. (1969). The listed *Leucosporidium* and *Mrakia* spp. have also been isolated from Arctic sources and even from areas of more temperate climate (summarized by Summerbell, 1983). More recently, Middelhoven et al. (1992) reported isolations of *L. scottii* from Dutch soil by enrichment with ferulic and vanillic acids at 10C. When subjected to supermaximal temperatures, psychrophilic strains may survive by sharing the dissociative Arrhenius curve reported for two strains of *M. frigida* (van Uden et al., 1968), i.e., thermal death became significant only at the same temperatures as for mesophilic yeasts.

The two remaining psychrophilic species are known only from lesser latitudes. *Tausonia pamirica* is from Tadjikistan, which lies at approximately 38° N but has temperatures affected by the altitude of the high plains (around 3.4 km alt.). *Sporobolomyces dracophyllus* is from an area of New Zealand roughly as far south, at ca. 39° S, but only ca. 1 km alt. However, the mean January (summer) temperature of this climatic area is only 16-17C, with at least 50 days when the temperature falls below freezing in the rainy winter season. The discovery of these species suggests that additional psychrophiles may occur in similar areas which so far have lacked the appropriate mycologists.

The psychrotrophic species listed, in contrast, were discovered and have been isolated, almost without exception, only outside of the Antarctic. Of the psychrotrophs listed in table 1, only *Trichosporon pullulans* has been isolated from Antarctic soil (Atlas et al., 1978), but so rarely as to be reasonably considered an aeolian deposit rather than an indigene (see Vishniac, 1996). This is somewhat surprising, since yeasts generally survive cold, particularly dry cold, very well (for example, lyophilization) and *T. pullulans* is a relatively common yeast in temperate climates.

3. Physiology

The *Mrakia* spp. are among the most easily manipulated subjects for physiological studies. While they have attracted some attention, little has been added to the work reviewed by van Uden (1984), here briefly summarized. The contribution of cell membrane stability was recognized in 1968, when Nash and Sinclair reported that *Mrakia* (as *Candida nivalis*) membranes leaked at 25C. Membrane lipid composition has been considered to account for the lower and upper limits of *Mrakia* growth (Arthur and Watson, 1976). Ninety percent of the mitochondrial membrane lipids of *Mrakia frigida* (as *Leucosporidium frigidum*) grown at -0.5C were unsaturated (27% $C_{18:2}$, 53% $C_{a18:3}$) (Arthur and Watson, 1976). Three strains of *Mrakia* (as *L. frigidum, L. gelidum,* and *L. nivale*) cells grown at -1C contained 84 to 89% unsaturated fatty acids, with linolenic again dominant but at 17C, while still containing 80 to 84% unsaturated fatty acids, adjusted viscosity with decreased percentages of linolenic acid and increases in linoleic and oleic acids (Watson et al., 1976). *Candida psychrophila* and *C. austromarina* grown at 0C contained (respectively) 92% and 86% unsaturated fatty acids and at 15C underwent similar changes in the proportions of linolenic vs linoleic and oleic acids (Watson, 1980). More recently, an analysis by MIDI found that the fatty acids of *Cryptococcus antarcticus* strains grown at 13C were 87 to 93% unsaturated and of the type of *C. vishniacii* 91% unsaturated. Neither species produced linolenic acid under these conditions (Vishniac and Kurtzman, 1992).

Most of the other work reviewed by van Uden involved exposure to temperatures too high to assess the contribution of other factors to thermosensitivity. However, at temperatures only 1 to 3 degrees above T_{max}, a suite of problems led by failure in DNA synthesis contributed to preventing reproduction. "*Leucosporidium stokesii*" (i.e., *Mrakia* sp.) showed impairment of DNA synthesis at 23C (Silver et al. 1977). While cell mass increased, atypical buds (produced at supramaximal temperatures up to 23C) were anucleate and aseptate (Silver and Sinclair, 1979). RNA synthesis became impaired only at 25C, while protein synthesis remained active at that temperature.

This does not leave the psychrophilic yeasts without recourse to the usual assistance of heat shock proteins (hsps) and trehalose accumulation after exposure to appropriately higher supramaximal temperatures. Deegenaars and Watson (1997) found that while "nearly 100%" of stationary phase *C. psychrophila* cells survived 60 minutes at 35C, but less than 0.1% of exponential phase cells did so. Exposed to a heat shock of 3 hours at 25C, "almost 100% " of exponential phase cells could tolerate 30 minutes at 35C. Heat shock proteins 60, 70, 80, 90, and a novel 110 were produced. Hsp 104, so important to thermotolerance in *Saccharomyces cerevisiae* (Sanchez et al., 1992), was not. Extending their studies to basidiomycetous psychrophilic yeasts, Deegenaars and Watson (personal communication), found that a heat shock of 3 h at 25C (i.e., 10 centigrade degrees above the growth temperature of 15C) produced greatest thermotolerance to 35C exposure and maximal trehalose accumulation in the psychrophilic yeasts *Mrakia* sp. (as *M. frigida, M. gelida, M.*

stokesii) and *Leucosporidium antarcticum*. The induction of several hsps in each of these yeasts was detected by ^{35}S-methionine labeling and their familial affinities established with monoclonal antibodies to *S. cerevisiae* hsps 60, 70, 90, and 104. All of the psychrophiles produced hsps 70 and 90, but none hsps 60 or 104. Since the psychrotrophic *Leucosporidium fellii* and *L. scottii* did produce hsp 104, the authors suggest that its absence distinguishes psychrophiles from psychrotrophs. Neither the psychrophilic *L. antarcticum* nor the psychrotrophic *L.* spp, however, produced hsp 110, while the *Mrakia* spp did. They suggest that induction of this hsp, though present in the ascomycetous *C. psychrophila*, may have a generic character.

4. Literature cited:

Arthur, H., and Watson, K. (1976) *J. Bacteriol.* **128**, 56-68.

Atlas, R. M., di Menna, M. E., and Cameron, R. E. (1978) *Antarctic Research Series* **30**, 27-34.

Babjeva, I. P. (1997) *Mikrobiologia* (Microbiology) (in Russian, in press; cited from *Yeast Newsletter* **66** (1): 2)

Baharaeen, S. and Vishniac, H. S. (1982) *Int. J. Syst. Bacteriol.* **32**, 229-232.

Deegenaars, M. and Watson K. (1997) *FEMS Microbiology Letters* **151**, 191-196.

di Menna, M. (1960) *J. gen. Microbiol.* **23**, 295-300.

di Menna, M. (1966) *Antonie van Leeuwenhoek* **32**, 29-38.

Fell, J. W. (1974) In R. R. Colwell and R. Y. Morita (eds), *Effect of the ocean environment on microbial activities*, Baltimore University Park Press, pp. 510-523.

Fell, J. W. (1993) In R. R. Reynolds and J. W. Taylor (eds), *The Fungal Holomorph: Mitotic, Meiotic and Pleomorphic Speciation in Fungal Systematics*, CAB International, Wallingford, pp. 247-251.

Fell, J. W. and Hunter, I. L. (1974) *Antonie van Leeuwenhoek* **40**, 307-310.

Fell, J. W., Statzell, A. C., Hunter, I. L., and Phaff, H. J. (1969) **35**, 433-462.

Goto, S., Sugiyama, U., and Iizuka, H. (1969) *Mycologia* **61**, 748-774.

Hamamoto, M. and Nakase, T. (1995) *Antonie van Leeuwenhoek* **67**, 151-171.

Larkin, J. M. and Stokes, J. L. (1968) *Can. J. Microbiol.* **14**, 97-101.

Martini, A. V. (1991) *Exp. Mycol.* **15**, 140-145.

Middlehoven, W. J., Koorevaar, M., and Schuur, G. W. (1992) *Plant and Soil* **145**, 37-43.

Morita, R. Y. (1975) *Bacteriological Reviews* **39**, 146-167.

Nash, C. H. and Sinclair, N. A. (1968) *Can. J. Microbiol.* **14**, 691-697.

Sanchez, Y., Taulien, J., Borkovich, K. A., and Lindquist, S. (1992) *EMBO J.* **11**, 2357-2364.

Silver, S. A. and Sinclair, N. A. (1979) *Mycopathologia* **67**, 59-64.

Silver, S. A., Yall, I., and Sinclair, N. A. (1977) *J. Bacteriol.* **132**, 676-680.

Summerbell, R. C. (1983) *Can. J. Bot.* **61**, 1402-1410.

van Uden, N. (1984) *Adv. Microb. Physiol.* **25**, 195-251.

van Uden, N., Abranches, P., and Cabeça-Silva, C. (1968) *Archiv für Mikrobiologie* **64**, 381-393.

Vancanneyt, M., van Lerberge, E., Berny, J.-F., Hennebert, G. L., and Kersters, K. (1992) *Antonie van Leeuwenhoek* **61**, 69-78.

Vidal-Leira, M., Buckley, H., and van Uden N. (1979) *Mycologia* **71**, 493-501.

Vishniac, H. S. (1985) *Int. J. Syst. Bacteriol.* **35**, 119-122.

Vishniac, H. S. (1995) *Microbial Ecology* **30**, 309-320.

Vishniac, H. S. (1996) *Biodiversity and Conservation* **5**, 1365-1378.

Vishniac, H. S. and Baharaeen, S. (1982 *Int. J. Syst. Bacteriol.* **32**, 437-445.

Vishniac, H. S. and Hempfling, W. P. (1979) *Int. J. Syst. Bacteriol.* 29: 153-158.

Vishniac, H. S. and Kurtzman, C. P. (1992) *Int. J. Syt. Bacteriol.* **42**, 547-553.

Watson, K. (1980) In M. Kates and A. Kuksis (eds), *Membrane Fluidity*, Humana Press, Clifton N.J., pp 349-363.

Watson, K., Arthur, H., and Shipton, W. A. (1976) *J. Gen. Microbiol.* **97**, 11-18.

Yamada, Y. and Nakagawa, Y. (1990) *J. Gen. Appl. Microbiol.* **36**, 63-68.

Note added in Proof.

Tausonia pamirica should not have been included in this article, as it does not produce blastospore.

BioData of **Dr. Silvno Onofri**
Contribution: ***Antarctic Microfungi***.

Silvano Onofri is a Professor of **B**otany at the Tuscia University in Viterbo (Italy). He gained his Doctor degree in Biology from the University of Rome, and serves currently as the head of the mycological program within the Italian National Program for Antarctic Research (PNRA). Among his publications are articles about fungal taxonomy, succession in mycocoenoses and adaptation of fungi to extreme environments.
<E-mail: ***onofri@unitus.it***>

J. Seckbach (ed.), Enigmatic Microorganisms and Life in Extreme Environments, 323–336.

ANTARCTIC MICROFUNGI

SILVANO ONOFRI

Dipartimento di Scienze Ambientali, Università della Tuscia
Via San Camillo de Lellis, 01100 Viterbo, Italy

1. Introduction

Antarctica is the continent of microorganisms, as they are the ecologically dominant organisms both in the ocean and on dry land (Friedmann, 1993). About 98% of Antarctica is covered by ice. Ice-free zones occur in the Antarctic Peninsula and in scattered coastal areas, such as along the Trans-Antarctic mountains (Claridge and Campbell, 1985), the most important being the Ross Desert in the McMurdo Dry Valleys. These ice-free areas have also been called "oases" (Pickard, 1986). Although some microorganisms are able to grow on ice, like some yeasts (Sinclair and Stokes, 1965), or survive in ice, like some microfungi and bacteria (Abyzov, 1993), the colonization of microorganisms is mainly limited to the ice-free zones. Factors affecting the possibility of microbial colonization are: viable propagules, suitable transport mechanisms, suitable micro-habitats, and successful growth and reproduction (Ellis-Evans and Walton, 1990). The suitability of micro-habitats depends mainly on climate which controls before everything water availability, and on the presence of suitable substrata. The macroclimate affects both the microclimate and the presence of primary producers, a fundamental source of substrata for fungi.

2. Climate

The Antarctic climate is the coldest and the dryest known on earth. It is not uniform all over the Continent and creates conditions which do or do not allow the microorganisms to survive. Antarctica can be divided into three climatical regions (Holdgate, 1977):

(a) The sub-Antarctic region (including most of the Antarctic islands north to 60° South) is characterized by an oceanic climate with a mean annual temperature above 0°C and precipitation over 1000 mm; vascular plants dominate vegetation near sea level.

(b) The climate in the maritime Antarctic region (consisting of South Sandwich Islands, South Orkney Islands, South Shetland Islands, Palmer Arcipelago and the west coast of Antarctic Peninsula) is cold maritime, with mean monthly temperatures above 0°C in the warmer period, rarely below -15°C. Precipitation varies from 250 to over 1000 mm.

(c) Coastal climate in continental Antarctic region is cold, with all monthly means below 0°C and ranging from -15° to -25°C during the winter; precipitaton above

100–150 mm. Moving to the interior the climate becomes harsher, reaching mean monthly temperatures well below -25°C on the ice plateau.

Differences can be detected in the maritime and continental Antarctic mycoflora. Besides the relatively milder climate, the presence in the former of phanerogamic and rich bryological vegetations (Longton, 1985) significantly affects the mycoflora, supplying a differentiated and abundant substrata and allowing the presence of a higher number of specialized fungal species. The sub-Antarctic region is not considered here.

3. Autochthonous, potential and expressed mycoflora

The long distance dispersal capability of the air-borne mycoflora, added to transportation by animal and man, make it difficult to distinguish between autochthonous and imported fungal species (Ellis-Evans and Walton, 1990). New species described for Antarctica could be considered autochthonous (Vishniac, 1993). In any case many Antarctic microfungi are air-borne species, therefore new species described for Antarctica are not endemic by definition.

Despite the importance of knowing autochthonous (i.e. native) and allochtonous (i.e. imported) microfungal species, in practical terms it is more effective to distinguish between potential and expressed mycoflora. Potential mycoflora is represented by species (or strains) which can express themselves in suitable conditions, being present only as viable propagules. This species group is very important, because its colonization could be linked to climatical changes. Expressed mycoflora is represented by actually growing fungi. Criteria in distinguishing between these two groups could be found in psychrophily or psychrotolerance, xerophily and capability to overcome extremely dry and cold conditions, at least as spores.

4. Total mycodiversity

In this chapter only organisms belonging to the Kingdom Fungi (Hawksworth *et al.*, 1995) are reported.

Since the beginning of the century, Antarctica attracted the interest of microbiologists (Bommer and Rousseau, 1900, 1905; Tsiklinsky, 1908; Gazert, 1912; McLean, 1918). The first new fungal species described from Antarctica was *Sclerotium antarcticum* Bommer and Rousseau - its teleomorph *Sclerotinia antarctica* was described as new species by Gamundi and Spinedi (1987) - from Damco Coast, maritime Antarctic Peninsula. From that pioneering period till now 27 new filamentous fungi and 19 yeast species have been described for Antarctica and about 190 species have been recorded for continental Antarctica. Mycodiversity in maritime Antarctica is probably higher and at least 12 differential species of *Ascomycota*, 13 species of *Basidiomycota* and many Imperfect Fungi, unknown for continental Antarctica, have been recorded in its relatively more suitable environment.

a. *Chytridiomycota* are represented only by a few species. *Phlyctochytrium recurvistomum* and *Rhizophydium proliferum* were described as new species by Knox and Paterson (1973), both from soils in several sites in Victoria Land, mainly in the McMurdo Dry Valleys. *Chytridium versatile* has been recorded as diatoms parasite,

Scherffeliomyces appendiculatus as *Chlamydomonas* sp. parasite in lake water, *Karlingia rosea* from soil and freshwater, and *Catenophlyctis variabilis* on penguin and skua feathers (Paterson and Knox, 1972; Knox and Paterson, 1973), from soil (Paterson and Rooney, 1972) and snake skin baits (Knox and Paterson, 1973), all from Victoria Land. Other species belonging to this group have been recorded also from maritime Antarctica (Harder and Persiel, 1962).

b. Many species in *Zygomycota* are considered psychrotolerant and their survival strategy is often ruderal (Pugh, 1980). Species belonging to *Mortierella* (including the new species *M. antarctica*), *Absidia, Mucor, Rhizopus, Syncephalastrum* have been isolated in continental Antarctica mainly from mosses, soil, and soil under mosses (Tubaki and Asano, 1965; Zycha *et al.*, 1969; Sun *et al.*, 1978; Greenfield, 1981; Del Frate and Caretta, 1990; Onofri *et al.*, 1994; Zucconi *et al.*, 1996), but also from the water of the Lake Vanda, Dry Valleys (Sugiyama *et al.*, 1967; Sugiyama, 1970), air (Sun *et al.*, 1978), green algae and soil (Fletcher *et al.*, 1985) and the core of an ice sample extracted from the dephts of the ice sheet, near Vostok Station (Abyzov and Belyakova, 1982; Abyzov, 1993). In maritime Antarctica *Cunninghamella antarctica* has been described as a new species and *C. echinulata* recorded from the west coast of the Antarctic Peninsula (Caretta and Piontelli, 1977), as well as *Mortierella gamsii* (Möller and Dreyfuss, 1996) and *M. parvispora* (Latter and Heal, 1971).

c. *Ascomycota* are pretty rare in continental Antarctica: the most common species is the psychrophilic or psychrotolerant - depending on the strains - and coprophilous *Thelebolus microsporus* (Wicklow and Malloch, 1971; Fletcher *et al.*, 1985; Ryan *et al.*, 1989; Del Frate and Caretta, 1990; Kerry, 1990a; Kerry, 1990b; Onofri *et al.*, 1991; Montemartini Corte *et al.*, 1993; Onofri *et al.*, 1994). *Gymnascella citrina* has been isolated from soil in McMurdo Station (Sun *et al.*, 1978) and some *Chaetomium* species from soils and pond mud (Ellis, 1980; Montemartini Corte, 1991). *Pseudogymnoascus roseus* and members of the *Trichocomaceae* (limited to teleomorphs, i.e. sexaul stages, of *Aspergillus*) are known in anamorphic (i.e. asexally reproducing) forms. Among Ascomycetes isolated as anamorphs, the presence of *Amorphotheca resinae* ('kerosene fungus') in soils contaminated with oil (Kerry, 1990b) and kerosene (Line, 1988) is remarkable. The new species *'Diplonevia' parmeliae* (Dodge and Baker, 1938) and *Orbicula buelliae* (Dodge, 1948) were reported on the lichens *Physcia caesia* and *Buellia muscicola*, respectively. Some new species and many records are reported on vascular plants, mosses and lichens (see 5.1, 5.2, 5.3) in maritime Antarctica (Pegler *et al.*, 1980). Gamundi and Spinedi (1988) published four new species and one new variety from the Damco Coast (maritime Antarctica): *Cistella polytrichi* var. *antarctica* epiphitic on *Polytrichum alpinum, Bryoscyphus aestivalis* parasitic on *Bryum dichotomum*, *Octospora ciervensis* parasitic on *Bryum dichotomum*, *Psilocistella deschampsiae* on *Deschampsia antarctica* and *Laetinaevia colobanthi* on *Colobanthus quietensis*.

d. Macrofungi are not part of this paper, but in any case they are represented in the maritime Antarctica, whereas *Basidiomycota* - excluding yeasts - are virtually absent from continental Antarctica. Only *Pycnoporus coccineus (Poriales)* was recorded (basing on an isolate identified by cultural characteristics) from a mummified penguin and associated soil, Taylor Dry Valley (Baublis *et al.*, 1991); its distribution elsewhere seems to be restricted to the Islands of Pacific and Indian Oceans and countries bordering on these waters.

The first basidiomycete described for maritime Antarctica was *Omphalina antarctica*, growing on or among mosses over rocky soil in Deception Island, South

Shetland Islands (Singer, 1957) and recorded also from South Orkney and South Sandwich Islands. Many other species of *Hymenomycetes* have been recorded from South Orkney Islands: *Exidia* sp., *Dacrymyces stillatus, Galerina antarctica, G. sideroides, G. moelleri, Simocybe antarctica, Gerronema fibula;* from South Shetland Islands: *Leptoglossum lobatum* var. *antarcticum;* from South Sandwich Islands: *Galerina longiqua, Gerronema fibula, G. schusteri* and *Tephricybe* sp.; and from Antarctic Peninsula: *Galerina perrara* (Pegler *et al.*, 1980). *Leptoglossum omnivorum* Agerer (1984) was described as new species from maritime Antarctic Islands, on timber and decaying feathers, nearby mosses and algae.

e. Imperfect (or Mitosporic, i.e. lacking of known sexal reproduction) Fungi are dominant in the Antarctic mycoflora. More than 100 species have been recorded and at least 10 new species have been described.

Besides the newly described species, many psychrotolerant Hyphomycetes could be considered actually growing in Antarctica. The survivor-escaper (Pugh, 1980) *Geomyces pannorum* var. *pannorum* is probably the most common Antarctic fungal species (Tubaki, 1961a, 1961b; Tubaki and Asano, 1965; Cameron, 1971; Greenfield, 1981; Pugh and Allsopp, 1982; Fletcher *et al.*, 1985; Heatwole *et al.*, 1989; Ryan *et al.*, 1989; Del Frate and Caretta, 1990; Kerry, 1990a, 1990b; Onofri *et al.*, 1991; Montemartini Corte, 1991; Finotti *et al.*, 1993; Mercantini *et al.*, 1993; Onofri *et al.*, 1994; Möller and Dreyfuss, 1996). It is considered psychrotolerant, but there exist strictly psychrophilic strains (Zucconi *et al.*, 1996) and it has been also isolated from the air (Sun *et al.*, 1978; Czarnecki and Bialasiewicz, 1987). Psychrotolerant are also: *Aureobasidium pullulans*, *Cladosporium cladosporioides*, *C. herbarum*, *Embellisia tellustris, Epicoccum nigrum*, *Penicillium aurantiogriseum*, *P. brevicompactum*, *P. crustusum*, *P. glabrum*, *Phialophora lagerbergii*, *P. malorum*, *Stachybotrys chartarum* (Tubaki and Asano, 1965; Sugiyama, 1970; Cameron *et al.*, 1970; Cameron, 1971; Cameron *et al.*, 1972; Atlas *et al.*, 1978; Sun *et al.*, 1978; Greenfield, 1981; Simmons, 1983; Fletcher *et al.*, 1985; Heatwole *et al.*, 1989; Ryan *et al.*, 1989; Del Frate and Caretta, 1990; Kerry, 1990b; Baublis *et al.*, 1991; Montemartini Corte, 1991; Möller and Dreyfuss, 1996; Zucconi *et al.*, 1996; Fenice *et al.*, 1997) and *Verticillium lecanii*, which Antarctic strains are good producers of chitinase (Fenice *et al.*, 1996, 1997), active at low temperature (Fenice, pers. comm.). *Ulocladium consortiale* (Montemartini Corte, 1991), recorded also from litter in Alaskan Arctic (Matsushima, 1975), is able to survive and grow after exposure to extreme (-94/+36°C) diurnal temperature changes and to resist to UV irradiation (Domsch *et al.*, 1980). Possessing an extremely wide range of temperatures requirements and psychrotolerant (Zucconi *et al.*, 1996), *Phoma herbarum* (Coelomycetes) is quite common in the soils in coastal sites. The numerous recorded species of common air-borne fungi, chiefly in the genus *Penicillium*, are possibly dormant in ice, soil, mosses etc.

f. Yeasts are quite common in Antarctic soils and many of them are probably endemic. Most of the about 60 recorded species are psychrophilic or psychrotolerant (Vishniac, 1993; Zucconi *et al.*, 1996). Several new species have been described, mainly from Ross Desert soils (di Menna, 1966a, 1966b; Fell *et al.*, 1969; Goto *et al.*, 1969; Vishniac and Hempfling, 1979a; Baharaeen and Vishniac, 1982; Vishniac and Baharaeen, 1982; Vishniac, 1985a; Vishniac, 1985b; Vishniac and Kurtzman, 1992).

5. Microfungi associated to different habitats

5.1. VASCULAR PLANTS

Colobanthus quitensis (Kunth) Bartl. *(Caryophyllaceae)* and *Deschampsia antarctica* Desv. *(Poaceae)* are the only vascular plants growing in the (maritime) Antarctic Continent (Greene and Holtom, 1971). Gamundi and Spinedi (1988) described new *Ascomycetes* species growing on them (see **4.**). Three new Hyphomycetes have been isolated from the rhizosphere of *C. quitensis* and *D. antarctica* in Danco Coast: *Acrodontium antarcticum, Chalara antarctica, Phialophora dancoii* (Cabello, 1989). Möller and Dreyfuss (1996) isolated from samples of both plant species collected on King George Island (maritime Antarctica) several microfungi not recorded for continental Antarctica, such as *Phaeoseptoria microscopica, Volucrispora graminea, Chromelosporium ollare, Pleospora vagans*. Some nematophagous species have been isolated from *D. antarctica* and *C. quitensis,* and *Monacrosporium lypsipagum* is known in Antarctica only from *D. antarctica* (Gray *et al.*, 1982).

5.2. MOSSES

Many species are recorded from mosses: *Mortierella antarctica* (Zycha *et al.*, 1969; Del Frate and Caretta, 1990; Zucconi *et al.*, 1996), *M. alpina, Alternaria alternata, Aspergillus versicolor* (Onofri *et al.*, 1994), *Arthrobotrys ferox* (Onofri and Tosi, 1992), *A. conoides* (Caretta *et al.*, 1994), *Cladosporium herbarum* (Möller and Dreyfuss, 1996; Zucconi *et al.*, 1996), *Embellisia tellustris* (Simmons, 1983; Onofri, *et al.*, 1991, 1994 [as *Dendryphiella salina*]), *Engyodontium album* (Cameron, 1971; Caretta *et al.*, 1994), *Verticillium lecanii* (Fenice *et al.*, 1997), *Phoma herbarum* (Kerry, 1990a; Zucconi *et al.*, 1996), *Phoma sorghina* (Zucconi *et al.*, 1996), and *Geomyces pannorum*, which is psychrotolerant from mosses in maritime Antarctica (Möller and Dreyfuss, 1996) and psychrophilic from mosses in continental Antarctica (Zucconi *et al.*, 1996). The foregoing species are recorded also from soils under mosses. *Mucor hiemalis* (Fletcher *et al.*, 1985), *Chaunopycnis alba, Paecylomyces variotii, Penicillium janthinellum, Phaeoseptoria eustoma, Phialophora hyalina, Phialophora melinii, Tolypocladium nubicola* (Möller and Dreyfuss, 1996) have been recorded only from mosses.

One is inclined to doubt the active colonization by the mesophilic strains of *Pseudogymnoascus roseus, Penicillium glabrum, Phialophora malorum, Penicillium brevicompactum*, and *Epicoccum nigrum* (Kerry 1990b; Möller and Dreyfuss, 1996)

5.3. LICHENS

Lichens represent one of the most important and diversified group in the Antarctic biota (Kappen, 1993). Mycoflora associated to lichen thallus is commonly rich (Hawksworth, 1979; Petrini *et al.*, 1990).

'Diplonevia' parmeliae and *Orbicula buelliae* are new Ascomycetes on lichens (see **4.c.**). *Polycoccum rugulosarium* (Pegler *et al.*, 1980) was found several times mainly in the apothecia or the thallus of*Caloplaca regalis, C. cirrochrooides* and *C. sublobulata*

on wood and on rock in sub-Antarctic and maritime Antarctic, and in *C. rugulosa* in Tasmania, Australia (Hawksworth and Diederich, 1988), recorded also in Deception Island (Aptroot and van der Knaap, 1993) and in Victoria Land (continental Antarctica) on *Xanthoria elegans* (Onofri *et al.*, 1991).

Hawksworth (1979) described the new Hyphomycete *Acremonium antarcticum* growing on *Caloplaca* sp.; Möller and Gams (1993) isolated the new Hyphomycetes *Chaunopycnis ovalispora* and *Acremonium psychrophylum* from *Caloplaca regalis* and *Turgidosculum complicatum*, respectively. Möller and Dreyfuss (1996) isolated many mitosporic fungal species, and *Zygomycota* as *Mortierella gamsii*, from lichens collected near Arctowski and Jubany Antarctic Bases in Antarctic Peninsula.

5.4. SOILS

The peculiar capability of the soil to store propagules, especially in cold climates, makes it difficult to distinguish dormant and growing microfungi. On the other hand, differences in mycoflora composition, detected in ornithogenic and colonized by mosses soils, can help in describing characteristic fungal communities possibly composed by actually colonizing microfungi.

Most of Antarctic *Chytridiomycota* are from soils (see **4.a.**), and *Catenophlyctis variabilis* particularly is from ornithogenic soil. Typical of ornithogenic soils are the Ascomycetes *Thelebolus microsporus* (see **4.c.**) and *Chaetomium gracile* (Ellis, 1980), and the Hyphomycetes *Chrysosporium merdarium* (Sun *et al.*, 1978; Ryan *et al.*, 1989), *Scopulariopsis brevicaulis* (Mercantini *et al.*, 1989) and *Penicillium aurantiogriseum* (Cameron, 1971; Fletcher *et al.*, 1985; Heatwole *et al.*, 1989; Kerry, 1990b), which is a psychrotolerant species. *Aspergillus fumigatus* has been isolated from soil (Greenfield, 1981) and from guano (Ellis, 1980); Kageruka (1967) observed that pulmonary aspergillosis, frequently occurring in imported penguins, did not exist in a natural state, but the reported records indicate the existence of this species in natural habitats of penguins and it can probably grow on the guano in the sun (Wicklow, 1968).

Among *Zygomycota* from soils, *Mortierella alpina, M. antarctica, Mucor flavus* and *M. mucedo* seem to be exclusive of soils under mosses, as the Ascomycete *Thielavia heterothallica* and the Imperfect Fungi *Cladosporium cladosporioides, Embellisia tellustris, Engyodontium album, Epicoccum nigrum, Verticillium lecanii* and *Phoma sorghina* (Cameron, 1971; Greenfield, 1981; Simmons, 1983; Heatwole *et al.*, 1989; Ryan *et al.*, 1989; Kerry, 1990b; Caretta *et al.*, 1994; Onofri *et al.*, 1994; Zucconi *et al.*, 1996; Fenice *et al.*, 1997). Some species have been isolated both from ornithogenic and colonized by mosses soils, as the Ascomycete *Pseudogymnoascus roseus* (Mercantini *et al.*, 1993) and the Imperfect Fungi *Cladosporium herbarum* (Greenfield, 1981; Fletcher *et al.*, 1985; Del Frate and Caretta, 1990; Zucconi *et al.*, 1996), *Phialophora fastigiata* (Greenfield, 1981; Del Frate and Caretta, 1990; Kerry, 1990b; Zucconi *et al.*, 1996), and *Phoma herbarum* (Sun *et al.*, 1978; Fletcher *et al.*, 1985; Del Frate and Caretta, 1990; Kerry, 1990a; Zucconi *et al.*, 1996).

5.5. GEOTHERMALLY HEATED SOILS

Thermophilic and thermotolerant species can grow in Antarctica in soils heated by volcanism. Thermotolerant Ascomycetes as *Chaetomium* spp. and thermophilic as *Thielavia heterotallica* [as the anamorph *Myceliophtora thermophila*] have been isolated from hot soils on Mt. Melbourne, as well as the thermophilic imperfect fungus *Malbranchea cinnamomea* (Broady *et al.*, 1987; Broady, 1993; Zucconi *et al.,* 1996). Thermophilic *Penicillium*, *Chaetomium* and *Mucor*, and thermotolerant *Paecilomyces* and *Aspergillus* have been isolated from hot soils on Mt. Erebus (Greenfield, 1983).

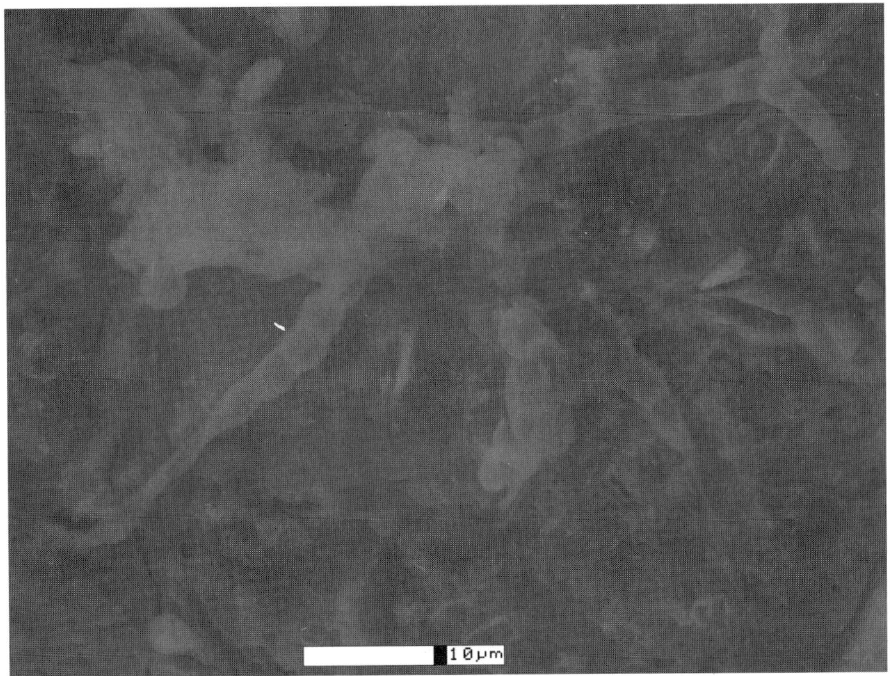

Figure 1. An undescribed imperfect fungus, that grows endolithically in pegmatite in Northern Victoria Land (Environmental Scanning Electron Micrograph).

5.6. ROCKS

Microfungi can grow in fissures and cracks in the rocks (chasmoendolithic habitat) and structural cavities within porous rocks (cryptoendolithic habitat) (Golubic *et al.*, 1981), as part of rock communities. *Acremonium strictum, Cladosporium herbarum, Geomyces pannorum* and *Phoma herbarum* group grow with algae in cracks in rocks and below translucent stones - quartz rock communities - (Kerry, 1990b). Cryptoendolithic microbial communities inhabit porous sandstones of the Beacon

Supergroup in the Ross Desert (Friedmann and Ocampo, 1976; Friedmann, 1982) and in other areas of Southern Victoria Land. The most common and most characteristic microbial cryptoendolithic community in the Antarctic desert is the "lichen dominated community" (Friedmann *et al.*, 1988; Nienow and Friedmann, 1993); it inhabits also microfractures in granite and pegmatite (volcanic rocks) (Onofri and Friedmann, 1998). Few data are available on microfungi living in Antarctic cryptoendolithic communities. None of these isolates is identified, out of some yeasts and *Tilletiopsis washingtonensis*, which have been isolated from sandstone in the Ross Desert (Vishniac and Hempfling, 1979a, 1979b). Their role in the community is uncertain, but some of them are antibiotic producers, possibly controlling the growth of other members of the community (Nienow and Friedmann, 1993). Most of them look somewhat like black yeasts (Fig. 1), some are probably new genera or new species possibly belonging or comparable to *Bispora, Capnocheirides, Coniosporium, Phaecoccomyces, Phaeotheca, Sarcinomyces, Sclerococcum, Scytalidium, Spilodochium, Taeniolella, Xylohypha*, and other genera.

6. Mycorrhizae

An undetermined mycorrhizogenous fungus of the liverwort *Cephaloziella exiliflora* (Tayl.) Steph., collected from Windmill Islands (maritime Antarctica) and from Botany Bay (Granite Harbour, continental Antarctica), has been observed in the rhizoids and isolated by Williams *et al.* (1994). It is an ascomycetous endophyte; possibly more than one fungus was present in the mycorrhiza.

7. Predaceous fungi

Predaceous (i.e. capturing nematodes or springtails) fungi are rather common on mosses and in barren Antarctic soils, mainly containing organic matter resulting from decomposition of mosses or lichens. At least 23 predaceous species have been recorded and Antarctic isolates are generally more psychrotolerant than European isolates (Gray, 1982). The first predaceous fungus from continental Antarctica was recorded by Maslen (1982) as an undetermined nematophagous hyphomycete, producing constricting rings. It was isolated from algal and moss material growing in a pond near Ablation Point (Alexander Island). Onofri and Tosi (1992) published the new species *Arthrobotrys ferox*, the first springtail-predaceous fungus in Antarctica and the second known. Nematode-trapping hyphomycetes from maritime Antarctica mainly belong to *Monacrosporium* and *Arthrobotrys,* and endozoic fungal parasites of nematodes mainly belong to *Harposporium* and *Verticillium* (Duddington *et al.*, 1973; Gray, 1982; Gray *et al.*, 1982; Gray and Lewis Smith, 1984).

8. Antarctic microfungi under UV-radiation

Although hyaline fungi are less protected against UV-radiation than dematiaceous ones (Wheeler and Bell, 1988), they are well represented in the Antarctic mycoflora.

Probably fast growth and abundant sporulation are the strategies allowing these fungi to live in the Antarctic environment. In addition there is evidence that spores of the probably endemic Antarctic species *Arthrobotrys ferox* possess substances - possibly mycosporines (Fig. 2) and pigments (not shown), evidenced by spectrofluorimetry - (Arcangeli *et al.*, 1996, 1997), that increase its ability to survive under UV-B radiation, as compared to a strain from temperate Europe of the very close species *A. oligospora:* this is supported by the higher germination percentage (36% and 19%, respectively) and the lower modification in the fluorimetric spectra of proteins in *A. ferox* (Fig. 3, after 60 min. of UV-B irradiation).

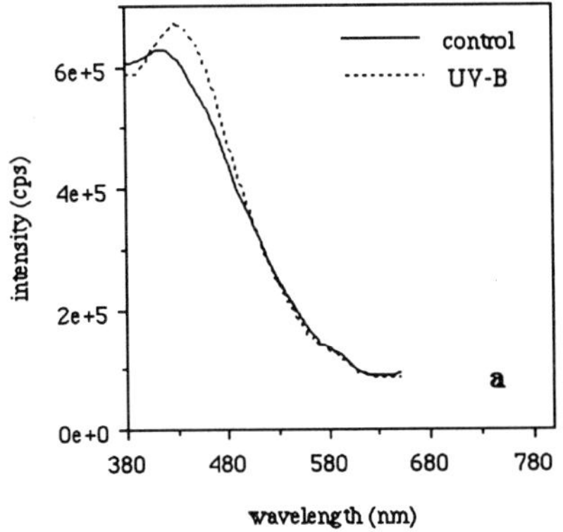

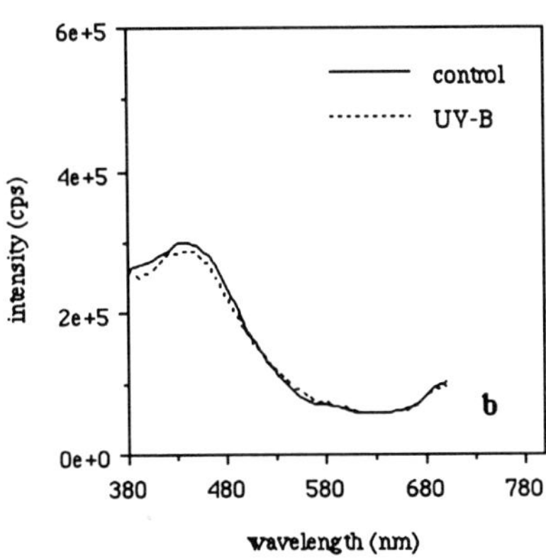

Figure 2. Comparison between emission spectra (λ_{ex} = 360 nm) obtained from UV-B irradiated and control spores of *A. ferox* (a) and *A. oligospora* (b); differences at 430 nm could be due to the presence in *A. ferox* of mycosporines, that are modified by irradiation.

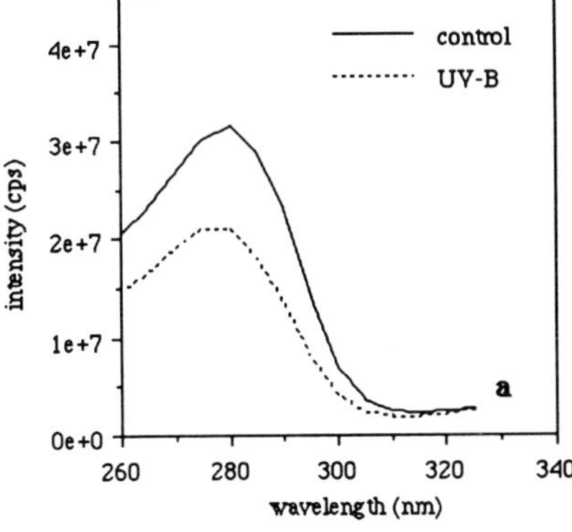

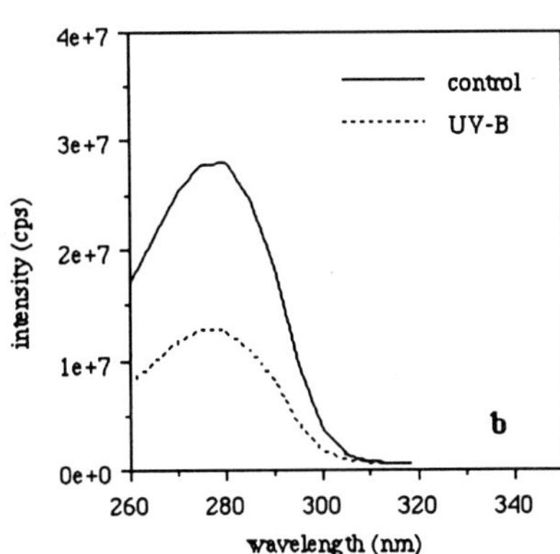

Figure 3. Comparison between excitation spectra (λ_{em} = 340 nm) obtained from UV-B irradiated and control spores of *A. ferox* (a) and *A. oligospora* (b); differences at 280 nm could be due to a more severe damage of proteins in irradiated spores of *A. oligospora.*

9. Temperature requirements

Mycoflora of Antarctica is mainly composed by psychrotrophic (i.e. mesophilic psychrotolerant) and few psychrophilic species (Latter and Heal, 1971; Kerry, 1990a; Möller and Dreyfuss, 1996; Zucconi *et al.*, 1996; Azmi and Seppelt, 1997). As observed by Baross and Morita (1978), psychrophiles are almost exclusively isolated from stable cold environments. The unstable conditions of the microhabitats of the Antarctic fungi justify such a numerous presence of psychrotrophic instead of psychrophilic strains. Generally, Antarctic strains are more psychrotolerant than corresponding strains from temperate areas (Latter and Heal, 1971; Kerry, 1990a). For instance, the Antarctic isolates of *Monacrosporium cionopagum* and *M. ellipsosporum* are psychrotolerant and show lower maximum, optimum and minimum temperatures for growth than the British isolates (Gray, 1982).

In addition to species growing in heated soils (see 5.5.), inexpectedly thermophilic *Thermomyces lanuginosus* (Del Frate and Caretta, 1990) and *Thielavia heterotallica,* and thermotolerant strains of *Aspergillus fumigatus* and *Chaetomium gracile* (Ellis, 1980) have been isolated from cold soils: it is not quite clear whether or not they are actually growing.

This work is carried out within the framework of the Italian National Program for Research in Antarctica (PNRA). The author wishes to thank Professor E.I. Friedmann (Florida State University), Professor A. Rambelli, Drs S. Tosi and L. Zucconi (Tuscia University in Viterbo, Italy) for critical advices, Professor S. Cannistraro (Tuscia University) for spectrofluorimetrical analyses, Mr R. Goddard (High Magnetic Field National Laboratory, U.S.A.) for technical assistance in Environmental Scanning Electron Microscopy, Dr M.A. Korovkin for the revision of the english-language version of the text and Italian National Council of Researches (CNR) for financial support.

10. References

Abyzov, S.S. (1993) in Friedmann E.I. (ed.), *Antarctic Microbiology*, Wiley-Liss, New York, pp. 265–295.

Abyzov, S.S. and Belyakova, L.A. (1982) *Izvestiya Akademii Nauk SSSR. Seriya Biologicheskaya* **1982(3)**, 432–435.

Agerer, R. (1984) *Trans. Brit. Mycol. Soc.* **82**, 184–186.

Aptroot, A. and van der Knaap, W.O. (1993) *Nova Hedwigia* **56**, 183–192.

Arcangeli, C., Zucconi, L., Onofri, S. and Cannistraro, S. (1996) *Physica Medica* **12**, 119–123.

Arcangeli, C., Zucconi, L., Onofri, S. and Cannistraro, S. (1997) *Journal of Photochemistry and Photobiology, B: Biol.* **39**, 258–264.

Atlas, R.M., di Menna, M.E. and Cameron, R.E. (1978) *Antarctic Research Series* **30**, 27–34.

Azmi, O.R. and Seppelt, R.D. (1997) *Polar Biol.* **18**, 128–134.

Baharaeen, S. and Vishniac, H.S. (1982) *Int. J. Syst. Bacteriol.* **32**, 229–232.

Baross, J.A. and Morita, R.Y (1978) in Kushner D.J. (ed.), *Microbial life in extreme environments*. Academic Press, London, pp. 9–71.

Baublis, J.A., Wharton, R.A. Jr. and Volz, P.A. (1991) *Journal of Basic Microbiology* **31**, 3–12.

Bommer, E. and Rousseau, M. (1900) *Bull. Acad. Roy. Sci. Belgique, Cl. Sci.*, 640–646.

Bommer, E. and Rousseau, M. (1905) in *Résultats du Voyage du S.Y. Belgica, Expédition Antarctique Belge 1897–1899, Botanique,* pp.1–5.

Broady, P.M. (1993) in Friedmann E.I. (ed.), *Antarctic Microbiology*, Wiley-Liss, New York, pp. 413–432.

Broady, P., Given, D., Greenfield, L. and Thompson, K. (1987) *Polar Biol.* **7**, 97–113.

Cabello, M.N. (1989) *Mycotaxon* **36**, 91–94.

Cameron, R.E. (1971) in Quam L.O. and Porter H.D. (eds.), *Research in the Antarctic*, American Association for the Advancement of Science Publications, Washington D.C., pp. 137–189.

Cameron, R.E., King, J. and David, C.N. (1970) *Soil Science* **109**, 110–120.
Cameron, R.E., Morelli, F.A. and Randall, L.P. (1972) *Antarctic Journal of the United States* **7**, 254–258.
Caretta, G., Del Frate, G. and Mangiarotti, A.M. (1994) *Boletín Micológico* **9**, 9–13.
Caretta, G, Del Frate, G and Tosi, S. (1995) *Boletín Micológico* **10**, 37–41.
Caretta, G. and Piontelli, E. (1977) *Sabouraudia* **15**, 1–10.
Claridge, G.G.C. and Campbell, I.B. (1985) in Bonner W.N. and Walton D.W.H. (eds.), *Key Environments, Antarctica,* Pergamon Press, Oxford, pp. 62–70.
Czarnecki, B. and Bialasiewicz, D. (1987) *Polish Polar Research* **8**, 153–158.
Del Frate, G. and Caretta, G. (1990) *Polar Biol.* **11**, 1–7.
di Menna, M.E. (1966a) *Antonie van Leeuwenhoek* **32**, 25–28.
di Menna, M.E. (1966b) *Antonie van Leeuwenhoek* **32**, 29–38.
Dodge, C.W. (1948) *Reports of B.A.N.Z. Antarctic Research Expedition 1929-31, Ser. B,* **7**, 276 pp.
Dodge, C.W. and Baker, G.E. (1938) *Annals of the Missouri Botanical Garden* **25**, 515–718.
Domsch, K.H., Gams, W. and Anderson, T.-H. (1980) *Compendium of soil fungi.* Academic Press, London, Vols 1-2.
Duddington, C.L., Wyborn, C.H.E. and Lewis Smith, R.I. (1973) *Br. Antarct. Surv. Bull.* **35**, 87–90.
Ellis, D.H. (1980) *Mycologia* **72**, 1033–1036.
Ellis-Evans, J.C. and Walton, D. (1990) *Proceedings of NIPR Symposium on Polar Biology* **3**, 151–163.
Fell, J.W., Statzell, A.C., Hunter, I.L. and Phaff, H.J. (1969) *Antonie van Leeuwenhoek* **35**, 433–462.
Fenice, M., Selbmann, L., Di Giambattista, R., Petruccioli, M., Federici, F. (1996) *Chitin Enzymology* **2,** 285–292.
Fenice, M., Selbmann, L., Zucconi, L. and Onofri, S. (1997) *Polar Biol.* **17**, 275–280.
Finotti, E., Moretto, D., Marsella, R. and Mercantini, R. (1993) *Polar Biol.* **13**, 127–130.
Fletcher, L.D., Kerry, E.J. and Weste, G.M. (1985) *Polar Biol.* **4**, 81–88.
Friedmann, E.I. (1982) *Science* **215**, 1045–1053.
Friedmann, E.I. (1993) *Antarctic Microbiology*, Wiley-Liss, New York.
Friedmann, E.I., Hua, M. and Ocampo-Friedmann, R. (1988) *Polarforschung* **58**, 251–259.
Friedmann, E.I. and Ocampo, R. (1976) *Science* **193**, 1247–1249.
Gamundi, I.J. and Spinedi, H.A. (1987) *Mycotaxon* **29**, 81–89.
Gamundi, I.J. and Spinedi, H.A. (1988) *Mycotaxon* **33**, 467–489.
Gazert, H. (1912) in *Deutsche Südpolar Expedition, 1901–1903*, G. Reimes, Berlin, **7**, 268–296.
Golubic, S., Friedmann, E.I., and Schneider, J. (1981) *Journal of Sedimentary Petrology* **51**, 475–478.
Goto, S., Sugiyama, J.E. and Iizuka, H. (1969) *Mycologia* **61**, 748–774.
Gray, N.F. (1982) *Plant and Soil* **64**, 431–435.
Gray, N.F. and Lewis Smith, R.I. (1984) *Mycopathologia* **85**, 81–92.
Gray, N.F., Wyborn, C.H.E. and Lewis Smith, R.I. (1982) *Oikos* **38**, 194–201.
Greene, D.M. and Holtom, A. (1971) *Br. Antarct. Surv. Bull.* **26**, 1–29.
Greenfield, L., (1981) in Greenfield L. and Wilson G. (eds.), *University of Canterbury, Antarctic Expedition No. 19.* Christchurch, University of Canterbury, pp. 4–22.
Greenfield, L. (1983) *New Zealand Antarctic Record* **4**, 10–11.
Harder, R. and Persiel, I. (1962) *Archives of Microbiology* **41**, 44–50.
Hawksworth, D.L. (1979) *Bull. Br. Mus. (Natural History), Botany Series* **6**, 183–300.
Hawksworth, D.L. and Diederich, P. (1988) *Trans. Brit. Mycol. Soc.* **90**, 293–312.
Hawksworth, D.L., Kirk, P.M., Sutton, B.C. and Pegler, D.N. (1995) *Ainsworth & Bisby's Dictionary of the Fungi*, CAB International, Wallingford.
Heatwole, H., Saenger, P., Spain, A., Kerry, E. and Donelan, J. (1989) *Antarctic Science* **1**, 225–234.
Holdgate, M.W. (1977) *Philosophical Transactions of the Royal Society of London* B **279**, 5–25.
Kageruka, P. (1967) *Acta Zoologica et Pathologica Antverpensia* **44**, 87.
Kappen, L. (1993) in Friedmann E.I. (ed.), *Antarctic Microbiology*, Wiley-Liss, New York, pp. 433–490.
Kerry, E. (1990a) *Polar Biol.* **10**, 293–299.
Kerry, E. (1990b) *Polar Biol.* **10**, 423–430.
Knox, J.S. and Paterson, R.A. (1973) *Mycologia* **65,** 373–387.
Latter, P.M. and Heal, O.W. (1971) *Soil Biology and Biochemistry* **3**, 365–379.
Line, M.A. (1988) *Polar Biol.* **8**, 421–427.
Longton, R.E. (1985) in Bonner W.N. and Walton D.W.H. (eds.), *Key Environments, Antarctica,* Pergamon Press, Oxford, pp. 73–105.
Maslen, N.R. (1982) *Br. Antarct. Surv. Bull.* **51**, 285–296.
Matsushima, T. (1975) *Icones microfungorum a Matsushima lectorum.* Published by the Author, Kobe, Japan.
McLean, A.L. (1918) *Nature* (London) **102**, 35–39.

Mercantini, R., Marsella, R. and Cervellati, M.C. (1989) *Mycopathologia* **106**, 47–52.

Mercantini, R., Marsella, R., Moretto, D. and Finotti, E. (1993) *Mycopathologia* **122**, 169–175.

Montemartini Corte, A. (1991) in Battaglia B., Bisol P.M. and Varotto V. (eds.), *Proceedings of the 1st Meeting on 'Biology in Antarctica', Scienza e Cultura*, Edizioni Universitarie Patavine, Padova, pp. 67-76.

Montemartini Corte, A., Caretta, G. and Del Frate, G. (1993) *Mycotaxon* **48**, 343–358.

Möller, C. and Dreyfuss, M.M. (1996) *Mycologia* **88**, 922–933.

Möller, C. and Gams, W. (1993) *Mycotaxon* **48**, 441–450.

Nienow, J.A. and Friedmann, E.I. (1993) in Friedmann E.I. (ed.), *Antarctic Microbiology*, Wiley-Liss, New York, pp. 343–412.

Onofri, S. and Friedmann, E.I. (1998) in Tamburrini M. and D'Avino R. (eds.), *Newsletter of the Italian Biological Research in Antarctica*, No. 2, *Camerino University Press*, Camerino (in press).

Onofri, S., Rambelli, A., Maggi, O., Persiani, A.M., Riess, S., Tosi, S. and Grasselli, E. (1991) in Battaglia B., Bisol P.M. and Varotto V. (eds.), *Proceedings of the 1st Meeting on 'Biology in Antarctica', Scienza e Cultura*, Edizioni Universitarie Patavine, Padova, pp. 55–65.

Onofri, S. and Tosi, S. (1992) *Mycotaxon* **44**, 445–451.

Onofri, S., Tosi, S., Persiani, A.M., Maggi, O., Riess, S. and Zucconi, L. (1994) in Battaglia B., Bisol P.M. and Varotto V. (eds.), *Proceedings of the 2nd Meeting on 'Biology in Antarctica', Scienza e Cultura*, Edizioni Universitarie Patavine, Padova, pp. 19–32.

Paterson, R.A. and Knox, J.S., (1972) in Parker B.C. (ed.), *Conservation problems in Antarctica. Proceeding of a Colloquium*, Backsburg, Virginia, pp. 185–192.

Paterson, R.A. and Rooney, H. (1972) *Antarctic Journal of the United States* **7**, 85.

Pegler, D.N., Spooner, B.M. and Lewis Smith, R.I. (1980) *Kew Bull.* **35**, 499–562.

Petrini, O., Hake, U. and Dreyfuss, M.M. (1990) *Mycologia* **82**, 444–451.

Pickard, J. (1986) *Antarctic Oasis*, Academic Press, Sidney.

Pugh, G.J.F. (1980) *Trans. Brit. Mycol. Soc.* **75**, 1–14.

Pugh, G.J.F. and Allsopp, D. (1982) *Br. Antarct. Surv. Bull.* **57**, 55–67.

Ryan, P.G., Watkins, B.P., Lewis Smith, R.I., Dastych, H., Eicker, A., Foissner, W., Heatwole, H., Miller, W.R. and Thompson, G. (1989) *South African Journal of Antarctic Research* **19**, 10–20.

Simmons, E.G. (1983) *Mycotaxon* **17**, 216–241.

Sinclair, N.A. and Stokes, J.L. (1965) *Can. J. Microbiol.* **11,** 259–269.

Singer, R. (1957) *Sydowia,* Ser. 2, Beih. **1**, 16–23.

Sugiyama, J. (1970) *Polar News,* Tokyo **6**, 17–24.

Sugiyama, J., Sugiyama, Y., Iizuka, H. and Torii, T. (1967) *Report of the Japanese Summer Parties in Dry Valleys, Victoria Land, 1963–1965*, 23–32.

Sun, S.H., Huppert, M. and Cameron, R.E. (1978) *Antarctic Research Series* **30**, 1–26.

Tsiklinsky, M. (1908) in Charcot J. (ed.), *Expedition Antartique Française 1903–1905*, Paris.

Tubaki, K. (1961a) *Antarctic Research* **11**, 161–162.

Tubaki, K. (1961b) *Biological Results of the Japanese Antarctic Research Expedition,* Special Publications from the Seto Marine Biological Laboratory, Sirahama, Wakayama-Ken, Japan, **14**, 1–9

Tubaki, K. and Asano, I. (1965) *Scientific Reports of the Japanese Antarctic Research Expedition 1956–1962 , Series E* **27**, 1–12.

Vishniac, H.S. (1985a) *Int. J. Syst. Bacteriol.* **35**, 119–122.

Vishniac, H.S. (1985b) *Mycologia* **77**, 149–153.

Vishniac, H.S. (1993) in Friedmann, E.I. (ed.), *Antarctic Microbiology*. Wiley-Liss, New York, pp. 297–341.

Vishniac, H.S. and Baharaeen, S. (1982) *Int. J. Syst. Bacteriol.* **32**, 437–445.

Vishniac, H.S. and Hempfling, W.P. (1979a) *Int. J. Syst. Bacteriol.* **29**, 153–158.

Vishniac, H.S. and Hempfling, W.P. (1979b) *Journal of General Microbiology* **112**, 301–314.

Vishniac, H.S. and Kurtzman, C.P. (1992) *Int. J. Syst. Bacteriol.* **42**, 547–553.

Wheeler, M.H. and Bell, A.A. (1988) *Curr. Top. Med. Mycol.* **2**, 338–387.

Wicklow, D.T. (1968) *Can. J. Microbiol.* **14**, 717–719.

Wicklow, D.T. and Malloch, D. (1971) *Mycologia* **63**, 118-131.

Williams, P.G., Roser, D.J. and Seppelt, R.D. (1994) *Mycological Research* **98**: 34–36.

Zucconi, L., Pagano, S., Fenice, M., Selbmann, L., Tosi, S. and Onofri, S. (1996) *Polar Biol.* **16**: 53–61.

Zycha, H., Siepmann, R. and Linnemann, G. (1969) *Mucorales. Eine Beschreibung aller Gattungen und Arten dieser Pilzgruppe*. J. Cramer, Lehre.

Biodata of **Aharon Oren** contributor of " ***The Enigma of Square and Triangular Halophilic Archaea.*** "

Dr. Aharon Oren is an Associate Professor in Microbiology and Molecular Ecology at the Institute of Life Sciences, The Hebrew University of Jerusalem. He earned his Ph.D. at the Hebrew University (1978) and spent his postdoctoral periods at the University of Illinois (Urbana-Campaign) with R.S. Wolfe and C.R. Woese. His research interests is in ecology, physiology and biochemistry of halophilic microorganisms and the microbiology of hypersaline environments. Dr. Oren wrote a few Hebrews microbial textbooks and has many publications in international journals concerning the microbial communities of the Dead Sea. Recently he edited a book on ***"Microbiology and Biogeochemistry of Hypersaline Environments"*** (published by CRC Press, Boca Raton).
E-Mail: **orena@shum.cc.huji.ac.il**

J. Seckbach (ed.), Enigmatic Microorganisms and Life in Extreme Environments, 337–355.

THE ENIGMA OF SQUARE AND TRIANGULAR HALOPHILIC ARCHAEA

A. OREN
Division of Microbial and Molecular Ecology, Institute of Life Sciences, and the Moshe Shilo Minerva Center for Marine Biogeochemistry, The Hebrew University of Jerusalem, Jerusalem 91904, Israel

1. Introduction

An overview of the prokaryotic world shows that the morphological variety among bacteria is limited, and most prokaroytic microorganisms, whatever their taxonomic position or their physiological properties might be, can be classified in a very limited number of morphological types: rods, cocci, and spirilla. Only occasionally are prokaryotic microorganisms found with more elaborate shapes, such as the stalked or prosthecate bacteria. A feature shared by all common bacterial shapes is the absence of sharp angles, and rounded corners and spherical shapes are the rule. Bernard Dixon (1994) states it clearly: "Look around the natural world and what do you see? ... What you do not see is squares. Whether in the shape of a building or a book, whether as the object of a chess game or a child's first geometry lesson at school, a quadrilateral composed of four equal sides and four right angles is a feature not of the biosphere but of civilization and the built environment".

In the light of the above, the discovery of bacteria shaped as almost perfectly square thin boxes or sheets came as a huge surprise. These square bacteria were first observed in 1980 by A.E. Walsby in a coastal brine pool (the Gavish sabkha) in the Sinai Peninsula, currently in Egypt (Walsby, 1980). In the course of a search for gas vesicle-containing bacteria in hypersaline lakes and pools, Walsby recognized the extremely thin, square-shaped objects as prokaryotes thanks to the presence of gas vesicles. These gas vesicles are visible within the cells as refractile bodies in the phase contrast microscope. A posteriori it may be assumed that many microbiologists have observed these square organisms before, but did not realize their true nature, and they were either overlooked or considered as some kind of abiotic structures. After the first description appeared, and scientists learned how to recognize these square bacteria, they were rapidly reported to occur in other hypersaline environments. They can be found in most thalassohaline (sea water-derived) hypersaline environments in which the salt concentration exceeds 200-250 g l^{-1} and up to NaCl saturation.

In a living world dominated by cells whose morphology is determined and stabilized by high turgor pressures exerted by the cytoplasm on the cell membrane, a square flat morphology would be precluded by osmotically generated internal hydrostatic pressure. However, the halophilic archaea, a group of organisms containing extremely high salt concentrations (mainly KCl) in their cytoplasm to balance the salinity of the outside medium, lack a measurable turgor pressure. The same is probably true for the square bacteria, and this may explain the stability of a flat square-shaped cell.

Today, more than 17 years after the discovery of the square bacteria, they remain enigmatic to a large extent. This is due in part to the fact that the species (possibly more than one) still defy the microbiologists' attempts to grow them. In spite of many efforts, no growing cultures of the square gas vesicle-containing cells are extant. Thus, all research on these organisms thus far has been performed using material collected from hypersaline field sites. Attempts to isolate the square bacteria have, however, led to the isolation of other interesting morphological types of halophilic archaea ("box-shaped bacteria"), and species producing triangular and rhomboid cells are now also known.

In this chapter I will review our present knowledge of the square bacteria. In addition, other halophilic archaea with unusual morphology, such as the triangular and box-shaped cells, will be discussed.

2. The square bacteria

2.1. OCCURRENCE OF SQUARE BACTERIA

Large flat square bacteria containing gas vesicles were first observed by Walsby in the salt-saturated brines of the Gavish sabkha in the Sinai Peninsula. They may be present in extremely high numbers: counts as high as $7x10^7$ ml^{-1} brine were reported (Walsby, 1980). These morphologically fascinating organisms often resemble small halite crystals, but they are easily recognizable by their thinness (less than 0.25 μm, possibly even as thin as 0.07 μm), and by the presence of refractile gas vesicles. In midsummer the square bacteria were found in massive numbers on the salt crust. In the winter months the sabkha brine is diluted by freshly introduced sea water and by flash floods, and the squares can only be found on salt crust on the edge of the water body (Kessel and Cohen, 1982).

Undivided squares were found in a wide range of sizes, from 1.5 up to 11 μm in cross section (Walsby, 1980); cells between 2-5 μm in diameter are most commonly found. Division planes are often seen, with an arrangement indicating that divisions occur in two planes alternating at right angles, so that each square grows to a rectangle which then divides into two equal squares (Parkes and Walsby, 1981). Thus, sheets of cells arranged like postage stamps are often formed. Sheets of eight squares were not uncommon, and sheets of 16 were found with individual squares tearing free (Walsby, 1980). In one case, even an arrangement was observed, resulting from six divisions of a single cell without separation of the daughter cells (Kessel and Cohen, 1982) (Fig. 1).

Since the square bacteria were first recognized, they have been reported to occur in many places throughout the world (Oren, 1994a), and it may be assumed that they are common inhabitants of thalassohaline hypersaline environments worldwide (Grant

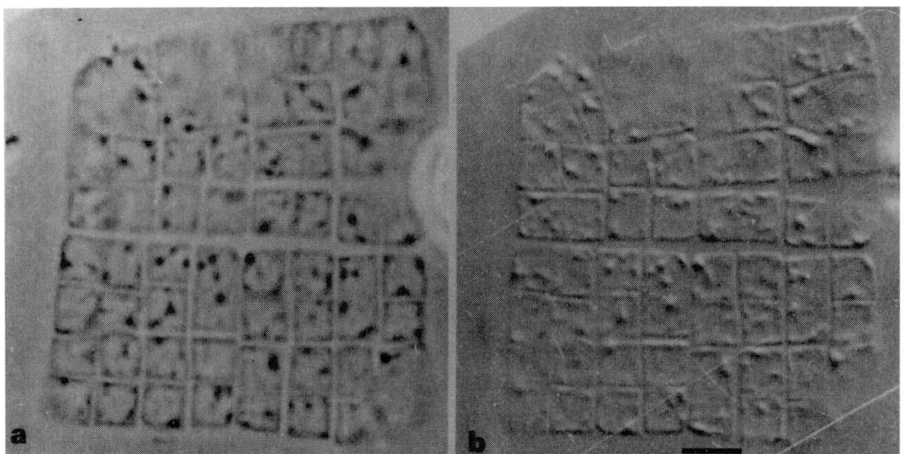

Figure 1. Phase contrast (a) and Nomarski interference (b) photomicrographs of square bacteria which have remained associated after a series of six divisions. Not all of the cells have completed the most recent cell division. The dark bodies seen in phase contrast are cellular inclusions bulging against the cell wall. Bar, 5 μm. From Kessel and Cohen (1982), reproduced by permission

and Larsen, 1989). The occurrence of the flat, perfectly square boxes has now been documented in hypersaline water bodies at or close to halite saturation, such as saltern crystallizer ponds in Eilat, Israel (Oren, 1994b; Oren *et al.*, 1996), Spain (Guixa-Boixareu *et al.*, 1996; Torrella, 1986), and the salterns on San Francisco Bay (USA) and Baja, California (Mexico) (Javor, 1989; Stoeckenius, 1981). They were further found in salt lakes in the Crimea (Ukraine) (Romanenko, 1981), and Senegal (West Africa) (Grant and Larsen, 1989).

Bacterial samples from the salterns of San Francisco Bay contained about 10 times the total number of cells found in the brines collected from the sabkha in Sinai and the salt flats of Guerrero Negro and La Paz, Baja California (Mexico), but square cells were rare (Stoeckenius, 1981). The smallest square cells observed at these sites measured 2 x 2 μm, and larger cells were often rectangular. It was noted that the real number of square cells may easily be underestimated in light-microscopic studies, as only cells with the flat side aligned in the plane of focus will be recognized as squares, while similar cells positioned at an angle will resemble rods (Stoeckenius, 1981; Oren *et al.*, 1996). Cells are extremely thin - about 0.25 μm only or even less - and grow from small squares to larger squares of the same thickness, and then undergo two divisions, with or without separation of the daughter cells (Stoeckenius, 1981)

Square, gas vesicle-containing cells similar to those described from the sabkha in Sinai (Walsby, 1980) were found in high numbers in saltern crystallizer ponds in a solar salt facility in Eilat, Israel (Fig. 2). The relative contribution of the square cells to the total community density in these crystallizer ponds was variable, but values as high as 22-32% were reported in Spanish salterns (Guixa-Boixareu *et al.*, 1996), and from 22% up to as high as 55% in salterns in Eilat, Israel (Oren, 1994b; Oren *et al.*, 1996). These values relate to cells with a recognizable square or rectangular morphology, and they may underestimate the true contribution of the square archaea to the community, as some of squares may mistakenly be counted as rod-shaped cells when seen in profile (see above). The high value of 55% flat square or rectangular, trapezoid, or triangular shaped bacteria was observed in brine collected in November 1995 from a saltern pond (water density 1.239 g ml^{-1}, 360 g l^{-1} total dissolved salts, $1.3x10^7$ microscopically recognizable bacteria ml^{-1}). The cells were highly variable in size, and measured between 2.5 and 7.5 μm in diameter. Direct microscopic examination of the square cells in the brine showed the presence of refractile gas vesicles in most of the cells. No gas vesicles are seen in Fig. 2, as the high-speed centrifugation employed in sample preparation caused their collapse. Incubation of the brine with 2-(*p*-iodophenyl)-3-(*p*-nitrophenyl)-5-phenyl tetrazolium chloride (INT) to obtain information on the number of actively respiring cells, showed the majority (about 83%) of the square cells to be active in respiration, as seen by the accumulation of red formazan granules within the cells (Oren *et al.*, 1996).

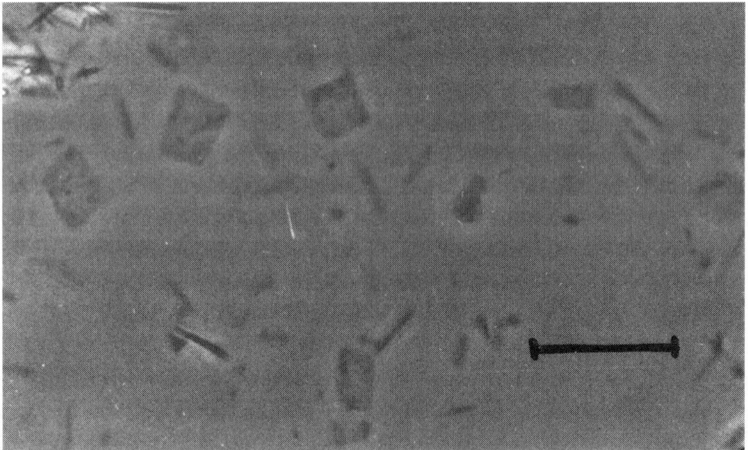

Figure 2. Square bacteria from a saltern crystallizer pond in Eilat, Israel (salinity about 350 g l^{-1}). No gas vesicles are visible, as high-speed centrifugation used to concentrate the cells caused collapse of the gas vesicles. Bar, 10 μm.

Square bacteria were also seen in abundance in different saltern sites in Spain (Guixa-Boixareu *et al.*, 1996; Torrella, 1986), and at salinities exceeding 250 g l^{-1} their numbers may exceed 25% of the total prokaryote assemblage (Torrella, 1986).

In the hypersaline Lake Saxkoye, Crimea Peninsula, Ukraine, a lake with a total salt content varying between 160 and 250 g l^{-1}, square bacteria were found in the upper layers (Romanenko, 1981). The size of the square cells found in this lake was 1.4-2.5 μm, being somewhat smaller than the square cells reported from other locations.

2.2. ATTEMPTS TO OBTAIN SQUARE BACTERIA IN CULTURE

Although more than 17 years have passed since square, gas vesicle-containing bacteria were discovered in 1980, they still have not been brought into culture. Thus, all studies performed on these unusual prokaryotes have been based on material collected from natural accumulations of the square bacteria in hypersaline environments.

It is relatively easy to enrich the square bacteria from brine samples collected from the field, exploiting the buoyancy of the cells resulting from their high content of gas vesicles. The cells will slowly float toward the surface meniscus in stationary cell suspensions, whereafter they can be collected from the water surface (Kessel *et al.*, 1985). Flotation can be aided by low speed centrifugation, taking care not to exceed the critical pressure at which gas vesicles start to collapse (e.g. centrifugation for 1 h at 700 rpm, so that the pressure inside the centrifuge tubes pressure does not exceed 20 kPa or 0.2 bar, see Parkes and Walsby, 1981). A similar approach to concentrate the square bacteria was used by Walsby (1980) and by Kessel and Cohen (1982). This procedure will of course also enrich for other types of bacteria with a high buoyancy due to the presence of gas vesicles.

In spite of the ease with which the square cells can be observed in brine samples, and their numbers can be enriched by flotation, thus far, they seem to have resisted isolation in pure culture. A single report of the successful isolation of this type of bacteria was published by Torrella (1986), who used brines from Spanish salterns as his source of the organisms. It was stated that the organisms were isolated using "specially designed enrichment cultures". However, no information was given on the composition of these media. Growth of the isolated strains was reported to be completely inhibited or retarded by complex media high in nutrients, especially in static non-shaken cultures. Two kinds of gas vesicle-containing cells were found to develop, depending on growth conditions. Small hypervacuolated forms that floated to the surface of the media mainly appeared upon exhaustion of the nutrients. Larger gas vesicle-containing squares developed under different conditions and colonized both the surface and the bulk of the liquid. The production of gas vesicles was found sensitive to nutrient enrichment, being inhibited above a nutritional threshold. Moreover, high nutrient concentrations inhibited the development of gas vesicles. Under certain conditions, the square cells were transformed into small sized motile cells. Growth was aerobic or microaerophilic. It is most unfortunate that the isolate has not been deposited in a culture collection, and the culture was lost many years ago (F. Torrella, personal communication).

Attempts by Javor and coworkers to isolate square bacteria from a variety of hypersaline environments (the Gavish sabkha, salterns of Guerrero Negro, Baja California, and San Diego, California) failed to yield the desired squares, but led to the isolation of "box-shaped" pleomorphic (i.e., having different forms at different stages of the life cycle) red halophilic archaea, belonging to the genus *Haloarcula* (see below).

An independent attempt to isolate extremely halophilic bacteria from the Gavish sabkha, the site from which the square bacteria were first documented, yielded cocci, pleomorphic rods and reportedly also square bacteria (Kessel *et al.*, 1985). A square isolate, lacking gas vesicles, and designated strain 81830-2, was used for study of carbon and ionic requirements. It was found to grow on single carbon sources such as acetate, lactate, malate, glycolate, succinate, oxalate, and glutamine. No or poor growth was obtained on glucose, sucrose, and starch (Kessel *et al.*, 1985). The strain is not available from culture collections, and it is not clear what its relationship is with the gas vesicle-containing square bacteria observed microscopically in samples from the site. The same may be true for the square, gas-vesicle-lacking isolate 801030/1, derived from the same site and used in a study by Alam *et al.* (1984).

2.3. GAS VESICLES AND TURGOR PRESSURE

The square bacteria were first identified as prokaryotic organisms based on their content of gas vesicles. In many of the cells the gas vesicles were found concentrated at the cell periphery and along the presumptive division lines (Walsby, 1980), but in other cells the gas vesicles were found to be randomly dispersed within the cells. Some of the cells appeared to be nearly filled with gas vesicles, and 120 vesicles at least were counted in one single cell (Walsby, 1980; Parkes and Walsby, 1981). The gas vesicles are morphologically similar to those found in other halophilic archaea such as *Halobacterium salinarum* or *Haloferax mediterranei*, and both spindle-shaped vesicles and cylindrical vesicles with conical ends occur (Parkes and Walsby, 1981) (Fig. 3).

The appearance of the refractile rod-shaped gas vesicles within the square cells led at least once to the erroneous interpretation that the square cells may be microcolonies of rod-shaped cells, embedded in a square matrix. Romanenko (1981) described the squares observed in brines of Lake Saxkoye, Crimea, as being variable in size (length 1.4-2.5 μm, surface area 1.5-3 μm^2) and as microcolonies consisting of not yet cultured rod-shaped bacteria embedded in a mucous matrix. In most cases, these "bacteria" were found to settle the edge of the "microcolonies" (consistent with the finding that the gas vesicles are often concentrated at the periphery of the cells). It was suggested that the square shape of the "microcolonies" might be determined by the fact that they are formed on the surface of square salt crystals. A simple experiment of pressurizing the "microcolonies" may have disclosed the real identity of the "bacteria" within the square structures as gas vesicles.

Gas vesicles are found in a limited number of halophilic archaea. Their primary function seems to be to provide buoyancy, enabling the cells to float toward the water surface, and thus toward oxygen (Larsen, 1984). The solubility of oxygen in salt-saturated brines is limited, and the presence of dense microbial communities in

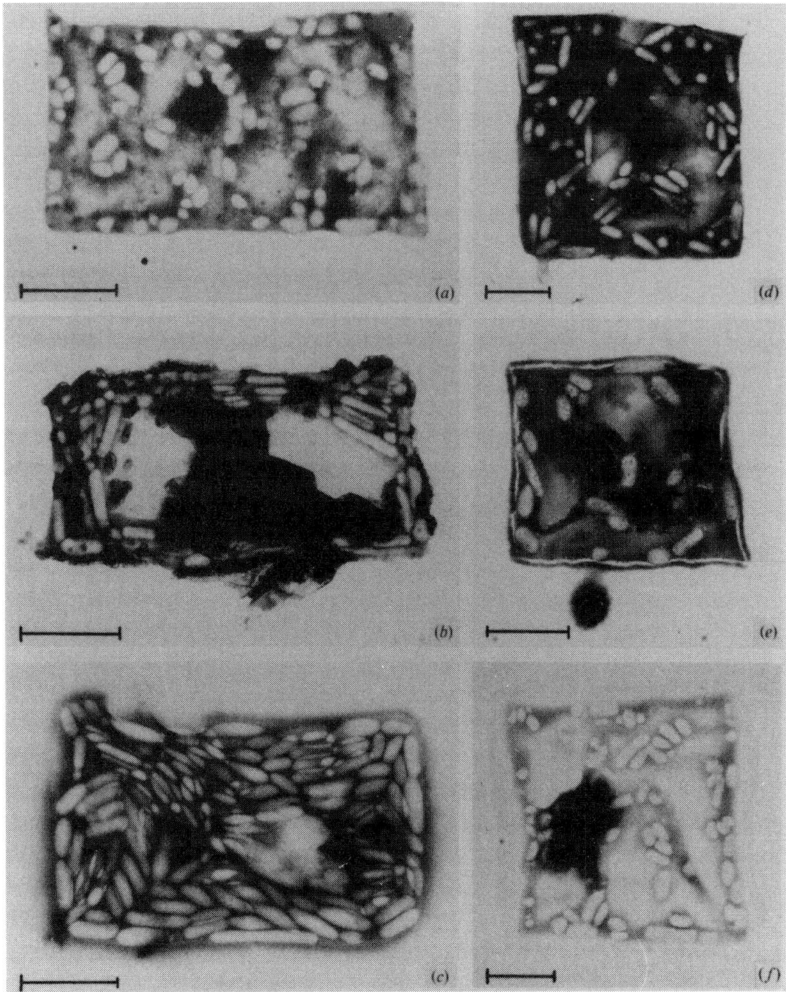

Figure 3. Electron micrographs of square bacteria showing the range of shape and gas vesicles: (*a-c*) rectangular form, (*a*) with short, spindle-shaped gas vesicles, (*b*) with peripheral gas vesicles and overlying salt crystals, and (*c*) packed with over 120 gas vesicles, mostly of the cylindrical form; (*d-f*) square form, (*d*) with no sheath, (*e*) with a thin sheath and (*f*) with extensive surrounding sheath. The bar markers represent 1 μm. From Parkes and Walsby (1981), reproduced by permission.

these brines may further reduce the availability of oxygen. However, the hypothesis that gas vesicles serve the halophilic organisms to obtain oxygen has yet to be substantiated by field observations (Walsby, 1980). The presence of light-driven proton pumps such as bacteriorhodopsin-like pigments in the square bacteria (see below) may present another incentive for the cells to float toward the surface where more light may be available. However, field studies to assess the importance of buoyancy to increase light availability have never been performed.

A characteristic property of gas vesicles is their sensitivity to pressure. The disappearance of the refractile gas vesicles in the square cells upon application of pressure led to the proper recognition of the nature of the square bacteria (Walsby, 1980). After pressure-induced collapse of the gas vesicles the cells sank to the microscope slide, rather than floating to the coverslip as native cells do (Walsby, 1980).

The maintenance of a square shape is hardly imaginable in cells possessing a turgor pressure. In the words of Bernard Dixon (1994): "Cells are spherical because osmotic pressure inside forces the enveloping membrane to form this shape, like a balloon inflated with air." The presence of gas vesicles presents the investigator with a convenient tool to assess the presence of a turgor pressure inside the cell. The pressure required to be applied to cause collapse of gas vesicles depends not only on the intrinsic strength of the gas vesicles (which is more or less species-dependent), but also on the turgor pressure inside the cells. In gas vesicle containing eubacteria, such as cyanobacteria, the pressure required for the gas vesicles to break is much higher in cells suspended in a concentrated solution of sucrose than in water, the difference in pressure required being equivalent to the turgor pressure in water-suspended cells. No such difference in critical pressure was found in gas vesicle-containing *Halobacterium* cells, which led to the conclusion that these halophilic archaea lack a turgor pressure (Walsby, 1971). The unusual absence of turgor in halobacteria is evidently related to the high salt concentration of their surroundings. In the square cells collected from the Gavish sabkha a definite reduction in the number of gas vesicles was seen after application of 1.5 bars, both in the presence and in the absence of sucrose, most of the vesicles had disappeared after 2.5 bars, and none remained after 3 bars (Walsby, 1980). The lack of difference in the sensitivity of the vesicles to pressure-induced collapse in the presence and absence of high sucrose concentrations suggests that the square bacteria, similar to other halophilic archaea, lack a turgor pressure.

2.4. CELL WALL STRUCTURE

In the absence of turgor, there is no obvious constraint on the shape of a bacterium, and thus the shape of the cell may be solely dependent on the shape and arrangement of the wall subunits.

In thin sections through the square bacteria a 40-nm thick cell wall can be seen, in which, especially in grazing sections, a subunit structure can be discerned, forming a hexagonal lattice with a lattice constant of 13.5 nm (Kessel and Cohen, 1982). Examination of the surface layer of the square bacteria in the electron microscope after metal shadowing showed the presence of a hexagonal lattice of particles, 20 nm across, very similar to the surface glycoprotein S-layer of *Halobacterium*, *Haloferax*, and other non-coccoid halophilic archaea (Parkes and Walsby, 1981; Walsby, 1980).

A more detailed structural analysis of the surface layer of the square bacteria was performed by examining debris of cells that had lysed in the brine by natural cell death or phage infection (Kessel, 1983; Kessel *et al.*, 1985). These debris provided especially good preparations of walls for analysis of periodic structures. Shadowed preparations and surface replicas showed an ordered array of subunits, the center to center spacing between the subunits being 16 nm. The particle arrangement was predominantly hexagonal, but occasionally patches with a square arrangement of subunits were observed. Optical diffraction patterns could be obtained from both sides of the cell wall (Kessel *et al.*, 1985). In carbon-platinum replicas of the cell wall at least 3 different types of subunit arrangement and subunit sizes could be discerned by optical diffraction. The most common pattern was a hexagonal arrangement of particles, with a lattice constant of 15.9 nm, but in addition patches with hexagonal patterns having a lattice constant of 22 nm, and tetragonal particle arrangements were found with a 14.6 nm lattice constant (Kessel and Cohen, 1982). It is not yet clear whether the different particle arrangements and lattice constants can be found in a single species of square bacteria, or whether a mixed population was present of different species, each with its characteristic arrangement of wall subunits. This question can only be answered using pure cultures of the organisms.

From measurements based on shadow angle and shadow length in shadow cast preparations in the electron microscope, the square bacteria from the Gavish sabkha were calculated to be a minimum of 0.07 μm along the edges. At sites where gas vesicles were present, the cells were thicker, with an apparent thickness of about 0.15 μm. The overall impression is that the cell envelope is flattened, but highly plastic and distended in places by granular contents within the cell (Parkes and Walsby, 1981). Shadowed replicas of whole cells showed bulges randomly distributed within the cells, which probably can be attributed to the presence of poly-β-hydroxyalkanoate granules (Kessel *et al.*, 1985), corresponding with the empty swollen regions seen in thin sections (Stoeckenius, 1981). These granules are also visible in the phase contrast microscope (Kessel and Cohen, 1982; Oren *et al.*, 1996, see also Fig. 1 and 3).

Electron microscopic examination of cells from the Gavish sabkha after staining with ruthenium red, which stains acid polysaccharides, suggested that more than one type of cells may occur, as some square cells appeared to possess a loose fibrillar sheath (Kessel and Cohen, 1982; Kessel *et al.*, 1985).

2.5. PIGMENTATION

The square halophilic bacteria are probably similar in pigmentation to *Halobacterium* and other halophilic archaea, and contain bacterioruberins and bacteriorhodopsin-like photoactive pigments. The occurrence of both bacterioruberins and photoactive pigments was reported in the (lost) pure culture of square gas vesicle-containing bacteria isolated by Torrella (1986) from a Spanish saltern.

Flash spectroscopy of material collected by precipitation of 6 liter brine from the Gavish sabkha showed a transient absorbance change indicative of the presence of bacteriorhodopsin: the difference spectrum 0.5 msec after an actinic flash at 500 nm showed a minimum near 560 nm and a maximum near 400 nm, indicating a bacteriorhodopsin like pigment (Stoeckenius, 1981). Only rarely were small ordered arrays of intramembrane particles found, and these did not show the typical appearance of the characteristic purple membrane of *Halobacterium salinarum*. The presence of bacteriorhodopsin-like pigments in square bacteria was also suggested by Kessel *et al.* (1985). Once more, however, the fact that a natural mixed community, rather than a pure culture of square bacteria was examined does not enable us to draw unequivocal conclusions on the presence of these pigments in the square bacteria. Photoactive, bacteriorhodopsin-like pigments were found in a variety of halophilic archaeal strains isolated from the site, including long and short rods, coccoid and flat-shaped bacteria (Stoeckenius *et al.*, 1985).

2.6. MOTILITY

Most reports on the square bacteria in natural brine samples do not mention the cells as being motile, and this author has never seen active movement of square cells in samples from the Eilat crystallizer ponds. The gas vesicle-containing square cells were described as non-motile by Grant and Larsen (1989).

However, an isolate from the Gavish sabkha, designated 801030/1, having a square morphology but lacking gas vesicles, was found to motile by flagella. The taxonomic status of strain 801030/1 has not been ascertained, and its relationship with the gas vesicle-containing square bacteria is as yet unknown. The mechanism of its flagellar movement had many features in common with that of *Halobacterium salinarum*: presence of right-handed helical flagella, which cause forward movement by clockwise rotation and backward movement by rotating counterclockwise. Cells may possess a single flagellum, or several flagella anchored from a single or from several locations on the cell surface. Several filaments may form a stable bundle that does not fall apart when changing from clockwise to counterclockwise movement and vice versa. However, instead of the three flagellar proteins present in flagella from *Halobacterium*, SDS-polyacrylamide gel electrophoresis of isolated flagella of strain 801030/1 showed the presence of a single flagellin band with an apparent molecular mass of 73 kDa (Alam *et al.*, 1984; Spencer, 1985).

2.7. VIRUSES OF SQUARE BACTERIA

Similar to other halophilic archaea, the square bacteria are subject to lytic attacks by viruses. Bacteriophage particles associated with lysed square bacteria were first observed by Kessel (1983). Electron-microscopic examination of samples from two Spanish salterns (Tarragona, Alacant) showed the presence of up to $2\text{-}3\times10^8$ prokaryotes, including up to 4×10^7 square cells, and up to 10^9 virus-like particles ml^{-1} at the highest salinities. One to ten percent of the square cells observed contained phage particles. A lemon-shaped phage, similar in morphology to those spindle-shaped phages described for other archaea such as *Sulfolobus*, was found infecting the squares. It abundance increased along the salinity gradient together with the abundance of the square archaea. In addition, many square cells were infected by viruses with other morphologies. Viral production was estimated to exceed 10^8 virus particles ml^{-1} day^{-1} above 250 g l^{-1} salinity, due to the large burst size (estimated between 100-380 viruses per cell, with an average value of 203) of the square archaea (Guixa-Boixareu *et al.*, 1996).

2.8. TAXONOMIC AFFILIATION

As long as the gas vesicle-containing square bacteria have not been isolated in pure culture and cannot be studied in laboratory cultures, their taxonomic and phylogenetic affiliation cannot unequivocally be determined. They undoubtedly belong to the extremely halophilic branch of the archaea (order *Halobacteriales*, family *Halobacteriaceae*), as based on their pigmentation, cell wall structure, and salt requirement, but assignment to one of the recognized genera or to a newly created genus is not possible at present. In the absence of any convincing laboratory cultures it is not even possible to tell whether more than one type exists.

New species names have tentatively been proposed for the squares. Parkes and Walsby (1981) suggested the name *Quadra* ("square"), while Stoeckenius (1981) proposed *Arcula* ("a small box"). The last name has already caused some confusion due to its resemblance to the recognized genus name *Haloarcula*, and at least in one article the square cells were called Haloarculae (Dixon, 1994). There is no a priori reason that they should be associated with the genus *Haloarcula* other than the observation that *Haloarcula* species do occasionally produce squares in culture. However, the flat sheets of Walsby's organism have not been seen in any laboratory cultures of representatives of the pleomorphic genera *Haloarcula* or *Haloferax* (Grant and Larsen, 1989).

One of the characteristic differences between the recognized genera within the family *Halobacteriaceae* is the types of polar lipids found in the membranes, and especially the glycolipid species present. The types and amounts of glycolipids present in the known types of halophilic archaea do not vary greatly under different growth conditions. Little evidence has been presented to date on the occurrence of phenotypic variations in the types of glycolipids present in halophilic archaea, though the relative amount of the different glycolipids may vary slightly according to growth conditions. As a result, specific glycolipids can be used as diagnostic signature compounds, enabling identification of halophilic archaea at the genus level (Tindall, 1992).

An attempt has been made to obtain information on the nature of the square bacteria in the Gavish sabkha based on the nature of the glycolipids present. Using fast atom bombardment mass spectrometry, Fredrickson and coworkers (1989) identified in addition to the diether derivatives of phosphatidylglycerol (PG), phosphatidylglycerophosphate (PGP), and phosphatidylglycerosulfate (PGS), a trisaccharide glycolipid and its sulfated derivative, identical to the TGD-1 and S-TGD-1 glycolipids of *Halobacterium salinarum*. However, some uncertainty may arise on the nature of the sample examined: the authors state that they analyzed "the extremely halophilic square archaebacterium originally isolated from the Gavish sabkha", but as noted above, this type of square bacterium has not yet been cultured.

Microscopic examination of brines from the Eilat saltern crystallizer ponds showed a mixed community, in which a substantial part of the cells showed a square to trapezoid flat shape with gas vesicles. Their contribution to total bacterial numbers was 20-23% in April-July, 1993 (Oren, 1994b), and as high as 55% in November 1995 (Oren *et al.*, 1996). The contribution of these square cells to the total lipid content of the biomass is expected to be much larger due to their large surface area. We calculated an average surface area of $31.4 \pm 15.6\ \mu m^2$ for the square cells, and $7.0 \pm 3.4\ \mu m^2$ for the observed rods. Assuming that the glycolipid content per unit of surface of the square cells is comparable to that of the rod-shaped archaea, we calculated that the square bacteria contributed about 85% of the total glycolipids in November 1995. This dominance of square archaea presented us with the opportunity to obtain information on their taxonomic position by lipid analysis of the total community, using polar lipids as chemotaxonomic markers.

Thin layer chromatography of extracts of the biomass collected from the Eilat saltern brines separated four types of polar lipids: PG, PGP, PGS, and a single glycolipid, chromatographically identical with S-DGD-1, the major glycolipid found in all *Haloferax* species. The same result was obtained in April-July, 1993 (20-23% square bacteria) and in November 1995 (55% square bacteria) (Oren, 1994b; Oren *et al.*, 1996). Therefore it may be tentatively concluded that the square bacteria contained S-DGD-1 as their (principal) glycolipid.

This finding alone does not justify the conclusion that the square cells may belong to the genus *Haloferax*. *Haloferax* species lack PGS, a lipid found in abundance in the extract of the saltern community. Moreover, the minor glycolipids characteristically found in extracts of representatives of the genus *Haloferax* were not detected. The polar lipid composition of the saltern brine biomass examined most closely resembled that of the recently created genus *Halorubrum*, whose members (including *H. saccharovorum*, *H. sodomense*, *H. lacusprofundi*, and *H. trapanicum*) are characterized by the presence of a sulfated diglycosyl diether and PGS. However, the single glycolipid found in the brine was chromatographically different from that of *H. saccharovorum*. Thus, a positive assignment of the square bacteria to one of the known genera within the *Halobacteriaceae* based on polar lipid composition is still not possible, and they may belong to a yet undescribed genus.

The results clearly showed that the square bacteria found in the Eilat salterns were unrelated to *Halobacterium salinarum*, as no trace of the triglycosyl diether lipids TGD-1 and S-TGD-1 could be detected in the lipid extract of the saltern biomass. Likewise, a possible taxonomic relationship with the genus *Haloarcula*, as suggested earlier (Stoeckenius, 1981) can be excluded, as TGD-2, the characteristic triglycosyl diether lipid of *Haloarcula* was not found either.

The possibility cannot be ruled out that the typical morphology of square flat cells with gas vesicles may be common to more than one type of organism, belonging to different genera. In that case, the sample analyzed by Fredrickson *et al.* (1989) and found to contain the *Halobacterium* signature glycolipids TGD-1 and S-TGD-1 may have contained a square bacterium with a taxonomic affiliation different from that of the square cells abundant in the Eilat saltern brine.

3. Other unusually shaped halophilic Archaea

The gas vesicle-containing square bacteria present just the most extreme case of unusual morphology within the halophilic archaea. In fact, the group of halophilic archaea contains many more unusually shaped organisms. The absence of turgor pressure in the cells releases the cells from the kinds of restraints acting on other types of bacterial cells with respect to the morphologies possible. Flat pleomorphic cells were found to dominate in the biota of the Dead Sea (Kaplan and Friedmann, 1980; Oren, 1988; Oren, 1994a), and this flat pleomorphic morphology is found in most representatives of the genera *Haloferax* and *Haloarcula*. Scanning electron micrographs of the Dead Sea isolate *Haloferax volcanii* show flat disk-shaped, 0.3-0.4 μm thick, and often slightly cupped cells (Grant and Larsen, 1989; Kessel *et al.*, 1988; Larsen, 1984). This type of cell morphology has aptly been compared to the shape of potato chips (Englert *et al.*, 1992).

Attempts to isolate halophilic archaea from hypersaline environments, including those rich in square bacteria, often yielded cultures of pleomorphic, flat or "box shaped" cells. Thus, the brines of the Gavish sabkha were the source of isolation of "*Haloarcula sinaiiensis*", and a somewhat different strain from a Californian saltern was tentatively described as "*Haloarcula californiensis*" (Javor *et al.*, 1982) (both species have not yet been validly described). The relationship of these pleomorphic "box-shaped" bacteria with Walsby's gas vesicle-containing square bacteria can not yet be ascertained, as long as the latter are still awaiting isolation. The isolates are smaller, more variable in appearance, and lack gas vesicles. As stated by Javor *et al.* (1982): "We do not believe that they are identical to Walsby's square cells". Additional pleomorphic strains were isolated by Kessel and coworkers (1985).

A new, extremely interesting organism from the point of view of cell morphology was isolated a few years ago in Japan. Rather than being square, many of the cells are triangular shaped, while rhomboid cells occur as well. In contrast to Walsby's gas vesicle-containing square bacterium, the triangular bacterium is available in pure culture, and it was assigned to the genus *Haloarcula* as a new species: *Haloarcula japonica* (Takashina *et al.*, 1990). In the following section, some of its properties will be discussed.

4. *Haloarcula japonica* - the triangular bacterium

If square bacteria can exist - why not triangular ones? In 1986, Koki Horikoshi and a team of Japanese microbiologists isolated the first known triangular bacterium. Once more we deal with an extremely halophilic microorganism, and it was isolated from soil surrounding a saltern on the Noto peninsula in western Japan (Anonymus, 1986; Hamamoto *et al.*, 1988; Otozai *et al.*, 1991; Takashina *et al.*, 1990).

The cells measure 2-5 x 0.3-0.5 μm, and are motile by means of flagella, attached to one apex of the cell. A large number of sugars can be used for growth. The NaCl concentration range for growth is between 150 and 300 g l^{-1} (optimum 200 g l^{-1}), and the temperature range is 24-45°C. Analysis of polar lipids by thin layer chromatography and 16S rRNA sequence analysis showed the new isolate to be affiliated with the genus *Haloarcula*, and it was described as the new species *Haloarcula japonica* (Takashina *et al.*, 1990).

The unusual triangular shape of a large proportion of the cells, both in liquid culture and on agar media, is a direct result of the asymmetric mode of cell division of *H. japonica*. Cell division was followed in time lapse microcinematographic experiments of cells on agar medium. The mean time between cell divisions was about 3.7 h. The cell plate formed either between the apex of a triangle and the middle of the opposite side, or between the middle of any two sides. The former, roughly symmetric manner of division, yielding two triangular cells from one parent triangular cell, was seen in 20% of the observed divisions. Rhomboid cells were generated, in addition to triangular cells, by division in the second, asymmetric manner (Fig. 4). On agar most of the divisions were asymmetric, generating rhomboid cells. In liquid medium most cells (over 80%) were triangular, compared to about half for cells grown on agar. It was suggested that on agar the cells lose their flexibility, so that the change to triangular shapes occurs more slowly (Hamamoto *et al.*, 1988).

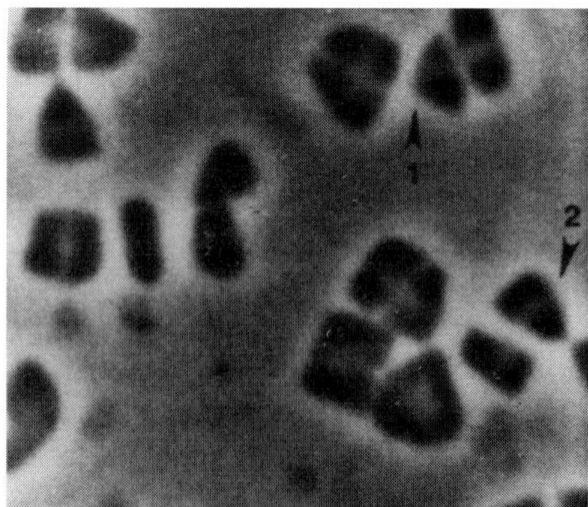

Figure 4. Haloarcula japonica, triangular and rhomboid cells. Reprinted from FEMS Microbiology Letters 56, T. Hamamoto et al., Asymmetric cell division of a triangular halophilic archaebacterium, pp. 221-224, 1988, with kind permission of Elsevier Science - NL, Sara Burgerhartstraat 25, 1055 KV Amsterdam, The Netherlands.

The cell wall of *Haloarcula japonica* is about 20-25 nm thick, and its appearance in the electron microscope after fixation or freeze-substitution is similar to that of *Halobacterium salinarum*. In grazing sections, a hexagonal array of particles is visible (Nishiyama *et al.*, 1992). The cell surface glycolipid of *H. japonica* was further characterized by Nakamura *et al.* (1992). Its properties were similar to those of other halophilic archaeal wall glycoproteins: its apparent molecular mass was 170 kDa, it could be stained with the periodate-Schiff stain for sugars, and it stained poorly with silver stains. The presence of high magnesium concentrations was required for structural integrity of the wall. Removal of magnesium, either by centrifugation of cells and resuspension in medium lacking magnesium, or by chelation of the magnesium with equimolar amounts of EDTA, led to shedding of part of the glycoprotein and converted the cells to spheroplasts (i.e., bacteria from which the cell wall has been removed but have not been lysed). Thus, when grown at low magnesium concentrations (40 mM and lower), cells were spherical, as compared to triangular and rhomboid at the optimal magnesium concentration of about 160 mM. A similar behavior has been reported for other, pleomorphic but not triangular, representatives of the genera *Haloarcula* and *Haloferax* (Cohen *et al.*, 1983).

In summary, with the exception of the unusual way daughter cells are formed in an asymmetrical way during cell division, *Haloarcula japonica* is not greatly different from other species within the genus *Haloarcula*, a genus anyhow characterized by the highly pleomorphic nature of its species.

5. Epilogue

The halophilic archaea can assume a great variety of shapes, including square or triangular morphologies, which are highly unusual in the living world. Direct microscopic examination of samples from hypersaline environments in which salt concentrations exceed 200-250 g l^{-1}, such as salt lakes, salterns, and salt pools, often show a tremendous variation in shapes.

It is not surprising that these unusually shaped bacteria are only found in environments characterized by high salt concentrations. The halophilic archaea do not possess a turgor pressure that would otherwise have forced the cells to assume a spherical shape. Absence of a turgor pressure has been documented in *Halobacterium salinarum*, in an elegant study in which its intracellular gas vesicles were used as pressure probes (Walsby, 1971). The lack of difference in the sensitivity of the gas vesicles to pressure-induced collapse in the presence and absence of high sucrose concentrations suggests that the square bacteria collected from the Gavish sabkha lack a turgor pressure as well (Walsby, 1980). By extrapolation it may be assumed that lack of turgor pressure is a common feature of all halophilic archaea of the family *Halobacteriaceae*. In the absence of turgor pressure there is less constraint on the shapes possible for bacterial cells, and it has been suggested that the shape of halophilic archaea is to a large extent determined by environmental factors (Grant and Larsen, 1989).

Prokaryotes with a square morphology may not be restricted to hypersaline environments. The recent discovery of small (about 0.3 μm) square non-halophilic bacteria in a Japanese forest soil sample (Whang and Hattori, 1989) may prove of great interest, and an investigation of their taxonomic position and their physiological properties will undoubtedly add new information on the enigma of the square bacteria.

The square, gas-vesicle-containing cells, which were not recognized as bacteria until 1980, but which now appear to be common inhabitants of hypersaline environments all over the world, are especially intriguing, as they are still waiting to be cultured, and their taxonomic affiliation is still unknown. The possibility cannot be ruled out that the typical morphology of square flat cells with gas vesicles may be common to more than one type of organism, possibly even belonging to different genera. The difference in glycolipid patterns found in the square bacteria analyzed by Fredrickson *et al.* (1989) and by Oren *et al.* (1996) suggest that the square morphology may occur in more than one organism. Positive assignment of the squares to one of the presently recognized or yet to be described genera within the *Halobacteriaceae* will have to wait until they can be studied in pure culture.

Acknowledgement

My present studies on halophilic archaea from saltern ponds are supported by grant no. 95-00027 from the United States - Israel Binational Science Foundation (BSF, Jerusalem, Israel).

References

Alam, M., Claviez, M., Oesterhelt, D. and Kessel, M. (1984) *EMBO J.* **3,** 2899-2903.

Anonymus (1986) *New Scientist* **July 1986,** 25.

Cohen, S., Oren, A. and Shilo, M. (1983) *Arch. Microbiol.* **136,** 184-190.

Dixon, B. (1994) *Power Unseen. How microbes rule the world.*, W.H. Freeman/Spektrum, Oxford.

Englert, C., Wanner, G. and Pfeifer, F. (1992) *Mol. Microbiol.* **6,** 3543- 3550.

Fredrickson, H.L., Rijpstra, W.I.C., Tas, A.C., van der Greef, J., LaVos, G.F. and de Leeuw, J.W. (1989) in Y. Cohen and E. Rosenberg (eds.), *Microbial Mats: Physiological Ecology of Benthic Microbial Communities*, American Society for Microbiology, Washington, DC, pp. 455-468.

Grant, W.D. and Larsen, H. (1989) Group III. in J.T. Staley, M.P. Bryant, N. Pfennig and J.G. Holt (eds.), *Bergey's Manual of Systematic Bacteriology, Vol 3,* Williams & Wilkins, Baltimore, pp. 2216-2233.

Guixa-Boizareu, N., Calderón-Paz, J.I., Heldal, M., Bratbak, G. and Pedrós-Alió, C. (1996) *Aquat. Microb. Ecol.* **11,** 215-227.

Hamamoto, T., Takashina, T., Grant, W.D. and Horikoshi, K. (1988) *FEMS Microbiol. Lett.* **56,** 221-224.

Horikoshi, K., Aono, R. and Nakamura, S. (1993) *Experientia* **49,** 497-502.

Javor, B. (1989) *Hypersaline Environments. Microbiology and Biogeochemistry*, Springer-Verlag, Berlin.

Javor, B., Requadt, C. and Stoeckenius, W. (1982) *J. Bacteriol.* **151,** 1532-1542.

Kaplan, I.R. and Friedmann, A. (1980) *Israel J. Chem.* **8,** 513-528.

Kessel, M. (1983) in G.W. Bailey (ed.), *Proceedings of the 41st Annual Meeting of the Electron Microscopy Society of America*, San Francisco Press, San Francisco, pp. 746-747.

Kessel, M., Buhle, E.L., Cohen, S. and Aebi, U. (1988) *J. Ultrastr. Mol. Struct. Res.* **100,** 94-106.

Kessel, M. and Cohen, Y. (1982) *J. Bacteriol.*. **150,** 851-860.

Kessel, M., Cohen, Y. and Walsby, A.E. (1985) in G.M. Friedman and W.E. Krumbein (eds.), *Hypersaline Ecosystems. The Gavish Sabkha*, Springer-Verlag, Berlin, pp. 267-287.

Larsen, H. (1984) in N.R. Krieg and J.G. Holt (eds.), *Bergey's Manual of Systematic Bacteriology, 9th ed, Vol 1,* Williams & Wilkins, Baltimore, pp. 261-267.

Nakamura, S., Aono, R., Mizutani, S., Takashina, T., Grant, W.D. and Horikoshi, K. (1992) *Biosci. Biotechnol. Biochem.* **56,** 996-998.

Nishiyama, Y., Takashina, T., Grant, W.D and Horikoshi, K. (1992) *FEMS Microbiol. Lett.* **99,** 43-48.

Oren, A. (1988) in K.C. Marshall (ed.), *Advances in Microbial Ecology, Vol. 10,* Plenum Publishing Company, New York, pp. 193-229.

Oren, A. (1994a) *FEMS Microbiol. Rev.* **13,** 415-440.

Oren, A. (1994b) *Int. J. Salt Lake Res.* **3,** 15-29.

Oren, A., Duker, S. and Ritter, S. (1996) *FEMS Microbiol. Lett.* **138,** 135-140.

Otozai, K., Takashina, T. and Grant, W.D. (1991) in K. Horikoshi and W.D. Grant (eds.), *Superbugs. Microorganisms in Extreme Environments*, Japan Scientific Societies Press, Tokyo/Springer-Verlag, Berlin, pp. 63-75.

Parkes, K. and Walsby, A.E. (1981) *J. Gen. Microbiol.* **126,** 503-506.

Romanenko, V.I. (1981) *Mikrobiologiya* **50,** 571-574 (in Russian).

Spencer, M. (1985) *Nature* **313,** 183.

Stoeckenius, W. (1981) *J. Bacteriol.* **148,** 352-360.

Stoeckenius, W., Bivin, D. and McGinnis, K. (1985) in G.M. Friedman and W.E. Krumbein (eds.), *Hypersaline Ecosystems. The Gavish Sabkha*, Springer-Verlag, Berlin, pp. 288-295.

Takashina, T., Hamamoto, T., Otozai, K., Grant, W.D. and Horikoshi, K. (1990) *System. Appl. Microbiol.* **13,** 177-181.

Tindall, B.J. (1992) in A. Balows, H.G. Trüper, M. Dworkin, W. Harder and K.H. Schleifer (eds.), *The Prokaryotes. A Handbook on the Biology of Bacteria: Ecophysiology, Isolation, Identification, Applications, 2nd ed.,* Springer-Verlag, New York, pp. 768-808.

Torrella, F. (1986) in *Abstracts of the Fourth International Symposium on Microbial Ecology,* Slovene Society for Microbiology, Ljubljana, p. 59.

Walsby, A.E. (1971) *Proc. R. Soc. B* **178,** 301-326.

Walsby, A.E. (1980) *Nature* **283,** 69-71.

Whang, K. and Hattori, T. (1989) *Bull. Jap. Soc. Microb. Ecol.* **5,** 9-11.

Biodata of **Antonio Ventosa** contributor (with D.R. Arahal) of *" Microbial Life in the Dead Sea."*

Dr. Antonio Ventosa is a Professor in the Department of Microbiology and Parasitology and is Dean of the Faculty of Pharmacy of the University of Sevilla, Spain. He received his Ph.D. in 1981 in Pharmacy from the University of Granada (Spain). Dr. Ventosa has published ~ 100 articles in books and scientific journals on halophilic microorganisms, and is a member of editorial boards. He gained the "Jaime Ferran" award of Microbiology (by the Spanish Society of Microbiology) in 1991.
E-mail: ventosa@cica.es

Biodata of **David R. Arahal** coauthor (with A. Ventosa) of *"Microbial Life in the Dead Sea".*

David R. Arahal is a postgraduate fellowship of the Spanish Ministry of Education and Science at the Department of Microbiology and Parasitology of the University of Sevilla, Spain. He spent some studies and research time abroad in Boston (USA), Munich (Germany) in Woodshole (USA).and in Athens (Greece). For his qualification in the degree of Pharmacy he gained the "Aguilar Ceballos"prize.

J. Seckbach (ed.), Enigmatic Microorganisms and Life in Extreme Environments, 357–368.

MICROBIAL LIFE IN THE DEAD SEA

A. VENTOSA and D.R. ARAHAL
Departament of Microbiology and Parasitology,
Faculty of Pharmacy, University of Sevilla, 41012 Sevilla, Spain.

1. Introduction

Located in the Syrian-African rift valley, the Dead Sea is a terminal lake showing some very unique properties from a geochemical point of view. It is recorded to be not only the lowest point on the continents (the surface is at 409 meters below sea level and the deepest part of it reaches -731 meters) (Hall, 1996), but also one of the saltier and larger hypersaline lakes in the world (close to 34% total dissolved salts) (Yechieli et al., 1995).

As a terminal lake the only outlet of water is evaporation, while the inflow comes mainly from the Jordan River. Until February 1979 the lake was permanently stratified (Steinhorn and Gat, 1983). The bottom layer was anoxic and contained sulfide. It was also more dense than the upper layer. It has been estimated that the meromictic state of the lake had existed for almost 300 years (Stiller and Chung, 1984). The main reason for the mixing of the two layers was the use of water from the Jordan River and its tributaries for irrigation in agriculture.

After the 1979 overturn, the lake has been stratified two more times following intense winter rainfloods in 1980 (Anati and Shasha, 1989) and 1992 (Oren, 1993) but none of these times the duration of the stratification was long enough to permit the formation of a reduced lower layer.

The Dead Sea is an athalassohaline lake with a very high ratio of divalent to monovalent cations (Oren, 1992). In contrast to most hypersaline environments the main cation is Mg^{2+} instead of Na^{+}. Cl^{-} is the predominant anion (99%), but Br^{-} (about 1%) is relatively more abundant than in thalassohaline waters (Nissenbaum, 1975).

The special features of this lake have been well known from ancient times by all the civilizations that populated its surroundings. Indeed it has been used as an economic resource for more than 10,000 years for obtaining salt and other chemicals, as well as therapeutics, cosmetics, etc. (Nissenbaum, 1993), and even today most of these resources are still under exploitation. Nevertheless, the lake was never used for fishery for one simple reason: there are no fishes living in its salty waters. As it is shown in a mosaic from the 6th century found in Madaba, fishes from the Jordan River coming into the Dead Sea try to avoid its salty brines (Nissenbaum, 1975). It is not suprising then that the lake was historically considered to be sterile or "dead", and this reputation did not change for the eyes of the world until 1936 (Volcani, 1936).

2. The pioneering microbiological studies on the Dead Sea

During the 19th century some chemical and geological expeditions were carried out at the Dead Sea, but attempts to find any sign of life were unsuccessful. Convinced of the sterility of the Dead Sea water and its applicability as an aseptic liquid, Lortet inoculated some sediment samples in appropriate media and to his surprise growth occurred consisting of anaerobic sporeforming rods of the genus *Clostridium* (Lortet, 1892). In spite of this finding the Dead Sea was still considered to be absent of life since these bacteria were unable to grow at high salt concentrations, and therefore they were considered to be contaminants whose endospores could resist the extreme salinity of the Dead Sea.

The discovery of extreme halophiles at the beginning of the 20th century brought reasonable doubts to the sterility of the Dead Sea, and finally in 1936 Benjamin Elazari Volcani (who also signed as B. Wilkansky and B. Elazari-Volcani) reported the isolation of microorganisms from this habitat, thus ending the reputation of lifelessness of the Dead Sea (Volcani, 1936). Volcani's first publication was a short article in Nature which was followed by several others (Volcani, 1940a, 1943a, 1943b, 1944a, 1944b) and his PhD thesis (Volcani, 1940b).

From water samples of the Dead Sea, he succeeded isolating extreme halophiles (*Halobacterium marismortui*, *Halobacterium trapanicum*, and *Micrococcus morrhuae*), moderate halophiles (*Chromobacterium marismortui*, *Pseudomonas halestorgus,* and *Flavobacterium halmephilum*) and ten sporulating Gram-positive halotolerant bacteria. In addition, the physiological groups of bacteria found in Dead Sea water include denitrifying bacteria, aerobic cellulose-decomposing bacteria, and fibrinolytic bacteria (Volcani, 1940b, 1944b).

From the bottom sediments Volcani isolated *Flavobacterium* (*Halobacterium*) *maris-mortui*, and described several physiological groups of aerobic and anaerobic bacteria (Volcani, 1943a).

Among the algae, *Dunaliella viridis* was described as a common inhabitant of the Dead Sea, but also some others were found, belonging to the genera *Aphanocapsa*, *Chlorella*, *Microcystis*, *Nostoc*, *Phormidium* and *Stichococcus* (Volcani, 1940a, 1944b).

Other organisms described were an amoeba of the genus *Dimastigomoeba* (Volcani, 1943b) and a ciliate (Volcani, 1944a).

3. Microbiological studies on the Dead Sea during the second half of the century

The work of Volcani was not followed by other authors for a period of almost 20 years. He himself devoted his research efforts to a different topic, and so it was not until the 1960's and 1970's that a few new microbiological studies were conducted on the Dead Sea. Kaplan and Friedmann (1970) and Kaplan and Baedecker (1970) measured quantitatively the biological productivity in the Dead Sea water column and in the sediments. Mullakhanbhai and Larsen (1975) described the isolation of a new species from the Dead Sea, which was named after Volcani as *Halobacterium volcanii*, and which is now placed in the genus *Haloferax* (Torreblanca et al., 1986). Additional studies were performed by Nissenbaum and coworkers (Nissenbaum, 1975; Nissenbaum and Kaplan, 1976).

Since the physico-chemical evolution of the Dead Sea is quite remarkable, and since important changes occurred in this period, the lack of a more complete information about the ecology of the lake is deplorable.

From 1980 onwards a large number of studies covering different microbiological aspects have been conducted. Most of these contributions were due to Aharon Oren and coworkers. Thanks to this intensive research activity detailed information on the microbial populations of the lake is now available (Oren, 1981, 1983b, 1985, 1988, 1993; Oren and Shilo, 1982). Some new species coming from the Dead Sea have been described (Huval et al., 1995; Oren, 1983c, 1983d; Oren et al., 1984, 1987, 1989, 1990, 1995). Other studies dealing with microorganisms from the Dead Sea focused on topics such as their biochemistry (Kessel et al., 1988; Pundak and Eisenberg, 1981; Pundak et al., 1981; Steinert et al., 1997), genetics (Gomes and Gupta, 1997, Kleman-Leyer et al., 1997; St Jean and Charlebois, 1996), physiology (Mojica et al., 1997; Oren, 1983a; Oren and Gurevich, 1994), etc.

Recently, our group has carried out a taxonomic study using some old enrichments prepared by Volcani at the time when he was working on his PhD thesis. These enrichments consisted of a peptone-brine medium inoculated with water samples collected at different points in the northern part of the lake. After having been stored in a refrigerated dark room for 57 years, they proved to contain viable microorganisms (Arahal et al., 1996). From these samples we isolated a total of 261 strains, which were classified according to their salt range and optimal salt concentration for growth: 158 strains were extreme halophiles, and 103 strains were moderately halophilic bacteria.

Phenotypic characterization of these strains, as well as phylogenetic studies based on the comparison of 16S rRNA sequences of some representative strains, showed that the extreme halophiles were members of the *Halobacteriaceae*, and clustered into six groups. Three of these were assigned to the genus *Haloarcula*, two to the genus *Haloferax*, and the last one to the genus *Halobacterium* (Arahal and Ventosa, unpublished results). In the case of the moderately halophilic strains, 12 were Gram-negative aerobic rods of the genus *Halomonas*, and the other 91 strains were Gram-positive endospore-forming rods that could represent a new species within the genus *Bacillus*, growing best at 10% salts (Arahal and Ventosa, unpublished results).

4. Microorganisms from the Dead Sea

4.1. ARCHAEA

Although Volcani (1940b) described the presence of several halobacteria in the Dead Sea, these cultures were lost, and are not available nowadays from any culture collection.

Some of these organisms were similar to previously described species, such as *Bacterium trapanicum* (lately renamed as *Halorubrum trapanicum*), *Micrococcus morrhuae* and *Sarcina morrhuae*; these two later organisms were considered by Volcani as two different species, but currently they are members of a single species of the genus *Halococcus*, named *Halococcus morrhuae* (Grant and Larsen, 1989). Since that time, this species has not been reported or isolated from the Dead Sea, and recent attempts to reisolate it from Volcani's enrichments have not been successful, probably because it could constitute a minor proportion of the total community, as has been reported in the case of other hypersaline environments (Rodriguez-Valera *et al.*, 1985).

In addition, Volcani (1940b) reported the isolation of a new species from the Dead Sea, named *Flavobacterium maris-mortui*, which was subsequently included in the genus *Halobacterium* as *H. marismortui* (Volcani, 1957). Unfortunately, this organism was lost. During the 1960's a new organism with characteristics quite similar to those of *H. marismortui* was isolated (named as the "*Halobacterium* of the Dead Sea") (Ginzburg *et al.*, 1970). The new name *Haloarcula marismortui* has been used to designate it (Oren *et al.*, 1990). Morphologically, the cells are pleomorphic, flat and disk-shaped, and require 20-25% NaCl; the optimal temperature for growth is 40 to 50°C. One interesting property of this species is that it is able to grow anaerobically with nitrate as an electron acceptor (Oren *et al.*, 1990). This organism has two genes encoding 16S rRNA with different nucleotide sequences (rrnA and rrnB genes) (Mylvaganam and Dennis, 1992) (Figure 1).

Another indigenous archaeon isolated from the bottom sediment of the Dead Sea is *Haloferax volcanii* (Torreblanca *et al.*, 1986), originally described as *Halobacterium volcanii* (Mullakhanbhai and Larsen, 1975). This organism shows a lower NaCl requirement (10-15%) than other halobacteria. It grows quite well at high $MgCl_2$ concentrations (optimal growth at 0.2 M, but 1.5 M is tolerated); these features reflect the different salt composition of the Dead Sea in comparison with other thalassohaline habitats. The cells of *H. volcanii* are extremely pleomorphic, from disk-shaped to spherical or irregular structures. This organism is unable to grow anaerobically with nitrate, but uses certain carbohydrates as single carbon sources (Mullakhanbhai and Larsen, 1975; Grant and Larsen, 1989).

In 1983 Oren reported the isolation of a slender, rod-shaped halobacterium, motile by polar flagella, from a water sample of the Dead Sea by using a medium containing starch. This organism, designated as *Halobacterium sodomense* (recently renamed as *Halorubrum sodomense* [McGenity and Grant, 1995]), shows some interesting features, especially concerning its salts requirements: its requirement for divalent cations (Mg^{2+} or Ca^{2+} ions) is higher than that of most other halobacteria. Optimal growth was reported at 0.6 to 1.2 M Mg^{2+} (in the presence of 2.1 M NaCl), and an increase of Na^+ concentration did not depress the requirement for Mg^{2+}. On the other hand, the specific requirement for Na^+ ions was exceptionally low in comparison with other halobacteria. Optimal growth was obtained at about 10-12% NaCl, but good growth was still observed at about 3% NaCl (Oren, 1983d). This species requires starch (0.2% for optimal growth), produces acid from glucose and other sugars, but is unable to grow anaerobically with nitrate (Oren, 1983d).

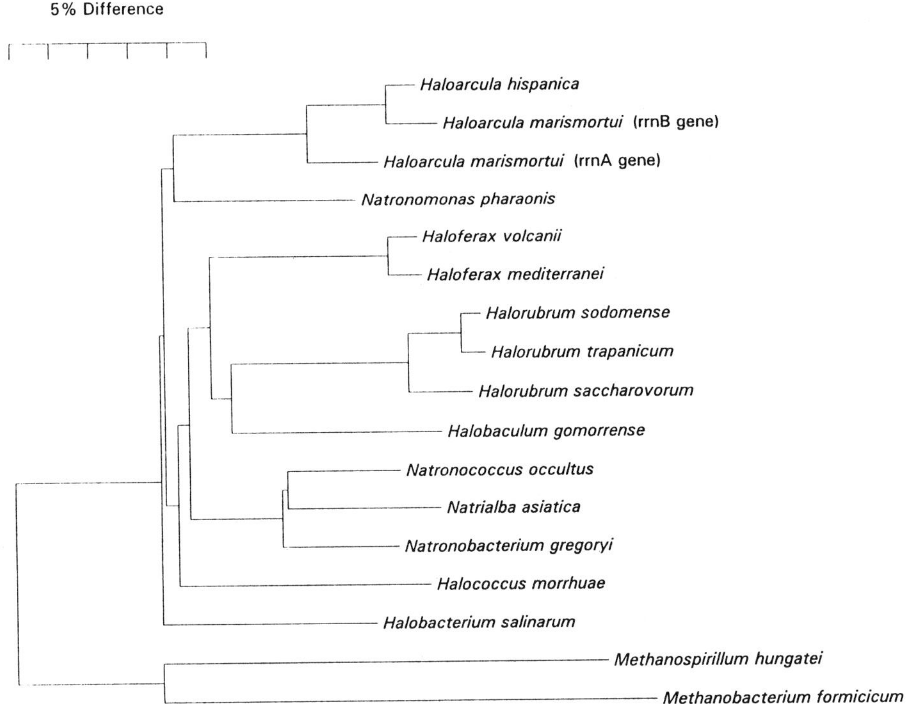

Figure 1. Phylogenetic tree, based on the 16S rRNA nucleotide sequence comparison, of some representatives of the *Halobacteriaceae*, showing the position of the extremely halophilic Archaeal species that have been isolated and described from the Dead Sea (marked with an asterisk). The methanogenic Archaea *Methanospirillum hungatei* and *Methanobacterium formicicum* were used as outgroups.

The halophilic archaeal species most recently described from the Dead Sea is *Halobaculum gomorrense* (Oren *et al.*, 1995). This organism was isolated from the water body of the Dead Sea in the summer of 1992, when a dense bloom of red Archaea developed, imparting a reddish color to the water. *Halobaculum gomorrense* resembles *H. sodomense* in many features, such as its high magnesium requirement (optimal growth at 0.6 to 1.0 M Mg^{2+}), and a relatively low requirement for sodium ions for growth (1 M NaCl, in the presence of Mg^{2+} ions). Moreover, the cells are rod shaped, and produce acid from glucose and other sugars. However, on the basis of its polar lipids composition, *H. gomorrense* was most similar to species of the genus *Haloferax*. Phylogenetic studies, based on the 16S rRNA sequence comparison, showed that *H. gomorrense* did not cluster with species of *Haloferax*, and occupied an intermediate position between the species of the genera *Halorubrum* (*H. sodomense, H. saccharovorum, H. lacusprofundi*, etc.) and

Haloferax (Figure 1). This finding, together with other phenotypic differences, supported the placement of this Dead Sea Archaeon in a separate genus, which is currently represented by a single species (Oren *et al.*, 1995).

4.2. BACTERIA

In addition to extremely halophilic Archaea, several halophilic Bacterial species have been described from the Dead Sea. These include heterotrophic aerobic and strictly anaerobic, as well as photosynthetic representatives. Among the three new species described by Volcani (1940b), *Pseudomonas halestorgus* is the only one which is not currently accepted as a valid species, since no culture is available. *Chromobacterium marismortui*, is a Gram-negative aerobic species which produces a yellow and a violet blue pigments and whose cells are rods, motile by peritrichous flagella (Volcani, 1940b). However, it could not any longer be classified as a member of the genus *Chromobacterium*, because the pigment is not violacein. For many years it was included as a "incertae sedis" species of this genus (Sneath, 1974). In 1982, Ventosa and coworkers (1982) isolated a group of strains from Mediterranean salterns located near Alicante (Spain), which showed phenotypic features similar to those described by Volcani for *C. marismortui*. They were able to grow in media containing 1 to 30% salt, with an optimum at 10%. Later studies showed that they could be classified in a single species, in spite of the fact that they were inhabitants of hypersaline environments with different salt compositions. It was proposed to place them in a new genus as *Chromohalobacter marismortui* (Ventosa *et al.*, 1989). Recent studies, based on 16S rRNA sequence comparisons, supported this designation, and showed that *C. marismortui* is closely related to other moderately halophilic Bacteria belonging to the genus *Halomonas* which have been isolated from quite different hypersaline habitats, and are included in the family *Halomonadaceae*. Phylogenetically they constitute a branch of the gamma subclass of the *Proteobacteria* (Mellado *et al.*, 1995) (Figure 2). With respect to the Gram-negative rod *Flavobacterium halmephilum*, the third new bacterium that was described by Volcani (1940b), it was suggested that this organism was not a member of the genus *Flavobacterium*, because unlike the species of *Flavobacterium*, it contained ubiquinones and not menaquinones as the major respiratory quinones. More recent studies determined that it belongs to the genus *Halomonas* (Figure 2), and currently it is classified as *Halomonas halmophila* (Franzmann *et al.*, 1988; Dobson and Franzmann, 1996).

Another interesting organism that was isolated from the Dead Sea is a Gram-negative rod, designated as Ba_1 (Rafaeli-Eshkol, 1968). This strain has been used for many biochemical and physiological studies concerning osmoregulation, ions transport or the effect of salt on respiratory enzymes. Recent studies showed that it represents a new species, and it was designated as *Halomonas israelensis* (Huval *et al.*, 1995). This species produces acid, but no gas, from sucrose and other sugars. The genus *Halomonas* was proposed in 1980 by Vreeland *et al.* for a single species, *H. elongata*, but presently a large number of species are classified within this genus, which is widely distributed in hypersaline environments.

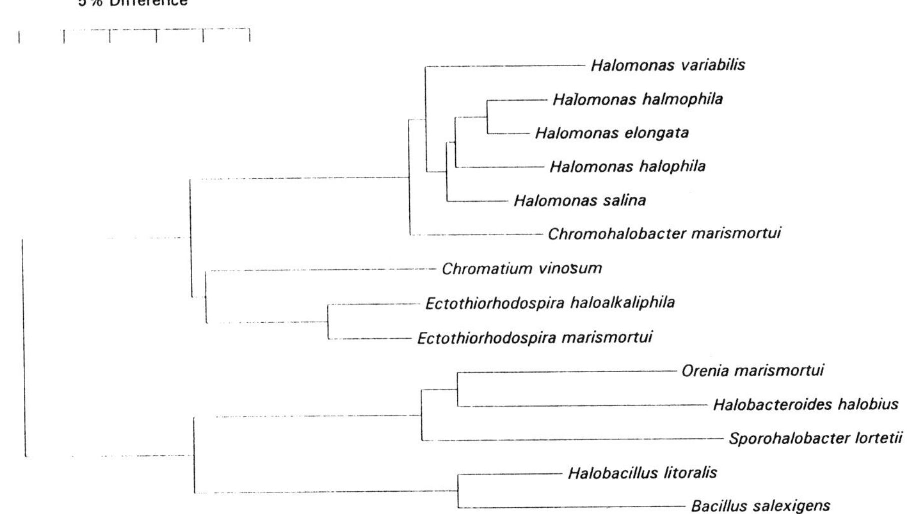

Figure 2. Phylogenetic tree based on 16S rRNA sequence analysis, indicating the position of the bacterial species described from the Dead Sea (marked with an asterisk) among the representatives of the *Proteobacteria* and the Gram-positive bacteria.

Three species of heterotrophic strictly anaerobic Bacteria have been described from the Dead Sea: *Sporohalobacter lortetii* (formerly designated *Clostridium lortetii*), *Halobacteroides halobius*, and *Orenia marismortui* (formerly named as *Sporohalobacter marismortui*) (Rainey *et al.*, 1995).

Sporohalobacter lortetii was isolated from Dead Sea sediment from a depth of 60 m, at about 2 km off the lake's shore near Massada. This species requires 5 to 15 % NaCl for growth. It is a Gram-negative, rod-shaped bacterium, able to produce endospores (Oren, 1983c). One interesting feature of this moderate halophile is that gas vacuoles are observed in sporulating cells, generally near the developing terminal endospores and they remain attached to the mature endospore after degeneration of the vegetative cell. This species produces the following fermentation compounds: acetate, propionate, isobutyrate, n-butyrate, isovalerate, and hydrogen. Glucose and some other carbohydrates stimulate the growth (Oren, 1983c).

Halobacteroides halobius is a long rod-shaped, Gram-negative, non-sporulating bacterium that was isolated from Dead Sea sediment. This organism requires about 8 to 15% NaCl for growth, and is able to ferment several mono- and oligosaccharides, starch and pyruvate. The products of glucose fermentation are ethanol, acetate, hydrogen and carbon dioxide (Oren et al., 1984).

Orenia marismortui (formerly named *Sporohalobacter marismortui*) is an anaerobic, rod-shaped, motile, Gram-negative, sporulating bacterium isolated from Dead Sea sediment, that requires from 3 to 18% NaCl for growth. The fermentation products on glucose are ethanol, acetate, butyrate, hydrogen and carbon dioxide (Oren et al., 1987; Rainey et al., 1995).

Phylogenetically, the species *Halobacteroides halobius*, *Sporohalobacter lortetii* and *Orenia marismortui* are closely related (Figure 2), and they are placed in the family *Halobacteroidaceae* within a recently proposed order *Haloanaerobiales* (Rainey et al., 1995).

Another obligately anaerobic organism isolated from a hypersaline sulfur spring located on the shore of the Dead Sea is the purple sulfur bacterium *Ectothiorhodospira marismortui*. This microorganism can grow photoautotrophically with sulfide as electron donor or photoheterotrophically with acetate and other compounds as carbon source. Its photosynthetic system contains bacteriochlorophyll *a* and carotenoids of the spirilloxanthin series, and is located on stacks or lamellar membranes in the cell cytoplasm. With respect to its salts requirements, it grows in the range 1 to 20 % NaCl (optimal growth at 3 to 8 % NaCl), but high divalent cation concentrations are not required (Oren *et al.*, 1989). Species of the genus *Ectothiorhodospira* have been obtained from marine, hypersaline or haloalkaline environments, and are classified, together with the members of the genus *Halorhodospira*, in the family *Ectothiorhodospiraceae,* which is phylogenetically closely related to the *Chromatiaceae*, within the gamma subclass of the *Proteobacteria* (Imhoff and Süling, 1996) (Figure 2).

4.3. OTHER MICROORGANISMS

Since the studies of Volcani, the presence of the unicellular green alga *Dunaliella* in the Dead Sea has been reported many times; although originally clasified as *Dunaliella viridis*, recent studies replaced this name by *Dunaliella parva* (Oren, 1988). Besides, other algae, cyanobacteria and protozoa have been observed in the Dead Sea (Volcani, 1940b; Oren, 1988). Studies on saline soils of the Dead Sea area (but not the water body) have permited the isolation of a variety of fungi belonging to the ascomycetes, basidiomycetes, zygomycetes and deuteromycetes (Steiman et al., 1994, 1995, 1996; Guiraud et al., 1995).

Virus-like particles have also been reported recently from the Dead Sea, and it was postulated that they were probably derived from the lysis of Archaea (Oren et al., 1997).

5. Epilogue

The Dead Sea can be considered as one of the most extreme environments on our planet, and this is reflected by its low microbial biodiversity. As we have reviewed, only a very limited number of species have been isolated and described as normal inhabitants of this peculiar habitat. One factor that may influence the dynamics of the microbial communities that develop in the Dead Sea is the changes in the salinity resulting from freshwater inflow after rainfloods periods. Recently developed molecular ecology techniques should be used for the characterization of the microbial communities as well as the biogeochemical activities that they carry out in this hypersaline environment. It may be expected that new microorganisms that have not yet been isolated and cultured in laboratory media with the currently available techniques will be described in the near future.

6. Acknowledgements

We thank A. Oren for critical reading of the manuscript. Our studies on halophiles are supported by grants from the European Commission (Generic Project Extremophiles as Cell Factories grant BIO4-CT96-0488), from the Ministerio de Eduación y Ciencia, Spain (grants BIO94-0846-CE and PB93-0920), and from the Junta de Andalucía.

7. References

Anati, D.A. and Shasha, S. (1989) *Israel J. Earth Sci.* **38,** 33-35.
Arahal, D.R., Dewhirst, F.E., Paster, B.J., Volcani, B.E. and Ventosa, A. (1996) *Appl. Environ. Microbiol.* **62,** 3779-3786.
Dobson, S.J. and Franzmann P.D. (1996) *Int. J. Syst. Bacteriol.* **46,** 550-558.
Franzmann, P.D.,Wehmeyer, U. and Stackebrandt, E. (1988) *Syst. Appl. Microbiol.* **11,** 16-19.
Ginzburg, M., Sachs, L. and Ginzburg, B.Z. (1970) *J. Gen. Physiol.* **55,** 187-207.
Gomes, I. and Gupta, R. (1997) *Biochem. Biophys. Res. Commun.* **237,** 588-594.
Grant, W.D. and Larsen, H. (1989) in J.T. Staley, M.P. Bryant, N. Pfennig and J.G. Holt (eds.), *Bergey's Manual of Systematic Bacteriology*, vol. 3, Williams & Wilkins, Baltimore, pp. 2216-2233.
Guiraud, P., Steiman, R., Seigle-Murandi, F. and Sage, L. (1995) *Syst. Appl. Microbiol.* **18,** 318-322.
Hall, J.K. (1996) *Tectonophysics* **266,** 177-185.
Huval, J.H., Latta, R., Wallace, R., Kushner, D.J. and Vreeland, R.H. (1995) *Can. J. Microbiol.* **41,** 1124-1131.
Imhoff, J.F. and Süling, J. (1996) *Arch. Microbiol.* **165,** 106-113.
Kaplan, I.R. and Baedecker, M.J. (1970) *Isr. J. Chem.* **8,** 529-533.
Kaplan, I.R. and Friedmann, A. (1970) *Isr. J. Chem.* **8,** 513-528.
Kessel, M.; Buhle, E.L. Jr, Cohen, S. and Aebi, U. (1988) *J. Ultrastruct. Mol. Struct. Res.* **100,** 94-106.
Kleman-Leyer, K., Armbruster, D.W. and Daniels, C.J. (1997) *Cell* **89,** 839-847.
Lortet, M.L. (1892) *Palestine Exploration Fund* **1892,** 48-50.
McGenity, T.J. and Grant, W.D. (1995) *Syst. Appl. Microbiol.* **18,** 237-243.
Mellado, E., Moore, E.R.B., Nieto, J.J. and Ventosa, A. (1995) *Int. J. Syst. Bacteriol.* **45,** 712-716.
Mojica, F.J., Cisneros, E., Ferrer, C., Rodr□guez-Valera, F. and Juez, G. (1997 (*J. Bacteriol.*5471-5481 ,179 .
Mullakhanbhai, M.F. and Larsen, H. (1975) *Arch. Microbiol.* **104,** 207-214.
Mylvaganam, S. and Dennis, P.P. (1992) *Genetics* **130,** 399-410.
Nissenbaum, A. (1975) *Microb. Ecol.* **2,** 139-161.
Nissenbaum, A. (1993) *Hydrobiologia* **267,** 127-141.
Nissenbaum, A. and Kaplan, I.R. (1976) in Nriagu, J.O. (ed.), *Environmental Biochemistry*, vol. 1. Ann Arbor Science Publishers, Ann Arbor, Michigan, pp. 309-325.
Oren, A. (1981) *Kieler Meeresforsh. Sonderh.* **5,** 416-424.
Oren, A. (1983a) *Limnol. Oceanogr.* **28,** 33-41.
Oren, A. (1983b) *Limnol. Oceanogr.* **28,** 1094-1103.
Oren, A. (1983c) *Arch. Microbiol.* **136,** 42-48.
Oren, A. (1983d) *Int. J. Syst. Bacteriol.* **33,** 381-386.
Oren, A. (1985) *Limnol. Oceanogr.* **30,** 911-915.
Oren, A. (1988) in K.C. Marshall (ed.), *Advances in Microbial Ecology*. Plenum Press, New York, pp. 193-229.

Oren, A. (1992) *Int. J. Salt Lake Res.* **1,** 7-20.
Oren, A. (1993) *Experientia* **49,** 518-522.
Oren, A., Bratbak, G. and Heldal, M. (1997) *Extremophiles* **1,** 143-149.
Oren, A., Ginzburg, M., Ginzburg, B.Z., Hochstein, L.I. and Volcani, B.E. (1990) *Int. J. Syst. Bacteriol.* **40,** 209-210.
Oren, A. and Gurevich, P. (1994) *FEMS Microbiol. Ecol.* **14,** 147-156.
Oren, A., Gurevich, P., Gemmell, R.T. and Teske, A. (1995) *Int. J. Syst. Bacteriol.* **45,** 747-754.
Oren, A., Kessel, M. and Stackebrandt, E. (1989) *Arch. Microbiol.* **151,** 524-529.
Oren, A., Pohla, H. and Stackebrandt, E. (1987) *Syst. Appl. Microbiol.* **9,** 239-246.
Oren, A. and Shilo, M. (1982) *Limnol. Oceanogr.* **27,** 201-211.
Oren, A., Weisburg, W.G., Kessel, M. and Woese, C.R. (1984) *Syst. Appl. Microbiol.* **5,** 58-70.
Pundak, S. and Eisenberg, H. (1981) *Eur. J. Biochem.* **118,** 463-470.
Pundak, S., Aloni, H. and Eisenberg, H. (1981) *Eur. J. Biochem.* **118,** 471-477.
Rafaeli-Eshkol, D. (1968) *Biochem. J.* **109,** 679-685.
Rainey, F.A., Zhilina, T.N., Boulygina, E.S, Stackebrandt, E., Tourova, T.P. and Zavarzin, G.A. (1995) *Anaerobe* **1,** 185-199.
Rodriguez-Valera, F., Ventosa, A., Juez, G. and Imhoff, J.F. (1985) *Microb. Ecol.* **11,** 111-115.
Sneath, P.H.A. (1974) in R.E. Buchanan and N.E. Gibbons (eds.), *Bergey's Manual of Determinative Bacteriology*, Williams & Wilkins, Baltimore, pp. 354-357.
Steiman, R., Guiraud, P., Sage, L. and Seigle-Murandi, F. (1994) *Syst. Appl. Microbiol.* **17,** 620-624.
Steiman, R., Guiraud, P., Sage, L., Seigle-Murandi, F. and Lafond, J.L. (1995) *Syst. Appl. Microbiol.* **18,** 310-317.
Steiman, R., Guiraud, P., Seigle-Murandi, F. and Sage, L. (1996) *Syst. Appl. Microbiol.* **19,** 182-190.
Steinert, K., Wagner, V., Kroth-Pancic, P.G. and Bickel-Sandkotter, S. (1997) *J. Biol. Chem.* **272,** 6261-6269.
Steinhorn, I. and Gat, J.R. (1983) *Sci. Am.* **249,** 102-109.
Stiller, M. and Chung, Y.C. (1984) *Limnol. Oceanogr.* **29,** 574-586.
St Jean, A. and Charlebois, R.L. (1996) *J. Bacteriol.* **178,** 3860-3868.
Torreblanca, M., Rodríguez-Valera, F., Juez, G., Ventosa, A., Kamekura, M. and Kates, M. (1986) *Syst. Appl. Microbiol.* **8,** 89-99.
Ventosa, A., Gutiérrez, M.C., García, M.T. and Ruiz-Berraquero, F. (1989) *Int. J. Syst. Bacteriol.* **39,** 382-386.
Ventosa, A., Quesada, E., Rodríguez-Valera, F., Ruiz-Berraquero, F. and Ramos-Cormenzana, A. (1982) *J. Gen. Microbiol.* **128,** 1959-1968.
Volcani, B.E. (Wilkansky, B.) (1936) *Nature* **138,** 467.
Volcani, B.E. (Elazari-Volcani, B.) (1940a) *Nature* **145,** 975.
Volcani, B.E. (Elazari-Volcani, B.) (1940b) *Studies on the microflora of the Dead Sea*. Ph.D. Thesis. The Hebrew University, Jerusalem.
Volcani, B.E. (Elazari-Volcani, B.) (1943a) *Nature* **152,** 274-275.
Volcani, B.E. (Elazari-Volcani, B.) (1943b) *Nature* **152,** 301-302.
Volcani, B.E. (Elazari-Volcani, B.) (1944a) *Nature* **154,** 335-336.
Volcani, B.E. (1944b) in *Papers Collected to Commemorate the 70th Anniversary of Dr. Chaim Weizmann*. Daniel Sieff Institute, Rehovoth, pp. 71-85.
Volcani, B.E. (Elazari-Volcani, B.) (1957) in R.S. Breed, E.G.D. Murray and N.R. Smith (eds.), *Bergey's Manual of Determinative Bacteriology*, 7th Ed., Williams & Wilkins, Baltimore, pp. 207-212.
Vreeland, R.H., Litchfield, C.D., Martin, E.L. and Elliot, E. (1980) *Int. J. Syst. Bacteriol.* **30,** 485-495.
Yechieli, Y., Ronen, D., Berkowitz, B., Dershowitz, W.S. and Hadad, A. (1995) *Water Resources Research* **31,** 893-902.

Biodata of Professor H. Nakamura

see his article on ***From Bacteria to Protista*** in this volume

J. Seckbach (ed.), Enigmatic Microorganisms and Life in Extreme Environments, 369–385.

SALT SENSITIVITY OF CELLS

HAKOBU NAKAMURA
Biological Institute, Faculty of Science
Konan University, Higashinada. Kobe 658, Japan

1. Introduction

The first life form, a proto-cell, is believed to have evolved in the primitive sea. The sea nurtured early life supplying both inorganic and organic materials. However, it is not clear whether all the minerals positively supported proto-cells, although the mineral elements required by present-day cells are abundant in seawater. Geological evidence suggests that sea water accumulated upon the cooling of dense vapor in the atmosphere which derived from active volcanoes on the primitive earth. Therefore, we can assume that very little or no sodium chloride was present in primitive seawater when the proto-cells evolved. However, it is a fact that seawater currently contains on average, 3.5% sodium chloride. The increase in the sodium chloride concentration likely resulted from:(i) extraction from the solid earth, which has been estimated to contain 460 sodium atoms per 10,000 atoms of silicon (Gymer,1973). (ii) evaporation of water into the atmosphere. Water evaporation from the surface of lakes is known to increase the sodium chloride concentration. For example, the salt concentration reaches saturation (about 35% at 20°C) in some area of the Dead Sea and Great Salt Lake. However, halophilic bacteria and certain other organisms flourish in this type of salt water.

2. Sodium / Potassium Content Of Cells

Sodium is an essential element for life, acting as a co-factor of certain enzymes, an osmotic balance of cell sap, a chelator of various macromolecular assemblies and, though indirectly, an antibiotic co-factor in the halophilic organisms. It seems reasonable to assume that the proto-cell preferentially utilized sodium as it was the most abundant metal ion in primitive seawater and its efflux from the cell consumes energy. However, an excess of sodium ions is toxic to a cell and even present-day living systems consume much energy to prevent their accumulation. It is not clear why an excess of sodium, but not potassium ions, threatens survival, although it likely involves some physico-chemical effect.

Table 1 shows the sodium and potassium balance in the living cells of various organisms (Nakamura, 1980). The sodium concentrations in the cells are always significantly lower than those in the environmental media, whereas the reverse is true for potassium. A characteristic balance of the concentrations of Na^+ and K^+ is regularly observed not only in sea but also terrestrial organisms. This suggests that normal cells efflux sodium ions into the medium or suppress accumulation of the ions. In contrast, the potassium concentrations of cells are balanced against the sodium gradient of the inner and outer sides of the cells.

It has been demonstrated that cells living in environments with a high concentration of sodium ions are osmotically compensated by intracellular accumulation of non-toxic, low molecular weight substances, such as free amino acids and potassium ions. However, there is a tendency for the cells to uptake the sodium ions and, conversely, to efflux them through systems in the plasma membrane: Na^+ / K^+ antiporters[1], Na^+ / H^+ antiporters and Na^+ / specific amino acid antiporters. The antiporters require ATP. For example, the Na^+ / K^+ -ATPase transports three sodium ions out of and simultaneously two potassium ions into the cell, consuming one molar of ATP in the process (Nakamura, 1979b).

The bacterial mechanism for sodium efflux may differ from the eukaryotic mechanism because the former is insensitive to the drug quabain while the latter is quite sensitive to it. However, both organisms have a similar Na^+ / K^+ antiporter which functions by consuming ATP. According to Nikaido (1994), the active efflux mechanisms of bacteria including *Escherichia coli* play an important role in resistance to various antibiotic and basic compounds and are determined by specific genes. Further, evidence has been accumulating that the efflux pumps are present within the plasma membrane.

Table 1 shows that marine bacterial cells contain much more sodium ions than marine green algal or animal cells, although the intracellular concentration of sodium ions is significantly low as compared to the extracellular concentration in all of the organisms. As discussed later, an excess of sodium ions is required for cell growth and for preventation of lysis of marine bacteria. Such bacteria are referred to as halophiles, while bacteria like *E. coli* which can grow well in very low concentrations of sodium ions, are called non-halophiles. Bacteria are classified into four classes based on the level of halophism: (i) non-halophilic (requiring up 0.3% NaCl), (ii) slightly halophilic (requiring 1 to 4% NaCl), (iii) moderately halophilic (requiring 5 to 10% NaCl), and (iv) extremely halophilic (12 to 35% NaCl). *Pseudomonas* 101 in Table 1 is moderately halophilic, while *Halobacterium salinarium* and *Sarcina morrhuae* in Table 2 are extremely halophilic. Furthermore, certain organisms are classified as euryhalines. These are aquatic and resistant to a wide range of salt concentrations. Euryhalines live in river mouths and migrate between the sea and the river.

Sodium pumps are located in the plasma membranes in all cells, and thus were probably acquired during the cellular evolution of bacteria as an adaptation to the accumulation of salt in oceans and lakes.

3. Salt Sensitivity of *E. Coli* Cells

Figure 1 shows the sensitivity of *E. coli* cells to an excess of sodium ions and cationic

---

[1] **Antiporter** is a membranes (protein) which transports two different ions or molecules in opposite directions, across a lipid bilayer

Table 1. Sodium and Potassium Balance Between Intra- and Extra-cellular Fluids

organism or cell	ion concentration (m mol/l)			
	intracellular		extracellular	
	Na^+	K^+	Na^+	K^+
Marine bacteria				
Pseudomonas 101	1040	670	3000	5.5
Marine green algae				
Chara	66	65	460	10
Valonia	35	576	460	10
Animal				
Squid nerve	50	400	440	20
Frog muscle	16	127	106	2.6
Human erythrocyte	11	91	138	4.2

Table 2. Efflux of Sodium Ions in the Low, Moderate, and Extreme Halophiles

microorganism	sodium ion concentration (molar)		
	medium	**cell sap**	**uptake (%)**
Low halophile			
Marine *Pseudomonad* B-16	0.3	0.12	41
Moderate halophile			
Vibrio costicola	1.0	0.684	68.4
	1.6	1.09	68.1
	2.0	0.895	44.9
Paracoccus halodenitificans	1.0	0.311	31.1
Pseudomonas 101	1.0	0.90	90.0
Extreme halophile			
Pseudomonas sp.	0.6	0.05	8.3
	4.4	0.62	14.1
Halobacterium cutirubrum 1	3.3	0.86	24.0
Halobacterium salinarium	4.4	1.39	34.0
Sarcina morrhuae	4.0	3.17	70.0

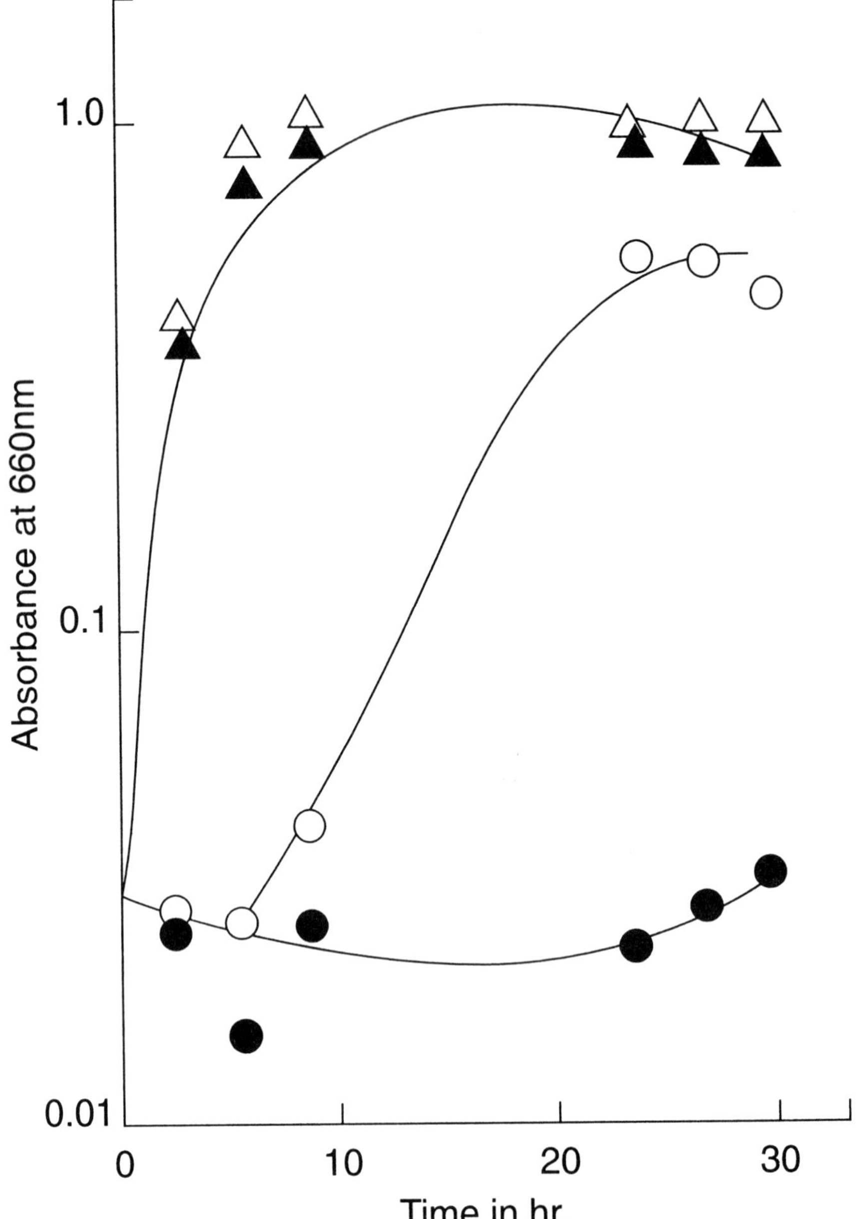

Fig. 1. Sensitivity of the *acrA* mutant and wild type *acrA*$^+$ cells in *Escherichia coli* to excess salt. Freshly grown cells of the *acrA* mutant (closed symbols) and *acrA*$^+$ (open symbols) cells were inoculated in normal (▲, Δ) and 5% (●, O)-NaCl broth.

dyes, such as acriflavine, was determined using the genetic locus *acrA* at min 10.6 on the chromosome (Nakamura,1965, 1868; Nakamura et al., 1975). We have demonstrated using biochemical and genetic techniques that the *acrA* mutant plasma membrane has a defect of protein of molecular weight 56,000, referred to as *acrA*$^+$ protein (Nakamura et al., 1981 and unpublished data) and confirmed this with electron microscopy using the freeze-fracture technique. The *acrA* mutant plasma membrane is missing a large class of proteinous particles, which are normally distributed in the *acrA*$^+$membrane (Nakamura et al., 1981, 1982). Furthermore, the reverse mutation of allele *acrA* to *acrA*$^+$ restores the levels of *acrA*$^+$ proteinous particles in the plasma membrane (Nakamura et al., 1981, 1982). Therefore, it is apparent that the *acrA*$^+$ protein is responsible for resistance to an excess sodium ions and to cationic dyes.

Fig. 2 shows that the *acrA* mutant strain is significantly sensitive to 5% NaCl in broth as compared to its wild type (*acrA*$^+$) strain. We have found that, when incubated in 5% sodium chloride-broth, the free proline pool of *acrA* mutant cells increases remarkably while that of *acrA*$^+$ cells increases only slightly (Nakamura, 1977, 1979a). It has been reported that the free amino acid pool increases when halophilic bacteria are incubated in the presence of high concentrations of sodium chloride (Britten and McClure, 1962: Measures, 1972: Hayzer and Leisinger, 1980: Wood, 1981: Csonka, 1983: Chen et al., 1985). We examined which amino acids are the most effective in increasing the resistance to an excess of sodium ions. One of eleven amino acids was added to the media at 20 μg per ml and cell growth was followed in viable cells. Proline and glutamate were the most effective in increasing the resistance to an excess sodium in both the *acrA* and *acrA*$^+$ strains. The effect may be due to the direct synthesis of proline through the three reactions from glutamate. The results indicate that the effect of proline on sodium resistance increases proportionally with the concentrations of the amino acid added. Why does an increase in the proline pool occurs in the presence of an excess of sodium ions?. How does the proline supply to the cells increase the resistance to the excess?. Possible mechanisms are as follows: (1). An increase in the proline pool has an osmotic effect to suppress dehydration from the cells in so-called hypertonic media. Therefore, low molecular weight compounds, such as amino acids and sugars, which accumulate in the cell sap play an important role. (2).Incorporation of proline activates a certain antiporter to efflux sodium ions, and thus addition of proline to the media decreases the intracellular sodium ion concentration. Thus, the behavior of the proline molecule in the cell sap is the most important theme in the resistance of *E. coli* cells to excess sodium.

Next, we examined the effect of mutations in proline biosynthesis, *proA* and *proB*, on the cellular tolerance to excess sodium. These *proA* and *proB* mutations involve defects in the enzymes α-glutamyl phosphate reductase and α-glutamyl kinase, respectively. Growth of the mutant cells is completely dependent on an external supply of proline. Fig. 2 shows the effect of *proA*, *proB*, or *acrA* mutation on sodium sensitivity. The combinations of mutations shown in the figure were prepared by genetic recombinations of alleles of *proA*, *proB*, and *acrA*. The sensitivities were determined by growth in 3% NaCl-synthetic media containing 20 μg of proline per ml. Findings were as follows: (A) the *acrA*$^+$ *proA*$^+$ strain is significantly less sensitive to an excess of sodium than is the *acrA*$^+$ *proA* mutant;(B) the *acrA proA*$^+$ mutant is less sensitive than the double acrA proA

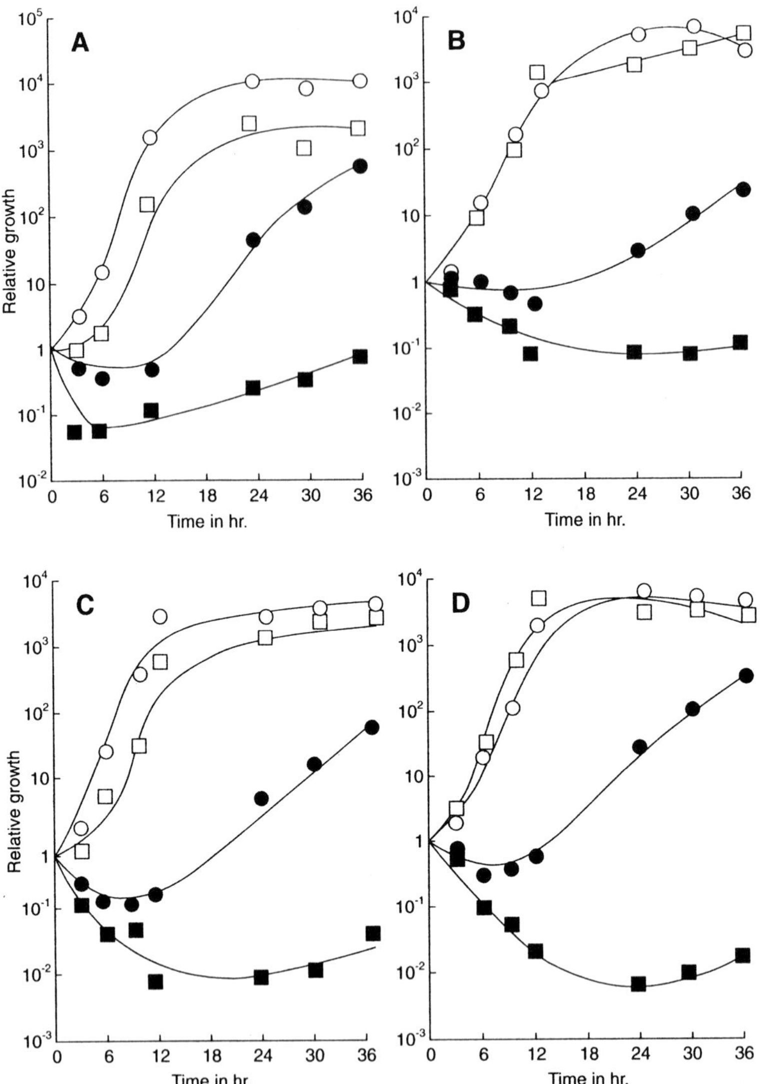

Fig. 2. Effect of proline-requiring (*pro*) mutations on sensitivity to excess salt of various recombinants between *acrA* and *pro* (*A* or *B*). Symbols in A: circles, *acrA*$^+$ *proA*$^+$; squares, *acrA*$^+$ proA; open, normal synthetic medium; closed, 3% NaCl-medium. Symbols in B: circles, *acrA proA*$^+$; squares, *acrA proA*. Open and closed symbols are as those in A. Symbols in C: circles, *acrA*$^+$ *proB*$^+$; squares, *acrA*$^+$ *proB*. Open and closed symbols are as those in A. Symbols in D: circles, *acrA proB*$^+$; squares, *acrA proB*. Open and closed symbols are as those in A. Proline was added always to 20 μg/ml.

mutant; (C) the *acrA*$^+$ *proB*$^+$ strain is less sensitive than the *acrA*$^+$ *proB* mutant; and (D) the *acrA proB* + mutant is less sensitive than the double *acrA proB* mutant. As a whole, the results show that a *proA* or *proB* mutation increases the sensitivity of cells to sodium while double mutations of *acrA* and *pro(A* or *B*) significantly increase sensitivity. The results suggest strongly that the proline metabolism closely correlates with a function of the *acrA* gene product to increase the resistance to an excess of sodium ions.

Fig. 3 shows sodium uptake of recombinants of *acrA* and *proA* genes (in A) and *acrA* and *proB* genes (in B). The cells were cultured in 11% NaCl-synthetic media containing 20 μg of proline per ml at 37°C. (For effective uptake of sodium ions by cells, relatively high concentrations of sodium chloride are used.) The figure shows that uptake was significantly higher in the *acrA* than in the *acrA*$^+$ strains. Furthermore, *proA* or *proB* mutation accelerates the rate of uptake. The double mutants *acrA proA* and *acrA proB* took up sodium at the highest rates. In other words, the phenotypic expressions of the response to the excess in the various strains tested coincided well with the amount of sodium ions taken up by the cells. Therefore, it seems reasonable to conclude that the sensitivities of the cells to an excess of sodium can be determined by their sodium contents, and that the wild type allele of the *acrA* gene functions to suppress any increase in sodium content.

We originally found the *acrA* gene in the *E. coli* chromosome as a determinant of sensitivity to acriflavine (Nakamura, 1965). Our laboratory (Nakamura and Shinya, 1985) and Midgley (1987) have demonstrated that the *acrA* gene is responsible for the effect of acriflavine and other cationic dyes. So, we examined whether or not this is also the case with sodium ions, and demonstrated clearly that the *acrA*$^+$ cells actively pump out more sodium ions than the *acrA* mutant cells. Therefore, the *acrA* and *pro(A* and *B*) genes also are responsible for sodium efflux. Fig. 4 shows the relative efflux of sodium ions by the *acrA* cells in a nutritionally rich (glucose) (Br) and a non-glucose (Mi) media, indicating that the sodium ion efflux requires a supply of energy, ATP. However, a specific inhibitor, quabain, of $Na^{+\cdot}$ K^+. - ATPase was again not effective in this experimental system.

4. Adaptation of Extreme Halophiles to Excess Sodium Ions

It has been clarified that all cells possess efflux systems in the plasma membrane regardless of low (or non-) and high halophiles. Therefore, the mechanism for efflux of excess sodium ions is assumed to have developed at a comparatively early stage of cellular evolution. Table 2 indicates characteristics common to low, moderate, and extreme halophiles; many of the halophiles take up rather large amounts of sodium ions from media. However, accumulation of an excess of sodium ions is avoided in three ways: (1). The cells actively incorporate sodium ions even from medium containing high concentration of sodium chloride. A very active system(s) is needed for efflux of sodium because the cell saps contain relatively low concentrations of the ions. (2). Cell saps of the extreme halophiles *Sarcina morrhuae* and *Halobacterium salinarium* contain very high concentrations of sodium, about 1.4 and 3.2 molar, respectively. Thus, the intracellular systems of these species are very resistant as a whole to an extreme of sodium ions. (3). In cells of the extreme halophile *Pseudomonas* sp., the sodium concentration is relatively low, suggesting that the permeability of the cell's plasma membrane to sodium ions is reduced, in order to reduce the concentration of sodium in the cell sap. Therefore, there

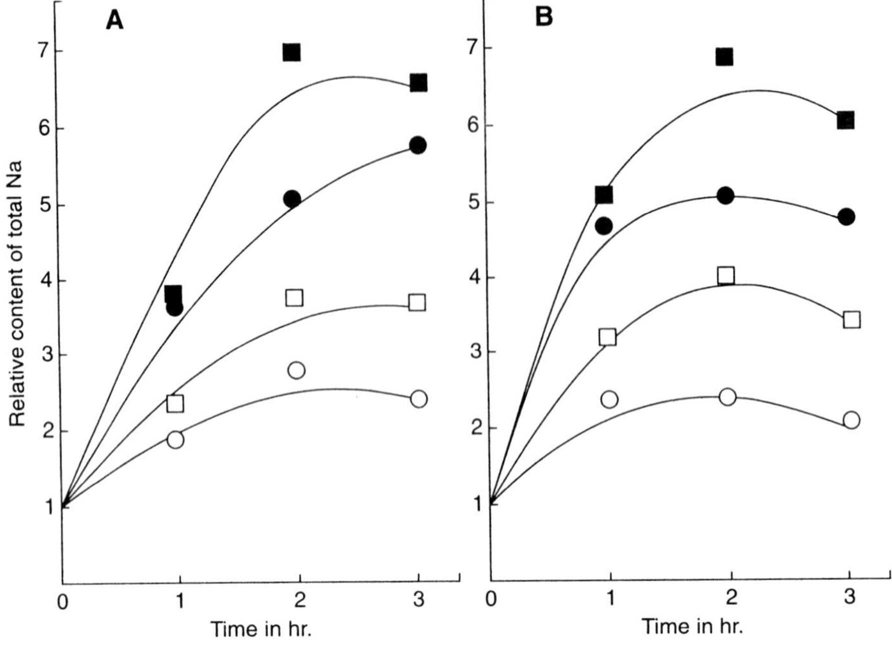

Fig. 3. Uptake of sodium ions by the *acrA* and *proA* (A) and the *acrA* and *proB* (B) mutant cells in an 11% NaCl-synthetic medium that contained 20 μg/ml proline. Symbols in A: *acrA*$^+$ *proA*$^+$; , *acrA proA*$^+$; , *acrA*$^+$ *proA*; , *acrA proA*. Symbols in B: ,*acrA*$^+$ *proB*$^+$; *acrA proB*$^+$; , *acrA*$^+$ *proB*; , *acrA proB*.

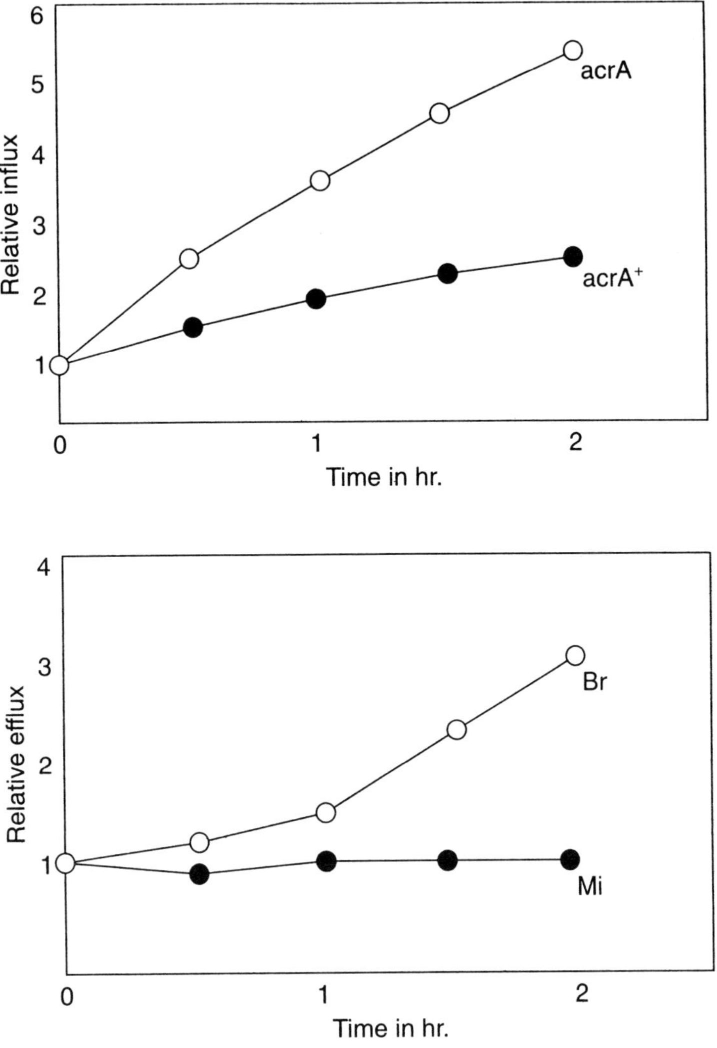

Fig. 4. Influx and efflux of Na^+ ions in the *acrA*$^+$ (wild type) and *acrA* (mutant) cells A: Relative influx of Na^+ ions in the *acrA*$^+$ (○) and acrA (●) cells. B: Relative efflux of Na^+ ions in a complete (Br) and a minimal (Mi) media. The data are shown as the ratio of *acrA*$^+$ to *acrA* cells. When incubated in a nutritionally rich medium (Br), the *acrA*$^+$ cells are much more active in the efflux activity than the *acrA* cells. However, under nutritionally poor conditions (Mi), the efflux is not very active.

is little doubt that present-day cells have multiple strategies for maintaining homeostasis.

Larsen (1962, 1967), Dondas (1977), and Kushner (1978) have demonstrated that not only the membranous functions of the extremely halophilic cells but also the enzymes themselves have acquired resistance to sodium excess. Fig. 5 shows activities of various isolated enzymes in a reaction mixtures containing excess sodium (Larsen, 1962). The enzyme sources are halotolerant *H. saliratum* and *S. morrhuae*, which grow optimally in a range of 20 to 30% NaCl, *Micrococcus halodenitrificans*, which shows optimal growth in a range of 1 to 4% NaCl, and non-halophilic *E. coli* strain. The enzymes from the extreme halophiles are more resistant in the presence of excess sodium ions than those of the moderate halophile, while the latter's enzymes are more resistant than those from the non-halophile. Two points should be emphasized: (1) although the enzymes of extreme halophiles have adapted to environments of excess salt, their substrate specificities have been strictly conserved during evolution for geological periods and (2) their levels of resistance to an excess of sodium is largely differential. All membranes of the family, Halobacteriaceae including genera *Halobacterium* and *Sarcina*, are extremely resistant to sodium ions. However, the gram-positive genus *Micrococcus* is not sodium resistant, while *M. halonitrificans* is an extreme halophile (Krieg and Holt, 1984). This means that sodium resistance is not a taxonomic criterion even for extreme halophiles.

Therefore, it is suggested that resistance to sodium as a characteristic of bacteria must not be determined by a specific gene but expressed synthetically by many loci on the chromosome. On the other hand, the substrate specificity of an enzyme is strictly determined by the amino acid sequence of its active center, independent of the sodium resistance of cells. Point (2) is also relevant to the present discussion. For example, succinate dehydrogenase of *S. morrhaue* is more resistant to an excess of sodium than *H. salinarium* and this relationship is reversed in the case of catalase and glutamate-aspartate transaminase.

Fig. 6 indicates different sensitivities to excess sodium of three kinds of enzymes isolated from *H. salinarium* cells (Larsen, 1967). A much higher sodium concentration is needed for activation of cytochrome oxidase than NADH dehydrogenase. Further, cysteine desulfhydrase is very sensitive to excess of sodium as compared to the former two. These results suggest strongly a subcellular differentiation, and that the enzymes are homeostatically arranged in the protoplasmic space depending on their sodium ion sensitivity. A possible explanation for the localization of these enzymes in the cell is that cysteine desulfhydrase must be stationed in the central area and thus is protected against the sodium ions from the cell surface. Halobacteriaceae is a kind of archaebacteria, and its cells are composed of a plasma membrane closely bounded by an outer proteinous layer, unlike ubiquitous bacterial walls, in which the proteins are strongly acidic and the lipids are ether-linked isoprenyl phosphoglycerides. Further, the cytoplasm of some strains contains gas vacuoles and differentiates complexly. Therefore, sodium ion-sensitive enzymes in particular areas may escape the toxicity of an excess of sodium ions. On the other hand, cytochrome oxidase seems to be integrated in the plasma membrane because its activity increases slowly with rising salt concentration in the medium. NADH dehydrogenase can be embedded in a cytoplasmic surface of the plasma membrane but its active center is opened to the plasma membrane; it is much more sensitive to sodium ions than cytochrome oxidase. Current molecular cytology has suggested such relationships between these enzymes and the plasma membrane (Alberts et al., 1994).

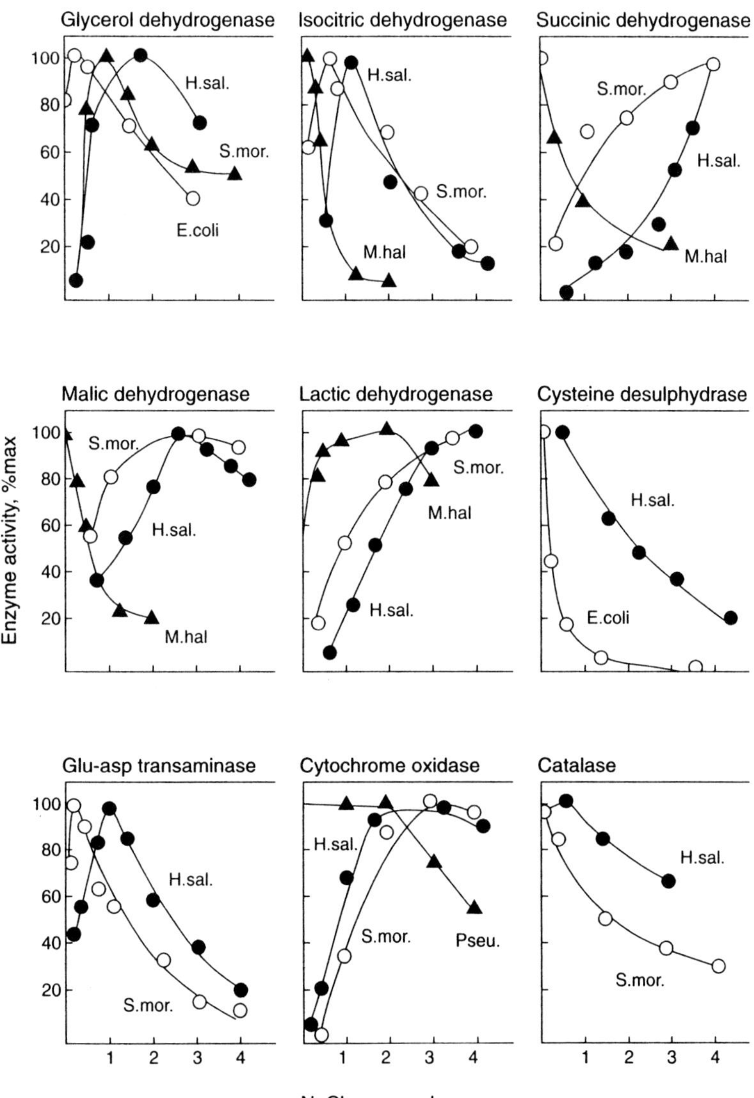

Fig. 5. Effect of salt hypertonicity on the *in vitro* activity of enzymes of extreme halophiles (*Halobacterium salinarium* and *Sarcina morrhuae*), moderate halophile (*Micrococcus halodenitrificans* and *Pseudomonas sp.*) and non-halophile (*E.coli*) (Larsen, 1967)

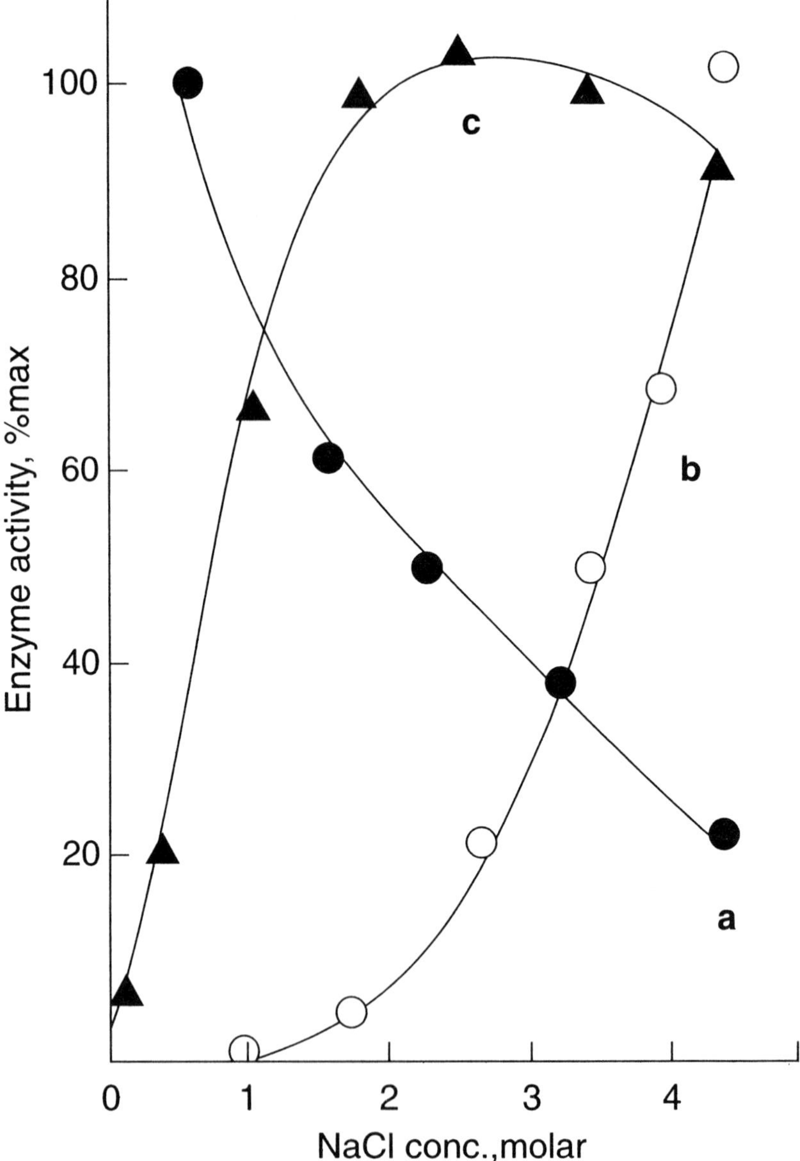

Fig. 6. Effect of salt hypertonicity on the *in vitro* activities of of enzymes cysteine desulfhydrase (a), cytochrome oxidase (b), and NADH dehydrogenase (c) of *H. salinarium* (Larsen, 1967)

4.1. HOMEOSTASIS OF THE UNICELLULAR LEVEL IN EXCESS SALT

The natural environment imposes severe stresses on organisms. For marine organisms, this includes temperature, osmotic pressure, and salinity. The evolution of an organism is the history of adaptation to virulent conditions. Perhaps, the most important and comprehensive concept of adaptation is homeostasis. Historically, the idea was first proposed by C. Bernald (1865), and then by W. B. Cannon (1932) who used the mild inner environments of the body of higher vertebrates as a typical example. The term homeostasis refers to the tendency of a system, in particular, the physiological system of higher animals, to maintain internal stability owing to the coordinate response to any situation or stimulus tending to disturb normal condition or function (after Webster's Dictionary, Random House, New York, 1991). Therefore, the mechanisms of homeostasis interact to achieve transduction. Homeostatic systems are found in all living forms, namely in unicellular, multicellular, and ecological levels.

The three-dimensional space of a cell can be divided into two parts, an outer surface, and an interior, relative to the environment. In this case, environment refers to all the conditions surrounding and affecting a cell. Morphological differentiation brings about certain differential effects influenced by environmental factors. That is, direct effects occur on the surface of the cell while indirect ones occur in the interior. Therefore, any three-dimensional body essentially has a basic, though weak, mechanism of homeostasis regardless of whether it is alive. It is likely that the proto-cells generated in the primitive sea survived and reproduced using just such a weak homeostatic mechanism. In other words, the cell had little active homeostatic ability and, fortunately, its environment favored this. Bernald and Cannon used extant higher vertebrates as examples of organisms to which concept of homeostasis was applicable. However, current phylogeny shows clearly that extent bacteria are a better example, as they have survived over four billion years. There is no doubt that their vitality is beyond comparison with higher organisms.

As shown in Figs. 5 and 6, even extremely halophilic bacteria contain some enzymes which are sensitive to excess sodium. Such bacterial cells must possess certain organizations to defined an enzyme's territory within the cytoplasm against invasion of sodium ions. According to our experiments on the supersensitivity of *E. coli acrA* and *pro(A* and *B*) mutants to excess sodium, the proto-cells had no active mechanism to exclude the toxic ions. Therefore, if the water of the primitive oceans contained as much as is present today, life could not emerge. Bacterial cells evolved sodium ion-resistant mechanism(s) as the salt concentration of the ocean increased and were able to adapt to even salt-saturated water.

5. Conclusion

Bacteria resistant to excess sodium can be divided into three classes: (i) non-halophiles, (ii) moderate halophiles, and (iii) extreme halophiles. This classification is clear, corresponding with specific resistance. As explained in the foregoing sections, it is apparent that there are several mechanisms by which organisms display resistance to excess sodium. These mechanisms are as follows: (1) efflux of sodium ions from the cell, (2) reduced permeability of the plasma membrane to sodium ions, (3) high tolerance to excess sodium ions. Cellular resistance involves all three.

(1). The efflux of sodium ions is via a sodium pump present in the plasma membrane. All types of cells, from bacteria to higher organisms and regardless of sensitivity to excess sodium, possess this mechanism, indicating that it was acquired at a very primitive stage of evolution. Furthermore, the pumping activity can adaptively increase as some non-halophilic cells become tolerant to an excess of sodium ions. Such examples can be seen in low, moderate, and extreme halophilic bacteria as shown in Table 2.

(2). In principle, the phospholipid bilayer, which forms the skeleton of membranes including the plasma membrane, is impermeable to ions. Therefore, the permeable capacity of a plasma membrane to sodium and other ions are due to specific channels. It has been widely accepted that a membranous mechanism imparts drug resistance to bacterial cells. As shown in Fig. 3, the more sensitive a cell to excess sodium, the greater the accumulation of sodium ions. This can be explained by two mechanisms, efflux (1) and/or the reduced permeability (2) to sodium ions.

(3). Cellular enzymes of extreme halophiles are highly tolerant to excess sodium. These bacteria actually require large amounts of sodium ions to maintain vitality. *In vitro* experiments have demonstrated that an excess of sodium ions is required to maintain enzyme activities and macro-molecular conformations (Kushner, 1978: Morishita and Masui, 1980). However, some enzymes of extremely halophilic cells, as shown in Fig. 6, are sensitive to excess sodium. Therefore, a certain amount of protection must be provided even in these cells.

6. References

Alberts, B., Bray, D., Lewis, J., Raff, M., Roberts, K., and Watson, J. D. (1994) Molecular Biology of the Cell (3rd Ed.), Garland Publishing, Inc., New York.

Britten, R. J. and McClure, E. T. (1962) Bacteriol. Rev. 26, 22-335.

Chen, C. C., Tsuchiya, Yamane, Y., Wood, J. M. and Wilson, T. H. (1985) J. Membrane Biol. 84, 157-164.

Csonka, L. N. (1983) J. Bacteriol. 151, 1433-1443.

Dundas, I. E. D. (1977) In: Adv. Microbial Physiol. 15, 85-120.

Gymer, R. G. (1973) Cited in De Witt (1977) Biology of the Cell - An Evolutionary Approach, W. B. Saunders Company, Philadelphia.

Hayzer, D. S. and Leisinger, T. (1980) J. Gen. Microbiol. 118, 287-293.

Krieg, N. R. and Holt, J. G. (eds.) (1984) Bergey's Manual of Systematic Bacteriolgy vol. 1, Williams & Wilkins, Baltimore, pp. 261, 262, and 402.

Kushner, D. J. (1978) In: D. J. Kushner (ed.) Microbial Life in Extreme Environments, Academic Press, London.

Larsen, H. (1962) Halophilism, In I. C. Gunsalus and R. Y. Stanier (eds.)The Bacteria : A Treatise on Structure and Function, vol. 4, Academic Press, New York, pp. 297-342.

Larsen, H. (1967) Adv. Microbial Physiol. 1, 97-132.

Measures, J. C. (1975) Nature 257, 398-400.

Midgley, M. (1987) Microbial Science 4, 125-127.

Morishita, H. (1980) In H. Morishita and M. Masui(eds.) Salt Environment, Japanese Conference on Halophilic Microbiology, pp. 93-109.

Nakamura, H. (1965) J. Bacteriol. 90, 8-14.

Nakamura, H. (1968) J. Bacteriol. 96, 987-996.

Nakamura, H. (1977) J. Gen. Microbiol.102, 221-222.

Nakamura, H. (1979a) J. Gen. Microbiol. 113, 425-427.

Nakamura, H. (1979b) Viva Origino 8, 1-9.

Nakamura, H. (1980) Microbial Evolution, Baifukan Publisher, Tokyo.

Nakamura, H., Tojo, T. and Greenberg, J. (1975) J. Bacteriol. 122, 874-879.
Nakamura, H., Yokomura, E., Hase, A. and Suganuma, A. (1981) Plant Cell Physiol. 23, 1141-1148.
Nakamura, H., Yokomura, E. and Hirayoshi, K. (1982) J. Gen. Microbiol. 128, 3067-3070.
Nakamura, H. and Shinya, T. (1985) J. Gen. Microbiol.131, 1639-1647.
Nikaido, H. (1994) Science 264, 382-388.
Wood, J, M. (1981) J. Bacteriol. 146, 895-901.

Dr. Russell H. Vreeland is an Assistant Professor at West Chester University, West Chester, PA. His PhD in microbial physiology is from the University of Nebraska. He has studied halophilic bacteria since 1976 and isolated the first species of *Halomonas*. His current research is directed toward isolation of halophilic bacteria from ancient salt crystals. He is head editor for the CRC press series on Microbiology of extreme and usual environments. E-mail: < rvreeland@wcupa.edu>

Dr. William D. Rosenzweig is an Assistant Professor at West Chester University (West Chester, PA). His PhD in Microbiology is from New York University. His interest is in microbial ecology/environmental microbiology. Special areas of interest include the ecology and physiology of nematode-trapping fungi, the occurrence and role of fungi in potable water, and the ecology and physiology of extremely halophilic bacteria. He is on the editorial board of Standard Methods for the Examination of water and wastewater.

J. Seckbach (ed.), Enigmatic Microorganisms and Life in Extreme Environments, 387–398.

SURVIVAL OF HALOPHILIC BACTERIA IN ANCIENT SALTS: POSSIBILITIES AND POTENTIALS

RUSSELL H. VREELAND and WILLIAM D. ROSENZWEIG
Department of Biology, West Chester University
West Chester, PA. 19383

1. Introduction

The Earth currently contains a large number of underground salt formations of significant size. These formations range in age from Jurassic (60,000,000 Million year of age [Ma]) to Cambrian (570,000,000 Ma). Most are Permian in age (230 - 280 Ma) (Javor 1989). Salt formations are distributed across nearly every continent, with the largest number in the northern hemisphere (Zharkov 1981)

While there are presently a large number of hypersaline terminal lakes located on the surface of the Earth there are currently no depositional environments of the scale that produced the vast underground salt deposits. Some of these smaller lakes have been studied extensively by numerous researchers with the resulting isolation of a significant number of aerobic and anaerobic halophilic organisms,(i.e., Oren 1983, 1992, Post 1977, Tindall *et al.*. 1980, 1984, Vreeland *et al.*. 1980, 1995). Many of these present day hypersaline environments are of interest because they provide a glimpse into the conditions that probably existed in the vast evaporating areas of the early Earth.

One of the most important characteristics of these terminal lakes is a high level of productivity. Javor (1989) has provided evidence that primary productivity in terminal lakes fixes on the order of 150 to 520 mg C/m^2/yr. In addition, even during the latter stages of crystallization, many of these lakes support large bacterial populations with viable counts ranging from 10^6 to as many as 10^9 cells per ml of brine. Many of these lakes support such large populations of extreme halophiles, the water actually takes on a distinct red color. Even terminal lakes that do not receive large influxes of nutrients (such as those that are landlocked or in desert regions) often have relatively stable, albeit, smaller bacterial populations. Figure 1 shows the distribution of microorganisms in three small hypersaline lakes in a desert area in southeastern New Mexico, United States. The bacterial populations within these three lakes remained relatively stable over the course of a one year study period with numbers ranging between 1000 and 100,000 viable cells per milliliter of brine (Figure 1). The red halophilic Archaea represented between 90 and 100% of the total population at all times. From the standpoint of the current discussion, this type of standing microbial pool would represent a continuous source of cells for entrapment in newly forming crystals.

As the crystals form within these environments the organisms can become trapped inside brine inclusions that form along the crystal faces or within the intra-crystalline matrix (Vreeland and Powers 1998). The condition of the cells within the inclusions, has been examined in elegant detail by Norton and Grant (1988) who demonstrated that the bacteria not only survive in the crystal but that they are often so well adapted that motility actually continues for several weeks following final crystallization. The fact that halophilic bacteria can remain viable within salt crystals has been known for nearly a century. In fact, halophilic bacteria were isolated from materials such as cod fish (Poulsen 1879/80, Klebahn 1919) and hides (Lochhead 1934) that had been preserved using solar salt crystals before they were isolated from any natural terminal lakes (Elazari-Volcani 1940).

Figure 1. Distribution of halophilic bacteria within selected hypersaline lakes of Nash Draw, New

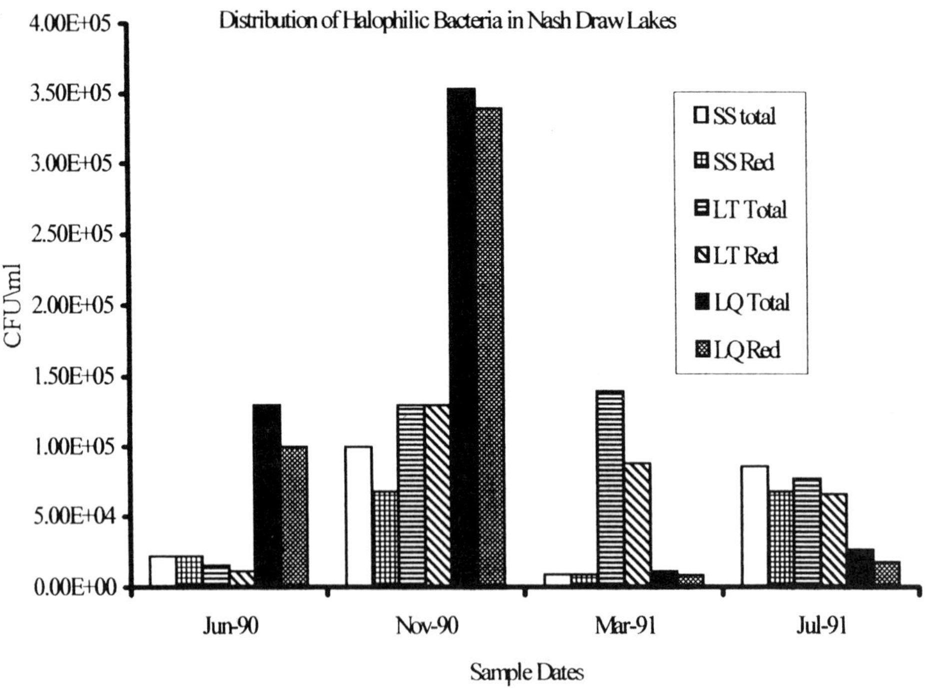

Mexico, United States. Lake brines were asceptically spread onto nutrient media containing 20% NaCl and incubated at 35C. Both total and red colony forming units were determined following 21 days of incubation. SS - Surprise Spring (a main feeder spring of Laguna Grande de la Sal, LT - Laguna Tres, LQ - Laguna Quatro.

Preservation within a crystal for a couple of years is of course a far cry from surviving several hundreds of millions years. Survival for hundreds of millions of years would require a complex interplay of forces that would serve to both protect and to stabilize the organisms and their macromolecules. Long term survival is not unknown in the bacterial world. Bacterial survival strategies include cryptobiotic spores, cysts, akinetes (Prescott *et al.* 1995) and even a state of starvation survival (Morita, 1990). Survival of spores from the inside of an amber encased, extinct bee has recently been demonstrated and has pushed the known limit of survival back to nearly 40 million years (Cano and Borucki 1995). However, long term survival inside of a salt crystal has not yet been demonstrated.

2. Geological factors impacting long term survival

As the evaporative environments continue to function over long periods of time a thick impenetrable salt crust develops. If the evaporative area is extensive and exists for a geologically significant period of time this crust can cover thousands of square miles and become thousands of meters thick (Zharkov 1981). Furthermore, once these formations become buried, the effects of geologic forces can alter the potential survival of trapped organisms, or macromolecules. In some cases these formations can even be penetrated by additional brines forming dissolution pipes that can ultimately trap more organisms as the fresher water recedes and the salt bed reforms (Powers and Hassinger 1985). Each of these occurrences can have implications on any experiments directed at recovering trapped microorganisms.

Vreeland and Powers (1998) have recently reviewed the issues surrounding experiments directed at isolating bacteria from these ancient salt crystals. The authors concluded that certain key data is needed to support claims that microorganisms, isolated from ancient geologic materials were trapped during the original deposition. These data should include geologic parameters to show that the formation being sampled has been isolated from the surface environment for a significant number of years, as well as data on the level of microbiological control used to minimize potential contamination (Vreeland and Powers 1998). But the real question in all of this is simply if it is even theoretically possible that an organism trapped inside a crystal could survive for thousands of millennia? Naturally the key to this survival would lie in the overall stability of the system and the degree of stress that would be imposed on the trapped organisms.

3. Ancient Salt Formations and Stability

The endurance of microorganisms trapped within these formations would ultimately be determined by the conditions experienced by the formation during the years following the initial deposition of the formation. Individual salt crystals that remain unaffected by powerful geologic and environmental forces would provide conditions that could preserve biological materials within the crystal structure. Over the last several years a wide variety of inclusion material from numerous salt formations have been sampled and chemically analyzed (Das *et al.* 1990, Horita *et al.* 1991, Horita *et al.* 1996, Lazar & Holland 1988). A summary of the data that has come from these studies is presented in Table 1. These data indicate, quite clearly, that the chemical conditions within the crystals would provide an excellent environment for the survival of halophilic organisms. In addition to the chemical analyses, examination of the fluid inclusions has shown that these too maintain a morphology that is almost identical to inclusions formed in crystals precipitating from modern day lakes. Data such as these indicate several things about the environment within a salt crystal. First the presence of these primary crystals suggest that formations such as the Permian age Salado have been isolated from surface influence since the original deposition occurred. Second, the fact that these primary inclusions have maintained their original morphology indicates that the crystals have not been subjected to the type of extreme conditions that could destroy trapped life forms or biological materials. The fact that conditions within the concentrated inclusion brines would be very stable can be inferred from the physical chemical research conducted by Brass (1980). In his studies, Brass has shown that water trapped in NaCl/KCl crystals would not freeze until temperatures reached below 22.9°C. Addition of even small amounts of alkali minerals to this mixture lowers the freezing point to -53°C. There would, of course, be a corresponding increase in the boiling temperatures of these brines. Consequently, liquid water would remain present within the inclusions of a salt crystal even under conditions that would normally cause ice formation or perhaps boiling. Vreeland and Powers (1998), in describing the considerations and constraints that should be imposed on studies designed to extract biological molecules from ancient geological materials discussed the need for careful selection of properly analyzed ancient formations. Based upon the descriptions and extensive analyses by Roedder (1984), Vreeland and Powers (1998) recommended that researchers concentrate only on crystals that contain primary inclusions as defined by Roedder (1984). Figure 2, is an example of the type of primary morphology recommended by Vreeland and Powers (1998). That is, inclusions being sampled must be square or cubic, they should not be penetrated by fractures or contain dissolution channels, all of which represent indications that the sample had been affected by intense geologic forces. The inclusion shown in Figure 2 is an excellent sample. While not shown in this figure, inclusions being targeted for sampling should also be completely filled with fluid without visible air bubbles. This is primarily due to the fact that while bubbles could arise from cooling of the trapped solution there is also the possibility that the bubbles entered the inclusion through a crack that could have allowed organisms to enter after the crystal originally formed.

4. Survival of Biological Material

The theoretical possibility that either biological remnants or an intact, viable, life form

Table 1

Chemical Composition of Inclusion Brines from Sodium Chloride Samples (mmol/kg of water)

Formation	Na	K	Mg	Ca	Li	Cl	SO_4	Br	Reference
Michigan Salina	2191	456	1781	1048	N.D	7447	<2	42	Das *et. al*. (1990)
Salado Inclusion	4770	124	808	16*	3	6068	193	548	Horita *et al*.. (1991)
Devonian Prairie	2365	165	1346	710	N.D	6667	N.D	23	Horita *et al*.. (1996)
Messinian (1988)	2810	307	1900	<7	N.D	6340	337	37	Lazar & Holland
Subsurface Brine	1544	236	517	2358	32	7290	N.D	48	Das et. al. (1990)
Permian Seawater	460 - 630	11	55	5 - 20	N.D	550 - 730	20 - 45		Horita *et al*.. (1991)
Modern Seawater	485	11	55	11	N.D	565	29	0.87	Horita *et al*.. (1991)

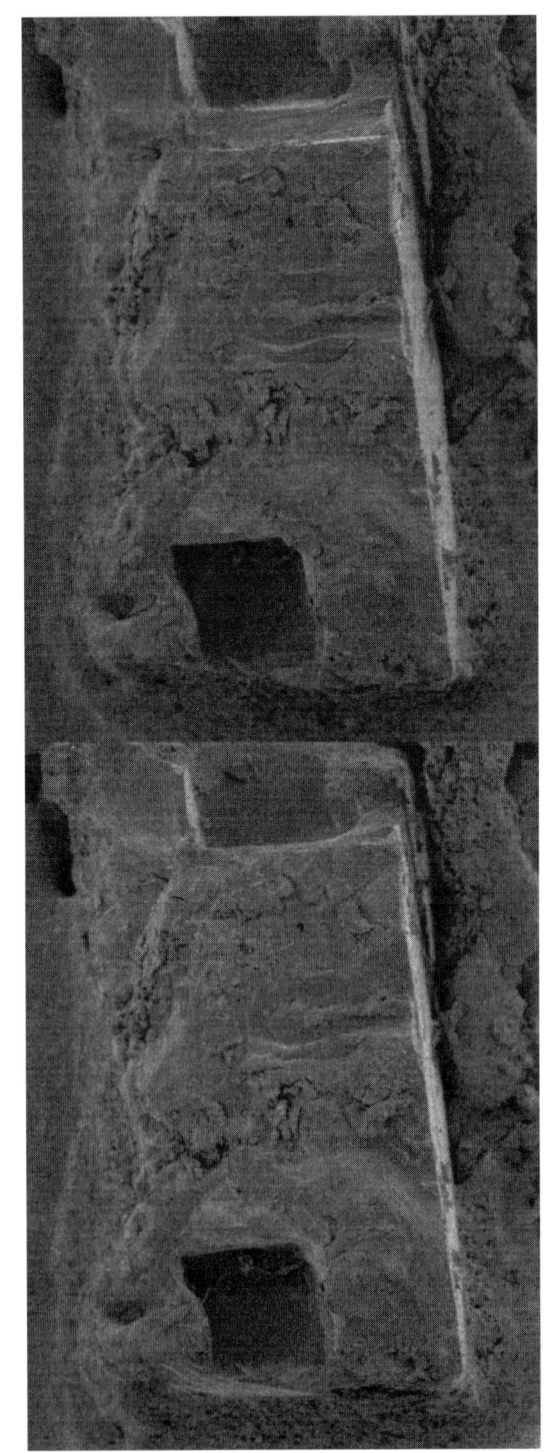

Figure 2. Stereo image of an unaltered brine inclusion from a primary crystal taken from the 250 ma Salado formation. This figure illustrates a high quality sample that would be a good candidate for isolating organisms from the inclusion brine. The crystal face was exposed, immediately coated with gold and viewed in a Joel 840A scanning electron microscope.

could be adequately preserved within an ancient material can no longer be summarily dismissed (Cano and Borucki 1995, Kennedy *et al...* 1994).

At this writing, living microorganisms have been found in everything from 100 year old beer bottles (an event which restored Flag Porter ale to beer drinkers) (Kennedy *et al..* 1994) to 40 million year old amber (Cano and Borucki 1995). Several groups have also reported finding microorganisms associated with 200 to 250 million year old rocks (Norton *et al.* 1993, Denner *et al.* 1994, Kennedy *et al.* 1994). Unfortunately for reasons pointed out by Vreeland and Powers (1998) these bacteria can not be considered as proof that the isolates had been trapped in the rocks since the time of deposition. Vreeland and Powers (1998) have pointed out the need to integrate both geologic and microbiological considerations that show, first the integrity of the formation being sampled and the integrity of the microbiological procedures used in isolating the microorganisms. This type of intense integration has not been addressed in any of the studies that have been conducted in salt formations to this time. Further, Vreeland *et al.* (1997) have demonstrated that while halophilic bacteria can be found associated with salt samples the microorganisms are heterogeneously distributed throughout the salts. Consequently, single sample type isolations, such as those that have been conducted to date, may yield negative results more often than they do positive isolations (Vreeland *et al.* 1997). One primary reason for the need to closely integrate these types of studies is that underground salt formations contain numerous different lithofacies within the same region. The different sedimentary layers arise at different periods during the entire depositional cycle. Holt and Powers (1990) have also demonstrated the fact that some areas of these formations were actually formed by the incursion of brines into already formed layers during subsequent flooding events. While these pit or pipe forms represent salt that may harbor microorganisms they are not necessarily identical (e.g. age or composition) to the sedimentary layer in which they exist. Because of this, microbiological sampling and analysis without regard to the geologic story told by the sedimentary layering could easily lead to erroneous conclusions regarding which portions of a formation are most likely to contain surviving microorganisms. The diversity of the sample types that are available in a single well characterized salt formation (i.e., the Permian aged Salado Formation) can be seen from the data in Table 2.

The cores listed in Table 2 were obtained directly from the Salado Formation (near Carlsbad, New Mexico) during a recently completed sampling trip. The cores listed between South 2180 and South 2200 represent a series of cores taken across a single depostional sequence (vertical distance approximately 1.5 meters). Within this sequence there were several different sedimentary layers all containing individual primary salt crystals. The layers included pressed mud (with obvious drying cracks), polyhalite, pure halite, interbedded gray clays and green clays reminiscent of current hypersaline mud containing green sulfur bacteria. Examination of selected unaltered crystals taken from the Salado formation has already shown the existence of bacteria like forms inside a Permian aged crystal. One of these forms is shown in Fig 3. This feature was found by the authors on a freshly cut crystal face. While being rather long for a typical bacterium (at least 30 microns) it is certainly the correct width (0.5 microns), has rounded ends which are not typical of crystals and appears to be either slightly bent or possibly collapsed (arrow Fig 3). If this form is indeed a preserved bacterial cell the degree of preservation would make it easy to believe that this bacterium could still be viable.

Table 2
Composition of Core samples from the Salado Formation

Air Intake Shaft Depth (ft)	Composition
1848	Primary halite
1850	Primary halite (inclusions with gas bubbles)
1852	Primary halite with Polyhalite
1854	Primary halite with Polyhalite
1857.5	Primary halite with Polyhalite
1858	Primary halite with Polyhalite
1858.5	Primary halite with Polyhalite
1858.7	Primary halite with Polyhalite
1859.7	Primary halite with Polyhalite
Mine Location	
W30 S2000 Lower	Primary halite with clay
W30 S2000 Upper	Primary halite
W30 S2180	Polyhalite, primary halite, clay
W30 S2209	Polyhalite with clay
W30 S2209.5	Polyhalite with clay
W30 S2210	Primary halite,
W30 S2211	Polyhalite with mud
W550 N90	Primary halite with trace clay

All mine locations have been given in terms of the normal coordinates showing the distance (in ft) from the main exit shaft. The samples listed from W30 S2180 to W30 S2211 represent a sampling of a complete depositional sequence. These samples were separated by a vertical distance of approximately 1.5 M.

5. Conclusions

While it would appear that survival of bacteria inside of ancient salt crystals is at least theoretically plausible this has not yet been proven. Despite many claims of isolations from ancient crystals no researcher has as yet obtained a living organism directly from the inside of a properly surface sterilized, methodically dated, unaltered crystal. If long term survival can be demonstrated to the satisfaction of at least a majority of the scientific community the value of the organism and the techniques employed would be tremendous. This would open up many important areas of future research on salt formations and on other ancient rocks that may also contain trapped stable fluids. In addition, there would wonderful experiments that could be conducted on different types of sedimentary layers to determine if particular types of sediments are better at preserving life. Numerous questions about the environment of the early earth, which are now only hinted at, could be addressed directly by examination of the physiology and genetics of the isolates. The evolutionary tree might begin to find it's roots. Finally, we would have developed the techniques, knowledge and confidence to hunt for life's remains in all sorts of ancient materials and environments.

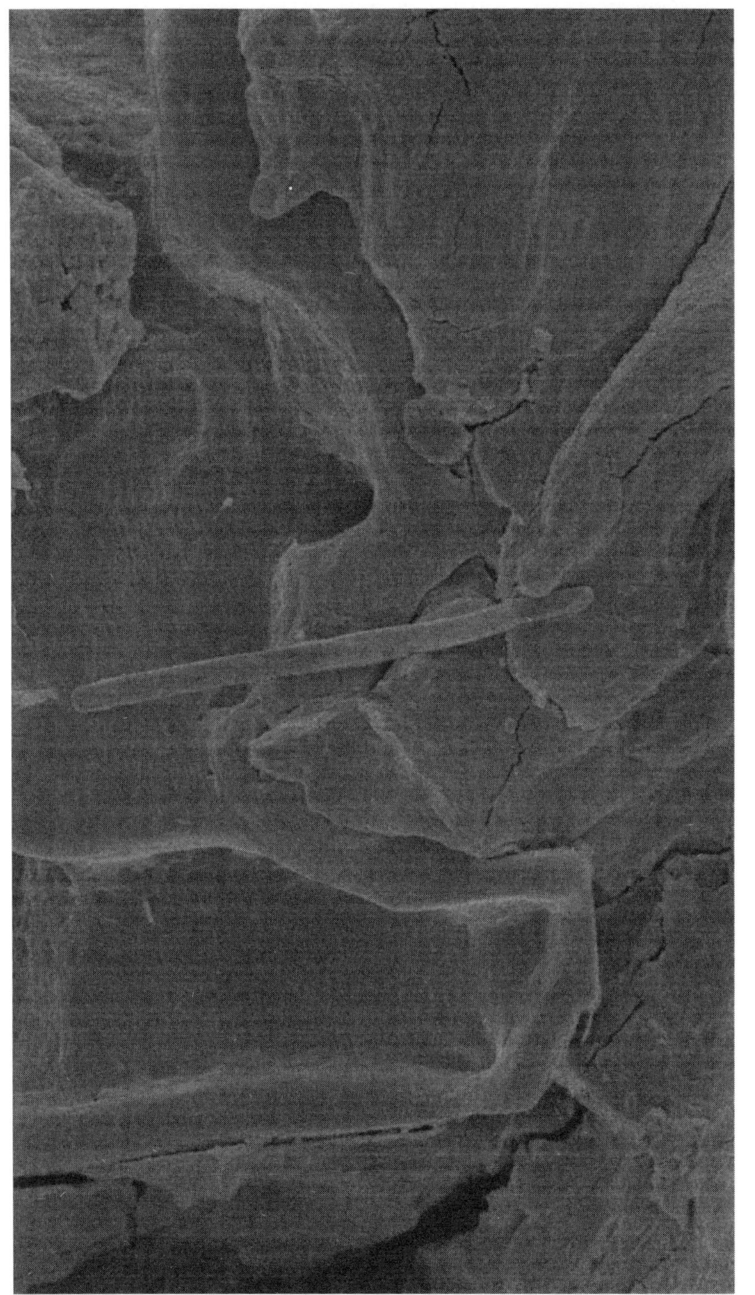

Figure 3. Bacterial cell form found on the freshly exposed face of an ancient salt crystal (Permian age, 250 ma). This form was found in another region of the crystal shown in Figure 2.

6. Acknowledgements

The authors wish to acknowledge the expert help of the people who helped to obtain the core samples from the Saldo Formation. These include the shaft crew chief Teddy Garcia, hoistmen David Avecedo, Stacy Britain, Edgar Keyser, Bill Kinsler, Bobby Modrall, Andy Morin, and Troy Padilla. Special thanks are also extended to Mr Dave Belski who spent the day handling the drill and to our honorary microbiologist Dr. Dennis Powers (Consulting Geologist) who led us through the formation. This work was supported by grant # EAR-9714203 from the US National Science Foundation LExEn Program.

7. References

Elazari-Volcani, B. (1943) Nature 152:274-275
Cano, R.J. and Borucki. M.K. (1995). Science. 268:1060-1064
Das, N., Horita, J., Holland, H.D. (1990). Geochim. Cosmochim. Acta 54:319-327
Denner, E.B.M., McGenity T.J.,. Busse H.J, Grant W.D., Wanner G. and Stan-Lotter. H. (1994). Int. J. Syst. Bacteriol. 44:774-780
Fredrickson, J.K. and Onstott, T.C. (1996) Sci. Amer. 275:68-73
Horita, J., Friedmann, T.J., Lazar, B., and Holland, H.D. (1991) Geochim. Cosmochim. Acta 55:417-432
Horita, J. Weinberg, A., Das, N. and Holland, H.D. (1996) J. Sediment. Res. 66:956-964
Javor, B. (1989) Hypersaline Envrionments, Springer-Verlag New York
Kennedy, M.J., S.L. Reader, and L.M. Swierczynski. (1994). Microbiology 140:2513-2529
Klebahn. (1919) Mitt. Inst. Allg. Bot. Hamburg. 4:11-69
Lazar,B. and Holland, H.D. (1988) Geochim. Cosmochim. Acta 52:485-490
Lochhead, A.G. (1934) Can. J. Res. 10:275-286
Morita, R.Y. 1990. Experentia 46:813-817
Norton, C. and Grant, W.D. (1988) J. Gen. Microbiol. 134:135-137
Norton, C.F., McGenity T.J., and Grant W.D.. (1993) J. Gen. Microbiol. 139:1077-1081.
Ollivier, B. Caumette, P., Garcia, J-L., and Mah, R.A. (1994) Microbiol. Rev. 58:27-38
Oren, A. (1983) Arch. Microbiol. 136:42-48
Oren, A. (1992) In. Balows, A., Truper, H.G., Dworkin, M., Harder, W. and Sclicfer, K.H. (eds) The Prokaryotes vol. 2 Springer Verlag, New York
Oren, A. (1992) Int. J. Salt Lake Res. 1:7-20
Post, F.J. (1977) Microbial Ecology. 3:143-165
Poulsen, V.A. (1879/1880) Vidensk. Medd, Dan. Naturhist. Foren. Khobenhavn pgs 231-254
Powers, D.W., and B.W. Hassinger. 1985. J. Sed. Petrol. 55:769-773.
Prescott, L.M., Harley, J.P. and Klein, D.A. (1996) Microbiology 3rd ed. Wm. C. Brown, Dubuque, IA.
Tindall, B.J., Mills, A.A. and W.D. Grant. (1980). J. Gen. Microbiol. 116:257-260
Tindall, B.J., Ross, H.N.M., Grant, W.D. (1984) Syst. Appl. Microbiol. 5:41-57
Vreeland, R.H. and D.W. Powers. (1998). In Oren A. (ed) Microbiology and Biogeochemistry of Hypersaline Environments CRC Press Boca Raton FL.
Vreeland, R.H., Piselli A.F., McDonnough S., and Myers S.S.. (1997) Submitted to Extremophiles (submitted).
Zharkov, M.A. (1981) In Yanshin. A.L. (ed) History of Paleozoic Salt Accumulation. Springer-Verlag, New York.

Biodata of **Ami Ben-Amotz** author of ***"Dunaliella β=Carotene from Science to Commerce."***

Dr. Ami Ben-Amotz is a Research Professor of Marine Biology at the National Institute of Oceanography, Israel Oceanographic & Limnological Research, Haifa, Israel, and a Coordinator of the Biotechnological *Dunaliella* Research, a joint activity between the Weizmann Institute of Science, Rehovot, Israel, NBT Ltd., Eilat, Israel, and Nikken Sohonsha Co., Gifu, Japan. Dr. Ben-Amotz obtained his Ph.D. degree in 1973 from the Department of Biochemistry, the Weizmann Institute of Science on the features of halotolerance in *Dunaliella* and since then his research activities are devoted to studies on the osmoregulation and β-carotene biosynthesis in *Dunaliella*.

E-mail: **<amiba@netvision.net.il> & <ami@ocean.org.il>**

J. Seckbach (ed.), Enigmatic Microorganisms and Life in Extreme Environments, 399–410.

DUNALIELLA β-CAROTENE
From Science to Commerce

AMI BEN-AMOTZ
The National Institute of Oceanography, Israel Oceanographic and Limnological Research
Tel-Shikmona, P.O. Box 8030, Haifa 31080, Israel

1. Introduction

The biflagellate unicellular alga *Dunaliella* thrives in media with extreme concentrations of salt in desert lands exposed to high solar radiation. *Dunaliella* produces and accumulates large amounts of intracellular glycerol to counter high extracellular osmotic concentration in the surrounding medium, and a high content of a mixture of β-carotene stereoisomers to defend the cell from the *damaging* effect of excessive radiation. Cultivation of autotrophic *Dunaliella* in large-scale open ponds composed of salt, inorganic nutrients and carbon dioxide in either extensive or intensive mode yields β-carotene rich algae ready for concentration by harvesting for use as high β-carotene algal powder or for extracted β-carotene in oil. The algal β-carotene differs from the synthetically available β-carotene in its stereoisomeric composition exhibiting equal amounts of *all-trans* and 9-*cis* isomers, differing in their physicochemical features and anti-oxidative activity. The high biotechnological production capability of natural β-carotene by *Dunaliella* and its unique stereoisomeric controlled composition has the potential to provide high quality β-carotene for a wide variety of marketing applications.

1.1. *DUNALIELLA*

1.1.1 *Biology and Halotolerance*

The biflagellated alga *Dunaliella* is classified under Chlorophyceae, Volvocales, which includes a variety of ill-defined marine and fresh water unicellular species (Avron and Ben-Amotz, 1992). *Dunaliella,* like *Chlamydomonas,* is characterized by an ovoid cell volume usually in the shape of a pear, wider at the basal side and narrow at the anterior flagella top. The cellular organization of *Dunaliella* is no different than that of other members of the Volvocales presenting one large chloroplast with single-centered starch surrounded by pyrenoid, a few vacuoles, a nucleus and a nucleolus. However, unlike other green algae, *Dunaliella* lacks a rigid polysaccharide cell wall, and is enclosed by a thin elastic plasma membrane covered by a mucous surface coat. The lack of a rigid cell wall permits rapid cell volume changes in response to extracellular changes in osmotic pressure. Commonly osmotically treated *Dunaliella* vary from round to eight shape. Under extreme salt concentrations greater than 4 M *Dunaliella* loses its flagella and the surrounding mucous; as the cell rounds with a buildup of a thick surrounding wall to form a dehydration resistant cyst.

Dunaliella occurs in a wide range of marine habitats such as oceans, brine lakes, salt marshes and salt water ditches near the sea, predominantly in water bodies containing more than 2M salt and high levels of magnesium. The effect of magnesium on the distribution of *Dunaliella* is not clear, but in many "bittern" habitats of marine salt producers, *Dunaliella* usually flourishes. The phenomenon of orange-red algal bloom in such marine environments is usually related to combined sequential growth of *Dunaliella* first, and then halophilic bacteria, as commonly observed in concentrated saline lakes such as the Dead Sea in Israel, the Pink Lake in Western Australia, the Great Salt Lake in Utah, U.S.A, and in many other places around the globe. *Dunaliella* is the most halotolerant eukaryotic organism known, showing a remarkable degree of adaptation to a variety of salt concentrations from as low as 0.1 M, to salt saturation.

Dunaliella uses a special mechanism of osmoregulation to adapt to the various salt concentrations in the surrounding media by varying the intracellular concentration of glycerol in response to the extracellular osmotic pressure. The intracellular concentration of glycerol is directly proportional to the extracellular salt concentration and is sufficient to account for all the required cytoplasmic and most of the chloroplastic osmotic pressures. Glycerol biosynthesis and elimination occur in the light or dark by a novel glycerol cycle which involves several specific enzymes (Avron, 1992).

1.1.2 *β-Carotene Production*

Among the many known strains of *Dunaliella*, only a few sub species of *salina* Teod. have been shown to produce and accumulate large amounts of β-carotene. In hypersaline lakes, which are generally low in available nitrogen and exposed to high solar radiation, these β-carotene producing strains of *Dunaliella* predominate over all other organisms to a seasonal bloom of about 0.1 mg β-carotene per liter. Under such stressful environmental conditions more than 12% of the algal dry weight is β-carotene, usually associated with a sharp decline in the thylakoids chlorophyll. The β-carotene in *Dunaliella* accumulates within distinctive oily globules in the interthylakoid spaces of the chloroplast periphery. Analysis of the globules showed that the β–carotene of *Dunaliella* is composed mainly of two stereoisomers: *all-trans* and 9-*cis*, with the rest a few other mono-*cis* and di-*cis* stereoisomers of the β-carotene. Both the amount of the accumulated β-carotene and the 9-*cis* to *all-trans* ratio depend on light intensity and on the algal division time which is determined by the growth conditions. Thus, any growth stress which will slow down the rate of cell division under light will in turn increase β-carotene production in *Dunaliella*. In fact, high light and many environmental stress conditions such as high salt, low temperature, extremes of pH, nutrient deficiencies, and others affect the content of β-carotene in *Dunaliella*. It was previously suggested that the equation of the amount of light absorbed by the cell during one division cycle integrates the effect of all growth ariables on the content and isomeric ratio of β-carotene in *Dunaliella* (Ben-Amotz and Avron 1990). The exceptions to this integration are nitrogen deficiency and low growth temperatures, both of which induce extreme intracellular accumulation of β-carotene under any light intensity. Taking into account that nitrogen starvation inhibits chlorophyll production in algae as part of its inhibitory effect on protein biosynthesis, the prolonged nitrogen independent biosynthesis of carotenoids will then protect the chlorophyll-reduced cells against the lethal damage of light. The effect of low temperatures on *Dunaliella* can be analyzed by measuring the specific stimulation effect of chilling on the biosynthesis of 9-*cis* β-carotene. The physicochemical properties of 9-*cis* β-carotene differ from those of *all-trans* β-carotene. *All-trans* β-carotene is practically insoluble in oil and is easily crystallized at low temperatures, while 9-*cis* β-carotene is much more soluble in hydrophobic-lipophilic solvents, very difficult to crystallize and generally oily in its concentrated form. To avoid cellular crystallization of *all-trans* β-carotene and to survive at low temperatures, *Dunaliella* produces a higher ratio of 9-*cis* to *all-trans* β-carotene, where the 9-*cis* stereoisomer functions *in vivo* as an oily matrix for the *all-trans* form(Ben-Amotz, 1996).

1.1.3 *Caroteneogenesis*

Although 272 geometric isomers of β-carotene can exist theoretically, 12 *cis* forms in total have been noted and recorded. Two *cis* forms, neo-β-carotene B and U, *all-trans* and 9-mono-*cis*, respectively were reported by Zechmeister (1962). Physico-chemical methods for stereomutation of β-carotene involve heat, light with no catalyst, and iodine catalysis under light yielding between one-third to one-half of the pigment in *cis* configuration. *Dunaliella* follows the same biosynthetic pathway of carotenoids with the same substrates and same intermediates as found in other eukaryotic organisms and plants. Mevalonic acid through isopentenyl diphosphate, geranyl diphosphate and geranylgeranyl diphosphate forms phytoene which undergoes a few desaturation steps and cyclization to β-carotene (Goodwin, 1988). The isomerization reaction, which produces 9-*cis* β-carotene, is not identified as yet. The observation of two stereoisomers of phytoene, 9-*cis* and *all-trans*, suggests that the biosynthesis of 9-*cis* β-carotene initiates probably early in the pathway of carotene biosynthesis at or before the formation of phytoene. Thereafter, all the intermediates occur in two isomeric forms. The induction of β-carotene in *Dunaliella* as described above was successfully applied to accumulate any choice of stereoisomeric intermediates in the caroteneogenesis pathway, yielding large quantity of the twin stereoisomers, *all-trans* and 9-*cis*: phytoene, phytofluene, neurosporene, ξ-carotene, β-zeacarotene, γ-carotene and β-carotene (Shaish et al., 1990, 1991). Subsequently, Ebenezer and Pattenden (1993), used ^{1}H and ^{13}C NMR on purified phytoene and phytofluene from *Dunaliella bardawil* extracts to verify the HPLC stereoisomeric analysis. Their NMR analysis showed 15-*cis* phytoene and 9-*cis* phytofluene as an indication that the phytoene is the branch point for the formation of 9-*cis* β-carotene. The contradiction between the HPLC and NMR analyses left the intriguing site of stereoisomerization open at any site between isopentenyl diphosphate to phytoene. Furthermore, the induced enzyme, enzymes or enzyme complex responsible for the high accumulation and isomerization of β-carotene in *Dunaliella* have not been identified as yet.

A few speculative hypotheses have been postulated for the function of the β-carotene globules in *Dunaliella*. One suggests that β-carotene accumulation is related to a protecting effect against chlorophyll catalyzed singlet oxygen production under high light radiation. However, the electron micrographs of *Dunaliella bardawil* (Ben-Amotz and Avron, 1990) which indicate that the massive accumulation of β-carotene occurs in globules located in the interthylakoid space of the chloroplast is in disagreement with this hypothesis. The large distance between the β-carotene globules and the thylakoid located chlorophyll would not allow efficient quenching of singlet oxygen, or any other chlorophyll generated product. Other hypotheses concerning the possible function of the globules such as carbon storage for use under limiting growth conditions were eliminated by the observation that β-carotene rich *Dunaliella* cannot utilize or metabolize the β-carotene in the dark nor under CO_2-deficient conditions. The most accepted hypothesis suggests that the β-carotene globules protect the cell against injury by high intensity radiation under limiting growth conditions by acting as a screen to absorb excess radiation. The harmful effect of the blue region of the spectrum is screened by the algal peripheral located globules thus preventing any cellular damage. Strains of *Dunaliella* and other algae unable to accumulate β-carotene die when exposed to high levels of radiation while the β-carotene-rich *Dunaliella* flourishes. Moreover, protection against photoinhibition by the massively accumulated β-carotene is observed only when the photoinhibitory light is composed of wavelengths absorbed by β-carotene, light in the blue region. No photoprotection at all is observed when red light, which is not absorbed by β-carotene, serves as the photoinhibitory agent. This is in agreement with the observation on the location of the β-carotene globules, distant from the thylakoid located chlorophyll, and with the above-mentioned hypothesis that the mode of action of the massively accumulated carotene is mostly a screening effect. It was shown previously (Ben-Amotz et al., 1989b) that a series of sequential events lead to the photodestruction of chlorophyll in *Dunaliella bardawil* . 9-*cis* β-carotene is destroyed first, then the *all-trans* β-carotene, and later the chlorophylls, indicating higher sensitivity of 9-*cis* β-carotene to direct elevated levels of light or to damaging chlorophyll generated free radicals.

The induction of β-carotene biosynthesis in *Dunaliella* is light dependent. Photosynthetic Active Radiation (PAR) is absorbed by the algal chlorophylls triggering the massive synthesis of β-carotene in *Dunaliella*. The question regarding the involvement of chlorophyll generated active oxygen species was studied kinetically by adding promoters of oxygen radicals and azide, an inhibitor of catalase and superoxide dismutase, to *Dunaliella* during a selective induction period of β-carotene (Shaish et al., 1993). The actual action of active oxygen species in triggering the biosynthesis of β-carotene remains open for further confirmation.

1.1.4 *Biotechnology of β-Carotene Production by Dunaliella*

Dunaliella is a most suitable organism for mass cultivation outdoors in open ponds. The ability to thrive in media with high sodium, magnesium, calcium and the respective anions, chloride and sulfate, in desert high solar irradiated land with access to brackish water or sea water at extreme temperatures from around -5° to above 40°C, make *Dunaliella* most attractive for the biotechnologists and venture capitalists. In fact, since 1980 several firms, government authorities, and industries have invested capital in the application of *Dunaliella* for the production of natural β-carotene.

Large-scale *Dunaliella* production is based on autotrophic growth in media containing inorganic nutrients with carbon dioxide as exclusive carbon sources. Attempts to commercially develop heterotrophic strains or mutants of *Dunaliella* for growth on glucose or acetate, such as *Chlorella* or *Chlamydomonas*, respectively, were not successful. Due to the demand for high light intensity for maximal β–carotene production beyond that required for normal growth, production facilities are located in areas where solar output is maximal and cloudiness is minimal. Most of the present *Dunaliella* production plants are located close to available sources of salt water, e.g., sea-salt industries usually in warm and sunny areas where the rate of water evaporation is high and non-agricultural land is abundant.

Four modes of cultivation have been used in large-scale production of *Dunaliella*. The first, termed extensive cultivation, uses no mixing and minimal control of the environment. To decrease attacks by zooplanktonic predators, such as certain types of ciliates, amoebae, or brine shrimp, the growers employ very high salt concentrations. *Dunaliella* grows slowly in shallow lagoons in nearly saturated brine and predators are largely eliminated. The naturally selected strain of *Dunaliella* is well adapted to nearly salt saturation conditions, partially loses its flagella and produces a thick cell wall on the transformation to a cyst form. Extensive cultivation productivity is low, and the area needed for commercial production is very large, however, the low operating costs of such facilities has led to the development of two commercial plants in Australia and China. The second, termed intensive cultivation, uses high biotechnology to control all factors affecting cell growth and chemistry. The ponds are usually oblong lined constructed raceways varying in size up to a production surface area of approximately 3000m^2. The use of long arm, slow revolution paddle wheels is presently common in the large-scale facilities in Israel, USA, China, Chile, and Portugal. One production-sized shallow water pond of 20 cm on an area of 3000m^2 (600m^3) containing 15g β-carotene/m^3 yields 9 kg β-carotene on total harvest. The current large scale production of β-carotene under intensive cultivation is around 200 mg β-carotene/m^2/day on a yearly average, thus, a modern intensive plant of 50,000m^2 produces 10 kg β-carotene/day. Between the extensive and intensive modes there are examples in Australia and China of third type, semi-intensive mode, where the ponds are enlarged ten times, to about 50,000m^2 each with partial control and no mixing. The fourth is highly intensive cultivation in closed photobioreactors. Many trials have been initiated in the last decade to grow *Dunaliella* in different models of closed photobioreactors with attempts to design the best sunlight harvesting unit for β-carotene optimization. The different designs include, narrow very long plastic tubes, plastic bags, trays and more. However, as of today, none of these trials have taken production beyond the laboratory or small pilot plant volume, mainly due to economic limitations and non-feasible large scale development. The few industrial ventures of high intensive-closed photobioreactors became insolvent and no longer exist.

Generally, large-scale optimization of β-carotene production is achieved in all modes of cultivation by high salt stress and by nitrogen deficiency. For pragmatic reasons, the first is applied in the extensive mode, while nitrogen starvation controls the intensive mode. Most species of *Dunaliella* grow optimally in a medium containing 1-2 M NaCl in accordance with the medium temperature, exhibiting closely similar growth rates at moderate temperatures of 25°C in 2M NaCl, and at low temperatures of 15°C in 1M NaCl. The algal composition changes respectively by selective accumulation of glycerol and starch. This unique environmental adaptation of *Dunaliella* allows successful intensive outdoor growth in cold seasons and in cold areas such as the Ukraine and Inner Mongolia. Most commercial *Dunaliella* ponds employ evaporated concentrated seawater, or seawater augmented with dry salt to reach the desired concentration in the medium. Favored sites for *Dunaliella* cultivation are along the sea shore or close to salt lagoons and salt producing industries for use of a mixture of seawater and concentrated seawater in order to obtain the desired salt concentration by season and temperature. The one exception is the Nutrilite Division of Amway, Corporation, located far from the sea in Imperial Valley, CA, USA, which employs freshwater augmented with commercial salt. The use of recycled high salt medium is common in a few plants after harvesting of the algae by oxidative treatment of the organic load and filtration. The use of recycled medium enriches the medium with higher concentrations of magnesium, calcium and sulfate. *Dunaliella* was found to grow well in seawater based media containing around 1.5M NaCl, more than 0.4 M $MgSO_4$ and 0.1 M $CaCl_2$ under strict pH control.

Dunaliella adapts to a broad range of pH, from 5.8 to 11. In natural fresh water or seawater several calcium salts, most notably carbonates, sulfates and phosphates tend to precipitate causing algal flocculation and a reduction in growth rate. It is, therefore, important to maintain the pH in the *Dunaliella* culture within the neutral range (pH ~7) and to avoid an increase of pH above the salting out range of the divalent cation salts. In autotrophic algal cultures the pH rises mostly due to photosynthesis. Uncontrolled autotrophic algal cultures show an increase in pH during the day which can exceed pH 10. Intensive algal growers use pure liquid CO_2 and commercial pH controllers to maintain the pH at a preset level. The CO_2 bubbles diffuse through commercial gas diffusers from outlets in the bottom of the pond, dissolve to carbonic acid, lower the pH, and then equilibrate to bicarbonate and carbonate as a function of the hydrogen-ion concentration in the medium and finally may be used in photosynthesis. CO_2 consumption in commercial algal plants is higher in the summer and minimal in the winter due to the lower solubility of the carbon dioxide gas in the medium at elevated temperatures varying from about 6 g CO_2/g dry *Dunaliella* to 3 g CO_2/g, respectively. Today, large-scale intensive algal cultivation employs modern computerized pH control of many ponds simultaneously by pH sensors in each pond transmitting amplified outputs to a central control room. Supply of CO_2 is monitored and controlled continuously by the central pH computer, thus allowing large-scale intensive cultivation. Alternatively, HCl and bicarbonate can be used to maintain the required pH and photosynthesis in the pond. Bicarbonate may avoid the above-mentioned diffusion-solubility and may replace CO_2, however, the market cost of bicarbonate together with HCl and the safety issue of the latter are prohibiting factors in industrial operation.

1.1.5 *Extraction and Concentration*

Separation and harvesting of *Dunaliella* out of the growth medium from around 0.1% biomass to a concentrated algal paste of more than 15% should take into consideration the specific features of the alga: (1) lack of cell wall and enclosure of the cell by fragile cell membrane; (2) an outer cell layer, termed a surface coat which possesses mucus properties; (3) elasticity of the cell shape; and (4) high extracellular salinity. Most of the previous approaches to harvest *Dunaliella* by different modes of filtration were not successful and have not reached the level of scale up, suffering mainly from the presence of glutinous surface coat and from the high elasticity of the cell membrane that block permeation through many filters and screens. Industrial extensive producers of *Dunaliella* are presently using flocculation and cell surface absorption while the intensive growers employ centrifugation and flocculation. Flocculation combined with sedimentation or flotation is being used extensively to harvest *Dunaliella*. The unicellular algae flocculate on the addition of inorganic or organic flocculants such as aluminum sulfate (alum), ferrous, ferric chloride, ferric sulfate, lime, polyelectrolytes, guar, polysaccharides, etc. Alum is extensively used to flocculate *Dunaliella* in high salt media at variable concentrations in relation to pH, temperature, dissolved organic load, bacteria and mineral concentration. Flocculated *Dunaliella* cannot be used directly for the food market unless the flocculant is safe or is completely released from the algae prior to utilization. Different plants extract the β-carotene from the flocculated *Dunaliella* leaving behind the aluminum contaminated cell debris. Centrifugation has been widely used for harvesting *Dunaliella* and other unicellular algae. The most common design is continuous-flow, stainless steel, batch-type centrifuge, or de-sludge-type centrifuge (self discharging), operating at large scale at about 1 m^3/hr, at 15,000 rpm, or 15 m^3/hr, at 5,000 rpm, respectively. Initial investment is high and operation energy is costly, but the high efficiency and the clean product attract many industries to employ such centrifuges. The discharged paste varies from 40% algal dry weight in the batch-type, down to 20% in the de-sludge type. Another method of harvesting *Dunaliella,* hydrophobic binding, is used by Betatene Ltd., Australia. *Dunaliella* grown at salinity higher than 4M converts to cysts and acquires the above-mentioned hydrophobic surface coat. These coated cells absorb to hydrophobic surfaces and are removed out of the growth suspension to yield algal paste ready for successive β-carotene extraction. The method is most suitable for extensive cultivation where the concentration of *Dunaliella* is below 0.1 g/L and the volume for daily harvest exceeds 20,000m^3.

β-Carotene can be extracted directly from the wet *Dunaliella* paste and this is the process employed in several production facilities by different extraction processes using edible oil with or without organic solvents, liquid carbon dioxide extraction, crystallization, and other common food technology methods. The need for dry *Dunaliella* powder as a product for the health food market has led to the development of several techniques for processing and dehydrating *Dunaliella* paste while protecting its oxygen-sensitive stereoisomers of β-carotene. Several methods have been employed to dry microalgae such as *Chlorella*, *Scenedesmus*, and *Spirulina* (Ben-Amotz and Avron, 1989), and all are based on conventional food technology for drying different food products such as milk powder, egg powder, tomato powder, baby food powder and coffee. Most common are spray dryers that work with input of the paste at high temperature (>150°C) for a short holding time, and on output temperature below 100°C yielding dry algal powder of low moisture content (<5%). The higher sensitivity of 9-*cis* β-carotene to oxidation and to thermal destruction may lead to rapid loss of total β-carotene in association with a reduction in the 9-*cis* to *all-trans* β-carotene ratio during the drying process. Other techniques of drying like drum drying, freeze drying or sun drying have been tested but not employed with *Dunaliella* due to the lack of product uniformity and the subsequent need to grind the dry material. Similar problems are encountered on drying different food and feed products, but the high sensitivity of *Dunaliella* and β-carotene to oxidation enforces a specific technological approach.

1.1.6 *Natural Versus Synthetic β-Carotene*

β-Carotene has been used for many years as a food coloring agent, as pro-vitamin A (retinol) in food and animal feed, as an additive to cosmetics, multivitamin preparations, and in the last decade as a health food product under the "antioxidant" claim. Many epidemiological and oncological studies suggest that humans fed on a diet high in carotenoid-rich vegetables and fruits, who maintain higher than average levels of serum carotenoids, have a lower incidence of several types of cancer and cardiovascular disease (Krinsky, 1989). Recently, the Alpha-Tocopherol, β-Carotene (ATBC) Cancer Prevention Study Group, and the β-Carotene and Retinol Efficacy Trial (CARET) clearly showed that not only did β-carotene fail to reduce the incidence of cancer and cardiovascular disease, but in fact it increased it in smokers and workers exposed to asbestos (Omenn et al., 1996; Hennekens et al., 1996). These two USA National Cancer Institute supported studies and numerous earlier trials which dealt with the question of the protective role of β-carotene against chronic diseases, used only synthetically formed *all-trans* β-carotene.

Despite the fact that the most convincing reports support a direct connection between high intake of fruits and vegetables and low incidence of cancer and cardiovascular disease, the literature lacks specific information on the possible medical contribution of natural carotenoids, isomers of carotenoids, and carotenoid fatty acid esters. Experimental nutrition and medical studies with natural carotenoids originating from different plants, fruits, vegetables and algae, have been very limited and such research is in its infancy.

β-Carotene is present in most plants and algae in small amounts of ~0.2% of the dry weight with approximately 1/3 as 9-*cis* β-carotene. The observations of the high β-carotene content of *Dunaliella* containing more than 50% 9-*cis* β-carotene gave impetus to studies on the metabolism, storage and function of 9-*cis* β-carotene in animals and humans with emphasis on the possible role of the 9-*cis* stereoisomer in scavenging reactive oxygen species. The first dietary studies showed that low doses of the algal β-carotene are as potent as doses of synthetic *all-trans* β-carotene in providing retinol in rats and chicks (Ben-Amotz et al., 1989a). Later, preferential and selective uptake of either *all-trans* or 9-*cis* β-carotene was noted in different animals and humans. 9-*cis* β-Carotene is not detected in serum of chicks and rats nor in humans fed a diet rich in *Dunaliella* (Ben-Amotz and Levy, 1996) or oil extract of *Dunaliella* (Stahl et al., 1993, Gaziano et al., 1995). The β-carotene detected in any of these sera was *all-trans* β-carotene. Rats and chicks that generally do not accumulate β-carotene in their tissues showed uptake of β-carotene into the tissues when the diet was supplemented with β-carotene-rich *Dunaliella.* The lack of 9-*cis* β-carotene in mammals serum and the different response to CNS toxicity (Bitterman and Ben-Amotz, 1994) and to *in vitro* oxidation (Levin and Mokady, 1994) led to the hypothesis that the isomeric structure of 9-*cis* β-carotene acts as a quencher of singlet oxygen and other free radicals serving as effective *in vivo* antioxidant. Generally, carotenoids exert an antioxidative effect by a mechanism that results in the formation of new products that are more stable and more polar. Different carotenoids exhibit different antioxidative-anti peroxidative capacities (Krinsky, 1989). Opening of the β-ionone ring, the addition of a chemical group on the ring, or replacement of the ring by various groups can modify the antioxidative activity. Therefore, structural variables besides the length of the polyene chain may direct the scavenging properties of the carotenoids. Insertion of any *cis* position along the conjugated double bond chain may lead to modified antioxidative properties. Assuming that the *cis* conformational change leads to a higher steric interference between the two parts of the carotene molecule, the *cis* polyenic chain will be less stable and more susceptible to low oxygen tension in *in vivo* oxidation. Recent studies strongly supported this hypothesis by measuring different antioxidative activity (Levy et. al., 1995, 1996) and much lower concentration of low-ultraviolet dienes in the serum of humans supplemented with β-carotene rich-*Dunaliella* compared with *all-trans* β-carotene (Ben-Amotz and Levy, 1996). However, the possibilities of low intestinal absorption, isomerization or tissues uptake have to be considered as well (Johnson et al., 1996). The possibility of synergistic effects and the possible beneficial potency of the different plant nutrients and carotenoids are still obscure and warrant further research.

1.1.7 *Dunaliella Market Products*

Dunaliella natural β-carotene is widely distributed today in many different markets under two different categories, β-carotene extracts, and dry *Dunaliella.* Extracted purified β-carotene is sold mostly in vegetable oil in bulk concentrations from 1% to 20%, or for personal use in soft gels, 5 mg β-carotene each gel. Due to the lack of fine purification during extraction, the β-carotene is generally accompanied by the other carotenoids of *Dunaliella,* predominantly: lutein, neoxanthin, zeaxanthin, violaxanthin, cryptoxanthin, α-carotene, each with its different stereoisomers, comprising approximately 15% of the β-carotene concentration. Under the same category of extracts, one can find granular powder of microencapsulated β-carotene (2-7.5%), and crystals of β-carotene packed under vacuum or nitrogen atmosphere. A variety of such formulations of β-carotene are presently found and distributed in the world under the market sections, "health food" or "food supplement."

The second category covers a line of dried *Dunaliella* powders, intact algae, harvested and dried as described above for marketing in the form of tablets or hard capsules containing between 3 to 20 mg β-carotene per unit. To avoid β-carotene oxidation the tablets are coated with sugar and the capsules are packed separately in aluminum/ polyethylene blisters. Dry *Dunaliella* is distributed popularly in Japan and in other Far East countries where the consumers are more familiar with edible algae and accustomed to *Chlorella* and *Spirulina*.

1.1.8 *Commercial Producers*

At present, the following companies are actively engaged in cultivating *Dunaliella* for commercial purposes. These are and their mode of cultivation is:

1. Betatene Ltd., Cheltenham, Victoria 3192, Australia, a division of Henkel Co., Germany. Extensive mode.
2. Cyanotech Corp., Kailua-Kona, HI 96740, USA. Intensive mode.
3. Inner Mongolia Biological Eng. Co., Inner Mongolia, 750333, P. R. China. Semi-intensive mode.
4. Nature Beta Technologies (NBT) Ltd., Eilat 88106, Israel, a subsidiary of Nikken Sohonsha Co., Gifu, Japan. Intensive mode.
5. Nutrilite, Calipatria, CA 92233, USA, a division of Amway Corporation, USA. Intensive mode.
6. Photo Bioreactors, PBL Ltd., Murcia 30003, Spain. Highly intensive mode, closed photobioreactors
7. Tianjin Lantai Biotechnology, Inc. Nankai, Tianjin, in collaboration with the Salt Scientific Research Institute of Light Industry Ministry, P. R. China. Intensive mode.
8. Western Biotechnology Ltd. Bayswater, W. A. 6053, Australia, a subsidiary of Coogee Chemicals Pty. Ltd., Australia. Semi-intensive mode.

2. References

Avron, M. (19920 In: M. Avron and A. Ben-Amotz (eds.) *Dunaliella: Physiology, Biochemistry, and Biotechnology*, pp.135-164, Boca Raton: CRC Press.

Avron, M., and Ben-Amotz, A. (eds.), (1992) *Dunaliella: Physiology, Biochemistry, and Biotechnology*, Boca Raton: CRC Press, pp. 240.

Ben-Amotz, A. (1996) *J. Phycol.*, **32**, 272-75.

Ben-Amotz, A., and Avron, M. (1990) *Trends Biotechnol.*, **8**, 121-26.

Ben-Amotz, A., and Levy, I. (1996) *Am. J. Clin. Nutr.*, **63**, 729-34.

Ben-Amotz, A., Mokady, S., Edelstein, S., and Avron, M. (1989a) *J. Nutr.*, **119**, 1013-19.

Ben-Amotz, A., Shaish, A., and Avron, M.(1989b) *Plant Physiol.*, **91**, 1040-43.

Bitterman, N., and Ben-Amotz, A. (1994) *J. Appl. Physiol.*, **76**, 1073-76.

Ebenezer, W. J., and Pattenden, G. (1993) *J.Chem. Soc. Perkin Trans.*, **1**, 1869-1873.

Gaziano, J. M., Johnson, E. J., Russel, E. M., Manson, J.E., Stampfer, M.J., Ridker, P.M., Frei, P.M., Hennekens, C.H. and Krinsky, N.I. (1995) *Am. J. Clin. Nutr.*, **61**, 1248-52.

Goodwin, T. W. (ed.) (1988) *Plant Pigments* London: Academic Press, pp. 362.

Hennekens, C. H., Buring, J.E., Manson, J.E. Stampfer, M.J., Rosner, B., Cook, N.R., Belanger, C., LaMomtte, F., Gaziano, J.M., Willett, W. and Peto, R. (1996) *N. Engl. J. Med.*, **334**, 1145-49.

Johnson, J. E., Krinsky, N. I. and Russell, R. M., (1996) *Circulation*, **15**, 620-4.

Krinsky, N. I. (1989) *Free Radic. Bio. Med.*, **7**, 617-35.

Levin, G., and Mokady, S. (1994) *Free Radic. Biol. Med.*, **17**, 77-82.

Levy, Y., Ben-Amotz, A., and Aviram, M. (1995) *J. Nutr. Environm. Med.*, **5**, 13-22.

Levy, Y., Kaplan, M., Ben-Amotz, A., and Aviram, M. (1996) *Isr. J. Med. Sci.* **32**, 473-78.

Omenn, G. S., Goodman, G.E., Thornquist, M.D., Balmes, J., Cullen, M.R., Galss, A., Keogh, J.P., Meyskens, F.R., Malanis, B., Williams, J.H., Barhart, S., Cherniack, M.G., Brodkin, C.A. and Hammar, S. (1996) *N. Engl. J. Med.*, **334**, 1150-55.

Shaish, A., Avron, M., and Ben-Amotz, A. (1990) *Plant Cell Physiol.*, **31**, 689-96.

Shaish, A., Avron, M., Pick, U., and Ben-Amotz, A. (1993) *Planta*, **190**, 363-68.

Shaish, A., Ben-Amotz, A., and Avron, M. (1991) *J. Phycol.*, **27**, 652-56.

Stahl, W., Schwartz, W., and Sies, H. (1993) *J. Nutr.*, **123**, 847-51.

Zechmeister, L., 1962, *cis-trans Isomeric Carotenoids, Vitamin A, and Arylpolyenes,* Vienna: Springer-Verlag

Biodata of Dr. **Richard F. Shand** author (with Abril Perez) of ***"Haloarchaeal Growth Physiology."***

Dr. Richard F. Shand received his Ph.D. in Microbiology from the University of California at Davis in 1986. After a post-doctoral fellowship at the University of California at San Francisco he joined the faculty of Northern Arizona University (in 1990) where he is an Associate Professor of Microbiology in the Department of Biological Sciences. His research interest include elucidating the mechanisms of stationary phase gene regulation in extreme halophiles using halocins as a model system, and recovering extreme halophiles from ancient halite deposits.

E-Mail: **richard.shand@nau.edu**

Biodata of **Abril M. Perez** co-author with Dr. Richard F. Shand of: ***"Haloarchaeal Growth Psysiology""***.

Abril M. Perez received her B.S. in Biology from Northern Arizona University in 1996. She is currently a graduate student with R.F. Shand and is supported through the National Institute of Health's Minority Biomedical Research Support program. She hopes to use the techniques she is developing for halocins to pursue a career dedicated to the discovery of novel plant extracts for the treatment of disease. Her other interestes include botany, dancing, reading, and making life that much more interesting for her family, friends, and colleagues.

J. Seckbach (ed.), Enigmatic Microorganisms and Life in Extreme Environments, 411–424.

HALOARCHAEAL GROWTH PHYSIOLOGY

RICHARD F. SHAND AND ABRIL M. PEREZ
Department of Biological Sciences
Northern Arizona University
Flagstaff, AZ 86011 USA

1. Introduction

Of all of the archaea, the extreme halophiles (the "haloarchaea") are the easiest to culture. Most can be grown in a buffered, complex medium containing peptone, yeast extract and/or tryptone, a few salts, and between 2 and 4.5 Molar NaCl (DasSarma et al., 1995). All are aerobic (so there is no need for anaerobic chambers, as with the methanogens) and temperature requirements are usually between 37° and 42°C, (so there is no need for specialized equipment and procedures for growth at high temperatures, as with the hyperthermophiles). Moreover, many can be grown in a defined medium (DasSarma et al., 1995). Consequently, one can take advantage of all of the growth physiology paradigms developed over the years for members of the domain bacteria. One of these paradigms is the concept of steady-state, balanced growth as described by Campbell (1957). Balanced growth is a physiological state whereby "...all cell constituents begin to increase by the same proportion over the same interval. In other words, the time it takes to double the number of cells will be the same time it takes for these cells to double their content of DNA, ribosomes, individual enzymes and so on." (Neidhardt et al., 1990). Since balanced growth is the most reproducible physiological state of a bacterial culture, experiments performed in different laboratories can be compared directly. We hasten to add that balanced growth is a mostly artificial state, rarely existing for any period of time in the environment. Nevertheless, since most physiology studies are done in the laboratory, cultures should be placed in balanced growth before an experiment begins. This provides a reproducible starting point, even if the aim of the experiment is to follow a particular parameter throughout the growth curve where there are periods of unbalanced growth (e.g., observing the expression of a particular gene as the culture enters stationary phase).

Unfortunately, the importance of understanding the physiological state of a bacterial culture and how it relates to various processes in the cell (e.g., gene regulation) has become lost in a blizzard of studies at the molecular level. The frustration with this state of affairs has been articulated by Neidhardt et al. (1990):

> Reference is often made in the literature to cultures harvested in "early" or "late" log phase, [or] even in "mid-log"; such arbitrary statements are disturbing because they give the impression that no great care was taken to establish balanced growth conditions. There is no assurance that the growth rate was not changing at the time of sampling. Unless growth is monitored throughout a physiological experiment, the results may not be reproducible, whether in your own or in somebody else's laboratory. In fact, a physiological experiment done with a poorly characterized culture is *all but useless.*

To facilitate duplicating the physiological state of a culture, Neidhardt and Umbarger (1996) provide a list of six parameters that should be specified: (i) identification of the organism by strain and source, (ii) a description of the culture medium, (iii) a description of how the culture was aerated (including the ratio of culture volume to flask size, and type of closure), (iv) the temperature, (v) the growth phase (or the physiological state of the culture at the beginning of the experiment and how it was achieved; e.g., "balanced exponential growth, achieved by serial subcultivation through five mass doublings"), and (vi) the growth rate.

A practical approach to growing haloarchaea has been provided by Rodriguez-Valera (1995) in which suggestions for optimizing growth conditions and for avoiding various problems when working with media containing very high NaCl concentrations have been included. The following sections expand upon some of those suggestions, add new suggestions, and provide examples of the usefulness of understanding haloarchaeal growth physiology.

2. Practical Considerations for Growing Haloarchaea in Batch Cultures

2.1. BALANCED GROWTH IN PRACTICE

The most effective way for generating cultures in balanced (steady-state) growth is to use a chemostat or turbidostat, the mathematics of which are well understood (e.g., see Neidhardt et al., 1990). Extreme halophiles have an advantage in these systems as contamination is rarely a problem, but the high salt levels can be hard on some of the probes and on the components made of stainless steel. In practice, balanced growth is often approximated by repetitive subculturing from exponentially growing batch cultures. Many haloarchaea demonstrate growth kinetics very similar to those for *Escherichia coli* and *Salmonella*: repetitive subculturing produces cultures with no lag phase, and which grow exponentially to optical densities of about 1 before they become unbalanced (e.g., *Haloferax mediterranei*; Cheung et al., 1997). However, other haloarchaea depart from exponential growth and begin growing slower at much lower cell densities (optical densities of about 0.1; see Figures 3 and 4). In order to approximate steady-state conditions in batch for organisms with these types of growth kinetics, the culture must be subcultured at low optical densities before it becomes unbalanced. The only way to determine at what optical density a particular haloarchaeal culture departs from its initial growth rate is to generate growth curves in which samples are taken very closely together.

2.2. OXYGENATION

One of the most important aspects of obtaining maximal growth rates for haloarchaea involves oxygenation. Oxygen solubility depends on salt concentration, total pressure, organics concentration and temperature (oxygen concentration decreases as temperature increases), and can vary significantly, depending upon the nature of the solution: 21.9 mg/L in 50% methanol-water, 9.2 mg/L in water, and 2 mg/L in 4 M KCl (Gary, 1989). The solubility of oxygen is decreased further in complex haloarchaeal media due to the presence of amino acids and peptides, and to the elevated incubation temperatures.

The most common way to grow haloarchaeal cultures for physiological studies is in batch using Erlenmeyer flasks in orbital shakers. Erlenmeyer flasks that have baffles at the bottom and that are closed with Morton-type closures provide superior aeration over non-baffled Erlenmeyer flasks closed with cotton plugs. The baffles in the flasks are so efficient at increasing the surface to air ratio that shaker speeds can be reduced from 25 to 50%. The increased efficiency in aeration allows cultures to be grown at higher temperatures, and the reduction in shaker speed generates significantly less foaming by media components such as peptone, yeast extract and tryptone (see below).

2.2.1. *The Effects of Different Forms of Oxygenation may be Subtle*

Two batch cultures of haloarchaea may appear to exhibit identical growth kinetics, even though they are being oxygenated in different ways. An example of how oxygenation levels can affect a physiologically dependent parameter is shown in Figure 1. Halocins are protein antibiotics that are excreted into the environment by the producer strain and kill or inhibit other halobacterial cells. Several liters of culture supernatant are routinely prepared for purification and characterization of various halocins. Figure 1 correlates growth and halocin activity in two cultures of *Halobacterium* species GN101. One culture was grown in a 4 L baffled Erlenmeyer flask containing 2 L of media, while the second parallel culture was grown in a 2.8 L unbaffled Fernbach flask containing 1 L of media. The cultures were grown side-by-side at 37.5°C in a large, orbital air shaker at 200 rpm. Throughout the experiment, the two cultures grew in parallel. However, despite the apparent similarities in growth, there was a significant effect on the kinetics of halocin activity. Onset of halocin activity in the baffled flask was rapid as the culture entered stationary phase, with activity reaching a maximum value and then leveling off. These kinetics were identical to those in much smaller batch cultures grown in shaking water baths. In contrast, halocin kinetics in the unbaffled Fernbach flask were greatly slowed, but halocin activity eventually reached the same maximal level. In this case, the growth curve does not reveal the subtlety in the oxygenation levels between the two cultures. Note that oxygen levels may not be affecting halocin kinetics directly; other alternatives such as differences in energy production may be the cause.

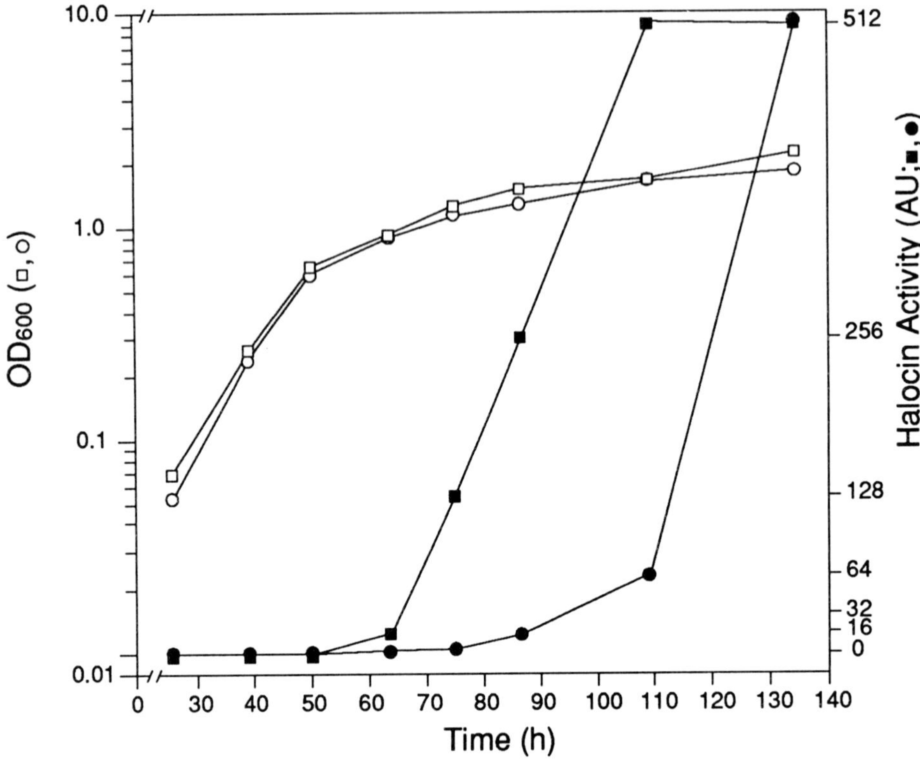

Figure 1. Growth curves (open symbols) and halocin activity (closed symbols) for *Halobacterium* species GN101. Cultures were grown in a complex medium (Rdest and Sturm, 1987) in either a baffled flask (squares; g = 6 h) or in an unbaffled Fernbach flask (circles; g = 5.5 h). Halocin activity was quantified by serial two-fold dilutions to extinction and reported as arbitrary units (AU).

2.3. FOAMING

Complex media containing nutrients such as peptone, tryptone, and yeast extract are prone to foaming when shaken rapidly. In contrast, foaming is greatly reduced in defined media, even if heavily supplemented with amino acids and nucleosides (see Table 1). Although using baffled flasks at reduced shaker speeds helps to reduce foaming, it does not always eliminate it. Foaming in batch cultures presents a variety of problems. First, foam is a barrier to oxygen transfer to the culture. Second, "...cells partition at the air/water interface with a higher concentration of cells in the foam ... than in the liquid." (Koch, 1997). Third, foaming often results in air denaturation of proteins, especially those that are secreted into the growth medium. Silicon-based antifoams such as Antifoam A and its derivatives precipitate at high salt concentrations, can clog filters and can generate erroneous spectrophotometric measurements. Antifoam 204 (Sigma, St. Louis, MO) is a non-silicon defoamer and is very effective in haloarchaeal media at very high dilutions (1:10,000). As with any additive, parallel cultures should be grown with and without antifoam prior to the experiment to determine if there is any effect on the growth of the cells or the parameter(s) one is monitoring.

2.4. BUFFERS AND HUMIDITY

Another area of concern is the buffering capacity of the medium. For neutrophilic haloarchaea, growth media is often titrated to pH 7.0-7.2 using Tris-HCl (tris[Hydroxymethyl]aminomethane). However, the useful pH range of this buffer is 7.0 to 9.0. Growth of neutrophilic haloarchaea in the presence of carbohydrates can result in the production of large amounts of acid. Since Tris is being used at the lower end of its useful range, MOPS (3[N-Morpholino]propanesulfonic acid), with a useful range of 6.5 to 7.9, is a better buffer for cultures whose pH optima is about 7.0, although it is more expensive than Tris. In any case, buffers should be selected so that the pH at which the culture is to be grown is near the buffer's pKa and in the middle of the buffer's useful range.

Finally, haloarchaeal growth curves often take from several days to weeks to complete, depending upon the growth conditions. Consequently, wherever possible, batch cultures in baffled Erlenmeyer flasks should be grown in covered, shaking water baths (not in air shakers) so that the atmosphere is humidified and the amount of evaporation from the culture minimized.

2.5. EFFECTIVENESS OF THE ABOVE CULTURING CONDITIONS AND THE ARRHENIUS PLOT

A practical demonstration of the effect of the above procedures on haloarchaeal physiology is shown in Figure 2. This plot relates k, the

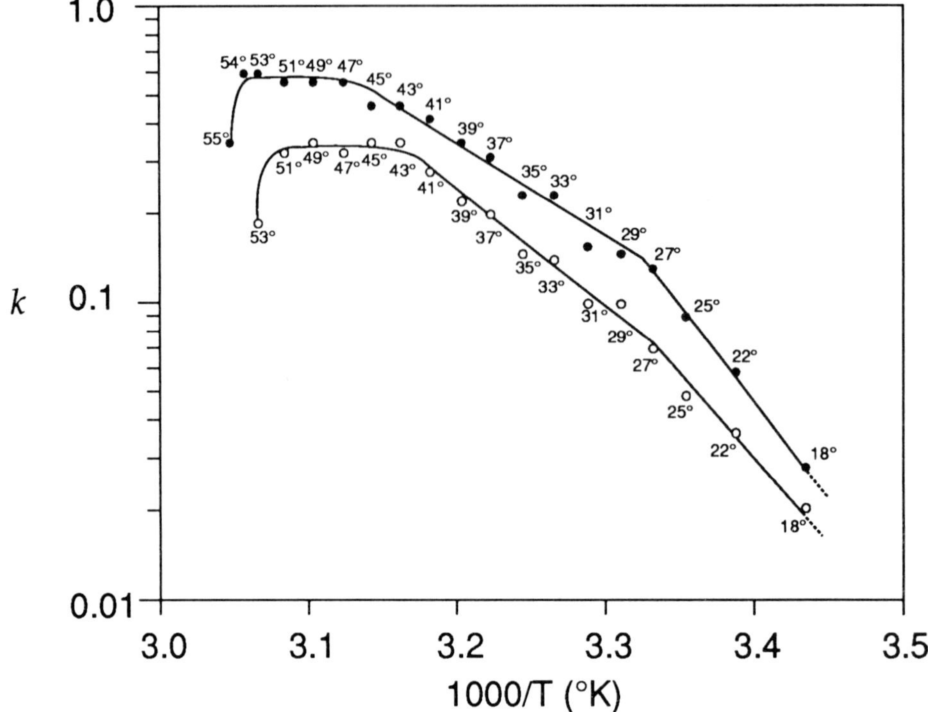

Figure 2. An Arrhenius plot for *Hf. mediterranei* R4. Complex medium (closed circles) was ATCC medium 1176 (American Type Culture Collection, 1989) and defined medium (open circles) was minimal medium #1 for *Hf. mediterranei* (DasSarma et al., 1995). Cultures (25 ml) were grown in 125 ml baffled Erlenmeyer flasks as described in the text.

specific growth rate constant (plotted logarithmically) to reciprocal temperature (in °K) for *Haloferax mediterranei* R4 (ATCC 33500). k (h^{-1}) is related to generation time (g, in h) by

$$k = \frac{0.693}{g}. \tag{1}$$

This type of plot, called an "Arrhenius plot", is named after the 19th century Swedish chemist Arrhenius, who discovered how temperature affected the rate of chemical reactions. The form of the plot is based on the following equation

$$\ln v = (-\Delta E^*/R)(1/T) + S \tag{2}$$

where v is the velocity of the reaction, S is a constant, ΔE^* is the activation energy of the reaction, R is the gas constant, and T is the temperature in °K (for a detailed description, see Neidhardt et al., 1990, and Herenden et al., 1979). In an analogous manner, bacterial physiologists use Arrhenius plots to show how temperature affects the rate of bacterial growth. When *Hf. mediterranei* R4 was initially characterized (Rodriguez-Valera et al., 1983), the investigators determined that the maximum and minimum temperatures for growth were 45°C and 25°C, respectively, with an optimum at 35°C (Rodriguez-Valera et al., 1983). The procedure used to establish these temperatures involved following growth "...in screw-capped spectrophotometer tubes (outside diameter 13 mm) containing 3 ml of medium and shaken laterally in a near-horizontal position at 100 strokes per min" (Rodriguez-Valera et al., 1983). With respect to growth rates, the authors reported that, "In both media the growth rate is very high, approximately double the highest growth rate described for other *Halobacterium* species..." (Rodriguez-Valera et al., 1983). From the growth curves provided in the manuscript, we calculated the generation times to be approximately 3.25 h in yeast extract-glucose medium and 4.25 h in glucose minimal medium (at 38°C). In comparison, the data in Figure 2 show that the maximal growth temperature for *Hf. mediterranei* R4 is 55°C in complex medium (10°C higher than previously reported) and 53°C in defined medium. Note that the fall off from the "optimal temperature plateau" in both media is very steep, and that this organism does not simply grow slower and slower at higher temperatures (as it does at the lower temperatures), but rapidly reaches a threshold temperature above which it will not grow. The optimal growth temperature is not defined by a single temperature, but by a range of temperatures: 47°C to 54°C (average g = 1.22 h) in complex medium, and 43°C to 51°C (average g = 2.07 h) in defined medium; the optimal growth temperature initially reported was 35°C. Similarly, *Hf. mediterranei* R4 grew at 18°C in both complex (g = 25 h) and defined media (g = 34 h), which was the lowest temperature we were able to achieve. This organism almost certainly will grow at temperatures below 18°C, which is 7°C lower than the minimum temperature of 25°C reported previously.

In addition to the large range of temperatures for optimal growth, the Arrhenius plot shows two other interesting features. First, the maximal growth temperatures (55°C *vs* 53°C) and the temperature ranges for optimal growth (47-54°C *vs* 43-51°C) are only slightly affected by the composition of the medium. Second, there is a significant affect on growth rate between 27°C (g = 10 h) and 25°C (g = 27 h) in minimal medium. What types of physiological adjustments the cell must make to grow at or below 25°C in minimal medium are unknown.

3. Complex Growth Physiology of *Halobacterium salinarum* NRC817.

Our focus on the growth kinetics of *Halobacterium salinarum* NRC817 was first brought to our attention as we were consistently unable to predict when cultures

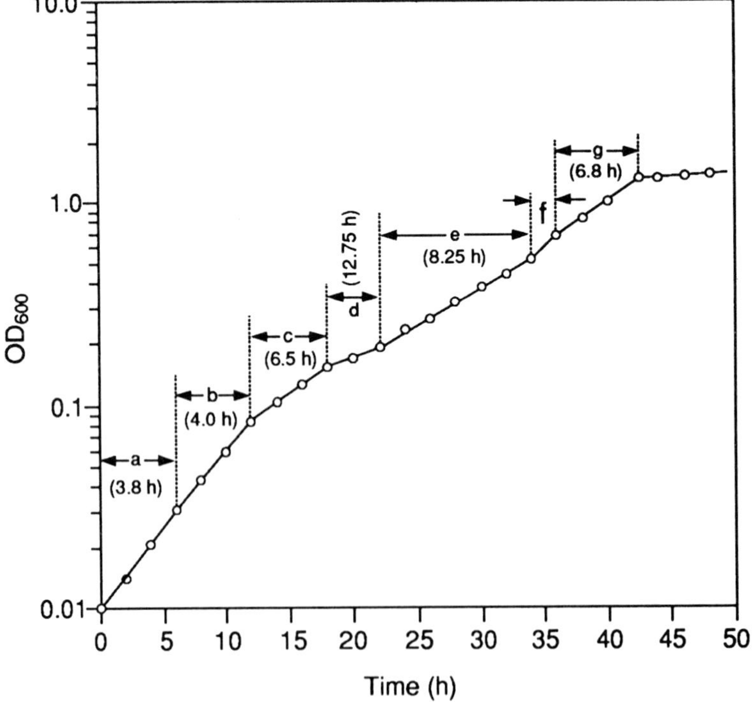

Figure 3. Growth curve for *Hb. salinarum* NRC817 in complex medium (Cheung et al., 1997). The culture (25 ml) was grown in a 125 ml baffled flask at 200 rpm and 41°C. Generation times are indicated in parentheses.

would reach a specific optical density after they had been subcultured to an initial OD_{600} of 0.01. Figure 3 is a growth curve of *Hb. salinarum* NRC817 in a complex medium containing yeast extract and typtone and grown under the conditions described above. The culture was sampled at 2 h intervals. Between 0 h and 22 h, the culture appeared to have four different generation times, each one slower than the previous with the culture growing very slowly at "d." During the period indicated by "e" in Figure 3, the culture resumed growth, with a second slight increase at "g" just before entrance into stationary phase. For *Hb. salinarum*, four amino acids are required for growth, and five others are stimulatory (Grey and Fitt, 1976). Given that the amino acid content of different lots of yeast extract/tryptone vary, one interpretation of the growth pattern in Figure 3 is that between each change in growth rate, a nutrient (or nutrients) supplied exogenously in the medium was exhausted, and the cells in the culture needed to initiate endogenous synthesis and thus grow somewhat slower. The most noticeable of these adjustments occurred during period "d", where the culture grew very slowly (g = 12.75 h), but after recovery, grew faster during period "e" (8.25 h)

To approach this issue with an eye to reducing or eliminating these periodic decreases in growth rate, we prepared a supplemented defined medium based on the synthetic media of Onishi et al. (1964), Ducharme et al. (1972), Grey and Fitt (1976), and Oesterhelt and Krippahl (1983). The supplemented defined medium (Table 1) contained (in excess) all of the amino acids, 5 nucleosides, 3 vitamins, and 4 trace elements. Figure 4 shows the growth curve of *Hb. salinarum* NRC817 in this highly supplemented, defined medium. This medium supported a very similar initial growth rate (g = 3.5 h) compared to the initial growth rate in complex medium (g = 3.8 h). Although there seemed to be some improvement at the beginning of the growth curve (compare period "a" in Figures 3 and 4), periods of slower growth remained and were seen during "b", "d" and "f", with periods of adjustment at "c" and "e." The nutritional requirements of this organism appear to be complex, and we continue to experiment with media of varying composition to determine what additives are most affective in extending exponential growth at a constant rate.

4. Summary and Recommendations

Although extremely halophilic members of the domain archaea are easy to culture, they may or may not exhibit classical growth rate kinetics similar to those seen for *E. coli* or *Salmonella*. Some haloarchaea have very complex growth kinetics (and complex nutritional requirements?) and close attention to

Table 1. Composition of supplemented defined medium for *Hb. salinarum*

Basal Salts	g/L	mM	Amino Acids	mg/L	mM
NaCl	235	4.02 M	L-alanine	509	5.71
$MgSO_4 \cdot 7H_2O$	10	40	L-arginine·HCl	400	1.90
MOPS	8.37	40	L-asparagine	127	0.96
KCl	2	27	L-aspartic acid	250	1.88
$Na_3C_6H_5O_7 \cdot 2H_2O$	0.5	1.7	L-cysteine	245	2.02
$NaH_2PO_4 \cdot H_2O$	0.276	2.0	L-glutamic acid	1,565	10.64
KNO_3	0.1	1.0	L-glutamine	730	5.00
			L-glycine	75	1.00
Nucleosides	g/L	mM	L-histidine·HCl	50	0.239
Adenosine	0.104	0.389	L-isoleucine	440	3.35
Cytidine	0.109	0.448	L-leucine	800	6.10
Guanosine	0.068	0.240	L-lysine·HCl	200	1.04
Thymidine	0.077	0318	L-methionine	90	0.603
Uridine	0.109	0.446	L-phenylalanine	50	0.303
			L-proline	46	0.400
Trace Elements	mg/L	µM	L-serine	305	2.902
$CuSO_4 \cdot 5H_2O$	0.05	0.20	L-threonine	500	4.20
$Fe(NH_4)_2(SO_4)_2 \cdot 6H_2O$	4.55	11.6	L-tryptophan	20	0.098
$MnSO_4 \cdot H_2O$	0.3	1.78	L-tyrosine	112	0.618
$ZnSO_4 \cdot 7H_2O$	0.44	1.53	L-valine	250	2.135
			L-cystine	45	0.187
Vitamins	mg/L	µM			
Folic Acid	5	11.33	Glycerol	ml/L	mM
Thiamine·HCl	5	14.82		1	13.68
Biotin	0.5	2.05			

their growth physiology is needed to understand exactly how these organisms grow. Furthermore, studies of cellular regulation are essentially physiological and well defined physiology data are important when correlating growth with studies on gene regulation. Therefore, we make the following recommendations:

- prior to the start of an experiment, cultures should be placed in balanced growth by serial subcultivation through 10 to 15 mass doublings, wherever possible

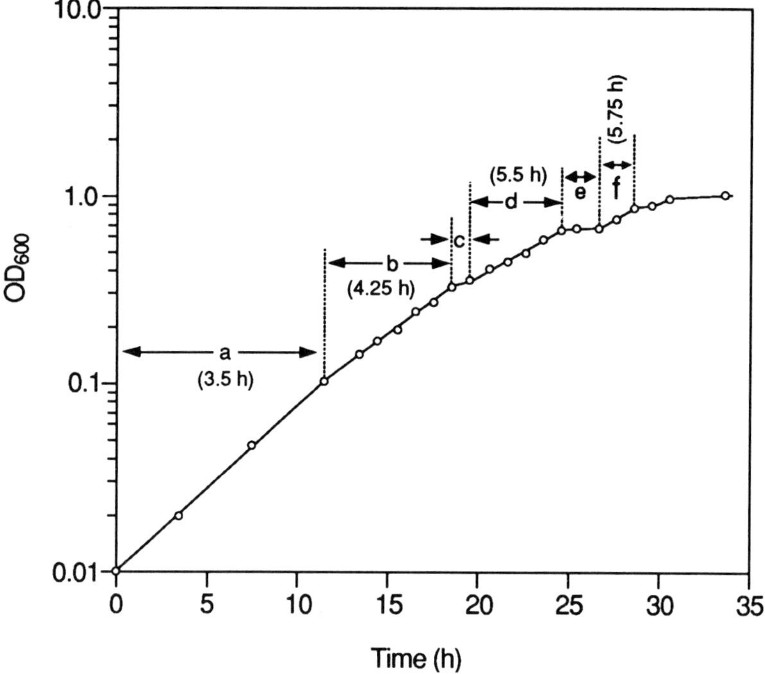

Figure 4. Growth curve of *Hb. salinarum* in supplemented defined medium (see Table 1). Conditions are as described in the legend to Figure 3, except baffled flasks were not used (no foaming occurs in this medium) and the shaker speed was 300 rpm. Generation times are shown in parentheses.

- an initial growth curve where samples are taken closely together should be done to determine where the culture falls off from its initial growth rate

- oxygenation of cultures should be optimized by using baffled flasks, antifoam (where needed), appropriate buffers and shaking water baths [Note: the experiments presented here were done at an altitude of 7,000 ft (2,133 m)]

- an Arrhenius plot should be developed during the initial characterization of new haloarchaeal isolates

Finally, in all of the studies presented here, growth was monitored spectrophotometrically using a Shimadzu UV160U dual beam spectrophoto-meter at 600 nm. Caution must be used when follow growth using optical density as some haloarchaea can generate compounds or structures that can interfere with spectrophotometric measurements. For example, some haloarchaea can produce large amounts of exopolymers (Rodriguez-Valera, 1995), while others can synthesize large numbers of refractile gas vesicles resulting in erroneously high spectrophotometric readings compared to viable counts (Shand and Betlach, 1991). Under these conditions, an alternative parameter such as viable counts should be used for following growth.

5. Acknowledgments

We would like to acknowledge the contributions of the following students who over the years willingly became sleep-deprived in order to carry out many detailed growth experiments: Kristen Bastron, Michael Braswell, Martín Casado, Joyce Cheung, Casey Lehman, Elizabeth O'Connor and Stacey Silveira. This work was supported by grants to RFS from the Arizona Disease Control Research Commission, the National Institutes of Health, and by Organized Research of Northern Arizona University.

6. References

American Type Culture Collection (1979) *Catalogue of Bacteria and Bacteriophages* (17th ed), ATCC, Rockville, MD. p. 336.

Campbell, A.M. (1957) Bact. Rev. **21**, 263-272.

Cheung, J., Danna, K.J., O'Connor, E.M., Price, L.B. and Shand, R.F. (1997) J. Bacteriol. **179**, 548-551.

DasSarma,S., Fleishmann, E.M. and Rodriguez-Valera, F. (1995) Appendix 2: Media for halophiles, in S. DasSarma, E.M. Fleischmann (eds.), *Archaea: a laboratory manual: halophiles*, Cold Spring Harbor Laboratory Press, Cold Spring Harbor, New York, pp. 225-230.

Ducharme, L., Matheson, A.T., Yaguchi, M. and Visentin, L.P. (1972) Can. J. Microbiol. **18**, 1349-1351.

Gary, K. (1989) American Biotech. Lab. **7**, 26-33.

Grey, V.L. and Fitt, P.S. (1976) Can. J. Microbiol. **22**, 440-442.

Herenden, S.L., VanGogelen, R.A. and Neidhardt, F.C. (1979) J. Bacteriol. **139**, 185-194.

Koch, A.L. (1997) Microbiol. and Molec. Biol. Rev. **61**, 305-318.

Neidhardt, F.C., Ingraham,J.L. and Schaechter, M. (1990) *Physiology of the Bacterial Cell*, Sinauer Associates, Inc., Sunderland, Massachusettes.

Neidhardt, F.C. and Umbarger, H.E. (1996) Chemical Composition of *Escherichia coli*, in F.C. Neidhardt (ed.) *Escherichia coli* and *Salmonella*, ASM Press, Washington, D.C., pp. 13-16.

Oesterhelt, D. and Krippahl, G. (1983) Ann. Microbiol. (Inst. Pasteur) **134B**, 137-150.

Onishi, H., McCance, M.E. and Gibbons, N.E. (1965) Can. J. Microbiol. **11**, 365-373.

Rdest, U. and Sturm, M. (1987) Bacteriocins from halobacteria, in *Protein Purification: Micro to Macro*, Alan R. Liss, Inc. pp. 271-278.

Rodriguez-Valera, F. (1995) Cultivation of halophilic bacteria, in S. DasSarma, E.M. Fleischmann (eds.), *Archaea: a laboratory manual: halophiles*, Cold Spring Harbor Laboratory Press, Cold Spring Harbor, New York, pp. 13-16.

Rodriguez-Valera, F., Juez, G. and Kushner, D.J. (1983) System. Appl. Microbiol. **4**, 369-381.

Shand, R.F. and Betlach, M.C. (1991) J. Bacteriol. **173**: 4692-4699.

Biodata of **Joseph Seckbach,** editor of this volume, and author of ***"The Cyanidiaceae: Hot Spring Acidophilic Algae"*** and co-author (with M.M Walsh) of " ***The Versatility of Microorganisms."***

Dr. Joseph Seckbach, served as the first selected occupant (1997/1998) of the John P. Laborde Endowed Chair for Louisiana Sea Grant Research and Technology Transfer, and as visiting Professor in the Department of Life Sciences at Louisiana State University (LSU), Baton Rouge, LA. He earned his Ph.D. from the University of Chicago (Chicago, IL) in 1965, and spent his postdoctoral years in the Division of Biology at Caltech (Pasadena, CA). Then he headed at the University of Californian at Los Angeles (UCLA) a team for searching of extraterrestrial Life. Dr. Seckbach has been appointed to the faculty of the Hebrew University (in Organic Chemistry and later in the School for Over Seas Students) performing algal research and teaching (resp.). He spent his sabbatical periods in Tübingen (Germany), UCLA (Los Angeles), and Harvard University (Cambridge, MA). Among his publications are three books and numerous scientific articles concerning plant ferritin, Cellular evolution, and acido-thermophilic algae. His recent interest is in the field of enigmatic microorganisms and Life in extreme environments.

E-Mail: **seckbach@cc.huji.ac.il**

J. Seckbach (ed.), Enigmatic Microorganisms and Life in Extreme Environments, 425–435.

THE CYANIDIOPHYCEAE: HOT SPRING ACIDOPHILIC ALGAE[1]

JOSEPH SECKBACH[2]
Hebrew University of Jerusalem
Mount Scopus, 91905 Israel

1. Introduction

The Cyanidiaceae are exceptional organisms among the microalgal communities, these cells thrive in extreme ecological conditions (see Seckbach, 1992, 1994, 1996 and 1997). They are considered very primitive eukaryotes and possess features of prokaryotic algae. Although basically they are autotrophic, they also tolerate anaerobic conditions (Lafraie and Betz, 1985), have very low pH ranges, flourish at elevated temperature ranges, and thrive under pure carbon dioxide (Seckbach et al. 1970). More research has been concentrated on these inconspicuous and controversial algae in recent years than on any other organism (see Fredrick, 1998). These unicellular microalgae appear similar to *Chlorella* (Chlorphyta) in size and external morphology under the light microscope. The blue-green coloration of these eukaryotic algae is due to their phycocyanin and chlorophyll-**a** photo-pigments. Despite their coloration they are currently classified in the lower Rhodophyta (red algal division).

Modern investigations in the domain of *Cyanidium caldarium (*Cyanidiaceae*)* began to interest scientists over 40 years ago with the reports by Fukuda (1958) and Mary Bell Allen (1959). Currently these acido-thermophilic algae (Cyanidiaceae) consist of three different but related genera, namely, *Galdieria, Cyanidium* and *Cyanidioschyzon,* each genus having its own species. They are found in the same hot, acidic environments. Another microalgal "cousin" of *Cyanidium caldarium* has been discovered in some lime caves in Chile, France, Spain and Israel (Hoffmann, 1994; Hoffmann and Mayele, 1993). This ambient alga grows under a regime of dim light and neutral pH conditions, and has been labeled "cave Cyanidium." We will not treat this ambient alga in this chapter (for its features, see Hoffmann and Mayele, 1993; Hoffmann, 1994; Seckbach, 1996). The first cell isolated from acidophilic hot springs and to have its natural history described was *Cyanidium caldarium (*Fukuda, 1958, Allen, 1959). Publications on this alga and its cohorts have multiplied to a large extent (see list of references in Seckbach, 1994).

The taxonomy, systematics and phylogeny of *Cyanidium* have puzzled many phycologists. For decades biologists hesitated over how to name this alga (with

1 Dedicated to Professor **Raphael Ikan**, a friend, colleague and coauthor, upon his retirement from the Hebrew University of Jerusalem.

2 This article was initiated at Louisiana State University (Baton Rouge) while the author served as a visiting Professor and the John P. Laborde Chair for Louisiana Sea Grant (1997/1998).

uncetain taxonomic position) or where to place these enigmatic algae among all the algal Divisions. At one time or another *Cyanidium* has been given various genera and species' names (see Seckbach, 1994). The nomenclature, taxonomy, and phylogeny of *Cyanidium* [and its cohorts] has caused confusion since the early investigations on it. Lewin (1974) commented that "Each new analysis... serv[es] only to complicate the picture further." He further remarked that "Certain despairing phylogenists are inclined to assign it [*Cyanidium* cell] to limbo or to a subclass by itself." In that way this alga has been placed in various taxons, the blue-green algae (cyanobacteria) (see Seckbach, 1994), Chlorophyta (Allen, 1959), Cryptophyta (Lewin, 1961), Glaucophyta (Kremer, 1982), with it setting finally in the red algal Division (Seckbach, 1972; Seckbach et al. 1983). Others workers (Kremer and his associates 1978, 1982, 1983; Castenholz, 1979) considered *Cyanidium* cells, on the basis of their ecology and biochemistry, as a symbiotic association between a cyabobacterium and a protistan (colorless) host cell (similar to the glaucophytan, *Cyanophora paradoxa).* Because of their primitiveness and reduced rhodophytan features there has been a proposal to place the *Cyanidium caldarium* and its members in a new taxon**, Prerhodophyta**, (see Seckbach, 1987, 1992, Seckbach et al., 1983, Seckbach and Ott, 1994). Ohta et al., q.v., and other molecular biologists (Bhattacharya and Medlin, 1995) support the primitive status of the Cyanidiaceae among the other rhodophytans. They noticed that *Cyanidium caldarium* is located on the phylogenetic tree lower than the rest of the red algae and form a cluster distinct from that defined by other rhodophytes (Bhattacharya and Medlin, 1995; Ragan and Gutell, 1995).

Seckbach and Ott (1994), as well as Ott and Seckbach (1994) summarized the taxonomy and systematics of these hot spring algae and proposed a new system of their systematics and taxonomy (see comments by Gross, q.v.). Currently, the cyanidiacean members are placed among the lower rhodophytes.

The earlier investigators dealing with these algae noticed differences in the morphology and biochemical and fine structural features which caused them to distinguish between two "formae" or subspecies of *Cyanidium caldarium (*Brock, 1978, Merola et al. 1981, Seckbach et al. 1981*).* Indeed, reports throughout the earlier literature contained contradictory morphological, physiological and biochemical observations among these microorganisms. It turned out that in their natural habitat there exist two different genera and species (and sometimes even an additional alga) all named as "*Cyanidium caldarium*" taxon (see Merola et al., 1981).

Seckbach and his associates (1983, 1994) studied the characteristics of these algae and noticed a gradual progress of evolution within this algal group in the domains of: **(1) morphology**: – cellular size, mode of cell division and number of the autospores; **(2) ultrastructure**: – see Kuroiwa et al 1994, Seckbach, 1992, 1994; Seckbach et al., 1993; Itoh and Kuroiwa, q.v.). – The cellular shape and the intracellular location of plastid DNA indicates an elevation (Kuroiwa et al, 1994, for general fine cellular fine structure see Seckbach, 1992, 1994; Seckbach et al. 1993, Kuroiwa et al. 1994; Itoh and Kuroiwa, q.v.); **(3) biochemical features:** – including the halo-tolerance to NaCl, storage glucans types and presence or absence of linolenic acid, phytoglycogen, the ability to metabolize nitrates, and so on.

Cyanidioschyzon merolae seems to be the most primitive eukaryotic cell (Seckbach et al. 1992) based on morphological aspects, it divides by binary fission similar to cyanobacteria (Seckbach, 1994; Itoh and Kuroiwa, q.v.) and as far as molecular sequence is concerned (Ohta et al., q.v.), this alga has the smallest known amount of nuclear DNA among all eukaryotes (Ohta et al, 1993; Terui et al. 1995). *Cyanidium caldarium* is the best known alga in this group and serves as an intermediate (divides into four endospores). *Galdieria* is larger, than the other two genera, in size, possesses higher values of nuclear DNA, and multiplies into 8, 16 and 32 endospores). It shows biochemical and molecular characteristics of higher taxons and is closely related to the other unicellular red algae (see Seckbach, 1992; 1994 [see diagram of "composite theory of evolution" in Fredrick, 1998]; Seckbach and Walsh, 1998; Seckbach et al. 1983).

In spite of the early reports revealing the existence of more than one organism within this group, many investigators have continued to confuse their laboratory reports. Some workers have been investigating *Galdieria sulphuraria* but published results about "*Cyanidium caldarium.*" Recently, Seckbach (1991) has set new criteria for distinguishing between those thermophilic algae. The last publication noted drew definite lines between both cells, and (most) cyanidiacean investigators are currently able to identify their specific organism.

Gross (q.v.) recently discussed biochemical features obtained from new isolates of *Galdieria* and *Cyanidium caldarium.* His observations are different from the "classical" known characteristics as revealed first by the Neapolitan group (Merola et al., 1981). For example, Gross observed linolenic acid in *Cyanidium caldarium* isolates (although it is known that alpha linolenic acid is presence only in *Galdieria*). Twenty years earlier, Bedord et al. (1978) found an abundance of this tri-en unsaturated fatty acid only when these cells were cultured in 20^0C but they did not detect it in 53^0C. It seems therefore very likely that some enzymatic systems are turned on and off following external culturing conditions. When we (Seckbach, unpublished) cultured *Cyanidium caldarium* cells in an iron-deficient medium, the cells exhibited new ulltrastructural features, many of them exposed prominent vacuoles ("normally" *Cyanidium* cells do not possess them) and under ironless condition we observed heavily stained and reduced-in-size thylakoids in the chloroplasts. Since Gross' (q.v.) findings were performed in 20^0C, one should be cautious in determining "new" biochemical features when culturing thermophilic algae at room temperature, rather than at elevated temperatures.

2. Habitat

The Cyanidiophyceae thrive ubiquitously in warm (45 - 56^0C) and acidic (pH ~2) environments (they have been isolated from sites in Europe, Asia and America). Allen (1959) reported that the "anomalous" *Cyanidium caldarium* can tolerate even a solution of 1 N H_2SO_4. This should be a practical method to clean and isolate pure cultures. The internal pH of *Cyanidium caldarium* is ~ neutral (Mattoon, 1990; see, Pinto, 1993). A similar large pH gradient exists in other obligate acidophiles (grown in optimal pH ranges of 2-4), they have the ability like *Cyanidium* and *Dunaliella acidophila* to maintain a neutral internal pH (Pick, q.v., Matin, 1990; Pinto, 1993). *Cyanidium* was first isolated from sulphuric hot springs (such as in Yellowstone National Park, Wyoming USA). Later on they were observed in warm acidic soils (Brock, 1978), or even covered by a soft layer of lime stone (Seckbach, 1994). *Cyanidium caldarium*-like algal cells have been also discovered in caves in Chile and in some Mediterranean zones (France, Israel, and Italy) growing under ambient conditions (see, Hoffman 1994; Seckbach, 1996). Gross et al. (1998) observed these cells growing under dim light regime as cryptoendolithic populations. Such *Galdieria sulphuraria* cells have been covered by silica and some survive as heterotrophs by consuming organic compounds released by other dead cells (Gross et al. 1998).

Seckbach and his coworkers (1970) succeeded in showing a vigorous growth of *Cyanidium caldarium* cultures and higher photosynthetic rates under a constant stream of pure carbon dioxide in acidic media. The lower pH level of the growing media is not affected by the CO_2 and this gas just gives an additional promotion of carbon sources to the algal culture.

3. Morphology and Ultrastructure

All Cyanidiophycean are unicellular eukaryotes, containing one nucleus, a single chloroplast and one mitochondrion (Suzuki et al., 1994) and the major other [slightly reduced] eukaryotic cellular constituents. Baily and Stehaelin (1968) analyzed the cell wall of *Cyanidium caldarium* and reported that it is primarely proteinaceous without significant amounts of cellulose. This wall composition protects the two larger cyanidian against the warm and acidic hostile environment. On the other hand, *Cyanidioschyzon* (the smallest alga in this series) has a very fragile cellular wall and sometimes is completely absent in the electron-micrographs. Itoh and Kuroiwa (q.v.) present the special ultrastructural features and the dividing method of *Cyanidioschyzon* and *Cyanidium.* They demonstrate the "dividing cellular rings" in these algae. *C. merolae* is the only organism in which both plastid dividing (PD) and mitochondrial dividing (MD) rings have been observed (Itoh and Kuroiwa, q.v.; Itoh et al. 1996, 1997).

The size of the cyanidian members varies from 1x2 mμ (*Cyanidioschyzon*) to 3 mμ (*Cyanidium*) to larger than 6 mμ (*Galdieria*). Today we recognize several additional species of *Cyanidium* and *Galdieria* (for further details, see Seckbach 1994; Sentsova, 1994). The cellular cross sections of *Galdieria sulphuraria* (viewed electron microscopically) exhibit several contours of chloroplasts and profiles of mitochondria, giving the impression of several organelles within this cell. (Seckbach, 1992; Seckbach et al. 1981). It turns out that both these organelles are multi-lobed and their cross-ultra-sectioning plans reveal figures of several chloroplasts and mitochondria branching (all actually the proliferation of a single organelle). Although there are still some technical difficulties in dealing with these algae in the electron microscopic preparations, we now almost have a complete picture of their fine cellular structrue. The major electronmicroscopic cellular morphology (Itoh and Koruiwa, q.v.) and molecular biology (Ohta et al., q.v.) has been recently investigated by Kuroiwa and his group in Japan, who discovered new fine structural appearances and the special mechanism of organellar and nuclear divisional ring in these cells.

4. Biochemistry and Physiology

The cyanidiacean store their photosynthetically assimilated products (polyglucans) in the cytosole (like the rest of the chromophytes and rhodophytans) while in thechlorophytes (and in higher plants) the storage (amylopectin and amylose, or the starch) is accumulated inside the chloroplasts (Fredrick 1978, 1993, 1998; Fredrick and Seckbach, 1986). *Cyanidioschyzon merolae* and *Cyanidium caldarium* accumulate a highly branched polyglucan molecule, i.e., phytoglycogen (similar to the cyanobacteria), while *Galdieria sulphuraria* stores floridian starch that is a less branched molecule and resembles red algal products. *Galdieria* (in contrast to the other two algae) can grow autotrophic and heterotrophic on a wide variety of organic compounds (see Gross et al 1997 and chapter by Gross, q.v., for his new approach). This alga possesses (like most of the eukaryotes) linolenic acid although it is unable to metabolize nitrates. Thus *Cyanidium caldarium* and *Cyanidioschyzon merolae* are both only autotrophic and are among the only known eukaryotes lacking linolenic acid. Both microalgae are considered very primitive eukaryotes and are on a lower evolutionary level than *Galdieria.* Heimann et al. (1997) detected, in the plasma membrane vesicles of *Cyanophora paradoxa* (Glaucocystophyta) only traces of <1% α-linolenic acid, similar to these cyanidean algae. Recently Gross and Schnarrenberger (1995) and Stein et al. (1997) reported that *Galdieria sulphuraria* can grow heterotrophic on at least ten different polyols (sugar alcohols), and this team has analyzed the metabolism of these polyols. Seckbach et al. (1992) discusses the activity of catalase in the Cyanidiaceae and reports that the electronmicrographs revealed microbodies only in *Galdieria* and in *Cyanidium.* Recent ultrastuctural publications by Itoh et al (1997), Kuroiwa et al. (1994) and by Okuwaki at al. (1996) demonstated the presence of microbodies in these algae, which might be related to mitochondrial division.

5. Molecular studies with the Cyanidiophyceae

Currently most of the investigations with these algae are focused on the molecular biological aspects studying various gene sequences (see chapter by Ohta et al. q.v., Zetsche, 1994, Troxler, 1994). The nuclear genomes of the Cyanidiophyceae show an increase from *C. merolae* (~ 8 mbp), to *Cyanidium caldarium* (13 mbp), and to Galdieria sulphuraria (48.7 mbp) [see Toda et al (1995), Okuwaki et al. (1996) and Suzuki et al. (1995)]. Similar increase in genomic values has been detected in the plastid-DNA of these algae. This sprouting in size of the plastid and nuclear DNA values is also reflected by the cellular size of all three genera (see above). Tanaka et al (1996) reported the nuclear encoding of a chloroplast RNA polymerase Sigma subunit in *C. caldarium (*see, Troxler et al., 1994). Itoh et al. (1996) demonstrated that Aphidicolin (a specific inhibitor of DNA polymerase α) uncouples the chloroplast division cycle from the mitotic cycle in *C. merolae.* In these experiments they succeeded in separating cytokenesis (that stopped dividing) from chloroplast that proceeded to divide and generated four or more chloroplasts per cell (instead of the single organelle) see Itoh et al. 1996 and Terui et al. 1995). These investigators also determined the mitotic division synchronized cycle (in *C. merolae* it is 12 hrs.) and the order of the cellular division (nuclear → chloroplast → mitochondrion), all divide with Divisional Rings.

6. Phylogeny

These Cyanidia are considered primitive members of rhodophytans. These thermophiles first arose, most probably among the earliest eukaryotes as we are able to judge from the "simple" morphological, biochemical features and molecular sequences (Seckbach, 1992). It is well accepted among the microbial evolutionists that the first living forms originated from ancient hot springs (Seckbach, 1995, Rachel, q.v.). Others, think that even the most primitive eukaryote, *C. merolae* (Rhodophyta) diverged after the division of Protozoa, but before the divisions of Fungi, Animalia and Chlorophyta (Takahashi et al. 1995).

Hori et al (1990) studied the rRNA sequences and indicated that the red algae (Rhodophyta) were the first evolved eukaryotic microorganisms. Several diagrams of evolutionary trees based on various molecular sequences insert members of the Rhodophyta close to the base (Van den Eynde, 1988, cf. Riding, 1992).

It is interesting that the Cyanidiophyceae (*Cyanidium/Galdieria*) are positioned even earlier than the red algae on these trees of evolution (Loiseaux-de Goër, 1994). The cyanobacteria are often placed on these evolutionary trees (by molecular biologists) at a lower level than the rhodophytes and are positioned close to the cyanidiaceaen members. Zesche and his team studied the molecular sequences of various algal plastid genes and concluded that the origin of the rhodoplasts (red algal plastids) points to a large phylogenetic distance between the plastid of *Cyanidium* (rhodoplasts), cyanobacteria (cyanelles) and other plastid types (Zetsche, 1994, Kostrzewa et al., 1990; Valantin et al., 1992). Similar studies were conducted by Lüttke (1991) who sequenced proteins (of the small and large subunits of Rubisco and ferredoxin) of various algal groups and noticed, likewise, that their structures suggest separate origins for chloroplasts of green algae (*Chlamydomonas, Scenedesmus*, or *Euglena*) and rhodoplasts (*Cyanidium, Porphyra, Porphyridium*).

However, regardless of the initial dating period for the appearance of *Cyanidium* and its cohorts, these enigmatic algae show some archaic features (*Cyanidioschyzon merolae* and *Cyanidium caldarium* resemble the Cyanobacteria) while *Galdieria sulphuraria* is more closely related to the Rhodophyta unicellular members. When analyzing all the characteristics of the cyanidiaceaen members, we (Seckbach et al. 1983; Seckbach, 1992, 1994) placed them in a progressive order: *Cyanidioschyzon merolae* → *Cyanidium caldarium* → *Galdieria sulphuraria* → rhodophytes → Chromophyta. All this cluster of cells may serve as a transition group between the cyanobacteria and the eukaryotic rhodophytes (Seckbach et al. 1983). Kuroiwa et al. (1994) shows an evolutionary algal tree based on ultrastructure and placed the Cyanidian tribe in the same order as our model. Seckbach and Fredrick (1981) considered *Cyanidium caldarium* the "missing link" in the eukaryogenesis and considered these algae as the only relic of the bridge between prokaryotes and eukaryotes. Also, the enzyme chemistry of the glucan biosynthesis indicate these algae as the evolutionary bridge of pro-eukaryotes (Fredrick, 1998).

Seckbach and his coauthors (1993, 1994, 1996, 1997) have proposed a model for the cyanidiophyceaen phylogeny in the following paths:

An Archaean cell (*Thermoplasma acidophilum*-like organism [known to live in acido thermophilic habitat similar to the Cyanidiaceae] incorporated a thermophilic **cyanobacterium** [thriving in neutral or alkaline pH ranges], and this dual association resulted in the first cell of the Cyanidian member (i.e., ***Cyanidioschyzon*** *merolae*). The engulfed cyanobacterium is now protected against the hostile acidic environment by finding inside its host a neutral surrounding (Pinto, 1993). Following this association, the incorporated cyanobacterium began to function as a photosynthetic organelle (chloroplast) of *Cyanidioschyzon.* Some of these *Cyanidioschyzon* cells evolved autogenously (non-symbiotically) further into ***Cyanidium caldarium*** cells, while certain of the later cells continued to develop into more advanced algae, i.e., ***Galdieria,*** that already resemble the first **rhodophytes.** However there are other views against Eukaryogenesis that indicate a direct autogenous pathway (compartmentation theory) from the bacteria or cyanobacteria to the first eukaryotes (see, Nakamura, q.v., Jensen, 1994; Seckbach et al., 1998).

In summary we can declare that *Cyanidium caldarium* which has been considered just an ugly duckling among the microalgae has turned out to be a beautiful swan. For further information on this algal group, see the special volume devoted to *Cyanidium caldarium* and its cohorts (Seckbach, 1994) and articles by Seckbach and Walsh (1998) or Seckbach et al. (1998).

7. References

Allen (1959) Archiv. fur Mikrobiol. 32, 270-277.

Baily, R. W. and Staehelin, L. A. (1968) J. Gen. Microbiol. 54, 269-276.

Bedord, C.J., McMahon, V. and Adams, A. (1978) Arch. Biochem. Biophys. 185, 15-20.

Bhattacharya, D. and Medlin, L. (1995) J. Phycol. 31, 489-498.

Brock, T.D. (1978) The genus *Cyanidium.* in: *Thermophilic Microorganisms and Life at High Temperatures.* Springer, New York. pp 255-302.

Castenholz, R.W. (1979) In: M. Shilo (ed.) Strategies of Microbial Life in Extreme Environments. Verlag Chemie, Weinhein, N.Y,. pp. 373-392.

Fredrick, J.F. (1978) J. Theorm. Biol. 3, 1-4.

Fredrick, J.F. (1993) In: S. Sato, M. Ishida and H. Ishikawa (eds.) *Endocytobiology V,* Tübingen University Press, pp.475-479

Fredrick, J.F. (1998) In: E. Wagner (ed.) Endocytobiology VII. Freiburg conference (April 1998) In press.

Fredrick J.F. and Seckbach J. (1986) Phytochem. 25, 363-365,

Fukuda, I. (1958) Bot. Mag. Tolyo 71, 79-86.

Gross W. and Schnarrenberger, C. (1995) Plant. Cell, Physiolo. 36, 633-638.

Gross, W, Küver, J., Tischchendorf, G., Bouchaala, N. and Büsch W. (1998). Eur. J. Phycol. 33, 25-31.

Gross, W., Seipold, P and Schnarrenberger, C. (1997) Plant Physiol. 114, 231-136.

Heimann, K., Becker, B., Harnisch, H., Mukherjee, K. D. and Melkonian, M. (1997) Bot. Acta 110, 401-410.

Hoffmann, L. (1994) In: J. Seckbach (ed.) *Evolutionary Pathways and Enigmatic Algae: Cyanidium caldarium and Related Cells.* Kluwer Academic Publishers, Dordrecht, The Netherlands. pp 175-182.

Hoffmann, L. and Mayele, B. (1993) Phycologia 32, 307-309.

Hori, H., Satow, Y., Inoue, I. and Chihara, M. (1990) In: P. Nardon, V.Gianinazzi-Pearson, A.M. Grenier, L. Margulis and D.C. Smith (eds.) Endocytobiology IV, INSA, Paris, pp. 557-559.

Itoh, R., Takahashi, H., Toda, K., Kuroiwa, H. and Kuroiwa, T. (1996) Europ. J. Cell Biol. 71, 303-310.

Itoh, R., Takahashi, H., Toda, K., Kuroiwa, H. and Kuroiwa, T. (1997) Protoplasma 196, 135-141.

Jensen, T.E. (1994) In: J. Seckbach (ed.) *Evolutionary Pathways and Enigmatic Algae: Cyanidium caldarium and Related Cells.* Kluwer Academic Publishers, Dordrecht, The Netherlands. pp.53-66.

Kostrzewa, M., Valentin, K., Maid, U., Radetsky, R. and Zetsche, K. (1990) Curr. Genet. 18, 465-469.

Kremer, B.P. (1982) Br. phycol. J. 17, 51-61.

Kremer, B.P. (1983) In: H.E.A. Schenk and W. Schwemmler (eds.) *Endocytobiology II,* W. deGruyter and Co. Berlin. New York. pp. 963-970.

Kremer, B.P., Feige, B. and Schneider Hj. A. W. (1978) Naturwissenschft 65, 157.

Kuroiwa, T., Kuroiwa, H., Mita, T. and Ohta, N. (1994) In: J. Seckbach (ed.) *Evolutionary Pathways and Enigmatic Algae: Cyanidium caldarium (Rhodophyta) and Related Cells.* Kluwer Academic Publishers, Dordrecht, The Netherlands. pp.239-253.

Lafraie, M.A. and Betz,, A. (1985) Planta 163, 38-42.
Lewin R.M. (1961) New Bull. Phycol. Soc. Am. 14, 6-7.
Lewin R.M. (1974) In: W.D.P. Steward (ed.) *Algal Physiology and Biochemistry*, University of California Press, berkeley and Los Angeles.
Loiseaux-de Goër, L.S. (1994) Recent Progress in Phycol. Research 10, 137-177.
Lüttke, A. (1991) Endocytobiosis & Cell Research. 8, 75-82.
Matin, A. (1990) In: H.P. Durre (ed.), FEMS Microbiol Rev. (Amsterdam) 75, 307-318.
Merola, A., Castaldo, R., De Luca, P., Gambardella, R., Mauacchio, A. and Taddei, R. (1981) Giorn. Bot. Ital. 115, 189-195.
Okuwaki, T., Takahashi, H., Itoh, R., Toda, K., Kawazu, T., Kuroiwa, H. and Kuroiwa, T. (1996) Cytologia 61, 69-74.
Ott and Seckbach (1994) In: J. Seckbach (ed.) *Evolutionary Pathways and Eingmatic Algae: Cyanidium caldarium (Rhodophyta) and related Cells.* Kluwer Academic Publisher, Dordrecht, The Netherlands. pp. 113-132 and 145-152.
Pinto, G. (1993) Giorn. Bot.Ital. 127, 400-406.
Ragan, M.A. and Gutell, R.R. (1995) Bot. J. Linn. Soc. 118, 81-105.
Riding, R. (1992) Eurp. J. Biochem. 201, 343-349.
Seckbach, J. (1972) Microbios 5, 133-142.
Seckbach, J. (1987) Ann. NYAcSc. 503, 424-437.
Seckbach, J. (1991) J. Phycol. 27, 794-796.
Seckbach, J. (1992) In: W. Reisser (ed.) *Algae and Symbioses: Plants, Animals, fungi, Viruses, Interactions Explored,* Biopress Lmt. Bristol, England, pp. 399-426.
Seckbach J. (1994) editor, *Evolutionary Pathways and Enigmatic Algae: Cyanidium caldarium (Rhodophyta) and Related Cells.* Kluwer Academic Publishers, Dordrecht, The Netherlands, pp. 99-112.
Seckbach, J. (1995) J. Biol. Physics 20, 335-345.
Seckbach, J. (1996) In: J. Chela-Flores and F. Raulin (eds.) *Chemical Evoluion: Physics of the Origin and Evolution of Life.* Kluwer Academic Publishers, The Netherlands. pp. 197-213.
Seckbach, J. (1997) In: C.B. Cosmovici, S. Bowyer, and D. Werthimer (eds.) *Astronomical and Biochemical Origins and the Search for Life in the Universe.* Editrice Compositori, Italy. pp. 511-523.
Seckbach, J., Baker, F.A. and Shugarman, P.M. (1970) Nature 227, 744-745.
Seckbach, J. and Fredrick, J.F, (1981) In: Y. Wolman (ed.) *Origin of Life.* D. Reidel Publishing Comp. Dordrecht. The Netherlands. pp. 567-574/
Seckbach, J., Fredrick, J.F. and garbary, D.J. (1983) In: H.E.A. Schenk and W. Schwemmler (eds.) *Endocytobiology II*, W. deGruyter and Co. Berlin. New York. pp. 947-962.
Seckbach, J., Gonalez, E., Wainwright, I.M. and Gross, W. (1992) Nova Hedwigia 55, 99-109.
Seckbach J., Hammerman, I.S. and Hanania, J. (1981) Ann. N.Y.Ac.Sc, 361, 409-425.
Seckbach, J., Ikan, R., Nagashima, H. and Fukuda, I. (1993) In: S. Sato, M. Ishida and H. Ishikawa (eds.) *Endocytobiology V*, Tübingen University Press, pp. 241-254.
Seckbach, J., Jensen, T.E., Matsuno, K., Nakamura, H., Walsh, M.M. and Chela-Flores, J. (1998) In: J.Chela-Flores and F.Raulin (eds.) V. Trieste Coference on *Chemical Evolution. Exobiology: Matter, Energy, and Information in the Origin and Evolution of Life in the Universe.* Kluwer Academic Publishers, The Netherlands. (in press)
Seckbach, J. and Ott, F.D. (1994) In: J. Seckbach (ed.) ***Evolutionary Pathways and Eingmatic Algae: Cyanidium caldarium (Rhodophyta) and Related Cells.*** Kluwer Academic Publishers, Dordrecht, The Netherlands. pp.133-143.
Seckbach, J. and Walsh, M.M. (1998) In: E. Wagner (ed.) *Endocytobiology VII.* Freiburg conference (April 1998) In press.
Sentsova, O. (1994) In: J. Seckbach (ed.) *Evolutionary Pathways and Eingmatic Algae: Cyanidium caldarium (Rhodophyta) and related Cells.* Kluwer Academic Publisher, Dordrecht, The Netherlands. pp. 167-174.
Stein, R. Gross, W. and Schnarrenberger C. (1997) Characterization of xylitol dehydrogenase and a D-arabitol dehydrogenase from the thermo- and acidophilic red alga *Galdieria sulphuraria.* Planta202: 487-493.
Suzuki, K., Kawano, S. and Kuroiwa, T. (1994) Phycologia 33, 298-300.
Suzuki, K., Kawazu, T., Mita, T., Takahashi, H., Itoh, R., Toda, K. and Kuroiwa, T. (1995) Eurp . J. Cell Biol. 67, 170-178.
Takahashi, H., Takano. H., Yoloyama, A., Hara, Y. Kawanu, S., Toh-e, A. and Kuroiwa, T. (1995) Curr. Genet. 28, 484-490.
Tanaka, K., Oikawa, K., Ohta, N., Kuroiwa, H., Kuroiwa, T. and Takahashi, H. (1996) Science 272, 1932-1935.

Terui, S., Suzuki, K., Takahashi, H., Itoh, R. and Kuroiwa, T. (1995) J. Phycol. 31, 958-961.
Toda, K., Takahashi, H., Itoh, R. and Kuroiwa, T. (1995) Cytologia 60, 183-188.
Troxler, R.F. (1994) In: J. Seckbach (ed.) *Evolutionary Pathways and Enigmatic Algae: Cyanidium caldarium (Rhodophyta) andRelated Cells*. Kluwer Academic Publishers, Dordrecht, The Netherlands. pp. 263-282.
Troxler, R.F., Zhang, H. and Bogorad, L. (1994) Plant Physiol. 104, 753-759.
Valntin, K., Maid, U., Emich, A. and Zetsche, K. (1992) Plant Mol. Biol. 20, 267-276.
Van den Eynde, H., De Baere, r., De Roeck, e., Van de Peer, Y., Vandenberghe, A., Willekens, P. and De Wachter, R. (1988) J. Mol. Evol. 27, 126-132.
Zetsche K. (1994) In: J. Seckbach (ed.) *Evolutionary Pathways and Enigmatic Algae: Cyanidium caldarium(Rhodophyta) andRelated Cells*. Kluwer Academic Publishers, Dordrecht, The Netherlands. pp. 255-262.

Biodata of Dr. Wolfgang Gross, author of ***"Revision of Comparative Traits for the Acido- and Thermophilic Red Algae Cyanidium and Galdieria."***

Dr. Wolfgang Gross is a research associate at the Free University of Berlin. His interest is on the growth and metabolism of unicellular red algae. Dr. Gross received his Diploma and D.Sc. from the University of Oldenburg (Germany) and spent his postdoctoral in USA at the Biology Departments in Universities of California (Santa Cruz) and in North Carolina State University (Raleigh). Dr. Gross publication list include Journal articles about the physiology, biochemistry and ecology of Cyanidium caldarium and Galdieria sulphuraria, as well as chapters in books.

E-Mail: galdi@zedat.fu-berlin.de

J. Seckbach (ed.), Enigmatic Microorganisms and Life in Extreme Environments, 437–446.

REVISION OF COMPARATIVE TRAITS FOR THE ACIDO- AND THERMOPHILIC RED ALGAE *CYANIDIUM* AND *GALDIERIA*

WOLFGANG GROSS
Institut für Pflanzenphysiologie und Mikrobiologie, Freie Universität Berlin, Königin-Luise-Str. 12-16, 14195 Berlin, Germany

1. Introduction

The red algae *Galdieria sulphuraria* (Galdieri) Merola (*Cyanidiaceae*) and *Cyanidium caldarium* (Tilden) Geitler (*Cyanidiaceae*) inhabit hot acidic sulphur springs and volcanic calderas with pH values from 0.5 to 3 and temperatures up to 56 °C (Doemel and Brock, 1970; De Luca and Taddei, 1976). Although both genera occur world-wide, their habitat is restricted to volcanic areas (Brock, 1978). Both algae usually occur in mixed populations (De Luca *et al.*, 1974; 1979). Dispersal of the algae by wind or water fowl seems to be very unlikely, because of the low desiccation tolerance of the algae and the inhospitableness of these sites for wild life. Aquatic dispersal is also not conceivable because the algae will not grow at neutral pH. These observations prompted Brock (1967) to speculate on the possibility that today's acidic hot springs are inhabited by relict populations from aeons when volcanism was much more widespread. The absence of *Galdieria* and *Cyanidium* from the geological young Hawaiian Islands - despite favorable habitats - might be a result of the low dispersal frequency of these algae (Brock, 1978). However, as suggested by Round (1981), this would imply that the algal population of Iceland survived several ice ages. At present, it is not clear how new habitats could be colonized by the two genera or how an exchange between separate populations might be possible. If the natural habitats of *Galdieria* and *Cyanidium* are indeed extremely isolated, the various populations should have developed some individual traits.

TABLE 1. Comparative traits of the type strains of *Galdieria*-species and *Cyanidium caldarium* used in the past for taxonomic purposes

Feature	*C. caldarium*	*G. sulphuraria*	*G. partita*	*G. maxima*	*G. daedala*
Size (µm)	4 - 6	3 - 11	3 - 11	6.5 - 16.5	3 - 11
Plastid morphology	cup-shaped	cup-shaped	cup-shaped	irregular to oval	irregular
Number of endospores	4	4 - 32	4 - 32	4 - 16	2 - 8
Heterotrophy	no	yes	yes	yes	yes
NO_3-utilization	yes	no	n. d.	n. d.	n. d.
Presence of linolenic acid	no	yes	n. d.	n. d.	n. d.

n. d.: not determined

Several attempts have been made to establish a taxonomic concept for *Galdieria* and *Cyanidium*. Because *Galdieria* and *Cyanidium* have a very similar morphology, it proved to be difficult to assign proper morphological characteristics to the two genera. In addition, morphological markers are often affected by culture conditions. Therefore, biochemical and physiological markers, such as heterotrophy, utilization of nitrate, and the occurrence of linolenic acid, have been used as well to distinguish between *Galdieria* and *Cyanidium* (Tab. I). The first to notice significant biochemical differences between two strain of *Cyanidium caldarium* were Merola *et al.* (1981). These authors established a new genus called *Galdieria* with only one species, *G. sulphuraria*. Reluctantly, this new classification was adopted by the scientific community (see Seckbach, 1991). However, the more strains were isolated the more difficult it was to assign newly isolated strains to one of the two taxons. This has recently led to the replacement of *Galdieria sulphuraria* by three new species (*G. daedala*, *G. maxima*, *G. partita*) (Sentsova, 1994) on the one hand. This reclassification was based mainly on morphological characteristics. On the other hand, Ott and Seckbach (1994) renamed *Galdieria* to *Cyanidium* because in their eyes different biochemical and ultrastructural characteristics would not justify a separation into two genera. These authors erected a new classification with seven species of *Cyanidium*. The main problem with biochemical markers appears that they have been established almost exclusively for strains isolated in Japan (Nagashima, 1994) and Italy (Merola *et al.*, 1981). Therefore, it is questionable whether these traits are also applicable for strains isolated from other locations. This has led me to investigate three common quoted biochemical characteristics (heterotrophy, nitrate utilization, fatty acid profile) in 20 different isolates of *Cyanidium* and *Galdieria*.

2. Evaluation of Biochemical Markers

2.1 HETEROTROPHIC GROWTH

The most notable difference between *Galdieria* and *Cyanidium* is the ability of the former to grow heterotrophically. Therefore, I studied the heterotrophic growth of 20 different isolates of *Galdieria* and *Cyanidium* obtained from various parts of the world. All strains tested derived from single colonies taken from the original stock cultures. Only the type strain of *C. caldarium* was unable to grow under heterotrophic conditions (at least within a 4-month period) (Tab. II). Most other strains - whether they were originally assigned to the genus *Galdieria* or not - grew on all eight different carbon sources supplied. Strain Y.G did not grow heterotrophically on mannitol, ribose, adonitol, and dulcitol. The latter polyol was also not a substrate for the strains Y.W and 500. The substrate spectrum of these strains did not change when cells were cultivated at higher temperatures (37 °C and 45 °C). The preference regarding the carbon source was very similar in all isolates. The doubling times were usually shortest for D-glucose, D-galactose, and D-mannitol (ranging from 40 to 100 h) and usually longest for L-sorbose and adonitol. For most substrates and isolates a lag-phase of 2 to 10 days was observed when cells were transferred from autotrophic to heterotrophic conditions. With dulcitol, however, the length of this lag-phase was 62 days for strain 002. As it has been shown previously the lag-phases were not due to a selection of heterotrophic mutants (Gross and Schnarrenberger, 1995). The observed length of the lag-phases differed considerably for some substrates, e. g. mannitol and dulcitol, even for strains isolated from the same location, e. g. strain 074W and 074G. Overall, no apparent consensus pattern was

TABLE 2. Heterotrophic growth and lag phases after transfer from auto- to heterotrophic growth conditions of strains of *Galdieria* and *Cyanidium* on eight different sugars and polyols

Isolate[a]	D-Glucose		D-Galactose		D-Ribose		D-Mannitol		Adonitol		Xylitol		Dulcitol		L-Sorbose	
	D[b]	lag[c]	D	lag	D	lag	D	lag	D	lag	D	lag	D	lag	D	lag
G. sulphuraria 074 G	43	2	47	2	83	10	112	15	248	3	55	3	56	38	150	6
G. sulphuraria 074 W	43	2	45	2	73	21	81	4	268	3	56	2	65	45	114	3
G. sulphuraria 002*	40	2	46	1	83	9	48	2	82	6	77	5	159	62	129	4
G. sulphuraria 002S	40	2	46	0	159	6	48	2	73	2	79	0	103	48	120	3
G. sulphuraria 002P	38	2	47	1	127	7	46	6	109	2	84	2	73	55	149	0
G. sulphuraria 063	104	0	112	0	95	8	99	0	216	20	100	3	102	31	213	46
G. sulphuraria 107.79	38	0	45	2	92	8	47	2	194	0	48	2	102	11	221	0
G. sulphuraria 108.79	34	0	39	0	115	6	52	2	145	3	60	2	123	26	191	4
G. sulphuraria Y.G	73	0	48	0	n.g.	-	n.g.	-	n.g.	-	230	0	n.g.	-	390	0
G. sulphuraria Y.W	45	8	49	9	70	23	62	9	128	11	79	9	n.g.	-	216	9
G. sulphuraria 21.92	69	3	70	3	370	21	69	11	350	1	76	8	183	42	295	8
G. partita K	34	3	28	5	166	4	44	6	150	4	77	4	79	10	169	6
G. partita 500*	42	10	86	6	165	9	46	6	192	0	94	6	n.g.	-	218	7
G. maxima 507*	86	2	69	31	349	10	109	1	250	48	179	0	620	35	222	0
G. daedala 508*	+++[d]	-	202	0	411	0	++[d]	-	++	-	++	-	+[d]	-	202	0
C. caldarium J	55	6	49	10	200	13	60	7	174	12	65	7	126	18	213	35
C. caldarium 182*	n.g.[e]	-	n.g.	-	n.g.	-	n.g.	-	n.g.	-	n.g.	-	n.g.	-	n.g.	-
C. caldarium 510	43	9	43	9	75	20	44	13	139	17	77	11	164	78	115	20
Cyanidium spec. 509	38	1	40	1	189	6	40	5	210	0	71	0	113	9	243	0
Cyanidium spec. 511	43	12	39	11	152	17	77	10	++	-	++	-	+	-	170	17

*: type strain; [a]: Source and geographic origin of the strains are listed in Tab. 3; [b]: Doubling time (h) at 25 °C; [c]: Duration of lag-phase (d); [d]: Cells adhere to the walls of the culture vessel, therefore growth was estimated visually (+++: D = approx. 50 h; ++: D = approx. 100 h; +: D = approx. > 150 h); [e]: no visible growth within a 4-month period.

noticed for the growth rates and lag-phases among the strains examined. Keeping the diversity of the substrates tested in mind, it is surprising that the majority of the strains can use even very rare sugars or polyols for growth. It appears that for all strains of *Galdieria* the ability to use organic compounds is of prime importance for the survival in their hostile environment. However, the questions arises under what conditions heterotrophy might be important. Field studies in the vicinity of Naples (Italy) revealed that the concentration of dissolved organic compounds is extremely low in the soils and waters in these volcanic areas (Gross *et al.*, 1998). Similar results were obtained by Smith and Brock (1973) in their studies done at Yellowstone National Park (Montana). These authors estimated that the concentration of organic material is too low to support significant heterotrophy and speculated that bacteria and fungi are the most likely consumers of organic matter. However, an examination of the organisms inhabiting the sulfur springs around Naples showed that *Galdieria* is without competitors for organic substrates. The key for the understanding of the role of heterotrophy might be found in the cryptoendolithic habitat of *Galdieria*. At two sites near Naples we found that a considerable part of the *Galdieria* population grows in 2 - 3 mm thick mats on rocks covered by a thin layer of silica sinter. Only 0.1 to 1 % of the sunlight can penetrate this upper layer but would still permit at least limited photosynthesis by the cells in the underlying mat. Under insufficient light conditions, however, cells at the underside of the mats would die and release soluble material into the surrounding. These compounds could then be used by the neighboring cells to prolong their viability until conditions are more favorable. In this scenario almost all compounds that are released would be taken up by other algal cells and the overall concentration of organic substrates remains

TABLE 3. Utilization of nitrate by strains of *Galdieria* and *Cyanidium* under autotrophic and heterotrophic (D-galactose) conditions at 25 °C.

Taxon	Strain	Geographic origin	Source	Doubling time (h) autotroph	Doubling time (h) heterotroph
G. sulphuraria	074 G	Mt. Lawu, Java	FUB	195	437
G. sulphuraria	074 W	Mt. Lawu, Java	FUB	191	104
G. sulphuraria	002[a]	Pisciarelli, Naples	IBN	-[b]	-
G. sulphuraria	002S	La Solfatara, Naples	FUB	-	-
G. sulphuraria	002P	Pisciarelli, Naples	FUB	-	-
G. sulphuraria	063	Comitini, Sicily	IBN	+[c]	125
G. sulphuraria	107.79	Sonoma Co., California	SAG	246	352
G. sulphuraria	108.79	Yellowstone Natl. Park	SAG	215	234
G. sulphuraria	Y.G	Yellowstone Natl. Park	FUB	+	355
G. sulphuraria	Y.W	Yellowstone Natl. Park	FUB	+	319
G. sulphuraria	21.92	Yangmingshan, Taiwan	SAG	+	151
G. partita	K	Kodakarajima Island, Japan	MBI	+	131
G. partita	500[a]	Mt. Uzon, Kamchatka, Russia	IPPAS	+	119
G. maxima	507[a]	Kunashir Island, Russia	IPPAS	+	177
G. daedala	508[a]	Kunashir Island, Russia	IPPAS	315	-
C. caldarium	J	Japan	MBI	+	248
C. caldarium	182[a]	Mt. Lawu, Java	IBN	85	-
C. caldarium	510	unknown	IPPAS	90	-
Cyanidium spec.	509	Kamchatka, Russia	IPPAS	+	220
Cyanidium spec.	511	Kamchatka, Russia	IPPAS	+	83

[a]: type strain; [b]: no growth; [c]: growth observed; IBN: Culture Collection of the University of Naples, Italy; IPPAS: Culture Collection of Microalgae of the Institute of Plant Physiology, Russia; MBI: Marine Biotechnology Institute, Kamaishi Laboratories, Japan; SAG: Sammlung von Algenkulturen at the University of Göttingen, Germany; FUB: Freie Universität Berlin, Germany

low in this environment. Because I found that strains obtained from very different locations all show heterotrophy, the endolithic growth of *Galdieria* should be very widespread if the above hypothesis is true.

2.2 UTILIZATION OF NITRATE

The ability to use not only ammonium but also nitrate as nitrogen source has been suggested as a physiological marker for *Cyanidium caldarium* (De Luca and Taddei, 1972). Compared with the ability to grow heterotrophically the use of nitrate appears to be a weak characteristic because the necessary metabolic machinery is much simpler. When I tested 20 different strains of *Cyanidium* and *Galdieria*, only the three isolates from Naples (Italy) were not capable to use nitrate, regardless of autotrophic or heterotrophic conditions (Tab. III). Interestingly, most other isolates grew on nitrate even in the dark, except for the obligate autotrophic type strain of *C. caldarium* as well as *G. daedala* (Kunashir Island) and *C. caldarium* strain 510. The growth rates under these conditions were usually slower than observed with ammonium as nitrogen source. Thus, the lack of nitrate metabolism appears to be a characteristic not for *Galdieria* but for strains isolated in the vicinity of Naples. It is a coincidence that the only unequivocal report in the literature (De Luca and Taddei, 1972) on nitrate utilization was conducted with strain 002 from Naples.

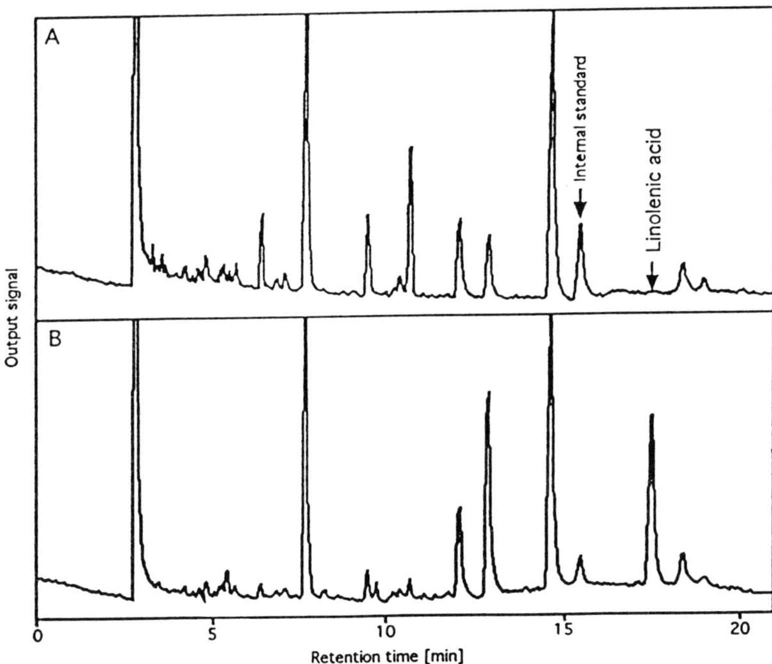

FIGURE 1. GC chromatograms showing the fatty acid profile of *Cyanidium caldarium* strain J (A) and *Cyanidium* spec. strain 509 (B) .

2.3 OCCURRENCE OF LINOLENIC ACID

The fatty acid profile has often been used as a taxonomic marker for algae (Volkman *et al.*, 1993). Boenzi *et al.* (1977) and Nagashima *et al.* (1986) reported the absence of linolenic acid ($C_{18:3}$) in *C. caldarium* isolates and suggested to use this as a marker to distinguish between *Galdieria* and *Cyanidium*. When I analyzed the fatty acid profile of 20 different isolates (listed in Tab. 3) I found that linolenic acid was absent in the type strain of *C. caldarium* as well as in strain 510 and strain J of *C. caldarium* (Fig. 1a). All other isolates contained significant amounts of linolenic acid including the two isolates from Kamchatka termed *Cyanidium* spec. strain 511 and strain 509 (Fig. 1b). These results indicate that the absence of linolenic acid is indeed a characteristic of *C. caldarium* although isolates from other habitats have to be analyzed before a generalization can be made.

3. Geographical Races of the Cyanidiaceae

A taxonomic classification has to meet two very different requirements: It should reflect the natural relations between species, and the markers used to erect this classification should be practical. This is very difficult for algal groups with few morphological features such as coccoid forms. Ultrastructural, biochemical, and genetic markers may help to escape this dilemma but they are often not feasible for practical usage. For the Cyanidiaceae the whole problem becomes even bigger because these algae have apparently been isolated for a very long time in a number of subpopulations and might have undergone parallel developments with little morphological changes. Therefore, taxonomic markers have to be established covering a reasonable number of strains from different locations. I examined twenty strains of *Cyanidium* and *Galdieria* representing the algal population of roughly eight natural habitats for three biochemical markers used in the past to distinguish between the two genera. With regard to heterotrophy the results obtained indicate that the type strain of *C. caldarium* (Java) is the only member of this taxon. In contrast, putting nitrate utilization to the test revealed that all strains, except those from Naples, would fit into the taxon *C. caldarium*. The marker which showed the best agreement with the taxonomic grouping is the absence of linolenic acid from the three strains classified as *C. caldarium*.

The most valuable taxonomic markers should be those directly linked to a special ecological niche. In this respect the ability to grow heterotrophically should be regarded as a highly significant characteristic of a taxon because it indicates a different way of survival. The results that 19 of the 20 strains examined can use a wide variety of substances for heterotrophic growth is an impressive example. If heterotrophy were not be of prime importance for the alga this feature would be lost very soon. Although the advantage of heterotrophy for these algae remains obscure, our field studies at Naples (Italy) indicate that cryptoendolithic growing cells make use of the remains of decaying cells by prolonging the survival under detrimental conditions. This might reflect special conditions at the examined site and it might not be the only reason for heterotrophy but constellations that demand heterotrophy for survival have to be present at all sites where heterotrophic strains have been found. It has to await further field studies in other continents to find possible other circumstances.

Biochemical markers such as nitrate metabolization may be useful to characterize geographic strains as indicated by the finding that only the strains isolated at Naples (Italy) are unable to use nitrate as nitrogen source. The absence of linolenic acid in the one obligate autotroph and the two facultative heterotroph strains of *C. caldarium* appears of secondary taxonomic importance although it is in agreement with the current classification. This marker is a reflection of a single missing enzymatic reaction and, therefore, might be not very consistent during evolution. Taken together it appears that a separation of the type strain of *C. caldarium* from *Galdieria* is justified because of the inability of the former to grow heterotrophically. All other strains examined should be regarded as members of the genus *Galdieria.* In combination with morphological characteristics the other two biochemical markers examined might be useful to further dissect the genus *Galdieria.* In view of a possible long isolation of the different habitats the occurrence of geographic races would be expected. Although, the occurrence of distinct races of *Galdieria* and *Cyanidium* might be proven only by DNA sequence analysis, the currently proposed reclassifications of the *Cyanidiaceae* (Sentsova, 1994; Ott and Seckbach, 1994) have to consider this possibility.

4. Conclusions

The red algae *Galdieria* and *Cyanidium* occur world-wide, but exclusively in areas with hot acidic sulphur springs. Twenty different strains of both genera were tested with regard to three comparative markers that have been used in the past to distinguish between *Galdieria* and *Cyanidium.* The following results were obtained: (i) Only the type strain of *Cyanidium caldarium* was an obligate autotroph. (ii) The inability to use nitrate is a specific trait of *Galdieria*-isolates from Naples (Italy). (iii) Linolenic acid was absent in the type strain and in two facultative heterotrophic strain of *C. caldarium* from Japan and Kamchatka. Therefore, with regard to the occurrence of linolenic acid as a taxonomic marker, one has to reject the ability for heterotrophic growth as specific for *Galdieria.* In view of the heterotrophic growth, the absence of linolenic acid is no longer a trait of *Cyanidium.* The disability to use nitrate appears to be a special trait of the strains from Naples. These results indicate that the occurrence of geographical races has to be taken into account for a taxonomic reorganization of these red algal genera.

The question remains: What markers can be used to distinguish *Galdieria* and *Cyanidium*? The strongest impact on the physiology and ecology of the two genera has the ability to grow heterotrophically. It is very unlikely that the obligate autotrophic *Cyanidium* could aquire an uptake system and enzymatic machinery for heterotrophy by simple mutations. For *Galdieria* a deletion in the uptake system for organic compounds is much more likely but this mutation apparently weakens the chances of survival as outlined above. Thus, at present heterotrophy appears to be the best marker to distinguish between the two genera, although a combinaton with other new markers would be much more desirable.

5. Acknowledgement

I thank I. Heilmann and R. Pausewein for technical assistance, H.-P. Haschke for the fatty acid analysis, and C. Schnarrenberger for valuable discussions.

6. References

Boenzi, D., De Luca, P., and Taddei, R. (1977) Fatty acids in *Cyanidium*, *Giorn Bot Ital* **111**, 129-134.

Brock, T.D. (1967) Life at high temperatures, *Science* **158**, 1012-1019.

Brock, T.D. (1978) The genus *Cyanidium*. in T.D. Brock (ed.), *Thermophilic microorganisms and life at high temperatures*, Springer-Verlag, New York, pp 255-302.

De Luca, P., Gambardella, R., and Merola, A. (1979) Thermoacidophilic algae of north and central america, *Bot Gaz* **140**, 418-427.

De Luca, P., Moretti, A., and Taddei, R. (1974) Nuove stazioni di *Cyanidium caldarium* nell'Italia meridionale ed in Sicilia, *Delpinoa* **14/15**, 49-60.

De Luca, P. and Taddei, R. (1972) Crescita comparata di due forme di *Cyanidium caldarium* dei Campi Flegrei (Napoli) in presenza di diverse fonti di azoto, *Delpinoa* **12/13**, 3-8.

De Luca, P. and Taddei, R. (1976) On the necessity of a systematic revision of the thermal acidophilic alga *Cyanidium caldarium* Tilden Geitler, *Webbia* **30**, 197-218.

Doemel, W.N. and Brock, T.D. (1970) The upper temperature limit of *Cyanidium caldarium*, *Arch Microbiol* **72**, 326-332.

Gross, W., Küver, J., Tischendorf, G., Bouchaala, N., and Büsch, W. (1998) Cryptoendolithic growth of the red alga *Galdieria sulphuraria* in volcanic areas, *Eur J Phycology* **33**, 25-31.

Gross, W. and Schnarrenberger, C. (1995) Heterotrophic growth of two strains of the acido- thermophilic red alga *Galdieria sulphuraria*, *Plant Cell Physiol* **36**, 633-638.

Merola, A., Castaldo, R., De Luca, P., Gambardella, R., Musachio, A., and Taddei, R. (1981) Revision of *Cyanidium caldarium*. Three species of acidophilic algae, *Giorn Bot Ital* **115**, 189-195.

Moretti, A. and Musacchio, A. (1982) Chemotaxonomic notes on *Porhyridium* Nag. (Rhodophyta, Porphyridiales), *Giorn Bot Ital* **116**, 269-274.

Nagashima, H. (1994) Natural products of the Cyanidiophyceae, in J. Seckbach (ed.), *Evolutionary Pathways and Enigmatic Algae: Cyanidium caldarium (Rhodophyta) and Related Cells*, Kluwer Academic Publishers, Dordrecht, pp 201-214.

Nagashima, H., Matsumoto, G.I., and Fukuda, I. (1986) Hydrocarbons and fatty acids in two strains of the hot spring alga *Cyanidium caldarium*, *Phytochem* **25**, 2339-2341.

Ott, F.D. and Seckbach, J. (1994) New classification of the genus *Cyanidium* Geitler 1933, in J. Seckbach (ed.), *Evolutionary Pathways and Enigmatic Algae: Cyanidium caldarium (Rhodophyta) and Related Cells*, Kluwer Academic Publishers, Dordrecht, pp 113-132.

Round, F.E. (1981) *The ecology of algae*, Cambridge University Press, Cambridge.

Seckbach, J. (1991) Systematic problems with *Cyanidium caldarium* and *Galdieria sulphuraria* and their implications for molecular biology studies, *J.Phycol.* **27**, 794-796.

Sentsova, O. Y. (1994) The study of Cyanidiophyceae in Russia. Algae of genus *Galdieria*: Diversity, characterization and occurrence in mixed populations with *Cyanidium caldarium*. in J. Seckbach (ed.), *Evolutionary Pathways and Enigmatic Algae: Cyanidium caldarium (Rhodophyta) and Related Cells*, Kluwer Academic Publishers, Dordrecht, pp 167-174.

Smith, D.W. and Brock, T.D. (1973) The water relations of the alga *Cyanidium caldarium* in soil, *J Gen Microbiol* **79**, 219-231.

Volkman, J.K., Brown, M.R., Dunstan, G.A., and Jeffrey, S.W. (1993) The biochemical composition of marine microalgae from the class Eustigmatophyceae, *J Phycol* **29**, 69-78.

Biodata of **Tsuneyoshi Kuroiwa** coauthor with R. Itoh of the chapter on *" **Mechanism and Evolution of Organelle Division.**"*

Dr. Tsuneyoshi Kuroiwa is a Professor of Biology and the Vice-Dean of the School of Science, University of Tokyo (Tokyo, Hongo). He received his ScD degree in 1971 from the University of Tokyo. Dr. Kuroiwa main interest is in the mechanism of organelle division and inheritance. His group is known to produce very high qualified electron-micrographs. Among their identifications is the chromosomal behavior during interphase (in *Crepis*), sexuality of mitochondria by plasmids, chromosomes of yeast and first identified the division apparatus of plastids and mitochondria by dividing rings. In 1998 he received the Japanese Society for Electron Microscopy prize for his (and associates) discovery of the plastid and mitochondrion dividing apparatus.

E-Mail: **kuroiwa@biol.s.u-tokyo.ac.jp**

Biodata of **Ryuuichi Itoh** contributor (with T. Kuroiwa) of *"**Mechanism and Evolution of Organelle Division.**"*

Ryuuichi Itoh is a graduate student (The 5th Grade in the Doctoral Course) in T. Kuroiwa's group at the Department of Biological Sciences , University of Tokyo (Tokyo, Japan). His main interest is in morphogenesis and proliferation of organelles, especially plastids and mitochondria. Currently he studies the chloroplast and mitochondrial FtsH proteins of the red alga *Cyanidioschyzon merolae*. Mr. Itoh has a few publications (with other member of kuroiwa's group) on the most primitive eukaryotic alga *Cyanidioschyzon merolae*.

E-Mail: **rgitoh@biol.s.u-tokyo.ac.jp**

J. Seckbach (ed.), Enigmatic Microorganisms and Life in Extreme Environments, 447–463.

MECHANISM AND EVOLUTION OF ORGANELLE DIVISION[1]

R. ITOH and T. KUROIWA
Department of Biological Sciences, Graduate School of Science, University of Tokyo, Hongo 113-0033, Japan

1. Introduction

Faithful inheritance of organelles by daughter cells is an essential component of cell division in eukaryotes. The mechanism of inheritance of the nucleus has been studied in more detail than that of any other organelle. Briefly, the chromosome DNA, the main cargo of the nucleus, duplicates at a specific point in the cell cycle (S phase). In the next stage (M phase), the nuclear envelope fragments into vesicles and the chromosomes are separated and transported to the opposite poles of the cell by the spindle, an apparatus consisting mainly of microtubules. After the chromosomes have separated, a nuclear envelope is reconstructed from the dispersed vesicles surrounding the chromosomes in each daughter cell (Warren and Wickner 1996). This process of nuclear inheritance has been described in detail at the molecular level. On the other hand, inheritance of the cytoplasmic organelles, such as the plastids and mitochondria, is not well understood at the present, and the molecular basis of the proliferation (or division) and distribution of the organelles remains largely unknown.

Plastids and mitochondria are generally called "semiautonomous organelles" because (1) it is believed that they originated from free-living blue-green algae and aerobic bacteria, respectively, (2) they still retain their own genetic system, and (3) they proliferate by the division of preexisting organelles with the transmission of their own genomic DNA. Such semiautonomous characteristics have stimulated many studies, which have increased our knowledge of the activities of these organelles tremendously - including the organization and replication of the genome, gene expression, protein synthesis, differentiation, energy conversion, and protein targeting and import. Nevertheless, one central problem remains unsolved - How do the organelles proliferate?

The key to solving this problem may be the "**plastid-dividing (PD) ring**" and the "**mitochondrion-dividing (MD) ring**". Both of these organelle-dividing rings were first identified in our laboratory in two species of **Cyanidiophyceae**: the PD ring in *Cyanidium caldarium* (Mita et al. 1986, Mita and Kuroiwa 1988) and the MD ring in *Cyanidioschyzon merolae* (Kuroiwa et al. 1993, 1995). In this chapter, we survey the recent progress in the study of plastid and mitochondrial division, with special focus on the organelle-dividing rings and the use of Cyanidiophyceae as model organisms for studying organelle division.

---

[1] Abbreviations: DiOC6, 3,3'-dihexyloxacarbocyanine iodide; PD, plastid-dividing; MD, mitochondrion -dividing

2. Organelle Division from the Perspective of the Endosymbiotic Theory

It is widely accepted that plastids and mitochondria originated from free-living cyanobacteria and aerobic eubacteria, respectively (Margulis 1981). As a matter of course, these ancestors of organelles possessed the molecular apparatus for their own division. Do the organelles retain any rudiments of the division apparatus of their ancestors? plastids have. In *Arabidopsis thaliana*, a nuclear gene encoding the FtsZ protein, which is localized in the plastid stroma, was discovered (Table 1; Osteryoung and Vierling 1995). FtsZ is a GTPase with tubulin-like motifs, and is ubiquitous in all prokaryotes studied, including cyanobacteria (RayChaudhuri and Park 1992, de Boer et al. 1992). Mutation of the *ftsZ* gene causes a defect in cytokinesis at certain restrictive temperatures, resulting in filamentous elongation of the bacteria. The FtsZ protein forms a ring structure beneath the cell membrane in the plane of cell division (Bi and Lutkenhaus 1991, Ma et al. 1996). *In vitro*, FtsZ is capable of filament formation (Bramhill and Thompson 1994, Erickson et al. 1996, Erickson and Stoffler 1996). From these characteristics, FtsZ is believed to be the bacterial division apparatus (Erickson 1995, 1997, Rothfield and Justice 1997). Plastid FtsZ (designated as cpFtsZ) in *A. thaliana* seems to associate with the inner envelope of plastids and to pinch off the inner membrane and thylakoids. This is logical, considering the topological correlation between the bacterial cell membrane and the plastid envelope. A second *A. thaliana* sequence FtsZ was deposited into the GenBank database as part of an expression sequence tag (Erickson et al. 1996). Although this second FtsZ has not yet been characterized, it is possible that it encodes a mitochondrial isotype. However, a search of the complete Saccharomyces cerevisiae nucleotide sequence has failed to reveal an FtsZ sequence in its genome. So at the least, FtsZ is not a necessary requirement for mitochondrial division in all eukaryotes.

Traces of the bacterial division apparatus are also found in the plastid genome. The complete nucleotide sequence of the chloroplast genome from the green alga *Chlorella vulgaris* (Wakasugi et al. 1997) revealed the existence of genes homologous to *minD* and *minE*, which are components of the system for determining the plane of cell division (Rothfield and Zhao 1996). A homolog of the bacterial cell division protein FtsH is also encoded in the chloroplast genome of *C. vulgaris* and red algae such as Porphyra purpurea (Reith and Munholland 1995), *Cyanidium caldarium*, and *Cyanidioschyzon merolae* (Ohta, personal communication). FtsH is an ATP-dependent metalloprotease in which mutations cause defective cell division and produce filamentous cells similar to those seen with the *ftsZ* mutation (Santos and Almeida 1975). In seed plants, plastid FtsHs are encoded in the nuclear genome, not the plastid genome, and have been characterized biochemically (Hugueney et al. 1995, Lindahl et al. 1996, Ostersetzer and Adam 1997). The role of FtsH in the process of bacterial cell and plastid division is largely unknown, but a role is suggested since the plastid FtsH of the red pepper *Capsicum annuum* was isolated as the vesicle fusion factor in the stroma during plastid differentiation (Hugueney et al. 1995). A similar role for FtsH *in C. annuum,* designated as Pftf (for **p**lastid **f**usion and/or **t**ranslocation **f**actor), implies that plastid FtsH is required for membrane fusion and the reorganization of the inner envelope and thylakoid that occurs during plastid division. In addition, plastid division of cultured *Physcomitrella patens* cells is inhibited by β-lactam antibiotics such as ampicillin, cefotaxime, and penicillin (Kasten and Reski 1997). The target of β-lactam

Table 1. Bacterial cell division genes, whose homologue(s) have been found in eukaryotes.

Gene	Phenotypes of mutants	Features of its product	Eukaryotic homologue(s)	Refs.
ftsZ	Inhibition of cytokinesis Filamentous elongation of cell	GTPase that self-assembles into filaments Formation of ring beneath the cell membrane surrounding the cell division plane	*cpFtsZ* , encoded on nuclear genome of *Arabidopsis thaliana* The product is localized in plastid stroma	1
ftsH	Inhibition of cytokinesis Filamentous elongation of cell	ATP-dependent metalloprotease Elimination of unassembled and misassembled proteins	Encoded on nuclear genomes of higher plants (plastid-localized type) and yeasts (mitochondrial inner membrane-localized type), and on plastid genomes of red- and green algae	2-16
minD	Minicell formation	Determination of cell division plane Inhibition of cell division at all potential division sites	Encoded on plastid genome of *Chlorella vulgaris*	2
minE	Formation of nonseptate filaments	Determination of cell division plane Permission of cell division at the midcell site by inhibition of MinCD division inhibitor	Encoded on plastid genome of *Chlorella vulgaris*	2

References: 1. Osteryoung and Vierling (1995), 2. Wakasugi *et al.* (1997), 3. Ohta, personal communication, 4. Reith and Munholland (1995), 5. Hugueney *et al.* (1995), 6. Lindahl *et al.* (1996), 7. Ostersetzer and Adam (1997), 8. Schnall *et al.* (1994), 9. Guélin *et al.* (1994), 10. Tauer *et al.* (1994), 11. Tzagoloff *et al.* (1994), 12. Thorsness *et al.* (1993), 13. Nakai *et al.* (1995), 14. Campbell *et al.* (1994), 15. Thorsness and Fox (1993), 16. Thorsness *et al.* (1993)

antibiotics is the bacterial cell division protein FtsI, which is required for division-septum formation and bringing FtsZ into contact with the cell membrane (Rothfield and Justice 1997). The inhibition of plastid division by β-lactam antibiotics indicates the existence of a FtsI-like protein involved in plastid division that acts in concert with FtsZ, although neither this protein nor its corresponding gene has yet been reported.

To date, no gene related to bacterial cell division has been reported to be encoded by the mitochondrial genome. In the nuclear genome, there are only three reported FtsH homologues, Afg3p/Yta10p (Schnall et al. 1994, Guélin et al. 1994, Tauer et al. 1994), Yme1p/Yta11p (Thorsness et al. 1993, Nakai et al. 1995), and Rca1p/Yta12p (Schnall et al. 1994, Tzagoloff et al. 1994), which are localized on the inner mitochondrial membrane of *S. cerevisiae*. These proteins have been reported to function primarily as factors required for protein assembly on the mitochondrial membrane and for the degradation of unassembled proteins. Therefore, the role of mitochondrial FtsH in mitochondrial division remains unclear. However, the mitochondria of strain *yme1* are punctate and swollen, unlike the tubular, reticular shape of the mitochondria of wild type cells (Campbell et al. 1994). In addition, the *yme1* mutation causes **DNA** to escape from the mitochondria into the nucleus (Thorsness and Fox 1993, Thorsness et al. 1993). These observations demonstrate the necessity of mitochondrial **FtsH** for maintaining mitochondrial morphology and integrity, and it probably also plays a role in mitochondrial division.

In conclusion, mitochondria retain fewer rudiments of the division apparatus of their ancestral bacteria than plastids. A large part of the molecular apparatus for organelle division was replaced by novel mechanisms with the evolution of eukaryotic cells. Then, what constitutes the new division apparatus, which evolved after endosymbiosis?

3. Organelles Division Apparatus

The central enigma of organelle division is the gap between the division apparatus of organelles and that of bacteria, which are the ancestors of organelles. This gap can be seen from two different, but interrelated perspectives. One concerns the topology of the membrane and the division apparatus, and the other concerns the biochemical composition of the division apparatus. The division apparatus of plastids is called the "plastid-dividing (PD) ring" and was first identified in the red alga *Cyanidium caldarium* (Fig. 1; Mita et al. 1986, Mita and Kuroiwa 1988). The PD ring is a closed ring, which consists of a bundle of many thin filaments, that encircles the cytoplasmic side of the outer envelope of the isthmus where the plastid will divide. As well as in the red algae, the PD ring is also found in green algae (Chida and Ueda 1991, Ogawa et al. 1995), brown algae (Hashimoto 1997), bryophytes (Tewinkel and Volkmann 1987, Vaughn et al. 1992, Duckett and Ligrone 1993a), ferns (Duckett and Ligrone 1993b), and seed plants (Hashimoto 1986, Oross and Possingham 1989, Modrusan and Wrischer 1990, Robertson et al. 1996). These findings strongly suggest the universality of the PD ring throughout the plant kingdom. The PD ring is often observed as a "**doublet**", with a cytoplasmic "**outer**" ring and an additional stromal "**inner**" ring backing the inner envelope (Hashimoto 1986, Kuroiwa et al. 1998). A counterpart of the PD ring was discovered on the single mitochondrion of the red alga *Cyanidioschyzon merolae* and

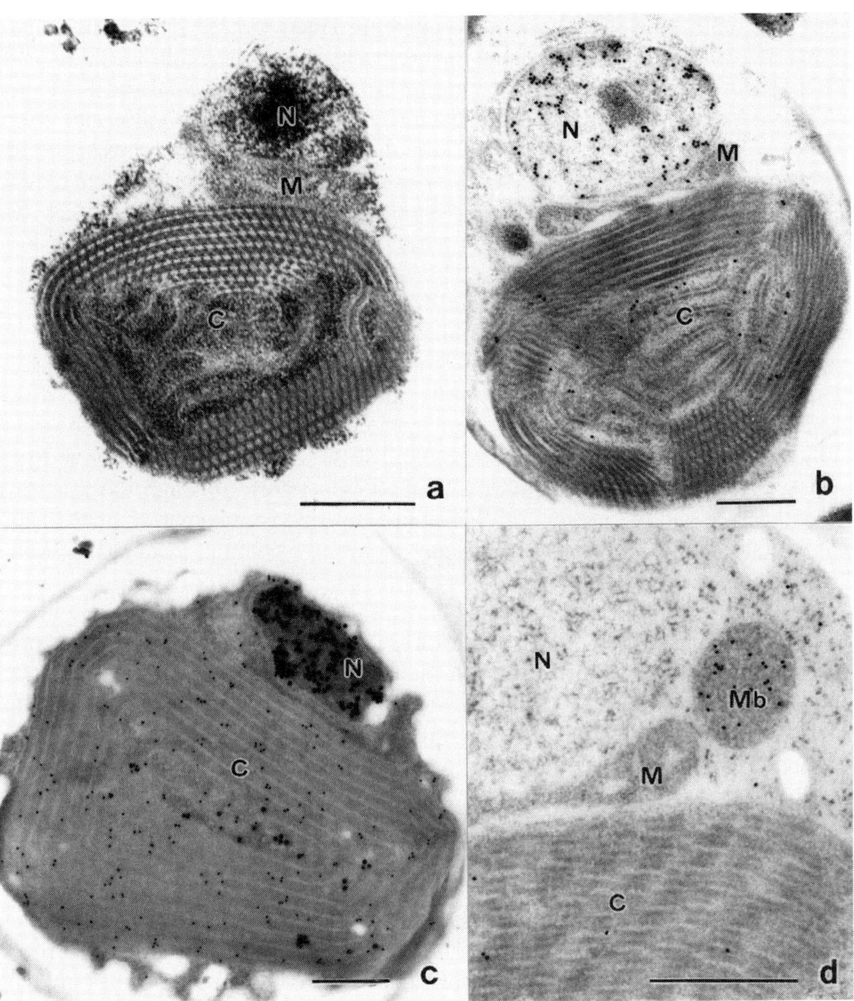

Fig. 1. Electron micrographs of *Cyanidium caldarium*, showing the accessibility of this cell for cytological and immunocytochemical studies. (a) Compact alignment of nucleus (N), mitochondrion (m), and chloroplast (C) in a cell in that order. (b) Immunogold localization of DNA on these three organelles. (c) Double immunogold localization of DNA (larger particle) and SigA (chloroplast RNA polymerase s subunit [Tanaka *et al.* 1996]; smaller particle), confirming the specific localization of SigA on chloroplast. (d) Immunogold localization of catalase on microbody (Mb). Scale bars represent 0.5 μm.

called the "mitochondrion-dividing (MD) ring" (Figs. 2, 3; Kuroiwa et al. 1993, 1995). The MD ring has the same general topology as the PD ring. It is a double closed ring that consists of a cytoplasmic ring on the outer membrane and a matrix ring beneath the inner membrane. These structures are different from the bacterial cell division ring composed of FtsZ protein, currently called the "**Z-ring**". The Z-ring is only seen beneath the cell membrane in the plane of cell division with immunogold labeling, although FtsZ filaments cannot be observed in vivo (Bi and Lutkenhaus 1991, Erickson 1997). No division structure has been detected outside the cell membrane in the periplasmic space. If there is a correspondence between the membrane topology of dividing organelles and their bacterial progenitors, then we can hypothesize that the inner **PD** and **MD** rings are rudiments of the bacterial cell division apparatus Z-ring. This hypothesis is supported by the recent discovery in *A. thaliana* of an FtsZ protein that localizes in plastids (cpFtsZ) and is nuclear-encoded (Osteryoung and Vierling 1995). However, bacteria do not possess any division structures that correspond to the outer **PD** or **MD** rings. These are novel division apparatus, which must have evolved with the rise of the eukaryotic cell. What are the molecular components of these rings? How did they evolve?

When observed under the electron microscope, both the outer PD and MD rings are composed of many thin filaments with a diameter of about 7 nm, which is the same diameter as actin filaments. Rhodamine-phalloidin staining of the green algae *Closterium ehrenbergii* (Hashimoto 1992) and *Nannochrolis bacillaris* (Ogawa et al. 1995) revealed that actin filaments girdle the constricted isthmus of the dividing chloroplasts. Similar actin rings surrounding the chloroplasts have also been observed in cultured tobacco BY-2 cells (Hasezawa et al. 1988) and protonemal cells of the fern *Adiantum capillus-veneris* (Kadota and Wada 1989, 1992), although the relationship between this ring and the process of chloroplast division is unknown. Cytochalasin, an inhibitor of actin polymerization, inhibits chloroplast division in *C. caldarium* (Mita and Kuroiwa 1988) and *C. ehrenbergii* (Hashimoto 1992), and mitochondrial division in *Physarum polycephalum* (Kuroiwa and Kuroiwa 1980). These facts suggest that both the outer PD and MD rings are probably composed of actin or an actin-like filament-forming protein. Remember that the cytoplasmic division ring (the contractile ring) is also composed of cytoplasmic actin filaments and is located immediately beneath the plasma membrane. Considering the endosymbiotic origin of plastids and mitochondria, it is easy to suppose that the cytoplasmic face of the outer membrane of these organelles and the cytoplasmic face of the plasma membrane are "equal" from a topological perspective. Therefore, we hypothesize that the outer PD and MD rings are derivatives of the contractile ring, and evolved at the same time as eukaryotes. Very early in the evolution of endosymbiosis, eukaryotic cells were probably extremely simple and contained one plastid, one mitochondrion, and one of every other organelle per cell, just like today's unicellular red alga *Cyanidioschyzon merolae* (see the next section for a detailed discussion). During cell division, these organelles may have moved into the plane of cell division and undergone binary division synchronously with cytoplasmic division, brought about by a single contractile ring. As the eukaryotic cells evolved a more complicated architecture, the plastids and mitochondria obtained their own outer dividing rings so that organelle division became independent of cytoplasmic division. Of course, bacteria do not possess a contractile ring that consists of actin filaments. However, an actin-like protein FtsA is found in bacteria. **FtsA** is involved in

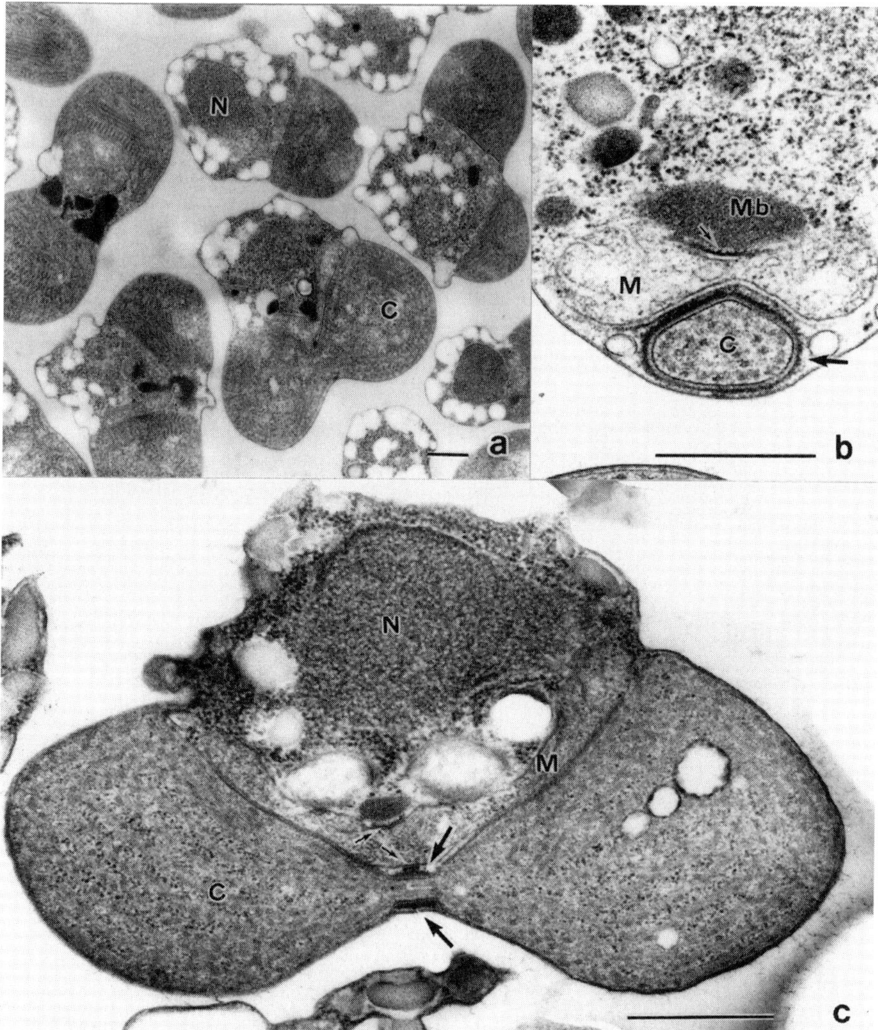

Fig. 2. Electron micrographs of the dividing profiles of *Cyanidioschyzon merolae.* (a) Highly synchronized division of cell and organelles, obtained by light-dark cycles. Most cells contain a dumbbell-shaped dividing chloroplast (C) or two small chloroplasts after its division. (b, c) Plastid-dividing (PD) ring (large arrow) and mitochondrion-dividing (MD) ring (small arrow). In b, as the section was cut parallel to the division plane, PD ring is observed as a closed ring surrounding the chloroplast. Closure of MD ring looks incomplete in this image. In c, as the section was cut vertically to the division plane, both PD ring and MD ring are seen as a couple of belt-like deposits, flanking the constricted isthmus on both sides. N; nucleus, M; mitochondrion, Mb; microbody. Scale bars represent 0.5 μm.

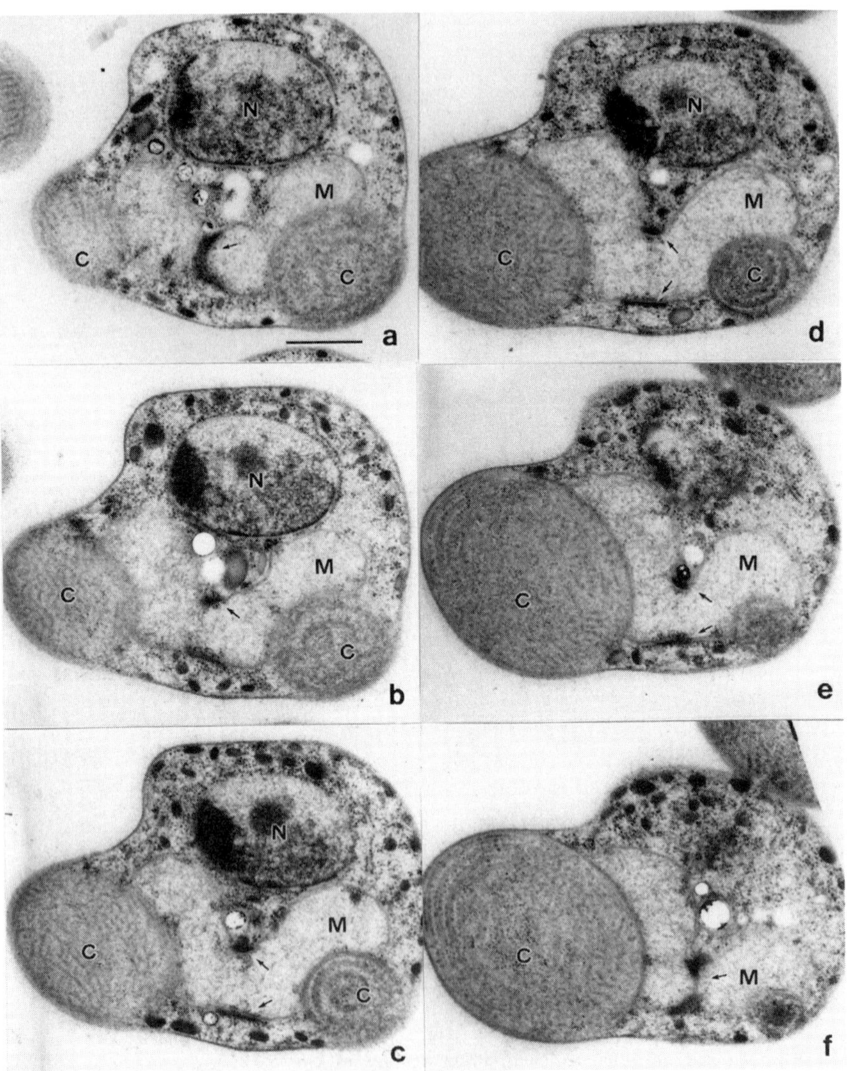

Fig. 3. Serially-sectioned images of a dividing mitochondrion (M) in a *Cyanidioschyzon merolae* cell, representing a closed mitochondrion-dividing (MD) ring (small arrow). On the top plane (a), the surface of MD ring is seen as an electron-dense band on the division plane of the mitochondrion. On the lower planes (b-e), this structure splits into two belt-like deposits flanking the constricted isthmus on both sides. On the bottom plane (f), two deposits are about to merge again into a single band like that seen in a, indicating the closure of MD ring. N; nucleus, C; chloroplast. Scale bars represent 0.5 μm.

bacterial cell division and co-localizes with the Z-ring (Ma et al. 1996). Hence, the FtsA ring in the host cells of protoeukaryotes may be the progenitor of the eukaryotic contractile ring and, if we accept the aforementioned hypothesis, the outer PD and MD rings. FtsZ in the host cells of protoeukaryotes may have evolved into the mitotic division apparatus, namely the spindle. It is interesting to imagine how the original single FtsZ/A ring for bacterial cell division split into five rings: the inner PD and MD rings made of FtsZ, and the outer PD and MD rings and a contractile ring made of FtsA (Fig. 4).

The doublet structure of the PD and MD rings seems to be problematic. How do the inner and outer rings cooperate to produce organelle division?. The inner ring forms before the outer ring, as revealed in electron microscopic studies of *C. merolae* (Miyagishima et al. submitted). This observation seems quite reasonable, because the plane of division is likely to be determined on the matrix side of organelles, possibly utilizing the **minCDE system**, the division plane determinant system of bacteria (Rothfield and Zhao 1996, Wakasugi et al. 1997). The inner ring may have a dual role, producing invagination of the inner membrane and acting as the "guide" that instructs the outer ring protein where to form the outer ring. Recently, a third ring was discovered in the intermembrane space between the inner and outer envelopes of chloroplasts in *C. merolae* (Miyagishima et al. submitted). This structure may be an "anchor" or "information transmitter" between the inner and outer rings, rather than the source of the contractile force. After the plane of division is determined and the inner ring has formed, the outer ring is built. Formation of the outer PD ring seems to involve vesicle transport to the plane of plastid division. Unidentified vesicles associated with the outer PD ring have been observed in *C. caldarium* (Mita and Kuroiwa 1988), *C. merolae* (Miyagishima et al. submitted), and the green alga *Trebouxia potteri* (Chida and Ueda 1991). It is important to identify these vesicles associated with the PD ring in order to understand the mechanism for the construction of the outer PD ring.

After the gap between the division apparatus of organelles and that of bacteria, the second most enigmatic feature of organelle division is the fact that the MD ring cannot be observed in organisms other than *C. merolae*. This is very strange, because it is believed that mitochondria proliferate in all organisms by the division of preexisting organelles, not by de novo synthesis. How do the mitochondria divide in organisms other than *C. merolae*?. The observation of mitochondrial division in the green alga *Scenedesmus acutus* may provide a clue to this problem (Ueda and Noguchi 1993). The mitochondrial division apparatus in *S. acutus* appears as an electron-dense region in the plane of division, not as a bundle of filaments surrounding the cytoplasmic side of the dividing isthmus as seen in *C. merolae*. In other words, while the mitochondrial division apparatus of *C. merolae* is "cytoskeletal", on the outside of the organelle, in *S. acutus* it appears to be a "mitochondrion-skeletal" apparatus that is intrinsic to the mitochondria themselves. These two organisms are the only ones in which the existence of a microscopically-visible mitochondrial division apparatus has been reported. However, we dare to hypothesize that the visibility or invisibility of the mitochondrial division apparatus in different species may reflect the evolution of the mitochondrial division apparatus. In the very early stages of the evolution of endosymbiosis, mitochondria may have been divided by a division ring, which was derived from the contractile ring, on the cytoplasmic side of organelles. Naturally, this type of mitochondrial division occurred at the same time as cytokinesis. The primitive asexual unicellular red alga *C.*

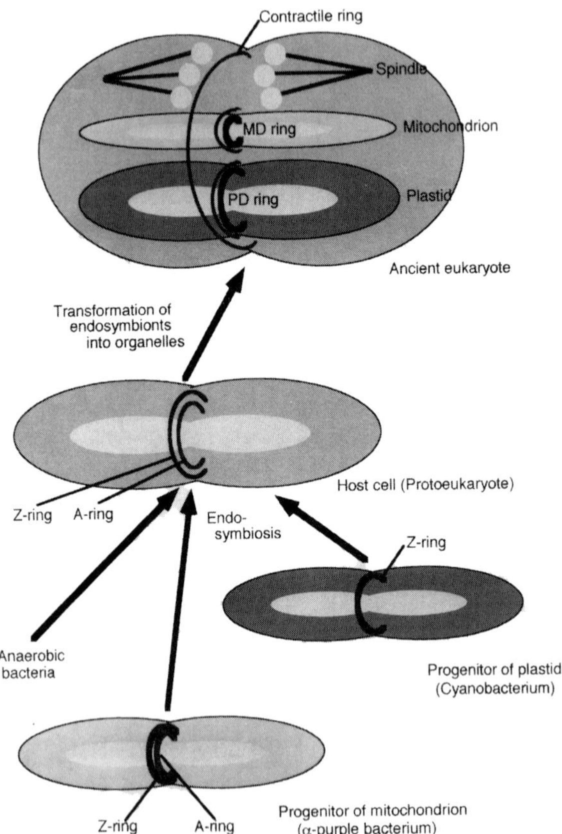

Fig. 4. **Hypothesis for the evolution of the division apparatuses of plastids and mitochondria. In this hypothesis, inner plastid-dividing (PD) ring is derived from the FtsZ ring (Z-ring), the general division apparatus of prokaryotes, of the progenitor of plastids (cyanobacterium). Similarwise, inner mitochondrion-dividing (MD) ring is derived from the Z-ring of the progenitor of mitochondria (a-purple bacterium). Three cytoplasm-facing rings - outer PD ring, outer MD ring, and cytoplasmic division ring (contractile ring) - are derived from the FtsA ring (A-ring) of protoeukaryotic host cell. The Z-ring of host cell may have converted into the mitotic division apparatus, spindle. Division apparatuses of host cell and their derivatives are drawn by narrower line, while division apparatuses of endosymbionts and their derivatives are drawn by thicker line. Structures composed of FtsZ or tubulin are drawn by dark color, while structures composed of FtsA or actin are drawn by brighter color. Whitish regions in cells are DNA-containing regions, *i. e.* nucleoids and chromosomes.**

merolae still retains this mode of mitochondrial division. As sexual reproduction evolved in eukaryotes, with the evolution of multicellular organisms and cell differentiation, mitochondria were required to assume different shapes, to move in the cytoplasm, and even to fuse with each other (Bereiter-Hahn and Vöth 1994, Köhler et al. 1997b), besides simply proliferating. To enable this, mitochondria evolved a highly ordered, sophisticated "mitochondrial skeleton". Gradually, the mitochondria evolved their own division apparatus, which is so sophisticated that we cannot visualize it in higher organisms with current techniques. The mitochondrial division apparatus *of S. acutus* may represent an intermediary in the evolution of mitochondrial division.

4. Cyanidioschyzon Merolae: The Simplest Plant Cell

C. merolae is a model organism for investigating the control mechanism of organelle division for the following reasons: (1) *C. merolae* is a unicellular red alga that contains one chloroplast, one mitochondrion, one microbody, one Golgi apparatus, and one nucleus per cell (Kuroiwa et al. 1994, Okuwaki et al. 1996). Before cytokinesis, these organelles divide in a specific order (Suzuki et al. 1994). (2) The mitotic and organelle division cycles can be synchronized by subjecting the cells to a 12-h light/dark cycle (Fig. 2a; Suzuki et al. 1994, Terui et al. 1995). (3) *C. merolae* is the only organism in which both PD and MD rings have been observed by electron microscopy (Figs. 2b, 2c, 3; Kuroiwa et al. 1993, 1995, Suzuki et al. 1994). (4) Its genome size is only 14 Mbp, which is as small as that of *S. cerevisiae* (Toda et al. 1995, Takahashi et al. 1995). (5) The complete nucleotide sequences of both the chloroplast and mitochondrial genomes have already been determined (Ohta, personal communication). Finally, (6) a molecular phylogeny based on the nucleotide sequences of actin genes indicates that *C. merolae* is quite primitive (Takahashi et al. 1995). Therefore, we can assume that *C. merolae* retains at least some of the simple plant cell architecture seen in the very early stages of the evolution of eukaryotic cells.

Using *C. merolae*, we performed a series of inhibitor experiments to elucidate both the nuclear controlling mechanism of organelle division and the differences between the division mechanisms of chloroplasts and mitochondria. Since the organelle division cycles of *C. merolae* appear to be tightly coupled with the mitotic cycle, we assumed the existence of a checkpoint control that inhibits organelle division prior to the completion of nuclear DNA synthesis. To test this assumption, we arrested the mitotic cycle in the S phase by adding aphidicolin, an inhibitor of DNA polymerase α, to a synchronous culture, and observed the mode of chloroplast division by both epifluorescence and electron microscopy (Itoh et al. 1996, 1997a). These experiments clearly distinguished between the controlling mechanisms of chloroplast and mitochondrial division. When cells in the S phase in synchronous cultures were treated with aphidicolin, mitosis and cytokinesis did not occur, while chloroplast division did. Moreover, the chloroplasts in these cells continued to divide, ultimately generating four or more chloroplasts per cell. In addition, microfluorometry revealed that the total amount of DNA in the chloroplasts in aphidicolin-treated cells remained constant during uncoupled chloroplast division. Consequently, the size and amount of DNA per chloroplast decreased in a stepwise manner with chloroplast division. Electron-microscopic examination of aphidicolin-treated cells showed that the second division of chloroplasts involved a PD ring similar to that of cells undergoing normal chloroplast division. These results suggest that chloroplast division brought about by the PD ring is free from the checkpoint control that inhibits the progression of mitosis and cytokinesis before the completion of nuclear DNA synthesis, and that chloroplasts lack a checkpoint control mechanism, which inhibits their division without chloroplast DNA synthesis or growth. On the other hand, epifluorescence microscopy after staining with the mitochondrion-specific fluorochrome DiOC6 revealed that mitochondrial division was completely inhibited. Nevertheless, electron-microscopic examination of the aphidicolin-treated cells clearly revealed the presence of an MD ring in the mitochondria of all cells examined, in spite of the absence of mitochondrial division. [It is worth noting that the organelle phenotype is quite similar to that of the dividing chloroplast of the *A. thaliana arc5* mutant (Robertson et al. 1996).] The microbodies also failed to divide. These results imply that a checkpoint control mechanism inhibits division of the mitochondria and microbodies in the absence of the synthesis of nuclear DNA. This checkpoint control mechanism allows formation of the MD ring, but not its contraction if the S phase is not completed. Consequently, mitochondrial division seems to involve two independent processes, the formation and the contraction of the MD ring. These two events can be separated experimentally. From these results, we can conclude that chloroplast and mitochondrial division are regulated independently in the cascades of cell cycle regulation. These observations - chloroplasts that proliferate independently of the nucleus and mitochondria that cease to proliferate under nuclear control - remind us of the amounts of rudimentary bacterial cell division apparatus that are found in the organelles. There are fewer rudiments in the mitochondria than in the plastids, as pointed out in Section 2.

A difference between plastids and mitochondria is also found in the division of their nucleoids. In *C. merolae*, each of these organelles has a simple centrally located nucleoid. Prior to mitosis and cytokinesis, division of the organelle nucleoid and the organelle itself occurs simultaneously. We investigated the effects of nalidixic acid, an inhibitor of DNA gyrase, on the replication and division of organelle nucleoids (Itoh et al. 1997b). DNA gyrase is the bacterial topoisomerase II required for DNA replication and subsequent chromosome segregation (Wheeler and Shapiro 1997). Using microfluorometry, we monitored the synthesis of organelle DNA (organelle S phase) in synchronous cultures. In both the chloroplast and the mitochondrion, DNA synthesis commenced at the start of the light period and ceased within 3 h. The addition of nalidixic acid at the start of the light period inhibited DNA synthesis in both organelles. Next, we added nalidixic acid to the synchronous culture at the beginning (just before the organelle S phase) or the end (organelle G2 phase) of the light period, and monitored the division of organelle nucleoids during the dark period. Approximately 50-70% of the chloroplast nucleoids divided unequally in both experiments. Despite the unequal division of the chloroplast nucleoids, electron microscopy revealed that the chloroplasts were divided into equal sized daughter chloroplasts by the PD ring. With the same treatment, all the mitochondrial nucleoids underwent normal and equal division. Our results suggest the involvement of DNA gyrase in the replication of chloroplast and mitochondrial DNA and in the division of the chloroplast nucleoid, as well as a difference in the mechanisms for the division of the chloroplast and mitochondrial nucleoids. It is likely that the chloroplast employs DNA gyrase in its nucleoid division, while the mitochondrion utilizes eukaryotic topoisomerase II for its nucleoid division, as does the cell nucleus during chromosome segregation. Again, we are reminded that plastid division has more prokaryotic features than mitochondrial division.

It should be noted that these results were obtained by simple observation of *C. merolae* cells treated with inhibitors. This demonstrates how useful *C. merolae* is as a model organism for studies of organelle division.

5. Future prospects

How do plastids and mitochondria divide? In order to answer this question, we must discover a protein key to organelle division, analogous to p34/*cdc2* in cell cycle research or FtsZ in bacterial cell *division research. Since 1991, when FtsZ protein was seen to localize in the Escherichia coli* cell division plane (Bi and Lutkenhaus 1991), progress in bacterial cell division research has accelerated, with a focus on the function of FtsZ. From this, we can surmise that the identification of a protein localized in the plane of organelle division would be a significant breakthrough for research in this field. Since it is not easy to identify the plane of division of organelles in higher organisms, discovering such proteins in these organisms seems difficult. *C. merolae* is ideal for this task. Proteins isolated from organisms other than *C. merolae* can be screened easily using immunolocalization in *C. merolae* cells in order to identify proteins localized in the plane of organelle division. Once such a protein is found, we should be able to visualize the organelle division apparatus in higher organisms, in which the organelle division apparatus has not presently been visualized, by immunolocalization or introduction of the chimeric gene fused to the GFP (green fluorescent protein) gene into living cells (Köhler et al. 1997a, b). We hope that by studying C. merolae, proteins localized in the plane of organelle division can be isolated.

6. Acknowledgments

R.I. was supported by a research fellowship (no.3641) for young scientists from the Japan Society for the Promotion of Science.

7. References

Bereiter-Hahn, J. and Vöth, M. (1994) Microsc. Res. Tech. 27:198-219.
Bi, E. and Lutkenhaus, J. (1991) Nature 354:161-164.
Bramhill, D. and Thompson, C. M. (1994) Proc. Natl. Acad. Sci. USA 91:5813-5817.
Campbell, C. L., Tanaka, N., White, K. H., and Thorsness, P. E. (1994) Mol. Biol. Cell 5:899-905.
Chida, Y. and Ueda, K. (1991) Ann. Bot. 67:435-442.
de Boer, P., Crossley, R., and Rothfield, L. (1992) Nature 359:254-256.
Duckett, J. G. and Ligrone, R. (1993a) G. Bot. Ital. 127:318-319.
Duckett, J. G. and Ligrone, R. (1993b) Ann. Bot. 72:619-627.
Erickson, H. P. (1995) Cell 80:367-370.
Erickson, H. P. (1997) Trends Cell Biol. 7:362-367.
Erickson, H. P. and Stoffler, D. (1996) J. Cell Biol. 135:5-8.
Erickson, H. P., Taylor, D. W., Taylor, K. A., and Bramhill, D. (1996) Proc. Natl. Acad. Sci. USA 93:519-523.

Guélin, E., Rep, M., and Grivell, L. A. (1994) Yeast 10:1389-1394.
Hasezawa, S., Nagata, T., and Syono, K. (1988) Protoplasma 146:61-63.
Hashimoto, H. (1986) Protoplasma 135:166-172.
Hashimoto, H. (1992) Protoplasma 167:88-96.
Hashimoto, H. (1997) Protoplasma 197:210-216.
Hugueney, P., Bouvier, F., Badillo, A., d'Harlingue, A., Kuntz, M., and Camara, B. (1995) Proc. Natl. Acad. Sci. USA 92:5630-5634.
Itoh, R., Takahashi, H., Toda, K., Kuroiwa, H., and Kuroiwa, T. (1996) Eur. J. Cell Biol. 71:303-310.
Itoh, R., Takahashi, H., Toda, K., Kuroiwa, H., and Kuroiwa, T. (1997a) Protoplasma 196:135-141.
Itoh, R., Takahashi, H., Toda, K., Kuroiwa, H., and Kuroiwa, T. (1997b) Eur. J. Cell Biol. 73:252-258.
Kadota, A. and Wada, M. (1989) Protoplasma 151:171-174.
Kadota, A. and Wada, M. (1992) Protoplasma 167:97-107.
Kasten, B. and Reski, R. (1997) J. Plant Physiol. 150:137-140.
Köhler, R. H., Cao, J., Zipfel, W. R., Webb, W. W., and Hanson, M. R. (1997a) Science 276:2039-2042.
Köhler, R. H., Zipfel, W. R., Webb, W. W., and Hanson, M. R. (1997b) Plant J. 11:613-621.
Kuroiwa, T., Kawazu, T., Takahashi, H., Suzuki, K., Ohta, N., and Kuroiwa, H. (1994) Cytologia 59:149-158.
Kuroiwa, T. and Kuroiwa, H. (1980) Experientia 36:193-194.
Kuroiwa, T., Kuroiwa, H., Sakai, A., Takahashi, H., Toda, K., and Itoh, R. (1998) Int. Rev. Cytol. 181:1-41.
Kuroiwa, T., Suzuki, K., Itoh, R., Toda, K., O'Keefe, T. C., and Kuroiwa, H. (1995) Protoplasma 186:12-23.
Kuroiwa, T., Suzuki, K., and Kuroiwa, H. (1993) Protoplasma 175:173-177.
Lindahl, M., Tabak, S., Cseke, L., Pichersky, E., Andersson, B., and Adam, Z. (1996) J. Biol. Chem. 271:29329-29334.
Ma, X., Ehrhardt, D. W., and Margolin, W. (1996) Proc. Natl. Acad. Sci. USA 93:12998-13003.
Margulis, L. (1981) Symbiosis in cell evolution. W. H. Freeman and Company.
Mita, T., Kanbe, T., Tanaka, K., and Kuroiwa, T. (1986) Protoplasma 130:211-213.
Mita, T. and Kuroiwa, T. (1988) Protoplasma Suppl. 1:133-152.
Modrusan, Z. and Wrischer, M. (1990) Protoplasma 154:1-7.
Nakai, T., Yasuhara, T., Fujiki, Y., and Ohashi, A. (1995) Mol. Cell. Biol. 15:4441-4452.
Ogawa, S., Ueda, K., and Noguchi, T. (1995) J. Phycol. 31:132-137.
Okuwaki, T., Takahashi, H., Itoh, R., Toda, K., Kawazu, T., Kuroiwa, H., and Kuroiwa, T. (1996) Cytologia 61:69-74.
Oross, J. W. and Possingham, J. V. (1989) Protoplasma 150:131-138.
Ostersetzer, O. and Adam, Z. (1997) Plant Cell 9:957-965.
Osteryoung, E. W. and Vierling, E. (1995) Nature 376:473-474.
RayChaudhuri, D. and Park, J. T. (1992) Nature 359:251-254.
Reith, M. and Munholland, J. (1995) Plant Mol. Biol. Rep. 13:333-335.
Robertson, E. J., Rutherford, S. M., and Leech, R. M. (1996) Plant Physiol. 112:149-159.
Rothfield, L. I. and Justice, S. S. (1997) Cell 88:581-584.
Rothfield, L. I. and Zhao, C.-R. (1996) Cell 84:183-186.
Santos, D. and Almeida, D. F. (1975) J. Bacteriol. 124:1502-1507.
Schnall, R., Mannhaupt, G., Stucka, R., Tauer, R., Ehnle, S., Schwarzlose, C., Vetter, I., and Feldmann, H. (1994) Yeast 10:1141-1155.
Suzuki, K., Ehara, T., Osafune, T., Kuroiwa, H., Kawano, S., and Kuroiwa, T. (1994) Eur. J. Cell Biol. 63:280-288.
Takahashi, H., Takano, H., Yokoyama, A., Hara, Y., Kawano, S., Toh-e, A., and Kuroiwa, T. (1995) Curr. Genet. 28:484-490.
Tauer, R., Mannhaupt, G., Schnall, R., Pajic, A., Langer, T., and Feldmann, H. (1994) FEBS Lett. 353:197-200.
Terui, S., Suzuki, K., Takahashi, H., Itoh, R., and Kuroiwa, T. (1995) J. Phycol. 31:958-961.
Tewinkel, M. and Volkmann, D. (1987) Planta 172:309-320.
Thorsness, P. E. and Fox, T. D. (1993) Genetics 134:21-28.
Thorsness, P. E., White, K. H., and Fox, T. D. (1993) Mol. Cell. Biol. 13:5418-5426.
Toda, K., Takahashi, H., Itoh, R., and Kuroiwa, T. (1995) Cytologia 60:183-188.
Tzagoloff, A., Yue, J., Jang, J., and Paul, M. F. (1994) J. Biol. Chem. 269:26144-26151.
Ueda, K. and Noguchi, T. (1993) Ann. Bot. 72:591-594.
Vaughn, K. C., Ligrone, R., Owen, H. A., Hasegawa, J., Campbell, E. O., Renzaglia, K. S., and Monge-Najera, J. (1992) New Phytol. 120:169-190.
Wakasugi, T., Nagai, T., Kapoor, M., Sugita, M., Ito, M., Ito, S., Tsudzuki, J., Nakashima, K., Tsudzuki, T., Suzuki, Y., Hamada, A., Ohta, T., Inamura, A., Yoshinaga, K., and Sugiura, M. (1997) Proc. Natl. Acad. Sci. USA 94:5967-5972.

Warren, G. and Wickner, W. (1996) Cell 84:395-400.
Wheeler, R. T. and Shapiro, L. (1997) Cell 88:577-579.

Biodata of **Uri Pick** contributor of "***Dunaliella acidophila - a Most Extreme Acidophilic Alga.***"

Dr. Uri Pick is an associate Professor in the Department of Biological Chemistry at the Weizmann Institute in Israel. He earned his Ph.D. in 1976 in the Weizmann Institute of Science, and spent his postdoctoral period at Cornell University (Ithaca, NY) and as a visiting Professor at the Imperial College in London, and in UC Berkeley (CA). Dr. Pick's main interests are: Plant stress, Algae, Ion transport, Photosynthesis and Bioenergetics.
E-Mail: **bcpick@weizmann.weizmann.ac.il**

J. Seckbach (ed.), Enigmatic Microorganisms and Life in Extreme Environments, 465–478.

DUNALIELLA ACIDOPHILA - A MOST EXTREME ACIDOPHILIC ALGA

URI PICK
Department of Biological Chemistry, The Weizmann Institute of Science, Rehovot 76100, Israel

1. Introduction

In the last few years there has been an increasing interest in organisms thriving at extreme environmental conditions, which have been termed extremophiles. They include mainly prokaryotes which survive at extreme high or low temperature, high salinity, high hydrostatic pressure and extreme acidic or alkaline pH. The interest in extremophiles stems not only from the keen desire to discover mechanisms of adaptation to extreme environments but also from practical considerations. Extremophiles prove to be a valuable source of unusual enzymes and chemicals for scientific, medical and industrial applications (Adams and Kelly, 1995; Pennisi, 1997).

The unicellular green algal genus *Dunaliella* includes several species which challenge the most extremophilic bacteria in their ability to adapt to harsh environmental conditions: many *Dunaliella* species adapt to the widest known range of salinities, from 0.1M to 5M NaCl (Avron, 1992); *D. bardawil* withstands exceptionally high irradiance by accumulation of β-carotene, a commercially valuable pigment (Ben-Amotz and Avron, 1990); *D. antarctica* survives at sub-zero temperatures (Franzmann, 1991) and *D. acidophila* is one of the most extreme acidophiles on earth, surviving at a narrow pH range between pH 0 and pH 3 (Gimmler and Weis, 1992).

Adaptation to such low pH necessitates modifications of basic transport mechanisms in order to maintain an intracellular neutral pH and to accumulate essential organic and inorganic elements. Moreover, the fact that at pH 0 or 1 chemical bonds in biopolymers become unstable requires special structural adaptations of extracellular components at the plasma membrane surface. Thus, study of extreme acidophiles such as *D. acidophila* is a challenge not only for ecologists and physiologists but also for biochemists and structural biologists.

2. *Dunaliella acidophila*

2.1 GEOGRAPHICAL DISTRIBUTION AND POSSIBLE ORIGIN.

D. acidophila was isolated from several acidic waters or soils in Italy and Czechslovakia, characteristically having pH values between 0.5 -2.0 (Kalina, 1965;

Albertano et al., 1981). Such natural environments of low pH typically originate by activity of chemotrophic bacteria, which oxidize sulfur to sulfuric acid. It is not surprising, therefore, that the preferred growth medium for *D. acidophila* around 0.5M sulfuric acid, similar to its natural habitat. The evolutionary origin of this species is not clear, but it is conceivable that *D. acidophila* evolved from halotolerant mesophilic species of *Dunaliella* by adaptation to acidic aquatic niches created by bacterial sulfur oxidation. In the past *D. acidophila* has been mistakably defined as *Spermatozopsis*, and was only later identified as *Dunaliella*, based on similarities in morphology, physiology and biochemistry to other *Dunaliella* species (Masyuk, 1971).

Here we concentrate on the special features related to adaptation to acidic pH in this species with reference to acidophilic bacteria. An extensive review of the general ecology, physiology and biochemistry of this species has already been published (Gimmler and Weis, 1992).

2.2 GROWTH CONDITIONS

Like other *Dunaliella* species, *D. acidophila* is a halotolerant alga and proliferates in the presence of up to 1.2 M Na_2SO_4 or 1.9 M NaCl. Optimal growth was observed at about 0.5 M of either salt (Fuggi et al.,1988). The pH range for growth of *D. acidophila* is 0-3 with a sharp optimum close to pH 1.0 (Gimmler and Weis, 1992).

The only other algal species reported to grow at pH 0 is the red alga *Cyanidium caldarum* (Seckbach, 1994). The latter species has a broad pH range from pH 0 to pH 8. The two species differ also in their temperature requirements: *C. caldarium* is a thermophilic alga that was isolated from acidic hot springs (Allen, 1959; Seckbach, 1994), whereas *D. acidophila* is thermosensitive and grows at the temperature range 18-26°C (Gimmler and Weis, 1992).

2.3 METABOLISM AND OSMOREGULATION

Photosynthetic and respiration rates in *D. acidophila* are similar to the activities in neutrophilic species (Gimmler and Weis,1992). Also the dependence of photosynthesis on CO_2 concentration is not unusual, although calculations of CO_2 permeability indicate the existence of a light-induced catalyzed uptake pathway for CO_2 in this species (Gimmler and Slovik, 1995).

The osmoregulatory mechanism of *D. acidophila* is typical to other *Dunaliella* species: At high salinity it accumulates glycerol as the sole internal osmotic element, and the glycerol concentration is proportional to the extracellular osmotic pressure (Fuggi et al., 1988).

2.4 pH HOMEOSTASIS

2.4.1 *Cytoplasmic pH*

Several lines of evidence suggest that *D. acidophila* maintains a neutral cytoplasmic pH. Firstly, it was found that enzymes of *D. acidophila* have similar pH optima as homologous enzymes from neutrophilic species, suggesting that the cytoplasmic pH is closer to neutrality than to the extracellular acidic pH (e.g. Sekler and Pick, 1993).

Direct measurements of the cytoplasmic pH of *D. acidophila*, using ^{31}P-NMR or distribution of a weak acid, showed that within the external pH range of 0.5-3 the cells maintain a constant neutral pH of 7.2 or 6.2, respectively (Gimmler et al., 1988). Thus, *D. acidophila* maintains a pH gradient of up to 6.5 pH units (!) across its plasma membrane.

How does *D. acidophila* resist cytoplasmic acidification? A combination of factors contribute to reduce the influx of protons and to actively extrude protons out of the cell. They include the positive inside transmembrane potential, the positive surface charge, and the existence of a potent plasma membrane H^+-ATPase.

pH HOMEOSTASIS IN *DUNALIELLA. ACIDOPHILA*

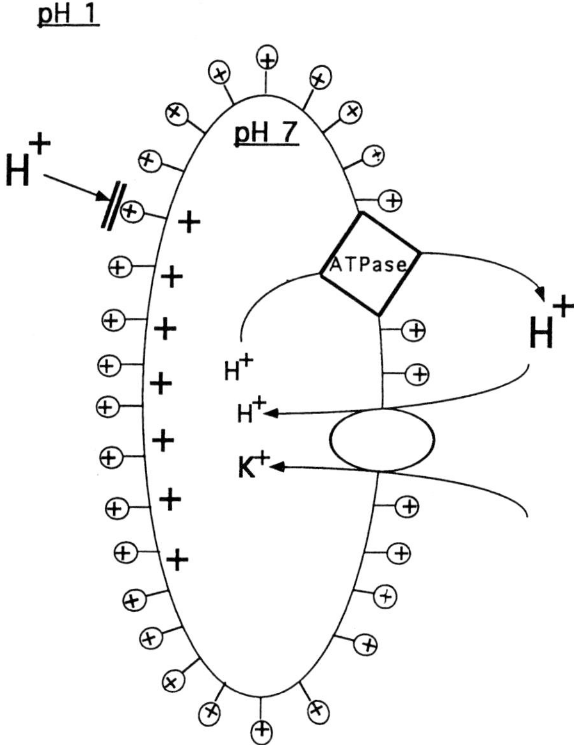

Figure 1. pH homeostatic mechanisms in *Dunaliella acidophila*.

One of the consequences of the large ΔpH across the plasma membrane is that the cells are expected to be particularly vulnerable to weak acids. At pH 0 or 1 most weak acids are protonated, and in this form can readily permeate biological membranes, liberate protons inside, and lead to cytoplasmic acidification. Indeed, acetic acid induces cytoplasmic acidification in *D. acidophila* at pH 1 (Gimmler et al., 1990). However, as noted previously (Gimmler and Weis, 1992), *D. acidophila* is surprisingly resistant to lipophilic substances, including weak acids. The calculated permeability coefficients of *D. acidophila* for different weak acids are 10^3-10^4 times lower than the corresponding permeability constants for a mesophilic *Dunaliella* species (Gimmler and Weis, 1992). This may provide an important protective adaptation against acid stress by common acids such as acetic acid and hydrochloric acid.

2.4.2 *The positive $\Delta\Psi$*

A most unusual feature in the bioenergetics of extreme acidophiles is their positive-inside electrical potential ($\Delta\Psi$) across the plasma membrane. $\Delta\Psi$ in *D. acidophila* at pH 1 ranged from +40 to +80 mV. In contrast, at pH 7 $\Delta\Psi$ was reversed to negative-inside (Gimmler et al., 1989b; Remis et al., 1992). The positive-inside $\Delta\Psi$ is not unique to *D. acidophila.* A similar phenomenon was reported in extreme acidophilic bacteria at pH <3 (reviewed by Bakker, 1990; Matin, 1990). Deenergized bacteria retain for several days a considerable protonophore-resistant pH gradient (1.5 units) at acidic pH, which is offset by the large positive $\Delta\Psi$. These results indicated that a positive $\Delta\Psi$ is a physical property of the cells at acidic pH and does not depend on energy-consuming reactions.

What is the origin of the positive-inside $\Delta\Psi$? The major contributor in deenergized acidophilic bacteria and probably also in *D. acidophila* is the H^+ diffusion potential. In the absence of active proton extrusion and of considerable transport of other ions, H^+ diffusion into the cells is expected to generate a positive $\Delta\Psi$ which will slow down the collapse of the ΔpH (Matin, 1990).

However, in metabolically-active cells under steady-state condition passive H^+ diffusion alone cannot exceed the active H^+ extrusion rate, and therefore cannot be the sole generator of the positive $\Delta\Psi$ as discussed by Bakker (1990).

An ion transporter which may contribute to the positive $\Delta\Psi$ in metabolically active *D. acidophila* cells is a plasma membrane H^+/K^+ cotransporter (section 3.2). Thus, both H^+ diffusion and membrane transporters may contribute to the positive $\Delta\Psi$.

In *D. acidophila* inhibition of proton extrusion by the H^+-ATPase inhibitor omeprazole (see below) increased the magnitude of the positive $\Delta\Psi$. This suggests that $\Delta\Psi$ in this alga is dominated by the H^+ diffusion potential (Remis et al., 1992).

The positive $\Delta\Psi$ may have a crucial role for survival at extreme acidic pH for several reasons: As found in deenergized bacteria, a large positive $\Delta\Psi$ slows down the rate of cytoplasmic acidification and therefore provides a safeguard against rapid inactivation. Another possible function may be to facilitate active H^+ extrusion by decreasing the overall proton motive force (PMF) across the plasma membrane. In the absence of a positive $\Delta\Psi$ the overall PMF may be so large as to retard proton extrusion from the cell via the H^+ pump. A pH gradient of 6 to 7 pH units is equivalent to about 360-420 mV, close to the estimated maximal ΔG_{ATP} of about 480 mV (Remis et al., 1992). Thus, in the absence of a positive $\Delta\Psi$ the rate of H^+ extrusion may slow down because H^+ pumping through the ATPase would approach thermodynamic equilibrium.

2.4.3 *Surface potential*

Another factor which has been suggested to decrease the permeability of *D. acidophila* cells to protons is their positive surface charge. Most living cells have a negative surface charge at neutral pH, due to the excess of acidic over basic amino acids in most proteins and the presence of acidic lipids at the extracellular surface of the membrane. However, at pH 1 most of the anionic groups are expected to be protonated (e.g. histidines, carboxyl groups, phosphates), leaving a net positive surface charge (Donnan potential).

Electophoretic measurements of intact *D. acidophila* cells at pH 2 show that the cells indeed possess a positive surface charge, whereas at pH 7 the direction of electrophoretic mobility is reversed, reflecting a negative surface charge, similar to neutrophilic species (Gimmler et al., 1991). Another indication for the positive membrane surface charge of *D. acidophila* is its high resistance to polyvalent cations.

The positive Donnan potential may have important consequences for survival at pH 0-1: The electrostatic positive charge should exclude protons from the membrane surface and therefore reduce proton influx. Proton exclusion would also shift the dissociation constants of acidic residues and increase the local pH at the membrane interface, thus protecting proteins and lipids against denaturation. However, the effectiveness of a Donnan potential will be decreased at high salinity and therefore may not be of major significance at highly saline growth media.

2.4.4 *Internal buffering capacity*

Acidophilic bacteria have higher internal buffering capacity than neutrophilic bacteria (Matin, 1990). Also *D. acidophila* has a high internal buffering capacity of about 100 mM l^{-1} pH unit^{-1}, significantly higher than neutrophilic *Dunaliella* species (Gimmler and Weis, 1992). It has been suggested that the high internal buffering capacity in acidophiles provides another safety device against rapid cytoplasmic acidification which may arise under sudden stress conditions.

2.4.5 *H^+ export*

A light-induced H^+ export was demonstrated in *D. acidophila* by pH measurements utilizing a special antimony electrode which detects pH changes at very acidic pH (Remis et al., 1994). Illumination of *D. acidophila* cells at pH 1 was found to induce a reversible acidification of the medium followed by a slow alkalinization. The proton export activity was sensitive to inhibitors of plasma membrane H^+-ATPases (e.g. vanadate, dicyclohexylcarbodiimide) and was enhanced by fussicoccin, a specific activator of plant plasma membrane H^+-ATPases. Furthermore, H^+ export was inhibited by omeprazole, an acid-activated thiol reagent, in parallel with inhibition of the plasma membrane ATPase activity (Sekler et al., 1993; Remis et al., 1994). These studies indicated that the light-activated proton extrusion activity in *D. acidophila* is mediated by a plasma membrane H^+-ATPase.

3. Plasma Membrane Transport Systems

A strict extreme acidophile is expected to possess unusual and/or structurally modified transport systems. First, it should have a potent H^+ extrusion mechanism capable of pumping protons at high rates against a large electrochemical proton gradient. Second, because of the positive-inside membrane potential, essential cations such as K^+ and Mg^{++} are maintained very far from electrochemical equilibrium, and therefore can not be taken up by passive mechanisms (e.g. channels). Third, because of the extremely low extracellular pH it may be expected that extracellular domains of transporters which face pH 0-1 should be structurally modified. Extensive analysis of the structure, function and regulation of the plasma membrane (PM) H^+-ATPase from *D. acidophila* reveals some unusual features which may be associated with adaptation to low pH. K^+ transport studies in plasma membrane vesicles and in whole cells reveal that K^+ is taken up via a K^+/H^+ cotransporter.

3.1 THE PLASMA MEMBRANE H^+-ATPase: THE ACTIVE PROTON EXPORT MECHANISM

P-type H^+-ATPases are the major primary pumps in plasma membranes of most non-animal eukaryotes including plants and yeast (Serrano, 1989; Sussman and Harper, 1989).

Until recently it was not clear whether algae possess plasma membrane H^+-ATPases. Early biochemical characterizations indicated that plasma membrane preparations from *D. parva, D. salina* and *D. bioculata* possess a vanadate-sensitive ATPase activity which was partially purified and identified as a 100 kD polypeptide. Based on sensitivity to inhibitors and on immunological properties it was suggested to function as a H^+-pump (Gimmler et al., 1989a; Smahel et al., 1990; Weiss et al., 1989). The presence directly of an ATP-dependent H^+ uptake activity in algae was first demonstrated in plasma membrane preparations from *D. acidophila* (Sekler et al., 1991) and from the marine alga *Platymonas viridis* (Popova and Balnokin, 1992).

Extensive characterization of the kinetic properties of the *D. acidophila* H^+-ATPase revealed clear resemblance to plant PM H^+-ATPases: H^+ uptake and ATPase activities were inhibited by vanadate, diethylstilbestrol, dicyclohexylcarbodiimide and by erythrosine, characteristic inhibitors of P-type ATPases. Fussicoccin, a fungicide which stimulates the plant PM H^+-ATPase via a distinct plasma membrane receptor protein, also stimulated the *D. acidophila* ATPase and H^+ uptake activities. Subsequent purification of the ATPase showed that it consisted of a single polypeptide of about 100 kDa which cross-reacted with antibodies against the *Arabidopsis. thaliana* enzyme (Sekler and Pick, 1993). However, the *D. acidophila* enzyme differed from typical plant H^+-ATPases in several kinetic and structural properties: (i) ATPase and H^+ transport were stimulated 2-3 fold by K^+ ions, in comparison to 20-30% stimulation in plants, (ii) the K_m for ATP of about 50 µM was considerably lower than for most known plant PM H^+-ATPases, (iii) the pH optimum of 6.0 was slightly lower than for plant enzymes and (iv) the apparent molecular weight of the ATPase polypeptide on polyacrylamide gels was higher than the plant PM H^+-ATPase (Sekler and Pick, 1993, Sekler et al., 1994).

The regulation of the H^+-ATPase from *D. acidophila* was investigated. It was shown that cleavage of a C-terminal fragment of *D. acidophila* PM H^+-ATPase by trypsin digestion stimulated H^+-ATPase activity, decreased the K_m(ATP) and shifted the pH optimum (Sekler et al., 1994) indicating that the basic mechanism of regulation resembled that in plants and yeast (Serrano et al., 1992). Interestingly, the cleaved C-terminal fragment contained an *in vivo* phosphorylation site which may be involved in regulation via a protein kinase as suggested for the plant and yeast enzymes (Chang and Slayman, 1991; Lino et al.,. 1998; Schaller and Sussman, 1988).

Extracellular inactivation of the ATPase by omeprazole, a membrane-impermeable acid-activated thiol reagent (Sachs et al., 1988) inhibited proton extrusion and photosynthetic activities (Sekler et al., 1993; Remis et al., 1994). These results demonstrated that the H^+-ATPase is crucial for survival of *D. acidophila* at pH 1.

What are the special features of the PM H^+-ATPase of *D. acidophila* which enable the cells to survive at pH 0?

Insights into this intriguing issue came from comparisons of the structure and expression of the enzyme in two related *Dunaliella* species: the ATPase polypeptide level (identified with specific antibodies) in *D. acidophila* was remarkably higher than in the neutrophilic halotolerant *D. salina,* indicating that the enzyme is highly overexpressed in *D. acidophila* (Sekler et al., 1991; Weiss and Pick, 1996).

The H^+-ATPase genes were cloned from *D. acidophila, D. bioculata* and *D. salina* (Wolf et al, 1995; Weiss and Pick, 1996). Comparative sequence analysis with other PM H^+-ATPases revealed the highest homology to plants (42-43% identity) and a lower homology to fungal genes (35-37% identity), but less than 20% identity between all three groups (plants, yeast and *Dunaliella,* Fig. 2, Weiss and Pick, 1996).

Sequence analysis of different domains revealed high homology in conserved motifs (the ATP binding and phosphorylation domains) and in the location of polar amino acids within putative trans-membrane helices (Fig. 11 in Weiss and Pick, 1996). Interestingly, the homology to a putative PM H^+-ATPase gene from the acidophilic red alga *C. caldarium* is only 40%, lower than the homology to higher plants, suggesting a divergent evolutionary origin (Ohta et al., 1997).

The only unusual feature of *Dunaliella* enzymes was their extended C-terminal region and the presence of two polar loops between the last putative membrane helices (Wolf et al., 1995, Weiss and Pick, 1996). Comparison of *D. acidophila* to *D. salina* gene sequences revealed an interesting difference: the two genes were 75% identical, but differed in the hydrophilic insert TM-VII/TM-VIII (47% identity). This domain, which presumably faces the cell exterior according to consensus organization of P-type ATPases, has excess negatively charged amino acids in *D. salina* and excess positively charged amino acids in *D. acidophila*. The significance of this observation is not clear. However, it was noted that two extracellular proteins which were cloned from *D. salina*, a carbonic anhydrase and an internally-triplicated transferrin-like protein, have excess negative/positive amino acids (Fisher et al., 1996, 1997), characteristic of halophilic bacterial proteins (Dym et al., 1995). Thus, the excess negative charge in *D. salina* TM-VII/TM-VIII insert reflects characteristics of halophilic proteins and may function in stabilization of the protein in hypersaline media. By analogy, the excess positive charge may stabilize proteins in highly acidic media.

PHYLOGENETIC ANALYSIS OF PLASMA MEMBRANE H^+-ATPases

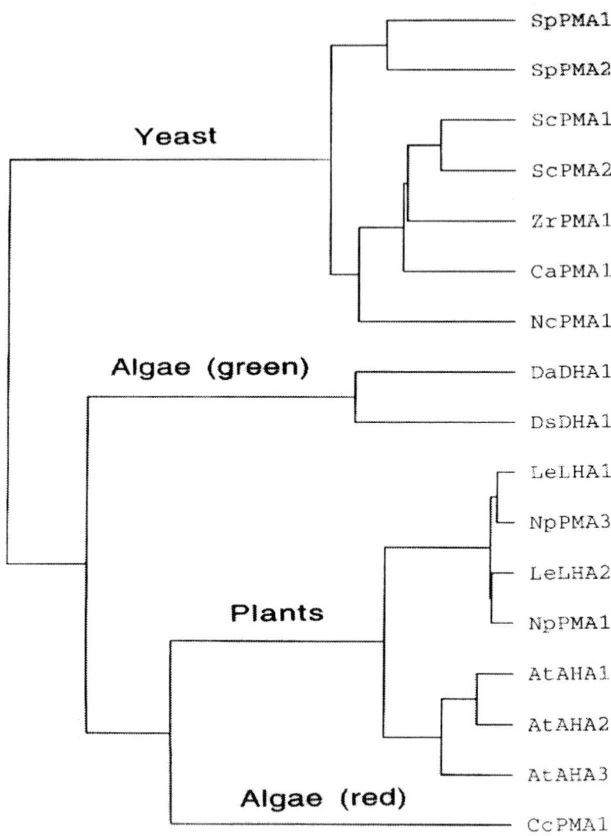

Figure 2: Graphical representation of relatedness of plasma membrane H^+-ATPases derived from amino acid sequences with a PILEUP program. Sp, *Schizosaccharomyces pombe*, Sc, *Saccharomyces cerevisiae*, Zr, *Zygosaccharomyces rouxii*, Ca, *Candida albicans*, Nc, *Neurospora crassa*, Da, *Dunaliella acidophila*, Ds, *Dunaliella salina*, Le, *Lycopersicon esculentum*, Np, *Nicotiana plumbaginifolia*, At, *Arabidopsis thaliana*, Cc, *Cyanidium caldarium*.

Another possible function of the excess positive charge in the extracellular domains of the H^+-ATPase may be to facilitate the extracellular release of protons. pH of 0 or 1 is lower than the pK values of all acidic amino acids. Decreasing the dissociation constants of proton leaving residues, may be crucial for extracellular release of protons from the ATPase (Weiss and Pick, 1996). Future studies are required to test these predictions.

The expression of the *D. salina* PM H^+-ATPase was found to be transcriptionally pH-regulated: the protein and mRNA levels were hardly detectable at alkaline pH, but significantly increased at pH 7. A partial correlation between pH and expression of the H^+-ATPase was observed also in *D. acidophila*, in which the overall expression was remarkably higher. Thus, in *D. acidophila* the expression of the enzyme is largely constitutive, probably as a means to ensure a high H^+ export capacity and for maintaining the large pH gradient across the plasma membrane. The close sequence homology between the H^+-ATPase genes suggests that the *D acidophila* may have evolved from a halotolerant neutrophilic *Dunaliella* species by acclimation to acidic environment through amplification and modification of the H^+-ATPase.

3.2 K^+/H^+ SYMPORT

K^+ plays a mayor role in volume and osmotic regulation in plant cells. Uptake of K^+ is driven by the negative-inside plasma membrane potential through special potential-regulated channels, and its elimination takes place via a distinct type of plasma membrane channels. In the halotolerant *D. salina* at steady-state conditions K^+ is maintained close to electrochemical equilibrium and its uptake is driven by the membrane potential probably through plasma membrane K^+ channels (Pick et al., 1986). Because of the positive membrane potential in *D. acidophila*, K^+ uptake cannot be driven by the membrane potential. Yet, *D. acidophila* maintains a large K^+ concentration gradient of >100-fold, indicating that it accumulates K^+ ions against a large electrochemical gradient. Studies in intact cells have shown that the cells possess a saturable K^+ uptake mechanism having a K_m of 0.3 mM and a V_{max} of 2 mol mg Chl^{-1} h^{-1} (Caradang et al., 1992). K^+ uptake was not affected by reduction of the ATP pool, but was sensitive to inhibitors of the PM H^+-ATPase (Omeprazole, dicyclohexylcarbodiimide), suggesting that it is not driven by a primary K^+ pump but rather by the pH gradient generated by the H^+-ATPase. Studies of Rb^+ transport in plasma membrane preparations indicated that *D. acidophila* cells possess a K^+/H^+ cotransport activity (Gläser et al., 1990). These results suggested that K^+ uptake is coupled to downhill H^+ influx and is driven by the large pH gradient across the plasma membrane. A similar K^+/H^+ cotransport activity has been described in membrane preparations of acidophilic bacteria, and was suggested to contribute to the generation of the positive-inside membrane potential (reviewed by Bakker, 1990). K^+/H^+ cotransport into the cells can explain the maintenance of a positive membrane potential at steady-state provided that the rate exceeds half the rate of H^+ extrusion. The measured rates of K^+ uptake in *D. acidophila* (2 μmol mg Chl^{-1} h^{-1}) were significantly lower than light-induced H^+ uptake rates (60-300 μmol mg Chl^{-1} h^{-1}) but in the same order as the dark passive H^+ leak (<6 μmol mg Chl^{-1} h^{-1}, Remis et al., 1994) which better represents steady-state rates of the H^+ extrusion. Therefore, it is consistent with the idea that K^+/H^+ cotransport contributes significantly to the positive $\Delta\Psi$ in *D. acidophila.*

4. Summary: Special Features Of Extreme Acidophiles

Although there are still many gaps in our knowledge about the molecular mechanisms enabling survival of organisms at very low pH, the results available from acidophilic bacteria and *D. acidophila* reveal several unusual features of extreme acidophiles which thrive at pH <3.

(i) Unusual bioenergetics. Extreme acidophiles maintain a neutral cytoplasmic pH by generating a large ΔpH across the plasma membrane. The pH gradient is generated in bacteria by electron transport reactions and in *D. acidophila* by a potent PM H^+-ATPase. A positive-inside $\Delta\Psi$ decreases the overall PMF across the plasma membrane to similar values as in neutophiles. The significance of lowering the PMF is probably to avoid kinetic limitations on electron transport (bacteria) and on H^+ extrusion (*D. acidophila*) and to decrease the driving force for H^+ influx.

(ii) Unusual permeability properties. Plasma membranes of extreme acidophiles have intrinsic low permeabilities to protons which is crucial to avoid cytoplasmic acidification.

(iii) Positive surface charge results primarily from protonation of most acidic groups at the membrane surface at low pH. Thus, the positive surface is not an adaptive property but a consequence of the low external pH.

(iv) High internal buffer capacity. Extreme acidophiles have higher internal buffering capacity for protons than mesophilic counterparts. The significance of the high buffering capacity may be to slow down the rate of cytoplasmic acidification under stress conditions.

(v) Overexpression of H^+ export enzymes. A major adaptive feature in *D. acidophila* is the high expression level of the PM H^+-ATPase, providing a high capacity for H^+-extrusion.

(vi) Unique transporters. Uptake of K^+ ions in extreme acidophiles is mediated primarily by K^+/H^+ cotransporters and not through K^+ channels. Acidophilic bacteria also possess high-affinity K^+ pumps (Kdp system) which are common also to mesophilic bacteria.

5. Open Questions

Although some common characteristics of extreme acidophilic microorganisms have been documented, our understanding of the molecular basis for their unusual acid tolerance is still vague. Some of the basic questions for which we have no answers are the following:

(i) The reason for the intrinsic low permeability of extreme acidophiles to protons and to lipophilic molecules is unknown. The composition of plasma membrane lipids and proteins have not been determined, and nothing is known about their organization.

(ii) It is not clear what protects proteins and other biomolecules at the membrane external surface against acid inactivation. If we learn what stabilizes proteins at low pH, it may be possible to engineer acid-resistant enzymes of commercial importance in the future.

(iii) Extreme acidophiles may be expected to possess unique ion transporters. The positive $\Delta\Psi$ and large ΔpH impose distinct energetic constrains in comparison to mesophiles. The low extracellular pH should affect the ionization state of most inorganic anions as well as acidic amino acids in proteins. This should in turn modify the binding of ions to transport proteins. However, our understanding of basic ion

transport mechanisms in extreme acidophiles is poor even at the kinetic level, and very few transporters have been characterized biochemically or cloned.
(iv) It is not clear what prevents obligatory acidophiles such as *D. acidophila* to thrive at neutral pH.

In summary, extreme acidophiles constitute an intriguing class of microorganisms, which have adapted to low pH by overexpression and/or modifications of plasma membrane proton and other ion transporters and by modifications of the plasma membrane which decrease its permeability to protons. The nature of these unusual features is just beginning to emerge.

6. References

Adams M.W.W. and Kelly R.M. (1995) Chem. Eng. News 73, 32-42

Albertano, P., Pinto. G., Santisi, S. and Taddei, R. (1981) G. Bot. Ital. 115, 65-76.

Allen, M.B. (1959) Arch. Microbiol. 32, 270-277.

Avron M. (1992) In: M. Avron and A. Ben-Amotz (eds) *Dunaliella*: Physiology, Biochemistryand Biotechnology, CRC Press, Boca Raton, Florida, pp. 135-164.

Bakker, E.P. (1990) FEMS Microbiol. Rev. 75, 319-334.

Ben–Amotz A. and Avron M. (1990) Trends Biotechnology, 8, 121-126.

Chang, A. and Slayman, C.W. (1991) J. Cell Biol. 115, 289-295.

Dym O.D., Mevarech M. and Sussman J.L. (1995) Science 267, 1344-1346.

Caradang, J.S., Pick, U., Sekler, I. and Gimmler, H. (1992) J. Plant Physiol. 139, 413-421.

Fisher, M., Gokhman, I., Pick, U. and Zamir, A. (1996) J. Biol. Chem. 271, 17718-17723.

Fisher, M., Gokhman, I., Pick, U. and Zamir, A. (1997) J. Biol. Chem. 272, 1565-1570.

Franzmann P.D. (1991) In: F. Rodriguez-Valera (ed) General and applied aspects of halophilic microorganisms, Plenum Press, New-York, pp.14.

Fuggi, A., Pinto, G., Pollio, A. and Taddei, R. (1988) Phycologia 27, 439-446.

Gimmler, H., Kugel, H., Leibfritz, D. and Mayer, A. (1988) Physiol. Plant. 74, 521-530.

Gimmler, H., Schneider, L and Kaaden, R. (1989a) Zeitschr. Naturforsch. 44,128-138.

Gimmler, H.,Weis, U., Kugel, H. and Treffny, B. (1989b) New Phytol., 113, 175-184.

Gimmler, H., Bental, M. Degani, H., Avron, M. and Pick, U. (1990) In: M, Baltchefsky (ed.) Current Research in Photosynthesis, vol. IV, Kluwer Academic Publishers, The Netherlands, pp. 773-776.

Gimmler, H., Schieder, M. Kowalski, M., Zimmermann, U. and Pick, U. (1991) Plant, Cell and Environment 14, 261-269.

Gimmler, H. and Weis, U. (1992) In: *M.* Avron and A. Ben-Amotz (eds) *Dunaliella*: Physiology, Biochemistry and Biotechnology, CRC Press, Boca Raton, Florida, pp. 99-134.

Gimmler, H. and Slovik, S. (1995) Plant Physiol. Biochem. 33, 655-664.

Gläser, H.U., Sekler, I. and Pick, U. (1990) Biochim. Biophys. Acta 1019, 293-299.

Kalina, T. (1965) Preslia 37, 9-12.

Lino, B., Baizab al- Aguirre, V.M. and Gonzales de la Vara, L.E. (1998) Planta 204, 352-359.

Masyuk. N.P (1971) Ukr. Bot. Zh., 26, 87-90.

Matin, A. (1990) Biochim. Biophys. Acta 1018, 267-270.

Ohta, H. Shirakawa, H., Uchida, K., Yoshida, M., Matuo, Y. and Enami, I. (1997) Biochim. Biophys. Acta 1319, 9-13.

Pennisi, E. (1997) Science 276, 705-706.

Pennisi, E. (1997) Science 276, 705-706.

Pick, U., Ben-Amotz, A., Karni, L., Seebergts, C.J. and Avron, M. (1986) Plant Physiol. 81, 875-881.

Popova, L.G. and Balnokin, Y.V. (1992) FEBS Lett. 309. 333-336.

Remis, D., Simonis, W. and Gimmler, H. (1992) Arch. Microbiol., 158, 350-355.

Remis, D., Treffny, B. and Gimmler, H. (1994) Plant Physiol. Biochem. 32, 75-84.

Sachs, G., Carlsson, E., Lindenberg, P. and Wallmark, B. (1988) Ann. Rev. Pharmacol. Toxicol. 28, 269-284.

Schaller, G.E. and Sussman, R.E. (1988) Planta 173, 509-518.

Seckbach, J. (1994) editor, *Evolutinary Pathways and Enigmatic Algae: Cyanidium caldarium (Rhodophyta) and Related Cells,* Kluwer Acad. Publisher. The Netherlands.

Sekler, I., Gläser, H.U. and Pick, U. (1991) J. Membr. Biol. 121, 51-57.

Sekler, I., Remis, D. Gimmler, H. and Pick, U. (1993) Biochim. Biophys. Acta 1142, 88-92.

Sekler, I. and Pick, U. (1993) Plant Physiol. 101, 1055-1061.

Sekler, I., Weiss, M. and Pick, U. (1994) Plant Physiol. 105, 1125-1132.

Serrano, R. (1989) Ann. Rev. Plant Physiol. Plant Mol. Biol. 40, 61-94.

Serrano, R., Portillo, F., Monk, B.C. and Palmgren, M.G. (1992) Acta Physiol. Scand. 146, 131-136.

Smahel, M., Hamann, A. and Grandmann, D. (1990) Planta, 181, 496-504.

Sussman, M.R. and Harper, J.F. (1989) Plant Cell 1, 953-960.

Weiss M., Sekler, I. and Pick, U. (1989) Biochim. Biophys. Acta 974, 254-260.

Weiss M. and Pick, U. (1996) Plant Physiol. 112, 1693-1702.

Wolf, A.H., Slaymann, C.W. and Grandmann, D. (1995) Plant Mol. Biol. 28, 657-666.

Biodata of **Masahiro Kamekura** (contributor of ***"Alkaliphilic Micro-organisms)"***

Dr. **Masahiro Kamekura** is a Senior Researcher at Noda Institute for Scientific Research at Noda-shi, Chiba-ken, Japan. He earned his Ph.D. in 1980 at the University of Tokyo. Dr. Kamekura has a research interest in taxonomy, enzymology and in gene engineering of halophilic Archaea.

E-Mail: mkamekur@supernig.nig.ac.jp

J. Seckbach (ed.), Enigmatic Microorganisms and Life in Extreme Environments, 479–485.

ALKALIPHILIC MICROORGANISMS

MASAHIRO KAMEKURA
Noda Institute for Scientific Research
399 Noda, Noda-shi, Chiba-ken 278, Japan
Phone: +81-471-23-5573 Fax: +81-471-23-5959
Electronic mail address: mkamekur@supernig.nig.ac.jp

1. Introduction

The microorganisms growing optimally in excess of pH 8 are defined as alkaliphiles. Obligate alkaliphiles are unable to grow at neutral pH values, while alkalitrophic (or alkalitolerant) microorganisms have pH optima at neutral side but are capable of growth at alkaline side (Grant,1992).

During an attempt to improve the enrichment medium for *Vibrio cholera*, Vedder (1934) isolated aerobic, endospore forming-bacteria from human and animal feces. He demonstrated that the isolates were obligate alkaliphiles, which he designated *Bacillus alcalophilus*. This is the first description of alkaliphilic microorganism. After a long lag period, Chislett and Kushner (1961) reported another isolate of alkaliphile, *Bacillus circulans* strain RU38 as a contaminant on an agar plate with a high pH. Soon after this, K.Horikoshi started his systematic survey of alkaliphilic microorganisms which produced extracellular enzymes (Horikoshi, 1971). Of the extremely alkaliphilic bacteria which have been isolated and characterized since then, most are members of the genus *Bacillus*. Other alkaliphilic eubacteria include *Pseudomonas*, *Micrococcus*, *Corynebacterium* etc. Alkaliphilic halophilic archaebacteria have also been isolated from alkaline saline environments. Alkaliphilic yeast and fungi have also been isolated, but little has been known about these eukaryotes (Horikoshi, 1991).

2. Alkaline Environments

Soda lakes represent the most alkaline naturally occurring environments on earth, with pH values generally higher than 10, occasionally reaching 12 (Grant *et al.*, 1990; Grant, 1992). The nature of alkalinity of a number of the soda lakes of the Kenyan-Tanzanian Rift Valley has been known to be derived from the presence of unusually high levels of sodium carbonate. The alkaline saline lakes are distributed on almost all major continents, and detailed explanation of the genesis of alkaline saline lakes is given by Grant (1992). In summary, the critical point is the topology of a drainage basin and the low levels of magnesium/calcium and high sodium ions leached from the surrounding rocks by carbonic acid contained in rain fall. Abundance of these divalent cations would result in the precipitation of magnesium or calcium carbonate. Arid and semi-arid tropical or semi-tropical climates where intense evaporative concentration rates exceeds inflow rates are also requisite for the formation of soda lakes.

These lakes are also somewhat saline due to the concomitant increase in chloride ions. The salinity of these lakes range from 5% to saturation (Jones *et al. 1994.*; Duckworth *et al.*,1996).

3. Isolation of Aerobic Alkaliphiles

Alkaliphiles or alkalitrophs are readily isolated from the soda lakes and other natural environments. Indeed, some alkaliphiles have been isolated from soil samples of neutral or even acidic pH (Hamamoto and Horikoshi,1992).

In order to isolate obligate alkaliphiles, it is preferable to use culture media of pH 10 or above, which is achieved by adding 1-5% sodium carbonate (Na2CO3) to commonly used media. In practice, maintenance of the initial pH of 10.5 of media is difficult due to the absorption of atmospheric carbon dioxide, particularly in the case of agar plates. Automated addition of NaOH is required to maintain the pH at high alkalinity of media in glass flasks or jar fermentor.

In the 1970s the physiologically unique bacilli became interesting because of their ability to produce unusual enzymes that are applicable in industrial and other processes at high pH range. The increased industrial interests in alkaliphilic organisms, especially members of the genus *Bacillus*, have resulted in the isolation of numerous alkaliphilic and alkalitolerant strains. A considerable number of these were deposited with culture collections in connection with patent applications and were designated only as *Bacillus* species. Some beared species names, which have not been validated. To date, a number of approaches have been done to define this heterogeneous group taxonomically.

3.1. TAXONOMY OF ALKALIPHILIC BACILLI

Fritze et al.(1990) initiated an extensive taxonomic study by measuring the extent of the guanine-plus-cytosine (G+C) content variation among 78 strains of alkaliphilic *Bacillus* and the levels of relatedness of these strains to the type strain of *Bacillus alcalophilus* as determined by DNA-DNA hybridization experiments. They ranged from 34 to 44 mol %G+C, suggesting that these must be assigned to numerous different species. In 1993 the second alkaliphilic species of the genus *Bacillus*, *B.cohnii* was described (Spanka and Fritze,1993), incorporating the strain RU38 isolated by Chislett and Kushner (1961).

These studies were followed by comparative 16S rDNA sequence analysis of some alkaliphilic bacilli (Nielsen *et al.*,1994). This study revealed that the alkaliphilic bacilli were phylogenetically heterogeneous, falling into three distinct groups.

Further detailed examination of physiological and biochemical characters were done for 119 strains, and 11 clusters comprising more than 3 strains were revealed by numerical analysis. Representatives of 9 clusters, excluding the two valid species,

B.alcalophilus and *B.cohnii*, were proposed as new species: *B.agaradhaerens*, *B.clarkii*, *B.clausii, B.gibsonii, B.halmapalus, B.halodurans, B.horikoshii*, *B.pseudalcaliphilus*, and *B.pseudofirmus* (Nielsen *et al.*,1995). The most recent proposal of a new species is on *B.haloalkaliphilus* (Fritze,1996), originally isolated from Wadi Natrun of Egypt by Weisser and Truper (1985).

Recently an alkaliphilic new species of the genus *Halomonas*, *H.pantelleriense* was proposed (Romano *et al.*,1996). Optimal growth occurred in media containing 10% NaCl at pH 9.0.

4. Facultatively Anaerobic Alkaliphilic Eubacteria

Niimura et al. (1990) isolated three strains of gram-positive, facultatively anaerobic, spore-forming, xylan-degrading organisms from alkaline composts of manure with grass and rice straw. These microorganisms were able to grow both under strict anaerobic condition (with titanium (III) citrate as a reducing agent) and in aerobic shaking culture, at pH 8-10 but not at pH 7. Its lack of cytochromes, quinones or catalase and its ability to form spores under aerobic and strict anaerobic conditions clearly distinguished this organism from the genera *Bacillus*, *Clostridium*, and *Sporolactobacillus*. They designated the isolates as *Amphibacillus xylanus* gen.nov., sp.nov.

5. Alkaliphilic Archaebacteria

The family *Halobacteriaceae*, the extremely halophilic archaebacteria have also genera comprising alkaliphilic strains, *Natronobacterium*, *Natronomonas*, and *Natronococcus*, in addition to the neutrophilic genera *Halobacterium*, *Halococcus*, *Haloarcula*, *Haloferax*, *Halorubrum*, *Halobaculum*, *Natrialba* (Grant and Larsen, 1989; McGenity and Grant, 1995; Kamekura and Dyall-Smith, 1995; Oren *et al.*, 1995; Kamekura *et al.*, 1997). The alkaliphilic extreme halophiles are unique in the *Halobacteriaceae* in having no detectable amount of glycolipids in their cell membranes, while neutrophilic halophiles contain diether glycerolipids with sugar moieties, sometimes sulfated, attached to *sn*-1 of the glycerol (Kates, 1993).

Though most of the methanogenic archaebacteria are neutrophiles, a few alkaliphilic methanogens have been reported, *Methanobacterium alcaliphilum, Methanonacterium thermoalcaliphilum,* and *Methanosalsus zhilinae*, with optimal growth at pH 8.5 to 9.2 and salinity of 0.3 to 0.7 M NaCl (Boone *et al.*, 1993).

6. Comprehensive Survey of Soda Lake Alkaliphiles

A group of W.D.Grant recently performed a comprehensive survey of diversity of alkaliphilic microorganisms in a range of East African soda lakes (Duckworth *et al.*,

1996). 16S rRNA genes amplified from a range of chemoorganotrophic, alkaliphilic isolates have been sequenced and subjected to phylogenetic analysis. Gram-negative aerobic alkaliphiles were found to be confined to the gamma3 subdivision of the proteobacteria, with many isolates related to *Halomomas*. Gram-positive alkaliphiles were found in both high G+C and low G+C divisions of the Gram-positive lineage, with many isolates related to the *Bacillus* group. An anaerobic isolate from alkaline hot spring was assigned to a new genus within the order *Thermotogales*.

The dilute soda lakes of the East African Rift Valley are regarded as one of the most productive aquatic environments, and the primary productivity in most of the lakes is due to the presence of dense populations of alkaliphilic cyanobacteria, notably *Spirulina* spp. with some contribution from alkaliphilic anoxygenic phototroph of the genus *Ectothiorhodospira* (Grant and Tindall, 1986).

7. Bioenergetics

When oxygen is admitted to an anaerobic suspension of neutrophilic bacteria, the pH of the suspension declines. The chemiosmotic theory by Mitchell of energy generation postulates that the free energy of electron transport is conserved by pumping out protons so as to create an electrochemical proton gradient across the membrane. The inside of respiring neutrophilic bacteria is electrochemically negative and alkaline relative to the suspending medium. The electrochemical potential of this gradient (or proton motive force) (Delta P) consists of the electrical membrane potential (Delta Phi) and the pH difference (Delta pH) in the following equation, Delta P = Delta Phi - Z Delta pH(mV), where Z is 2.3RT/F, namely 59mV at 25 C. The Delta P is harnessed for the synthesis of ATP, solute transport, flagella movement etc.

Intracellular pH was measured for a few isolates of alkaliphilic *Bacillus* by assaying the distribution of weak bases across the cell membrane, and there is a consensus that alkaliphiles regulate their internal pH within the range of 7-9 (Grant, 1989). The challenge of alkaliphiles to the chemiosmotic theory is that reversed pH difference, eg. 2.5 unit in a growth at pH11, generates a reverse force of around 150mV (Grant, 1989). The pH homeostasis clearly requires that the protons should be returned into the cytoplasm across the membrane. A sodium/proton antiporter is strongly implicated in the pH homeostasis in alkaliphiles. Membrane vesicle preparations from a alkaliphilic *Bacillus* sp. show a sodium dependent movement of proton (Kitada and Horikoshi, 1992).

8. Conclusion

In view of the wide range of physiological groups as described above, alkaliphile phenotype is not monophyletic in eubacteria and also in archaebacteria. The very diverse alkaliphiles may continue to attract peoples' interest in basic science and in industry.

9. References

Boone,D.R., Whitman,W.B., and Rouviere,P. (1993) In: J.G. Ferry (ed.) *Methanogenesis – Ecology, Physiology, Biochemistry & Genetics,* Chapman &Hall, New York, pp. 35-80.

Chislett,M.E. and Kushner,D.J. (1961) *J. Gen. Microbiol.* 24, 187-190.

Duckworth,A.W., Grant,W.D., Jones,B.E., and van Steenbergen,R. (1996) *FEMS Microbiol. Ecol.* 19, 181-191.

Fritze,D. (1996) *Int. J. Syst. Bacteriol.* 46, 98-101.

Fritze,D., Flossdorf,J., and Claus,D. (1990) *Int. J. Syst. Bacteriol.* 40, 92-97.

Grant,W.D. (1989) Alkaliphiles, in M.S. DaCosta, J.C. Duarte, R.A.D.Williams (eds.) *Microbiology of Extreme Environments and its Potential for Biotechnology,* Elsevier Applied Science, London, pp.346-366.

Grant,W.D. (1992) Alkaline environments, In: J. Lederberg (ed.) *Encyclopedia of Microbiology Vol.1.* Academic Press, New York, pp.73-80.

Grant,W.D. and Larsen,H. (1986) Group III. Extremely halophilic archaeobacteria, In: J.T.Staley, M.P.Bryant, N.Pfennig, and J.G.Holt (eds.) *Bergey's Manual of Systematic Bacteriology, Vol. 3*, Williams & Wilkins, Baltimore, pp. 2216-2233.

Grant,WD, and Ross,H.N.M. (1986) *FEMS Microbiol. Rev.* 39, 9-15.

Grant,W.D. and Tindall,B. (1986) The alkaline saline environment, In: R.A.Herbert and G.A.Codd (eds.) *Microbes in Extreme Environments,* Academic Press, London, pp.25-54.

Grant,W.D., Mwatha, W.E., and Jones,B.E. (1990) *FEMS Microbiol. Rev.* 75, 255-270.

Hamamoto,T. and Horikoshi,K. (1992) Alkaliphiles, In: J.Lederberg (ed.) *Encyclopedia of Microbiology Vol.1,* Academic Press, New York, p.81-87.

Jones,B.E., Grant,W.D., Collins,N.C., and Mwatha,W.E. (1994) In: F.Priest, A.Ramos-Cormenzana, and B.J.Tindall (eds). *Bacterial Diversity and Systematics.* Plenum Press, New York., pp.195-230.

Horikoshi,K. (1971) *Agric. Biol. Chem.* 35, 1407-1414.

Horikoshi,K. (1991) General view of alkaliphiles and thermophiles, In: K.Horikoshi and W.D.Grant (eds.), *Superbugs - Microorganisms in Extreme Environments*, Japan Scientific Societies Press, Tokyo, pp.3-14.

Kamekura,M. and Dyall-Smith,M.L. (1995) *J. Gen. Appl. Microbiol.* 14, 333-350.

Kamekura,M., Dyall-Smith,M.L., Upasani,V., Ventosa,A., and Kates,M. (1997) *Int. J. Syst. Bacteriol.* 47, 853-857.

Kates,M. (1993) *Experientia* 49, 1027-1036.

Kitada,M. and Horikoshi,K. (1992) *J. Bacteriol.* 174, 5936-5940.

McGenity,T.J. and Grant,W.D. (1995) *Syst. Appl. Microbiol.* 18, 237-243.

Nielsen,P., Fritze,D., and Priest,F.G. (1995) *Microbiology* 141, 1745-1761.

Nielsen,P., Rainey,F.A., Outtrup,H., Priest,F.G., and Fritze,D. (1994) *FEMS Microbiol. Lett.* 117, 61-65.

Niimura,Y., Koh,E., Yanagida,F., Suzuki,K-I., Komagata,K., and Kozaki,M. (1990) *Int. J. Syst. Bacteriol.* 40, 297-301.

Oren,A., Gurevich,P., Gemmell,R.T., and Teske,A. (1995) *Int. J. Syst. Bacteriol.* 45, 747-754.

Romano,I., Nicolaus,B., Lama,L., Manca,M.C. and Gamabacorta,A. (1996) *Syst. Appl. Microbiol.* 19, 326-333.

Ross,H.N.M. and Grant,W.D. (1985) *J. Gen. Microbiol.* 131, 165-173.

Spanka,R. and Fritze,D. (1993) *Int. J. Syst. Bacteriol.* 43, 150-156.

Weisser,J. and Truper,H.G. (1985) *Syst. Appl. Microbiol..* 6, 7-11.

Vedder,A. (1934) *Antonie van Leeuwenhoeck* J.Microbiol.. Serol. 1, 143-147.

Biodata of **Dr. Onstott et al.** This group contributed the chapter on ***"A Global Perspective on the Microbial Abundances and Activity in the Deep Subsurface."*** All the below written authors were principal investigators in the Subsurface Science Program of the U.S. Department of Energy and collaborated on studies of the deep subsurface microbiology.

Dr. Tullis C. Onstott is a geologist in the Department of Geosciences, Princeton University and is interested in the biogeochemical cycling of subsurface microbial ecology. He earned his Ph.D. in 1981 at Princeton University (in Geology).
E-Mail: tullis@princeton.edu

Dr. T.J. Phelps (Oak Ridge Natl. Lab.) is a microbiologist specializing in anaerobic microbial processes of the deep subsurface. He received his Ph.D. in 1985 at the University of Wisconsin, Madison (in Bacteriology).

Dr. Tom Kieft (Department of Biology, New Mexico Institute of Mining and Technology, Socorro, NM) is a microbiologist concerned with the starvation effects and survival of bacteria in subsurface habitats. He earned his Ph.D. in 1983 at the University of New Mexico (in Biology).

Dr. F.S. Colwell (Biotechnological Department, Idaho National Engineering and Environmental Lab. Idaho Falls, ID) is a microbiologist interested in the microbial activity and biomass of deep subsurface environments. He received his Ph.D. in 1986 from Virginia Tech. (in Microbiology).

Dr. D.L. Balkwill (Department of Biologic Sciences, Florida State University at Tallahassee, FL.) is a microbiologist specializing in the molecular phylogeny of deep subsurface microbial communities. He earned his Ph.D. in 1977 from Penn State University (in Microbiology).

Dr. J.K. Fredrickson (Battelle Pacific Northwest Laboratories, Richland, WA.) is a microbiologist interested in microbial ecology of the deep subsurface. He received his Ph.D. in 1984 from the Washington State University (in Soil Science).

Dr. Fred J. Brockman is a microbiologist interested in deep vadoze zone environments and the application of nucleic acid-based approaches for studying subsurface microbial ecology. Dr. Brockman received his Ph.D. in 1992 at the University of Idaho.
E-mail: fred.brockman@pnl.gov

J. Seckbach (ed.), Enigmatic Microorganisms and Life in Extreme Environments, 487–500.

A GLOBAL PERSPECTIVE ON THE MICROBIAL ABUNDANCE AND ACTIVITY IN THE DEEP SUBSURFACE

T.C. ONSTOTT
Dept. of Geosciences
Princeton University
Princeton, NJ 08544

T.J. PHELPS
Environmental Sciences Division
Oak Ridge National Laboratory
Oak Ridge, TN 37831-6036

T. KIEFT
Dept. of Biology
New Mexico Institute of Mining and Technology
Socorro, NM 87801

F.S. COLWELL
Biotechnologies Department
Idaho National Engineering and Environmental Laboratory
P.O. Box 1625
Idaho Falls, ID 83415-2203

D. L. BALKWILL
Department of Biological Sciences
Florida State University
Tallahassee, FL 32306

J. K. FREDRICKSON and F.J. BROCKMAN
Battelle Pacific Northwest Laboratories
Richland, WA 99352

1. Introduction

Reports that microorganisms and even microbial communities exist at great depths below the continental surface and below the sea floor have steadily increased in the past several years. The best evidence for the presence and diversity of microorganisms at hundreds to thousands of meters below the land surface (mbls.) lies with cores taken using tracers to detect contamination and processed by aseptic methods (Griffin et al., 1997). Other sources of information on the subterranean biosphere include ground water samples and microbial investigations of recently excavated mine faces. This review compares results from these studies with those of the marine subsurface in order to place constraints on the global subsurface biomass and rate of biogeochemical cycling.

2. Microbial biomass

Microbial processes in deeply buried marine sediments were first investigated by Claude Zobell (Zobell, 1947) and has been explored in greater detail during several recent Ocean Drilling Program (ODP) drilling legs (Parkes et al., 1994). One series of microbial samples collected during drilling in the Japan Sea was recovered from depths ranging up to 450 meters below the sea floor (J in Fig. 1).

Microbial samples have also been extracted from the continental subsurface by drilling at more than a dozen localities in the coterminous United States. Most of these subsurface investigations were supported by the Subsurface Science Program, U.S. Dept. of Energy. The maximum depth of microbial sampling ranged from 200 meters below land surface (mbls.) in the Eastern Coastal Plain sequences (M in Fig. 1; Chapelle et al., 1987; and S in Fig. 1, Balkwill, 1989) to 2800 mbls. of Virginia in Taylorsville basin (T in Fig. 1; Boone et al., 1995; Lehman et al., 1995; Tseng et al., 1995; Liu et al. 1997).

Investigations into the microbiology of mine workings were first mentioned by Humboldt in the late 18th century. More recent concerns about the disposal of radioactive waste have motivated microbial studies of tunnels into solidified volcanic tuff approximately 400 mbls. at the Nevada Test Site (N in Fig. 1; Haldeman and Amy, 1993). Preliminary results on the microbial biomass for gold mining addits 3400 mbls. of South Africa in the Witwatersrand basin (W and W in Fig. 1) have recently been presented by Onstott et al. (1997).

Oil field production water from 400 mbls. of Illinois was the source of samples for an early subsurface microbial study by Bastin (1926). More recently, Stetter et al. (1993) has reported hyperthermophilic archaea in water drawn from oil reserves located 3000 mbls. of the North Sea (NS in Fig. 1). Ground water samples from 150 mbls. in Snake River Plains basalt (I in Fig. 1; Colwell and Lehman, 1996) to up to 1200 mbls. in the Columbia River basalt (C in Fig. 1; Stevens et al., 1993) and from underground workings within granite up to 1200 mbls. of Sweden (S in Fig. 1; Pedersen and Ekendahl, 1990; Ekendahl and Pedersen, 1994) have revealed the presence of active subsurface microbial communities in these igneous rock environments.

A plot of the microbial biomass (cells/g) as determined by acridine orange direct counts (AODC) versus depth (Fig. 1) from which the samples were collected reveals the following:

1) The relationship of decreasing biomass with increasing depth for the continental subsurface is consistent with the relationship observed by Parkes et al. (1994) for the marine subsurface (lines plotted in Fig. 1).2) Some of the continental subsurface sites containing low permeability, consolidated sediments or fractured rocks yield biomass concentrations slightly below the marine trend. For example, the AODC-based biomass

values for subsurface shale and sandstone near the Cerro Negro volcanic intrusive in northwestern New Mexico (C in Fig. 1) and Piceance basin in western Colorado (P in Fig. 1) are less than what would be predicted from the trend observed in marine sediments. This may be related to the tectonic history of these settings, which indicates that they were buried to greater depths until geologically recent times. If the AODC-based biomass for these two sites are plotted at their maximum paleodepth (open arrows in Fig. 1), then they would compare more closely with deep marine sediments.
3) The concentration of bacteria in ground water samples do not obviously exhibit a diminishing trend with depth. At shallow depths the bacteria are preferentially concentrated on the solid phases relative to being suspended in water. At greater depths the magnitude of bacterial partitioning on to the solid mineral from the water is unknown and depends in detail on the pore structure and fracture network of the rock matrix. Another complication is that the habitat in and around wells represents a disturbed environment and thus microorganisms obtained from well samples may not represent the complete subsurface microbial community indigenous to the rock strata.
4) Microbial biomass values for depths greater than 800 m exist for only a few samples (C, P, T, and W in Fig. 1) and these may not be representative for the earth as a whole.

To estimate a global subsurface biomass, we have assumed an upper temperature limit for life. A value of 110°C for the maximum temperature for subsurface bacterial survival and the geothermal gradient constrains the maximum depth of the global subsurface biosphere. The 110°C maximum temperature is the observed maximum temperature in laboratory cultures of hyperthermophilic microorganisms. For the oceanic crust the average geothermal gradient is approximately 26 °C/km and for the continental crust it is approximately 19°C/km (Sclater et al., 1980). For these geothermal gradients the predicted average depths to the 110°C isotherm are 4 and 6 km for the oceanic and continental crusts, respectively (horizontal dashed line in Fig. 1).

Integrating the biomass versus depth relationship based upon AODC, as described by Parkes et al. (1994), the continental subsurface biosphere from 1 mbls. to the 110°C isotherm would be composed of approximately 5.4×10^{16} grams of microbial carbon (assuming 10^{-13} grams of carbon per bacterium). A similar calculation for the marine subsurface biosphere yields 1.5×10^{17} g of carbon. These quantities are significant when compared to 4.5×10^{17} g of biomass carbon for the continental surface biosphere and actually exceeds the 7×10^{15} g of biomass carbon for the marine biosphere (Lasaga, 1981).

In near surface environments, the phospholipids and nucleic acids from deceased cells are metabolized relatively quickly by the microbial community. In subsurface environments, the rate of turnover of cellular material from expired microorganisms is probably much slower. Consequently, the AODC may overestimate the viable or living biomass for subsurface samples, if remnant DNA are still stained and counted as cells by AODC.

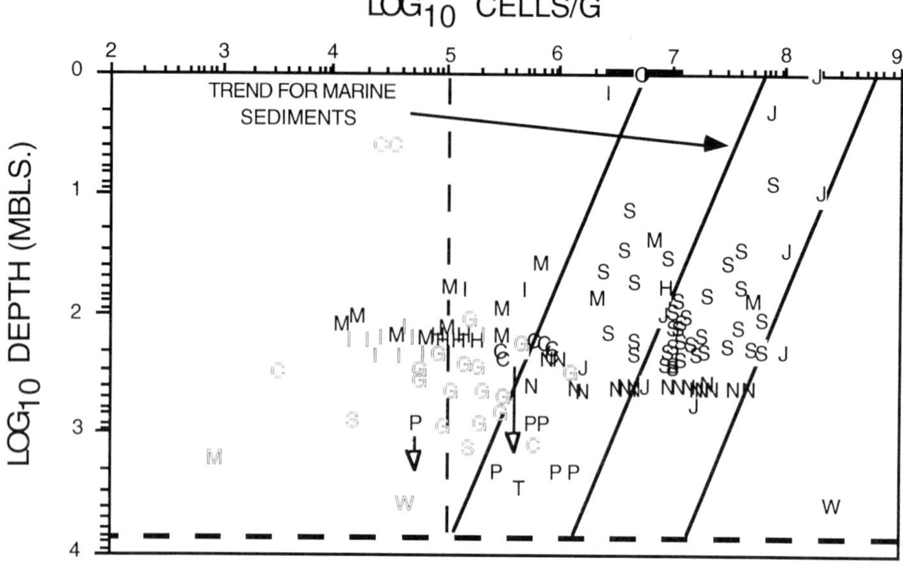

Figure 1. Biomass (bacterial cells/g of rock or cells/ml of water) as determined by AODC versus depth. Rock or Sediment: O = Oklahoma (Balkwill and Ghiorse, 1985), J = Japan Sea (Parkes et al., 1994), S = Savannah River Site (Balkwill, 1989; Fredrickson et al., 1991), H = Hanford Site (Kieft et al., 1995; Brockman et al., 1992), M = Maryland Coastal Plain (Chapelle et al., 1987), N = Nevada Test Site (Haldeman and Amy, 1993; Haldeman et al., 1993), I = Snake River Plains soils (Colwell, 1989), C = Cerro Negro (Fredrickson et al., 1997), P = Piceance Basin (unpublished data), T = Taylorsville basin (Lehman et al. 1995) and W = Witwatersrand (Onstott et al. 1997). Ground water: CC = Cape Cod aquifer (Harvey et al. 1994), I = Snake River Plains basalt (Colwell and Lehman, 1996), C = Columbia River basalt aquifer (Stevens et al., 1993), G and S = Stripa ground water (Pedersen and Ekendahl, 1990; Ekendahl and Pedersen, 1994), M = Madison formation (Olson et al., 1981), and W = Witwatersrand (Onstott et al. 1997). Trend for marine subsurface sediment is from Parkes et al. (1994) and represents the mean and 95% confidence levels. Vertical dashed line represents the limit of detection for rock samples. The detection limit for water samples is much lower. The horizontal dashed line represents the approximate maximum depth for bacterial survival based upon an average geotherm and a temperature of 110°C.

The actively metabolizing biomass concentrations, therefore, have been estimated for subsurface settings by a variety of techniques. For the rock formations sampled at Savannah River Site (S in Fig. 1), Hanford (H in Fig. 1), Cerro Negro (C in Fig. 2), Piceance basin (P in Fig. 2), the Nevada Test Site (N in Fig. 2) and for basaltic volcanic flows at Idaho National Engineering and Environmental Laboratory (I in Fig. 2), the most probable number (MPN) or colony forming units (CFU) per gram of sample were the measures used to define the proportion of bacteria that could be cultured relative to total populations determined by AODC.

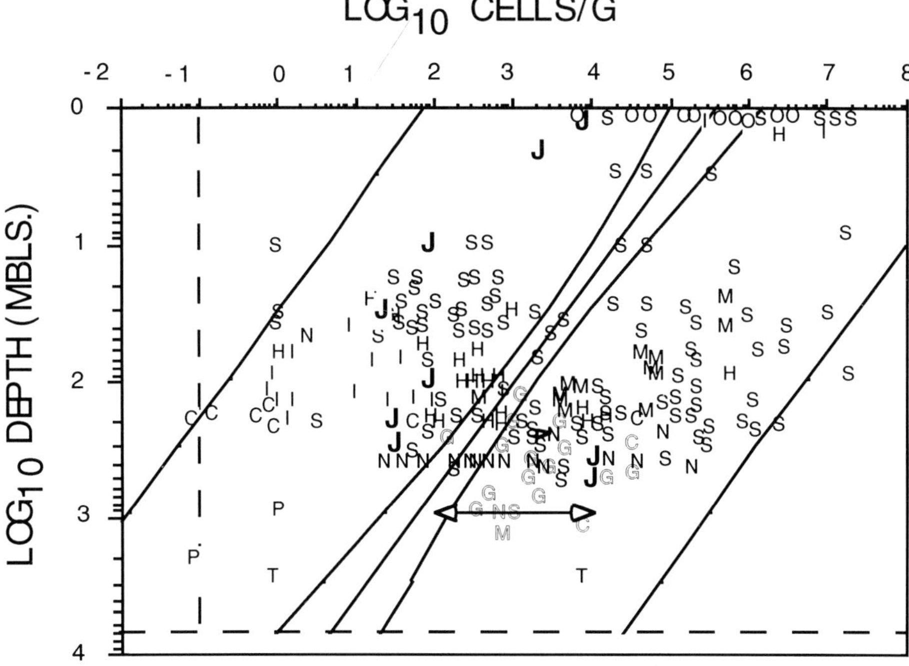

Figure 2. Biomass (bacterial cells/g or cells/ml), as determined by MPN or CFU, versus depth. Rock or Sediment: O = Oklahoma (Balkwill and Ghiorse, 1985), J = Japan Sea (Parkes et al., 1994), S = Savannah River Site (this paper), H = Hanford Site (this paper; McKinley, 1997), I = Snake River basalt (this paper), M = Maryland Coastal Plain (Chapelle et al., 1987), N = Nevada Test Site (Haldeman and Amy, 1993; Haldeman et al., 1993), C = Cerro Negro (Fredrickson et al., 1997), P = Piceance Basin (this paper), and T = Taylorsville basin (this paper). Ground water: C = Columbia River basalt aquifer (Stevens et al., 1993), G and S = Stripa ground water (Pedersen and Ekendahl, 1990; Ekendahl and Pedersen, 1994), M = Madison formation (Olson et al., 1981), and NS = North Sea production fluids (Stetter et al., 1993). The vertical dashed line represents the limit of detection for rock samples. The solid lines represent the least squares best fit to the data, the 95% confidence interval (hyperbolic lines) and the 95% variance.

Since only a small percentage of cells may be cultured on a given medium, microbial populations estimated from aerobic and anaerobic culturing techniques probably represent the minimum viable biomass for a given sample. A plot of the cultured biomass versus depth for the continental subsurface samples yields a least squares best fit relationship of viable cells/g = $4.38 \times 10^5 \times Z^{-1.28}$ (Fig. 2), where Z is the depth in mbls. The MPN's for the Japan Sea marine sediments (J in Fig. 2) overlap those of the continental subsurface, suggesting that this relationship holds for both the marine and continental subsurface biosphere. Using the same assumptions made in the AODC calculation, the integral of this viable cell relationship yields a continental subsurface biomass of $1x10^{14}$ g of carbon and a marine sub-sea floor biomass of $3x10^{14}$ g of carbon. This represents approximately 0.1% of the total living surface biosphere carbon on Earth (Lasaga, 1981).

The actual "living" subsurface biomass likely resides between the estimates provided by the AODC and MPN or CFU determinations. MPN and CFU methods are known to underestimate microbial populations in near surface sediments by one to two orders of magnitude. The MPN values relative to the AODC estimates also diminish more rapidly with depth. For example, at depths of 50 to 500 meters the population of culturable bacteria for all Savannah River Site samples (S in Figs. 1 and 2) ranged down to 10^2 cells/g; whereas, the total population of cells as determined by AODC was consistently greater than 10^6 cells/g.

For the Savannah River Site the percentage of the total population represented by culturable microorganisms (%MPN/AODC) ranged from 0.1 to 80% (S in Fig. 3) with the higher percentage found in the more permeable sandy sediment layers. This observation is also consistent with the results of Chapelle et al. (1990), who found for the Maryland Coastal Plain sediments, that direct counts of growing bacteria relative to the AODC were lower in the less permeable clay layers than in the sandier aquifers. The saturated zone of the Hanford Site (H in Fig. 3) yields %MPN/AODC values comparable to that of the clay layers at Savannah River. The vadose zones at both the Hanford site (data not shown in Fig. 3) and the vadose zone in the Snake River Plains basalt at Idaho National Engineering and Environmental Laboratory (I in Fig. 3), yielded much lower values of culturable bacteria. Because the permeability at the latter site is high, 100 times that of Savannah River clay layers, the low %MPN/AODC value of 0.01% probably reflects the arid environment, thus low water availability and low nutrient flux. For the low permeability, deep subsurface sites at Cerro Negro, Piceance Basin, and Nevada Test Site (C, P, and N in Fig. 3), the %MPN/AODC ranged from 0.1% to 0.001%.

The difference in the rate at which the culturable biomass and direct counts diminish with increasing depth may reflect the rapid reduction in porosity in sediments with increasing depth due to compaction and cementation. Sedimentary rock formations typically exhibit exponentially decreasing porosity as a function of increasing burial depth. This declining porosity is often accompanied by increasing cementation resulting in lower permeability and smaller pore throats. The changes in these two properties will conspire to reduce the nutrient flux and to occlude cellular material in deep subsurface sedimentary formations. The lower nutrient flux could reduce the culturable biomass; whereas the lower turnover of cellular material could help preserve the direct counts. The same behavior applies, though to a lesser extent, for igneous formations, such as basalt or granite, where increasing pressure with depth and mineralization processes reduce the fracture permeability and porosity. Opportunities for large scale microbial transport and the consequent colonization of rock formations is necessarily limited at greater depths.

These processes may result in moribund cells being included in the cell counts from the AODC. Although the MPN and plate count methods underestimate viable biomass, the viable subsurface biomass at one kilometer may be hypothetically closer to the 10^2 cells/g, predicted by the culturable biomass relationship, than the 10^7 cells/g predicted by the AODC relationship.

Another method used to measure viable biomass is the concentration of phospholipid fatty acid (PLFA). All bacteria contain polar, phospholipid in their membranes and that is generally considered an excellent measure of bacteria biomass. In water-saturated sediments where microorganisms are active, the phospholipid fatty acids of non-viable bacteria are rapidly recycled (White et al., 1979). In such environments the metabolically active microbial biomass, as estimated by viable counts, generally agrees with the biomass estimate by PLFA and AODC methods (Balkwill et al., 1988).

PLFA values were compared to viable and total bacterial population estimates for samples from the Hanford Site, Piceance basin, and Cerro Negro sites (H, P, and C in

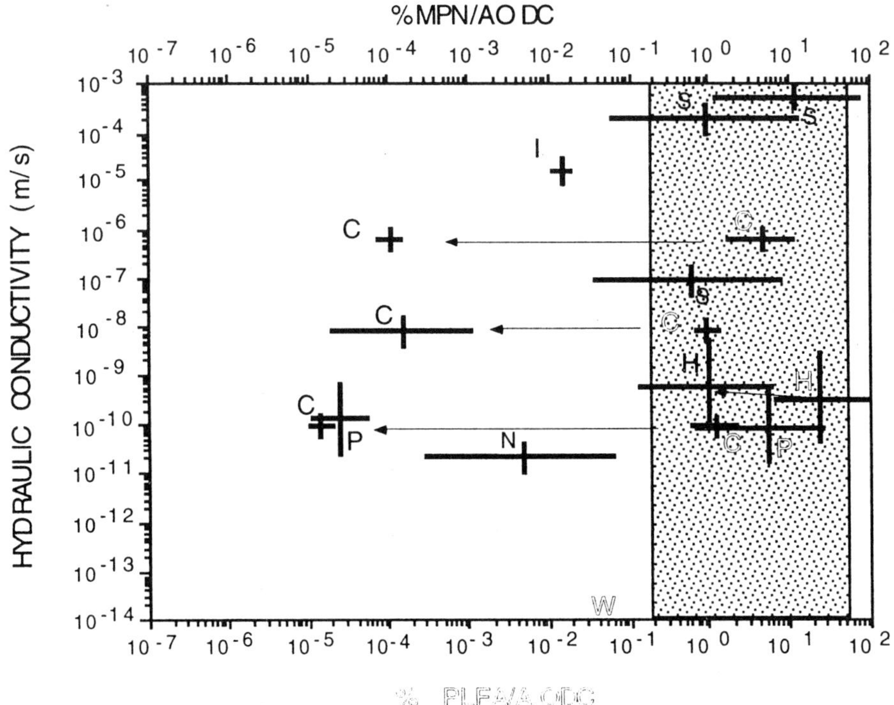

Figure 3. The %MPN/AODC (plain symbols) and the %PLFA/AODC (outline symbols) versus hydraulic conductivity. Letters refer to same sites as figure 1. Stippled pattern covers range of %MPN/AODC for ground water samples.

Fig. 3). These phospholipid fatty acid concentrations (pmole/g) were converted to cells/g using a factor of 2.5×10^4 cells/pmoles; a value for subsurface environments derived by Balkwill et al. (1988). With this conversion factor, the %PLFA biomass/AODC ranged from 1 to 100%. Some of this variation could reflect the dependence of the conversion factor on the physiological state of the microorganisms. For Piceance basin and the Cerro Negro sites the biomass derived from the PLFA was higher than that measured by viable counts and significantly less than the total determined by AODC. Furthermore, for these two sites, the differences between PLFA and MPN values were dramatic, a relationship that was not altered when hydraulic conductivities were $\leq 10^{-6}$ m/s versus as low as 10^{-10} m/s (C and P in Fig. 3).

In the case of Taylorsville basin samples no AODC values were available. The %MPN/PLFA, however, was 0.0003% indicating an enormous difference in the viable versus PLFA biomass estimates. The cause of this discrepancy is uncertain. It could be due to the inability to culture 99.999% of the bacteria in the core samples collected from these sites. Alternatively, it could be caused by preservation of the polar phospholipid membranes in nonviable bacteria in the low permeability rock and low living biomass.

3. Microbial activity

One method for estimating heterotrophic microbial activity in the subsurface is to determine the rate of organic carbon oxidation by estimating CO_2 production. A common approach is to augment samples with small quantities of ^{14}C-labeled organic substrate and to measure the $^{14}CO_2$ released. This method, however, typically overestimates *in situ* rates. A more accurate assessment of the *in situ* microbial respiration rate is to use geochemical mass balance estimates of CO_2 production (Phelps et al., 1994a). For fermenters and sludge digestors these rates are typically greater than a mole/liter-year (Fig. 4). For subsurface environments the CO_2 production rate ranges from nanomoles to picomoles/liter-year. The decrease in the rate of CO_2 production from fermenters to the subsurface environments may span 15 orders of magnitude, reflecting not only the number of metabolically active cells, but also the metabolic rate of each cell. Phelps and Zeikus (1985) estimated a 3.6% growth yield for bacteria in the sediments of Lake Mendota. If this same growth yield is applied to the deep subsurface, then a picomole/liter-year of CO_2 production would correspond to the growth of 10 cells/liter-year. If all this metabolic activity was associated with biomass estimated by viable counts (i.e., only the bacteria that could be cultured or about 10^5 cells/L for sediment one km below the surface), then the average doubling time for the subsurface microbial community would be 10^4 years (10^{11} seconds in Fig. 4). If the metabolic activity was associated with the biomass estimated by PLFA or AODC methods (i.e., 10^{10} bacteria/L), then the average community doubling would be on the order of 1000 million years (10^{16} seconds in Fig. 4).

Given these slow growth rates is likely that yield coefficients in deep subsurface sediments would be substantially lower than in eutrophic lake sediments. Decreasing or

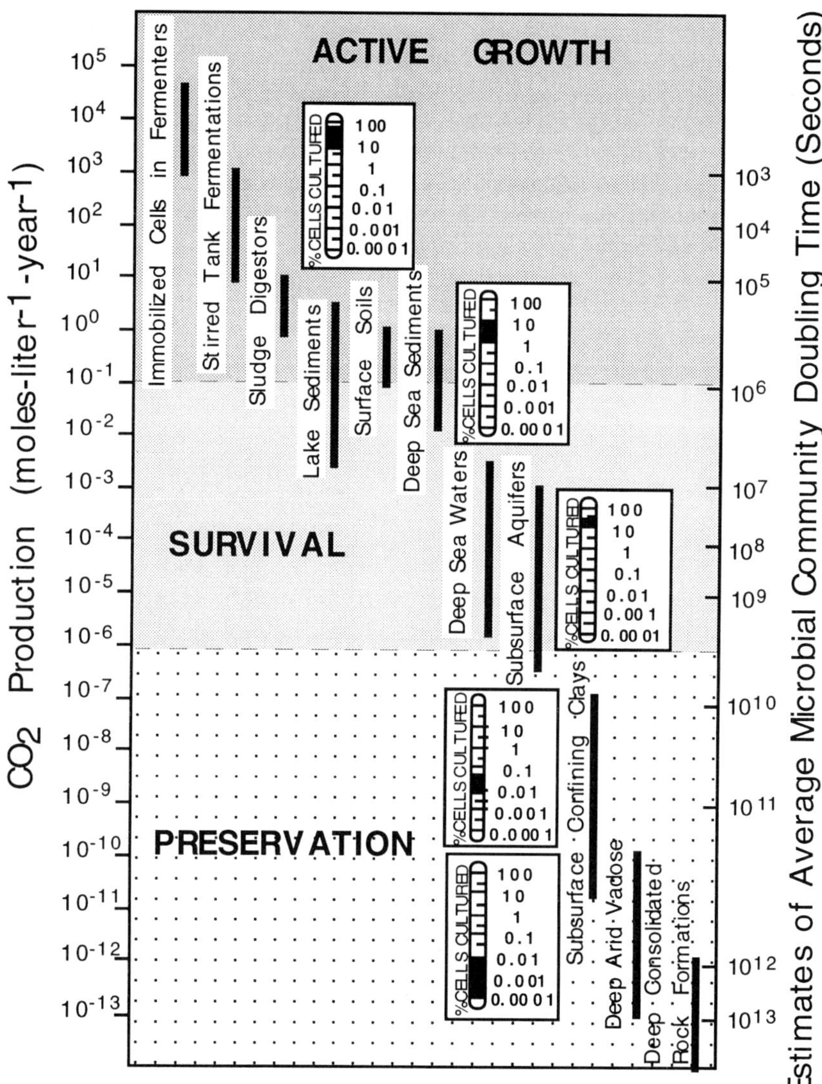

Figure 4. The rate of CO_2 production in moles/L-year and corresponding estimates of time for cell turnover in seconds for different microbial habitats. Immoblized cells in fermenters--(Webb et al., 1995; Berreto et al., 1991). Stirred tank fermentation--(Gschaedler et al., 1994; Atlas, 1988; Audet et al., 1989). Sludge digestors--(Conrad et al., 1985; Phelps, 1991). Lake sediments--(Phelps and Zeikus, 1984; Phelps and Zeikus, 1985; Phelps, 1991). Soils--(Phelps et al., 1994a; Siegrist et al., 1994; Chapatwala et al., 1996). Deep sea sediments--(Jahnke et al., 1982; Jorgensen, 1982). Deep Sea--(Williams and Carlucci, 1976). Deep subsurface--(Chapelle and Lovley, 1990; Lovley and Chapelle, 1995; McMahon and Chapelle, 1991; Murphy et al., 1992; Phelps et al., 1994a; this paper).

increasing the yield coefficient, however, does little to alter the doubling times, which must be glacially slow. Bacteria capable of efficiently repairing the radiation-induced DNA damage would have a higher probability of survival in such an environment.

Despite the slow CO_2 production rate that typifies the subsurface, the globallyintegrated CO_2 production from the subsurface would be 55×10^{13} grams of carbon/year. This value is comparable to the CO_2 exchange fluxes between the atmosphere and the continental surface biosphere (Lasaga, 1981). The subsurface CO_2 production may even be higher since autotrophic processes, which may be significant in some environments, fix CO_2.

4. Conclusions

In subsurface environments the biomass values determined by viable counts << PLFA biomass < total direct counts (AODC). Regardless of which procedure is used to estimate the microbial biomass in the subsurface or which technique is used to estimate microbial activities, the microbial biomass and activity in the subsurface are globally significant and influence the geology, geochemistry, and hydrology of these environments. Until the rise in O_2 concentrations in the Earth's atmosphere during the Proterozoic and the consequent development of an ozone shield against UV radiation, most of the Earth's biomass may have resided in the subsurface.

Microbial biomass and activity tends to decrease with depth and isolation from surface-derived nutrients and energy sources, although important exceptions to this relationship may exist. Microbial communities in low permeability, low porosity, deep subsurface environments are struggling for survival, with the average microorganisms apparently striving to maintain their machinery for biosynthesis in order to forestall death. Life processes in the deep subsurface, therefore, are by necessity less focused on cellular reproduction and more dedicated to maintaining cellular integrity, wherein a very small fraction of the community exhibits successful procreation in a given month or year.

5. Acknowledgments

This research for this article was supported by grant DE-FG02-94ER61821 from the Subsurface Science Program, U.S. Department of Energy to T.C.O. and by grants from Subsurface Science Program, U.S. Department of Energy to the coauthors.

6. References

Atlas, R. (1988) Macmillan Publishing Co. New York. P. 4640534.

Audet, P., Paquin, C. and Lacroix, C. (1989) *Appl. Environ. Microbiol.* **55**, 185-189.

Balkwill, D.L. and Ghiorse, W.C. (1985) *Appl. Environ. Microbiol.* **50**, 580-588.

Balkwill, D.L., Leach, F.R., Wilson, J.T., McNabb, J.F., and White, D.C. (1988) *Microb. Ecol.* **16**, 73-84.

Balkwill, D.L. (1989) *Geomicrobiol. J.* **7**, 33-52.

Barreto, M.T.O., Melo, E.P., Moreira, J.L. and Carondo, M.J.T. (1991) *J. Ind. Microbiol.* **7**, 63-70.

Bastin, E.S. (1926)*Science* **63**, 21-24.

Boone, D.R., Liu, Y., Zhao, Z., Balkwill, D.L., Drake, G.R., Stevens, T.O. and Aldrich, H.C. (1995) *Inter. Jour. System. Bacteriol.* **45**, 441-448.

Brockman, F.J., Kieft, T.L., Fredrickson, J.K., Bjornstad, B.N., Li, S. W., Spangenburg, W. and Long, P.E. (1992) *Microb. Ecol.* **23**, 279-301.

Chapatwala, K.D., Babu, G.V., Armstead, E., Palumbo, A.V., Zhang, C. and Phelps, T.J. (1996) *Appl. Biochem. Biotechnol.* **57/58**, 827-835.

Chapelle, F.H., Zelibor Jr., J.L., Grimes, D.J., and Knobel, L.L. (1987) *Water Resour. Res.* **23**, 1625-1632.

Chapelle, F.H. and Lovley, D.R. (1990) *Appl. Environ. Microbiol.* **56**, 1865-1874.

Colwell, F.S. (1989) *Appl. Environ. Microbiol.* **55**, 2420-2423.

Conrad, R., Phelps, T.J., and Zeikus, J.G. (1985) *Appl. Environ. Microbiol.* **50**, 595-601.

Ekendahl, S. and Pedersen, K. (1994) *Microbiology* **140**, 1565-1573.

Fredrickson, J.K., Balkwill, D.L., Zachara, J.M., Shu-Mei, W. Li., Brockman, F.J. and Simmons, M.A. (1991) *App. Environ. Microbiol.* **57**, 402-411.

Fredrickson, J.K., McKinley, J.P., Bjornstad, B.N., Long P.E., Ringelberg, D.B., White, D.C., Krumholz, L.R., Suflita, J.M., Colwell, F.S., Lehman, R.M., Phelps, T.J., and Onstott, T.C. (1997) *Geomicrobiol.J.* **14**, 183-202.

Griffin, W.T., Phelps, T.J., Colwell, F.S. and Fredrickson, J.K. (1997) in P.S. Amy and D.L. Haldeman (eds.), *The Microbiology of the Terrestrial Deep Subsurface*, CRC Lewis Publishers, Boca Raton, pp. 23-44.

Gschaedler, A., Thi Le, N., and Bourdant, J. (1994) *J. Ind. Microbiol.* **13**, 225-232.

Haldeman, D.L. and Amy, P.S. (1993) *Microb. Ecol.* **25**, 183-194.

Haldeman, D.L., Amy, P.S., Ringelberg, D. and White, D.C. (1993) *Microb. Ecol..* **26**, 145-159.

Harvey, R.W., Smith, R.L., and George, L. (1994*Appl. Environ. Microbiol.* **48**, 1197-1202.

Jahnke, R.A., Emerson, S.R., and Murray, J.W. (1982)*Limnol. Oceanogr.* **27**, 610-623.

Jorgensen, B.B. (1982) *Nature* **296**, 643-645.

Kieft, T.L., Fredrickson, J.K., McKinley, J.P., Bjornstad, B.N., Rawson, S.A., Phelps, T.J., Brockman, F.J. and Pfiffner, S.M. (1995) *App. Environ. Microbiol.* **61**, 749-757.

Lasaga, A.C. (1981) *Reviews in Mineralogy*, Volume 8, Mineralogical Society of America, 69-110.

Lehman, R.M., Colwell, F.S., Ringelberg, D.B. and White, D.C. (1995) *Microbiol. Methods* **22**, 263-281.

Liu, S.V., Zhou, J., Zhang C., Cole, D.R., Gajdarziska-Josifovsak, M., and Phelps, T.J. (1997) *Science* **277**, 1106-1109.

Lovley, D.R. and Chapelle, F.H. (1995) *Rev. Geophysics.* **33**, 365-381.

McKinley, J.P., Stevens, T.O., Fredrickson, J.K., Zachara, J.M., Colwell, F.S., Wagnon, K.B., Smith, S.C., Rawson, S.A., Bjornstad, B.N. (1997) *Geomicrobiol. J.* **14**, 23-39.

McMahon, P.B. and Chapelle, F.H. (1991) *Nature* **349**, 233-235.

Murphy, E.M., Schranke, J.A., Fredrickson, J.K., Bledsoe, H.W., Francis, A.J., Sklarew, D.S. and Linehan, J.C. (1992*Water Resour. Res.* **28**, 723-740.

Olson, G.J., Dockins, W.S., McFeters, G.A., and Iverson, W.P. (1981)*Geomicrobiol. J.* **2**, 327-340.

Onstott, T.C., Tobin, K., Dong, H., DeFlaun, M.F., Fredrickson, J.K., Bailey, T., Brockman, F., Kieft, T., Peacock, A., White, D.C., Balkwill, D., Phelps, T.J. and Boone, D.R. (1997) *SPIE 42nd Annual Mtg. Proceedings* **3111**, 344-357.

Parkes, R.J., Cragg, B.A., Bale, S.J., Getliff, J.M., Goodman, K., Rochelle, P.A., Fry, J.C., Weightman, A.J., and Harvey, S.M. (1994) *Nature* **371**, 410-413.

Pedersen, K., and Ekendahl, S. (1990) *Microb. Ecol.* **20**, 37-52.

Phelps, T.J. and Zeikus, J.G. (1984) *Appl. Environ. Microbiol.* **48**, 1088-1095.

Phelps, T.J. and Zeikus, J.G. (1985) *Appl. Environ. Microbiol.* **50**, 1285-1291.

Phelps, T.J. (1991) *J. Microbiol. Meth.* **13**, 243-254.

Phelps, T.J., Murphy, E.M., Pfiffner, S.M. and White, D.C. (1994a) *Microb. Ecol.* **28**, 335-350.

Phelps, T.J. Siegrist, R.L., Korte, N.E., Pickering, D.A., Strong-Gunderson, J.M., Palumbo, A.V., Walkerk J.F., Morrissey, C.M. and Mackowshi, R. (1994b) *Appl. Biochem. Biotechnol.* **45/46**, 835-846.

Tseng, H-Y, Onstott, T.C., Burruss, R.C. and Miller, D.S. (1995) *Chem. Geol.* **127**, 269-294.

Sclater, J.G., Jaupart, C. and Galson, D. (1980) *Rev. Geophys. Space Physics* **18**, 269-311.

Siegrist, R.L., Phelps, T.J., Korte, N.E., Pickering, D.A., Mackowski, R., and Cooper, L.W. (1994) *Appl. Biochem. Biotechnol.* **45/45**, 757-774.

Stevens, T. O., McKinley, J.P., and Fredrickson, J.K. (1993) *Microb. Ecol.* **25**, 35-50.

Stetter, K.O., Huber, R., Blochl, E., Kurr, M., Eden, R.D., Flelder, M., Cash, H. and Vance, I. (1993) *Nature* **365**, 743-745.

Webb, O.F., Davison, B.H., Scott, T.C. and Scott, C.D. (1995) *Appl. Biochem. Biotechnol.* **51/52**, 559-568.

White, D.C., Davis, W.M., Nickels, J.S., King, J.D. and Bobbie, R.J. (1979)*Oecologia* **40**, 51-62

Williams, P.M. and Carlucci, F. (1976) *Nature* **262**, 810-811.

Zobell, C.E. (1947) *Bull. Amer. Assoc. Petrol. Geol.* **31**, 1709-1751.

BioData of **Douglas H. Bartlett**
Contributor of: ***Membrane-based Adaptations of deep-sea Piezophiles.***

Dr. D.H. Bartlett is an Associate Professor at Scripps Institution of Oceanography, University of California, San Diego, CA. He earned his PhD from the University of Illinois (1985) and then spent several years in postdoctoral training in Agouron Institute, La Jolla, CA. His principal interest is high pressure and low temperature adaptation by dee-sea bacteria.
E-mail: <dbartlett@ucsd.edu>

J. Seckbach (ed.), Enigmatic Microorganisms and Life in Extreme Environments, 501–512.

MEMBRANE-BASED ADAPTATIONS OF DEEP-SEA PIEZOPHILES

D. H. BARTLETT AND K. A. BIDLE
Center for Marine Biotechnology and Biomedicine
Scripps Inst. of Oceanography
University of California, San Diego
La Jolla, CA 92093-0202 U. S. A.

1. Introduction

Pressure is an important thermodynamic variable of particular significance when comparing biological processes that take place at different depths. Most of what is known about high pressure habitats has come from studies of deep-sea environments. When one considers the tremendous diversity of deep-sea benthic invertebrates (Rex *et al.*, 1993), and the propensity of marine microorganisms to establish symbiotic associations, the combined diversity of free-living and host-associated deep-sea microorganisms must surely be immense. Add to these microorganisms those dwelling in the deep-earth (Kerr, 1997), and high pressure environments are seen to encompass a diverse biota as well as represent the vast majority of the biosphere by volume. Within both marine and terrestrial environments pressure usually increases by about 10 megapascal (MPa, 1 MPa = 10 bar ≈ 9.87 atmospheres ≈ 145 pounds per square inch) per kilometer depth, however, great variation in this rule is found within the deep earth. In contrast to the deep sea where an isothermal temperature near freezing (2°C) predominates, subterranean temperatures generally increase by 20-30°C per kilometer. The highest pressure environment yet sampled for microorganisms is within the Challenger Deep of the Mariana Trench at a depth of 10,898 m and a pressure of 110 MPa. The current upper pressure limit for microbial growth is about 130 MPa (Yayanos, 1986).

Microorganisms obtained from the deep-sea which exhibit greater growth rates at high pressure than atmospheric pressure have been defined alternatively as barophiles ("weight loving") or more accurately as piezophiles ("pressure loving"; Yayanos, 1995). Molecular phylogenetic studies derived from comparisons of 16S rRNA sequences indicates multiple distinct lineages of piezophilic and piezotolerant bacteria, including *Shewanella*, *Moritella*, *Colwellia*, *Photobacterium*, an unnamed genus within the Proteobacteria γ-subgroup and a piezotolerant *Bacillus* strain (Kato *et al.*, 1995; DeLong *et al.*, 1997). A piezophilic *Desulfovibrio* species has also recently been described (Bale *et al.*, 1997). It would appear that a great diversity of piezophiles probably exists in the deep sea and that the current collection of strains has been limited by the types of environmental samples used and the nature of the enrichments employed.

In this article we will address the effects of high pressure on membranes and membrane-localized processes, drawing upon information obtained from model lipids, and membrane and membrane protein studies of shallow-water and deep-sea organisms. Unlike many of the extreme environments described in this volume, even the deepest deep-sea environments contain both macroorganisms as well as microorganisms. Therefore, while our emphasis will be on the microbial world, some description of biochemical adaptation in deep-sea animals will also be provided.

Pressures within the range found in the known biosphere can affect many cellular processes (Bartlett, 1992). At the biochemical level such pressures are documented to perturb the macromolecular interactions among proteins and between proteins and nucleic acids. Nevertheless, lipids appear to be particularly sensitive to pressure effects. They are typically an order of magnitude more compressible than proteins (Weber and Drickamer, 1983). Thus far, experiments with deep-sea microorganisms have only identified membrane composition and membrane-localized functions as critical to high pressure growth. For information on other aspects of deep-sea microbiology (Bartlett, 1992; DeLong, 1992; Prieur, 1992; Yayanos, 1995) and high pressure biochemistry (Silva and Weber, 1993; Gross and Jaenicke, 1994; Robinson and Sligar, 1995; Royer, 1995; Moshaev *et al.*, 1996) additional reviews are cited.

The energetic consequences of a change in pressure on chemical equilibria and rates are related to the following free energy equations:

$$\Delta G = (\Delta E + P\Delta V) - T\Delta S \text{ and } \Delta G^{\ddagger} = (\Delta E^{\ddagger} + P\Delta V^{\ddagger}) - T\Delta S^{\ddagger}$$

where ΔG is the free energy change, ΔE is internal energy change, P is the pressure, ΔV is the volume change, T is the temperature, ΔS is the entropy change, and the double dagger symbol ($\ddagger$) represents the activation terms. These equations illustrate the connection between pressure and volume, namely that increased pressure perturbs both the equilibria and rates of chemical reactions by promoting decreased system volume changes and volume changes of activation, respectively. Because most molecular motions within the membrane are accompanied by a positive volume change, increased pressure favors a reduction in membrane conformational changes and thus promotes a decrease in fluidity. In this regard, elevated pressure shares with increased osmolarity (in cases where there is cytoplasmic water loss) and decreased temperature the promotion of increased membrane packing density.

2. Pressure Effects on Model Membranes

Studies with defined phospholipids indicate that lamellar lipid bilayers can assume a variety of structures depending on pressure, and that pressure also influences the formation of nonlamellar lipid structures including inverted hexagonal and cubic structures which may also play a functional role in biological processes (Mariani, Luzzati *et al.*, 1988; Wong *et al.*, 1988; Seddon, 1990). The membrane states most familiar to biologists are the fluid-like liquid crystal. In the liquid crystalline state of the membrane, the hydrocarbon chains are more disordered as opposed to the more extended and ordered gel state (which can assume a number of different structures). A minimal proportion of the bilayer in the liquid-crystal form is believed to be required *in vivo* for cellular functions such as electron transport, nutrient transport, and ion pumping. Increasing pressure enhances the order of hydrocarbon chains and raises the temperature of membrane phase transition from gel state L_β to the liquid crystalline state L_α (Macdonald, 1987; Wong *et al.*, 1988; Driscoll *et al.*, 1991). Neutron diffraction experiments indicate that pressure increases bilayer thickness by reducing the "kinking" of acyl chains (Braganza and Worcester, 1986). For many phospholipids and natural membranes the transition slope in a temperature, pressure-phase diagram for the L_β - L_α

main transition is approximately 20°C/100 MPa at pressures lower than 100 MPa. Based on this relationship the combined effect of pressure and temperature on the phase state of a membrane from a deep-sea organism existing at 100 MPa and 2°C is equivalent to an identical membrane at atmospheric pressure and ⁻18°C! This temperature-pressure relationship is particularly striking when one realizes that the lower temperature limit for microbial growth is currently ~ ⁻12°C. Fortunately for the residents of deep-sea environments changes in phospholipid head groups, acyl chain length, and extent and nature of unsaturation can dramatically alter temperature and pressure effects on membrane phase and fluidity changes (Wong *et al.*, 1988; Winter, 1996).

3. *In Vivo* Low Temperature Phospholipid Adaptations

Before tackling the membrane adaptations of deep-sea organisms it is worthwhile to consider organismal responses to a drop in temperature. As noted above temperature decreases have a similar effect on membranes to pressure increases, and there is a larger body of information regarding membrane thermal acclimation processes. All organisms appear to regulate their membrane composition in response to temperature. Homeoviscous adaptation is the theory espoused by Sinensky (1974) whereby organisms adjust their membrane composition in response to changes in temperature in order to maintain a nearly constant membrane "fluidity". For our purposes fluidity will be used as a general description of microviscosity within the membrane. It has also been suggested that homeophasic adaptation, maintaining a certain percentage of membrane phospholipids within a liquid crystalline state, is more physiologically relevant than the actual level of membrane fluidity (McElhaney, 1982).

Microorganisms modulate many facets of their membranes in response to temperature, although at low temperature changes in fatty acyl chain unsaturation appear to be particularly significant (Russell, 1989). Unsaturation can be introduced into fatty acids during their elongation or subsequent to their incorporation into phospholipids and membranes. Temperature regulation of unsaturated fatty acid synthesis can occur at the levels of transcription, enzyme activity, or enzyme turnover. In *E. coli* a temperature downshift leads to increased enzyme activity of β-ketoacyl-ACP synthase II (Garwin and Cronan, 1980). By contrast, in *B. subtilis* lower temperatures increase unsaturated fatty acid (UFA) synthesis as a result of increased desaturase gene transcription and decreased desaturase enzyme turnover (Fulco and Fujii, 1980; Grau and de Mendoza, 1993). Some bacteria can even utilize both fatty acid synthetases and desaturases for unsaturated fatty acid production. Strains of *Pseudomoans*, *Flexibacter*, and *Vibrio* have been demonstrated to employ both systems for the biosynthesis of unsaturated fatty acids (Wada *et al.*, 1989; Wada *et al.*, 1991; Intriago, 1992; Morita *et al.*, 1992). One of the clearest lines of evidence that UFAs can influence the survival of microorganisms at low temperatures is that the introduction of a desaturase gene (*desA*) from the cold temperature-resistant cyanobacterium *Synechocystis* PCC6803, into the cold-sensitive cyanobacterium *Anacystis nidulans* conferred increased chilling resistance concomittantly with the appearance of 16:2 and 18:2 polyunsaturated fatty acids (PUFAs; Wada *et al.*, 1990).

Other types of low temperature-inducible phospholipid changes have also been observed. For example, in psychrophilic microorganisms changes in utilization of primer molecules for the initiation of fatty acid biosynthesis (Nichols and Russell, 1996) and *cis/trans* isomerization of the double bonds in fatty acids (Okuyama *et al.*, 1991) have been noted. The protozoan *Tetrahymena pyriformis* exhibits yet another regulatory pattern in response to decreased temperature. In this organism temperature downshift is accompanied by an increase in phospholipase A activity and an immediate re-shuffling of the lipid fatty acyl chains so that a lower melting temperature lipid is generated (Ramesha and Thompson, 1984).

4. *In Vivo* High Pressure Phospholipid Adaptations

Many deep-sea organisms regulate their membrane fatty acid composition in response to pressure changes as well as temperature changes. This work was first pioneered by Cossins and Macdonald who both predicted and then demonstrated homeoviscous adaptation in membranes from deep-sea fish as a function of pressure (Macdonald and Cossins, 1985). This was accomplished first by measuring the fluidity of brain myelin samples from shallow-water and deep-sea fish using fluorescence polarization (Cossins and Macdonald, 1984), and later by correlating an increased proportion of unsaturated fatty acids in liver mitochondrion membranes with fish capture depth (Cossins and Macdonald, 1986).

As with deep-sea fish, piezophilic and piezotolerant bacteria generally produce increased proportions of unsaturated fatty acids. DeLong and Yayanos opened up this window into piezophily by examining the fatty acid profile of the piezophile CNPT3 grown at various pressures (DeLong and Yayanos, 1985). A correlation was noted between growth pressure and the ratio of unsaturated to saturated membrane fatty acids (Table 1). In particular a marked increase in the amount of palmitoleic acid (16:1) and oleic acid (18:1) was observed as the growth pressure increased above 17 MPa. Since a non-piezophilic microorganism has been examined and found not to exhibit an increase in unsaturated fatty acids with pressure (MacDonald, 1984), it seems plausible to suggest that piezophiles have evolved regulatory processes (ie., for genes encoding enzymes involved in unsaturated fatty acid synthesis) attuned to the hydrostatic pressure of the environment.

DeLong and Yayanos also examined fatty acid profiles of 11 different deep-sea bacteria whose depth of origin ranged from 1,200 meters to 10,476 meters (DeLong and Yayanos, 1986). A major finding of this study was the detection of PUFAs as part of the bacterial fatty acid profile. Production of PUFAs were generally observed to increase in reponse to increased growth pressure. Previous to this study, it was thought that bacteria only synthesized mono-unsaturated fatty acids (Shaw, 1966; Goldfine, 1972), with the notable exception of *Flexibacter* sp. (John and Perry, 1977). Based on data gathered from this study, it was hypothesized that deep-sea bacteria incorporated PUFAs into their membranes to facilitate growth at high pressures. A small amount of a highly polyunsaturated fatty acid can exert a disproportionately large effect on the melting temperature and pressure of a lipid bilayer. For example, the melting temperature of eicosapentaenoic acid (EPA, 20:5) at ¯54°C is far below that of the monunsaturated fatty acid oleic acid (18:1) which is 13.4°C. Similar studies performed by Wirsen *et al.* (1987) and Kamimura *et al.* (1993) have generally confirmed the findings of DeLong and Yayanos. In addition, these later reports noted pressure increases of 17:1 or iso-17:1 monunsaturated fatty acids.

TABLE 1. Proportion of fatty acids in CNTP3 membrane phospholipids as a function of pressure at 2°C. Values are percentages by weight (means ± standard errors for five separate experiments). Data presented with permission from A. A. Yayanos (DeLong and Yayanos, 1985).

Fatty acid type	Pressure (bars)				
	1	172	345	517	690
	Generation time (hours)				
	17.2	10.1	7.9	8	17.2
14:1	20.0 ± 0.8	19.6 ± 0.8	17.4 ± 1.0	13.7 ± 0.7	11.5 ± 0.1
14:0	7.5 ± 0.2	5.7 ± 0.3	5.2 ± 0.2	4.8 ± 0.5	3.5 ± 0.4
16:1	40.2 ± 0.7	44.0 ± 1.2	47.9 ± 1.1	46.8 ± 1.6	56.3 ± 0.9
16:0	25.2 ± 0.6	22.2 ± 0.6	20.7 ± 0.3	18.7 ± 1.0	18.5 ± 0.8
18:1	5.5 ± 0.4	5.9 ± 1.8	6.6 ± 1.8	11.0 ± 1.6	7.3 ± 0.5
18:0	1.5 ± 0.2	2.5 ± 0.5	2.1 ± 0.4	4.9 ± 1.1	2.8 ± 0.4

It is important to note that not all piezophiles display an increase in the unsaturation index of their fatty acids with pressure, nor is there a clear trend between the unsaturation index of bacteria and their degree of piezoadaptation (Yayanos, 1995). Measurements of the fluidity and phase of membranes from piezophiles need to be assessed for possible trends associated with these parameters, and if such trends are found, then it may be possible that alternative membrane fluidizing modifications exist.

Recently, Yano and colleagues examined the fatty acid profiles of bacterial isolates from deep- and shallow-sea fish, and shallow-sea invertebrates (Yano *et al.*, 1997). They reported that many isolates from deep-sea fish contained docosahexaenoic acid (DHA; 22:6) and/or EPA in high percentages of the total fatty acids observed. No DHA and less EPA was produced by the isolates from the shallow-sea fish and invertebrates. EPA may be important for adaptation to low temperature as well as high pressure environments. Higher percentages of polyunsaturated fatty acids, including EPA have been noted in Antarctic versus temperature bacteria (Nichols *et al.*, 1993).

Both EPA and DHA are considered important chemicals for biotechnology because of their activities in reducing the risk of cardiovascular disease and certain cancers, their use as dietary supplements in mariculture and poultry farming, and their application as substrates for the synthesis of certain hormone-like molecules (Simopoulos, 1991). As a result of this interest, PUFA-producing bacteria have been sought from a variety of sources. In addition to the deep-sea bacteria, EPA producers have been isolated from Antarctica, sea water, invertebrates and fish intestines (Yazawa *et al.*, 1988; Yazawa *et al.*, 1988; Ringo *et al.*, 1992; Nichols *et al.*, 1993; Henderson *et al.*, 1995; Iwanami *et al.*, 1995).

Much progress has been made in recent years towards understanding the genetic basis of EPA synthesis. Yazawa isolated an EPA-producing *Shewanella putrefaciens* species from the intestines of Pacific Mackerel (Yazawa *et al.*, 1988). From this *Shewanella* a 38 kbp genomic DNA fragment was isolated which conferred EPA production ability to *E. coli* (Yazawa, 1996). The nucleotide sequence of this DNA has been determined and based upon deletion analysis and deduced amino acid homology to other known enzymes involved in fatty acid synthesis, seven open reading frames have been identified which are important to EPA production (Yazawa, 1996).

Bacterial EPA biosynthesis has been proposed to depend on both fatty acid synthetases and desaturases (Nichols et al., 1993; Iwanami *et al.*, 1995). This is based upon assessments of total fatty acid compositions, together with radiotracer uptake experiments, the use of a fatty acid elongation inhibtor, and analyses with and without oxygen. Since EPA production can occur in the absence of oxygen it has been hypothesized that in some bacteria desaturases may be able to utilize one or more alternate electron acceptors under anaerobic conditions (Nichols *et al.*, 1992). Bacterial EPA is primarily associated with the *sn*-2 position of phosphatidylethanolamine and phosphatidylglycerol (Henderson *et al.*, 1995; Yazawa, 1996).

Much less is known about pressure adaptation in the membranes of piezothermophilic deep-sea Archaea compared with piezopsychrophilic deep-sea Bacteria. However, modulation of membrane structure with pressure has been observed in one such piezothermophile. The deep-sea hydrothermal vent archaeon *Methanocccus jannaschii* produces an increased proportion of the lipid core macrocyclic archaeol (2,3-di-O-cyclic-biphytanyl-sn-glycerol) and reduced levels of archaeol (2,3-di-O-phytanyl-sn-glycerol) and caldarchaeol (2,2',3,3'-dibiphytanyl-diglycerol) with increased pressure (Kanishiro and Clark, 1995). Although the loss of the more fluidizing archaeol at high pressure would exacerbate the structuring effect of high pressure on the membrane, electron paramagnetic resonance studies indicated no well-defined phase transition for the lipids from *M. jannaschii* from from 35 - 100°C and from 1 to 50 MPa. Thus, even in the absence of lipid core modification, the membranes of this organism would appear to be well suited for stability over a broad range of temperatures and pressures.

5. Membrane Protein Systems at High Pressure

Membrane-localized proteins have been found in many instances to be highly sensitive to elevated pressure, so much so that high pressure treatment of membranes has been proposed as a general method for extracting membrane proteins (Orr *et al.*, 1990). Although numerous examples of pressure-sensitivity among membrane proteins exist (Marquis and Bender, 1987; Chi and Bartlett, 1995; Kato *et al.*, 1996a and b), for the sake of brevity, only a few examples will be described below.

One of the first classes of membrane proteins to be examined from piezophiles were outer membrane porin-like proteins. Porins are channel proteins located within the outer membranes of gram negative bacteria which facilitate the diffusion of nutrients into the cell. In the case of the moderately piezophilic isolate *Photobacterium* species strain SS9 one porin-like protein, designated OmpL, is abundant at atmospheric pressure while another porin-like protein, designated OmpH, dominates at the pressure optimum for SS9, 20 - 30 MPa. (Bartlett *et al.*, 1989; Bartlett *et al.*, 1993; Welch and Bartlett, 1996) . Why should a deep-sea bacterium produce different porins as a function of pressure? Moderate pressures have recently been found to close porins from *E. coli* (A. G. Macdonald; personal communication), so it may be necessary for marine microorganisms which experience substantial vertical displacements in the water column to produce channels of the appropriate size, selectivity and activity for the pressure of their environment. Although neither *ompL* or *ompH* mutants exhibit altered growth as a function of pressure (Bartlett and Chi, 1994; Welch and Bartlett, 1996), other porins could be compensating for the loss of channels encoded by either of these genes. Uptake experiments suggest that OmpH may provide a larger channel than OmpL (Bartlett and Chi, 1994), a property that could be important in the deep sea where nutrients are particularly scarce.

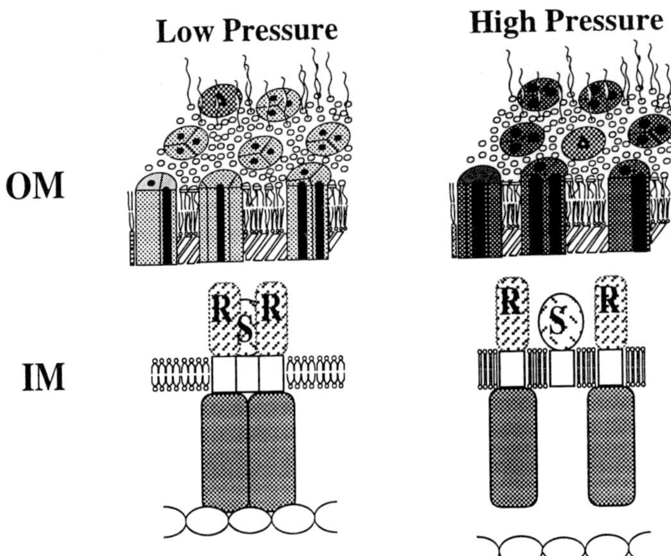

Figure 1. Pressure regulation of OmpL and OmpH production in *Photobacterium profundum* SS9. Above is a schematic of the outer membrane at low and high pressure, showing the replacement of OmpH over OmpL as the major protein at high pressure. Below is a hypothetical model depicting changes in the transcription factor regulating *ompL/H* gene expression; ToxR and ToxS quaternary structure and DNA binding ability are shown to change as a function of pressure. See text for details.

The pressure regulation of the OmpH and OmpL proteins is controlled at the transcriptional level (Bartlett *et al.*, 1989; Welch and Bartlett, 1996). Disruption of a gene bearing high homology to *toxR*, a gene previously identified in other bacteria within the family *Vibrionaceae*, resulted in the loss of OmpL production and constituitive OmpH levels (Welch and Bartlett, 1997). ToxR has been most thoroughly studied in the human pathogen *V. cholerae*, where it is documented to be a unique transmembrane DNA binding protein which extends from the periplasmic space across the inner membrane and into the cytoplasm (DiRita, 1992). There it coordinates the expression of a number of genes associated with virulence. The *toxR* gene has also been found in the marine bacterium*V. fischeri* (Reich and Schoolnik, 1994), where it may function to control the expression of genes involved in the symbiotic interactions of this species. Both *V. cholerae* and SS9 ToxRs control the inverse expression of a pair of *omp* genes. However, only in SS9 does ToxR appear to be a component of a pressure-responsive gene regulation system (Welch and Bartlett; unpublished results). We speculate that in certain deep-sea bacteria, ToxR has evolved the ability to sense (directly or indirectly) and respond to levels of hydrostatic pressure and thereby modulate porin gene expression (Fig. 1). Pressure controls both the abundance and activity of SS9 ToxR. ToxR abundance is down-regulated at high pressure, but even in strains which over-express a plasmid-encoded ToxR at high pressure, normal OmpH/L regulation occurs.

Another intriguing aspect of high pressure ToxR-mediated OMP regulation is its reversiblity by the addition of the local anesthetics (Welch and Bartlett, 1997). Even at high pressure OmpH/L levels can be reversed by the addition of small amounts (3 mM) of procaine or phenethyl alcohol. Since these concentrations of drug can fluidize cell membranes (Killian *et al.*, 1992), the simplest explanation is that pressure/anesthetic effects on ToxR-signaling are the result of changes in membrane fluidity, although protein effects can not yet be excluded. The SS9 ToxR system provides a simple model system in which to explore pressure/anesthetic antagonism, a phenomenon associated with a variety of high pressure neurological effects observed in higher organisms (Johnson and Miller, 1970).

6. Concluding Remarks

Pressures of the magnitude present in the biosphere exert a profound influence on the viscosity and phase properties of biological membranes. Deep-sea organisms appear to compensate for the membrane-ordering effects of high pressure by possessing unusually fluid fatty acids. Modification of some membrane proteins also appears to be required for piezophily. Additional characterization of the membrane and its connection to the growth and energetics of microbial life at high pressure will go a long way towards characterizing this intriguing group of extremophiles.

7. References

Bale, S., Goodman, K., Rochelle, P., Marchesi, J., Fry, J.C., Weightman, A.J., and Parkes, R.J. (1997) *Int. J. Syst. Bacteriol. 47*, 515-521.

Bartlett, D., Wright, M., Yayanos, A., and Silverman, M. (1989)*Nature 342*, 572-574.

Bartlett, D.H. (1992) *Sci. Progress 76*, 479-496.

Bartlett, D.H. and Chi, E. (1994) *Arch. Microbiol. 162*, 323-328.

Bartlett, D.H., Chi, E., and Wright, M.E. (1993) *Gene 131*, 125-128.

Braganza, L.F. and Worcester, D.L. (1986) *Biochem. 25*, 7484-7488.

Chi, E. and Bartlett, D.H. (1995) *Mol. Microbiol. 17*, 713-726.

Cossins, A.R. and Macdonald, A.G. (1984) *Biochim. Biophys. Acta 776*,144-150.

Cossins, A.R. and Macdonald, A.G. (1986) *Biochim. Biophys. Acta 860*, 325-335.

DeLong, E.F. (1992) In J. Lederberg (ed.), *Encyclopedia of microbiology, volume 2*, Academic Press, San Diego, pp. 405-417.

DeLong, E.F., Franks, D.G., and Yayanos, A.A. (1997) *Appl. Environ. Microbiol. 63*, 2105-2108.

DeLong, E.F. and Yayanos, A.A. (1985) *Science 228*, 1101-1103.

DeLong, E.F. and Yayanos, A.A. (1986) *Appl. Environ. Microbiol. 51*, 730-737.

DiRita, V.J. (1992) *Mol. Microbiol. 6*, 451-458.

Driscoll, D.A, Jonas, J., and Jonas, A. (1991) *Chem. Phys. Lipids 58*, 97-104.

Fulco, A.J. and Fujii, D.K. (1980) In M. Kates and A. Kuksis (eds.), *Membrane fluidity: biophysical techniques and cellular regulation*, Humana, New Jersey, pp. 77-98.

Garwin, J.L. and Cronan, J.E. (1980) *J. Bacteriol. 141*, 1457-1459.

Goldfine, H. (1972). *Adv. Microbiol. Physiol 8*, 1-58.

Grau, R. and de Mendoza, D. (1993) *Mol. Microbiol. 8*, 535-542.

Gross, M. and Jaenicke, R. (1994) *Eur. J. Biochem. 221*, 617-630.

Henderson, R.J., Millar, R.-M., and Sargent, J.R. (1995) *Lipids 30*, 181-185.

Intriago, P. (1992) *J. Gen. Micro. 138*, 109-114.

Iwanami, H., Yamaguchi, T. and Takeuchi, M. (1995) *Nippon Suisan Gakkaishi 61*, 205-210.

John, R.B. and Perry, G.J. (1977) *Arch. Microbiol. 114*, 267-271.

Johnson, S.M. and Miller, K.W. (1970) *Nature 228*, 75-76.

Kamimura, K., Fuse, H., Takimura, O. and Yamaoka, Y. (1993) *Appl. Environ. Microbiol. 59*, 924-926.

Kanishiro, S.M., and Clark, D.S. (1995) *J. Bacteriol. 177*, 3668-3672.

Kato, C., Ikegami, A., Smorawinska, M., Usami, R. and Horikoshi, K. (1997) *J. Mar.Biotech. 5*, 210-218.

Kato, C., Sato, T. and Horikoshi, K. (1995) *Biodiversity and conservation 4*, 1-9.

Kato, C., Smorawinska, M., Sato, T. and Horikoshi, K. (1996) Biosci. Biotech. *Biochem. 60*, 166-168.

Kato, C., Suzuki, S., Hata, S., Ito, T. and Horikoshi, K. (1995) *JAMSTEC Res. 32*, 7-13.

Kato, C., Tamegai, H., Ikegami, A., Usami, R. and Horikoshi, K. (1996) *J. Biochem. 120*, 301-305.

Kerr, R.A. (1997) *Science 276*, 703-704.

Killian, J.A., Fabrie, C.H., Baart, W., Morein, S. and de Kruijff, B. (1992) *Biochim. Biophys. Acta. 603*, 63-69.

MacDonald, A.G. (1984) *Phil. Trans. Soc. Lond. B304*, 47-68.

Macdonald, A.G. (1987) In H. W. Jannasch, R. E. Marquis, and A. M. Zimmerman (eds.), *High pressure biology*, Academic Press Inc., London, pp. 207-223.

Macdonald, A.G. and Cossins, A.R. (1985) *Soc. Exp. Biol. Symp. 39*, 301-322.

Mariani, P., Luzzati, V. and Delacroix, H. (1988) *J. Mol. Biol. 204*, 165-189.

Marquis, R.E. and Bender, G.R. (1987) In H. W. Jannasch, R. E. Marquis, and A. M. Zimmerman (eds.) *High pressure biology*.Academic Press Inc., London, pp. 65-73.

McElhaney, R.N. (1982) *Curr. Top Membr. Transp. 17*, 317-380.

Morita, N., Gotoh, M., Okajima, N., Okuyama, H., Hayashi, H., Higashi, S. and Murata, N. (1992) *FEBS lett. 297*, 9-12.

Moshaev, V.V., Heremans, K., Frank, J., Masson, P. and Balny, C. (1996) *Proteins: Struct. Funct. Genet. 24*, 81-91.

Nichols, D.S., Nichols, P.D. and McMeekin, T.A. (1992) *FEMS Microbiol. lett. 98*, 117-122.

Nichols, D.S., Nichols, P.D. and McMeekin, T.A. (1993) *Antarctic Science 5*, 149-160.

Nichols, D.S. and Russell, N.J. (1996) *Microbiol. 142*, 747-754.

Okuyama, H., Okajima, N., Sasaki, S., Higashi, S. and Murata, N. (1991) *Biochim. Biophys. Acta 1084*, 13-20.

Orr, N., Yavin, E., Shinitsky, M. and Lester, D.S. (1990) *Anal. Biochem. 191*, 80-85.

Prieur, D. (1992) In R. A. Herbert and R. J. Sharp (eds.), *Molecular Biology and Biotechnology of extremophiles*, Chapman Hall Inc., New York, pp. 163-202.

Ramesha, C.S. and Thompson, G.A. (1984) *J. Biol. Chem. 259*, 8706-8712.

Reich, K.A. and Schoolnik, G.K. (1994) *J. Bacteriol. 176*, 3085-3088.

Rex, M.A., Stuart, C.T., Hessler, R.R., Allen, J.A., Sanders, H.L. and Wilson, G.D.F. (1993) *Nature 365*, 636-745.

Ringo, E., Jostensen, J.P. and Olsen, R.E. (1992) *Lipids 27*, 564-566.

Robinson, C.R. and Sligar, S.G. (1995) *Methods Enzymol. 259*, 395-427.

Royer, C.A. (1995) *Methods Enzymol. 259*, 357-377.

Russell, N.J. (1989) In C. Ratledge and S. G. Wilkinson (eds.), *Microbial lipids*, Academic Press Inc., London, pp. 279-365.

Saitou, N. and Nei, M. (1987) *Mol. Biol. Evol. 4*, 406-425.

Schmehl, M., Jahn, A., Meyer zu Vilsendorf, A., Hennecke, S., Masepohl, B., Schuppler, M., Marxer, M., Oelze, J. and Klipp, W. (1993) *Mol. Gen. Genet.241*, 602-615.

Seddon, J.M. (1990) *Biochim. Biophys. Acta 1031*, 1-69.

Shaw, R. (1966) *Adv. Lipid Res. 4*, 107-174.

Silva, J.L and Weber, G. (1993) *Ann. Rev. Phys. Chem. 44*, 89-113.

Simopoulos, A.P. (1991) *Am. J. Clin. Nutr. 54*, 438-463.

Sinensky, M. (1974) *Proc. Natl. Acad. Sci. U. S. A. 71*, 522-526.

Wada, H., Gombos, Z. and Murata, N. (1990) *Nature 347*, 200-203.

Wada, H. and Murata, N. (1989) *Plant Cell Physiol. 30*, 971-978.

Wada, M., Fukunaga, N., and Sasaki, S. (1989) *J. Bacteriol. 171*, 4267-4271.

Wada, M., Fukunaga, N., and Sasaki, S. (1991) *J. Gen. Microbiol. 37*, 355-362.

Weber, G. and Drickamer, H.G. (1983) *Q. Rev. Biophys. 16*, 89-112.

Welch, T.J. and Bartlett, D.H. (1996) *J. Bacteriol. 178*, 5027-5031.

Welch, T.J. and Bartlett, D.H. (1998) *Mol. Microbiol.27*, 977-985.

Winter, R. (1996) In R. Hayashi and C. Balny (eds.) *High pressure bioscience and biotechnology, volume 13*, Elsevier Science B. V, Amsterdam, pp.21-28.

Wirsen, C.O., Jannasch, H.W., Wakeham, S.G. and Canuel, E. A. (1987) *Current Microbiol. 14*, 319-322.

Wong, P.T.T., Siminovitch, D.J., and Mantsch, H.H. (1988) *Biochim. et Biophys. Acta 947*, 139-171.

Yano, Y., Nakayama, A., and Yoshida, K. (1997) *Appl. Environ. Microbiol. 63*, 2572- 2577.

Yayanos, A.A. (1986) *Proc. Nat. Acad. Sci. U. S. A. 83*, 9542-9546.

Yayanos, A.A. (1995) *Annu. Rev. Microbiol. 49*, 777-805.

Yazawa, K. (1996) *Lipids 31*, S-297-S-300.

Yazawa, K., Araki, K., Okazaki, N., Watanabe, K., Ishikawa, C., Inoue, A., Numao, N., and Kondo, K. (1988) *J. Biochem. 103*, 5-7.

Yazawa, K., Araki, K., Watanabe, K., Ishikawa, C., Inoue, A., Kondo, K., Watabe, S., and Hashimoto, K. (1988) *Nippon Suisan Gakkaishi 54*, 1835-1838.

Biodata of R.W. Lee contributor of "***Chemoautotrophic Bacteria – Marine Invertebrate Symbiosis*** ."

Dr. Raymond W. Lee is an Assistant Professor in the Department of Zoology and Program in Biology, Washington State University (Pullman, WA). He received his Ph.D. in 1994 at the University of California, Santa Barbara and performed research during his postdoctoral periods in the University of Alabama at Birmingham and at Harvard University. His research interests are: comparative physiology, Chemoautotrophic bacteria–Invertebrate symbiosis, deep-sea hydrothermal vents and cold seep biology, carbon and nitrogen assimilation, sulfide utilization and detoxification by marine organism, stable isotopes in biological system.
E-mail: **rlee@mail.wsu.edu**

J. Seckbach (ed.), Enigmatic Microorganisms and Life in Extreme Environments, 513–523.

CHEMOAUTOTROPHIC BACTERIA - MARINE INVERTEBRATE SYMBIOSES

Adaptations for autotrophic carbon and nitrogen assimilation

R.W. LEE
Department of Zoology, Washington State University
Pullman, WA 99164-4236

1. Introduction

Symbiosis, the living together of dissimilar-named organisms, is a widespread phenomenon that can result in mutualistic benefit for both partners. In symbiosis, the metabolic capabilities of symbiotic partners are integrated to create a new organism. Thus symbiosis is a powerful means by which organisms can solve adaptive problems such as living in extreme environments.

Chemoautotrophic symbioses between thiotrophic bacteria and marine invertebrates were first discovered at deep-sea hydrothermal vents in 1977. This was a surprising finding since the deep-sea is generally a food-poor environment that does not support high biomass. At vents, hot, anoxic vent fluids rich in hydrogen sulfide, a potent inhibitor of aerobic respiration, emerge from the seafloor and mix with ambient bottom water. Despite this seemingly hostile environment, dense communities based on symbiosis-derived primary production thrive at these sites (see Lutz and Kennish 1993; Tunnicliffe 1991 for recent ecological reviews) with high biomass (Lutz and Kennish 1993) and, in some cases, high growth rates (Lutz et al. 1994).

Chemoautotrophic symbioses predominate in these environments because they can overcome low food availability and sulfide toxicity. The symbionts provide an internal source of nutrition and also consume sulfide which might otherwise poison aerobic metabolism of the host. Living within the host tissues enables the symbionts to gain simultaneous access to sulfide and oxygen. The exchange surfaces of the invertebrate hosts can contact both oxygenated seawater and sulfide-rich waters, and coupled with physiological mechanisms such as hemoglobin that binds sulfide and oxygen (Arp et al. 1985), are able to provide the symbionts with an environment suitable for sulfide-based chemoautotrophy. See Childress and Fisher (1992) and Somero et al. (1989) for recent reviews on sulfide utilization by chemoautotrophic symbioses.

These symbioses are now known to be widespread in areas where sulfide and oxygen are present (see Cavanaugh 1994; Fisher 1990 for recent reviews) and are found in a variety of habitats including eelgrass beds, mangrove swamps, anoxic basins, whale carcasses, pulpmill effluent sites, and sewage sludge outfalls. Like hydrothermal vent symbioses, their nutrition is based on symbiont chemoautotrophy. An additional type of chemosynthetic bacteria - invertebrate symbiosis between methanotrophic bacteria and deep-sea mussels has also been characterized (Cavanaugh et al. 1987; Cavanaugh et al.

1992; Childress et al. 1986).

These symbioses appear to be highly specific. Individual members of the same host species have the same symbiont and each host species harbors a particular symbiont phylotype (Cavanaugh 1994). For recent reviews of symbiont phylogeny, morphology, and enzymology see Cavanaugh (1994); Felbeck and Distel (1992); Fisher (1990); Nelson and Fisher (1995). In general, the endosymbionts fall into the gamma Proteobacteria group based on 16S rRNA sequences, and *Thiomicrospira* L-12, an isolate from hydrothermal vents, is the closest known free-living relative (Cavanaugh 1994). So far, no attempts to culture the symbionts have been successful and as a result, all studies must be conducted on tissues of the intact association or freshly-isolated symbionts. The symbionts are generally intracellular and housed in vacuoles within host cells. Cell densities can be as high as 10^9 per gram wet weight of tissue (Cavanaugh et al. 1981; Powell and Somero 1986)

Chemoautotrophic symbioses have been found in a variety of invertebrate phyla, including Vestimentifera, Pogonophora, Mollusca, Bivalvia, Annelida, Nematoda, Arthropoda, and Ciliophora with new symbioses continuing to be discovered. The most extensively studied symbioses are those involving vestimentiferan tubeworms, vesicomyid clams, bathymodiolid mussels, and solemyid clams (reviewed in Cavanaugh 1994; Fisher 1990; Nelson and Fisher 1995) which represent the two major body plans for this type of symbiosis. A well studied example of the vestimentiferan body plan is the unusual and highly specialized vent tubeworm symbiosis *Riftia pachyptila*. These worms can grow to lengths in excess of 1 meter and lack mouth, gut, and anus as adults (Jones 1981). The symbionts are housed in a large organ within the trunk of the worm that is densely supplied with capillaries of the worm's closed circulatory system. Respiratory exchange of sulfide, oxygen and other metabolites occurs across the plume which is the only part of the worm in direct contact with the environment. The other major body plan is that of the bivalve molluscs, such as the vent clam *Calyptogena magnifica*, vent mussel *Bathymodiolus thermophilus*, and coastal clam *Solemya reidi*, in which the symbionts are housed intracellularly within the gill filaments (Felbeck 1983; Fiala-Médioni and Métivier 1986). Since surrounding water is pumped through the gill filaments during ventilation, access of the symbionts to substrates from the environment is potentially enhanced. The degree of nutritional reliance on symbiosis is variable among bivalve molluscs. For example, *B. thermophilus* is a competent suspension feeder (Page et al. 1991) whereas *S. reidi* lacks both mouth and gut (Reid and Bernard 1980).

In order to grow autotrophically, inorganic carbon and nitrogen must be assimilated, and the acquisition of these capabilities was likely a driving force in the evolution chemoautotrophic symbiosis. This review will focus on what is presently understood about adaptations with respect to these key metabolic capabilities. In general, these animals and their symbionts are awash in high levels of inorganic carbon and nitrogen substrates and it is therefore not surprising that they are dominant primary producers forming the basis for entire ecosystems.

2. Carbon

Some of the first evidence for a novel form of nutrition of organisms at hydrothermal vents came from studies of stable carbon isotope ratios (Rau and Hedges 1979). These findings indicated early on that these organisms are not reliant on photosynthetically derived carbon from the surface, and that novel nutritional mechanisms are involved. The stable isotope approach has been very useful in understanding carbon flow and trophic interactions at deep-sea vents since extensive direct observations and laboratory studies are not feasible (Fisher et al. 1994; Kennicutt et al. 1992; VanDover and Fry 1989). In general, vent, seep and sediment dwelling symbioses have stable carbon isotope values that are very depleted in ^{13}C compared with marine photosynthetic primary producers which is consistent with the symbionts as the major source of organic carbon (reviewed in Kennicutt et al. 1992)

2.1. ENVIRONMENTAL AVAILABILITY

Most marine autotrophs are potentially limited by the availability of CO_2 (Raven 1991; Riebesell et al. 1993) since dissolved inorganic carbon (ΣCO_2) in seawater (pH 8.1) is present primarily (>99%) as bicarbonate ion (HCO_3-) rather than the freely-diffusing form, CO_2. However, the environment of chemoautotrophic symbioses differs from normal seawater in that there are elevated levels of both ΣCO_2 and dissolved CO_2. This was first recognized in the hydrothermal vent tubeworm *Riftia pachyptila* (Childress et al. 1993). Seawater in the vicinity of tubeworms is enriched in ΣCO_2 (2.7 to 7.1 mM) and low in pH, ranging from 5.6 to 7.2 (Childress et al. 1993; Goffredi et al. 1997). Given an apparent equilibrium dissociation constant (pK_{app}) of around 6 (Boutilier et al. 1984), these values correspond to CO_2 levels ranging to over 100 times normal seawater. Thus elevated inorganic carbon appears to be a previously unrecognized environmental parameter that may enhance autotrophy. Although inorganic carbon availability has not been as extensively characterized for other chemoautotrophic symbioses, high levels are probably typical. Symbioses living in reducing sediments likely encounter elevated ΣCO_2 and reduced pH conditions due to the proximity to sediment porewater. Indeed sediment from the Hyperion sewage sludge outfall where *Solemya reidi* are collected contains 4-60 mM ΣCO_2 (Lee et al. 1992). The pH of sediment porewater from these sites is around 7.0 to 7.5 (Lee unpublished results).

2.2. UPTAKE AND ASSIMILATION

Mechanisms of inorganic carbon uptake and supply have received the greatest attention in the tubeworm *Riftia pachyptila.* Carbon dioxide limitation is potentially a more serious problem for this symbiosis. There is a longer diffusion path length resulting from the symbionts being housed in an organ remote from the site of respiratory exchange and higher demands for carbon dioxide due to fast growth rates. This problem appears to be solved in part by the ability to accumulate ΣCO_2 in the circulatory fluids to levels several times higher than the environment (Childress et al. 1993). ΣCO_2 accumulation is facilitated by unusual properties of the blood. Rather than undergoing acidosis in response to elevated carbon dioxide levels, blood of *R. pachyptila* is maintained at around pH 7.4 (Toulmond et al. 1994). Since the pH of the surrounding waters is lower than 7.4 and elevated in ΣCO_2, equilibrium between dissolved CO_2 in the environment and blood results in an elevated blood ΣCO_2 pool.

Assuming that CO_3^- does not diffuse across the respiratory surface, the ratio of external/internal $\Sigma CO_2 = (10^{pHe-pK} + 1) / (10^{pHi-pK}+1)$ (Roos and Boron 1981). Thus for a worm with an internal pH (pHi) of 7.5 and an environmental pH (pHe) of 6.0 and an apparent pK of 6.1 (calculated in Goffredi et al. 1997), internal ΣCO_2 is predicted to be 15 times greater than external ΣCO_2. Typical values of ΣCO_2 from freshly collected worms of around 30-35 mM are common (Goffredi et al. 1997). Dissolved CO_2 in such fluids are around 2 mM. Since the K_m for carbon fixation of the symbionts is between 0.02–0.04 mM dissolved CO_2 (Scott et al. 1994), the symbionts are likely saturated with respect to inorganic carbon.

Intracellular and body fluid levels of inorganic carbon in bivalve symbioses have been much less well studied, but elevated levels appear to also be present. In *Calyptogena magnifica*, hemolymph of freshly collected clams exhibits a ΣCO_2 concentration of 3.3 mM, a pH of 6.9, and a dissolved CO_2 concentration of 0.4 mM (Childress et al. 1991). This CO_2 concentration is above the Km for dissolved CO_2 observed in *Riftia* symbionts and is 20 times that of normal seawater. Thus *C. magnifica* symbionts may also not be limited by carbon dioxide availability.

Fluxes of ΣCO_2 in symbiotic associations across multiple compartments is likely enhanced by the enzyme carbonic anhydrase (CA). The uncatalyzed interconversion of CO_2 and bicarbonate ion in solution is slow but is increased dramatically by the presence of CA which can reduce boundary layer depletion of CO_2. CA has been implicated in facilitating CO_2 supply in marine symbioses between algae and invertebrates (Weis et al. 1989; Yellowlees et al. 1993). CA may serve a similar function in chemoautotrophic symbioses. In a survey of chemoautotrophic symbioses, Kochevar et al. (Kochevar and Childress 1996) showed that CA activity can vary considerably among different species, with particularly high activities in plume and trophosome of tubeworms and symbiont-containing gill tissue of several species of *Calyptogena* clam. In *R. pachyptila*, two forms of CA are present. One is specific to the respiratory plume and the other is localized to the symbiont-containing trophosome (Kochevar et al. 1993). Carbon fixation rates of *R. pachyptila* symbiont preparations is reduced by CA inhibitors (Scott et al. 1994).

An additional mechanism that has been proposed for CO_2 transport in *R. pachyptila* is initial fixation of CO_2 into C4 compounds in the plume with subsequent decarboxylation at the site of CO_2 fixation in the trophosome. This mechanism is analogous to the CO_2 trapping mechanism of terrestrial C4 plants. Consistent with such a mechanism is the finding that high rates of $^{14}CO_2$ fixation are observed in the plume with fixation into malate and succinate (Felbeck 1985). It is hypothesized that malate and succinate are transported to the symbionts where they are decarboxylated. The importance of this mechanism as opposed to ΣCO_2 concentrating mechanisms is currently disputed. It should be noted that this issue may soon be resolved by the development of a new experimental preparation in which catheterized tubeworms can be maintained in pressurized aquaria, allowing time course sampling of ^{14}C labeled products in the blood (see "Farming tubeworms" Science V279 p. 663)

Fixation of CO_2 by the symbionts occurs via the Calvin-Benson cycle. Activities of the enzyme ribulose-1,5-bisphosphate carboxylase/oxygenase (Rubisco) have been measured extensively in chemoautotrophic symbioses (e.g., (Cavanaugh 1983; Cavanaugh et al. 1981; Felbeck et al. 1981) and are often used as a diagnostic indicator of the capability for autotrophic metabolism (Fisher 1990). Rubisco activities vary between different species (reviewed in Nelson and Fisher 1995) and may reflect differences in autotrophic potential. It has recently been shown that the chemoautotrophic symbionts have diverse types of Rubisco enzymes. The bivalves, *B. thermophilus* and *S. velum* appear to have the common form I Rubisco that is also found in green plants whereas the vent vestimentiferans *R. pachyptila* and *Tevnia jerichonana* have a less well studied type, termed form II, which has generally been observed in only a few species of autotrophic bacteria (Robinson and Cavanaugh 1995). The adaptive significance of form II Rubisco is not known, but it is postulated (Haygood 1996) that the form expressed is a function of intracellular environment since form II Rubiscos generally have higher catalytic efficiency but require lower oxygen tensions and higher CO_2 to reduce competing oxygenase activity. Thus form II Rubisco may allow high rates of carbon fixation in symbioses that can maintain high internal CO_2/low oxygen environments around the symbionts.

Whole organism physiological studies of autotrophic carbon fixation have also been conducted, but only on a few species. This is in part due to the difficulty involved in maintaining moderate concentrations of both sulfide and oxygen in respiration studies. In the case of vent symbioses, experiments must be conducted at high hydrostatic pressure to simulate natural conditions. Two published studies have documented net autotrophic fixation of carbon dioxide (assimilation in excess of respiration) in *R.pachyptila* (Childress et al. 1991) and *S. reidi* (Anderson et al. 1987). Recent experiments directed at developing the coastal clam *Solemya velum* as a model symbiosis have documented net autotrophy under laboratory conditions using a simple apparatus (Lee unpublished results). The potential to study a readily collected symbiosis under conditions of net autotrophy may lead to rapid advancement of our understanding of autotrophic functioning in these organisms.

3. Nitrogen

Like inorganic carbon, inorganic nitrogen assimilation appears to be an important metabolic capability of chemoautotrophic symbioses, and the availability of nitrogen nutrients is high. Early evidence that these symbioses were growing on inorganic nitrogen came from stable nitrogen isotope measurements. Symbioses living at deep-sea hydrothermal vents and cold seeps are generally depleted in ^{15}N compared with elsewhere in the deep-sea (Fisher et al. 1994; Rau 1981; Rau 1985; VanDover and Fry 1989), indicating that organic nitrogen of these communities is assimilated from inorganic sources locally.

The availability of inorganic nitrogen sources can often limit primary production by marine autotrophs (Vitousek and Howarth 1991). Concentrations of sources such as nitrate and ammonia are often vanishingly low in surface seawater where there are photosynthetic primary producers. In contrast, ammonia and nitrate are often abundant where chemoautotrophic symbioses are found (Lee and Childress 1994). Deep-sea symbioses encounter nitrate levels typical of oceanic bottom water, ranging as high as 40 μM (e.g., Johnson et al. 1988; Karl et al. 1988; Tunnicliffe et al. 1986). In general, nitrate is the most abundant source at unsedimented deep-sea hydrothermal vents since ammonia concentrations are low. In some environments inhabited by these symbioses ammonia levels are also high. Porewater of sediment inhabited by *Solemya reidi* contains 60 μM ammonia (Lee et al. 1992) and high ammonia availability is likely typical of other symbioses that live in reducing sediments. Recently high ammonia has been discovered in vent fluids of hydrothermal vents on the Juan de Fuca ridge (Lilley et al. 1993) and ammonia is also high at vents in the Guaymas basin (VonDamm et al. 1985). Symbioses living at hydrocarbon seeps in the Gulf of Mexico also encounter high levels of ammonia in sediment porewater and hypersaline seep effluent (Kennicutt II et al. 1989; Lee and Childress 1994).

It is not clear how nitrate and ammonia are taken up by these symbioses since marine invertebrates do not normally metabolize nitrate and ammonia is usually excreted. Nitrate is present in seawater as an anion, while the bulk of ammonia (_NH_3, the sum of NH_4^+ and NH_3) at pH 8 is present as NH_4^+ (92% given a pK of 9.25 at 25 °C). Symbiotic invertebrates may have evolved transporters or channels to facilitate uptake of nitrate and NH_4^+ but these have so far not been demonstrated. Ammonia uptake kinetics by *S. reidi* are linear (Lee et al. 1992), as is observed in algal-invertebrate symbioses (Wilkerson and Trench 1986), and is suggestive of a diffusion mechanism (D'Elia and Cook 1988; Muscatine 1980). Seep mytilid Ia, a methanotrophic symbiosis, exhibits saturable uptake kinetics indicating active transport may be involved (Lee et al. 1992).

Assuming that NH_3 readily diffuses between host tissues and the medium, and NH_4^+ is less permeable, it is likely that ΣNH_3 concentrations are elevated in host tissues of bivalve symbioses. This is due to a more acidic internal pH compared with the medium which acts to trap ΣNH_3, which is the opposite of what occurs with ΣCO_2. Thus, when pHe is greater than pHi, ΣNH_3 is concentrated internally. ΣNH_3 of *Solemya reidi* is considerably elevated compared with the medium and falls within the predicted range of a 5 to 31 fold concentration gradient predicted by an acid trapping mechanism (Lee et al. 1997). Similarly, elevated ammonia concentrations have also been observed in seep mytilid Ia (Lee and Childress 1996). Ammonia is also elevated in the blood of *R. pachyptila* (Lee unpublished results), but presumably by a different mechanism since internal pH in this case is higher than that of the environment. Overall, these findings indicate that the intracellular environment encountered by the symbionts is likely extremely rich in ammonia.

Chemoautotrophic symbioses appear to have the requisite assimilation pathways for ammonia and nitrate assimilation. Nitrate reductase (NR) activity has been documented in a variety of chemoautotrophic symbioses, e.g., Felbeck et al. (1981). Recent studies have demonstrated activities of glutamine synthetase (GS) as well as glutamate dehydrogenase (GDH) Lee and Childress 1996; Lee et al. submitted). Nitrite reductase (NiR) and glutamate synthase (GOGAT) activities have also been detected in some vent symbioses (Lee unpublished results).

Although assimilation steps catalyzed by GS and GDH could be performed by the host, perhaps as a means of controlling nitrogen assimilation as in some algal invertebrate associations (McAuley and Cook 1994), this does not appear to generally be the case. Experiments have been conducted to distinguish between host and symbiont pathways for the key ammonia assimilating enzyme GS (Lee et al. submitted).

The relative proportions of host and symbiont GS has been determined in several chemoautotrophic symbioses and in most cases GS activity is predominantly that of the symbionts (Lee et al. submitted). Thus ammonia assimilation generally appears to be localized to the symbionts.

Consistent with the enzymatic potential for ammonia and nitrate assimilation is the finding that labeled nitrate and ammonia are assimilated into organic compounds by these associations (Lee and Childress 1994). Comparison of the rates of nitrogen incorporation with carbon-dioxide incorporation indicate that assimilation of these sources contributes significantly to nitrogen needs in support of autotrophic carbon fixation.

Interestingly, these symbioses may use nitrate a terminal electron acceptor as well as nitrogen source (Hentschel et al. 1993; Hentschel and Felbeck 1993; Wilmot and Vetter 1992). The characteristics of nitrate respiration have been most extensively studied in the sediment dwelling bivalve *L. aquizonata* (Hentschel et al. 1993). Nitrate respiration appears to occur under both aerobic and anaerobic conditions (Hentschel et al. 1996), which contrasts with what is observed in free-living bacteria which generally respire nitrate under anaerobic conditions (Kuenen and Robertson 1988). Further studies are needed on the dual role of nitrate as potential electron acceptor as well as nitrogen source.

4. Conclusions

Chemoautotrophic symbiosis allows marine invertebrates to thrive in food-poor and potentially toxic environments of deep-sea hydrothermal vents and is a widespread adaptation to other marine reducing environments where sulfide and oxygen co-occur. How these autotrophs make their living by utilizing inorganic substrates to produce biomass is the focus of ongoing physiological and biochemical research. The best characterized capability, inorganic carbon assimilation, appears to involve a combination of high environmental levels, concentrating mechanisms of the host, and novel assimilation enzymes which likely facilitate high rates of carbon fixation. Nitrogen assimilation has only recently been investigated, and shares some features in common with carbon assimilation such as high environmental availability and host concentrating mechanisms. Thus both carbon and nitrogen assimilation may be enhanced in these symbioses. Elevated levels of inorganic carbon and nitrogen may be typical of all marine reducing environments. By exploiting this opportunity, and by using sulfide as an energy source, these symbioses appear to be successful because of, rather than despite, the extreme environments they live in.

5. References

Anderson, A. E. , Childress, J.J. and Favuzzi, J. (1987) *J. Exp. Biol.* **133**, 1-31.

Arp, A.J., Childress, J.J. and Fisher C.R. Jr. (1985) *Bull. Biol. Soc. Wash.* **6**, 289-300.

Boutilier, R.G., Heming, T.A. and Iwama, G.K. (1984) In: W. S. Hoar and D. J. Randall (eds.) *Fish Physiology,* Academic Press, New York, vol. XA, pp. 401-430.

Cavanaugh, C. (1983) *Nature* **302**, 58-61.

Cavanaugh, C.M. (1994) *Amer. Zool.* **34**, 79-89.

Cavanaugh, C.M., Gardiner, S.L, Jones, M.L., Jannasch, H.W. and Waterbury, J.B. (1981) *Science* **213**, 340-342.

Cavanaugh, C.M., Levering, P.R., Maki, J.S., Lidstrom, M.E. and Mitchell, R. (1987) *Nature* **325**, 346-348.

Cavanaugh, C.M., Wirsen, C.O. and Jannasch, H.W. (1992) *Appl. Envir. Microbiol.* **58**, 3799-3803.

Childress, J.J. and Fisher, C.R. (1992) *Oceanogr. Mar. Biol. Annu. Rev.* **30**, 337–441.

Childress, J.J. et al. (1986) *Science* **233**, 1306-1308.

Childress, J.J. et al. (1991) *Biol. Bull.* **180**, 135–153.

Childress, J.J.. Fisher, C.R., Favuzzi, J.A. and Sanders, N.K. (1991) *Physiol. Zool.* **64**, 1444-1470.

Childress, J.J. (1993) et al., *Nature* **362**, 147-149.

D'Elia, C.F. and Cook, C.B. (1988) *Limnol. Oceanogr.* **33**, 1153-1165.

Felbeck, H. (1983) *J. Comp. Physiol.* **152**, 3–11.

Felbeck, H. (1985) *Physiol. Zool.* **58**, 272-281.

Felbeck, H. and Distel, D.L. (1992) In: A. Balows, H. Trüper, W. Harder, K. H. Schleifer (eds.) *The prokaryotes,* Springer-Verlag, Berlin, pp. 3891-3906.

Felbeck, H., Somero, G.N. and Childress, J.J. (1981) *Nature* **293**, 291-293.

Fiala-Médioni, A. and Métivier, C. (1986) *Mar. Biol.* **90**, 215-222.

Fisher, C.R. (1990) *Crit. Rev. Aquat. Sci.* **2**, 399-436.

Fisher, C.R., Childress, J.J. and Macko, S.A. (1994) J. M. Brooks, *Mar. Ecol. Prog. Ser.* **103**, 45-55.

Goffredi, S.K, et al. (1997) , *J. exp. Biol.* **200**, 883-896.

Haygood, M.G. (1996) *Limnol. Oceanogr.* **41**, 370-371.

Hentschel, U., Cary, S.C. and Felbeck, H. (1993) *Mar. Ecol. Prog. Ser.* **94**, 35-41.

Hentschel, U. and Felbeck, H. (1993) *Nature* **366**, 338-340.

Hentschel, U., Hand, S.C. and Felbeck, C. (1996) *J. exp. Biol.* **199**, 427-433.

Johnson, K.S., Childress, J.J.., Hessler, R.R., Sakamoto-Arnold, C.M. and Beehler, C.L. (1988) *Deep-Sea Res.* **35**, 1723-1744.

Jones, M.L. (1981) *Proc. Biol. Soc. Wash.* **93**, 1295-1313.

Karl, D.M. et al.. (1988) *Deep-Sea Res.* **35**, 777-791.

Kennicutt II, M.C., Brooks, J.M. and Burke R.A. Jr. (1989) *Offshore Tech. Conf.* **Paper No. 5952**, 649–654.

Kennicutt, M.C. et al. (1992) *Chem. Geol.* **101**, 293–310.

Kochevar, R.E. and Childress, J.J. (1996) *Mar. Biol.* **125**, 375-383.

Kochevar, R.E., Govind, N.S. and Childress, J.J. (1993) *Molec. Mar. Biol. Biotechnol.* **2**, 10-19.

Kuenen, J.G. and Robertson, L.A. (1988) In: J. A. Cole and S. J. Ferguson (eds.) *The Nitrogen and Sulphur Cycles,* Cambridge University Press, Cambridge, pp. 161-218.

Lee, R.W. and Childress, J.J. (1994) *Appl. Environ. Microbiol.* **60**, 1852-1858.

Lee, R.W. and Childress, J.J. (1996) *Biol. Bull.* **190**, 367-372.

Lee, R.W., Childress, J.J. and Desaulniers, N.T. (1997) *J. exp. Biol.* **200**, 2797-2805.

Lee, R.W., Thuesen, E.V. and Childress, J.J. (1992) *J. Exp. Mar. Biol. Ecol.* **158**, 75-91.

Lee, R.W., Thuesen, E.V., Childress, J.J. and Fisher, C.R. (1992) *Mar. Biol.* **113**, 99-106.

Lilley, M.D. et al. (1993) , *Nature* **364**, 45-47.

Lutz, R.A. and Kennish, M.J. (1993) *Reviews of Geophysics* **31**, 211-242.

Lutz, R.A. et al. (1994) *Nature* **371**, 663-664.

McAuley, P.J. and Cook, C.B. (1994) *Mar. Biol.* **121**, 343-348.

Muscatine, L. (1980) In: C. B. Cook, P. W. Pappas and E. D. Rudolph (eds.) *Cellular Interactions in Symbiosis and Parasitism* Ohio State University Press, Columbus, pp. 229-244.

Nelson, D.C. and Fisher, C.R. (1995) In: *The Microbiology of Deep-Sea Hydrothermal Vents,* CRC Press Inc. pp. 125-167.

Page, H.M., Fiala-Médioni, A., Fisher, C.R. and Childress, J.J. (1991) *Deep-Sea Res.* **38A**, 1455-1461.

Powell, M.A. and Somero, G.N. (1986) *Biol. Bull.* **171**, 274-290.

Rau, G. H. (1981) *Nature* **289**, 484-485.

Rau, G.H. (1985) *Bull. Biol. Soc. Wash* **6**, 243-248.

Rau, G.H. and Hedges, J.I. (1979) *Science* **203**, 648-649.

Raven, J.A. (1991) *Plant Cell Environ.* **14**, 779-794.

Reid, R. G. B. and Bernard, F.R. (1980) *Science* **208**, 609-610.

Riebesell, U. , Wolf-Gladrow, D.A. and Smetacek, V. (1993) *Nature* **361**, 249-251.

Robinson, J.J. and Cavanaugh, C.M. (1`995) *Limnol. Oceanogr.* **40**, 1496-1502.

Roos, A. and Boron, W.F. (1981) *Physiol. Rev.* **61**, 296-434.

Scott, K.M. , Fisher, C.R., Vodenichar, J.S., Nix, E.R. and Minnich, E. (1994) **67**, 617-638.

Somero, G.N., Childress, J.J. and Anderson, A.E. (1989) *Rev. Aquat. Sci.* **1**, 591-614.

Toulmond, A. et al. (1994) *Deep-Sea Res.* **41**, 1447-1456.

Tunnicliffe, V. (1991) *Oceanogr. Mar. Biol. Annu. Rev.* **29**, 319-407.

Tunnicliffe, V. et al.(1986) *Deep-Sea Res.* **33**, 401-412.
VanDover, C.L. and Fry, B. (1989) *Mar. Biol.* **102**, 257-263.
Vitousek, P.M. and Howarth, R.W. (1991) *Biogeochemistry* **13**, 87–115.
VonDamm, K.L., Edmond, J.M., Measures, C.I. and Grant, B. (1985) *Geochim. Cosmochim. Acta* **49**, 2221-2237.
Weis, W.M., Smith, G.J. and Muscatine, L. (1989) *Mar. Biol.* **100**, 195-202.
Wilkerson, F.P. and Trench, R.K. (1986) *Mar. Biol.* **93**, 237-246.
Wilmot, D.B. and Vetter, R.D. (1992) *Biol. Bull.* **182**, 444-453.
Yellowlees, D. et al.(1993) , *Mar. Biol.* **115**, 605-611.

V. EFFECT OF SUBSTANCES, GASES AND IRRADIATIONS

Biodata of Julia Foght (author with D. McFarlane of the chapter *"Growth of Extremophiles on Petroleum)".*

Dr. Julia Foght is an assistant Professor of Microbiology at the University of Alberta, where she earned her PhD in 1985. Her field of research is oil biodegradation, including genetic analysis of bacteria which oxidize polycyclic aromatic hydrocarbons. She also has been involved in collaborative research analyzing sulfate reducing bacterial communities associated with Petroleum reservoirs and production facilities. Her current interests include bioremediation of crude oil and fuels at low temperatures.
E-Mail: Julia_foght@biology.ualberta.ca

Biodata of D. McFarlane (coauthor with J. Foght of the chapter *"Growth of Extremophiles on Petroleum)".*

Douglas McFarlane is a PhD candidate at the University of Alberta. He is currently examining substrate specificity of polycyclic aromatic hydrocarbon (PAH) biodegradation using molecular genetic techniques. His other research interests include anaerobic PAH degradation and reductive dechlorination.

J. Seckbach (ed.), Enigmatic Microorganisms and Life in Extreme Environments, 527–538.

GROWTH OF EXTREMOPHILES ON PETROLEUM

J.M. FOGHT and D.M. McFARLANE
Department of Biological Sciences
University of Alberta, Edmonton, Alberta, Canada

1. Introduction

Petroleum is a general term for mixtures of hydrocarbons and their analogues that occur naturally in the earth (Speight, 1980). In this review, we will discuss primarily liquid petroleum, commonly called crude oil, and some refined products of petroleum, such as diesel fuel.

Petroleum can be divided into four chemical classes: the aliphatics (e.g. *n*-alkanes); the aromatics (e.g. benzene, naphthalene); the resins (polar compounds); and the asphaltenes (high molecular weight, pentane-insoluble planar molecules). Of these, the first two classes include compounds which are readily oxidized by a wide variety of microbial genera, with the aliphatics in general being metabolized more rapidly than the aromatics (Leahy and Colwell, 1990). The latter two classes are more recalcitrant to microbial degradation. Some petroleum components serve as carbon sources for microbes under ideal conditions, whereas under sub-optimal conditions these and other compounds are only partially oxidized. For example, fatty acids may arise from incomplete alkane oxidation, and aromatic acids and alcohols from the aromatics. These products in turn can serve as carbon sources for microbes which cannot attack the petroleum directly.

Petroleum typically is found in reservoirs exhibiting extreme conditions, potentially providing long-term selection for extremophiles with hydrocarbon-degrading abilities and associated microflora. Additionally, petroleum and its refined products may impact other extreme habitats which then require bioremediation. Extremophiles can have adverse economic effects due to their deleterious actions in petroleum reservoirs, and conversely beneficial environmental and economic effects when used for mitigation of oil spills or for industrial reactions. This review will summarize the literature describing extremophilic bacteria isolated from petroleum-impacted environments, and will speculate on the present and potential use of extremophiles in petroleum bioremediation and biocatalysis.

2. Extreme Habitats Associated with Petroleum

By some definitions, petroleum itself could be considered an extreme environment for bacterial growth because it usually contains short chain *n*-alkanes ($\leq C_8$) and low molecular weight aromatics (e.g. benzene, toluene, ethylbenzenes and xylenes; BTEX) known to be toxic (Section 2.3.1). It is typically anhydrous because of its hydrophobicity which restricts bacterial growth to the oil-water interface or to the aqueous phase of water-in-oil emulsions, thus impeding exchange of oxygen and nutrients.

2.1. PETROLEUM RESERVOIRS

Gold (1992) reviewed evidence for the existence in the earth's subsurface of a "deep, hot biosphere" experiencing high temperatures and pressures, and pointed out that some of these environments, both marine and continental, are associated with gaseous and liquid hydrocarbons. In fact, the deep subsurface microflora have been implicated in petroleum diagenesis (Gold, 1992). It is reasonable to assume that some of these microbes have evolved the ability to utilize hydrocarbons *in situ* despite combinations of extreme temperature, pressure and salinity (e.g. Clark *et al.*, 1981; Bernard *et al.*, 1992), low organic nitrogen content (Adkins *et al.*, 1992), and low oxygen and water availability, and indeed hydrocarbon-degrading microbes have been isolated from petroleum reservoirs.

Evidence exists for bacterial alteration of petroleum *in situ*. For example, Bailey *et al.* (1973) found that the physico-chemical profiles of a suite of crude oils were correlated with freshwater incursion, and postulated that a combination of waterwashing, decreasing salinity, and biodegradation had produced crudes ranging from unaltered to severely biodegraded. Three oils from the Beaufort Basin obtained from increasing depths (and therefore increasing temperature) showed decreasing alteration of their aliphatic (Burns *et al.*, 1975) and aromatic (Westlake, 1983) profiles, such that the shallowest sample resembled biodegraded oil while the deepest oil was not considered microbially altered. Similarly, Rubinstein *et al.* (1977) presented evidence that the massive oil sand bitumen reserves of northern Alberta, Canada represent the residues of microbial action on conventional petroleum *in situ* over geological time. Therefore, the crude oil present in some petroleum reservoirs may be considered as evidence of long-term microbial activity in those reservoirs.

2.1.1. *Hydrocarbon-Degrading Extremophiles in Petroleum Reservoirs*

While the existence of various mesophilic bacteria in reservoirs has been known for decades, only relatively recently have oil-degrading extremophiles been cultured from this environment. For example, Nazina *et al.* (1995) isolated aerobic, thermophilic, spore-forming bacteria from a high-temperature petroleum reservoir being flooded with water at ca. 60°C. Pure cultures of these bacteria, identified as *Bacillus thermoleovorans* (Nazina *et al.*, 1993), were shown to grow on crude oil, utilizing a wide range of *n*-alkanes (Nazina *et al.*, 1992). Highly saline wells in the same field, however, failed to produce oil-degrading thermophiles. Conversely, Kulichevskaya *et al.* (1991) and Belyaev *et al.* (1993) isolated alkane-degrading halotolerant and halophilic Bacteria and Archaea from reservoir formation waters having salinities from 2.5 to 5.5 M NaCl, yet these were uniformly mesophilic. Adkins *et al.* (1992) have also commented on the apparent difficulty of isolating thermophilic halophiles from petroleum reservoirs.

Hydrocarbon-Degrading Anaerobes. It was once assumed that *in situ* biotransformation of petroleum occurred only in the presence of oxygen (provided by ground water or injection of surface waters for oil recovery by waterflooding) through sequential attack, with anaerobes utilizing the byproducts of aerobic degradation (Jobson *et al.*, 1979; Nazina *et al.*, 1985). However, recent reports suggest that anaerobic extremophiles can directly alter petroleum *in situ*. In one study (Stetter *et al.*, 1993) the reservoirs examined were undergoing waterflooding with seawater and it was postulated that the thermophilic hydrocarbon-degrading bacteria isolated from produced waters had been introduced with the injection waters; in another case (L'Haridon *et al.*, 1995), indigenous barotolerant thermophiles were implicated.

Aeckersberg *et al.* (1991) were the first to demonstrate the ability of a purified sulfate-reducing bacterium to oxidize alkanes anaerobically, and Rabus *et al.* (1996) later described anaerobes which degraded alkylaromatics. Although these isolates were not extremophilic, the reports revealed the potential for anaerobic petroleum degradation.

Stetter *et al.* (1993) in fact successfully enriched a mixed population of hyperthermophiles which utilized crude oil anaerobically under *in situ* conditions, and Rueter *et al.* (1994) purified a sulfate-reducing bacterium which utilized *n*-alkanes at 60°C. Thermophilic hydrocarbon-degrading nitrate-reducers also have been isolated (Ivanov *et al.*, 1994).

2.1.2. *Reservoir-Associated (Non-Degradative) Extremophilic Anaerobes*
Mesophilic anaerobic Bacteria and Archaea which cannot directly attack oil but rather utilize metabolites of aerobic hydrocarbon attack have been isolated frequently from petroleum reservoirs and described in the literature (e.g. Machel and Foght, in press). Reports of extremophilic anaerobes associated with reservoirs (but not directly capable of degrading petroleum) are more recent, and include thermophilic and halophilic fermenters (e.g. Adkins *et al.*, 1992; Grassia *et al.*, 1996) and respirers: sulfate- and sulfur-reducing Bacteria and Archaea (e.g. Stetter *et al.*, 1987; Ravot *et al.*, 1997); and iron- and manganese-reducers (Greene *et al.*, 1997).

These reports of extremophilic "associated flora" are significant because sulfate-reducing bacteria, in particular, have been shown responsible for adverse economic effects in reservoirs, including souring (accumulation of biogenic H_2S), plugging of formations with biomass or sulfide precipitates, and biofouling and corrosion of production facilities (Cord-Ruwisch *et al.*, 1987). Recent isolation of hyperthermophilic sulfate-reducing Archaea (e.g. Nilsen *et al.*, 1996) from reservoir fluids has revealed their deleterious potential at greater depths than previously thought possible. On the other hand, some reservoir extremophiles have potential commercial applications (Section 3.2).

2.2. TERRESTRIAL ENVIRONMENTS

2.2.1. *Soils*

Low-Temperature Soils. Atlas (1986) has exhaustively reviewed the fate of petroleum pollutants in the Arctic, including tundra, freshwater and marine ecosystems where slow rates of oil biodegradation typically are observed. Cold-adapted microflora were found in all Arctic ecosystems surveyed and generally proliferated in response to crude oil incursion. Atlas concluded that the slow rates of *in situ* biodegradation were not due to lack of competent microbes, but neither were they due solely to low temperature. Rather, the systems which experienced prolonged low temperatures usually exhibited additional limitations such as oligotrophy and limited oxygen availability, compounding the problem of microbial adaptation to and growth on petroleum.

Cold-adapted bacteria have been purified from diverse soils and subsoils, and recently molecular genetics has been applied to their study. For example, certain of the psychrotrophic aromatic-degrading bacteria isolated by Whyte *et al.* (1996) were found to possess genes which hybridized to archetypal mesophilic genes for toluene catabolism. In contrast, the isolated alkane-degraders hybridized only to putative catabolic genes from another psychrophile. In a related report, a plasmid encoding toluene degradation was transferred from a mesophile to a psychrotroph (Kolenc *et al.*, 1988), suggesting that cold-tolerant bacteria may become hydrocarbon-degraders if they acquire the appropriate catabolic genes. This has implications for bioremediation of low-temperature habitats.

Recently, attention has turned to Antarctic soils where low levels of available water and nutrients, moderate alkalinity (pH$\geq$8) and sometimes salinity from salt spray combine with low temperatures (Vishniac, 1993). In both pristine and fuel-impacted Antarctic soils indigenous microflora were shown capable of degrading fuel components *in situ* (Kerry, 1993; Section 3.1). It has been noted that hydrocarbon degraders isolated from Antarctic soils tend to be psychrotrophic rather than psychrophilic (Kerry, 1990; J. Aislabie and J. Foght, unpublished observations), although they equally must be able to withstand the harsh environment.

High-Temperature Soils. Desert soils have been a source of thermophilic hydrocarbon-degraders. For example, *in situ* degradation of spilled oil, presumably accomplished by indigenous microbes, was documented in desert soils of Kuwait after massive petroleum contamination (Al-Gounaim *et al.*, 1995), and Sorkhoh *et al.* (1993) isolated an alkane-utilizing strain of thermophilic *Bacillus stearothermophilus* from such soils. High ambient temperature is not a pre-requisite for these isolations, however, as Kvasnikov *et al.* (1971) and Sukatsch and Johnson (1972) were able to enrich thermophilic alkane-utilizing bacteria from temperate soils.

2.2.2. *Freshwaters, Hotsprings and Muds*

Thermophilic, Gram negative alkane-degrading bacteria isolated from thermal and non-thermal sources (reviewed by Perry, 1985) seem to fall into two groups: those that utilize a variety of hydrocarbon and non-hydrocarbon sources, and those that obligately utilize a narrow spectrum of *n*-alkanes (e.g. n-C_{13} to n-C_{20}). An example of the latter group is the novel genus *Thermoleophilum*, which includes species obligate for both thermophily and *n*-alkane degradation (Zarilla and Perry, 1984, 1986). In addition to these Gram negative isolates, Gram positive thermophiles have been isolated from water, soil and mud (Klug and Markovetz, 1967), including *Bacillus thermoleovorans*, an *n*-alkane-utilizing thermophile related to *B. stearothermophilus* (Zarilla and Perry, 1987).

Culture collection strains of *Thermus aquaticus* (ATCC 25104) and *Thermus scotoductus* (ATCC 27978) originally isolated from thermal waters recently were tested for their ability to degrade BTEX (Chen and Taylor, 1995). Interestingly, although these strains did not utilize the BTEX compounds as growth carbon sources, they did effect removal of the compounds during incubation at $\geq$60°C, presumably through co-metabolism. This suggests that petroleum transformation in high temperature environments may occur without hydrocarbon utilization.

2.2.3. *Hypersaline Terrestrial Environments*

Ward and Brock (1978) studied hydrocarbon degradation in samples from salt evaporation ponds and Great Salt Lake, USA with salinities ranging from $<$1 M to 5 M NaCl. They found a reciprocal relationship between aliphatic hydrocarbon oxidation and salinity, and speculated that hypersalinity was incompatible with hydrocarbon biodegradation. This general observation held for years until the isolation from a salt marsh of hydrocarbon-degrading obligately halophilic Archaeon (Bertrand *et al.*, 1990).

2.3. MARINE ENVIRONMENTS

Bacteria adapted to extremes of pressure, salinity and cold have been isolated from marine environments. However, to date only a few of these have been described which also degrade petroleum hydrocarbons.

2.3.1. *Deep Sea Muds and Sediments*

Bertrand *et al.* (1993) enriched halotolerant hydrocarbon-degrading mixed bacterial populations from an oil-contaminated marine sediment. They then showed that salinity greater than 0.4M NaCl (i.e. natural seawater) adversely affected oil degradation by these communities, although biodegradation remained significant up to 2M NaCl. The effect depended somewhat on the composition of the substrate itself, as a simple synthetic hydrocarbon mixture was degraded more completely than a complex crude oil.

A suite of halotolerant bacteria enriched from deep sea sediments incubated with hydrocarbons (Moriya and Horikoshi, 1993a,b) yielded novel isolates demonstrating solvent tolerance (Section 3.2.2). Solvent components of petroleum such as toluene and benzene are toxic to most microorganisms at concentrations as low as 0.1% (Horikoshi and Grant, 1991) and even organisms that utilize these compounds can only tolerate concentrations up to ca. 0.3% (Inoue and Horikoshi, 1989). For this reason the BTEX components of petroleum and its refined products can themselves be considered extreme factors. Two deep-sea benzene-tolerant isolates, *Flavobacterium* sp. strain DS711 and *Bacillus* DS 994 (Moriya and Horikoshi, 1993a,b) were shown to degrade, respectively, the *n*-alkanes in kerosene, and thiophene and dibenzothiophene, common aromatic sulfur heterocycles in petroleum.

2.3.2. *Hydrothermal Vents*

The Guaymas Basin hydrothermal vent in the Gulf of California is believed to be unique in that it exhibits natural seepage of gasoline-range aliphatic and aromatic hydrocarbons (Simoneit and Lonsdale, 1982). These hydrocarbons would be predicted to be a significant carbon source for the vent microflora, and indeed, oxidation of both aliphatic and aromatic substrates by sediment slurries was observed by Bazylinski *et al.* (1989) over a wide temperature range (≤4 to ≥55°C). However, Goetz and Jannasch (1993) remarked upon their unexpected inability to isolate hydrocarbon-degrading thermophiles or psychrophiles from the vent sediments.

2.3.3. *Marine Surface Environments*

A naturally hypersaline salt marsh sediment yielded the first known hydrocarbon-degrading extremely halophilic Archaeon (Bertrand *et al.*, 1990). Strain EH4, which had an optimum salinity of 3.8 M NaCl, degraded pure alkanes and aromatic hydrocarbons. Later, the extremely halotolerant Bacterium *Marinobacter hydrocarbonoclasticus* was isolated from Mediterranean seawater near an oil refinery (Gauthier *et al.*, 1992) and shown to utilize eicosane (n-C_{20}) at salinities up to 2.5 M NaCl. Increasing salinity was noted to modify its production of a biosurfactant but did not change its alkane biodegradation rate (Fernandez-Linares *et al.*, 1996).

Although cold marine surface waters, coastal sediments (e.g. *Exxon Valdez*, Bragg *et al.*, 1994; *Bahia Paraiso*, Karl, 1992), and sea ice (Delille *et al.*, 1997) can be impacted by petroleum spills, surprisingly little attention has been paid to isolating extreme (i.e. psychrophilic) hydrocarbon-degrading bacteria from these environments.

3. Environmental and Biotechnological Applications

There has been great interest in recent years in biotechnological exploitation of extremophiles. In the petroleum industry, potential uses range from bioremediation of oil spills in extreme environments to microbial enhancement of oil recovery in petroleum reservoirs. This section of the review will discuss various applications of petroleum-degrading extremophiles and touch upon the vast biotechnological potential for these organisms.

3.1. HYDROCARBON BIOREMEDIATION IN EXTREME ENVIRONMENTS

Effective oil spill bioremediation is influenced by a number of environmental factors including nutrient availability, pH, temperature, salinity, and water activity (Atlas, 1981). Bioremediation of spills in harsh environments requires either treatments to ameliorate the extreme conditions, or the *in situ* activity of extremophilic petroleum-degrading organisms. We present one current example of the latter application, hydrocarbon bioremediation of Antarctic soils.

Antarctica is the coldest, driest, most barren continent on Earth and presents a challenge to microbial life. Until relatively recently it was undisturbed by human contact, but exploration of the continent has already resulted in fuel spills totaling hundreds of thousands of litres, primarily of jet and diesel fuel (Kennicutt *et al.*, 1990; Tumeo and Wolk, 1994). Even small spills can have major ecological consequences because the contaminants persist due to extreme environmental conditions. Physical removal of oiled soil is undesirable because of the resulting alteration of the natural landscape, as well as the immense expense of excavation and transportation of soil for *ex situ* treatment (Wardell, 1995). Abiotic losses due to volatilization, which can be significant in temperate soils, are minimized by the low ambient temperatures in cold soils and result in retention of toxic BTEX components of fuel, compounding the adverse conditions.

An alternative approach is *in situ* bioremediation, which has been examined in Antarctic soils only in recent years. The presence of indigenous microorganisms capable of petroleum degradation in Antarctica is critical to successful *in situ* bioremediation because the Antarctic treaty forbids the introduction of non-indigenous life forms (Draggan, 1993). Also, indigenous microbes should be better adapted to the prevailing cold, dry conditions and low nutrient content of Antarctic soils than introduced microbes. Therefore, most studies to date have concentrated on assessing the biological potential for bioremediation by indigenous cold-adapted hydrocarbon-degrading xerophiles and oligotrophs. From preliminary experiments, it would appear that many Antarctic soils support indigenous populations of microorganisms capable of degrading the aliphatic components of jet and diesel fuels (Kerry, 1990; MacCormack and Fraile, 1997). Aislabie *et al.* (in press) have isolated Antarctic bacteria which degrade polycyclic aromatic hydrocarbons, giving hope that even recalcitrant fuel components can be degraded *in situ*. If some contaminated soils are found to lack an appropriate indigenous microflora, it may be possible to seed these areas with fuel-degrading extremophiles from other Antarctic sites. Notably, however, Chablain *et al.* (1997) observed that low temperature affected the ability of psychrotrophic bacteria to grow on toluene, suggesting that the mere presence of cold-adapted, hydrocarbon-degrading bacteria in a cold environment does not necessarily equate with their activity *in situ*.

Experiments to ameliorate the environment itself, using nutrient amendments to relieve oligotrophic conditions, and gum xanthan or plastic coverings to improve water activity and incidentally increase soil temperature (Kerry, 1993), have met with some success. There is a need for further investigation into practical methods to enhance bioremediation of cold soils by stimulating the indigenous microflora.

3.2. BIOTECHNOLOGICAL APPLICATIONS AND POTENTIAL

3.2.1. *Potential Applications of Extremophiles in the Petroleum Industry*

Biocatalytic Desulfurization. Combustion of high sulfur content fuels releases toxic gases such as sulfur dioxide which contribute to acid rain. Increasingly stringent emission regulations combined with dwindling reserves of low-sulfur crude oils have led to investigation of new, inexpensive methods for petroleum desulfurization. Presently, most organic sulfur is removed from crude oil by thermochemical hydrodesulfurization which is energetically expensive and incurs high capital and operating costs. Biocatalytic desulfurization (BDS) is a process by which sulfur is enzymatically excised from organic sulfur (e.g. aromatic heterocycles), releasing sulfate and leaving the hydrocarbon skeleton intact so that fuel value is not decreased (Shennan, 1996). Its potential advantages include reduced costs due to low temperature-low pressure reaction conditions combined with chemical specificity.

Of the several described strains of bacteria capable of petroleum desulfurization, only one is extremophilic (Konishi *et al.*, 1997). However, since the mesophilic genes involved have been cloned and sequenced (Denome *et al.*, 1993; Piddington *et al.*, 1995), the regulatory elements characterized (Li *et al.*, 1996), and the individual enzymes involved isolated (Gray *et al.*, 1996), genetic engineering may prove useful by shuttling the genes for mesophilic BDS into extremophiles. For example, thermophilic BDS hosts would permit higher reaction temperatures and shorter retention times, while solvent-tolerant hosts (Section 3.2.2) may prove useful in significantly reducing water phase volumes for BDS and in prolonging the life of the biocatalyst in the presence of toxic hydrocarbons.

It has been proposed (e.g. Foght *et al.*, 1990) that BDS could be performed *in situ* by inoculating reservoirs with suitable bacteria, then allowing extended incubation time for sulfur excision to occur. Since many reservoirs are extreme environments, the use of thermophiles and halophiles capable of BDS would be logical. These organisms might have the added benefit of lowering the viscosity of the crude oil through thioether cleavage (Monticello *et al.*, 1995).

Microbial Enhancement of Oil Recovery (MEOR). MEOR is the utilization of microorganisms to increase the efficiency of oil recovery from reservoirs. There are several mechanisms by which this can be effected: improved formation permeability through acid generation; biomass production to selectively plug high permeability regions; production of biosurfactants, biopolymers or organic solvents to enhance oil flow; or generation of gases to increase pressure and decrease oil viscosity (e.g. McInerney and Westlake, 1990; Yen, 1990; Jack, 1991).

There are two general strategies used in MEOR: stimulating indigenous organisms by adding appropriate nutrients, or seeding reservoirs with selected microbes (plus amendments) to establish a new microflora with the desired metabolic capabilities. Presently, extremophiles (thermophiles, barophiles and halophiles) generally fall into the biostimulation category (Bhupathiraju *et al.*, 1993) while non-extremophiles are usually used for bioaugmentation (Bryant *et al.*, 1990). However, Pfiffner *et al.* (1986) described a halotolerant, thermotolerant *Bacillus*-like species for bioaugmentation, and development of extremophiles for seeding is currently underway in a number of laboratories (e.g. Trebbau de Acevedo and McInerney, 1996).

In most MEOR applications, petroleum-degrading bacteria are considered undesirable because of their potential to devalue the oil during recovery. However, it has been suggested (Chakrabarty, 1985; Premuzic and Lin, 1991) that controlled petroleum degradation might be useful during MEOR so that costly nutrient amendments would not be required. In particular, extremophiles might be selected or genetically engineered to degrade only the less valuable components of crude oil such as high molecular weight paraffins, asphaltenes or sulfur heterocycles. These organisms could be doubly useful as their activities would not only enhance oil recovery, but could potentially upgrade the petroleum by decreasing viscosity, unplugging well bore areas or biodesulfurizing the crude *in situ* (Section 3.2.1).

Remediation of Petrochemical Wastewaters. Before petrochemical wastewaters from oil and gas production can be discharged, removal of most of the toxic components and decrease of the chemical oxygen demand is required. While this may be accomplished in bioreactors, certain petrochemical wastewaters are too saline to be used directly with conventional biocatalysts. Woolard and Irvine (1995) demonstrated the potential usefulness of a moderate halophile isolated from Great Salt Lake, USA to treat a simulated hypersaline oil field produced water without dilution. However, Oren *et al.* (1992) have pointed out the current limits to our understanding of organic pollutant biodegradation at high salt concentrations, and further research is needed in this area.

Potential also exists for the use of organic solvent tolerant bacteria (below) in petroleum industry wastes, since oil and gas production wastewaters usually contain relatively high concentrations of toxic compounds such as phenols and toluene. Judicious use of extremophiles may provide solutions to these difficult remediation problems.

3.2.2. *Solvent Tolerant Bacteria*

In recent years, microorganisms tolerant of relatively high concentrations of organic solvents such as benzene and toluene have been isolated, some of which utilize components of petroleum as carbon and energy source (Moriya and Horikoshi, 1993a,b; Fukumaki *et al.*, 1994; Ramos *et al.*, 1995). To our knowledge, all naturally solvent tolerant microorganisms described to date have been isolated either from the deep sea or from soil samples not contaminated by hydrocarbons. Future surveys for solvent tolerant bacteria should include petroleum reservoirs where microbes have been exposed to high concentrations of hydrocarbons for a long time.

The biotechnological applications of solvent tolerant bacteria include use as biocatalysts in two-phase bioreactors for BDS (Section 3.2.1), bioremediation of petrochemical wastewaters or BTEX-contaminated environments, and transformation of petroleum compounds to value-added compounds such as *cis*-diols, epoxides or indigo (Ramos *et al.*, 1995). Bacteria with increased tolerance to polar solvents in particular would expand the array of useful solvents for these applications.

4. Concluding Remarks

The existence of extremophiles capable of growth on individual components of petroleum (e.g. utilization of pure alkanes by thermophiles) has been known for decades. As well, there have been reports of *in situ* alteration of petroleum in extreme environments, and the activity of indigenous populations has been inferred without actual isolation of the degradative microbes. What has been lacking is systematic enrichment and culture of microbes from petroleum-impacted environments to isolate extremophiles with broad spectrum petroleum-degrading capabilities.

Undoubtedly there are many petroleum-degrading extremophiles as yet undiscovered. It is hoped that increased interest in the biotechnological potential of these organisms will stimulate comprehensive microbiological surveys of likely habitats and lead to new culturing techniques to isolate these interesting microbes. The challenge will be discovery of petroleum-degrading extremophiles with combined tolerances, such as thermophilic halophiles, or psychrophilic oligotrophs.

5. References

Adkins, J.P., Cornell, L.A., and Tanner, R.S. (1992) *Geomicrobiol. J.* **10**, 87-97.

Aeckersberg, F., Bak, F., and Widdel, F. (1991) *Arch. Microbiol.* **156**, 5-14.

Aislabie, J., McLeod, M., and Fraser, R. (in press) *Appl. Microbiol. Biotechnol.*

Al-Gounaim, M.Y., Diab, A., Al-Abdulla, R., and Al-Zamil, N. (1995) *Arab Gulf J. Scient. Res.* **13**, 653-672.

Atlas, R.M. (1981) *Microbiol. Rev.* **45**, 180-209.

Atlas, R.M. (1986) *Wat. Sci. Tech.* **18**, 59-67.

Bailey, N.J.L., Jobson, A.M., and Rogers, M.A. (1973) *Chem. Geol.* **11**, 203-221.

Bazylinski, D.A., Wirsen, C.O., and Jannasch, H.W. (1989) *Appl. Environ. Microbiol.* **55**, 2832-2836.

Belyaev, S.S., Borzenkov, I.A., Milekhina, E.I., Zvyagintseva, I.S., and Ivanov, M.V. (1993) in E.T. Premuzic and A. Woodhead (eds.) *Microbial Enhancement of Oil Recovery - Recent Advances*. Elsevier, Amsterdam. pp. 79-88.

Bernard, F.P., Connan, J., and Magot, M. (1992) 67th Annual Technical Conference and Exhibition of the Society of Petroleum Engineers, Washington, D.C., Society of Petroleum Engineers. pp. 467-476.

Bertrand, J.C., Almallah, M., Acquaviva, M., and Mille, G. (1990) *Lett. Appl. Microbiol.* **11**, 260-263.

Bertrand, J.-C., Bianchi, M., Al Mallah, M., Acquaviva, M., and Mille, G. (1993) *J. Exp. Mar. Biol. Ecol.* **168**, 125-138.

Bhupathiraju, V.K., McInerney, M.J., and Knapp, R.M. (1993) *Geomicrobiol. J.* **11**, 19-34.

Bragg, J.R., Prince, R.C., Harner, E.J., and Atlas, R.M. (1994) *Nature* **368**, 413-418.

Bryant, R.S., Burchfield, T.E., Dennis, D.M., and Hitzman, D.O. (1990) *SPE Res. Eng.* **1990**, 9-13.

Burns, B.J., Hogarth, J.T.C., and Milner, C.W.D. (1975) *Bull. Can. Petrol. Geol.* **23**, 295-303.

Chablain, P.A., Philippe, G., Groboillot, A., Truffaut, N., and Guespin-Michel, J.F. (1997) *Res. Microbiol.* **148**, 153-161.

Chakrabarty, A.M. (1985) *TIBS* **3**, 32-39.

Chen, C. and Taylor, R.T. (1995) *Biotechnol. Bioeng.* **48**, 614-624.

Clark, J.B., Munnecke, D.M., and Jenneman, G.E. (1981) *Dev. Ind. Microbiol.* **22**, 695-701.

Cord-Ruwisch, R., Kleinitz, W., and Widdel, F. (1987) *J. Petroleum Technol.* **1987**, 97-106.

Delille, D., Bassères, A., Dessommes, A. (1997) *Microb. Ecol.* **33**, 97-105.

Denome, S.A., Olson, E.S., and Young, K.D. (1993) *Appl. Environ. Microbiol.* **59**, 2837-2843.

Draggan, S. (1993) in E.I. Friedmann (ed.) *Antarctic Microbiology*. Wiley-Liss, New York. pp. 603-614.

Fernandez-Linares, L., Acquaviva, M., Bertrand, J.-C., and Gauthier, M. (1996) *System. Appl. Microbiol.* **19**, 113-121.
Foght, J.M., Fedorak, P.M., Gray, M.R., and Westlake, D.W.S. (1990) in H.E. Ehrlich and C.L. Brierley (eds.) *Microbial Mineral Recovery*. McGraw-Hill, New York. pp. 379-407.
Fukumaki, T., Inoue, A., Moriya, K., and Horikoshi, K. (1994) *Biosci. Biotech. Biochem.* **58**, 1784-1788.
Gauthier, M.J., Lafay, B., Christen, R., Fernandez, L., Acquaviva, M., Bonin, P., and Bertrand, J.-C. (1992) *Int. J. Syst. Bacteriol.* **42**, 568-576.
Goetz, F.E. and Jannasch, H.W. (1993) *Geomicrobiol. J.* **11**, 1-18.
Gold, T. (1992) *Proc. Nat. Acad. Sci. USA* **89**, 6045-6049.
Grassia, G.S., McLean, K.M., Glénat, P., Bauld, J., and Sheehy, A.J. (1996) *FEMS Microbiol. Ecol.* **21**, 47-58.
Gray, K.A., Pogrebinsky, O.S., Mrachko, G.T., Xi, L., Monticello, D.J., and Squires, C.H. (1996) *Nature Biotech.* **14**, 1705-1709.
Greene, A.C., Patel, B.K.C., and Sheehy, A.J. (1997) *Int. J. Syst. Bacteriol.* **47**, 505-509.
Horikoshi, K. and Grant, W.D. (1991) *Superbugs.* Springer Verlag, Berlin.
Inoue, A. and Horikoshi, K. (1989) *Nature* **338**, 264-266.
Ivanov, V.N., Kachur, T.L., Dulgerov, A.N., and Saljuk, A.I. (1995) *Microbiol. Zhurnal.* **57**, 85-92.
Jack, T.R. (1991) *Current Opinion Biotech.* **2**, 444-449.
Jobson, A.M., Cook, F.D., and Westlake, D.W.S. (1979) *Chem. Geol.* **24**, 355-365.
Karl, D.M. (1992) *Microb. Ecol.* **24**, 77-89.
Kennicutt II, M.C. (1990) *Environ. Sci. Technol.* **24**, 620-624.
Kerry, E. (1990) *Polar Biol.* **10**, 423-430.
Kerry, E. (1993) *Polar Biol.* **13**, 163-170.
Klug, M.J. and Markovetz, A.J. (1967) *Nature* **215**, 1082-1083.
Kolenc, R.J., Inniss, W.E., Glick, B.R., Robinson, C.W., and Mayfield, C.I. (1988) *Appl. Environ. Microbiol.* **54**, 638-641.
Konishi, J., Ishii, Y., Onaka, T., Okomura, K., and Suzuki, M. (1997). *Appl. Environ. Microbiol.* **63**, 3164-3169.
Kulichevskaya, I.S., Milekhina, E.I., Borzenkov, I.A., Zvyagintseva, I.S., and Belyaev, S.S. (1991) *Mikrobiologiya (transl.)* **60**, 596-601.
Kvasnikov, E.I., Zhuravel', A.M., and Klyushnikova, T.M. (1971) *Mikrobiologiya (transl.)* **40**, 443-449.
L'Haridon, S., Reysenback, A.-L., Glénat, P., Prieur, D., and Jeanthon, C. (1995) *Nature* **377**, 223-224.
Leahy, J.G. and Colwell, R.R. (1990) *Microbiol. Rev.* **54**, 305-315.
Li, M.Z., Squires, C.H., Monticello, D.J., and Childs, J.D. (1996) *J. Bacteriol.* **178**, 6409-6418.
MacCormack, W.P. and Fraile, E.R. (1997) *Antarctic Sci.* **9**, 150-155.
Machel, H.G. and Foght, J. (in press) in S.M. Awramik and R. Riding (eds.) *Microbial Sediments*. Springer Verlag, New York.
McInerney, M.J. and Westlake, D.W.S. (1990) in H.L. Ehrlich and C.L. Brierly (eds.) *Microbial Mineral Recovery*. McGraw-Hill, New York. pp. 409-445.
Monticello, D.J., Murphy, S., and Johnson, S. (1995) Proceedings 12th annual general meeting of Biominet, Calgary, Alberta, Canada.
Moriya, K. and Horikoshi, K. (1993a) *J. Ferment. Bioeng.* **76**, 168-173.
Moriya, K. and Horikoshi, K. (1993b) *J. Ferment. Bioeng.* **76**, 397-399.
Nazina, T.N., Rozanova, E.P., and Kuznetsov, S.I. (1985) *Geomicrobiol. J.* **4**, 103-130.
Nazina, T.N., Ivanova, A.E., and Blagov, A.V. (1992) *Microbiologiya (transl.)* **61**, 216-221.
Nazina, T.N., Ivanova, A.E., Mityushina, L.L., and Belyaev, S.S. (1993) *Mikrobiologiya (transl.)* **62**, 359-365.
Nazina, T.N., Ivanova, A.E., Borzenkov, I.A., Belyaev, S.S., and Ivanov, M.V. (1995) *Geomicrobiol.* **13**, 181-192.
Nilsen, R.K., Beeder, J., Thorstenson, T., and Torsvik, T. (1996) *Appl. Environ. Microbiol.* **62**, 1793-1798.
Oren, A., Gurevich, P., Azachi, M., and Henis, Y. (1992) *Biodegradation* **3**, 387-398.
Perry, J.J. (1985) *Adv. Aquatic Microbiol.* **3**, 109-139.
Pfiffner, S.M., McInerney, M.J., Jenneman, G.E., and Knapp, R.M. (1986) *Appl. Environ. Microbiol.* **51**, 1224-1229.
Piddington, C.S., Kovacevich, B.R., and Rambosek, J. (1995) *Appl. Environ. Microbiol.* **61**, 468-475.
Premuzic, E.T. and Lin, M.S. (1991) *Resources Conserv. Recycl.* **5**, 277-284.
Rabus, R., Fukui, M., Wilkes, H., and Widdel, F. (1996) *Appl. Environ. Microbiol.* **62**, 3605-3613.
Ramos, J.L., Duque, E., Huertas, M.-J., and Haidour, A. (1995) *J. Bacteriol.* **177**, 3911-3916.
Ravot, G., Magot, M., Ollivier, B., Patel, B.K.C., Ageron, E., Grimont, P.A.D., Thomas, P., and Garcia, J.-L. (1997) *FEMS Microbiol. Lett.* **147**, 81-88.
Rubinstein, I., Strausz, O.P., Spyckerelle, C., Crawford, R.J., and Westlake, D.W.S. (1977) *Geochim. Cosmochim. Acta* **41**, 1341-1353.
Rueter, P., Rabus, R., Wilkes, H., Aeckersberg, F., Rainey, F.A., Jannasch, H.W., and Widdel, F. (1994) *Nature* **372**, 455-458.
Shennan, J.L. (1996) *J. Chem. Technol. Biotechnol.* **67**, 109-123.
Simoneit, B.R.T. and Lonsdale, P.F. (1982) *Nature* **295**, 198-292.
Sorkhoh, N.A., Ibrahim, A.S., Ghannoum, M.A., and Radwan, S.S. (1993) *Appl. Microbiol. Biotechnol.* **39**, 123-126.
Speight, J.G. (1980) *The Chemistry and Technology of Petroleum*. Marcel Dekker, Inc., New York.

Stetter, K.O., Huber, R., Blöchl, E., Kurr, M., Eden, R.D., Fielder, M., Cash, H., and Vance, I. (1993) *Nature* **365**, 743-745.
Stetter, K.O., Lauerer, G., Thomm, M., and Neuner, A. (1987) *Science* **236**, 822-824.
Sukatsch, D.A. and Johnson, M.J. (1972) *Appl. Microbiol.* **23**, 543-546.
Trebbau de Acevedo, G. and McInerney, M.J. (1996) *J. Ind. Microbiol.* **16**, 1-7.
Tumeo, M.A. and Wolk, A.E. (1994) *Antarctic J.* **1994**, 375-377.
Vishniac, H.S. (1993) in E.I. Friedmann (ed.) *Antarctic Microbiology*. Wiley-Liss, New York. pp. 297-341.
Ward, D.M. and Brock, T.D. (1978) *Appl. Environ. Microbiol.* **35**, 353-359.
Wardell, L.J. (1995) *J. Soil Contam.* **4**, 111-121.
Westlake, D.W.S. (1983) International Conference on Microbial Enhancement of Oil Recovery, Shangri-La, OK, US Dept. Energy. pp. 102-111.
Whyte, L.G., Greer, C.W., and Inniss, W.E. (1996) *Can. J. Microbiol.* **42**, 99-106.
Woolard, C.R. and Irvine, R.L. (1995) *Water Res.* **29**, 1159-1168.
Yen, T.F. (1990) *Microbial Enhanced Oil Recovery: Principle and Practice*. CRC Press, Inc., New York.
Zarilla, K.A. and Perry, J.J. (1984) *Arch. Microbiol.* **137**, 286-290.
Zarilla, K.A. and Perry, J.J. (1986) *Int. J. Syst. Bacteriol.* **36**, 13-16.
Zarilla, K.A. and Perry, J.J. (1987) *System. Appl. Microbiol.* **9**, 258-264.

Biodata of **Dag Klaveness** author of "***Metallogenium - A Microbial Enigma.***"

Dr. Dag Klaveness is a Professor in the Department of Biology at the University of Oslo (Norway). He received his Ph.D. in 1971 from the University of Oslo. Dr. Klaveness main research interests are in eukaryotic microbiology and teaching duties within the aquatic sciences. He has published articles on components of the plankton communities in ponds, lakes and fjords and the open ocean. Dr. Klaveness outdoor life includes fishing and skiing with wife Mona and his two sons.

E-Mail: **dag.klaveness@bio.uio.no**

J. Seckbach (ed.), Enigmatic Microorganisms and Life in Extreme Environments, 539–548.

METALLOGENIUM - A MICROBIAL ENIGMA

D. KLAVENESS
University of Oslo
Department of Biology
P.O. Box 1066 Blindern
0316 Oslo, Norway

1. Introduction

The genus *Metallogenium* was described by Perfilev and Gabe (1961, transl. 1969) from glass capillaries ("peloscopes") inserted through the sediment surface in freshwater lakes. The *Metallogenium* structure is easily recognised from its characteristic trichospherical (Perfilev and Gabe's term), actinomorphic or asteriform morphology as seen in lakes (Fig. 1), in combination with the deposition of oxides of manganese, which appear dark brown to black in the conventional light microscope (cf. Fig 1). Before Perfilev and Gabe's description was published, similar structures had been seen for example by Beger (1935), who identified them as bacteria and named them *Leptothrix echinata*, and by Ruttner (1938), who was of the opinion that they might be "Anorganische Eisenausfällungen". Perfilev and Gabe (1961) chose the specific epithet "personatum" (lat. "masked") referring to the manganese oxide deposits which "mask" a possible biological structure. The scientific name is thus *Metallogenium personatum* Perfilev and Gabe (1961).

The involvement of the microbiologist G.A. Zavarzin led to the parallel description of a second species, *Metallogenium symbioticum* Zavarzin (1961, 1964), isolated from a swamp near Moscow and kept in culture together with an unidentified fungus. All attempts to separate the two partners in this alleged symbiosis failed. The fungal partner grew vigorously upon various fungal media, but only following inoculation of the fungus on medium containing manganese acetate did *Metallogenium* betray its presence as brown rings where the growth of the mycelium proceeded. According to Zavarzin (1964, 1989), the differences between *Metallogenium personatum* and *Metallogenium symbioticum* are slight.

Since the description, *Metallogenium* has been discussed in more than 50 scientific papers and books. Despite this effort, there is still no general agreement among scientists as to whether *Metallogenium* is a biological object, what kind, whether it should be treated as a taxon, or whether it is an inorganic precipitate which may or may not be associated with biological activity. Relevant reviews and points of views include those of Klaveness (1977), Gregory et al. (1980), Margulis et al. (1983), Ghiorse (1984), Schmidt and Overbeck (1984), Zavarzin (1989) and Miyajima (1992a-c). This paper discusses various aspects of the *Metallogenium* story as it stands at present.

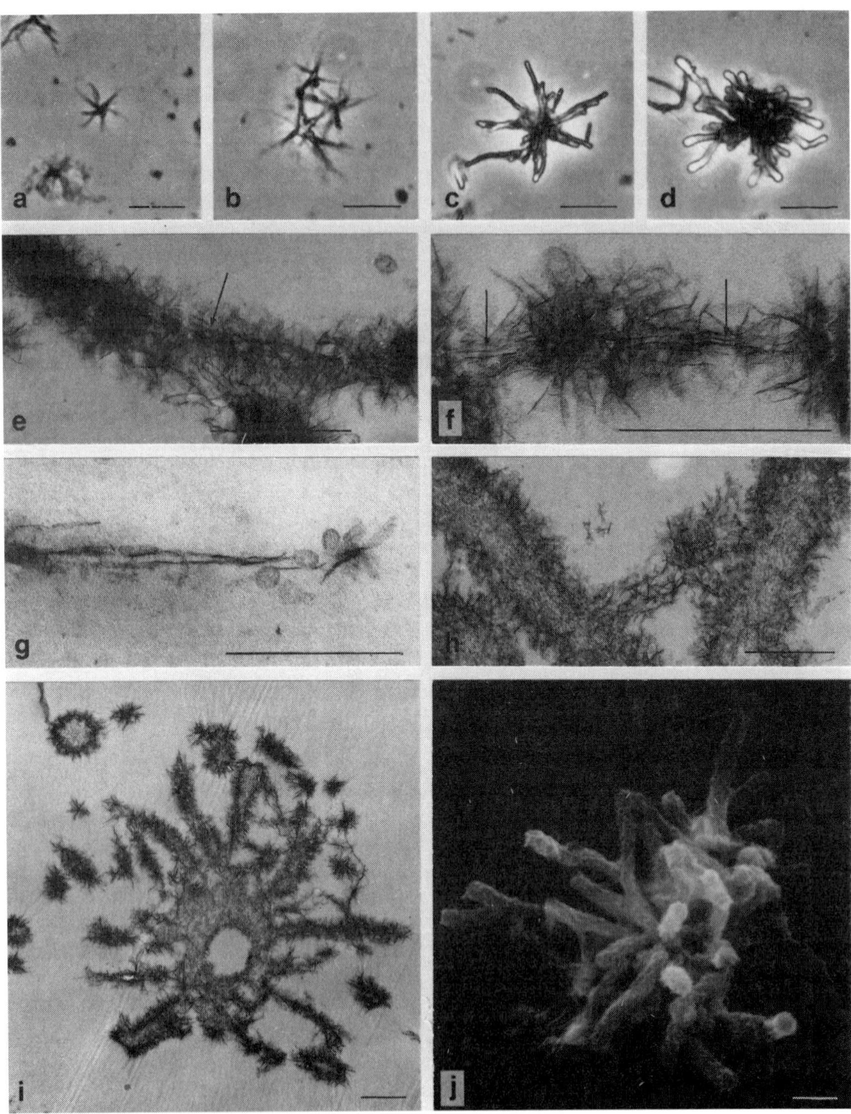

Figure 1. *Metallogenium personatum* Perfilev and Gabe, from Lake Nordbytjern in Norway. a-d) Light micrographs. Bar = 5 μm. Different degree of manganese oxide encrustation of *Metallogenium*. e-i) Transmission electron micrographs. Bar = 0.5 μm. e-f) Longitudinal sections of arai, showing central structures (arrows). g) Longitudinal section of the end of an arai of a small and less encrusted specimen, showing "vesicular" structures. h) Longitudinal section of arai of a heavily incrusted specimen. i) Transversal section of *Metallogenium* showing the central cavity. j) Scanning electron micrograph of *Metallogenium*. Bar = 0.5 μm. Reprinted from Klaveness (1977), Hydrobiologia 56 (1) Fig. 1 p. 27, with kind permission from Kluwer Academic Publishers.

2. Morphology and Taxonomy.

The original descriptions of both "species" were founded on morphological characters together with information about cell and/or "colony" (or "coenobium") size, tentative life-cycle information, and the presumed parasitic life-form. The characteristic morphology is easily recognised, and *"Metallogenium"* has been found in various environments spanning a wide range of environmental conditions (see Zavarzin 1989 for references).

Barghoorn and Tyler (1965) described a similar organism from the Gunflint formation, a hypersaline aqueous environment (Barghoorn 1977) formed about 1900 million years before the present and under very different atmospheric conditions (see Cloud (1965), Kasting (1993)). The authors were unaware of the extant *Metallogenium* morphotypes, and named their fossil specimens *Eoastrion simplex* Barghoorn and Tyler (1965). *Eoastrion* was considered a synonym of *Metallogenium* by Cloud (1965), and the taxonomic problem was later revisited by Barghoorn (1977) who discussed the nomenclature of these "actinomorphic, asteriform structures three-dimensionally preserved" and agreed to conspecificity with *Metallogenium. Eoastrion* is "so similar in some 14 different growth stages to the living, Cretaceous, and younger Proterozoic, budding bacterium *Metallogenium* that no grounds can be found for giving it a different name other than separation in time", said Cloud (1985). It is apparent that the fossilised *Metallogenium* occurs at several sites of different ages (e.g. Gunflint formation, Paradise Creek Formation, Duck Creek dolomite in Australia and the Seichal Cherts of Timor, see Crerar et al. (1980) for references). *Metallogenium* -like fossils occur at very high densities (7500/mm^3) in some Proterozoic deposits (Barghoorn 1977).

From studies of biocoenoses isolated from Laguna Figuerora, Baja California, Margulis et al. (1983) concluded that *"Metallogenium"* is a phenomenon of stellate colony growth and not a taxon of bacteria", since very different organisms give rise to stellate colony morphology of various sizes, from hundreds of m down to the limit of resolution of the light microscope. The "*Metallogenium* -like" colonies from this saline habitat may or may not be capable of manganese oxidation. Observations like this are certainly of great importance in interpreting fossil material - the context in which Margulis et al. made their observations. In my opinion, in the case of extant planktonic lake communities only, the lacustrine *Metallogenium* is a rather homogenous and well-defined morphotype occurring under particular sets of conditions, and possibly caused by a very specific kind of organism. Furthermore, I consider *Leptothrix echinata* Beger and *Metallogenium personatum* Perfilev and Gabe to be synonyms for one and the same "organism".

3. The Biology of Metallogenium

Following the studies of lacustrine communities using peloscopes, Perfilev and Gabe (1961, here from the transl. edition 1969) proposed four different stages in the development cycle of *Metallogenium:*

- a unicellular, motile and flagellate stage, cocci about 0.7 - 1.5 µm in diameter. A cell may give rise to thin, thread-like growths "at the limit of the resolving power of the microscope". The cell, when settled, may go through several cell divisions and give rise to a microcolony.

- the common centre of the microcolony thus gives rise to the "radial" or trichospherical stage, with "radially arranged, poorly developed, sometimes branching threads." The ends of the latter "swell into round structures, capable of becoming separate cells of the motile stage", or alternatively give rise to trichospherical colonies of the second order. "The young trichospherical microcolony (....) judging from its globular structure which consists of thin radial threads, ending sometimes with flagellate cells, and from its ability to live floating in the plankton (...) might perhaps be the reproductive and dispersal stage of the microbe (...)".

- "On further growth of the microcolony, the lobes fuse into one solid lump, in which a living organism can be recognised only with great difficulty. In this way there arises the most widespread third stage of microcolonies camouflaged by deposits of manganese and iron, and covered with round swellings. Sometimes the microcolony at this stage can grow into single thin filaments and give off buds with the secondary formation of trichospherical colonies (....)".

- "The fourth - branched zoogloeastage differs sharply from the preceding one by the thinness of the microcolonies which, on the substrate, exhibit a spreading tangle of more or less broad, flat, branching and intertwining cords, or sprawling zoogloea-like accumulations of irregular outlines".

In addition, the authors hypothesised a fifth, "quiescent" stage of larger spherical and refractile cells.

The presence of minute coccoid cells as part of a life cycle has also been implicated by Dubinina (1970) and Miyajima (1992b), from culture and field studies. Hard evidence for the different stages of a cycle, e.g. using light micrographs or electron microscopy, has been difficult to obtain. Using phase contrast light microscopy and electron microscopy of negatively stained preparations from mixed cultures of alleged "*Metallogenium*" and a fungus, Dubinina (1969) documented "polymorphic cells measuring 2.5-0.5 µm", "filamentous growth", "sprouting filament" and "formation of buds". Subsequent electron microscopy work on natural material of limnic origin (Klaveness 1977, Gregory et al. 1980, Jaquet et al. 1982, Schmidt and Overbeck 1984) failed to confirm the presence of what might have been living cells inside or in connection with the trichospheric stages of *Metallogenium*. Gregory et al. (1980) did find a low-contrast circular structure in the hollow core of *Metallogenium* collected from the lake, and what appeared to be a piece of a membrane-like structure in the central area of a *Metallogenium symbioticum* from a culture (obtained from G.A. Zavarzin). A fluorescent stain (Acridin orange) did not give results with *Metallogenium*. An important point, raised by Balashova and Dubinina (1989), is the possible interaction of manganese oxides with the osmium tetroxide used in the fixation procedure for electron microscopy, which might prevent adequate preservation of internal structures. Scanning electron microscopy of *Metallogenium* -like structures from rapid sand filters of water plants led Czekalla et al. (1985) to deduce a variant development cycle. He found different phases in the development of the trichospheric stages, starting with a single cell. In his view, a bacterial cell "attaches itself to growth material and forms, starting from the single filament stage, a number of filaments. Manganese oxidation then takes place around the cell so that a capsule is formed....". The cell leaves the encrusted capsule before division, "and the daughter cells again attach themselves" (Czekalla et al. 1985, p. 116 and 123).

The initial attempts to culture the lacustrine *Metallogenium* (=*Leptothrix echinata*) were successful, according to Beger (1935). Zavarzin (1961) isolated the second species (*M. symbioticum*) but in the presence of a sterile fungus only. Subsequent attempts to isolate either species in axenic culture have failed until recently (Balashova and Dubinina 1989). According to Dubinina (1969), Zavarzin suggested that *Metallogenium* might be related to the mycoplasmas. A mycoplasma-like organism was subsequently temporarily isolated into media for mycoplasmas (Dubinina 1969, 1970), and formed a sheath of manganic oxides when supplied with manganous salts. Experiments under various conditions (Dubinina 1984) proved that the organisms passed through 0.2 μm filters, could infect cultures of a wide range of other freshwater organisms, to influenced the growth of their "hosts" to a varying extent, and could induce oxidation of manganous salts to form the characteristic manganic oxide deposits of *Metallogenium*. The ultrastructure of *Metallogenium* grown in axenic culture was similar to that of mycoplasmas. Accordingly, the Russian scientists regard the organism as a free-living mycoplasma-like organism (see Dubinina (1969, 1984), Balashova (1974) and Balashova and Dubinina (1989)). The free-living mycoplasma *Acholeplasma laidlawii* (Sabin) Edward and Freund was also able to form a sheath of ferric oxides when tested for "biomineral" formation (biologically mediated mineral formation) by exposing it to ferrous salts (Balashova 1990, see also Bauminger et al. 1980). Alleged biomineral formation by an organism devoid of cell wall structures or extracellular polysaccharides require another explanation than conventionally invoked: binding of Mn^{++}-ions to extracellular polyanionic polymers followed by oxidation by factors present at the cell surface or in the polymer matrix (e.g. Ghiorse and Hirsch 1979). The factor responsible for oxidation of manganese has been shown to be a 110-kd protein in *Leptothrix discophora* (Schwers) Dorff (Boogerd and de Vrind 1987, Ghiorse 1987). Oxidation via metabolically formed peroxides has been suggested earlier (cf. Dubinina 1978, Zavarzin 1989).

By growing naturally occurring microorganisms from Lake Biwa in a semisolid medium, Miyajima (1992 c) established a mixed culture in an oxygen gradient, "a model of the redox boundary of lakes". In semi-solid media, different kinds of manganese-oxidizing bacteria made structures similar to the lacustrine *Metallogenium* from Lake Biwa. Miyajima did not succeed in isolating organisms parasitizing the manganese-oxidizing bacterial species, but could not exclude the possibility of that manganese-oxidizing mycoplasma-like organisms were present as proposed by the Russian scientists (loc. cit.).

In spite of the success of Beger (1935) in growing *Leptothrix echinata* some of us have tried and failed to produce transferable cultures, or living organisms, that are unequivocally identical to the *in situ Metallogenium* (Klaveness (unpublished), Gregory et al. 1980, Margulis et al. 1983, but see Miyajima 1992c).

The systematic positioning of *Metallogenium* has thus been the subject of a long-lasting debate. A phylogenetically founded systematics reflects significant biological knowledge by focussing on diacritical characters, those that create the synapomorphies in semic traits and distinguish the particular taxon from all others. In the case of *Metallogenium*, the morphological characters, the culture conditions, the fragmentary knowledge of a "developmental cycle" as reconstructed from natural material or observed in culture, and finally the ultrastructural features as presented by Balashova and Dubinina (1989), may suggest that the *Metallogenium* they are studying in their cultures is closely related to the mycoplasmas. But is this *Metallogenium* identical to the one regularly found in lakes or to extant or extinct *Metallogenium* -like structures found elsewhere ?

4. The Ecology of Metallogenium

The early survey and review by this author (Klaveness 1977) showed quite clearly that the distribution of the stage of *Metallogenium* that could be recognised in the plankton (the "trichospherical" stage) was limited to lakes with certain common properties. The typical *Metallogenium* lakes in Norway have medium to high concentrations of calcium, the Secchi disc transparency is 2-5 m and the colour (against the Secchi disc at ½ Secchi disc transparency depth) is mainly "yellowish green". The pH is generally above 6.50. Observations elsewhere (see Klaveness 1977, Giovanoli et al. 1981, Jaquet et al. 1982, Gorlenko et al. 1983, Stabel and Kleiner 1983, Schmidt and Overbeck 1984, Tipping et al. 1985, Eloranta 1987, Maki et al. 1987, Miyajima 1992a, Hupfer et al. 1995) support the idea that *Metallogenium* -like structures are a normal component of the plankton of lakes, most frequent near dysoxic gradients in the hypolimnion of small lakes or in the profundal zone of large lakes. However, it is still an open question whether *Metallogenium* -like structures from lakes of other trophic states, or differently influenced by humic matter, or with neutral or acid water (Walsh and Mitchell 1972) belong to the same "species". We do not have an explanation for the paucity of the lacustrine *Metallogenium* morph in recent dysoxic saline waters (Klaveness unpubl.) compared with its abundance in the fossil record (e.g. Barghoorn 1977).

The geochemical significance of biologically mediated manganese oxidation in various environments is undisputed (e.g. Dubinina 1973, Giovanoli et al. 1981, Diem and Stumm 1984, Hongve 1997). In addition to oxidation of manganese, absorption and chemical interactions with other ions, complexes and compounds take place. The interaction of biologically mediated manganese oxidation with the phosphate cycle in lakes has been suggested (Klaveness 1977, Hupfer et al. 1995, Hongve 1997). *Metallogenium* -like particles, or organisms, play a major role in these processes. The revelation of their true nature is therefore of more than academic interest.

5. Discussion

The *Metallogenium* enigma and its solution pose challenges within microscopy, molecular biology and palaeontology. There are currently no cultures of *Metallogenium personatum* and no new isolates of *Metallogenium symbioticum* except that of Zavarzin (see Zavarzin 1989). The isolation of new strains from natural material should therefore be given priority, particularly since the lacustrine *Metallogenium* appears to have a wide distribution. Today, methods exist to verify the identity of naturally occurring micro-organisms and those of allegedly identical species in cultures. This may permit us to answer questions such as the following:

- does the trichospheric stage form extracellularly in an organism not otherwise known to produce polysaccharides or cell wall structures otherwise (mycoplasmas), or in extracellular structures formed by "normal" bacteria?

- are the living part of the organism in the trichospheric stage inside the structure, at the tip of the radiating "arai", or both?

- under what physico-chemical conditions does biomineral formation take place: in connection with an organic "matrix" like in *Pedomicrobium* (Ghiorse and Hirsch 1979), and does the mineral of *Metallogenium* form as a result of oxidation by metabolically formed substances like proteins (cf. *Leptothrix*, Ghiorse 1986, Boogerd and de Vrind 1987) or peroxides (cf. Dubinina 1978, Zavarzin 1989, Miyajima 1992c)?

The emergence of molecular techniques and probes for detecting organisms and processes in situ may provide the solution to more than 35 years of bewildering search for "the true nature of *Metallogenium*".

6. References

Balashova, V.V. (1974) Mycoplasms and Iron Bacteria. Nauka, Moscow. 64 pp. (in Russ., cited from Zavarzin 1989).

Balashova, V.V. (1990) Mikrobiologiya 59, 418-424, transl. ed. 278-283.

Balashova, V.V. and Dubinina, G.A. (1989) Mikrobiologiya 58, 841-846, transl. ed. 681-685.

Barghoorn, E. (1977) In: C. Ponnamperuma (ed.), Chemical Evolution of the Early Precambrian, Academic Press, pp. 185-186.

Barghoorn, E.S. and Tyler, S.A. (1965) Science 147, 563-577.

Bauminger, E.R., Cohen, S.G., Labenski de Kanter, F., Levy, A., Ofer, S., Kessel, M.and Rottem, S. (1980) J. Bacteriol. 141, 378-381.

Beger, H. (1935) Zentralblatt Bakteriol., Abt. II, 92, 401-406.

Boogerd, F.C. and de Vrind, J.P.M. (1987) J. Bacteriol. 169, 489-494.

Cloud, P.E. Jr. (1965) Science 148, 27-35.

Cloud, P. (1985) In Drake, E.T. and Jordan, W.M. (eds.), Geologists and Ideas: A History of North American Geology, Geological Society of America, Centennial Special Volume 1, 151-156.

Crerar, D.A., Fischer, A.G. and Plaza, C.L. (1980) In: Varentsov, I.M. and Grasselly, Gy. (eds.): Manganese on the bottom of recent basins. Geology and Geochemistry of Manganese 3, 285-303. Proc. of the 2nd International Symposium on the Geology and Geochemistry of Manganese, E. Schweizerbart'sche Verlagsbuchh and lung (Naegele u. Obermiller), Stuttgart. Akad. Kiado, Budapest.

Czekalla, C., Mavius, W. and Hanert, H. (1985) Water Supply (Berlin) 3, 111-123.

Diem, D. and Stump, W. (1984) Geochim. Cosmochim. Acta 48, 1571-1573.

Dubinina, G.A. (1969) Doklady Akademii NAUK SSSR 184, 1433-1436, transl. ed. 87-90.

Dubinina, G.A. (1970) Z. allg. Mikrobiol. 10, 309-320.

Dubinina, G.A. (1973) Verh. Internat. Verein. Limnol. 18, 1261-1276.

Dubinina, G.A. (1978) Mikrobiologiya 47, 591-599, transl. ed. 471-478.

Dubinina, G.A. (1984) Curr. Microbiol. 11, 349-356.

Eloranta, P. (1987) Acta Academia Aboensis 47(2), 117-124.

Ghiorse, W.C. (1984) Ann. Rev. Microbiol. 38, 515-550.

Ghiorse, W.C. (1986) Biotechnol. Bioeng. Symp. 16, 141-148.

Ghiorse, W.C. and Hirsch, P. (1979) Arch. Microbiol. 123, 213-226.

Giovanoli, R., Br tsch, R., Diem, D., Osman-Sigg, G. and Sigg, L. (1981) Schw. Z. Hydrol. 42, 89-100.

Gorlenko, V.M., Dubinina, G.A. and Kuznetsov, S.I. (1983) Die Binnengew sser 28, IX + 1-252.

Gregory, E., Perry, R.S. and Staley, J.T. (1980) Microb. Ecol. 6, 125-140.

Hongve, D. (1997) Limnol. Oceanogr. 42, 635-647.

Hupfer, M., G chter, R. and Giovanoli, R. (1995) Aquatic Sciences 57, 305-324.

Jaquet, J-M., Nembrini, G., Garcia, J. and Vernet, J-P. (1982) Hydrobiologia 91, 323-340.

Kasting, J.F. (1993) Science 259, 920-926.
Klaveness, D. (1977) Hydrobiologia 56, 25-33.
Maki, J.S., Tebo, B.M., Palmer, F.E., Nealson, K.H. and Scaley, J.T. (1987) FEMS Microbiol. Ecol. 45, 21-29.
Margulis, L., Grozovsky, D.B.D., Stoltz, J.F., Gong-Collins, E.J., Lenk, S., Read, D. and Lopes-Cortes, A. (1983) Precambrian Res. 20, 443-477.
Miyajima, T. (1992a) Arch. Hydrobiol. 124, 317-335.
Miyajima, T. (1992b) Arch. Hydrobiol. 124, 411-426.
Miyajima, T. (1992c) Arch. of Microbiology 158, 100-106.
Perfil'ev, B.V. and Gabe, D.R. (1961) Capillary methods of investigating micro-organisms. Engl. transl. 1969, Oliver and Boyd, Edinburgh, 627 pp.
Ruttner, F. (1938) Arch. Hydrobiol. 32, 167-319.
Schmidt, W.D. and Overbeck, J. (1984) Zeitschr. f. allg. Mikrobiol. 24, 329-339.
Stabel, H-H. and Kleiner, J. (1983) Arch. Hydrobiol. 98, 307-316.
Tipping, E., Jones, J.G. and Woof, C. (1985) Arch. Hydrobiol. 105, 161-175.
Walsh, F. and Mitchell, R. (1972) J. gen. Microbiol. 72, 369-376.
Zavarzin, G.A. (1961) Mikrobiologiya 30, 393-395.
Zavarzin, G.A. (1964) Z. allg. Mikrobiol. 4, 390-395.
Zavarzin, G.A. (1989) Genus "Metallogenium" Perfil'ev and Gabe 1961, 50, in J.T. Scaley, M.P. Bryant, N. Pfennig and J.G. Holt (eds), Bergey's Manual of Systematic Bacteriology, Vol. 3, Williams and Wilkins, pp. 1986-1989.

Biodata of **Lynn J. Rothschild** author of ***"Microbes and Radiation."***

Dr. Lynn J. Rothschild received her Ph.D. in 1985 from Brown University. She spent her postdoctoral periods in Brown University and in NASA at Ames Research Center (California). Dr. Rothschild's current research focuses on the evolution and physiological ecology of protists (mainly algae), carbon metabolism and DNA synthesis patterns of microbial mats. Her study on the effect of UV radiation, has led to the first measurements of a form of UV damage repair in nature. Professional services include election as secretary to the International Society for Evolutionary Protistology and Chair of the nomination committee for the Society of Protozoologists, appointment to the Grants and Fellowship Committee of the Phycological society of America.
E-mail: **lrothschild@mail.arc.nasa.gov**

MICROBES AND RADIATION

L.J. ROTHSCHILD
Ecosystem Science and Technology
Mail Stop 239-12
NASA Ames Research Center
Moffett Field, CA 94035-1000 U.S.A.

1. Introduction

For life to exist, there must be an input of energy. Radiation, the movement of energy through space, is the critical source of energy for life because visible radiation in the 400 to 700 nm range (photosynthetically active radiation; PAR) powers photosynthesis. Other forms of radiation, for example UV and ionizing radiation, have effects on organisms that can range from beneficial (e.g., vitamin D synthesis) to lethal. Some forms of radiation have been a constant for organisms over geologic time, whereas others vary seasonally and diurnally. Exposure to radiation may depend on ecology, with, for example, organisms living near granite receiving more terrestrial radiation than those that do not. Man-made radiation can result in novel exposures or exposure levels. Some forms of radiation are blocked by the Earth's atmosphere and thus are newly relevant with respect to interplanetary travel or to potential extraterrestrial biota.

Much of the interest in radiation biology has focused on the effect of radiation on humans, with an emphasis on effects that are not relevant to microbiologists (e.g., central nervous system effects and cancer.) Yet, the effect of radiation on microbes is important from many perspectives. It has been known since 1877 that UV radiation can be lethal to bacteria, and since 1902 that it can be lethal to protozoa. The purpose of this chapter is to give a broad overview of what is known about radiation and microbes, bearing in mind that the effects vary among species. Mechanisms of radiation effects are presented, and the importance of these effects, from evolutionary biology to human health, and from commercial enterprises to exploring the space are discussed.

2. Radiation.

2.1. TYPES OF RADIATION

Radiation is energy in transit, either as particles or electromagnetic waves. Particulate forms of radiation include neutrons, electrons, protons, alpha particles (nuclei of helium atoms), and heavy ions which are the nuclei of any atom stripped of its planetary electrons and moving at high speed. Electromagnetic radiation includes gamma rays, x rays, ultraviolet

(UV) radiation, visible light, infrared radiation, microwaves and radiowaves. Radiation is either non-ionizing or ionizing. Non-ionizing radiation is radiation without sufficient energy to remove tightly-bound electrons from their orbits around atoms. Examples are microwaves and visible light. Ionizing radiation has sufficient energy to dislodge electrons from an atom, causing the atom to become ionized (charged). Ionizing radiation includes x and gamma radiation, electrons (beta radiation), neutrons, alpha particles and heavier charged atomic nuclei.

Cosmic radiation consists of very energetic particles that bombard the Earth from outer space. They are more intense at higher altitudes than at sea level where the Earth's atmosphere is most dense and gives the greatest protection. One form of cosmic radiation, a gamma ray burst, is among the most energetic events in the universe. In September 1997 NASA's Hubble Space Telescope announced observations from the Imaging Spectrograph suggesting that these explosions occur far away in other galaxies, possibly the result of the collision of neutron stars. Neutrons also come from outer space and from collisions with atoms in the atmosphere.

Table 1. Types of Radiation

	Radiation Type	Source	Wavelength
non-ionizing	ultraviolet radiation (UVR)	sun; divided into UVA (320-400 nm) UVB (280 - 320 nm) and UVC (1 -280 nm)	1-400 nm
	visible light	sun; photosynthetically active radiation (PAR) is defined as 400 to 700 nm	400-750 nm
	near infrared radiation	sun	750 nm - 2.5 μm
	infrared radiation	sun	2.5 - 25 μm
	microwaves		25 μm - 1 mm
	radio waves		> 1 mm
ionization	cosmic		
	gamma	electromagnetic waves or photos emitted by the nucleus of an atom	10^{-10}-10^{-12}m
	x-rays	electromagnetic waves of photons emitted by energy changes in electrons	
	alpha	helium nuclei (2 protons + 2 neutrons)	
	beta radiation	fast-moving electrons ejected from the nuclei of atoms	
	neutrons	neutral particles from nucleus of atom; ionize indirectly by colliding with atomic nuclei	
	protons	positive charged particles from nucleus of atom	
	heavier atomic nuclei	sun; outer galaxy	

2.2. MEASUREMENT OF RADIATION

Radiation can be measured using physical means (e.g., spectroradiometers, film detectors, nuclear emulsion, plastic nuclear track detectors) or by biological dosimetry (e.g., bacterial spores, uracil dosimeters). Each method has advantages and disadvantages, and the choice of method depends on the subsequent use of the data. For UV, visible and infrared radiation, spectrophotometry yields data that is of high resolution and easily comparable to existing data banks. Unfortunately, instrumentation is often expensive, large, requires power, and is not suitable for use in the field. Some equipment, such as broad band radiometers that measure photosynthetically active radiation, circumvent these problems but provide little spectral information. In contrast, biological dosimeters are often small, inexpensive to mass produce, and designed for field use both on Earth and in space. One popular biological dosimeter for UV radiation measures the survival of dried bacterial spores during exposure (e.g., Horneck et al., 1996; Horneck, 1997). The dosimeters are exposed, returned to the lab, and analyzed for spore inactivation. The disadvantages to this method include laborious and possibly expensive sample processing, difficulties in cross-calibration, and lack of spectral data.

The SI units for measuring radiation are the gray (Gy) and the sievert (Sv). A gray is the amount of radiation, or 'dose' received in terms of energy. Radiation does not necessarily result in the same biological effects in different organisms or tissues. Thus, the sievert is used to describe radiation effects, with one sievert of radiation producing a constant biological effect regardless of the type of radiation. The capacity of ionizing radiation to ionize is expressed as the linear energy transfer (LET).

3. Effects of Radiation on Microbes

3.1. MOLECULAR

One of the best-known consequences of radiation exposure is DNA damage. Ionizing radiation can cause direct damage by creating single or double strand breaks in DNA, but this is rare in nature. Indirect damage is more important, and is caused by free radicals, which also produce single and double strand breaks. UV-induced damage can result in the hydroxylation of cytosine and uracil, cytosine-thymine adducts, cross linking between DNA and protein, interstrand cross linking of DNA, and chain breakage and denaturation of DNA. The major UV-induced lesion is the production of photoproducts (cyclobutane pyrimidine dimers and the pyrimidine-pyrimidone (6-4) photoproduct) between adjacent bases (Harm, 1980). Repair of these lesions is discussed below (section 3.3).

The production of photoproducts has been assessed in laboratory studies and in natural communities. Jeffrey et al. (1996) quantified UVB-induced photoproducts in mixed natural communities of marine bacteria (<0.8 μM size) and eukaryotic phytoplankton (>0.8 μM) in the northern Gulf of Mexico. DNA damage was greatest near the surface on calm days, but was not detected in moderate seas even at the surface. Dimer induction per megabase of DNA was higher in the bacteria than in the eukaryotic plankton. In both cases dimer production increased starting at noon, increased to late afternoon (1500-1800 h), and then decreased to minimum levels after midnight. Karentz et al. (1991) showed a species-specific sensitivity of Antarctic microalgae to UV radiation.

3.2 BIOCHEMICAL AND PHYSIOLOGICAL

Radiation, except in the visible range, usually inhibits metabolism. For example, UV radiation has been shown to inhibit inorganic nitrogen assimilation in diatoms (Döhler et al., 1987), nitrogenase activity and heterocyst formation in cyanobacteria (Sinha et al., 1996). Well studied is the effect of UV radiation on photosynthesis where both UVA and UVB have harmful effects. The primary site of UVB inhibition of photosynthesis is photosystem II, specifically the D1/D2 polypeptide matrix (Renger et al., 1989). The carbon fixation enzyme, ribulose-1,5-bisphosphate carboxylase, is also damaged by UV radiation (Neale et al., 1993). Inhibition of photosynthesis is a balance between damage to the photosynthetic apparatus and repair of the damage. In nature, inhibition of photosynthesis depends on the time of day, which takes into account current UV fluxes, prior history, and metabolic state of the organism (Cockell and Rothschild, in press.)

3.3. AVOIDANCE AND REPAIR OF RADIATION DAMAGE

There are several classes of mechanisms that organisms use to avoid radiation-induced damage. One is to migrate away from the radiation, such as vertical migration within microbial mats (Garcia-Pichel and Castenholz, 1994). Another is the prevention of damage by attenuating UV before it reaches the interior of the cell. Water attenuates UV (Häder and Figueroa, 1997), but clear waters may transmit biologically significant amount of UV radiation to tens of meters depth. Gleason and Wellington (1993) reported signs of UV-induced bleaching in corals 12 m from the surface. Attenuation is enhanced if the water contains UV-absorbing materials such as iron (Olson and Pierson, 1986; Pierson and Olson, 1989), humic substances (Vincent and Roy, 1993), or nitrogenous salts (Rambler and Margulis, 1976). Organisms can live under a UV-absorbing medium such as an evaporite (Rothschild et al., 1994), $CaCO_3$, silica or sand and still obtain enough radiation to photosynthesize (Pierson, 1994; Rothschild, 1995; Garcia-Pichel and Bebout, 1996). Dead surface cells in microbial mats protect the lower layers (Margulis et al., 1976; Pierson, 1994). UV-absorbing pigments may provide protection, and could be an ancient UV-screening strategy (e.g., Proteau et al., 1993). Such pigments include scytonemin (e.g., Garcia-Pichel and Castenholz, 1991; Garcia-Pichel et al., 1992; Hill et al., 1994), mycosporine amino acid-like compounds (e.g., Garcia-Pichel and Castenholz, 1993), phycoerythrin (Tyagi et al., 1992) and carotenoids. Anti-oxidants neutralize the effects of radicals produced by UV reactions (Sies, 1985).

Alternatively, damage to DNA may be repaired. Repair is extremely important for survival. Escherichia coli and yeast strains deficient in repair are killed within minutes of exposure to sunlight (Nassim and James, 1978). The two major repair mechanisms for photoproducts are excision repair and photoreactivation. In nucleotide excision repair, the lesion and flanking nucleotides are removed from the damaged strand and the DNA is resynthesized using the undamaged strand as a template (Sancar, 1994). Excision repair is thought to make the greatest contribution to radiation resistance in microorganisms (Nasim and James, 1978). Photoreactivation is a light-dependent (UVA and/or short-wave visible to 480 nm) repair mechanism by which the pyrimidine dimers are monomerized by DNA photolyase, an enzyme found in most taxa but not in humans (Yasui et al., 1994). The 6-4 photoproducts do not seem to be repaired by photoreactivation. In postreplication (recombinational) repair, the undamaged double-stranded DNA formed after DNA replication is used for homologous recombination. The lesion is then removed by nucleotide excision repair. Microbes are known to repair DNA damage with all these mechanisms. Recently dark (excision) repair was demonstrated in two species of Halobacterium and Haloferax volcanii (McCready, 1996) and in mycoplasmas (Peterson et al., 1991). Bacteria subjected to potentially lethal conditions such as high UV fluxes may use error prone repair (Cooper, 1997). Since the lesion is used as a template to synthesize new DNA, this mechanism is inaccurate and often leads to mutations. There are protection and repair mechanisms specific to bacterial spores (rev. in Setlow, 1992).

3.4. SPECIFIC EXAMPLES OF MICROBIAL RESISTANCE

3.4.1. The Case of Deinococcus radiodurans

Deinococcus radiodurans is the most DNA-damage-resistant organism known (Battista, 1997). It is particularly well-studied with respect to its extraordinary ability to withstand ionizing radiation, probably by the efficient repair of the DNA damage. D. radiodurans also is extremely resistant to UV radiation, surviving doses as high as 1000J/m2. A shoulder of resistance to UV extends to over 500 J/m2. It is estimated that at this dose an average of one dimer forms for every 600 base pairs (Battista, 1997). The Deinococcaceae are non-spore-forming bacteria, but have the ability to survive prolonged desiccation. The selective pressure for extreme resistance to ionizing radiation is difficult to imagine since natural levels on the Earth are insignificant. Mattimore and Battista (1996) suggested that the ability of D. radiodurans to survive exposure to as high as 20 kGy of gamma radiation is a byproduct of an evolutionary adaptation to survive prolonged desiccation, a process that also results in DNA damage.

3.4.2. UV-Induced DNA Damage/Repair in Cyanobacteria

Cyanobacteria have several prevention and repair mechanisms for UV-induced mutations (see Herdman, 1984 for a review of early work). Strains of Anabaena photoreactivate UV-induced DNA damage efficiently, and are more resistant to UV damage than Synechocystis sp. strain PCC 6308 (Levine and Thiel, 1987). In Gloeocapsa, UV radiation is a very effective mutagen, increasing the spontaneous mutation frequency 4000-fold (Lambert et al., 1980). Most of the mutagenic effect is eliminated if photoreactivation is permitted, although cells are sensitive to mutation in the presence of caffeine, an inhibitor of excision repair (Lambert et al., 1980). Later work demonstrated that Gloeocapsa alpicola (Synechocystis PCC 6308) could perform excision repair (O'Brien and Houghton, 1982). Excision repair is extremely important for differentiation of vegetative cells into heterocysts in Anabaena aequalis (Blakefield and Harris, 1994).

3.4.3. Radiation Effects on Protists

As in other organisms, radiation effects on protists are species specific. The best general reviews of the effect of radiation on protozoa are Giese (1967), and Nassim and James (1978). Protozoa are constantly exposed to low levels of radioactive substances and cosmic rays, but natural exposures do not seem to have an effect (Giese, 1967). Similarly, radio waves and infrared radiation in themselves are not harmful to protozoa if the medium is kept from overheating (Giese, 1967).

Most work has focused on the effect of UV and visible radiation. Visible light is often, but not always, innocuous in non-photosynthetic forms, but is used as a cue in phototaxis and circadian rhythms. The former implies that visible light is used as a cue in distribution. Early work showed a negative phototaxis of ciliates, amoebae and flagellates in response to UV, and thus Giese (1967) speculated that the accumulation of protozoa under floating leaves, stones, pebbles etc. is in response to UV radiation. UV decreases growth rates, affects metabolism, decreases photosynthetic activity, impairs photo- and graviorientation in the water column, damages DNA and alters mobility. In protozoa, UV radiation can lower the lethal temperature, sensitize protozoa to heat, alter motility and shape, delay division and excystment of irradiated cysts, retard onset of mating reactions, change contractile vacuole activity, reduce respiration, delay regeneration of cut cells and change resistance to hydrostatic pressure. As in other microbes, desiccation during encystment or an increase in external osmotic pressure increases resistance to the effects of UV radiation.

The effect of radiation on motility has been studied in a variety of protists. Extremely low-frequency electromagnetic fields increase swimming velocity and directional change in several ciliates, a response that is mediated by changes in ion transport through the cell membrane (Hemmersbach et al., 1997). In contrast, UV radiation decreases swimming velocity in Blepharisma japonicum and Fabrea salina (Sgarbossa et al., 1995; Martini et al., 1997).

4. Commercial Applications

4.1. FOOD SAFETY

Approximately 25% of the global food supply is lost to spoilage, insects and rodents. Studies by the United States Center for Disease Control in early 1990's showed that even in the U.S. foodborne disease caused by pathogenic bacteria (e.g., Salmonella, Campylobacter) and by parasites (e.g., Trichinae) are responsible for 9000 deaths and cause 24-81 million cases of diarrhea annually. Ionizing radiation can damage or kill the bacteria, molds, parasites and insects that contaminate food in a very predictable and controlled way. Additionally, food can be irradiated after it is packaged and sealed, thus preventing re-contamination. The common method of food irradiation is to use a 60cobalt- or ^{137}cesium-based gamma ray irradiators. The irradiation process does not significantly increase the temperature because the distance between the source and the product being irradiated is significant or change the physical or sensory characteristics of most foods (exceptions include dairy products). Thus, irradiation is used to preserve or even sterilize food.

More than 40 countries have approved the use of irradiation on foods ranging from pork sausages (Thailand) to frog legs (France). The largest marketers of irradiated food are Belgium and France (~10,000 tons of food per year per country) and the Netherlands (~20,000 tons per year). Food irradiation has been endorsed by many organizations including the United Nations World Health Organization and the United States Food and Drug Administration. Yet, irradiation is not commonly used because of expense and consumer concern. Concerns that irradiated food actually becomes radioactive are not based on fact because the energies from the radiation sources used are too low to induce photoactivation. Worries about reduction in nutritional quality and production of hazardous side-products have also been shown to be unfounded. In the United States, approval was given in 1985 to irradiate pork to control trichinosis, and in 1990 to control Salmonella and other bacteria in fresh and frozen uncooked poultry. Poultry irradiation began commercially in 1993.

Astronauts have eaten irradiated food ever since the Apollo 17 moon shot in 1972, when they carried sandwiches made from irradiated ham, cheese, and bread. Space shuttle crews eat irradiated food, and it will undoubtedly continue to be used in the space program. Some hospitals and nursing homes serve irradiated chicken to people with weakened immune systems, including AIDS patients, burn victims, people undergoing chemotherapy, and patients who have just had a bone marrow or organ transplant.

4.2 OTHER COMMERCIAL USES

Irradiation is used to sterilize medical products such as surgical gloves, sutures, syringes, intravenous lines, tampons and contact lenses. It is also used to destroy bacteria in cosmetics.

Microbes are of interest in the field of nuclear waste disposal. For example the microbial program developed by Atomic Energy of Canada Limited focuses on several areas such as the survival of bacteria under relevant radiation and desiccation conditions; growth and mobility of microbes in compacted clay buffer materials and the potential consequences for container corrosion and microbial gas production; the presence and activity of microbes in deep granitic groundwaters; and the effects of biofilms on radionuclide migration in the geosphere (Stroes-Gascoyne and West, 1996).

5. Influence of Radiation on Evolution and Ecology of Microbes

5.1. EVOLUTIONARY HISTORY

Biological evolution consists of the generation of heritable variability and the differential survival of the resulting offspring (Darwin, 1859). Radiation has a dual role in evolution as both a potent mutagen and a selective agent. The role of UV radiation in evolution may include the evolution of DNA base composition (Boulikas, 1992) and sequence (Lesk, 1973), the ecology, biochemistry and evolution of the first life forms (e.g., Sagan, 1973) and the pattern of early evolution (e.g., Margulis et al., 1976).

UV radiation was probably an important factor in the early evolution of life, because life arose prior to the accumulation of UV-absorbing atmospheric ozone. Models suggest that pO2 was a relatively steady 10^{-13} bars from 3.5 to 2.5 Ga, then increased rapidly to $\sim 10^4$ bars by 2.4 Ga, followed by a relatively steady increase to the present atmospheric level at sea level of 0.212 bar (0.209 atm) by the beginning of the Cambrian, or possibly much earlier (e.g., Kasting, 1987; Holland et al., 1989; Kasting, 1993). We know early organisms were exposed to UV radiation because obligately and facultatively anaerobic bacteria show intrinsic resistance to UV radiation and repair DNA damage by photoreactivation (Rambler and Margulis, 1980; Pierson et al., 1993). Several mechanisms could have allowed an

Archean biosphere to thrive in the presence of UV. Overlying materials in the water column could have provided protection. Such substances include ferric iron, calcium carbonate and white sand which attenuate well in the UV region, but transmit significant amounts of radiation in the visible and near-infrared (400-1000 nm), wavelengths that are used in bacterial and oxygenic photosynthesis (Pierson, 1994). Photochemically-produced ferric iron may have been abundant in early Archean sediments, and because it absorbs strongly from 220-270 nm, ferric iron would have shielded organisms within millimeters below the water-sediment interface from surface UV (Olson and Pierson, 1986; Pierson et al., 1993, Pierson, 1994). Nitrogenous salts attenuate UV (Margulis et al;, 1976). If Earth's day length was significantly shorter during the Archean, perhaps only 15 h at 3.5 billion years (Walker et al., 1983), the shorter day length would have provided more frequent periods for dark damage-repair (Rambler and Margulis, 1976). Phototrophs could have formed mats to protect the lower layers of cells (Margulis et al., 1976) or used sunscreen pigments.

Perhaps most intriguing is a link between DNA repair and genetic recombination that has resulted in co-selection for sexual recombination and damage repair from UV (Margulis et al., 1976). Indeed, sex in general including natural genetic transformation in bacteria may have evolved as a DNA repair mechanism (Michod et al., 1988; Mongold, 1992). While some workers (Redfield, 1993) question whether DNA repair is the function of natural transformation systems, further transformation studies with homologous DNA demonstrated an increase in homologous recombination and transformation in response to DNA damage (Michod and Wojciechowski, 1994).

Sagan (1973) discussed the link between early UV fluxes and the evolution of cell repair mechanisms (e.g., pyrimidine dimer ligases and photoreactivation). He suggested that because early photosynthesizers lived near the surface, they protected themselves from UV by surrounding themselves with purines and pyrimidines. These organisms would have become large (10s to 100s of μm) and eventually evolved into eukaryotes.

Because radiation is energy in motion and organisms require energy for life, arguably the most important role of radiation in evolution has been as a source of energy for photosynthesis. With the exception of some prokaryotes, photosynthetic organisms use radiation in the range of 400 to 700 nm, although there are prokaryotes that can use energy at wavelengths into the near infrared. Why this window was selected is unknown. Interesting speculation includes the idea that this window was determined by the chemistry of the ancient oceans which may have contained high concentrations of Fe^{+2} (which absorb strongly at wavelengths >900 nm) and dissolved organic compounds (which absorb in the blue and near UV) (J. Berry, pers. commun.). Additionally, below 400 nm the energy of the photons makes photochemical damage increasingly likely, and wavelengths well beyond 700 nm may not contain sufficient energy to power photosynthesis (W. Briggs, pers. commun.).

5.2. GLOBAL CHANGE

There is active research on the effects of UV radiation as the result of concerns over the depletion of stratospheric ozone resulting in locally increased fluxes of UVB radiation (e.g., Nolan and Amanatidis, 1995). The interest in microbes has focused on the effect of elevated UV radiation fluxes on phytoplankton.

UV radiation affects many metabolic processes in microbes, but the most significant for global change are effects on algae, with an emphasis on photosynthesis, nitrogen metabolism, and structural damage to the nucleic acids. Several recent papers serve as an excellent review of the field, including those of Vincent and Quesada (1994), Booth et al. (1997), Häder (1993), Häder and Figueroa (1997). The reasons for concern about algae are because of their role in the global carbon cycle and their critical role as the base of aquatic food webs. Even coral bleaching in the Caribbean may, at least in part, be the result of elevated levels of UV radiation (Gleason and Wellington, 1993).

In addition to the primary effects of UV radiation, there are a few studies on secondary effects. Bothwell et al. (1994) showed that natural solar UV inhibits growth of benthic diatom communities in shallow freshwater, but that the inhibition of growth of the algal consumers (chironomid larvae) is greater. Thus, filtering out UV resulted in fewer diatoms because of the greater increase in herbivores. Another secondary example concerns soil microbes. These organisms are the base of the below-ground food web, providing functions such as mineralization, immobilization, decomposition and improvement in plant nutrition. Soil microbes rely on photosynthetic exudate from plant roots but, because the rate of photosynthesis is affected by solar radiation, soil microbes may be indirectly affected. Klironomos and Allen (1995) examined this possibility by exposing sugar maple (Acer saccharum) seedlings to the UV. The plants showed no visual responses to the UV treatments, but the microbial population associated with the rhizosphere was affected. Mycorrhizal fungi changed in morphology, and as a result, and probably did not acquire as much inorganic nutrients from the soil. Bacteria and non-mycorrhizal fungi increased in number significantly. In sum, the community shifted from a mutualistic-closed, mycorrhizal-dominated system to an opportunist-open, saprobe/pathogen dominated system.

6. Radiation, Microbes and Space

Once in space, microbes can be exposed to different radiation regimes than those on Earth. The amount and type of radiation depend on location, time (e.g., solar flares are sporadic events) and whether the location of interest is free space or inside a spacecraft. An older, but very comprehensive review, of microbes and space is given in Tobias and Todd (1974). Included are discussions of the effect of solar electromagnetic radiation, magnetic fields and heavy particle irradiation. Of particular interest to space radiation biology is the interaction of the effect of gravity and radiation, resulting in different radiation effects than may be experienced on Earth's surface.

6.1. INTERPLANETARY TRAVEL

Once off the surface of the Earth, the influence of radiation from the Earth's minerals vanishes, and exposure to cosmic radiation increases rapidly. Cosmic radiation comes from galactic radiation which itself comes from outside the solar system, from charged particles trapped in layers circling the Earth, and from the occasional bursts of radiation from the sun in termed solar particle events (principally protons and secondarily alpha particles) (Hall, 1984).

UV radiation would be a controlling factor in the type and distribution of life elsewhere, notably Mars because of its proximity to the Sun and lack of a UV-attenuating atmosphere (e.g., Rothschild, 1990). For this reason, UV effects influence the search for life on Mars, including site selection for potential life. UV radiation is relevant to assess planetary protection protocols (Nealson et al., 1992). Organisms undergoing forward transfer away from the Earth or back transfer from other bodies to the Earth are exposed to high fluxes of UV during interplanetary transfer. Research in these areas takes advantage of both ground-based facilities with elevated radiation fluxes, and flight experiments on the survival of microorganisms and biomolecules on board a satellite in Earth orbit containing the BIOPAN experiments (e.g., 1994 Committee on Space Research, COSPAR, meeting, Symposium F3.1). In one such experiment, Mancinelli et al. (in press) showed that two desiccation resistant prokaryotes were able to withstand two weeks exposure of the space environment. UV-induced DNA damage may have been the major cause of death in those organisms that did not survive.

When humans go into space, they are exposed to novel radiation regimes. Humans travel with their associated microbes, including biofilms on teeth, pathogenic and non-pathogenic microbes in the spacecraft, and microbes involved in waste recycling. Thus, the effect of radiation on microbes in the space environment is critical for human missions.

6.2. TERRAFORMING MARS

Mars may have had an early environmental history similar to that of the Earth; if not "warm and wet", perhaps "cold and wet" and suitable for life as we know it (McKay, 1997). Interest exists in creating (re-creating?) clement conditions on Mars in the future for the benefit of terrestrial life, including humans (McKay et al., 1991). One major environmental factor that precludes terrestrial life from living on the surface today is the solar UV flux (Rothschild, 1990). Mars is 1.5 times further from the sun than the Earth, so the overall incidence of solar radiation on the atmosphere of Mars is 43% that of the Earth. Martian atmospheric CO_2 protects that surface from radiation below 190 nm, but surficial fluxes rise rapidly above 200 nm (Kuhn and Atreya, 1979). Thus, considerations of terraforming Mars must include measures to shield life from such high UV fluxes, either by adding shielding mechanisms, or exploiting the physiology and ecology of terrestrial organisms that can withstand high levels of radiation (Rothschild, 1990). Shielding is the only option for establishing a human presence on Mars.

7. References

Battista, J.R. (1997) Ann. Rev. Microbial. 51: 203-24.

Blakefield, M.K. and D.O. Harris (1994) Photochem. Photobiol. 59: 204-208.

Booth, C.R., Morrow, J.H., Coohill, T.P., Cullen, J.J., Frederick, J.E., Häder, D.-P., Holm-Hansen, O., Jeffrey, W.H., Mitchell, D.L., Neale, P.J., Sobolev, I., van der Leun, J., and Worrest, R.C. (1997) Photochem. Photobiol. 65: 252-269.

Bothwell, M.L., Sherbot, D.M.J., and Pollock, C.M. (1994) Science 265: 97-100.

Boulikas, T. (1992) J. Mol. Evol. 35: 156-180.

Cockell, C. and Rothschild, L.J. J. Phycol. in press.

Cooper, G.M. (1997) The Cell. A Molecular Approach. ASM Press, Washington D.C.

Darwin, C. (1859) The Origin of Species By Means of Natural Selection or The Preservation of Favoured Races in the Struggle For Life. John Murray, London.
Döhler, G., Worrest, R.C., Biermann, I., and Zink, J. (1987) Physiol. Plantarum 70: 511-515.
Garcia-Pichel, F. and B.M. Bebout. (1996) Mar. Ecol. Progr. Ser. 131: 257-262.
Garcia-Pichel, F. And Castenholz, R.W. (1994) In: L.J. Stal and P. Caumette (eds.), Microbial Mats NATO ASI Series, Vol. G 35, Springer-Verlag, Berlin, pp.77-83.
Garcia-Pichel, F. and R.W. Castenholz (1991) J. Phycol. 27: 395-409.
Garcia-Pichel, F. and R.W. Castenholz (1993) Appl. Environ. Microbiol. 59: 163-169.
Garcia-Pichel, F., N.D. Sherry, and R.W. Castenholz (1992) Photochem. Photobiol. 56: 17-23.
Giese, A.C. (1967) In: T.-T. Chen (ed.) Protozoology, Vol. 2 Pergamon Press Inc., New York, pp. 268-356.
Gleason, D.F. and Wellington, G.M. (1993) Nature 365: 836-838.
Häder, D.-P. (1993) Progr. Phycol. Res. 9: 1-45.
Häder, D.-P. and F.L. Figueroa (1997) Photochem. Photobiol. 66: 1-14.
Hall, E. (1984) Radiation and Life, Second ed. Pergamon Press Inc., New York, 255 pp.
Harm, W. (1980) Biological Effects of Ultraviolet Radiation. IUPAB Biophysics Series I. Cambridge University Press, Cambridge, 216 pp.
Hemmersbach, R., Becker, E., and Stockem, W. (1997) Bioelectromagnetics 18: 491-498.
Herdman, M. (1982) In: N.G. Carr and B.A. Whitton (eds.) The Biology of the Cyabacteria. Botanical Monographs, vol. 19, pp. 263-305.
Hill, D.R., Peat, A., and Potts, M. (1994) Protoplasma 182: 126-148.
Holland, H.D., Feakes, C.R., and Zbinden, E.A. (1989) Am. J. Science 289: 362-389.
Horneck, G. (1997) Trends In Photochemistry & Photobiology 4: 67-78.
Horneck, G., Rettberg, P., Rabbow, E., Strauch, W., Seckmeyer, G., Facius, R., Reitz, G, Strauch, K., and Schott, J.-U. (1996) Photochem.Photobiol. B: Biology 32: 189-196.
Jeffrey, W.H., Pledger, R.J., Aas, P., Hager, S., Coffin, R.B., Von Haven, R., and Mitchell, D.L. (1996) Marine Ecol. Progr. Ser. 137: 283-291.
Karentz, D., Cleaver, J.E., and Mitchell, D.L. (1991) J. Phycol. 27: 326-341.
Kasting, J.F. (1987) Precambr. Res. 34: 205-229.
Kasting, J.F. (1993) Science 259: 920-926.
Klironomos, J.N. and Allen, M.F. (1995) Functional Ecology 9, 923-930.
Kuhn, W.R. and Atreya, S.K. (1979) J. Mol. Evol. 14: 57-64.
Lambert, J.A.M., Williams, E., O'Brien, P.A., and Houghton, J.A. (1980) J. Gen. Microbiol. 121: 213-220.
Lesk, A.M. (1973) J. Theor. Biol. 40: 201-202.
Levine, E. and Thiel, T. (1987) J. Bacteriol. 169: 3988-3993.
Mancinelli, R. L., White, M.R., and Rothschild, L.J. Adv. Space Research, In Press.
Margulis, L., Walker, J.C.G., and Rambler,M. (1976) Nature 264: 620-624.
Martini, B., Marangoni, R., Gioffré, D., and Colombetti, G. (1997) J. Photochem. Photobiol. B: Biology 39: 197-203.
Mattimore, V. and Battista, J.R. (1996) J. Bacteriol. 178: 633-637.
McCready, S. (1996) Mutation Res. 364: 25-32.
McKay, C.P. (1997) Origins Life Evol. Biosphere 27: 263-289/
McKay, C.P., Toon, O.B., and Kasting, J.F. (1991) Nature 352: 489-496.
Michod, R.E. and Wojciechowski, M.F. (1994) J. Evol. Biol. 7: 147-175.
Michod, R.E., Wojciechowski, M.F. and Hoelzer, M.A. (1988) Genetics 118: 31-40.
Mongold, J.A. (1992) Genetics 132: 893-898.
Nasim, A. and James, A.P. (1978)In: D.J. Kushner (ed.), Microbial Life in Extreme Environments. Academic Press, New York, pp. 409-439.
Neale, P.J., Cullen, J.J., Lessen, M.P., and Melis, A. (1993) in H.Y. Yamamoto and C.M. Smith, C.M. (eds.), Photosynthetic Responses to the Environment., American Society of Plant Physiologists, Rockville, Maryland, pp 61-77.
Nealson, K. et al. (1992) Biological Contamination of Mars: Issues and Recommendations. Task Group on Planetary Protection, National Academy Press, Washington, D.C.
Nolan, C.V. and Amanatidis, G.T. (1995) J. Photochem. Photobiol. B: Biol. 31: 3-7.
O'Brien, P.A. and Houghton, J.A. (1982) Photochem. Photobiol. 36: 417-422.
Olson, J.M. and Pierson, B.K. (1986) Photosynth. Res. 9: 251-260.
Peterson, S.N., Schramm, N., Hu, P.C., Bott, K.F., and Hutchison, C.A. III (1991) Nucleic Acids Res. 19:6027-.

Pierson, B.K. (1994) in Bengtson, S. (ed.), Early Life On Earth. Nobel Symposium No. 84. Columbia University Press, New York, pp. 161-180.

Pierson, B.K. and Olson, J.M. (1989) in Y. Cohen and E. Rosenberg (eds.), Microbial Mats. American Society for Microbiology, Washington, D.C. pp. 403-427.

Pierson, B.K., Mitchell, H.K., and Ruff-Roberts, A.L. (1993) Origins Of Life Evol. Biosphere 23: 243-260.

Proteau, P.J., Gerwick, W.H., Garcia-Pichel, F. and Castenholz, R. (1993) Experientia (Basel) 49: 825-829.

Rambler, M. and Margulis, L. (1976) Origins of Life 7: 73-74.

Rambler, M.B. and Margulis, L. (1980) Science 210: 638-640.

Redfield, R.J. (1993) Genetics 133: 755-761.

Renger, G., Volker, M., Eckert, H.J., Fromme, R., Hohm-Veit, S., and Graber, P. (1989) Photochem. Photobiol. 49 : 97-105.

Rothschild, L.J. (1990) Icarus 88: 246-260.

Rothschild, L.J. (1995) Adv. Space Res. 15(3): 223-228.

Rothschild, L.J., Giver, L.J., White, M.R. and Mancinelli, R.L. (1994) J. Phycol. 30: 431-438.

Sagan, C. (1973) J. Theor. Biol. 39: 195-200.

Sancar, A. (1994) Science 266: 1954-6.

Setlow, P. (1992) J. Bacteriol. 174: 2737-2741.

Sgarbossa, A., Lucia, S., Lenci, F., Gioffré, D., Ghetti, F,. and Checcucci, G. (1995) J. Photochem. Photobiol. B: Biology 27: 243-249.

Sies, H. (ed.).(1985) Oxidative Stress., Academic Press, San Diego, 507 pp.

Sinha, R.P., Singh, N., Kumar, A., Kumar, H.D., Häder, M., and Häder, D.P. (1996) J. Photochem Photobiol. 32 : 107-113.

Stroes-Gascoyne, S. and West, J. M. (1996) Can. J. Microbiol.42: 349-366.

Tobias, C.A. and Todd, P. (1974) Space Radiation Biology and Related Topics. Academic Press, New York, 648 pp.

Tyagi, R., Srinivas, G., Vyas, D., Kumar, A. and Kumar, H.D. (1992) Photochem. Photobiol. 55: 401-407.

Vincent, W.F. and Quesada, A. (1994)Ultraviolet Radiation in Antarctica: Measurements and Biological Effects, Antarctic Research Series, 62: 111-124.

Vincent, W.F. and Roy S. (1993) Environ. Rev. 1 : 1-12.

Walker, J.C.G., Klein, C., Schidlowski, M., Schopf, J.W., Stevenson, D.J, and Walter, M.R. (1983) in J.W. Schopf (ed.), Earth's Earliest Biosphere, Princeton University Press, Princeton, NJ, pp. 260-290.

Yasui, A., Eker, A.P.M., Yasuhira, S., Yajima, H., Kobayashi, T., Takao, M. and Oikawa, A. (1994) EMBO J. 13: 6142-6151.

Dr. W. Reisser - Airborne Algae.

Prof. Werner Reisser is teaching General and Applied Botany at the Institute of Leipzig University in Germany. His research interests center on soil and airborne algae, algal viruses and molecular ecology of soil ecosystems. Dr. Reisser is the editor and organizer of "**Algae and Symbioses: Plants Animals Fungi Viruses - Interactions Explored** (published by Biopress LMT, Bristol UK 1992).
E-Mail: < reisser@rz.uni-leipzig.de >

J. Seckbach (ed.), Enigmatic Microorganisms and Life in Extreme Environments, 563–572.

THE UNKNOWN LIFE OF AIRBORNE ALGAE

W. REISSER
University of Leipzig
Institute of Botany
Johannisallee 21 - 23
04103 Leipzig Germany

1. Introduction

In common textbooks the term "algae" usually designates both pro- and eukaryotic photoautotrophic organisms which live under aerobic conditions in aquatic or frequently humid habitats. They show a plethora of morphological and physiological features ranging from unicellular organisms with cell diameters of ca. 10^{-6} m to complex parenchymatous plants with length of thallus of about 10^2 m. In aquatic habitats algae form the basis of the food chain and represent the most important primary producers. Thus algae have been studied thoroughly with regard to morphology, physiology and ecology and are duly represented in standard textbooks. However, even among botanists there is surprisingly little awareness of the fact that algae can also be found in other than aquatic habitats such as in deserts and caves, on rocks and leaves, in the air and in human bones.

In comparison to our knowledge of algae of aquatic habitats, information on non-aquatic algae is rather limited. The reasons for this are manifold. One caveat might be that non-aquatic algae colonise much more diverse habitats than aquatic specimens and thus show a variety of morphological and physiological adaptations which do not make them ideal candidates for standard culture techniques in liquid media. The somewhat isolated situation of non-aquatic algae in the mainstream of phycological research is also reflected by the great variety of different and sometimes confusing names which are ascribed to them by various authors (for an overview see Ettl and Gärtner 1995). Therefore, an important first step towards a better understanding of non-aquatic algae would be for a consensus on a common nomenclature to be established. I propose to divide non-aquatic algae into a terrestrial and an aerial group. Terrestrial algae are subdivided into euterrestrial (living on and in soil), hypolithic and endolithic (living under and in rocks) specimens. Aerial algae are algae which are exposed to the air, mainly as covers of different kinds of substrata, i. e. they are not in contact with soil. According to the substrate they are growing on, such algae can be epiphytic, xylophytic, lithophytic, phycobiontic, etc.

In reports on terrestrial as well as on aerial algae, the term "airborne algae" is frequently used (e. g. Gregory et al. 1955), often as a synonym for non-aquatic specimens. Unfortunately, this term has never been defined unambiguously. Sometimes it is used in a broader sense to address algae which may have settled their habitat via airbound transportation mechanisms, while in other cases it is restricted only to those algae which can be isolated directly from the air itself. Whereas the former definition probably holds for the majority of terrestrial and aerial algae and presumably also for some part of the aquatic species, the more restricted definition applies to a group of algae on which very little information exists and which in that sense is rather enigmatic. Thus, in the following, I will try to give a critical overview of the present knowledge on those algae which can be isolated directly from air currents.

Probably the first report on algae isolated from samples of air is given by Ehrenberg in 1844, who isolated diatoms from "samples of atmospheric dust", which had been collected by Darwin "outside the Cap Verde Islands." A good survey on the history of the discovery of the aeroplankton which is also called "anemoplankton" (Geissler and Gerloff 1965) or "aerial flora" (Carson and Brown 1976) is given by Schlichting (1974b).

2. Methods and Data Base

There exist a great number of different methods for the isolation of algae directly from air. The simplest and most common method uses petri dishes and slides with different kinds of agar media which are exposed to the air. Exposure times vary from minutes to days. The location of petri dishes depends on the project in question and the imagination of the researcher. Petri dishes have been exposed in bedrooms and operating theatres, on trees and roofs, and have been attached to the windshields of cars and exposed from aeroplanes flying at different altitudes (Brown et al. 1964, McGovern et al. 1965, Schlichting 1961, 1969, 1986). Algal numbers were determined by light microscopy as impacts on the agar surface or, after some time of incubation, as colony numbers which indicate the number of viable algae. More sophisticated methods allow the quantitation of algae in an air flow by using a spore trap or by bubbling a defined amount of air through a water column (Schlichting 1974a, 1974b).

Regardless of the different methods applied, a common drawback of all data on numbers and taxonomic affiliation of airborne algae which have been obtained so far is the fact that the experimental basis was far too small to give any statistically reliable information. Authors used different nutrient media, exposure times and locations; only few studies report on exposure times longer than three days. Furthermore, the data show that numbers of airborne algae can vary tremendously, depending on the precise geographic location and to within a few meters or minutes, so that data given in the "List of airborne algae" (Chapt. 5) should not be taken as representative of airborne algae in general.

It is clear that airborne algae are truly enigmatic not only regarding the insufficient and scattered information about them, but also in view of the potential they offer for future research. In the following I shall try to give an overview of data collected on airborne algae and to outline some promising tendencies for future research.

3. Current Topics

The main questions being tackled in current research on airborne algae are:

3.1. HOW MANY AND WHICH CELLS AND TAXA ARE IN THE AIR? DO THERE EXIST AUTOCHTHONOUS SPECIES

Reports by different authors show that the species composition of airborne algae includes uni- and oligocellular specimens of all major algal taxa. Among eukaryotic algae, there seems to exist some dominance of chlorophycean species and diatoms. Cyanobacteria are nearly always present. In general, no formerly unknown taxa have been isolated from air. The existence of an autochthonous flora of airborne algae seems to be rather improbable although, it must be stressed again, the available database is very small and selective (see "List of airborne algae" and references).

The sizes of collected algae vary between 1 μm and 150 μm (Schlichting 1986), but sometimes aggregates of algae reaching 300 m in diameter and containing up to 50 cells have been observed (Mc Govern et al. 1965).

Although the reported numbers of algae in 1 m^3 air vary tremendously, it must again emphasized, that the database is far too small to draw general conclusions: Neither the type of locality, sampling time, nor the type of selective media is standardised. The majority of authors report of 0.1 - 10 algal cells in 1 m^3 air. Geissler and Gerloff (1965) counted a maximum of 15 diatom cells in 1 m^3. Gregory et al. (1955) observed up to 800 individuals of Gloeocapsa in 1 m^3 air, Brown et al. (1964) up to 3 000 algal cells in 1 m^3.

3. 2. HOW DO CLIMATIC AND GEOGRAPHIC FACTORS INFLUENCE NUMBER AND SPECIES COMPOSITION OF AIRBORNE ALGAE

There is some evidence that the number of airborne algae may depend on both climatic and geographic factors. In general, above arid areas more airborne algae are collected during dry than during moist weather (Schlichting 1974b). Driving by car through a dust cloud in Texas, Brown et al. (1964) counted up to 3 000 algal cells in 1 m^3 of air. According to Schlichting (1969) there are fewer airborne algae above the ocean than above the land and the same observation holds for areas above cities and the open landscape (Parswanath and Ramalingham 1979, Schlichting 1974b). Experiments carried out from aeroplanes flying at 3700 m demonstrated a decrease of algal counts with altitude (Brown et al. 1964). Gallagher et al. (1979) isolated about 60 000 algal cells from 1 g of house dust.

All such reports are rather enigmatic and are only useful for indicating some tendencies. Clearly the number of counted airborne algae depends on the duration of sampling time and the meteorological conditions during sampling, such as wind direction and velocity, temperature, moisture, etc. (Brown et al. 1964, Schlichting 1974b, Smith 1973). Usually after a rainfall, the number of airborne algae decreases tremendously. The influence of geographic factors should also be taken into account: Forests probably could serve as a sort of comb for airborne algae, particularly those in hilly areas. An interesting observation (Reisser, unpublished) which shows that airborne algae may also be filtered out by grassland areas was made on a grass lawn cut weekly during summertime from which about 50 000 algal cells per 1 m^2 (= 2,2 g dw of grass) were isolated from grass blades.

3.3. WHERE DO AIRBORNE ALGAE STEM FROM

Data do not support the hypothesis of an autochthonous flora of airborne algae but suggest a constant efflux from terrestrial and aquatic sources into the air. In terrestrial areas the main sources are probably algae growing on the soil surface and as covers on tree barks, rocks, fences, etc. which are carried away by wind erosion. In the upper layers of farmlands up to 10^8 algal cells per 1 g dw of soil were counted (Reisser, unpublished). The distances over which algae are carried may vary between a few meters to a continent wide and overseas distribution. However, for algae on tree barks, as Brand (1925) points out, most are firmly stuck to their substrate and have a wax-like composition of cell walls which makes them difficult to wet. He suggests that only dead cells are blown away by air currents and proposes that animals such as insects and squirrels are the main vectors of distribution of bark algae.

Another source of airborne algae are certainly aquatic algal populations. Geissler and Gerloff (1965) showed that the species composition of airborne diatoms above the city of Berlin is identical with the corresponding composition of diatoms in Berlin lakes and rivers. Interestingly, they also observed tropical species in the air. Obviously aquatic algae gain access to the atmosphere by aerosols formed by waterfalls and waves, mainly in the form of seafoam (Schlichting 1971). Marine microalgae have been detected at altitudes of 2700 m (Maynard 1968) and obviously could have been transported over long distances. Experiments by Schlichting (1974a) ascertained that air bubbled through a water column can indeed transport aquatic algae into the atmosphere. Aquatic algae have also been found among the feathers of ducks and splashing raindrops may also help to constantly supply the atmosphere with algae (Carson and Brown 1976, see also Ehresman and Hatch 1975).

In conclusion, there is convincing evidence that the pool of airborne algae is constantly fed from aquatic and terrestrial algal populations. Although members of all the major algal taxa have been observed among airborne specimens, it is a common observation that those algal taxa which are dominant among airborne algae are also dominant among algae which cover the surface of trees, fences, etc. and the upper soil layers, i.e. diatoms, chlorophyceae and cyanobacteria. Typical surface algae such as *Chlorella, Chlorococcum, Apatococcus, Anacystis, Nostoc, Gloecapsa* are also found among airborne algae which do not obviously show any autochtonous taxa.

3.4. DO AIRBORNE ALGAE SHOW SPECIAL FEATURES OF MORPHOLOGICAL AND PHYSIOLOGICAL ADAPTATION

As far as physiological features of airborne algae are concerned, hard data are not available and this topic, although fascinating, is largely a matter of speculation. There is no information available on the length of time that airborne algae usually stay in the air. Depending on local factors, this time may range from a few minutes to days or even weeks as might be the case with algae detected above the open ocean. It is also unknown what fraction of the airborne algal population is physiologically active, i. e. how many cells are already dead when dispersed into the air, how many die during their transportation or whether they are able to form special resistant cell forms. Since it is a well established observation that new habitats such as vulcano ash can be populated rapidly by airborne algae (Rayburn et al. 1982) it is conceivable that at least some algae could have established special adaptive features towards e. g. extreme temperatures, dryness, high light intensities and exposition to UV-light. Edlich (1936) showed that Pleurococcus sp. can endure -80 °C for 24 hours and that several chlorophycean species which colonise the barks of trees show a remarkable drought resistance and can still grow after exposure to UV-rays. Thus the physiology of airborne algae is a fascinating field still open to discovery.

3.5. DO AIRBORNE ALGAE PLAY A SPECIAL ECOLOGICAL ROLE AND DO THEY EXERT ANY INFLUENCE ON HUMAN HEALTH

The ecological impact of airborne algae is still a matter of speculation and indirect conclusions. Although there are no data available on the average time airborne algae stay in the air it is obvious that they must ultimately settle down somewhere. In general, one accepts that there is most probably a constant input of airborne algae into all types of aquatic and terrestrial ecosystems, whatever consequences this may have. Most algae are probably not able to grow or even survive in their new habitat but it has been shown that airborne algae play some role in the repopulating of artificial lakes and ponds, of devastated areas or of vulcano ash (Rayburn et al. 1982). Data indicate that a significant amount of airborne algae is washed out of the air through rainfall and that forests are able to comb out algae which might then be transported by rainwater elsewhere into the ecosystem. Another interesting aspect to the connection between airborne algae and weather comes from studies which suggest that airborne algae might directly interfere through the formation of condensation nuclei, or indirectly by the production of DMS with the formation of clouds (Breuer 1990, Vali 1979).

A most interesting aspect of the ecology of airborne algae is their interference with human health: There is abundant information on allergenic properties of airborne algae which probably act on the respiratory system in a similar way to pollen. Most strikingly, data indicate that algae could be a cause of more medical problems than pollen, since in contrast to pollen, algae are able to actively accumulate pollutants, such as heavy metals or radioactive dust material (Bergman et al. 1983, Bernstein and Safferman 1965, 1972, Lunceford 1968, McElhenney et al. 1965, McGovern et al. 1965, McGovern et al. 1966, Mittal et al. 1979, Schlichting 1974b, 1986, Tiberg 1989). The impact of airborne algae on human health should not be underestimated. Schlichting (1969) estimates that under normal conditions about 2900 algal cells are inhaled within 24 hours. Schellmann and Sperl (1979) could detect diatoms in the bone marrow of nondrowned bodies and Geissler and Gerloff (1965) found up to 210 diatoms per 1 cm^2 of lung tissue and in other internal organs of unborn foetuses.

4. Outlook

Taken published reports on airborne algae together there is an urgent need for further information and investigation. Insufficient data about physiological adaptations and ecological impact make them the least understood and most enigmatic group of algae.

5. List of airborne algae

CYANOBACTERIA

Anabaena helicoidea (4), A. circinalis (4), A. sp. (1, 4), Anacystis sp. (1, 4), Aphanocapsa delicatissima (4), A. sp. (4), Aphanothece castagnei (4), A. saxicola (4), Arthrospira sp. (1, 4)
Calothrix sp. (2, 4), Chroococcus minutus (4), C. turgidus (4), C. sp. (4), C. like (1)
Entophysalis sp. (2)
Fremyella sp. (1, 4)
Gloeocapsa magna (4), G. montana (4), G. sp. (1, 3, 4), Gloethece rupestris (4)
Lyngbia perelegans (4), L. vesicolor (4), L. sp. (1, 2, 4), Lyngbya like (4)
Merismopedia sp. (1, 4), Microcystis flos-aquae (4), M. sp. (4), M. like (4), Microcoleus vaginatus (4), M. sp. (1, 4), Myxosarcina sp. (1, 4)
Nostoc ellipsosporum (4), N. sp. (1, 4)
Oscillatoria subbrevis (4), O. sp. (1, 2, 4, 6)
Pelogloea bacillifora (4), Phormidium ambiguum like (4), P. foveolarum (4), P. inundatum like (4), P. luridum (4), P. minnesotense (4), P. orientale (4), P. sp. (1, 4), Plectonema carneum, (4), P. sp. (4), Pleurocapsa minor (4)
Schizothrix calcicola (4), S. sp. (1, 4), Scytonema hofmannii (4), S. sp. (1, 4), Symploca muscorum (4), S. sp. (2), Synechococcus sp. (1, 4), Synechocystis sp. (2)
Tolypothrix byssoidea (4), T. sp. (1, 4)
Xenococcus kerneri (4)

CHLOROPHYCEAE

Actinastrum sp. (4), Ankistrodesmus convolutus (4), Asterococcus superbus (4)
Borodinella sp. (1, 4), Bracteacoccus sp. (1, 4, 6)
Characium sp. (6), Chlamydomonas sp. (1, 4, 6), Chlorella ellipsoidea (4, 5), C. pyrenoidosa (4), C. vulgaris (4, 5) C. sp. (1, 2, 4, 6), Chlorococcum sp. (1, 2, 4, 5, 6), Chlorosarcina sp. (1, 4), Chlorosarcinopsis sp. (1, 4), Chlorosphaeropsis sp. (1, 4), Coccomyxa dispar (4), Coelastrum sp. (1, 4), Coleochaetae irregularis (4), Cosmarium sp. (1, 4), Cylindrocystis sp. (1, 4)
Dictyochloris sp. (1, 4)
Eudorina californica (4)
Friedmannia sp. (1, 4)
Gloeococcus schroeteri (4), Gloeocystis gigas (4)
Hormidium dissectum (4), H. flaccidum (4), H. sp. (1, 2, 4, 6), Hormotilopsis sp. (1, 4)
Microspora sp. (4), Myrmecia like (4), Monallantus sp.(2)
Nanochloris bacillaris (4), N. sp. (1, 4), Neochloris sp. (4, 6)
Oedogonium sp. (4), Oocystis sp. (1, 2, 4), Ourococcus sp. (1, 4),

Palmella sp. (1, 4), Palmellococcus protothecoides (4), P. sp. (1), Planktosphaeria sp. (1, 4), Pleurastrum sp. (1, 4) , Pleurococcus vulgaris (4), Prasiola like (4), Protococcus viridis (4), P. like (1), P. sp. (Pleurococcus) (2, 4), Protosiphon sp. (1, 4), Pseudoulvella like (1, 4)
Radiococcus sp. (1, 4), Radiosphaera sp. (4, 6), Rhizoclonium sp. (4), Roya sp. (1, 4)
Scenedesmus bijuga (4), S. denticulatus (4), S. obliquus (4), S. sp. (1, 4, 6), Sphaerocystis schroeteri (4), Spongiochloris sp. (1, 4, 6), Spongiococcum sp, (1, 4), Stichococcus bacillaris (1, 4), S. sp. (1, 4)
Tetracystis sp. (1, 4, 6), Tetraspora sp. (1, 4), Tetraedron bifurcatum (4), T. minimum (4), T. sp. (4), Trebouxia cladoniae (4) T. sp. (1, 4), Treubaria like (4)
Ulothrix sp. (1, 4)
Westella botryoides (4), W. sp. (1, 4)

CHRYSOPHYEAE

Botrydiopsis sp. (1, 4)
Chromulina sp. (4), Chrysocapsa sp. (1)
Heterococcus sp. (1, 4)
Monocilia sp. (1, 2, 4)
Tribonema sp. (1, 4)
Vaucheria sp. (4)

BACILLARIOPHYCEAE

Achnanthes spp. (4, 7), Amphora ovalis (7), A. sp. (4), Asterionella formosa (7)
Chaetoceros sp. (4), Cocconeis spp. (7), Coscinodiscus lacustris (7), C. like (4), Cyclotella spp. (7), Cymbella spp. (7)
Diatoma elongatum (7)
Epithemia sp. (7), Eunotia lunaris (7)
Fragilaria spp. (7), Frustulia spp. (7)
Gomphonema spp. (4, 7)
Hantzschia amphioxis (4, 7), H. sp. (1, 4)
Melosira spp. (7), Melosira like (1), M.. circulare (7)
Navicula minuscula (4), N. spp. (1, 4, 5, 7), Nitzschia frustulum (4), N. palea (4), N. spp. (7)
Pinnularia spp. (7)
Rhoicosphenia curvata (7)
Stephanodiscus spp. (7), Stauroneis sp. (7), Synedra spp. (7)

1: Brown et al. 1964, 2: Carson and Brown 1976, 3: Gregory et al. 1955, 4. Schlichting 1969, 5: Schlichting 1973, 6: Smith 1973, 7: Geissler and Gerloff 1965

6. References

Bergman, B., Wictorin, B., and Willén, T. (1983) Alger som allergener, Läkartidningen 80, 4792-4795.

Bernstein, I. L. and Safferman, R. S. (1965) Sensitivity of skin and bronchial mucosa to green algae, J. Allergy 38, 166-172.

Bernstein, I. L. and Safferman, R. (1972) Clinical sensitivity to green algae demonstrated by nasal challenge and in vitro tests of immediate hypersensitivity, J. Allerg. Clin. Immunol. 51, 22-28.

Brand, F. (1925) Analyse der aerophilen Grünalgenanflüge, insbesondere der proto-pleurococcoiden Formen, Arch. Protistenk. 52, 265-355.

Breuer, G. (1990) Algen und Wolken, Naturw. Rdsch. 43, 405.

Brown, R. M., Larson, D. A., and Bold, H. C. (1964) Airborne algae: Their abundance and heterogeneity, Science 143, 583-584

Carson, J. L. and Brown, R. M. (1976) The correlation of soil algae, airborne algae, and fern spores with meteorological conditions on the island of Hawaii, Pacific Sciences 30, 197-205.

Edlich, F. (1936) Einwirkung von Temperatur und Wasser auf aerophile Algen, Arch. Mikrobiol. 7, 62-104.

Ehresman, D. W. and Hatch, M. T. (1975) Effect of relative humidity on the survival of airborne unicellular algae, Applied Microbiology 29, 352-357.

Ettl, H. and Gärtner, G. (1995) Syllabus der Boden-, Luft- und Flechtenalgen, G. Fischer, Stuttgart, Jena, New York.

Gallagher, J. S., Enrione, M., Johnson, L., Safferman, R. S., Bernstein, I. L., and Arlian, L. G. (1979) Repetitive household sampling of mites and algae compared to clinical parameters of house dust sensitivity, J. Allergy Clin. Immunol. 63, 215.

Geissler, U. and Gerloff, J. (1965) Das Vorkommen von Diatomeen in menschlichen Organen und in der Luft, Nova Hedwigia 10, 565 -577.

Gregory, P. H., Hamilton, E. D., and Sreeramulu, T. (1955) Occurrence of the alga Gloeocapsa in the air, Nature 176, 1270.

Lunceford, T. M. (1968) Algae as an allergen, J. Kans. Med. Soc. 69, 466-467.

Maynard, N. G. (1968) Significance of air-borne algae, Zeitschrift für Allgem. Mikrobiologie 8, 225-226.

McElhenney, T. R., Bold, H. C., Brown, R. M., and McGovern, J. P. (1962) Algae: A cause of inhalant allergy in children, Annals of Allergy 20, 739-743.

McGovern, J. P., McElhenney, T. R., and R. M. Brown (1965) Airborne algae and their allergenicity. Part 1. Air sampling and delineation of the problem, Annals of Allergy 23, 47-50.

McGovern, J. P., Haywood, T. J., and McElhenney, T. R. (1966) Airborne algae and their allergenicity: II. Clinical and laboratory multiple correlation studies with four genera, Annals of Allergy 24, 145-149.

Mittal, A., Agarwal, M. K., and Shivpuri, D. N. (1979) Studies on allergenic algae of Delhi area: botanical aspects, Ann. Allerg. 42, 248-252.

Parswanath, H. V. and Ramalingham, A. (1979) Seasonal variation in the airborne algae over a rural and an urban area, Current Science 48, 956-957.

Rayburn, W. R., Mack, R. N., and Metting, B. (1982) Conspicuous algal colonization of the ash from Mount St. Helens, J. Phycol. 18, 537-543.

Schellmann, B. and Sperl, W. (1979) Diatomeen-Nachweis im Knochenmark (Femur) Nichtertrunkener, Z. Rechtsmed. 83, 319-324.

Schlichting, H. E. (1961) Viable species of algae and protozoa in the atmosphere, Lloydia 24, 81-88.

Schlichting, H. E. (1969) The importance of airborne algae and protozoa, APCA-Journal 19, 946-951.

Schlichting, H. E. (1971) A preliminary study of the algae and protozoa in seafoam, Botanica Marina 14, 24-28.

Schlichting, H. E. (1974a) Ejection of microalgae into the air via bursting bubbles, J. Allergy Clin. Immunol. 53, 185-188.

Schlichting, H. E. (1974b) Periodicity and seasonality of airborne algae and protozoa, in H. Lieth (ed.), Phenology and Seasonality Modeling, Springer-Verlag, New York, pp. 407-413.

Schlichting, H. E. (1986) Airborne algae and protozoa, VI. Coloquio de investigacion en ciencias de la salud el medio ambiente y la educacion, Universidad nacional Autonoma de Mexico, 398-409.

Smith, P. E. (1973) The effects of some air pollutants and meteorological conditions on airborne algae and protozoa, J. of the Air Pollution Control Association 23, 876-880.

Tiberg, E. (1989) Microalgae as allergens with special emphasis on the genus Chlorella Beij., Dissert. Univ. of Uppsala

Vali, G. (1979) The role of biological materials in atmospheric precipitation processes, in R. L. Edmonds (ed.), Aerobiology, Dowden, Hutchinson & Ross, Stroudsburg, pp. 255-278.

VI. LIVING TOGETHER: SYMBIOSIS

Biodata of **Douglas Zook** contributor of ***"Organisms Living Inside Others: the Symbiotic Wonders."***

Dr. Douglas Zook is an associate Professor of Science Education and Biology at Boston University. He received his Ph.D. in Biology from Clark University with special studies at the University of Tübingen, Germany. Dr. Zook teaches an intensive graduate course in Symbiosis. Previous research has focused on *Cyanophora paradoxa* as a model symbiosis of studying plastid evolution and bacterial populations within lichens. Dr. Zook also directs the international Microcosmos curriculum and professional development program and has recently served as a member of the National Academy of Sciences' Science Education Committee.
E-Mail: **dzook@seacoast.com**

J. Seckbach (ed.), Enigmatic Microorganisms and Life in Extreme Environments, 575–584.

ORGANISMS LIVING INSIDE OTHERS

THE SYMBIOTIC WONDERS

DOUGLAS P. ZOOK
Boston University
605 Commonwealth Avenue
Boston, MA 02215

1. Introduction

In the light of continuing research, the traditional symbiosis definitions of the "living together of two or more unlike organisms," based in large part on the discussions of A.B. Frank who first coined the term (Sapp, 1994) and A. DeBary who initially promoted the concept through lichen discussions (*Ibid.*), should perhaps be replaced by the "acquisition and active maintenance of one genome by another" (Zook, 1997) or the "acquisition of novel metabolic capability and structures" (Douglas, 1994). Such definitions avoid the oversimplifications inherent in such terms as "mutual benefit," help focus the discipline appropriately on the microbial world, and indirectly promote the linkage of symbiosis to horizontal gene transfer mechanisms.

Mitochondria have been well-established as endosymbiotically derived from ancient purple bacteria (Gray, 1989, Margulis, 1993) that entered another cell around two billion years ago. Plastids are symbiotically derived, likely from cyanobacteria (Whatley and Whatley, 1981). But, it is only recently that we have indications of the derivation of the host nucleocytoplasm. The discovery and elucidation of the Archaea domain offer strong candidates for the origin of the eukarya host. Indeed, research on lipids, genetic, and cytoskeletal features show Archaea more aligned with eukarya than with the eubacteria (Dyer and Obar, 1994).

Ongoing research reflects a duality: Symbiosis growth has been seeded by molecular biology, and yet, it is breaking away from reductionist analysis and evoking global holistic views. Symbiosis research will likely re-shape aspects of ecology, environmental management, biotechnology, geosciences and even medical research.

2. Boreal and Temperate Forests

2.1. ECTOMYCORRHIZAE

Vascular plants and even some bryophytes (Pocock and Duckett, 1984) are nearly universally associated with the common thread-like mycorrhizal fungi in soils.

Traditionally acknowledged as enhancing element uptake, such as phosphorus and ammonium ions, mycorrhizae are now recognized as being far more diverse and complex than previously realized. Ectomycorrhizae are fungi that penetrate roots but remain intercellular in a net arrangement. Its nutrient transfer is a key to the survival and growth of boreal and temperate forests, and they can even influence the diversity and distribution of plants (Grime *et al*, 1987).

Mycorrhizae can link from one root system to another. Studies indicate that photosynthate from one plant species may maintain a mycorrhizal individual that is in fact implicated in the survival of an adjacent (even different species) plant (Simard, 1995). Thus, ectomycorrhizae serve as a kind photosynthate-based ecological conduit.

At least 15% of the net primary production of forest trees are allocated to ectomycorrhizae (Vogt, *et al.*, 1982). The profound importance of ectomycorrhizae in the establishment and maintenance of forest systems is indirectly apparent through research that shows up to 3.5 kg of living mycelium per hectare (Smith and Read, 1997).

2.2. LICHENS

Lichens are microbial-based symbiotic associations among fungi and algae and occasionally with cyanobacteria. The intimate interrelationship among compatible fungi and algae result in a novel structure called a "thallus." Lichens cover vast areas of the substrate bark. While there are reports of some deleterious infections as the fungal hyphae penetrate into the vascular system of the tree (Avalos, *et al.* 1986), the lichen symbiosis contributes to the total carbon fixation of the forest community through its photosynthesizing phycobiont. Secondly, foliose and fruticose lichens hold substantial water, and the leachate of minerals which slowly seep out in rainfall and at senescence contribute to the bark and floor nutrient availability and to pH (Pike, 1978). Thirdly, some lichens have nitrogen-fixing cyanobacterial symbionts, usually in sequestered regions known as "cephalodia." These lichens increase nitrogen availability to the overall tree community (Esseen, 1985).

2.3. *FRANKIA*

Non-leguminous dicots of temperate regions also associate with nitrogen-fixing filamentous bacteria of the genus *Frankia. Frankia* exists in soils apart from the symbioses, but more than 200 angiosperms require its infection (Huss-Danell, 1997). There are two major modes of infection, one of which is similar to the *Rhizobium*-legume and *Azolla-Anabaena* symbioses in that plant hairs are involved. The root hairs become deformed and branch once they come in contact with the *Frankia* colony. The bacterial hyphae penetrate the cell wall and grow through the root hair, the epidermal root cells, and the into the cortical cells. The hyphae then become encapsulated within the plant root cells, while cortical cell differentiation and the formation of a nodule evolves. The bacteria infect the nodule and continue to grow, breaking through the root surface. A second mode is through intercellular penetration. The root cortex is directly invaded by *Frankia* as the nodule grows to the root surface. (Berry and Sunnell, 1990)

Most *Frankia*-plant symbiosis involve the development of vesicles where nitrogenase is isolated from high oxygen levels . Plant photosynthate is presumably transferred, but details of the transfer and its metabolic products are unknown (Huss-Danell, 1997).

2.4. TERMITE GUTS

At the forest soil/litter interface are important recyclers of cellulosic material and nitrogen, the termites. Within the paunch of xylophagous termites are up to 10^7 bacteria and 10^{11} protists, mostly anaerobic, which metabolize the ingested wood remnants and recycle waste nitrogen of the insect (Breznak, 1982). They also generate massive amounts of methane due to the presence of Archaean methanogenic bacteria in their guts. Because methane is highly reactive with oxygen, this may help stabilize an oxygenated atmosphere (Lovelock and Margulis, 1974).

3. Sub-tropical and Tropical Rainforests

3.1 ORCHID MYCORRHIZAE

Besides vesicular-arbuscular forms discussed below under grasslands, tropical forests feature a wide variety of orchids obligately dependent on mycorrhizae. There are approximately 20,000 orchid species and each require mycorrhizal infections at pre-germination stages. As with VA and ectomycorrhizae, orchid mycelia disperse within the soil, take up nutrients -- particularly phosphorous -- and transfer them to the plant.

However, orchid mycorrhizae are fast-growing saprophytes and also transfer carbohydrates to the plant. This is essential for the orchid, as all orchids pass through a prolonged seedling stage in which they are unable to photosynthesize (Smith and Read, 1997). While most orchids by adult stages become photosynthetic, approximately 200 species remain achlorophyllous and depend on the mycorrhizae for nearly all essential nutrients in its life cycle (Leake, 1994). There is mounting evidence that some orchids and associated fungi use photosynthate from adjacent plants through subterranean ectomycorrhizal linkages (Zelmer and Currah, 1995).

3.2 CYAN0-LICHENS

Epiphytes are as characteristic of rainforests as abundant rainfall and high humidity. The colonization of plant epiphytes and the nutrient flow that supports mid- and lower-canopy diversity appears to often depend, in turn, on the colonization of lichens.

Lichens are classic representatives of symbiosis. Most are an integrated relationship among a fungus and a chlorophyte alga. But, approximately 10% of the world's 18,000 lichens have cyanobacterial symbionts, either in a tripartite symbiosis or with the fungus alone (Friedl and Büdel, 1996). Although the lichen thallus is exclusively a result of symbiosis, creating foliose, fruticose and crustose morphologies, the symbionts are principally microbial.

As in the mycorrhizal symbiosis, a major acquisition is that of photosynthate, in this case from the phycobiont to the fungus. The selective advantage for the alga is less clear, but high light intensities have been found to adversely effect many of the free-living algal species that are normally phycobionts (Demmig-Adams, 1990). Through lichenization, these algae are able to expand their niche into previously inhospitable high light intensity environments.

In the rainforest, lichens absorb mineral rich water from leachates along the bark surface above the colony (Lang *et al.*, 1976). Lichens provide a mass upon which scattered plant materials can accumulate. Several arboreal-dwelling lichens are even susceptible to the trapping of organic particles and nutrients within their prolific intercellular spaces (Nieboer, *et al.*, 1978). Those lichens that likely have the greatest impact on rainforest community structure are cyano-lichens. Cyanobionts of lichens such as *Nostoc*, for example, have five times more nitrogenase-containing heterocysts than free-living cyanobacteria (Feige and Jensen, 1992). Nitrogen-fixing cyano-lichens are common in tropical rainforests where they substantially increase available nitrogen to the community (Green, *et al.*, 1980).

3.3 GUNNERA-NOSTOC

Because of the prominent cell wall of plants, symbiosis that directly involves non-root plant tissue is rare. However, a prolific herbaceous plant of the sub-tropics and tropical lowland tropical forests that demonstrates a profound symbiosis is the ancient, extant *Gunnera*. This broad, leafy plant features intracellular cyanobacterial colonies, *Nostoc punctiforme*, in the plant's red stem glands, which are also the site of initial infection (Silvester and McNamara, 1976). It is the only known angiosperm to be symbiotic with cyanobacteria (Bonnet and Silvester, 1981). The *Nostoc* fixes substantial nitrogen which is transferred to the *Gunnera*. Much like *Anabaena* within the *Azolla* symbiosis and *Nostoc* cyanobionts of lichens, photosynthesis in *Nostoc punctiforme* is greatly reduced once in this obligate relationship, as *Gunnera* transfers sugars (Silvester, 1975). This symbiosis is common in the South Pacific and South America, especially in forests of high humidity and where rainfall exceeds 2000 mm/yr. (Osborne *et al.*, 1991).

4. Global Grasslands and Agricultural Regions

4.1 AZOLLA-ANABAENA

Through its nitrogen fixing association with the cyanobacterium *Anabaena*, the small water fern *Azolla* has had a major influence on the survival of human populations. Its overall impact on world economy has not been inconsequential. *Oryza sativa* (rice) is one of the world's most grown and utilized grain crop. Its ability to proliferate is due in large part to the presence of a "green manure" (Wagner, 1997). The nitrogenous *Azolla-Anabaena* grows, dies off, and rice seedlings are then transplanted.

The relationship is tightly coordinated. After contact with a plant primary root hair and associated simple hairs within a cavity of the fern leaflets, the *Anabaena* transfers ammonium ions to the *Azolla*. (Shi and Hall, 1988). Ammonium uptake is partly due to the Azolla's ability at the transcription level to shut down the production of nitrogen-assimilating glutamine synthetase in the *Anabaena*, which then allows the nitrogenous compounds to be directly available to the *Azolla*. (Nierzwicki-Bauer and Haselkorn, 1986). Remarkably, nearly all of the required carbohydrate needs of *Anabaena* are met through the transfer of photosynthate from the increased photosynthetic activity of the fern. Cyanobacterium hydrogeneases recycle electrons and help generate the high amount of ATP-bound energy necessary for nitrogen fixation and photosystem reactions (Peters and Calvert, 1983).

4.2 RHIZOBIUM-LEGUMINOUS PLANTS

Another major agricultural-based symbiosis common among grassland and small arboreal communities is the bacterium *Rhizobium* in association with a wide range of plants in the leguminoseae family, including *Glycine max* (soy), *Medicago sativa* (alfalfa), and *Trifolium melilotus* (yellow clover). The bacteria -- including their nodulating genes – are activated through the secretion of flavonoids by the plant (Zaat, *et al.* 1988). The *Rhizobia* attach to and then infect root hairs. Simultaneously, root cortical cells divide and give rise to nodules that house the bacteria, which become known as "bacteroids." The bacteroids inhabit plant-derived membranous structures known as "symbiosomes" within the nodules. (Vance, 1983)

As in heterocysts of nitrogen-fixing cyanobacteria, the symbiosomes isolate the nitrogen fixation process. The *Rhizobium*-legume symbiosis produces a laeghemoglobin protein which both binds the available oxygen and, in effect, keeps it from nitrogenase, but yet still is able to shunt oxygen to the mitochondria (Appleby, 1984).

Extensive research has shown that the complex series of responses and reactions in the symbiosis is governed through a series of nodulation genes, which express before the uptake of nitrogenous compounds by the plant (Gloudemans and Bisseling, 1989). *Rhizobia* contain a mega-plasmid with specialized genes that code for the nitrogen fixation (David, 1988).

While fertilizers contribute to the fixation of nitrogen in soils worldwide, biomes are still largely dependent on the natural processes of symbiotic nitrogen fixation, with an estimated 35 x 10^6 tons of nitrogen fixed by legumes each year.

4.3 VA MYCORRHIZAE

Mycorrhizae, which branch and curl partially within plant root cells and feature rounded structures between the cells, are known as vesicular-arbuscular (VA). While they are found among temperate and tropical zone plants, they particularly dominate global grasslands (Smith and Read, 1997). As in ectomycorrhizae, the vesicular-arbuscular forms are the main conduit for the transfer of phosphorus into associated plants. However, growing evidence shows a wide range of benefits to host plants, with uptake of micronutrients ZN (Faber, *et al.*., 1990), CU (Li, *et al.*, 1991) and nitrate assimilation increased (Hamel, *et al.*, 1991).

VA mycorrhizae are more diverse than previously realized (Giovannetti and Gianinazzi-Pearson, 1994). Classification of the fungi is based upon the morphology and wall development of spores, which can exceed 500 microns and often contain more than 1000 nuclei (*Ibid*, 1997). Symbiotic specificity is low, and flavonoids may be involved in initiation (Gianinazii-Pearson, *et al.* 1989), much as in *Rhizobium*-legumes.

4.4 ENDOPHYTIC FUNGI-GRASSES

Besides infection by vescicular-arbuscular mycorrhizae into the root cells, a wide number of grasses from the family Clavicipitaceae show symbiotic intercellular infection by other endophytic fungi. Several of these endophytic fungi, which were previously considered pathogenic or transient, are ecologically obligate symbionts and directly protect host grasses from biotic stresses such as nematodes (Kimmons, *et al.*, 1990). Fungi such as Epichloe produce alkaloids which are active against insects within the plant tissue (Siegel, *et al.*, 1990).

Research indicates a kind of "balanced antagonism" or pleiotropic symbiosis (Schardl, *et al.*, 1997) among the microbial fungi and the grasses, for both produce substantial compounds that are directly toxic to the metabolism or reproduction of the other (Schultz, 1997). Because the fungi are maternally transmitted, produce anti-herbivore compounds, and improve fitness to hosts, the diversity and characteristics of endophytes likely play a major role in the long-term distribution of many grass species (Tsai, *et al.*, 1994).

4.5 RUMINANTS-RUMEN MICROORGANISMS

Bacteria, protists and fungi live extensively in symbiosis within megafauna, such as mammals, amphibians, reptiles, and birds. Previously regarded as either commensal or pathogenic, vertebrate-microbial symbioses are increasingly recognized as prolific and complex. Grasslands and agricultural areas can be especially abundant in this symbiosis.

Ruminants feature a large foregut sac, the rumen, among a multi-chambered digestive system. Most of the cellulosic food in the form of leaves and grasses is converted by microorganisms to vitamins and volatile fatty acids. Indeed, the nutritional requirements of ruminants, such as cattle, sheep, and deer are substantially met by the rumen microbes.

The rumen is anaerobic and includes astounding population densities of 10^{10} bacteria/ml, 10^6 heterotrophic protists/ml, and between 10^3 to 10^7 fungi/ml (Hespell, *et al.* 1997). The rumen stomach chambers can be thought of as churning fermentation chambers. The wide diversity of anaerobic bacteria assume several niches within the gut community. Generally, most of the smaller bacteria are digested each time that a portion of the rumen plant contents is transferred to the gastric region of the stomach. This digestion allows for the recycling of nitrogen within the animal (Douglas, 1994). Other bacteria such as *Fibrobacter succinogenes* efficiently degrade plant cell walls (Dehority and Scott, 1967). The rumen also includes fermenting Archaeans, including up to 10^8 methanogens/ml rumen contents in alfalfa hay-fed cattle (Miller, *et al.*, 1986). Gut protist feeding may help maintain appropriate bacterial population levels (Hespell *et al.*, 1997), while rumen fungi actively colonize and degrade the lignified cell walls of forage plant fiber (Bauchop, 1979). More significantly, rumen fungi are responsible for the break down of the protective surface of leaves and stems, the cuticle. This allows other gut microbiota to gain access to the internal portions of the plant for fermentation (Akin and

Rigsby, 1987). Thus, the ruminant symbiosis remarkably involves all three domains of life and all the major eukarya kingdoms -- animals, plants, protists, and fungi.

5. Marine

5.1 CORAL-ALGAL REEFS

Coral reefs are estimated to cover 6 x 10^5 km of the ocean photic zone and fix 3 to 4 times more carbon than the diatoms, dinoflagellates and other phytoplankton of the cooler, upwelling areas of the seas. Corals are second only to rainforests in the diversity of organisms among ecosystems. They are a nutrient cycling oasis in the relatively species-poor tropical waters.

Hermatypic corals are a symbiosis between the coral animal and encysted *Symbiodinium* dinoflagellate algae. The algae are either acquired directly from surrounding waters at each generation as is often in sexual-producers, while asexual corals transmit the algae through coral fragments (Muller-Parker and D'Elia, 1997).

Studies have consistently shown that the coral receive carbohydrates directly from the photosynthesis of the dinoflagellate (Muscatine, 1990). The symbiotic algae obtain inorganic nutrients via the recycling of coral waste and other nutrients from dissolved inorganic compounds in the zooplankton and seawater taken in by the coral (Muller-Parker and D'Elia, 1997). The need for CO_2 in photosynthesis by the algae drives the calcium deposition of the symbiosis and results in a limestone reef (Goreau, 1961).

These highly active *Symbiodinium spp.* were once thought to be a single species, but evidence is mounting that there may be a wide variety of symbiotic dinoflagellates, some more fit for certain environmental parameters than others (Rowan and Powers, 1991). There are even hypotheses which suggest that "coral bleaching" may be adaptive in that the coral replaces less adapted symbionts for "new" free-living dinoflagellates which are more fit in the changing conditions. (Buddemeier and Fautin, 1993).

References

Akin, D.E. and Rigsby, L.L. (1987) *Appl. Environ. Microbiol.* **53**, 1987-1995.
Appleby, C.A. (1984) *Annual Review of Plant Physiology* **35**, 443-468.
Avalos, A., Legaz, M.E., and Vincente, C. (1986) *Biochem Syst. Ecol.* **14**, 381-384.
Bauchop, T. (1979) *Appl. Environ. Microbiol.* **38**, 148-158.
Berry, A.M. and Sunell, L.A. (1990) in C.R. Schwintzer and J.D. Tjepkema (eds.), *The Biology of Frankia and Actinorhizal Plants*, Academic Press, San Diego, pp. 61-81.
Bonnett, H.T. and Silvester, W.B. (1981) *New Phytologist* **89**, 121-128.
Boucher V.L. and Nash, T.H. III (1990) *Bot. Gaz.* (Chicago) **151**, 114-118.
Bradshaw, A.D. and Chadwick, M.J. *(1980) The Restoration of Land*, Blackwell Scientific, Oxford.
Breznak, J.A. (1982) *Annual Review of Microbiology* **36**, 323-343.
Buddemeier, R.W. and Fautin, D.G. (1993) *BioScience* **43**(5), 320-326.
David, M., Davran, M.L., Batut, J., Dedieu, A., Domergue, O., Ghai, J., Hertig, C., Boistard, P., and Kahn, D. (1988) *Cell* **54**, 671-683.
Dehority, B.A. and Scott, H.W. (1967) *J. Dairy Science* **50**, 1136-1141.
Demmig-Adams, B., Magyasm, C., Adams, III, W.W., Meyer, A., Kilian,E., and Lange, O.L. (1990) Plantae **180**, 400-409.
Douglas, A.E. (1994) *Symbiotic Interactions*, Oxford: Oxford University Press, Oxford.
Dyer, B. and Obar, R. (1994) *Tracing the History of Eukaryotic Cells*, Columbia U. Press.
Esseen, P.A. (1985) *Can J. of Botany* **63**, 980-987.
Faber, B.A., Zasoski, R.J. Burau, R.G., and Uria, K. (1990) *Plant and Soil* **129**, 121-120.
Feige, G.B. and Jensen, M. (1992) in W. Weisser (ed.), *Algae and Symbioses*, Biopress, Bristol, pp. 227-299.

Friedl, T. and Büdel, B. (1996) in T. Nash,(ed.), *Lichen Biology*, Cambridge U. Press, Cambridge, pp. 8-23.
Gianinazzi-Pearson, V., Branzanti, B. and Gianinazzi, S. (1989) *Symbiosis* **7**, 243-255.
Giovannetti, M. and Gianinazzi-Pearson, V. (1994) *Mycol. Res.* **98**, 705-715.
Gloudemans, T. and Bisseling,T. (1989) *Plant Science* **65**, 1-14.
Gray, M.W. (1989) Origin and evolution of mitchondrial DNA. *Annual Review of Cell Biology* **5**, 25-50.
Green, T.G.A., Horstmann, J., Bonnett, Wilkins, H.A. and Silvester, W.B. (1980) *New Phytol.* **84**, 339-348.
Grime, L.P., Mackay, J.M.L., Hillier, S.H., and Read, D.J. (1987) *Nature* **328**, 420-422.
Goreau, T.F. (1961) *Endeavour* **20**, 32-39.
Hamel, C., Furlan, V. and Smith, D.L. (1991) *Plant and Soil* **133**, 177-185.
Hespell, R.B., Akin, D.E. and Dehority, B.A. (1997) Bacteria of the Rumen, in R.I. Mackie, B.A. Whyte, R.E. Isaacson (eds.), *Gastrointestinal Microbiology* Chapman-Hall, NY, pp. 59-141.
Huss-Danell, K. (1997) New Phytologist **136**, 375-405.
Kimmons. C.A., Gwinn, K.D. and Bernard, E.C. (1990) *Plant Dis.* **74**, 757-761.
Lang, G.E., Reiners, W.A. and Heier, R.K. (1976) *Oecologia* **25**, 229-241.
Leake, J.R. (1994) *New Phytol.* **127**, 171-216.
Li, X.L., Marschner, H. and George, E. (1991) *Plant and Soil* **136**, 41-48.
Lovelock, J. and Margulis, L. (1974) *Icarus* **21**, 471-489.
Margulis, L., (1993) *Symbiosis in Cell Evolution*, 2nd edition, W.H. Freeman, San Francisco.
Miller, T.L., Wolin, M.J., Zhao, H., Bryant, M.P. (1986) *Appl. Environ. Microbiol.* **51**, 201-202.
Muller-Parker, G. and D'Elia, C.F. (1997) Interactions between corals and their symbiotic algae, in C. Birkland (ed.), *Life and Death of Coral Reefs*, Chapman and Hall, New York. pp. 96-116.
Muscatine, L., Falkowski, P., Porter, J., and Dubinsky, Z. (1984) *Proc. R. Soc. Lond. B* **222**, 181-202.
Muscatine. L. (1990) The role of symbiotic algae in carbon and energy flux in reef corals, in Z. Dubinsky (ed.), *Coral Reefs*, Elsevier, Amsterdam, pp. 75-87.
Nash, T.H. III, (1996) *Lichen Biology*, Cambridge University Press, Cambridge,
Nieboer, E., Richardson, D.H.S. and Tomassini, F.D. (1978) *Bryologist* **81**, 226-246.
Nierzwicki-Bauer, S.A. amd Haselkorn, R. (1986) EMBO Journal **5**, 29-35.
Osborne, B., Doris, F., Cullen, A., McDonald, R., Campbell, G. and Steer, M. (1991) *BioScience* **41**(4), pp. 224-234.
Peters, G.A. and Calvert, H.E. (1983) in L. Goff (ed.), *Algal Symbiosis -- A Continuum of Interactive Strategies*, Cambridge University Press, New York, pp. 109-145
Pike, L.H. (1978) *Bryologist* **81**, 247-257.
Pocock. K. and Duckett, J.G. (1984) *J. of Bryology* 13, 227-233.
Rhoades, F.M. (1995) in M. Lowman and N.M. Nadkarni, *Forest Canopies*, Academic Press, London, pp. 393-397.
Rowan, R. and Powers, D.A. (1991) *Science* **251**, 1348-1351.
Sapp, J. (1994) *Evolution by Association*, Oxford University Press, Oxford.
Schardl, C.L. (1994) *Proc. Natl. Acad. Sci* 91, 2542-2546.
Schardl, C.L., Leuchtmann, A., Kuang-Ren Chung, Penny, D. and Siegel. M. (1997) *Mol. Biol. Evol.* **14**(2), 133-143.
Schultz, B. (1997) Presentation, "Endophyte-host relationship: a balanced antagonism," at Second International Congress on Symbiosis at Woods Hole, April 15.
Shi, Ding-Ji and Hall., D.O. (1988) *Botanical Rev.* **54**(4), 535-386.
Siegel, M.R., Latch, G.C.M., Bush, L.P. Fannin, F.F., Rowan, D.D., Tapper, B.A., Bacon, C.W. and Johnson, M.C. (1990) *J. Chem. Ecol.* **16**, 3301-3315.
Silvester, W.B. (1975) in P.S. Nutman (ed.), *Symbiotic Nitrogen Fixation in Plants*, Cambridge University Press, Cambridge, pp. 521-538.
Silvester, W.B. and McNamara, P.J. (1976) *New Phytol.* **77**, 135-141
Simard, S.W. (1995) PhD thesis, Dept. of Forest Science, Oregon State University, Corvallis, OR.
Smith, D.C. and Douglas, A.E. (1983) *Biology of Symbiosis*, E.A. Arnold Publishers, Baltimore
Smith, S.E. and Read, D.J. (1997) *Mycorrhizal Symbiosis*, 2nd edition, Academic Press, London.
Tsai, Huei-Fing, Liu, Jih-Shiou, Staben, C., Christensen, M.J., Latch, G.C.M., Siegel, M.R., and Schardl, C.L. (1994) *Proc. Natl. Acad. Sci. USA* **91**, 2542-2546.
Vance, C.P. (1983) *Ann Rev. Microbiol.* **37**, 399-424.
Veira, D.M., Ivan, M., and Jui, P.Y. (1983) *J. Dairy Science* **66**, 1015-1022.
Vogt.K.A., Grier, C.C., Meier, C.E., and Edmunds, R.L. (1982) *Ecology* 63, 370-380.
Wagner, G.M. (1997) Botanical Rev. **63**(1), 1-26.
Whatley, J.M. and Whatley, F.R. (1981) *New Phytologist* **87**, 233-247.
Zaat, S., Spaink, H.P., Wijfeelman, C.A., van Brussel, A., Okker, R., and Lugtenberg, B. (1988) in Scannerini, S. Smith, D. Bonfante-Facolo, P. and Gianinazzi-Pearson, V. (eds.), *Cell to Cell Signals in Plant, Animal, and Microbial Symbiosis*, Springer-Verlag, New York, pp. 189-205.
Zelmer, C.D. and Currah, R.S. (1995) *Can. J. of Botany* **73**, 862-866.
Zook, D.P. (1997) Curriculum book for graduate Symbiosis course, Boston University. Unpublished.

Biodata of **Kwang W. Jeon** (contributor of ***Symbiogenesis of Bacteria within Amoebae***).

Dr. **Kwang W. Jeon** is a Professor in the Department of Biochemistry and Cellular and Molecular Biology in the University of Tennessee (Knoxville). He received his PhD from the University of London, England (1964). His major research interests are to elucidate host-symbiont relationships at the cellular level and to formulate a comprehensive theory for the integration of newly acquired symbionts into host cells leading to the acquisition of new cell components and changes in cellular phenotypic characters. Dr. Jeon's experimental model is an amoeba-bacteria symbiosis which was recently established under laboratory conditions. The symbiotic bacteria were originally parasitic and harmful to the host but changed to required cell components within a few years and both the host and symbionts have changed to adapt themselves to stabilize the symbiotic relationship. His current aims are to determine the roles of symbiont-synthesized proteins in the symbiotic interactions, to clarify the roles of components of the symbiosome membranes which so not fuse with lyzosomes, to continue their search for a possible ancestral strain of bacteria in the environment from which the symbiotic bacteria arose and to determine the genetic basis for the parasite-to-symbiont transition, and to study the roles of plasmid DNAs of the symbionts in the establishment and maintenance of Endosymbiosis.

E-Mail: jeon-k@utk.edu

J. Seckbach (ed.), Enigmatic Microorganisms and Life in Extreme Environments, 585–598.

SYMBIOGENESIS OF BACTERIA WITHIN AMOEBAE

Symbiogenesis in amoebae

K. W. JEON
Department of Biochemistry
University of Tennessee
Knoxville, TN 37996, USA

1. Introduction

Intracellular symbiosis is a wide-spread and important biological phenomenon in terms of genetic novelty, bringing about genetic changes that may be greater in magnitude than those which may result from mutation, hybridization or ploidy changes [Margulis, 1993; Sato et al., 1994], because symbiosis draws genomes from the entire biosphere [Jeon & Danielli, 1971; Gray, 1992-3; Maynard Smith & Szathmary, 1995]. Bacterial infection and intracellular survival inside eukaryotic cells have been extensively studied, but much remains still unknown about the mechanisms underlying the phenomena [Galan, 1994; Bull & Fogarty, 1995; Jeon, 1995-6; Russell, 1995; Finlay & Cossart, 1997].

Once endosymbiosis has been established, host cells provide a suitable "shelter" and supply material needs for endosymbionts. Meanwhile, endosymbionts must overcome many difficulties to avoid destruction by the host cell, and hence the cytoplasm of a host cell is considered to be a hostile environment for intracellular symbionts or parasites [Moulder, 1985, 1991; Morioka & Ishikawa, 1992]. Most infecting microbes are digested within the host cell's phagolysosomes, but in rare occasions some microbes avoid digestion and survive inside host cells as intracellular symbionts. During and after symbiogenesis, the expression of symbionts' genes is often modulated by the host cell [Ishikawa, 1984; Abu Kwaik et al., 1993-4]. The host's proteins also change after infection. Recent studies have shown that gene transfers may occur between symbionts and hosts [Schenk et al., 1992; McFadden et al., 1994; Henze et al., 1995; Nardon, 1995; Keeling & Doolittle, 1997] as well as between cell organelles and cell nuclei [Brinkmann & Martin, 1996; Burgos et al., 1996; Horner et al., 1996; Liaud et al., 1997] or among organellar genomes.

In avoiding destruction by their hosts, infecting microbes employ various tactics. Some microbes are able to prevent lysosomal fusion with symbiosomes [Hart & Young, 1979; Ahn & Jeon, 1979,1982; Pfeifer, 1987; Holtzman, 1989; Ishibashi & Arai, 1990; Buchmeier & Heffron, 1991; Berger & Isberg, 1993; Russell, 1995; Bozue

& Johnson, 1996]. Symbionts also modulate acidification of symbiosomes and hence may indirectly prevent lysosome-symbiosome fusion [Horwitz and Maxfield, 1984; Eissenberg et al., 1993; Sturgill-Koszycki et al., 1994]. The macrophage infectivity potentiator (Mip) proteins of bacteria potentiate intracellular infection of eukaryotic cells [Engleberg et al., 1989; Cianciotto & Fields, 1992; Lundermose et al., 1993; Berger et al., 1994; Brand et al., 1994; Ludwig et al., 1994; Cianciotto et al., 1995; Mo et al., 1995; Wintermeyer et al., 1995], and some genes have been identified that control phagosome-lysosome fusion and organelle recruitment [Marra et al., 1992; Berger & Isberg, 1993; Hacker et al., 1993; Sadosky et al., 1993]. Other genes are also associated with the entry and survival of various pathogenic microbes [Arruda et al., 1993; O'Connell et al., 1995]. Many of the infecting bacteria are Gram negative and appear to have inherently digestion-resistant cell surfaces.

Together with several known molecules, the heat-shock proteins (HSP) appear to be involved in host-symbiont interactions, including host-parasite relationships [Hall & Joiner, 1991; Polla, 1991; Healy et al., 1992]. As molecular chaperonins, GroE analogs are thought to play a protective role for symbionts [Ohtaka et al., 1992; Morioka et al., 1993] as well as performing an essential function for normal cellular activities [Fayet et al., 1989]. An enhanced expression of selected *Salmonella* HS genes including *groEL* and *dnaK* may contribute to bacterial survival within their hosts, macrophages [Buchmeier & Heffron, 1990; Abshire & Neidhardt, 1993]. This chapter summarizes what is known about the process and consequences of symbiogenesis of a strain of bacteria, called X-bacteria (XB), in amoebae.

2. Survival of Symbionts in the Host

2.1. HOSTS AND SYMBIONTS

As free-living cells, amoebae are constantly exposed to other microbes present in their habitats and such microbes enter amoebae passively by phagocytosis. Amoebae are strongly phagocytic and digest all foreign organisms that enter them, but an exception was found in 1966 when Gram-negative XB spontaneously infected and survived inside amoebae. When XB first infected the D strain of *Amoeba proteus* [Jeon & Lorch, 1967], they were harmful to amoebae and killed most newly infected amoebae. However, a few infected amoebae survived and developed a close relationship with XB, becoming dependent on XB within a few years. Now, xD amoebae harbor about 42,000 XB per cell (about 10% of an amoeba's volume) and grow well. When XB are introduced into XB-free amoebae of the same or different strain, some XB survive and establish a stable association with their new host. Thus, XB are no longer virulent but influence the physiology and phenotypic characters of their newly infected host cells.

2.2. SYMBIOTIC X-BACTERIA

Roth and Daniels [1961] were among the first to confirm by electron microscopy the bacterial nature of previously reported bacteria-like particles in vacuoles of *A. proteus*

[Cohen, 1957]. Other reports followed to confirm the presence of bacterial inclusions in various strains of amoebae [Chapman-Andresen and Hayward, 1963; Drozanski, 1963; Wolstenholme and Plaut, 1964; Wolstenholme, 1966; Proca-Cibnau et al., 1975; Drozanski et al., 1984; Hall and Voelz, 1985; Fritsche et al., 1993; Yagita et al., 1995]. However, integration of invading bacteria into their host cells or the development of dependence on each other has not been reported in these cases.

X-Bacteria are Gram-negative rods (0.5 x 2 μm), enclosed in symbiosomes of varying sizes. Their ultrastructure is similar to that of other free-living or symbiotic bacteria and each XB appears to be embedded in a matrix of fibrous matter when observed in the electron microscope after freeze-fracture [Jeon & Hah, 1977]. A monoclonal antibody (mAb) against one component of the matrix, a 85-kDa protein, was obtained and used by Kim & Jeon [1995].

2.3. SYMBIONTS' AVOIDANCE OF DIGESTION

2.3.1. *Resistance to digestion*

X-Bacteria avoid digestion by amoebae first by being resistant to lysosomal enzymes when exposed to them within amoeba's phagolysosomes [Ahn & Jeon, 1979]. Isolated XB are also resistant to lysing agents such as enzymes and mild detergents (Triton-X 100 and Brij) [Han & Jeon, 1980]. X-Bacteria harbor two kinds of plasmid DNAs of about 21 and 59 kb, respectively, as determined by gel electrophoresis and by size measurements in the electron microscope [Han & Jeon, 1980]. It is suspected that one or both of the plasmids plays a role in protecting XB from digestion by coding for a digestion-resisting cell-surface component(s), although detailed mechanisms are not yet known. Any procedure such as treatment with antibiotics or raising the culture temperature which removes XB from xD amoebae causes the plasmids to disappear first [Jeon, unpub.], the findings suggesting that the plasmid DNAs are needed for infection by XB. X-Bacteria cured of their plasmids do not survive when introduced into other amoebae. Thus, XB's infectivity appears to be correlated with the presence or absence of plasmid DNAs in XB.

2.3.2. *Prevention of lysosome-symbiosome fusion*

X-Bacteria avoid digestion by amoebae also by preventing lysosomal fusion with symbiosomes [Jeon & Jeon, 1976; Ahn & Jeon, 1979,1982; Kim et al., 1994]. Symbiosome membranes (SM) do not fuse with lysosomes and it is suspected that SM-specific components play a role in the prevention of lysosomal fusion with symbiosomes [Ahn et al., 1990; Choi & Jeon, 1992; Kim et al., 1994]. For example, symbiont-derived lipopolysaccharides (LPS) are present on the cytoplasmic side of the SM [Choi & Jeon, 1992; Kim et al., 1994] and they seem to be responsible for the prevention of lysosomal fusion. If the SM components are masked with antibodies, symbiosome's ability to avoid fusing with lysosomes is lost [Kim et al., 1994]. Thus, when an anti-LPS mAb is injected into xD amoebae, some symbiosomes fuse with lysosomes, the results suggesting that unmasked LPS blocks lysosomal fusion. The sugar composition of purified XB's LPS is of a smoother type as compared with those of *Legionella* sp., *Yersinia* sp., *Shewanella* sp. and *Coxiella*, and XB's LPS produce a typical ladder-type banding pattern in SDS-PAGE [K. J. Kim & Jeon, unpub.; cf. Choi & Jeon, 1992]. They also contain the rare yersiniose A as do LPS of other microorganisms mentioned above. Purified XB LPS have been analyzed by SDS-PAGE and immunoblotting with anti-LPS mAbs after the separation of polysaccharides, core oligosaccharides, and lipid A fractions. The 3-hydroxy fatty acid composition is also similar to that found in lipid A of the above bacteria, but it is closest to the lipid A of *C. burnetii.* [cf. Sonesson et al., 1989; Kawahara et al., 1992; Phillips et al., 1993]. Since the anti-XB-LPS mAb does not cross-react with other LPS tested, it appears that XB LPS have unique epitopes. The SM also has an XB-produced 96-kDa protein [Ahn et al., 1990] and it may play a role in blocking lysosomal fusion as well.

2.4. MOLECULES INVOLVED IN SYMBIONGENESIS

It is not clear how XB survive inside the cytoplasm of amoebae, which is considered to be a hostile environment for them [cf. Moulder, 1985]. Meanwhile, there are several proteins that may be involved in the symbiogenesis of XB in amoebae.

2.4.1. *The S29x protein*

The cytosol of xD amoebae and of newly infected D amoebae contain a large amount of XB-produced 29-kDa protein (S29x) [Ahn & Jeon, 1983; Kim & Jeon, 1986-7; Park & Jeon, 1990; Pak, 1996]. The *s29x* gene has been sequenced, together with its putative promoter sequences that are different from those of known bacterial consensus promoter sequences [Pak & Jeon, 1994-6]. The *s29x* gene has an open reading frame (ORF) of 774 nucleotides, coding for 258 amino acids, equivalent to M_r of 29,968, the gene not sharing any sequence similarity with those in the GenBank.

Inside amoebae, S29x is synthesized by XB and is transported to the amoeba cytoplasm. S29x enters the amoeba's nucleus as shown by electron microscopic study using anti-S29x mAb and immunogold labeling [Pak & Jeon, 1997], although it is not clear if S29x is selectively accumulated inside the nucleus. While S29x moves freely from XB to the amoeba cytoplasm and enters the amoeba's nucleus, S29x does not have any potential signal sequences at the N-terminus or trans-membrane domains. In fact, amino acids of S29x are mainly hydrophilic except for a cluster of 15 hydrophobic amino acids at the C-terminal end. The hydrophobic region at C-terminus is not a secretory signal as in the ABC transporter systems of other bacteria. Thus, S29x appears to be transported across membranes by other mechanisms. The kinetics of synthesis of S29x was studied by using ^{35}S-methionine as a precursor and by following its incorporation into newly synthesized S29x under various conditions [Pak & Jeon, 1997]. The results indicated that S29x is exported to the amoeba cytoplasm soon after synthesis. Continued protein synthesis is not required for the secretion of S29x and hence its transport appears to be post-translational. In addition, secretion of S29x appears to be energy-independent. The function of S29x is not yet known but the protein may indirectly regulate a gene involved in bacterial infectivity or their intracellular survival since it is a prominent bacterial protein and the inhibition of its synthesis causes the death of host cells.

2.4.2. *Heat-shock proteins*

X-Bacteria contain much GroELx [Choi et al., 1991], its gene having a high degree of identity with those of other bacterial *groEL* genes and complementing mutations of *groESL* in *E. coli* [Ahn et al., 1991,1994]. The *groELx* gene is controlled by a novel and potent second promoter (*P2*) in addition to the heat-shock consensus promoter (*P1*). The *P2* promoter enhances the expression of *groELx* and other fusion genes in transformed *E. coli*. The *groES* gene of *Legionella pneumophila* also has nucleotide sequences corresponding to the whole region of *P2* of *groEx* [cf. Hoffman et al., 1990; Sampson et al., 1990]. However, no second promoter activity has been reported in the *groE* operon of *L. pneumophila*. The *P2* activity of the *groEx* gene was first detected accidentally by finding that the pAJX91 plasmid obtained as a result of automatic excision lacked *P1* but it still produced GroELx in transformed *E. coli* [Ahn et al., 1991]. *E. coli* cells transformed with DNA inserts containing no *P1* produced GroELx and it was evident that there was another promoter (putative *P2*) in the *groESx* portion of the gene. The activity of the putative *P2* was checked with a newly constructed expression vector pUXGPRM-*lac*. Transformed *E. coli* containing the whole *groEx* fragment in the fusion plasmid produced GroESx in addition to the fusion protein. Even when *P1* was removed, the putative *P2* alone promoted the production of a large amounts of ß-gal. This showed that the putative *P2* had a strong activity. The presence of the putative *P2* promoter was further demonstrated by primer-extension analysis, which revealed the presence of two transcription-initiation sites within the coding region of *groESx*.

E. coli transformed with a DNA insert containing *groEx* produced a large amount of the fusion protein and the cells could grow at their normally lethal temperature, 45° C. The results showed that the inserted gene was expressed and that the gene product was functional in protecting cells at high temperatures [T.I. Ahn & K.W. Jeon, unpub.]. It has been speculated that XB survive inside amoebae by having potent HS promoters in their *groEx* operon which enable XB to synthesize a large amount of GroESLx and use it as a chaperonin. Thus, possession of such potent promoters would be a useful adaptation for XB to survive within the potentially hostile environment of amoeba's cytoplasm.

2.4.3. *Mip-like protein*

XB contain a Mip-like protein that is recognized by a mAb against Mip of *L. pneumophila* [Cianciotto et al., 1995]. The *mipx* gene was cloned from a genomic DNA of XB [Oh & Jeon, 1996a], and the ORF had a 78.8% nucleotide-sequence identity with that of *mip* of *L. micdadei* [cf. Bangsborg et al., 1991].

2.5. EXPERIMENTAL INFECTION

Isolated XB may be transferred into other cells either by microinjection [Jeon & Jeon, 1976] or by induced phagocytosis [Ahn & Jeon, 1979; Kim & Jeon, 1986]. X-Bacteria do not multiply *in vitro* but stay viable for several days at 4° C. When infected with XB, new host amoebae become dependent on their symbionts within 200 cell generations (18 months) following experimental infection [Jeon & Ahn, 1978].

2.6. LOSS OF D-AMOEBA PROTEINS.

When D amoebae are infected with XB, they lose several proteins [Ahn & Jeon, 1983], among which a 45-kDa cytoplasmic protein that functions as S-adenosyl-L-methionine synthetase (SAMS) in symbiont-free D amoebae is prominent [Choi et al., 1997]. Once infected with XB, amoebae do not transcribe *sams* and hence stop producing SAMS as early as 50 days after infection with XB. Thus, the absence of the protein in xD amoebae is attributable to their failure to transcribe the corresponding gene as a result of harboring XB. However, xD amoebae possess about half the level of SAMS enzyme activity found in D amoebae, indicating that they use an alternative source for the enzyme. xD Amoebae exhibit irreversible nucleolar abnormalities when XB are removed and the findings suggest that XB supply the needed enzyme.

It is not likely that *sams* is deleted from infected xD amoeba's genome so soon with the observed regularity, but the gene is probably present and is not expressed. It is possible that a regulatory factor(s) produced by XB binds to a site on *sams* of xD amoebae and prevents its transcription. A 1.3-kb cDNA (GenBank, Acc. No. U91602) encoding the amoeba SAMS has been cloned with a monoclonal antibody (mAb) against the 45-kDa protein as a probe [Choi et al., 1997]. The encoded protein has a 37.1 - 42.9% homology with SAMS of *E. coli* (*metX*), *Saccharomyces cerevisiae, Arabidopsis thaliana,* mouse and tomato. The cDNA-encoded protein is shorter by three amino acids at its N-terminus compared with those of the above three organisms and it has an extra internal stretch of 28 amino acids that are not found in other SAMS. However, the highly conserved consensus sequences of SAMS from various other organisms are well maintained in the amoeba protein. The gene product of *E. coli* transformed with amoeba cDNA has SAMS activities [Choi et al., 1997]. *E. coli* cells containing *sams* of XB show a SAMS activity which is more than 25 times that of *E. coli* transformed with pBSK only. Thus, the cDNA-encoded protein is an active SAMS.

These results suggest that symbiotic XB might become essential cellular components of xD amoebae by supplementing what xD amoebae themselves no longer produce after the integration of XB (Fig. 1). This is the first reported example in which symbionts have been shown to alter the host's gene expression and to block the production of an essential protein. This is also the first viable clue on why symbiont-bearing amoebae become dependent on the symbiotic bacteria.

The enzyme SAMS mediates the formation of SAM from methionine and ATP, and SAM is known to be the principal methyl donor in methylation of cytosine or adenine of DNA, rRNA, tRNA, various proteins, and small molecules that are important in various organisms [Usdin et al., 1979]. Deficiency in DNA methylation causes partial differentiation in mammalian cells [Razin & Riggs, 1980] and *Chlamydomonas* [Sager et al., 1981]. In yeast, the level of SAM affects meiosis and sporulation [Varma et al., 1985]. Since SAM is the precursor for many cellular constituents such as spermidine, methylated bases of nucleic acids, and cyclopropane fatty acids, mutation causing an extreme deficiency of SAM would be lethal [Greene et al., 1973]. It could be assumed that an amoeba's inability to transcribe *sams* would be equally fatal to the cell.

It is not clear how the transcription of *sams* is suppressed in xD amoebae. However, it should be possible to elucidate the mechanism by examining the inhibition of gene expression in xD amoebae compared to that in D amoebae, for example, by genomic footprinting [Mueller & Wold, 1989; Ausubel et al., 1992; cf. Leblanc & Moss, 1994; Shaw & Stewart, 1994; Kim et al., 1996]. By comparing genomic footprints of D and xD amoebae, it should be possible to detect a protected site(s) on or near the gene. The XB-produced S29x protein is a potential candidate to be such a DNA-binding regulatory protein.

2.7. MUTUAL DEPENDENCE.

Established XB symbionts and host amoebae are mutually dependent for survival. The host's dependence on symbionts has been demonstrated in several different ways such as nuclear transplantation and selective removal of XB from the host. If XB are selectively removed from xD amoebae, the latter lose viability but XB-deprived amoebae may be rescued by reintroducing XB [Lorch & Jeon, 1980]. X-Bacteria cannot live outside amoebae and many attempts have been made to grow XB in vitro in vain [Jeon, unpub.]. X-Bacteria accumulate host-synthesized actin, spectrin and myosin on their surfaces [Kim & Jeon, 1987; Oh & Jeon, 1996b], but their roles are not known.

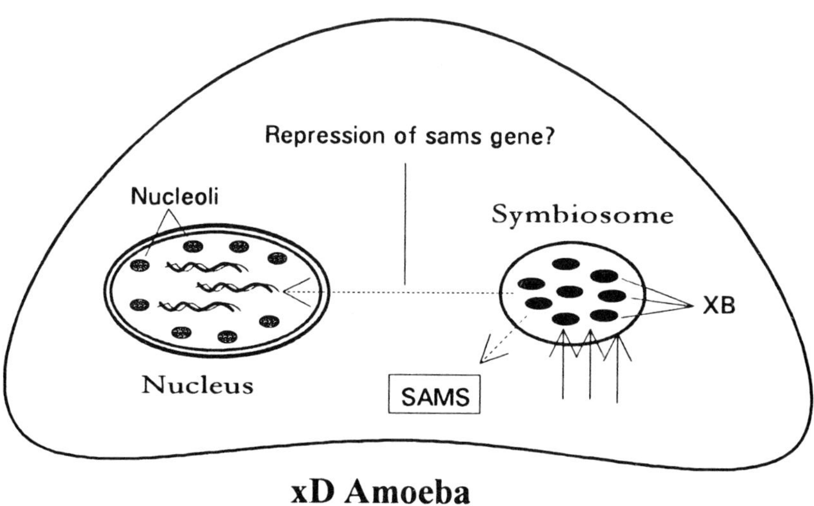

Figure 1. A schematic diagram to show possible molecular interactions between an amoeba and X-bacteria to cause the host's dependence on symbionts. The transcription of amoeba's *sams* gene is blocked but an alternative source for SAMS is provided by X-bacteria. Arrows represent the direction of possible molecular actions. (Adapted from Choi et al., 1997)

3. Host-Symbiont Interactions

3.1. EFFECT OF SYMBIOSIS ON THE HOST.

The presence of XB causes various physiological changes in host amoebae as listed below [Jeon & Jeon, 1976,1982; Ahn & Jeon, 1979,1983; Jeon, 1995-6].

3.1.1. *Accelerated Initial Growth.*
During the initial phase of infection for up to 12 months, newly infected amoebae grow faster than either established xD amoebae or XB-free D amoebae [Jeon & Ahn, 1978].

3.1.2. *Newly Acquired Temperature Sensitivity.*
xD Amoebae cannot grow at temperatures above 26° C and all die within 2 weeks [Jeon & Ahn, 1978]. The death of amoebae is preceded by the disappearance of XB, apparently by digestion [Lorch & Jeon, 1980]. Since amoebae are dependent on their symbionts for survival, they die after losing symbionts. Interestingly, the disappearance of XB is in turn preceded by the disappearance of their plasmids [Jeon, unpub.]. Thus, temperature sensitivity of xD amoebae may be traced to that of bacterial plasmids.

3.1.3. *Symbiont-Produced Proteins.*
Several new proteins are found in the cytoplasm of xD amoebae that are produced by XB. Among them, the most prominent is a 29-kDa S29x protein, and S29x has been suspected to be one of the components required for the survival of host amoebae. When protein synthesis by symbionts is selectively inhibited with chloramphenicol or rifampicin, xD amoebae die while symbiont-free D amoebae continue to grow under the same condition [Kim & Jeon, 1986].

3.1.4. *Altered Nucleocytoplasmic Compatibility.*
Nuclei of xD amoebae are not compatible with the cytoplasm of D amoebae, the original strain [Jeon & Jeon, 1976]. When D amoebae are infected with XB either by intracellular injection or by induced phagocytosis, their nuclei become incompatible with the cytoplasm of D amoebae within a few weeks [Lorch & Jeon, 1981-2]. Thus, XB cause nucleocytoplasmic incompatibility among amoebae quickly.

3.1.5. *Nuclear Lethal Effect.*
Nuclei of xD amoebae are not only incompatible with the cytoplasm of D amoebae but also exert a strong lethal effect on D amoebae [Lorch & Jeon, 1981]. When the nucleus of an xD amoeba is transplanted into D amoeba containing its own nucleus, the latter loses viability. If D amoebae are newly infected with XB, their nuclei acquire the ability to exert a lethal effect within a few cell cycles [Lorch & Jeon, 1982]. The nuclear lethal effect is due to a high-M_r protein of 200 kDa, newly synthesized by xD amoebae as a result of XB infection [Lorch et al., 1985].

3.1.6. *Specificity of Host-Symbiont Relationship.*
X-Bacteria are able to infect only amoeboid protozoa. Thus, they establish stable

association even with a different genus *Chaos carolinensis* [Jeon & Jeon, 1982], but they do not grow inside other cells such as ciliates or *Xenopus* eggs [Jeon, unpub.].

3.1.7. *Increased Sensitivity to Starvation, Overfeeding and Overcrowding.*
Symbiont-containing xD amoebae are more sensitive to starvation, overfeeding, crowding than are normal amoebae [Jeon, unpub.].

4. Concluding Remarks

The amoeba-XB symbiosis system is unique and novel, in which the host and symbionts became associated and developed a stable symbiosis within a few years under continuous observations. It is different from other known cases of endosymbiosis in that the history of the establishment of symbiosis is known, symbionts bring about the host's dependence for survival within a short period of time (200 amoeba generations or 18 months), and cellular character changes caused by symbiosis can be reproduced *ad libitum* under laboratory conditions. The symbionts, XB, are different from other bacterial inclusions [Roth & Daniels, 1961; Rabinovitch & Plaut, 1962; Drozanski et al., 1984; Springer et al., 1992] in their large number present in each host cell, their morphology including the membranes surrounding them, and their relationship to the host nuclear genome.

One could postulate the following scheme for the involvement of a molecule such as SAMS in host-symbiont interactions in xD amoebae (Fig. 1): Infection of amoebae by XB would initiate a change in the amoeba's *sams* gene by a DNA-binding protein(s), blocking its transcription. Then in the absence of amoeba's SAMS, XB would supplement the genetic defect by providing their hosts with an alternative source of SAMS. Thus, when XB are removed from xD amoebae, amoebae would be deficient in SAMS, and hence methylating reactions using SAM as the methyl-group donor would be blocked. If such interference occurred in the processing of pre-rRNA methylation, the result would be an abnormal morphology of amoeba's nucleoli as has been actually observed [cf. Lorch and Jeon, 1980]. Thus, xD amoebae would not recover unless live XB were reintroduced. Otherwise, the damaged nucleus would have to be transplanted into normal xD cytoplasm for recovery.

Symbiogenesis of XB in amoebae is a good model with which to study various steps involved in the initial establishment and maintenance of endosymbiosis. The system is well suited to study the process of symbiont integration into host cells and the consequences in general, leading to the acquisition of new cell components and changes in cellular phenotypic characters arising from the production of new macromolecules by both the host and symbionts.

Acknowledgments.

I thank my present and former graduate students and co-workers for their contributions to the project. Our own research work as been supported in parts by grants from the National Science Foundation, University of Tennessee Faculty Research Funds, and

American Philosophical Society. I also thank Dr. J. A. MacCabe for his critical reading of the manuscript.

References

Abshire, K.Z. and Niedhardt, F.C. (1993) *J. Bacteriol.* 175, 3734-3743.
Abu Kwaik, Y., Eisenstein, B.I. and Engleberg, N.C. (1993) *Infect. Immun.* 61, 1320-1329.
Abu Kwaik, Y., Fields, B.S. and Engleberg, C. (1994) *Infect. Immun.* 62, 1860-1866.
Ahn, G.S., Choi, E.Y. and Jeon, K.W. (1990) *Endocyt. Cell Res.* 7, 45-50.
Ahn, T.I. and Jeon, K.W. (1979) *J. Cell Physiol.* 98, 49-58.
Ahn, T.I. and Jeon, K.W. (1982) *Exp. Cell Res.* 137, 253-268.
Ahn, T.I. and Jeon, K.W. (1983) *J. Protozool.* 30, 713-715.
Ahn, T.I., Leeu, H.K., Kwak, I.H. and Jeon, K.W. (1991) *Endocyt. Cell Res.* 8, 33-44.
Ahn, T.I., Lim, S.T., Leeu, H.K. and Jeon, K.W. (1994) *Gene* 128, 43-49.
Arruda, S., Bomfim, G., Knights, R., Huima-Byron, T. and Riley, L.W. (1993) *Science* 261, 1454-1457.
Ausubel, F.M., Brent, R., Kingston, R.E., Moore, D.D., Smith, J.A., Seidman, J.G. and Struhl, K. (eds.) (1992) *Current Protocols in Molecular Biology*, John Wiley, New York.
Bangsborg, J.M., Cianciotto, N.P. and Hindersson, P. (1991) *Infect. Immun.* 59, 3836-3840.
Berger, K.H. and Isberg, R.R. (1993) *Mol. Microbiol.* 7, 7-19.
Berger, K.H. and Isberg, R.R. (1994) *Methods Cell Biol.* 45, 247-259.
Bozue, J.A. and Johnson, W. (1996) *Infect. Immun.* 64, 668-673.
Brand, B.C., Sadosky, A.B. and Shuman, H.A. (1994) *Mol. Microbiol.* 14, 797-808.
Brinkmann, H. and Martin, W. (1996) *Plant Mol. Biol.* 30, 65-75.
Buchmeier, N.A. and Heffron, F. (1990) *Nature* 248, 730-732.
Buchmeier, N.A. and Heffron, F. (1991) *Infect. Immun.* 59, 2232-2238.
Bull, L. and Fogarty, T.C. (1995) *Artf. Life* 2, 269-292.
Burgos, P.A., Velazquez, E. and Toro, N. (1996) *Mol. Plant Microbe. Interact.* 9, 843-849.
Chapman-Andresen, C. and Hayward, A.F. (1963) *Brit. Dutch Scand. Soc. Exp. Biol.*, Oxford England.
Choi, E.Y. and Jeon, K.W. (1992) *J. Protozool.* 39, 205-210.
Choi, E.Y., Ahn, G.S. and Jeon, K.W. (1991) *BioSystems* 25, 205-212.
Choi, J.Y., Lee, T.W., Jeon, K.W. and Ahn, T.I. (1997) *J. Eukaryot. Microbiol.* 44, 412-419.
Cianciotto, N.P. and Fields, B.S. (1992) *Proc. Natl. Acad. Sci. USA* 89, 5188-5191.
Cianciotto, N.P., Stamos, J.K. and Kamp, D.W. (1995) *Curr. Microbiol.* 30, 247-250.
Cianciotto, N.P., O'Connell, W., Dasch, G.A. and Mallavia, L.P. (1995) *Curr. Microbiol.* 30, 149-153.
Cohen, A. (1957) *J. Biophys. Biochem. Cytol.* 3, 859-866.
Drozanski, W. (1963) *Acta Microbiol. Pol.* 12, 3-8.
Drozanski, W., Drozanska, D. and Wicinska, M. (1984) *Acta Microbiol. Pol.* 33, 195-206.
Eissenberg, L.G., Goldman, W.E. and Schlesinger, P.H. (1993) *J. Exp. Med.* 177, 1605-1611.
Engleberg, N.C., Carter, C., Weber, D.R., Cianciotto, N.P. and Eisenstein, B.I. (1989) *Infect. Immun.* 57, 1263-1270.
Fayet, O., Ziegelhoffer, T. and Georgopoulos, C. (1989) *J. Bacteriol.* 171, 1379-1385.
Finlay, B.B. and Cossart, P. (1997) *Science* 276, 718-725.
Fritsche, T.R., Gautom, R.K., Seyedirashti, S., Bergeron, D.L. and Lindquist, T.D. (1993) *J. Clin. Microbiol.* 31, 1122-1126.
Galan, J.E. (1994) *Curr. Opin. Immunol.* 6, 590-595.
Gray, M.W. (1992) *Int. Rev. Cytol.* 141, 233-357.
Gray, M.W. (1993) *Curr. Opin. Genet. Dev.* 3, 884-890.
Greene, R.C., Hunter, J.S.V. and Coch, E.H. (1973) *J. Bacteriol.* 115, 57-67.
Hacker, J., Otto, M., Wintermeyer, E., Ludwig, B. and Fischer, G. (1993) *Int. J. Med. Microbiol. Virol.* 278, 348-358.
Hall, B.F. and Joiner, K.A. (1991) *Immunol. Today* 12, A22-27.
Hall, J.V. and Voelz, H. (1985) *J. Parasitol.* 71, 89-95.
Han, J.H. and Jeon, K.W. (1980) *J. Bacteriol.* 141, 1466-1469.
Hart, P.D. and Young, M.R. (1979) *Exp. Cell Res.* 118, 365-375.
Healy, A.M., Mariethoz, E., Pizurki, L. and Polla, B.S. (1992) *Aging and Cellular Defense Mechanisms* 663, 320-330.

Henze, K., Badr, A., Wettern, M., Cerff, R. and Martin, W. (1995) *Proc. Natl. Acad. Sci. USA* 92, 9122-9166.
Hoffman, P.S., Houston, L. and Butler, C.A. (1990) *J. Bacteriol.* 158, 3380-3387.
Holtzman, E. (1989) *Lysosomes*, Springer-Verlag, New York, 439 pp..
Horner, D.S., Hirt, R.P., Kilvington, S., Lloyd, D. and Embley, T.M. (1996) *Proc. R. Soc. Lond.* B263, 1053-1059.
Horwitz, M.A. and Maxfield, F.R. (1984) *J. Cell Biol.* 99, 1936-1943.
Ishibashi, Y. and Arai, T. (1990) *FEMS Microbiol. Immunol.* 2, 35-43.
Ishibashi, Y., Nobuta, K. and Arai, T. (1992) *Microb. Pathog.* 13, 317-323.
Ishikawa, H. (1984) *Insect Biochem.* 14, 417-425.
Jeon, K.W. (1995) *Trends Cell Biol.* 5, 137-140.
Jeon, K.W. (1996) *Jap. J. Protozool.* 29, 11-19.
Jeon, K.W. and Ahn, T.I. (1978) *Science* 202, 635-637.
Jeon, K.W. and Danielli, J.F. (1971) *Int. Rev. Cytol.* 30, 49-89.
Jeon, K.W. and Hah, J.C. (1977) *J. Protozool.* 24, 289-293.
Jeon, K.W. and Jeon, M.S. (1976) *J. Cell. Physiol.* 89, 337-347.
Jeon, K.W. and Jeon, M.S. (1982) *J. Protozool.* 29, 493A.
Jeon, K.W. and Lorch, I.J. (1967) *Exp. Cell Res.* 48, 236-240.
Kawahara, K., Dejsirilert, S., Danbara, H. and Ezaki, T. (1992) *FEMS Microbiol. Lett.* 96, 129-134.
Keeling, P.J. and Doolittle, M.F. (1997) *Proc. Natl. Acad. Sci. USA* 94, 1270-1275.
Kim, H.B. and Jeon, K.W. (1986) *Endocyt. Cell Res.* 3, 299-309.
Kim, H.B. and Jeon, K.W. (1987) *J. Protozool.* 34, 393-397.
Kim, K.J. and Jeon, K.W. (1994) *Mol. Biol. Cell* Suppl. 5, 452a.
Kim, K.J. and Jeon, K.W. (1995) *Molec. Biol. Cell* 6 Suppl., 108a.
Kim, K.J., Na, Y.E. and Jeon, K.W. (1994) *Infect. Immun.* 62, 65-71.
Kim, S.W., Ahn, I.M. and Larsen, P.R. (1996) *Mol. Cell Biol.* 16, 4465-4477.
Leblanc, B. and Moss, T. (1994) DNase I footprinting, in *DNA-Protein Interactions: Principles and Protocols*, G.G. Kneale (ed.), pp. 1-10.
Liaud, M.F., Brandt, U., Scherzinger, M. and Cerff, R. (1997) *J. Mol. Evol.* Suppl. 44, 1:S28-37.
Lorch, I.J. and Jeon, K.W. (1980) *J. Protozool.* 27, 423-426.
Lorch, I.J. and Jeon, K.W. (1981) *Science* 211, 949-951.
Lorch, I.J. and Jeon, K.W. (1982) *J. Protozool.* 29, 468-470.
Lorch, I.J., Kim, H.B. and Jeon, K.W. (1985) *J. Protozool.* 32, 745-746.
Ludwig, B., Rahfeld, J., Schmidt, B., Mann, K., Wintermeyer, E., Fischer, G. and Hacker, J. (1994) *FEMS Microbiol. Lett.* 118, 23-30.
Lundermose, A.G., Rouch, D.A., Penn, C.W. and Pearce, J.H. (1993) *J. Bacteriol.* 175, 3669-3671.
Margulis, L. (1993) *Symbiosis in Cell Evolution: Microbial Communities in the Archean and Proterozoic Eons*, 2nd Edn., W.H. Freeman, New York, 452 pp.
Marra, A., Blander, S.J., Horwitz, M.A. and Shuman, H.A. (1992) *Proc. Natl. Acad. Sci. USA* 89, 9607-9611.
Maynard Smith, J. and Szathmary, E. (1995) *The Major Transitions in Evolution*, Freeman, Oxford.
McFadden, G.I., Gilson, P.R., Hofman, C.J., Adcock, F.J. and Maier, U.G. (1994) *Proc. Natl. Acad. Sci. USA* 91, 3690-3694.
Mo, Y.Y., Cianciotto, N.P. and Mallavia, L.P. (1995) *Microbiology* 141, 2861-2871.
Morioka, M. and Ishikawa, H. (1992) *J. Biochem.* 111, 431-435.
Morioka, M. and Ishikawa, H. (1993) *J. Biochem.* 114, 468-472.
Morioka, M., Muraoka, H. and Ishikawa, H. (1993) *J. Biochem.* 114, 246-250.
Moulder, J.W. (1985) *Microbiol. Rev.* 49, 298-337.
Moulder, J.W. (1991) *Intracellular Parasitism*, CRC Press, Boca Raton.
Mueller, ,P. and Wold, B. (1989) *Science* 246, 780-786.
Nardon, P. (1995) *Bull. Soc. Zoo. Fr.* 120, 397-406.
O'Connell, W.A., Bangsborg, J.M. and Cianciotto, N.P. (1995) *Infect. Immun.* 63, 2840-2845.
Oh, S.W. and Jeon, K.W. (1996a) *Mol. Biol. Cell* Suppl. 7, 299a.
Oh, S.W. and Jeon, K.W. (1996b) *J. Eukaryot. Microbiol.* 43, 7A.
Ohtaka, C., Nakamura, H. and Ishikawa, H. (1992) *J. Bacteriol.* 174, 1869-1874.
Pak, J.W. (1996) *Characterization of a novel s29x gene and its product of symbiotic bacteria in Amoeba proteus*, Ph.D. Thesis, University of Tennessee, Knoxville.
Pak, J.W. and Jeon, K.W. (1994) *Mol. Biol. Cell* Suppl. 5, 375a.
Pak, J.W. and Jeon, K.W. (1995) *Mol. Biol. Cell* 6 Suppl., 398a.
Pak, J.W. and Jeon, K.W. (1996) *Gene* 171, 89-93.

Pak, J.W. and Jeon, K.W. (1997) *J. Eukaryot. Microbiol.*, in press.
Park, M.S. and Jeon, K.W. (1988) *Endocyt. Cell Res.* 5, 215-224.
Park, M.S. and Jeon, K.W. (1990) *Endocyt. Cell Res.* 7, 37-44.
Pfeifer, U. (1987) Functional morphology of the lysosomal apparatus, in *Lysosomes: Their Role in Protein Breakdown*, H. Glaumann, and E.J. Phillips (eds.), Academic Press, New York, pp. 3-59.
Phillips, N.J., Apicella, M.A., Griffiss, J.M. and Gibson, B.W. (1993) *Biochemistry* 32, 2003-2012.
Polla, B.S. (1991) *Immunoparasitol. Today* A38-A41.
Proca-Ciobanu, M., Lupascu, G.H. and Petrovic, A.L. (1975) *Int. J. Pasitol.* 5, 49-54.
Rabinovitch, M. and Plaut, W. (1962) *J. Cell Biol.* 15, 535-540.
Razin, A. and Riggs, A.D. (1980) *Science* 210, 604-610.
Roth, L.E. and Daniels, E.W. (1961) *J. Biophys. Biochem. Cytol.* 9, 317-323.
Russell, D.G. (1995) *Curr. Opin. Immunol.* 7, 479-484.
Sadosky, A.B., Wiater, L.A. and Shuman, H.A. (1993) *Infect. Immun.* 61, 5361-5373.
Sager, R.C., Grabowy, C. and Sand, H. (1981) *Cell* 24, 41-47.
Sampson, J.S., O'Connor, S.P., Holloway, B. and Plikaytis, B.B. (1990) *Infect. Immun.* 58, 3154-3157.
Sato, S., Ishida, M. and Ishikawa, H. (Eds.) (1994) *Endocytobiology V*, Tubingen Univ. Press, Tubingen, 586 pp.
Schenk, H.E., Bayer, M.G., Maier, T.L., Lutte, A., Gebhart, U. and Stevanovic, S. (1992) *Z. Naturforsch.* 47c, 387-393.
Shaw, P.E. and Stewart, A.F. (1994) Identification of protein-CNA contacts with dimethyl fulfate, in *DNA-Protein Interactions: Principles and Protocols*, G. G. Kneale (ed.), Humana Press, Totowa, NJ, pp. 79-87.
Sonesson, A., Jantzen, E., Bryn, K., Larson, L. and Eng, J. (1989) *Arch. Microbiol.* 153, 72-78.
Springer, N., Ludwig, W., Drozanski, W., Amann, R. and Schleifer, K.H. (1992) *FEMS Microbiol. Lett.* 96, 199-202.
Sturgill-Koszycki, S., Schlesinger, P.H., Charkraborty, P., Haddix, P.L., Collins, H.L., Fok, A.K., Allen, R.D., Gluck, S.L., Heuser, J. and Russell, D.G. (1994) *Science* 263, 678-681.
Usdin, E., Brochardt, R.T. and Creveling, C.R. (1979) *Transmethylation*, Elsevier/North-Holland Biomedical Press, New York.
Varma, A., Freese, E.B. and Freese, E. (1985) *Mol. Gen. Genet.* 201, 1-6.
Wintermeyer, E., Ludwig, B., Steinert, M., Schmidt, B., Fischer, G. and Hacker, J. (1995) *Infect. Immun.* 63, 4576-4583.
Wolstenholme, D.R. (1966) *Nature* 211, 652-653.
Wolstenholme, D.R. and Plaut, W. (1964) *J. Cell Biol.* 22, 505-513.
Yagita, K., Matias, R.R., Yasuda, T., Natividad, F.F., Enriquez, G.L. and Endo, T. (1995) *Parasitol. Res.* 81, 98-102.

Biodata of the **J. P. Hackstein et al.** authors of ***SYMBIOTIC ASSOCIATIONS BETWEEN METHANOGENIC ARCHAEA, PROTISTS AND METAZOA: EVOLUTIONARY IMPLICATIONS.***

Dr. Johannes P. Hackstein is an associate Professor at the Department of Microbiology and Evolutionary Biology at the University of Nijmegen, The Netherlands. He received his Ph.D. from the University of Cologne in 1976. Later he studied the genetics of spermatogenesis in *Drosophila melanogaster* and *D. hydei* at the Friedrich-Miescher-Laboratorium of the Max-Planck-Gesellschaft (Tübingen, Germany) and at the Department of Genetics at the University of Nijmegen. Since 1992 he is head of the research group "Anaerobic Protists." His main interests are the symbioses between animals and methanogenic archaea, the evolution of anaerobic protists, and spermatogenesis.
E-Mail: **Hack@sci.kun.nl**

J.H.P. Hackstein

Angela H.A.M. van Hoek and **Wander W. Sprenger** are graduate students at the Department of Microbiology and Evolutionary Biology. They are studying the physiology, ecology and evolution of anaerobic protists and methanogenic archaea.

Dr. Jörg Rosenberg is a senior scientist at the Department of animal physiology at the Ruhr-Universität Bochum, Germany. After receiving his Ph.D. (1974) at the University of Giessen, Germany, he became engaged in various projects concerning the morphology, ecology and ultrastructure of arthropods. His main interests are the ultrastructure and the development of arthropods.

A.H.A.M. Van Hoek

J. Rosenberg

W.W. Sprenger

J. Seckbach (ed.), Enigmatic Microorganisms and Life in Extreme Environments, 599–611.

SYMBIOTIC ASSOCIATIONS BETWEEN METHANOGENIC ARCHAEA, PROTISTS AND METAZOA: EVOLUTIONARY IMPLICATIONS

J.H.P. HACKSTEIN, A.H.A.M. VAN HOEK, W.W. SPRENGER AND J. ROSENBERG*
Dept. Microbiology and Evolutionary Biology, Fac. Science, Catholic University of Nijmegen, Toernooiveld 1, NL-6525 ED Nijmegen, The Netherlands
**** Dept. Animal Physiology, Ruhr-Universität Bochum, Postfach 10 21 48, D-44780 Bochum, Germany***

1. Introduction

Long-lasting associations between organisms that belong to different taxa are described as "symbiosis". They are very complex phenomena and, as we will emphasize in this essay, "symbiosis" can have a meaning beyond an exchange of nutrients and vitamins (Margulis 1976; Rennie 1992). Symbiosis requires a continuous cross-talk between organisms in the form of informational, physiological and behavioural interactions. Additionally, symbiotic associations have their peculiar evolutionary histories, involving co-evolution of host and symbiont (Douglas 1994; Moran and Baumann 1994; Page and Hafner 1996; Werren 1997; Hackstein 1997).

2. The Gastro-intestinal Tract of Animals: Complex Open Ecosystems

Herbivorous vertebrates are largely dependent on the metabolic activities of intestinal microorganisms. Bacteria, methanogenic archaea, anaerobic protists and chytridiomycete fungi co-exist in specialized compartments of the digestive tract such as rumina and caeca (Hungate 1966; Clarke and Bauchop 1977; Parra 1978; Hobson 1988; Langer and Snipes 1991; Langer 1994). The intestinal symbionts form a complex microbial ecosystem which degrades insoluble plant polymers that are resistant to the host's digestive enzymes. Such microbial communities are extremely complex and numerous. In the bovine rumen, for example, more than 100 different species of bacteria grow up to populations of about 10^{11} cells per ml of rumen fluid; some 50 species of ciliates occur at cell densities exceeding 10^5 cells per ml, and a few species of anaerobic fungi are not only present with a substantial vegetative biomass, they also contribute more than 10^3 flagellated spores per ml (Hungate 1966; Hobson 1988; Williams and Coleman 1991).

The intestinal microbiota of non-ruminant herbivorous animals form such highly complex open ecosystems that the apparent stability of these biota presents a great enigma (c.f. Finlay et al. 1997). This phenomenon becomes even more puzzling if one considers the fact that mammals are born with sterile gastro-intestinal tracts. All intestinal microorganisms are acquired after birth, and the "adult" gut microbiota are the result of a long-lasting colonization process. Substantial changes in the composition of the microbiota occur when the diet of the young mammal changes gradually from milk to solid food during the weaning period (c.f. Eisenberg 1993). It has been shown that the colonization by microorganisms and the differentiation of the intestinal tract requires a sophisticated cross-talk between the microorganisms and their host (Bry et al. 1996).

3. Host-dependent Constraints for the Gut Microbiota

Morphological adaptations of the gastro-intestinal tract to host symbiotic microbiota evolved several times, involving different parts of the gut (Langer 1988,1991; Langer and Snipes 1991). Apparently, there were a number of constraints that shaped the evolution of the gastro-intestinal tract. One obvious constraint is the size of the animal. Extant foregut fermenters, e.g. ruminants, range in size from approximately 3 kg to 1,500 kg fresh-weight (Fig. 1).

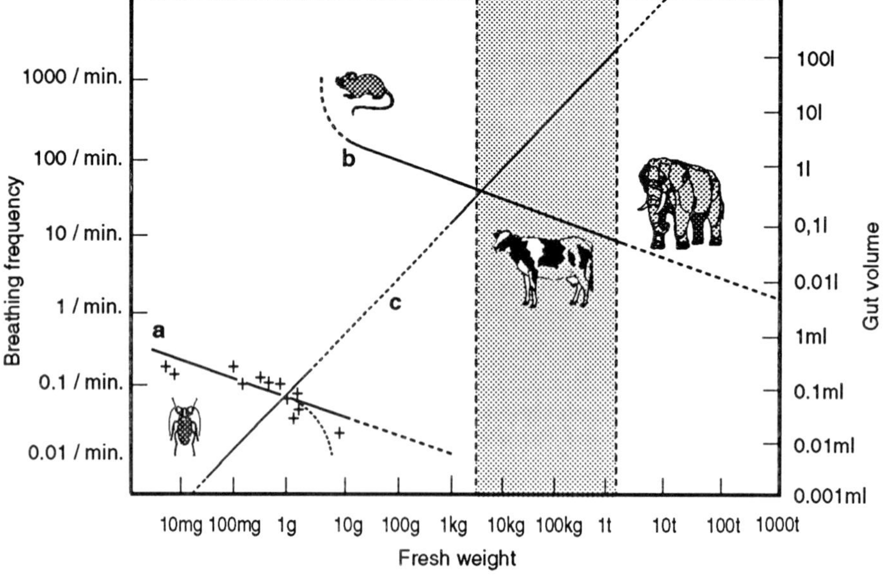

Fig. 1: 'Cockroach to elephant plot': Gut-volume, breathing frequencies and body weight are correlated in a logarithmic way. Whereas the correlation between body weight and gut-volume is constant over the whole range from cockroach larva to elephant (Hackstein and Cazemier, unpublished), the correlation between breathing frequency and body weight indicates that insects differ substantially in their metabolic activities (Bijnen et al. 1995, 1996). Foregut fermenters are restricted to body sizes between 3 to 1,500 kg.

It has been postulated that the metabolic reactions in the rumen might constrain the size of the ruminant, but the question of why hindgut fermenting animals can exist beyond these size limits remains unanswered (Prins and Kreulen 1990). Elephants and rhinos represent the largest extent hindgut-fermenting, land-living animals. Their adult fresh-weights far exceed 1,500 kg, and fossil evidence suggests that considerably larger animals that relied on alloenzymatic digestion also existed (Langer 1994). The possible existence of a lower weight-limit for hindgut-fermenting is even less clear: mice, for example, have a weight of less than 30 g and possess elaborate differentiations of the intestinal tracts (Snipes 1981). However, it remains unclear whether their intestinal biota digest plant polymers; there is some evidence that these microbiota might be involved in detoxifying the diet (Prins 1977; Perrin 1987).

On the other hand, studies on insects such as termites and cockroaches have revealed that tiny hindgut-fermenting insects are capable of maintaining complex microbiota which allow alloenzymatic digestion of insoluble plant polymers at a mg scale (Breznak 1982; Cruden and Markovetz 1987; Cazemier et al. 1997a,b; Brune 1998). Even more astonishing, scaling of the hindguts of cockroaches of various sizes reveals that the ratio between gut volume and fresh-weight is constant and nearly identical to the corresponding correlation in mammals (Fig. 1; Parra 1978).

4. The Rationale

The high complexity of the intestinal microbiota limits systematic studies on the specificity of the microbe host associations. Since many - if not the vast majority - of bacterial and eukaryotic symbionts have not yet been cultured, the functional analysis of these communities is seriously hampered. Only culture-independent techniques such as DGGE/TGGE analyses of PCR-amplified rDNA genes and FISH allow a straightforward analysis of such highly complex microbial communities (Ward et al. 1992; Amman et al. 1995; Wawer and Muyzer 1995). However, these techniques provide only rather indirect information about the potential metabolic functions of symbiotic microorganisms. Therefore, the use of indicator-organisms for fermenting microbiota is a reasonable alternative.

Methanogenic archaea are very suitable indicator-organisms for the detection of fermenting microbiota. Their presence requires anaeroby and the availability of fermentation products such as H_2, CO_2, methanol, formate, or acetate as substrates for methane biosynthesis (Miller and Wolin 1986). Methane production by intestinal methanogens is easily detected by gas chromatography of breath or faecal samples. Photoacoustic laser technologies even allow the recording of fast, cyclic releases of methane, CO_2, and water (Bijnen et al. 1996). Moreover, the characteristic blue autofluorescence of the cofactor F_{420} allows the detection of individual methanogenic bacteria in faeces or biopsies by epifluorescence microscopy (Doddema and Vogels 1978). Since these techniques are inexpensive and fast, large scale screening for methanogenesis in the intestinal tracts of animals is feasible.

Systematic screening of more than 250 vertebrate species for methane emissions via breath or faeces has revealed that most of the herbivorous animals hosted fermenting microbiota that included methanogens (Hackstein and van Alen 1996). Such animals frequently possess highly differentiated compartments in their gastro-intestinal tract. However, there were many exceptions: neither a vegetarian diet nor the presence of complex morphological adaptations guarantee the presence of methanogens in the gastro-intestinal tract. On the other hand, a protein-rich diet and a gut with a relatively simple structure do not prevent the presence of intestinal methanogenic archaea. One has to conclude that the availability of methanogenic substrates, anaerobic environments, and moderate temperatures, are not the only factors determining the presence of methanogens.

5. Evolutionary Aspects: Dollo's Law Rules Over the Symbiosis Between Intestinal Methanogens and Their Vertebrate Hosts

An analysis of the distribution of symbiotic methanogens among vertebrates reveals that certain taxa are completely devoid of species that host methanogens, irrespective of the nutritional preferences of the species (Hackstein and van Alen 1996; Hackstein 1997). In other taxa, all species (or at least the vast majority of species) share the ability to host symbiotic methanogens - again, without a consistent correlation with the feeding habits of particular hosts. Methanogenesis seems to depend on a character that obeys Dollo's law, i.e. if it is lost in a particular phylogenetic branch, "methanogenesis" will not reappear in any species of this branch - regardless of whether these species are herbivorous or not. The phylogenies of ostriches and rodents also strongly suggest that even the presence of highly differentiated digestive tracts and a vegetarian diet cannot prevent a loss of methanogenic symbionts (Hackstein and van Alen 1996; Hackstein 1997). These Dollo phylogenies strongly suggest that the presence of symbiotic methanogens depends on a genetic predisposition that - if lost in evolution - is unlikely to be restored by another mutational event.

The phylogenies of methanogenic animals strongly suggest that the presence of symbiotic methanogenic archaea in the gastro-intestinal tract was a necessary requirement for the evolution of foregut- and hindgut-fermenting structures. In taxa that lacked the predisposition to host intestinal methanogens before their radiation, e.g. insectivores, carnivores and chiropters, specialized adaptations of the digestive tract did not evolve (Langer 1988, 1991; Langer and Snipes 1991; Hackstein and van Alen 1996). In methanogenic taxa, on the other hand, the various foregut and hindgut fermenting structures evolved repeatedly from different parts of the gut (Fig. 2). In some taxa, such as the *Colobidae*, and the *Tayassuidae,* both fore - and hindgut- differentiations evolved.

A secondary loss of methanogens does not lead to a loss of the previously gained intestinal differentiations, but the phylogenetic trees suggest that the loss of methanogenesis was a crucial event in the evolution of small animals. Small-sized mammals in general have many generations per year and a large litter-size. Such an evolutionary strategy requires short lactation- and weaning periods that seem to be prohibitive for the establishment of a methanogenic intestinal microbiota (Hackstein et al. 1996; Hackstein and Langer 1997). Moreover, small-sized animals have high energy requirements and one might speculate whether anaerobic fermentations can satisfy this demand (Fig. 1; c.f. Prins and Kreulen 1990). Consequently, the exploitation of alternative diets, such as seeds, and the evolution of coprophagy or caecotrophy among rodents and caviomorphs might have been a necessity for small mammals. These arguments might explain how mammals can become smaller to occupy new ecological niches. However, all these hypotheses cannot explain why methanogenic intestinal microbiota did not reappear when a new evolutionary trend favoured again the rise of large herbivorous mammals such as giant panda. Moreover, these considerations fail to explain the presence of methanogenic fermenting microbiota in small insects.

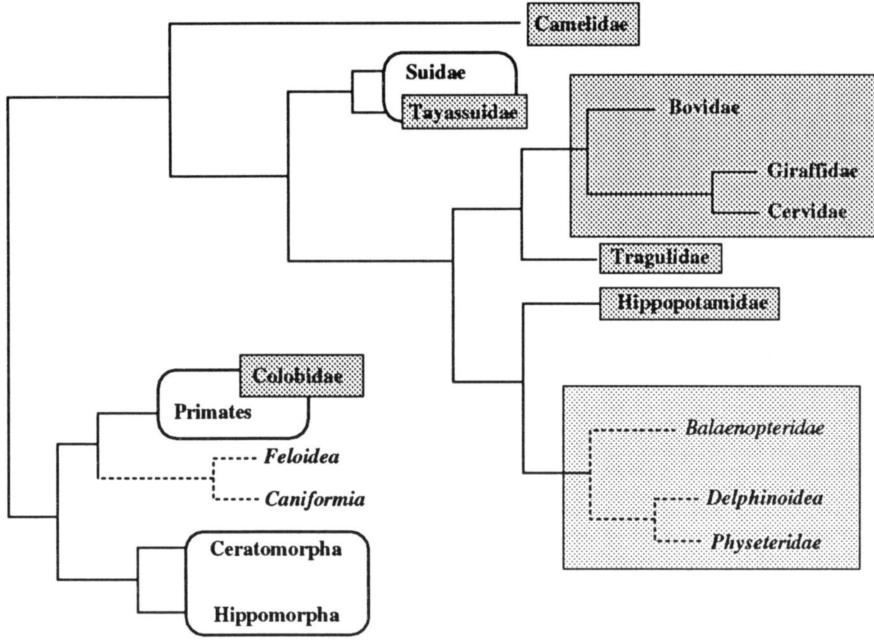

Fig. 2: Phylogenetic tree after Gatesy (1997). Foregut- and hindgut fermenting structures evolved independently and repeatedly in taxa that host methanogenic microbiota in their gastro-intestinal tract (solid lines). No such structures evolved in taxa that lack intestinal methanogenic archaea (broken lines). Rectangular boxes: foregut fermenting structures. Rounded boxes: hindgut-fermenting structures.

6. A Different World: Insects

The largest insects have a weight that equals that of the smallest rodents. Such insects have a hindgut with a volume well below 1 ml; in termites and larvae of cockroaches the hindgut can have a volume of less than 1 µl. In spite of their small size, such guts can host an extremely complex and dense population of bacteria, methanogenic archaea, yeasts, and - in certain hosts - also anaerobic protists (Breznak 1982; Cruden and Markovetz 1987; Cazemier et al. 1997a). The microbial consortia of such hosts produce methane, and they are capable of degrading plant-polymers to various degrees (Breznak 1982; Hackstein and Stumm 1994; Hackstein 1997; Cazemier et al. 1997b; Brune 1998)

The digestive tracts of insects represent extreme environments: the small size of the intestinal tract (below 1μl) causes surface/volume ratios of about 5000 m^2/m^3. Since the surface/volume ratio for the rumen of a cow is about 10 m^2/m^3, one can predict a substantially different composition of the microbiota of ruminants and hindgut fermenters on the one side, and termites and cockroaches on the other side (Brune 1998). Nevertheless, the evolution of the intestinal tracts of insects suggests an evolutionary strategy comparable to that seen in vertebrates (Fig. 1). One trend common to both phyla is the enlargement of the surface of the gastro-intestinal tracts. Among insects, this strategy could eventually lead to the evolution of a more or less aerobic gut. The cockroach *Nyctibora* might be an example of such an evolutionary strategy; the surface of the hindgut is substantially enlarged by invaginations of the hindgut epithelium (Fig. 3). Tracheae and tracheoles deliver oxygen nearly into the center of the hindgut. However, even under these conditions, methanogens exist and one can speculate whether efficient syntrophic associations with aerobic eubacteria guarantee an anaerobic micro-environment for the methanogens, or whether these methanogens are especially resistant against oxygen. Microsensor experiments provide evidence for the presence of extremely efficient sinks for O_2 and H_2 in the hindgut of termites and cockroaches (Brune 1998; Hackstein and Brune, unpublished). Thus, the extreme environment of insect guts provides very special ecological niches and, considering the antiquity of the insect phylum, it is not surprising that a long lasting evolution of host and gut microbiota led to a fascinating spectrum of morphological adaptations (Buchner 1965).

In clear contrast to the above mentioned *Nyctibora,* roze-chafer larvae evolved rather different adaptations to increase the efficiency of their gut. The most obvious adaptation is the high degree of compartmentalization of the gastro-intestinal tract of the latter organism. The midgut and the hindgut are rather voluminous and obviously, have rather different functions. The midgut is extremely alkaline (pH > 11) and hosts a microbial community that is rather different from the microbiota of the hindgut (Cazemier et al. 1997a,b). The hindgut has a near neutral pH and is populated by an extremely numerous and complex prokaryotic community of up to 10^{11} cells /ml. In the larval hindgut of all roze-chafers studied so far, bristle-like differentiations evolved (Hackstein and Stumm 1994). They increase the surface of the hindgut by a factor of 5 to 7 and serve for the attachment of complex prokaryotic biota, including methanogens. Electron microscopy has revealed that these structures represent a new type of epithelial differentiation that we have called "pseudosetae" (Rosenberg and Hackstein 1995; Hackstein et al. 1996). The cuticula that covers the pseudosetae is spongy, and in the central cylinder of the structure massive accumulations of microtubules and vacuoles are found (Fig. 4).

It is very tempting to assume that these microtubules have a function in the transport of intestinal fermentation products - presumably acetate and other short chain fatty acids - from the anaerobic lumen of the hindgut to an aerobic environment at the hindgut epithelium and the hemolymph. Interestingly, "pseudosetae" are lacking in the adult hindgut, that is also macroscopically clearly different. The adults have a diet rather different from the detrivorous larvae, and only low level methane emissions can be found. The microbiota of the adult hindgut seem to be rather different from the larval ones, and one might speculate that the larval microbiota may create a signal that causes the differentiation of the pseudosetae - in addition to their involvement in the degradation of food.

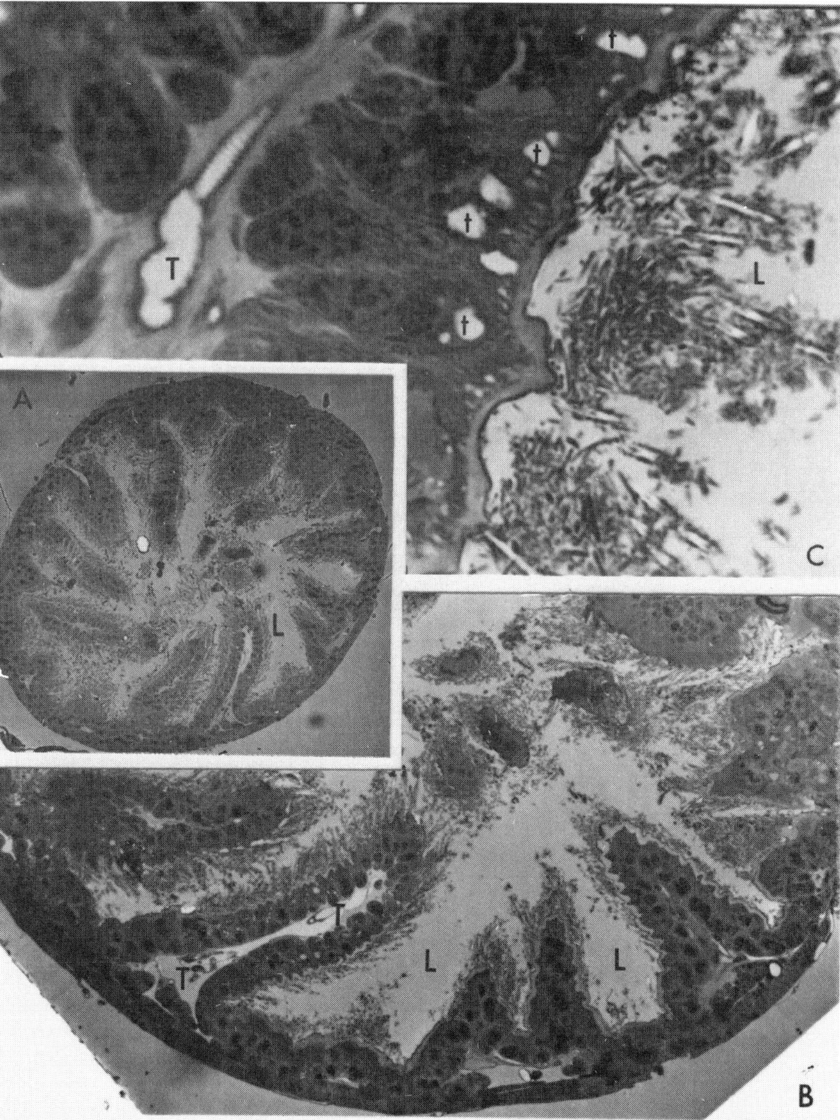

Fig. 3: Histological sections through the hindgut of the cockroach *Nyctibora spec.*. Numerous villus-like structures protrude into the lumen of the gut (A,B,C). Each villus is supported by at least one trachea (T) that branches into several tracheoles (t). They transport oxygen into the hindgut epithelium and the gut lumen (L).

7. 10 - 100 Millions of Insect Species: Niches for a Plethora of Highly Specialized Symbionts

It is beyond the scope of this paper to discuss the physiological implications of the symbiotic gut microbiota of insects and arthropods. There is evidence that both the host and the intestinal symbionts contribute to the digestion of the diet of insects (Buchner 1965; Cazemier et al. 1997a,b; Brune 1998). There is also evidence for the involvement of microorganisms in the detoxification of secondary plant substances, since many insects that consume a poisonous diet host intestinal microorganisms (Buchner 1965; c.f.

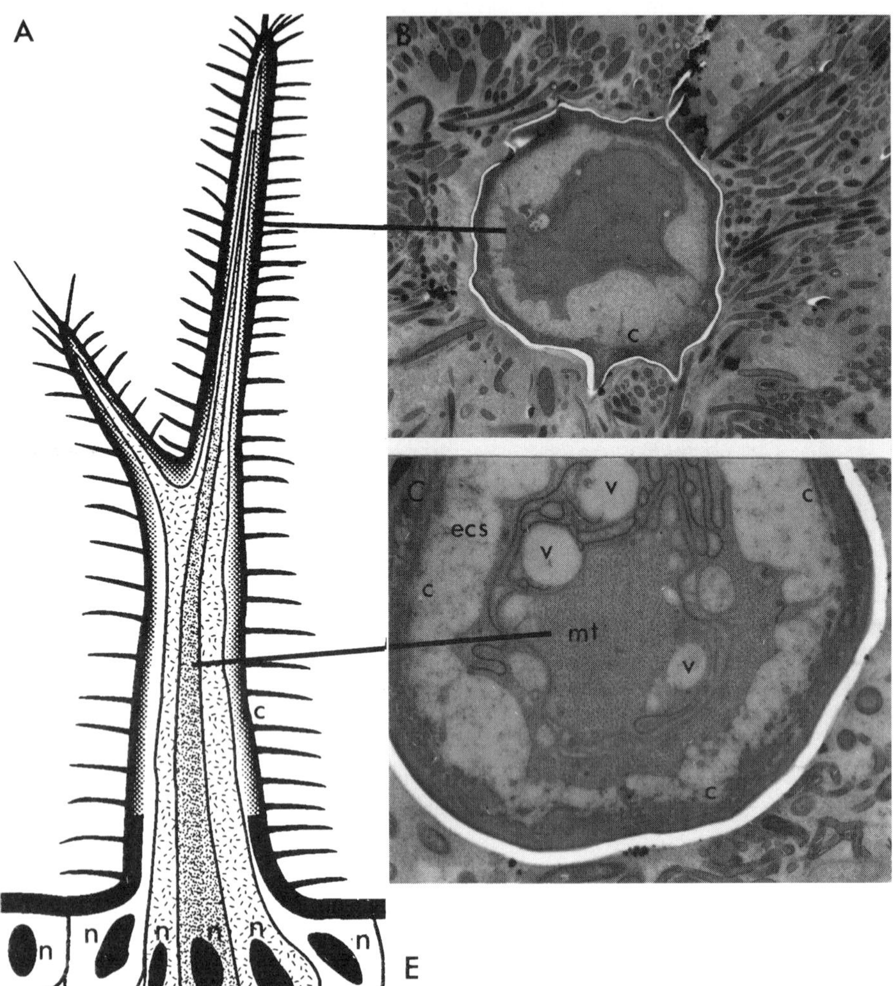

Fig. 4: Cartoon (A) and histological cross-sections (B,C) of a 'pseudoseta' from the hindgut of a larva of the rose-chafer *Pachnoda marginata*. Electron microscopy reveals thick bundles of microtubules (mt) that seem to be engaged into the transport of vacuoles (v) with fermentation products. The cuticle (c) of the pseudoseta is spongy; below the cuticle, extended extracellular spaces (ecs) seem to facilitate the uptake of fermentation products. n: nuclei of the gut epithelium. E: epithelium; N: nucleus

Hackstein and Stumm 1994; Hackstein et al. unpublished). The central question is whether the intestinal symbionts are host-specific or more or less accidentally acquired from the environment (Werren 1997). If they are host-specific, we might expect that a long lasting co-evolution occurred. Here, we will discuss the evolution of the symbiosis between methanogens, ciliates and cockroaches.

It has been known for a long time that protists are found in the intestinal tracts of termites and cockroaches (Leidy 1849; Cleveland 1926; Hoyte 1961a,b; Hackstein and Stumm 1994; Hackstein 1997). Phylogenetic aspects of the presence of intestinal flagellates in termites and the cockroach *Cryptocercus* have been discussed by Hennig (1981), Hackstein and Stumm (1994), and Grandcolas and Deleporte (1996). Systematic screening in 44 species of cockroaches has revealed that about 80% of the species emitted methane and hosted methanogens in the hindgut (Hackstein 1997). Symbiotic ciliates are found exclusively in species that also host intestinal methanogens. The non-producers - actually species lacking intestinal ciliates - clustered among the Blattellidae, suggesting that also among arthropods, the taxonomic position is more crucial for methanogenesis than feeding habits or intestinal differentiations.

Molecular studies revealed that the intestinal ciliates of cockroaches exhibit a high degree of host-specificity and only rarely do host switches occur (van Hoek et al. 1998a). The transmission of the ciliates is predominantly vertical, and the genetic isolation during the evolution of this symbiosis has been stringent enough to allow a substantial DNA sequence divergence. In addition, the methanogenic endosymbionts of the ciliates are specific for their protist-hosts and clearly different from their intestinal relatives (van Hoek et al. 1998b).

An intestinal methanogenic bacterium, *Methanomicrococcus blatticolus,* has been isolated from the intestinal tract of the American cockroach, *Periplaneta americana* (Sprenger et al. 1998). It differs substantially from the methanogenic endosymbionts of the ciliates (van Hoek et al. 1998b), and it occupies a niche similar to the recently described isolates of methanogenic archaea from the hindgut of the termite *Reticulitermes flavipes* (Leadbetter and Breznak 1996). However, *Methanomicrococcus blatticolus* has unique metabolic properties (Sprenger et al. 1998). It relies on the substrates methanol and hydrogen for growth and methane production. This specialism was rather unexpected for a gut-dwelling organism, since the methanol concentrations in the hindgut are below the detection level. *Methanomicrococcus blatticolus* is dependent on exogenous supply of acetate and HS-CoM, an essential co-factor in methane biosynthesis from methanol and hydrogen (Ferry 1993). These auxotrophies suggest that *Methanomicrococcus blatticolus* is highly adapted to its niche and that it has lost many biosynthetic capacities in the course of a long-lasting co-evolution with its host. Thus, the presence of highly specialized intestinal methanogenic archaea and ciliates seems to be the result of a co-evolution that might have lasted for more than 100 million years - comparable to the association between cockroaches and their flavobacterial endosymbionts that thrive in the mycetocytes of the host's fatbody (Bandi et al. 1994, 1995).

8. Concluding Remarks

Intestinal symbionts can have functions beyond degrading plant-polymers that are otherwise resistant to the hosts digestive enzymes. They were essential for the evolution of intestinal differentiations, and they also substantially influenced the evolution of their hosts. Notwithstanding the fact that intestinal tracts are open ecosystems, the available data suggest a high degree of host-specificity. Consequently, co-evolution of the gut microbiota and their hosts occurs, and the many potential hosts - especially among the arthropods - suggest that an incredible diversity of gut microbiota evolved since the radiation of insects occurred several hundred million years ago. The extant 10 to 100 million species of insects still provide myriads of tiny "Galapagos-Islands" on 6 legs - and it is reasonable to assume that they host a plethora of hitherto unknown microorganisms. Insects provide not only an unfathomed resource of enzymes, antibiotics and microbial biodiversity, they also host symbiotic microorganisms that bear the potential to evolve into life-threatening pathogens (Werren 1997; Min and Benzer 1997; Schilthuizen and Stouthamer 1997; Noda et al. 1997).

9. Acknowledgements

We are indebted to Prof. Dr. Allen Schwartz for improving the English phrasing and to Urs Hackstein for proof-reading.

10. References

Amann, R.I., Ludwig, W. & Schleifer, K-H. (1995). Microbiol. Rev. 59: 143-169

Bandi, C., Damiani, G., Magrassi, L., Grigolo, A., Fani, R. & Sacchi, L. (1994). Proc. R. Soc. London B 257: 43-48

Bandi, C., Sironi, M., Damiani, G., Magrassi, L., Nalepa, C.A., Laudani, U. & Sacchi, L. (1995). Proc. R. Soc. London B 259: 293-299

Bijnen, F.G.C., Kestler, P., Hackstein, J.H.P., Harren, F.J.M. & Reuss, J. (1995). In: *Refined CO-laser photoacoustic trace gas detection; observations of anaerobic processes in insects, soil, and fruit.* Thesis of F.G.C. Bijnen, Nijmegen, Chapter 8, pp. 141 - 172

Bijnen, F.G.C., Harren, F.J.M., Hackstein, J.H.P. & Reuss, J. (1996). Appl. Optics 35/27, 5357 - 5368

Breznak, J. A. (1982). Annu. Rev. Microbiol. 36, 323-343

Brune, A. (1998). Trends in Biotechnol. 16, 16-21

Bry, L., Falk, P.G., Midtvedt, T. & Gordon, J.I. (1996). Science 273, 1380 - 1383

Buchner, P. (1965) *Endosymbioses of animals with plant microorganisms.* Wiley, London

Cazemier, A.E., Hackstein, J.H.P., op den Camp, H.J.M., Rosenberg, J. & van der Drift, C. (1997a). Microbial Ecol. 33, 189-197

Cazemier, A.E., op den Camp, H.J.M., Hackstein, J.H.P. & Vogels, G.D. (1997b). Comp. Biochem. Physiol. 118 A, 101-109

Clarke, R.T.J. & Bauchop, T. (1977). *Microbial ecology of the gut.* Academic Press, New York

Cleveland, L.R. (1926). Quart. Rev. Biol. 1, 51-60

Cruden, D. L., & Markovetz, A. J. (1987). Annu. Rev. Microbiol. 41, 617-643

Doddema, H.J. & Vogels, G.D. (1978). Appl. Env. Microbiol. 36, 752 - 754

Douglas, A.E. (1994). *Symbiotic interactions.* Oxford University Press, Oxford, New York, Tokyo

Eisenberg, J. (1993). In: *Biology of the Heteromyidae.* Genoways, H.H. & Brown, J.H. (Edits.). American Society of Mammalogists, Special Publication 10. pp. 479 - 490

Ferry, J.G. (1993). *Methanogenesis, Ecology, Physiology, Biochemistry and Genetics.* Chapman and Hall, New York, London

Finlay, B.L., Maberly, S.C. & Cooper, J.J. (1997). Oikos 80, 209-213

Gatesy, J. (1997). Mol. Biol. Evol. 14, 537-543

Grandcolas, P. & Deleporte, P. (1996). Cladistics 12, 93-98
Hackstein, J.H.P. & Stumm, C.K. (1994). Proc. Natl. Acad. Sci. USA 91, 5441-5445
Hackstein, J.H.P. & van Alen, T.A. (1996). Evolution 50(2), 559 - 572
Hackstein, J.H.P., Langer, P. & Rosenberg, J. (1996). Env. Monit. Assessm. 42, 59 - 76
Hackstein, J. H. P. (1997). Antonie van Leeuwenhoek 72, 63-76
Hackstein, J.H.P. & Langer, P. (1997). In: *Eukaryotism and Symbiosis. Intertaxonic combination versus symbiotic adaptation.* Schenk, H.E.A., Herrmann, R.G., Jeon, K.W., Müller, N.E. & Schwemmler, W. (Edits.). Springer Verlag, Berlin. pp. 501-506
Hennig, W. (1981). *Insect Phylogeny.* John Wiley & Sons, Chichester
Hobson, P.N. (1988). *The rumen microbial ecosystem.* Elsevier Applied Science, London, New York
Hoyte, H.M.D. (1961a). Parasitology 51, 437-463
Hoyte, H.M.D. (1961b). Parasitology 51, 465-495
Hungate, R.E. (1966). *The rumen and its microbes.* Academic Press, New York
Langer, P. (1988). *The mammalian herbivore stomach. Comparative anatomy, function, and evolution.* Gustav Fischer Verlag, Stuttgart, New York
Langer, P. (1991). Verh. Dtsch. Zool. Ges. 84, 169-193
Langer, P. & Snipes, R.L. (1991). In: *Physiological aspects of digestion and metabolism in ruminants.* Tsuda, T., Sasaki, Y. & Kawashima, R. (Edits.). Academic Press, San Diego. pp. 349-384
Langer, P. (1994). In: *The digestive system in mammals.* Chivers, D.J. & Langer, P. (Edit.). Cambridge University Press, Cambridge, UK. pp. 9-23
Leadbetter, J.R. & Breznak, J.A. (1996). Appl. Env. Microbiol. 62, 3620-3631
Leidy, J. (1849). Proc. Acad. Natl. Sci. USA 4, 233
Margulis, L. (1976). Exp. Parasitol. 39, 277-349
Miller, T.L. & Wolin, M.J. (1986). Syst. Appl. Microbiol. 7, 223-229
Min, K-T. & Benzer, S. (1997). Proc. Natl. Acad. Sci. USA. 94, 10792-10796
Moran, N. & Baumann, P. (1994). Trends Ecol. Evol. 9, 15-20
Noda, H., Munderloh, U.G. & Kurtti, T. J. (1997). Appl. Env. Microbiol. 63, 3926-3932
Page, R.D.M. & Hafner, M.S. (1996). In: *New uses for new phylogenies.* Harvey, P.H., Leigh Brown, A.J., Maynard Smith, J. & Nee, S. (Edits.). Oxford University Press, Oxford, New York, Tokyo. pp. 255 - 270
Parra, R. (1978). In: *The ecology of arboreal foliovores.* G.G. Montgomery (Edit.). Smithonian Institution Press, Washington, D.C.. pp. 205-229
Perrin, M.R. (1987). S. Afr. J. Zool. 22: 67 - 76.
Prins, R.A. (1977): In: *Microbial ecology of the gut.* Clarke, R.T.J. & Bauchop, T. (Edits.) Academic Press, London, New York, San Francisco. pp. 73 - 183
Prins, R.A. & Kreulen, D.A. (1990). In: *The rumen ecosystem.* Hoshino, S., Minato, H. & Itabashi, H. (Edits.). Japan Scientific Societies Press, Tokyo. pp. 109-120
Rennie, J. (1992). Scientific American 266 (January) , 105-113
Rosenberg, J. & Hackstein, J.H.P. (1995). Verh. Westd. Entom. Tag 1994, Löbbecke-Mus., Düsseldorf. pp. 67-72
Schilthuizen, M. & Stouthamer, R. (1997). Proc. R. Soc. London B 264, 361-366
Snipes, R. L. (1981). Anat. Embryol. 162, 455 - 474
Sprenger, W. W., van Belzen, M., Rosenberg, J., Hackstein, J.H.P. & Keltjens, J.T. (1998).*submitted*
van Hoek, A.H.A.M., van Alen, T.A., Sprakel, V.I., Hackstein, J.H.P. & Vogels, G.D. (1998a). Mol. Biol. Evol., *i n press*
van Hoek, A.H.A.M., van Alen, T.A., Hackstein, J.H.P. & Vogels, G.D. (1998b). *submitted*
Ward, D.M., Bateson, M., Weller, R., & Ruff-Roberts, A.L. (1992). In: *Advances in microbial ecology.* Marshall, K.C. (Edit.). Plenum Press, New York. pp. 219-286
Wawer, C. & Muyzer, G. (1995). Appl. Env. Microbiol. 61, 2203-2210
Werren, J.H. (1997). Proc. Natl. Acad. Sci. USA. 94, 11154 - 11155
Williams, A.G., & Coleman, G.S. (1991). *The rumen protozoa.* Springer Verlag, New York

Biodata of **Birgitta Bergman, Kaye Bateman & Ulla Rasmussen,** authors of "***Cyanobacteria Symbiosis with Plants and Fungi.*** **"

Dr. Birgitta Bergman is a Professor of Plant Physiology at Stockholm University. She earned her Ph.D. from The University of Uppsala (Sweden) in 1977. Dr. Bergman's research interest concern both field and laboratory studies of cyanobacteria, particularly their importance in nitrogen fixation in symbioses with diverse hosts and in marine ecosystems.
E-Mail: bergmanb@botan.su.se

Dr. Kaye Bateman is a post-Doctoral Fellow at the Department of Plant Physiology in Stockholm University. She earned her Ph.D. from the University of Melbourne, Australia in 1994. Her research interest is on the symbiotic relationship between the cyanobacterial genus *Nostoc* and the plant host *Gunnera.*
E-Mail: bateman@botan.su.se

Dr. Ulla Rasmussen is a Research Associate who also works in the Department of Plant Physiology at Stockholm University. She earned her Ph.D. in Molecular Plant Pathology is from the University of Copenhagen, Denmark in 1992.
E-Mail: rasmussu@botan.su.se

J. Seckbach (ed.), Enigmatic Microorganisms and Life in Extreme Environments, 613–627.

CYANOBACTERIA IN SYMBIOSES WITH PLANTS AND FUNGI

BERGMAN, B., BATEMAN, K. & RASMUSSEN, U.
Department of Botany, Stockholm University,
S-106 91 Stockholm, Sweden

1. Introduction

In nature, interactions between prokaryotic and eukaryotic organisms are common. The prokaryote may form a pathogenic or symbiotic relationship with the eukaryotic organism. Symbiosis is here defined as an eukaryotic host that is living together with a prokaryote in a mutualistic rather than a pathogenic interaction. If pathogenic, eukaryotic defences may prevent or limit the infection while if symbiotic, the prokaryote occupies predestined eukaryotic cells or tissues.

Cyanobacteria are able to tolerate many different environments. One such environment is within eukaryotic host organisms. Cyanobacterial symbioses are highly mutualistic, and although only a few cyanobacterial genera are compatible, these interact with a variety of eukaryotic organisms (Rai, 1990a). The hosts are from all kingdoms: protists, animals, fungi and plants, and among plants there are representatives from each division. In plants, most organs may be infected and the cyanobacteria may be located extra- or intracellularly. Hence, symbiosis-compatible cyanobacteria must posses structural and metabolic flexibilities in order to tolerate the variety of conditions offered by the hosts. In addition, according to the endosymbiont theory, cyanobacteria are also progenitors of chloroplasts (Margulis, 1970; Douglas, 1994) and cyanobacteria-like cyanelles are found in some protists (Löffelhardt and Bohnert, 1994).

Although our knowledge about pathogenic and symbiotic interactions has widened considerably in recent years, certain areas are less well understood, for instance, the conditions offered to compatible prokaryotes on entering eukaryotes to form symbioses. In this chapter, this aspect as well as the range of challenges faced by both partners in order to establish cyanobacterial symbioses will be highlighted.

2. Cyanobacteria: Nature and Distribution

Cyanobacteria are a major and ecologically highly successful group of prokaryotes (Stanier and Cohen-Bazire, 1977). They have a world-wide distribution and colonize many terrestrial (neutral and alkaline) and aquatic (limnic and marine) habitats, from polar to tropical areas of the globe (Carr and Whitton, 1982; Dodds et al., 1995; Stal,

1995; Bergman, 1996; Capone et al., 1997). Many cyanobacteria can also adjust to a variety of extreme growth conditions such as salt marshes, hot springs or deserts (Carr and Whitton, 1982; Dodds et al., 1995) and can withstand many abiotic stresses (Tandeau de Marsac and Houmard, 1993; Grossman et al., 1994; Potts, 1996).

A key to their success is most likely their great structural and functional plasticities (Fay, 1992; Buikema and Haselkorn, 1993; Tandeau de Marsac and Houmard, 1993; Dodds et al., 1995). For instance, all cyanobacteria are photoautotrophs with a chlorophyll a based oxygenic photosynthesis and use light energy to capture carbon and generate ATP. In addition, some genera can live heterotrophically in the dark provided they are fed with appropriate exogenous carbohydrates. Cyanobacteria are also morphologically flexible, as they are capable of differentiating various cell types. Under nutrient limitations or other stresses many differentiate their vegetative cells into baeocytes (small reproductive cells) or into akinetes (thick-walled, perennial unicellular spores). Some cyanobacteria can differentiate their entire filaments into short and small celled filaments, hormogonia. These filaments are often motile and sometimes gas vacuolated and function as a means of dispersal. Furthermore, under nitrogen limiting conditions, differentiation of a subset of the vegetative cells into nitrogen-fixing cells, heterocysts, may occur in some species (Fay, 1992; Buikema and Haselkorn, 1993; Wolk et al., 1994). Heterocysts protect the oxygen labile nitrogen-fixing enzyme nitrogenase from inactivation and allow nitrogen fixation to occur even under aerobic conditions. Some cyanobacteria lacking heterocysts can also fix nitrogen but only under low ambient oxygen concentrations or in darkness (Fay, 1992; Bergman et al., 1997). Nitrogen fixation, together with an efficient ammonia assimilating machinery using glutamine synthetase (Fay, 1992), allows cyanobacteria to survive in nitrogen-poor areas. Furthermore, some cyanobacteria can withstand prolonged water stress due to the efficient water-holding capacity of their extracellular polysaccharides (Dodds et al., 1995; Potts, 1996). Cyanobacteria also show ultrastructural flexibility as they are capable of storing important macronutrients such as carbon, nitrogen and phosphate in distinct subcellular compartments (Shively, 1974). Under nutrient limitations, the cyanobacteria can mobilize these stored nutrients.

The adaptive nature of cyanobacteria is also seen in the ability of some to form symbioses with different eukaryotic organisms and to adjust their metabolism to conform to that of the host (Figure 1, Table 1).

3. Cyanobacterial-Eukaryotic Interactions: Integration and Evolution

As cyanobacteria are likely to be the progenitors of algal and plant chloroplasts, the intimacy of interactions with eukaryotes spans from total integration to form chloroplasts (Margulis, 1970; Giovannoni, 1988; Douglas, 1994), over the less integrated cyanelle-protist interactions (Löffelhardt and Bohnert, 1994), to the least integrated, those of present day cyanobacterial symbioses (Rai, 1990a; Bergman et al., 1992a; Bergman et al., 1996). As seen in Figure 1, eukaryotic hosts in cyanobacterial symbioses include only a few animals (some echiuroid worms and sponges) but several fungi as well as representatives from all plant divisions (Rai, 1990a). However, there are no known monocotyledonous hosts. Chloroplasts and cyanelles are obligate endosymbionts, while most fungal and plant symbioses are facultative and hence both partners can regain independence upon isolation. This has been shown for some lichen-, bryophyte-, cycad- and *Gunnera*-symbioses (Rai, 1990a; Bergman et al., 1992a; Johansson and Bergman, 1994). However, cyanobacteria in *Azolla* symbioses appear to be incapable of an independent life (Braun-Howland and Nierzwicki-Bauer, 1990).

The variation in the degree of intimacy suggests evolutionary differences among the interactions. The chloroplasts are probably the oldest being the most integrated (Douglas, 1994). These have lost part of their genome to the host nucleus. A smaller part of the cyanelle genome has also been integrated into the protist nucleus but the cyanelles are still surrounded by rudimentary cyanobacterial cell walls (Löffelhardt and Bohnert, 1994). In contrast, DNA transfer seems unlikely in the facultative cyanobacterial symbioses, except perhaps in the *Azolla* symbiosis. This suggests that cyanobacterial symbioses are the youngest, although the hosts given in Table 1 are among the oldest plants on earth (Kenrick and Crane, 1997). Hence, present day cyanobacterial symbioses "may represent relicts of ancient partnerships" (Sprent and Raven, 1985) rather than newly created associations.

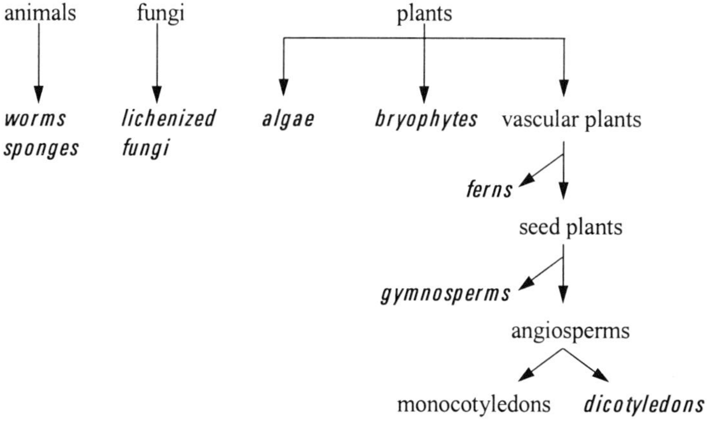

Figure 1. Eukaryotic organisms acting as hosts in present day cyanobacterial symbioses. Members within the italicized groups form symbioses with cyanobacteria.

4. Cyanobacteria in Symbioses

To be symbiotically competent, cyanobacteria have to fulfil a number of criteria, for instance: they must be present in the vicinity of and attracted to potential hosts, be able to avoid, suppress or overcome host defences and be capable of adapting their metabolism to the new niche offered. In addition, the cyanobacterial growth must be reduced to restrict the area of infection. The cyanobacteria must also alter their metabolism to exchange metabolites with the host. It is therefore not surprising that rather few cyanobacteria pass the transition from being free-living and self-supportive

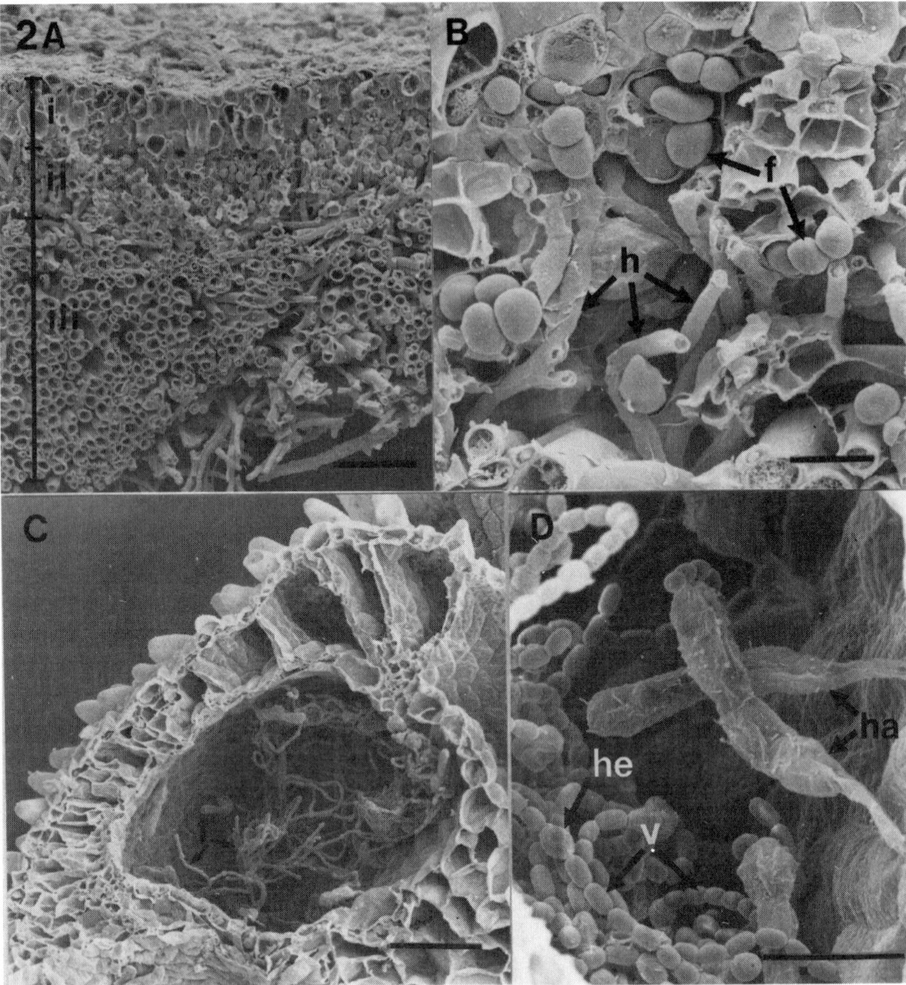

Figure 2. Scanning electron micrographs of cyanobacterial symbioses with fungal (A & B) and water fern (C & D) hosts.

A: A transverse section through a bipartite lichen (*Peltigera* sp.) thallus consisting of three distinct zones, upper cortex (i), cyanobacterial zone (ii) and lower medulla (iii). The cyanobacteria are extracellular and support the fungus with both nitrogen and carbon which is also the case in *Geosiphon*. Bar = 50 μm. B: The cyanobacterial zone showing fungal hyphae (h) intertwined amongst cyanobacterial filaments (f). Bar = 10 μm. C: An *Azolla* sp. leaf cavity containing numerous cyanobacterial filaments (f). As in bryophytes, the extracellularly located cyanobacteria are within a mucilage containing cavity in photosynthetic tissue and supply the host with nitrogen. Bar = 100 μm. D: The leaf cavity interior showing cyanobacterial filaments with vegetative cells (v) and heterocysts (he). The eukaryotic hair cells (ha) are involved in nutrient transfer between cyanobacteria and the plant. Bar = 25 μm.

to symbiotic and at least partially dependent on the host.

As seen in Table 1, by far the most common cyanobacteria in fungal and plant symbioses belong to one genus, the heterocystous filamentous genus *Nostoc* (Rai, 1990a; Bergman et al., 1992a; Dodds et al., 1995; Bergman et al., 1996). The diversity of cyanobacteria is largest in lichenized fungi. Besides *Nostoc*, the heterocystous genera *Fischerella* and *Scytonema* as well as some unicellular taxa, primarily *Gloeocapsa*, act as symbionts in lichens (Rai, 1990b). Moreover, representatives from the genus *Calothrix* are symbionts in the marine lichenized fungus, *Lichina* (Rai, 1990b; Janson et al., 1993) and in the cycad genus, *Encephalartos* (Lindblad and Bergman, 1990). However, the latter is alternatively infected by *Nostoc*. The cyanobacterial symbiont in the diatoms *Rhizosolenia* and *Hemiaulus* are short filaments with end heterocysts and are classified as *Richelia* spp. (Taylor, 1982; Janson et al. 1995).

As mentioned above, a geographic co-location is important for initiating symbiosis. *Nostoc* is both an ancient (late Precambrian, or older) (Schopf and Walter, 1982) and a very common (terrestrial and aquatic) cyanobacterial genus with a cosmopolitan distribution (Dodds et al., 1995; Potts, 1996). The likelihood that *Nostoc* is located close to a potential host is therefore high. The same is true for *Calothrix* which typically inhabit marine coastal zones where the host fungus *Lichina* occurs. The cyanobacterium, *Richelia* is also co-located with diatom hosts in marine oceans (Bergman, 1996).

The targeting and attraction of a cyanobacterium to its host probably involve host derived chemotactic signals, while the tighter binding may involve surface structures like, for example, carbohydrates and lectins (Kobiler et al., 1982; Meeks, 1990; Bergman et al., 1996; Knight and Adams, 1996; Schüßler et al., 1997). To successfully enter the host, the prokaryote must have evolved means to avoid, suppress or overcome host defences. This may involve masking cyanobacterial surfaces with host-like molecules or continuously changing surface structures in order to avoid defences. Another possibility is that the host recognises the cyanobacterial partner as mutualistic, or, as has recently been suggested for some *Rhizobium*- and mycorrhiza-plant interactions, some defence reactions are elicited in the hosts but the symbioses are still established (Vasse et al., 1993; Gianinazzi-Pearson et al., 1996). Specificity is yet another prerequisite of compatible cyanobacteria. For instance, although bacteria, fungi and non-compatible cyanobacteria may inhabit the same mucilage-filled channel leading into the *Gunnera* stem, only compatible cyanobacteria penetrate the host cells (Bergman et al., 1992b; Bergman et al., 1996). On the other hand, in the leaf cavity of *Azolla*, both bacteria and cyanobacteria occur together (Braun-Howland and Nirezwicki-Bauer, 1990).

Another prerequisite of symbiotic cyanobacteria is their great structural and functional plasticities (Cohen et al., 1994; Bergman et al., 1996). For instance, it is necessary for most plant compatible cyanobacteria to differentiate motile hormogonia, a capacity which however is not exclusive to symbiotic cyanobacteria. At least cyanobacteria infecting bryophytes, cycads and *Gunnera* are known to use hormogonia as the "infective units" and their differentiation is elicited by plant exudates (Meeks, 1990; Lindblad and Bergman, 1990; Gantar et al., 1993; Rasmussen et al., 1994; Cohen and Meeks, 1997).

Compared to when free-living, almost all cyanobacteria in symbioses exhibit enhanced frequencies of heterocysts and fix higher amounts of nitrogen, most of which is released for the benefit of the host (Rai, 1990a; Bergman et al., 1992a; Bergman et al., 1996). In these symbioses, the hosts are typically self supportive in terms of carbon (for example, in algae, bryophytes, *Azolla*, cycads and *Gunnera;* Table 1) or contain an additional photosynthetic symbiont (for example, the green algae *Coccomyxa* in *Peltigera aphthosa*, a tripartite lichen). As many other nitrogen-fixing cyanobacteria do not enter symbioses it is obvious that neither heterocyst differentiation nor nitrogen fixation are sufficient for a cyanobacterium to become compatible. However, if the main role of the cyanobacterium is to support a non-photosynthetic heterotrophic host with carbon, the cyanobacterium retains a low heterocyst frequency (Rai, 1990b; Mollenhauer et al., 1996). This is the case in some bipartite lichen symbioses (Table 1). Akinete differentiation is uncommon in symbioses, and are only known to play a role in *Azolla* in which they function as cyanobacterial inoculants for the next sporophyte generation (Braun-Howland and Nierzwicki-Bauer, 1990).

The great adaptability of compatible *Nostoc* strains is also reflected in the ability of some to infect not only different hosts, organs and tissues, but also in their ability to be located extra- or intracellularly depending on the host (Table 1) (Enderlin and Meeks, 1983; Johansson and Bergman, 1994). Obviously, such cyanobacteria are capable of changing their gene expression (transcription/translation) as an adaptive means. Recently, a new type of genome rearrangement which may be linked to the capacity of cyanobacteria to adapt to changeable and extreme environments has been described (Robinson et al., 1997). It is also known that many cyanobacteria posses mechanisms for genome rearrangement/s in relation to heterocyst differentiation (Golden et al., 1985; Matveyev et al., 1994). The potential involvement of gene rearrangements in symbiotic cyanobacteria, is therefore an interesting aspect to explore.

5. Conditions Offered by Fungi and Plants

It is apparent that highly variable environments are experienced by cyanobacteria in symbioses depending on the host and the organ, tissue or cell infected (Table 1, Figures 2 and 3). The cyanobacteria may for instance be internalized in highly oxygenic photosynthetic plant cells or tissues. This is the case for cyanobacteria of the genus *Richelia* occupying the interior of marine diatoms and for *Nostoc* in both the thalli of the bryophytes (for example *Anthoceros*) and in the dorsal leaves of *Azolla* (Figure 2 C, D). However, in these symbioses the cyanobacteria are exposed to light filtered through light-absorbing chlorophyll-containing host cells and have reduced CO_2 fixation rates compared to their free-living state. Under these conditions, the cyanobacteria are supported by carbon from the host (Kaplan and Peters, 1988; Meeks, 1990). A requirement for nitrogen-fixation in highly oxygenic tissues is the protection of the oxygen sensitive nitrogenase. An efficient protection is provided by the heterocysts but in addition, oxygen tensions are lower than ambient in the cavities of *Azolla* and *Anthoceros* (Grilli Caiola et al., 1989; Meeks, 1990), the significance of which is not known.

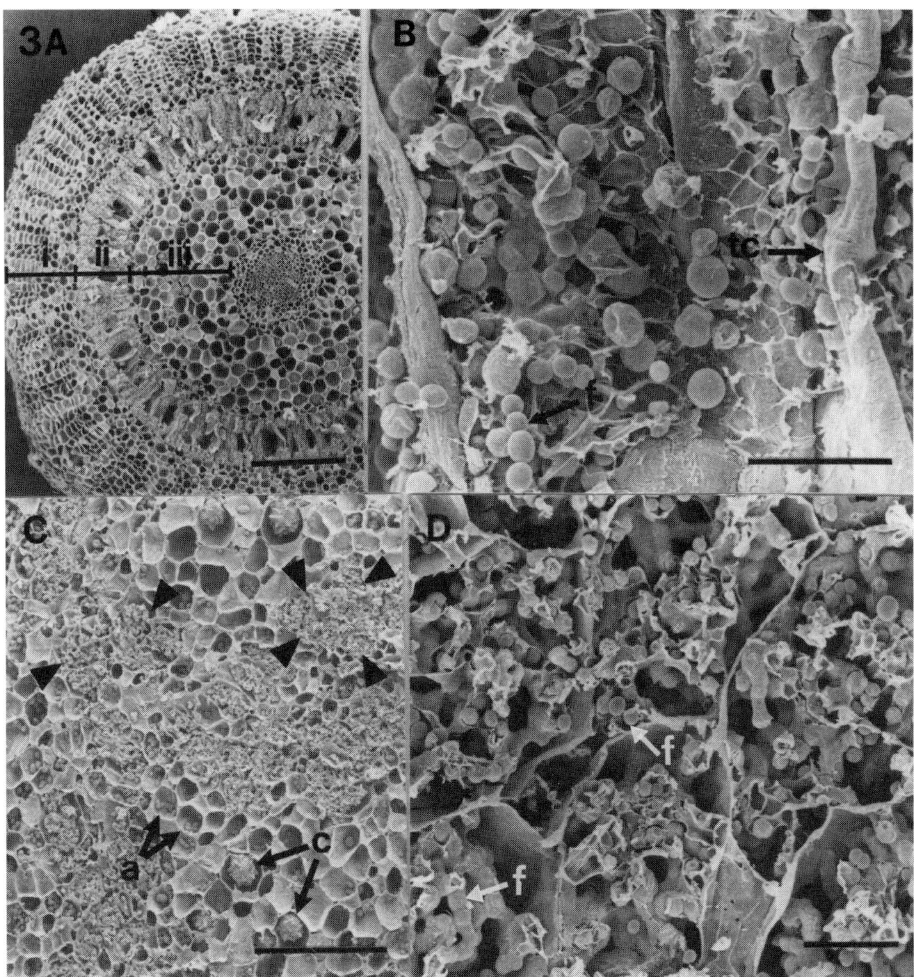

Figure 3. Scanning electron micrographs of cyanobacterial symbioses with cycad (A & B) and *Gunnera* (C & D) hosts.

A: A cross section through a coralloid root of *Bowenia* sp. showing distinct cylindrical zones, outer cortex (i), cyanobacterial zone (ii) and inner cortex (iii). The cyanobacteria are extracellular, occur in darkness in non-photosynthetic tissues and support the host with nitrogen. Bar = 500 μm. B: The cyanobacterial zone showing cyanobacterial filaments (f) and elongated cycad cells (tc) that transverse this zone to interconnect the two adjacent cortical layers. Bar = 25 μm. C: A cross section through a *Gunnera* sp. stem gland showing clusters of host cells infected by cyanobacterial filaments (▼). The intracellularly located cyanobacteria are in non-photosynthetic stem tissue and support *Gunnera* with nitrogen. Crystals (c) and amyloplasts (a) can be seen in non-infected cells. Bar = 250 μm. D: Cyanobacterial filaments (f) located within host stem cells. Bar = 25 μm.

A different scenario is experienced by cyanobacteria infecting other plants such as the non-photosynthetic root tissue (below ground) of cycads (Figure 3 A, B) and the chlorophyll-free inner stem tissues of *Gunnera* (Figure 3 C, D). In these symbioses, the cyanobacteria are not reached by light and have abandoned their typical photoautotrophic mode of life. Instead, they have adjusted to a heterotrophic metabolism that relies on host supplied carbon (Table 1) (Lindblad and Bergman, 1990; Söderbäck and Bergman, 1993). It is perhaps therefore not surprising that the frequency of heterocysts is highest (50-70%) in these symbioses (Bergman et al., 1992a). Fructose is apparently a suitable carbon source for heterotrophic or mixotrophic growth of cyanobacterial isolates from *Azolla* and *Gunnera*, and in addition it stimulates heterocyst differentiation (Rozen et al., 1986; J. Wouters and B. Bergman, unpublished data).

In most symbioses, the cyanobacteria are extracellular in relation to their hosts (Table 1). The cyanobacteria are either embedded by fungal hyphae as in lichens (Figure 2 A, B), or enter into preformed extracellular spaces or cavities, as in bryophytes and in *Azolla* (Figure 2 C, D). Yet others pass through the eukaryotic cell walls and are enclosed intracellularly, although still outside the cell plasmalemma. Intracellular symbioses are formed with the fungus *Geosiphon* (Mollenhauer and Kluge, 1993/94), with the marine diatoms (Taylor, 1982; Janson et al., 1995), with the angiosperm *Gunnera* (Figure 3 C, D) (Bergman et al., 1996) and possibly with some cycads (Lindblad and Bergman, 1990). In *Geosiphon,* the cyanobacteria are located in hyphal bladders (Mollenhauer and Kluge, 1993/94), in the unicellular diatoms they are found between the plasmalemma and the frustule (Janson et al., 1995), whereas in *Gunnera* they enter stem cells (Bergman et al., 1996). In *Geosiphon* and *Gunnera,* the cyanobacteria occur inside chloroplast-free cells, while in diatom symbioses the nitrogen-fixing cyanobacterium and the photosynthetic chloroplasts share the same cell.

The conditions experienced inside the plant cells and the reason/s why some hosts allow cell wall penetration by the cyanobacterium while others do not (Table 1), are still open questions. However, it has been shown that in the bryophyte *Anthoceros* and in the angiosperm *Gunnera*, the host plant dictates the position of the infecting cyanobacteria (Bergman et al., 1996). A cyanobacterium isolated from a lichen or from *Anthoceros* may pass through the cell walls to obtain an intracellular position within the *Gunnera* stem. Finally, as cyanobacteria in many plant symbioses migrate through mucilage (for example, in *Gunnera*) or end up in mucilage containing "openings" (for example, in bryophytes, *Azolla* and cycads) (Bergman et al., 1996), osmotic stress tolerance (Close and Lammers, 1993; Potts, 1996) may be another vital character needed for competence.

6. The Role of Cyanobacteria and of Symbioses

Depending on their host, the role of the cyanobacterium varies (Table 1). An important key role in symbioses is apparently to fix nitrogen and to supply the host plant with its total nitrogen requirement. This is even the case in the *Gunnera* symbiosis where the host is one of the largest herbs on earth and where the cyanobacteria constitute only

TABLE 1. Characteristics of present day fungal and plant symbiotic cells/tissues containing cyanobacteria. The cyanobacteria are capable of forming functional symbioses in a diverse range of uni- and multicellular hosts [39]

Symbiotic host	Organ/tissue/ cell infected	Intra- or extracellular	Conditions offered by the host	Typical cyanobacteria and their role in symbiosis
fungi	lichen thallus	extracellular among the fungal hyphae	bipartite: non-photosynthetic fungal tissues	*Nostoc;* donate C and N to the fungal host
	lichen thallus	extracellular among the fungal hyphae in cephalodia	tripartite: photosynthetic tissues due to a second phototroph (alga)	*Nostoc, Calothrix, Fischerella*; donate N to the fungal host
	fungal bladders	intracellular	inside non-photosynthetic fungal cells	*Nostoc*; donate C and N to the fungal host
algae (diatoms)	unicellular hosts	intracellular	photosynthetic cells	*Richelia*; role unknown
bryophytes mosses	gametophytic thallus	usually epiphytic or intracellular within haline cells	photosynthetic tissues	*Nostoc*; role unknown
liverworts	gametophytic thallus	epiphytic or extracellular in mucilage containing cavities	photosynthetic tissues; cb reached by filtered light and receive C from the host	*Nostoc*; donate fixed N to the host
hornworts	gametophytic thallus	extracellular in mucilage containing cavities	photosynthetic tissues; cb reached by some filtered light and receive C from the host	*Nostoc*; donate fixed N to the host

Azolla (fern)	leaf cavities	extracellular, enclosed by an envelope of plant origin and contained in mucilage	photosynthetic tissues; cb reached by some filtered light and receive C from the host	*Nostoc;* donate fixed N to the host
cycads (gymnosperm)	coralloid roots	usually extracellular in a mucilage containing zone transversed by elongated host cells	non-photosynthetic tissues; cb in darkness and receive C from the host	*Nostoc, Calothrix;* donate fixed N to the host
Gunnera (angiosperm)	stem-tissues	intracellular, surrounded by the plant plasmalemma	non-photosynthetic tissues, cb in darkness and receive C from the host	*Nostoc;* donate fixed N to the host

(cb=cyanobacteria; C=carbon; N=nitrogen)

about 1% of the total biomass. The importance of nitrogen-fixation in symbioses is also reflected in enhanced nitrogen fixation rates and dramatic upshifts in the numbers of heterocysts. Compared to heterocyst frequencies typical for free-living *Nostoc* (5-7% heterocysts), the frequencies increase from several fold in three-component lichens and bryophytes to 10-fold in cycad and *Gunnera* symbionts (Bergman et al., 1992a). It is evident that as the dependency on the photosynthetic vegetative cells decreases (i.e. more carbon is derived from the host) the heterocyst frequencies increase. The release of the nitrogen fixed by the cyanobacterium is probably of vital importance for maintaining symbiosis (Rai, 1990a; Bergman et al., 1992a; Stock and Silvester, 1994; Silvester et al., 1996). The key-regulatory event in this context is a down-regulation of the synthesis and/or activity of the cyanobacterial glutamine synthetase (Rai, 1990a; Bergman et al., 1992a). In return, the hosts must provide the symbiotic cyanobacteria with all other macro- and micronutrients required (Table 1). However, the role of some competent cyanobacteria is to feed the hosts with both nitrogen and carbon. This is the case in bipartite lichens (Table 1) (Rai, 1990b; Wastlhuber and Loos, 1996) and in the *Geosiphon* symbiosis (Mollenhauer and Kluge, 1993/94) where the heterocyst frequencies remain low.

There is also the possibility that nutrient exchange may involve other plants in the community. For example, nutrient exchange between nitrogen-fixing *Gunnera* and other nearby plants is possibly networked via a third symbiotic partner, mycorrhiza (Koske et al., 1992), the mycelia of which may extend from plant to plant.

For a symbiosis to persist through evolution there must be a tight co-evolution and a mutual benefit to both organisms. Cyanobacterial-fungal/plant symbioses probably have evolved in order to expand the ecological niche for both hosts and microsymbionts. For example, the host plant can sustain growth and reproduction in nitrogen deficient environments and the cyanobacterium gains a new habitat inside another organism (above or below ground) which, at least to some extent, is sheltered from adverse external conditions. In symbiosis with lichenized fungi, cyanobacteria may even expand their distribution to include acidic forest soils. This diversity in habitat will provide security to both partners and they may be more protected from competitors and/or predators.

7. Concluding Remarks and Future Perspectives

In the absence of exact data on the physical and chemical conditions offered by eukaryotic hosts it is difficult to fully understand what is required by cyanobacteria to form functional symbioses with these hosts. However, judging from the behaviour of cyanobacteria in symbioses it appears that the eukaryotic hosts offer some unique conditions and to a large extent control the behaviour of the cyanobacteria.

Our knowledge of cyanobacterial symbioses is limited but these symbioses deserve attention because of their evolutionary and ecological significance and their potential economic importance in future agriculture and biotechnology. In this review, we have highlighted what is known about the conditions offered by eukaryotic hosts and pointed to the remarkable adaptivity of symbiotic cyanobacteria. We suggest that to be able to fully understand these highly interesting symbioses it is important to: i) understand the mechanism/s involved in the establishment of balanced symbioses, ii) identify symbiotic genes and gene products in cyanobacteria and hosts to further understand aspects of communication, adaptation and co-ordination, iii) determine the physical and chemical conditions present in symbiotic eukaryotic tissues, iv) study the evolution, nature and prevalence of cyanobacterial symbioses and v) create new symbioses with cyanobacteria and agriculturally important plants.

8. Acknowledgements

The financial support from The Swedish Natural Science Research Council and The Swedish Forestry and Agricultural Research Council is acknowledged. We thank Susanne Lindwall for preparing the electron micrographs.

9. References

Bergman, B., Rai, A.N., Johansson, C. and Söderbäck, E. (1992a) *Symbiosis* **14**, 61-81.
Bergman, B., Johansson, C. and Söderbäck, E. (1992b) *New Phytol.* **122**, 379-400.
Bergman, B., Matveyev, A. and Rasmussen, U. (1996) *Trends in Plant Science* **1**, 191-197.
Bergman, B. (1996) In M. Björk, A.K. Semesi, M. Pedersén and B. Bergman (eds.), *Current Trends in Marine Botanical Research in the East African Region*, Sida, Stockholm, pp. 39-59.
Bergman, B. Gallon, J.R., Rai A.N. and Stal, L.J. (1997) *FEMS Microbiol. Rev.* **19**, 139-185.
Braun-Howland, E.B. and Nierzwicki-Bauer, S.A. (1990) In A.N. Rai (ed.) *Handbook of Symbiotic Cyanobacteria*, CRC Press, Boca Raton, pp. 65-118.
Buikema, W.J. and Haselkorn, R. (1993) *Ann. Rev. Plant Physiol. & Plant Mol. Biol.* **44**, 33-52.
Capone, D.G., Zehr, J.P., Paerl, H.W., Bergman, B. and Carpenter, E.J. (1997) *Science* **276**, 1221-1229.
Carr, N. G. and Whitton, B. A. (1982) *The Biology of Cyanobacteria*, Blackwell Scientific Publications, Oxford.
Close, T.J. and Lammers, P.J. (1993) *Plant Physiol.* **101**, 773-779.
Cohen, M.F., Wallis, J.G., Campbell, E.L. and Meeks, J.C (1994) *Microbiol.* **140**, 3233-3240.
Cohen, M.F. and Meeks, J.C. (1997) *Mol. Plant-Microbe Interact.* **10**, 280-289.
Dodds, W.K., Gudder, D.A. and Mollenhauer, D. (1995) *J. Phycol.* **31**, 2-18.
Douglas, S.E. (1994) In D.A. Bryant (ed.), *The Molecular Biology of Cyanobacteria*, Kluwer Academic Publishers, Dordrecht, pp. 91-118.
Enderlin, C.S and Meeks, J.C. (1983) *Planta* **158**, 157-165.
Fay, P. (1992) *Microbiol. Rev.* **56**, 340-373.
Gantar, M., Kerby, N.W. and Rowell, P. (1993) *New Phytol.* **124**, 505-513.
Gianinazzi-Pearson, V., Dumas-Gaudot, E., Gollotte, A., Tahiri-Alaoui, A. and Gianinazzi, S. (1996) *New Phytol.* **133**, 45-57.
Giovannoni, S.J., Turner, S., Olsen, G.J., Barns, S., Lane, D.J. and Pace, N.R. (1988) *J. Bacteriol.* **170**, 3584-3592.
Golden, J.W, Robinson, S.J. and Haselkorn, R. (1985) *Nature* **314**, 419-423.
Grilli Caiola, M., Canini, A. and Moscone, D. (1989) *FEMS Microbiol. Lett.* **59**, 283-288.
Grossman, A.R., Schaefer, M.R., Chiang, G.G. and Collier, J.L. (1994) In D.A. Bryant (ed.), *The Molecular Biology of Cyanobacteria*, Kluwer Academic Publishers, Dordrecht, pp. 641-675.
Janson, S., Rai, A.N. and Bergman, B. (1993) *New Phytol.* **124**, 149-160.
Janson, S., Rai, A.N. and Bergman, B. (1995) *Marine Biology* **124**, 1-8.
Johansson, C. and Bergman, B. (1994) *New Phytol.* **126**, 643-652.
Kaplan, D. and Peters, G.A. (1988) *Symbiosis* **6**, 53-68.
Kenrick, P. and Crane, P.R. (1997) *Nature* **389**, 33-39.

Knight, C.D. and Adams, D.G. (1996) *Physiol. Molec. Plant Pathol.* **49**, 73-77.
Kobiler, D., Cohen-Sharon, A. and Tel-Or, E. (1982) *Israel J. Bot.* **31**, 324-328.
Koske, R.E., Gemma, J.N. and Doyle, M.F. (1992) *Pacific Science* **46**, 480-483.
Lindblad, P. and Bergman, B. (1990) In A.N. Rai (ed.) *Handbook of Symbiotic Cyanobacteria*, CRC Press, Boca Raton, pp. 137-160.
Löffelhardt, W. and Bohnert, H.J. (1994) In D.A. Bryant (ed.), *The Molecular Biology of Cyanobacteria*, Kluwer Academic Publishers, Dordrecht, pp. 65-89.
Margulis, L. (1970) *Origin of Eukaryotic Cells*, Yale University Press.
Matveyev, A.V., Rutgers, E., Söderbäck, E. and Bergman, B. (1994) *FEMS Microbiol. Lett.* **116**, 201-208.
Meeks, J.C. (1990) In A.N. Rai (ed.) *Handbook of Symbiotic Cyanobacteria*, CRC Press, Boca Raton, pp. 43-64.
Mollenhauer, D. and Kluge, M. (1993/94) *Endocytobiosis & Cell Res.* **10**, 29-34.
Mollenhauer, D., Mollenhauer, R. and Kluge, M. (1996) *Protoplasma* **193**, 3-9.
Potts, M. (1996) *Physiol. Plant.* **97**, 788-794.
Rai, A.N. (1990a) *Handbook of Symbiotic Cyanobacteria*. CRC Press, Boca Raton.
Rai, A.N. (1990b) In A.N. Rai (ed.) *Handbook of Symbiotic Cyanobacteria*, CRC Press, Boca Raton, pp. 9-42.
Rasmussen, U., Johansson, C. and Bergman, B. (1994) *Mol. Plant-Microbe Interact.* **7**, 696-702.
Robinson, P.J., Cranenburgh, R.M., Head, I.M. and Robinson, N.J. (1997) *Mol. Microbiol.* **24**, 181-189.
Rozen, A., Arad, H., Schönfeld, M. and Tel-Or, E. (1986) *Arch. Microbiol.* **145**, 187-190.
Schopf, J.W. and Walter, M.R. (1982) In N.G. Carr and B.A Whitton (eds.) *The Biology of Cyanobacteria*, Blackwell Scientific Publications, Oxford, pp. 543-564.
Schüßler, A., Meyer, T., Gehrig, H. and Kluge, M. (1997) *Eur. J. Phycol.* **32**, 233-239.
Shively, J.M. (1974) *Ann. Rev. Microbiol.* **28**, 167-187.
Silvester, W.B., Parsons, R. and Watt, P.W. (1996) *New Phytol.* **132**, 617-625.
Sprent, J.I. and Raven, J.A. (1985) *Proc. Royal Soc. Edinburgh* **85B**, 215-237.
Stal, L.J. (1995) *New Phytol.* **131**, 1-32.
Stanier, R.Y. and Cohen-Bazire, G. (1977) *Ann. Rev. Microbiol.* **38**, 1-25.
Stock, P.A. and Silvester, W.B. (1994) *New Phytol.* **126**, 259-266.
Söderbäck, E. and Bergman, B. (1993) *Physiol. Plant.* **89**, 125-132.
Tandeau de Marsac, N. and Houmard, J. (1993) *FEMS Microbiol. Rev.* **104**, 119-190.
Taylor, F.J.R. (1982) *Ann. Inst. Oceanogr.* **58** (suppl.), 61-90.
Vasse, J., de Billy, F. and Truchet, G. (1993) *The Plant J.* **4**, 555-566.
Wastlhuber, R. and Loos, E. (1996) *Lichenologist* **28**, 67-78.
Wolk, C.P, Ernst, A. and Elhai, J. (1994) In D.A. Bryant (ed.), *The Molecular Biology of Cyanobacteria*, Kluwer Academic Publishers, Dordrecht, pp. 769-823.

Biodata of **Maria Grilli Caiola**, contributor of "***The hard life of Prokaryotes in leaf cavities of Azolla.***"

Dr. Maia Grilli Caiola is a Professor of Botany at the University of Rome "Tor Vergata", Rome, Italy. Her main interests are in biology of cyanobacteria, cyanobionts of Cycads and *Azolla* and in extreme environments. Dr. Grilli investigating the reproductive biology of *Crocus* and the role of calcium in the plant cell. She coordinates the Project of *Azolla-Anabaena* symbiosis and Plant reproduction sponsored by the Italian National Council of Research and the Italian Ministry of the University and Scientific and Technological Research (resp.). She is a member of various societies and acts as the Vice President of the Italian Botanical Society.
E-Mail: **grilli@utovrm.it; travaglini@tovvx1.ccd.utovrm.it**

Biodata of **Cinzia Forni**, coauthor with M. Grilli on the chapter: "***The hard life of Prokaryotes in leaf cavities of Azolla.***"

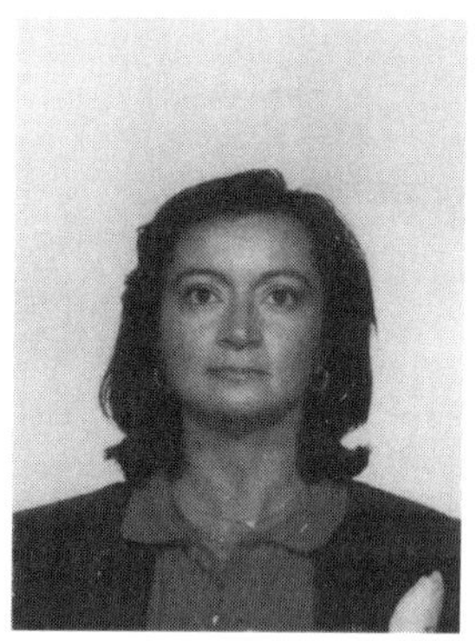

Dr. **Cinzia Forni** is an Assistant Professor at the Department of Biology, University of Rome "Tor Vergata", Rome, Italy. She spend (as a researcher fellowship award) a few years in Rome, San Carlos, California (USA) and in Rehovot, Israel. Her current interests are in *Azolla-Anabaena* association, nitrogen fixation activity of free-living and symbiotic cyanobacteria, polysaccharides production by cyanobacteria, phytoremendiation and cell differentiation in strawberry cell culture.
E-Mail: **Cinzia.Forni@UniRoma2.It**

J. Seckbach (ed.), Enigmatic Microorganisms and Life in Extreme Environments, 629–639.

THE HARD LIFE OF PROKARYOTES IN THE LEAF CAVITIES OF *AZOLLA.*

M. GRILLI CAIOLA AND C. FORNI
Dipartimento di Biologia, Università di Roma "Tor Vergata", Via della Ricerca Scientifica, 00133 Roma, Italy.

1. Introduction

Life of cyanobionts and bacteriobionts inside the leaf cavities of *Azolla* cannot be so easy. Fern shows many features of remedation plant, able to adapt itself to the change of the environmental conditions such as high light intensity, high temperature or water composition. Cyanobionts seem to modify their metabolism according to that of the fern, and whenever the host life conditions become stressing, the symbiont can be eliminated (Hill, 1975; Kitoh et al., 1991). Therefore speaking in term of symbiosis, it may be not so easy for the symbionts, since *Azolla* behaves at its convenience, forcing them to follow its benefit. Anyway, some authors consider the *Azolla-Anabaena* relationship like an happy marriage between lucky partners, where each of them contributes to the menage. Some considerations can be made on the different aspects of such unusual association.

2. *Azolla* Association.

The host plant is a small aquatic fern, *Azolla* (Fam. Azollaceae) (Fig. 1) a genus containing seven species, mainly typical of tropical areas, but now spread also in temperate regions. Whatever is their origin, all the *Azolla* species contain in their leaf cavities cyanobacteria and bacteria. The first description of this association was made by Strasburger in the 1873, and since then the importance of this symbiosis has caught the attention of several scientists, in particular for its great interest to tropical agronomy, especially with regard to lowland rice cultivation (Lumpkin and Plucknett, 1980).

The fern sporophyte, normally floats freely on the water surface and multiplies vegetatively. The floating stem (rhizome) bears two lateral rows of alternately arranged,deeply bilobed leaves on the dorsal surface and adventitious roots at nodes on the ventral surface. The thin ventral leaf lobe is nearly achlorophyllous and in contact with the water surface. The aerial dorsal leaf lobe at maturity contains an ellipsoid cavity, an extracellular space formed by an infolding of the adaxial epidermis during development, where the cyanobionts and bacteriobionts are included. The leaf cavity is also formed in *Azolla* freed of *A. azollae* (Forni et al., 1991). Like *Anabaena*, the bacteria are also a constant component of this association (Grilli, 1964; Forni and Grilli Caiola, 1993).

3. *Azolla* Leaf Cavity Organization

Azolla leaf cavity (Fig. 2) may be considered as the basic physiological unit of the association (Canini and Grilli Caiola, 1995). In fact it is the site where the nitrogen fixation and the metabolic exchanges of nitrogenous and carbon compounds occur. Therefore the leaf cavity is structured in such a way to meet both the host and symbionts needs. The form is ellipsoidal, lined by inner epidermal cells, on which simple or branched hairs are formed (Calvert and Peters, 1981). Two branched hairs are sited near the foliar trace, while numerous single hairs are randomly distributed on the inner epidermal surface. The cavity contains mucilage which acts to adhere symbionts to the host cells (Robins et al., 1986), and as a compartment in which many exchanges of metabolites occur.The mucilage originates from the activities of *Azolla*, cyanobacteria and bacteria (Forni et al., 1992). This content can be isolated by cellulolytic digestion of the fern, as packets limited by a very thin envelope about 10-20 nm thick, which completely lines the leaf cavity . Present in both *Azolla* containing and endophyte free plants, this envelope was shown to be resistant to various enzymes and chemicals (Uheda, 1986; Uheda and Kitoh, 1991), and recently evidence has been reported about the probable cutinic nature of the envelope (de Roissart et al., 1994). In the young leaves mucilage fills the cavity, but in the old ones it is mainly peripheral and central part of the cavity is occupied by liquid or gas. However, there is no common opinion on the algal packets content; in fact some authors detected the presence of gas in the cavity (Peters and Meeks, 1989), while others, by microeletrodes, revealed the presence of liquid containing many ions (Canini et al., 1992). The mucilage composition varies from the apex towards the basis of the stem. In the cavity of upper leaves mainly basic polysaccharides have been found and the pH was 6.5-6.7. Viceversa in the basal leaves, the mucilage is acidic and the pH decreases at 6.5. The cyanobiont filaments are embedded in the mucilage layer (Klein et al. 1992).

4. Cyanobionts

With few exceptions, all the *Azolla* species contain in their leaf cavities cyanobacteria with similar life cycle (Fig. 3). However, the morphology of the cyanobiont is different depending on the leaf age (Canini et al., 1992). In the apical leaves, there are mainly short filaments of small vegetative cells similar to hormogonia; they are dispersed inside the mucilage filling the cavity and are poor of cyanophycin and phycobiliproteins (Fig. 4). These characters suggest a mixotrophic behaviour of the symbiont. In the median leaves, the cyanobacteria have long filaments of vegetative cells and heterocytes which frequency can reach up to 30%. Heterocytes are ovoidal or roundish shaped, pale yellow, while the vegetative cells are green and rich in cyanophycin granules. In these leaves the nitrogen fixation reaches the highest values. In the basal leaves, heterocyst frequency decreases to 12%, and nitrogen fixation too. Many heterocytes show degenerative aspects as membrane vesciculation, cell wall breaks, and absence of glycogen. The akinetes rarely form.

Anabaena azollae classification has not been modified until recently, when the new approaches on cyanobacterial systematics induced to a revision of the *Anabaena* taxonomy. So the Azolla cyanobiont has been proposed as *Nostoc* (Vagnoli et al., 1988) or *Trichormus* (Komàreck and Anagnostidis, 1989). A real problem of the *Azolla*

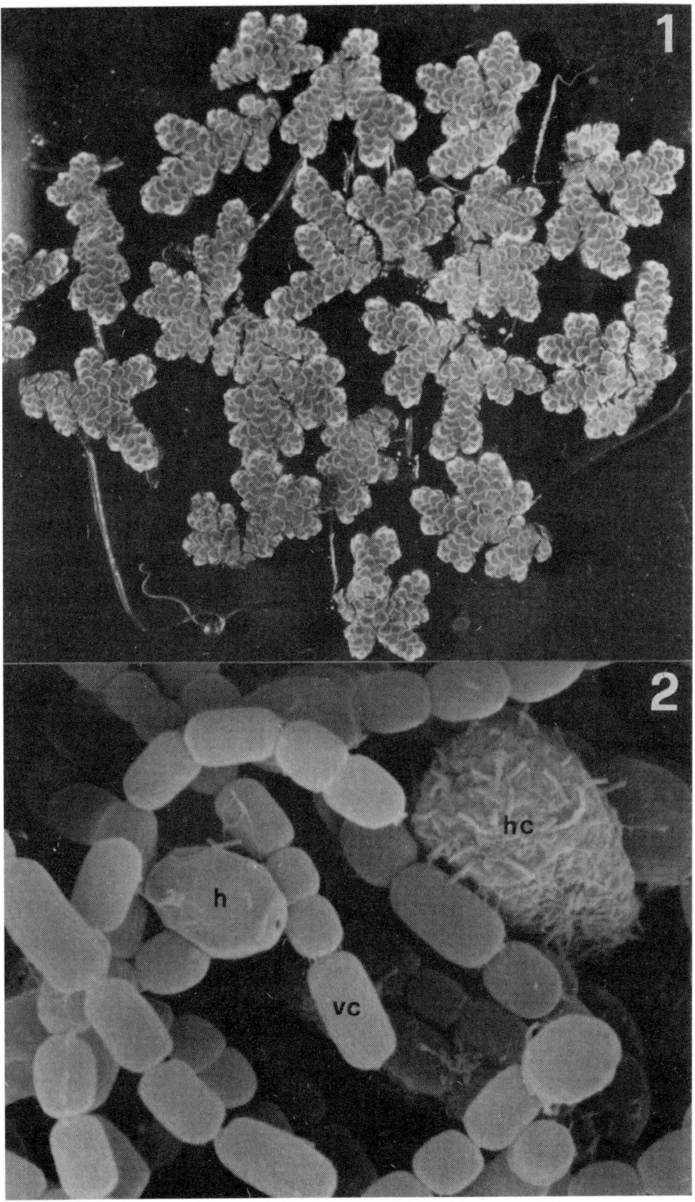

Figure 3. Cyanobiont packets with vegetative cells (vc) and heterocytes (h) in the *Azolla* median leaf cavity (650 x). Fig 4. Cyanobiont vegetative cells from apical leaves of *Azolla*. TEM micrograph (17,000 x). A large nucleoplasm (n) and the photosynthetic apparatus (p) are evident. Fig. 5. Immunogold labeling of the bacteriobionts within the leaf cavity of *Azolla*. Bacteria are labeled with antibodies raised against the type strain Arthrobacter globiformis ATCC 8010. TEM micrograph (63,000 x).

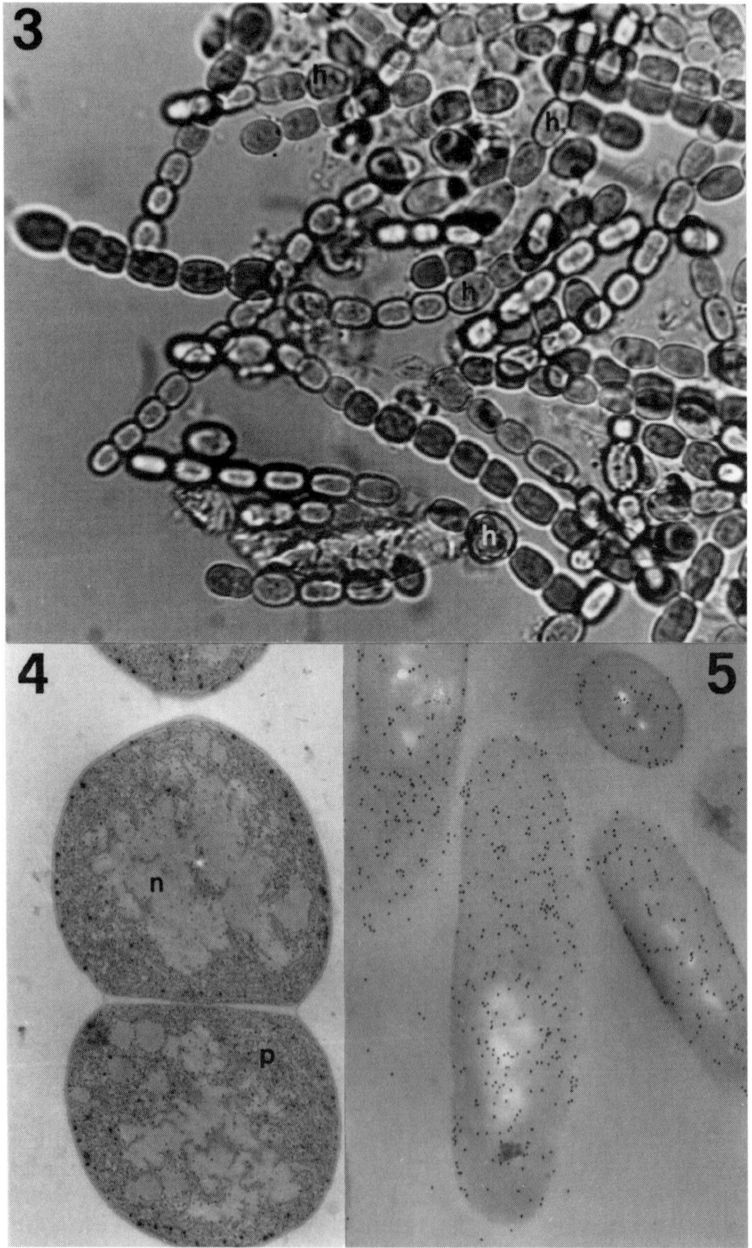

Figure 1. *Azolla filiculoides* Lam. L.m. (385 X); Fig. 2. *Azolla* leaf cavity, where cyanobacteria filaments of vegetative cells (vc) and heterocytes (h) are visible. Among the filaments a hair cell (hc) with attached numerous bacteria is visible. SEM micrograph (7600 x).

cyanobiont is the failure of the temptative of isolation and growth in pure culture, and the consequently lack of reinoculation of the *Azolla* cyanobiont, to satisfy Kock' spostulate. According to some authors, the maintenance of the symbiosis throught the life cycle of the fern might exclude the possibility of free-living stage of *Anabaena*. Another unresolved problem concerns the presence of one or more cyanobionts in the leaf cavities. Attempts to solve the controversy on the cyanobiont taxonomy were made on the basis of akinetes differentiation (Grilli Caiola et al., 1993), biochemical (Caudales et al., 1992; Canini et al., 1992), immunological (McCowen et al., 1987) and molecular (Gebhardt and Nierzwicki-Bauer, 1991) data, but the results were not conclusive.

Literature reports stated changes in the morphology of the cultured cyanobiont if compared to the one *in vivo*, i.e. the heterocyte frequency (Grilli Caiola, 1992), higher in symbiosis, and akinetes differentiation, which mostly lacks in *in vivo*.However, the cyanobiont differentiates akinetes under the indusium of micro and megasporocarps. These cells are the resting form of the *Anabaena*, and after germination they represent the new inoculum attached to the apex of the sporophyte (Becking, 1987). However, during the vegetative multiplication, the cyanobiont is already present in the young leaves and there is no need of new infection.

5. Bactobionts

The presence of bacteria, in the fern leaf cavities of *Azolla* was first reported by Grilli (1964) and then cofirmed by several other authors (Wallace and Gates, 1986; Forni and Grilli Caiola, 1993). However, there is no common opinion about the nature of this microbial population and its role in symbiosis. TEM observations (Grilli Caiola et al., 1988) of the leaf cavities of *A. caroliniana* reveal a heterogenous population consisting of long rods and cocci. In the young leaves as well as in the old ones, the bacteria appear either close to the hair cells or localized nearby the epidermal cells of *Azolla*. Nierzwicki-Bauer and Aulfinger (1990, 1991) also reported the occurence of ultrastructurally distinct eubacteria in the leaf cavitites of *A.caroliniana* and *A.mexicana*.

The tripartite association is maintained throughout the sexual cycle of the fern (Forni et al., 1990; Carrapico et al., 1991). Akinetes and mainly rod shaped bacteria occur under the indusium of the mega and microsporocarps. In the sporocarps, the bacteria are localized mainly among the akinetes; they are mostly rod shaped, about 2μm long and 0.5μ m diameter, but some larger cocci are also visible (Grilli Caiola et al., 1993).

According to Becking (1987), the infection of the leaf by the cyanobacterium symbiont occurs when the cotyledon leaf passes the megaspore indusium cap containing the akinetes of the symbiont. It is possible that the bactobiont uses the same way to become associated with the fern apical meristem (Forni et al., 1990).

The presence of coryneform bacteria has been detected in the leaf cavity of *Azolla caroliniana* (Gates et al., 1980). According to these authors, only one type of bacterium was isolated from the leaf cavities. The isolated bacteria are aerobic, non-motile, and Gram-positive, becoming Gram variable with increased culture age. Plate counts indicate bacterial-*Anabaena* cells ratio of about one. Further studies of Gates and coworkers (Wallace and Gates, 1986; Petro and Gates, 1987) confirm the presence of coryneform bacteria in *Azolla* and they have identified thegenus *Arthrobacter* Conn and Dimmick.There is a direct linear relationship between numbers of these bacteria and leaf age in *A.caroliniana*, *A.filiculoides* and *A.mexicana* (Petro and Gates, 1987), but no such relationship has been found in *A.pinnata* which contains fewer numbers of bacteria.

The presence of arthrobacters in different species of *Azolla* has been confirmed by Forni et al. (1989), and this genus is also present together with the *Anabaena* akinetes in the mega and microsporocarps of *A. filiculoides* (Forni et al., 1990; Grilli Caiola et al., 1993). Immunogold labeling studies of the bacteria residing in the leaf cavities of *A.filiculoides* (Fig. 5) confirm the presence of *Arthrobacter* in the fern (Leonardi et al., 1993). *Arthrobacter* sp. strains, isolated from *A.filiculoides*, are morphologically similar to the type strains *A.globiformis* ATCC 8010 (Forni and Grilli Caiola, 1992). They are straight or bent rods in the log phase and cocci in the stationary growth stage. The cocci derive from the fragmentation of rods within an electron transparent mother cell.

In a recent study, Plazinski et al. (1990) characterized the eubacteria from *A. filiculoides* as members of the genus *Agrobacterium* on the basis of bacteriologic multitest results and DNA/DNA hybridization analysis. The bacterial strain isolated by these authors possesses an indigenous plasmid of molecular weight of 100 MDa, with DNA regions that hybridized with *Rhizobium nodL*, and weakly with *nodABC* operon. *Nod* genes are involved in the *Rhizobium*-legume symbiotic interactions (Hirsch, 1992), and if such genes are functional in the bacterial strain isolated from *Azolla*, they may affect the development of the *Anabaena-Azolla* symbiosis or help the bacterium to persist in the *Azolla* leaf cavity.

The presence of different biotypes of *Arthrobacter* and perhaps even unidentified genera of eubacteria within a leaf cavity of a single *Azolla* species has been suggested by Shannon et al. (1993).

6. *Azolla* Leaf Cavity as a Functional Unit for Nitrogen Fixation.

Azolla leaf cavity is an efficient machinery, where each partner plays a precise role to optimize N_2-fixation and the utilization of carbon and nitrogenous compounds. The nitrogenase activity occurs only in the heterocytes of the cyanobiont, where the enzyme is synthetized and it has the microaerobic environment necessary for its activity. So far, it has not been yet established if the bacteria can also fix nitrogen. Preliminary tests on bacterial nitrogenase activity were negative (Forni et al., 1989); but nitrogenase has been immunocytochemically localized in some bacteria associated with *Azolla* (Lindblad et al., 1991). Therefore, the presence of both the Fe and MoFe proteins of the nitrogenase complex in some of the bactobionts of *Azolla* reopens the possibility on a contribute of these microorganisms to the nitrogen-fixing activities of the association.

The respiration is the main source for the reducing power needed for the reduction of dinitrogen; so, there is an intense metabolic exchanges of carbon compounds among the fern, the cyanobionts and bacteriobionts. The role of the epidermal simple hairs is to import photosynthates in the cavity, while the branched hairs export the fixed nitrogen; the mucilage may act as buffer against abrupt changes of pH and nitrogen compound accumulation, that inhibits the nitrogenase activity.

A gradient of nitrogen fixation and nitrogen and carbon utilization occurs from the median region towards the apex and the basis of the stem. In such scenary, the cyanobacteria do not appear to be in an ideal environment. They have to work for the host and to live in a limited space, i.e. the leaf cavity, more similar to a golden prison than to a comfortable house.

The presence of bacteria may be significant for the *Azolla-Anabaena* symbiosis. The aerobic bacteria can utilize several carbon compounds for their growth (Forni et al., 1989), and they may compete with the cyanobiont for the nutrients, but in the meantime they may protect *Anabaena azollae* by diminishing intracavity ammonium accumulation (Okoronkwo and Van Hove, 1987).

A recent growing body of evidence suggests that various species of arthro-bacters associated with *Azolla* may act analogously to other plant-associated microorganisms. In fact, several strains of *Arthrobacter*, isolated from different species of *Azolla*, produce in pure culture large quantity of mucilagenous substances, which main components are polysaccharides. (Forni et al., 1992). In pure culture *A. globiformis* produce exopolysaccharides (EPS) and lower amounts of capsular polysaccharides (CPS). EPS may play a structural role in the symbiosis by enabling attachment of the bacteria to the fern surface, improving nutrient acquisition, or providing protection from environmental stresses and host defenses in the leaf cavity. Alternately, EPS may function as a signaling molecule, like in the *Rhizobium*-Legume interaction, where EPS is required for establishment of the nitrogen-fixing symbiosis (Gonzàlez et al., 1996).

Further analogies of arthrobacters with other plant-associated microorganisms are concerning the production of the phytohormone indole-3-acetic acid (IAA) (Forni et al., 1992 a). *Arthrobacter* species, isolated from *A.pinnata* and *A.filiculoides*, produce IAA when the precursor tryptophan is added to the medium. Part of the tryptophan is transformed to Na-acetyl-L-tryptophan and other related tryptophan compounds; the production of these compounds and of IAA, may be a mechanism by which the bacterial cells eliminate the possible toxic effects of high concentrations of tryptophan (Forni et al., 1992 a). In *Azolla* leaf cavity tryptophan is present (Yun-Lu et al., 1983), and it is likely that the bactobionts utilize this tryptophan to produce IAA, which is excreted back into the leaf cavity, where the hair cells might transfer it to the plant. However, the mechanism of IAA synthesis in the association seems to be rather complex, in fact all the three components are able to produce IAA (Forni et al., 1994; Forni, 1995), and the role of phytohormone biosynthesis by bactobionts has to be fully elucidated.

Why cyanobionts and bactobionts remain in the cavity ? One possibility is the beneficial protection from unfavorable environment, and it is likely that the cyanobiont has a free living stage, which is not yet discovered. This free-living stage may start from the germination in the water of the akinetes contained in the microsporocarps. Various questions on the *Azolla* prokaryotic partners remain to be answered, including the mechanism of invasion or capture of the bacteria in the leaf cavity, and the presence of specific signaling molecules interacting with either *Azolla* or the *Anabaena*. The recognition mechanisms of the three partners have not been clarified yet. Lectins (McCowen et al., 1987), polyphenols (Leizerovich et al., 1988) and other molecules (Ladha and Watanabe, 1984) have been involved in the recognition processes. They have been likely evolved long time ago, before the endophytes became obligate symbionts.

We still have to clarify if the bacteriobionts are selected inside the cavity for the sake of the host. Finally, it is possible that a bacterium which lives in the fern leaf cavities could exploited as a vehicle for introducing genes and gene product into the host plant. Plazinski et al. (1990) have explored this possibility with their *Agrobacterium* strain, isolated from *A.filiculoides*, and they have obtained a successful plasmid transfer into this strain. These data might be an indication about the possibility to deliver cloned genes into the bactobionts, and potentially to deliver gene products to *Azolla* cavities.

7. References

Becking, J. (1987). *Plant Soil* **100,** 183-212.

Calvert, H.E. and Peters, G.A. (1981). *New Phytologyst* **89,** 327-335.

Canini, A., Grilli Caiola, M., Bertocchi, P., Lavagnini, M.G. and Mascini, M. (1992). *Sensors and Actuators* **B 7,** 431-435.

Canini, A. and Grilli Caiola, M. (1995). In, F.E. Round and D.J. Chapman (eds.), *Progress in Phycological Research,* **vol. 11,** Biopress Ltd, Bristol, pp. 154-186.

Carrapiço, F. (1991). In: M. Polsinelli, R. Materassi, M.Vincenzini (eds.), *Nitrogen fixation*, Kluwer Academic Publishers, Dordrecht, pp.453-456.

Caudales, R., Wells, J.M. and Antoine, A.D. (1992). *J. General Microbiology* **138,** 1489-1494.

de Roissart, P., Jacques, C., Waterkeyn, L., Berghmans, P. and Van Hove, C. (1994). In, N.A. Hegazi, M. Fayez e M.Monib (eds.), *Nitrogen fixation with non-Legumes*, The American University in Cairo Press, pp.133-138.

Forni, C., Grilli Caiola, M. and Gentili, S. (1989). In, F.A. Skinner, R.M. Boddey and I. Fendrik (eds.), Nitrogen fixation with Non-Legumes, Kluwer Academic Publishers, Dordrecht , pp. 83-88.

Forni, C., Gentili S., Van Hove, C. and Grilli Caiola, M. (1990). *Annali di Microbiologia* **40,** 235-243

Forni, C., Tel-Or, E., Bar, E. and Grilli Caiola, M. (1991). *Plant and Soil* **137,** 151-155.

Forni, C. and Grilli Caiola, M. (1992). *Microbiologica* **15 (3),** 271-279.

Forni, C., Haegi, A., Del Gallo, M. and Grilli Caiola, M. (1992). *FEMS Microbiology Letters* **93,** 269-274.

Forni, C., Riov, J., Grilli Caiola, M. and Tel-Or, E. (1992 a). *J. General Microbiology* 138, 377-381

Forni, C. and Grilli Caiola, M. (1993). *Giornale Botanico Italiano* **127 (3)** , 422-427.

Forni, C., Grilli Caiola, M. and Tel-Or, E. (1994). In, N.A. Hegazi, M. Fayez e M.Monib (eds.), *Nitrogen fixation with non-Legumes*, The American University in Cairo Press, pp.117-125.

Forni, C. (1995). *Archiv fur Hydrobiologie, Algological Studies* **83,** 207-214.

Gates, J.E., Fisher, R.W. and Candler, R.A. (1980). *Archiv Microbiology* **127,** 163-165.

Gebhardt, J.S. and Nierzwicki-Bauer, S.A. (1991). *Applied Environmental Microbiology* **57,** 2141-2146.

Gonzàlez, J.E., York, G.M. and Walker, G.C. (1996). *Gene* **179,** 141-146.

Grilli, M. (1964). *Annali Microbiologia Enzimologia* **40,** 235-243.

Grilli Caiola, M. (1992). In, W. Reisser (ed.), Algae and symbioses: plants, animals, fungi, viruses, interactions explored, Biopress Limited, Bristol, pp.231-254.

Grilli Caiola, M., Forni, C. and Castagnola, M.(1988). *Symbiosis* **5,** 185-198.

Grilli Caiola, M., Forni, C. and Castagnola, M. (1993). *Symbiosis* **14,** 247-264.

Hill, D.J. (1975). *Planta* **122,** 179-184.

Hirsch, A.M. (1992). *New Phytologist* **122,** 211-223.

Kitoh, S., Shiomi ,N. and Uheda, E. (1991). *Soil Science Plant Nutrition* **37,** 323-329.

Klein,E., Bar, E., Forni, C., Malkin, S. and Tel-Or, E. (1992). *J. Microscopy*: **167,** 273-278.

Komàrek, J.S. and Anagnostidis, K. (1989). *Archiv fur Hydrobiologie, Algological Studies* **82,** 247-345.

Ladha, J.K. and Watanabe, I. (1984). *New Phytologist* **98,** 295-300.

Leonardi, D., Canini, A. and Forni, C. (1993). *Symbiosis* **15,** 269-283.

Leizerovich, I., Fleminger, G., Kardish, N., Freusdorff, A. and Galun, M. (1988). *Symbiosis* **5,** 209-222.

Linblad, P., Bergman, B. and Nierzwicki-Bauer, S.A. (1991). *Applied Environmental Microbiology* **57,** 3637-3640.

Lumpkin, T.A. and Plucknett, D.L. (1980). *Economic Botany* **34,** 111-153.

McCowen, S.W., MacArthur, L. and gates, J.E. (1987). *Current Microbiology* **14,** 329-333.

Nierzwicki-Bauer, S. A. and Aulfinger, H. (1990). *Current Microbiology* **21,** 123-129.

Nierzwicki-Bauer, S. A. and Aulfinger, H. (1991). *Applied Environmental Microbiology* **57**, 3629-3636.
Okorowkuo, N. and Van Hove, C. (1987). *Microbios* **49,** 39-45.
Peters, G.A.and Meeks, J.C. (1989). *Annual Review Plant Physiology Plant molecualr Biology* **40,** 193-210.
Petro, M.J. and Gates, J.E. (1987).*Symbiosis* **3,** 41-48.
Plazinski, J., Taylor, R., Shaw, W., Croft, L., Rolfe, B.G. and Gunning, B.E.S. (1990). *FEMS Microbiology Letters* **70,** 55-60.
Robins, R.J., Hall, D.O., Shi, D.-J., Turner, R.J. and Rhodes, M.J.C. (1986). *FEMS* Microbiology Letters **133,** 157-160.
Shannon, B.T., Gates, J.E. and McCowen, S.M. (1993). *Symbiosis* **15,** 165-175.
Strasburger, E. (1873). Uber *Azolla*. Jena.
Uheda, E. (1986). *Plant Cell Physiology* **27,** 1187-1190.
Uheda, E. and Kitoh, S. (1991). *Canadian J. Botany* **69**, 1418-1419.
Vagnoli, L., Margheri, M.C., Allotta, G. and Materassi, R. (1988). *New Phytologist* **120,** 243-249.
Wallace, W.H. and Gates, J.E. (1986). *Applied Environmental Microbiology* **52,** 425-429.
Yun-Lu, X., Ke-Zhi, B., Sai Ling, Y., Cheng, C., Bao-Zhen, H. and Xiu-Zhang, H. (1983). *Acta Botanica Sinica* **25**, 82-86.

Glossary

Akinete: cyanobacterial resting cell. It is characterized by a thick envelope, and numerous and large cyanophycin granules.

Arthrobacter: Gram-positive bacterial genus

Coryneform bacteria: pleomorphic bacteria with cocci, short curved rods and irregular rods in the same culture. Usually the coccoid forms prevail in older cultures.

Heterocyte: specialized cyanobacterial cell, carrying nitrogen fixation. It has a multilayered envelope and lacks of photosystem II, thus creating a microaerobic environment to protect nitrogenase

Megasporocarp: spororocarp containing one or more megaspore, from which female gametophyte arises.

Microsporocarp: sporocarp containing several microsporangia from which microspores are released. The germinating microspores originate the male gametophyte.

Sporocarp: fern sporangia enveloped in a sterile leaf (indusium).

Biodata of Dr. Volker Huss

Contribution: ***Fresh water Symbioses in Protozoa and Invertebrate***

Volker A. Huss is at the Institute for Botany and Pharmaceutical Biology of the University Erlangen in Germany. He received his Ph.D. from the Technical University of Munich, and spent some time as a postdoctor with Mitchell Sogin at the National Jewish Center for Immunology and Respiratory Medicine in Denver, CO. and with Stephen Giovannoni at the Oregon State University in Corvallis, OR.

His main interest lies in the molecular systematic of green algae, especially of *Chlorella* and related taxa *(Protheca, Scenedesmus, Ankistrodesmus* etc.). Dr. Huss studies include also endosymbiotic *Chlorella* from different hosts such as *Hydra virides* and *Ophrydium versalite.* In addition, he has been investigating the origin and evolution of land plants by comparison of 18S rRNA sequences of charophytes, bryophytes and pteridophytes. Also he characterizes the plastid genome of coccolithophorid prymnesiophytes.

E-Mail: <**Varhuss@biologie.uni-erlangen.de**>

J. Seckbach (ed.), Enigmatic Microorganisms and Life in Extreme Environments, 641–650.

FRESHWATER ALGAL SYMBIOSES IN PROTOZOA AND INVERTEBRATES

V.A.R. HUSS
Institut für Botanik und Pharmazeutische Biologie
Staudtstr. 5; D-91058 Erlangen; Germany

1. Introduction

A wide range of aquatic protists and invertebrates harbour endosymbiotic microalgae within their cells. Regardless of the nature of the algae, phycobionts in marine habitats, mostly dinoflagellates or diatoms, are traditionally called „zooxanthellae", while algae involved in freshwater endosymbiotic associations are referred to as „zoochlorellae". Although most of the freshwater symbionts have been assigned to the genus *Chlorella* (cf. Droop, 1963; Reisser and Widowski, 1992), their accurate taxonomic and phylogenetic position is largely unclear. One reason is that the genus *Chlorella* consists of an artificial assemblage of morphologically similar algae which are scattered over two different classes within the chlorophytes according to phylogenetic trees based on molecular data (Huss and Sogin, 1990; Friedl, 1995; Kessler *et al.*, 1997). Thus, symbiont morphology and even ultrastructure are of limited value as diagnostic markers for a taxonomic designation (Karakashian, 1970; Pardy, 1976a). Moreover, some of the phycobionts appear to be obligately symbiotic, i.e. cannot be grown outside of their host (Park *et al.*, 1967; Rahat, 1992). This prevents the application of chemotaxonomic studies which have been shown to be useful for the classification of *Chlorella* strains (Kessler, 1982, 1984).

One of the many interesting issues in symbiotic associations is the question of host/symbiont specificity. Why can some algae live inside the cells of a specific host, e.g. *Hydra viridis* (green hydra) or *Paramecium bursaria*, but not in other species of *Hydra* and *Paramecium*, and why are only some specific algae found as symbionts and not others, even closely related ones? How are suitable algae „recognised" by the host cell and finally retained while others are expelled or digested? These are some of the questions that have been studied intensively using the *Hydra/Chlorella* symbiosis as a model system. Apparently, several features are characteristic for the establishment and maintenance of such symbioses: (i) the algae are taken up by phagocytosis; (ii) the algae are enclosed in special compartments, the „perialgal vacuoles", where they are protected against the host's lytic enzymes; (iii) permanent enclosure of algae in the perialgal vacuoles is specific; (iv) nutrients are transferred between host and symbionts in one or both directions; and (v) growth and proliferation of algae must be somehow 'regulated' (cf. Smith, 1979; Muscatine and McNeil, 1989).

2. Host/Symbiont Partnerships found in Nature

The variety of zoobionts that live together with algae in close symbiotic associations is impressive (Table 1). Most of them are represented by protists such as amoeba and ciliates, but also several phyla of invertebrates are involved. These include sponges, coelenterates, molluscs, tube worms, and flatworms. In contrast, the phycobionts appear much more uniform. Algae other than *Chlorella* have only rarely been described as symbiotic partners. Among them are species of *Oocystis*, *Chlamydomonas*, *Graesiella* (= *Scenedesmus*), *Ankistrodesmus* and *Pleurococcus* (see Reisser [1992] and references therein). Most of these reports are based on field observations and therefore are hard to prove. More recently, the phycobiont of a Japanese *Hydra magnipapillata* was described as a new genus, *Symbiococcum hydrae*, and placed in the order Chlorosarcinales (Rahat and Reich, 1989).

TABLE 1. Selection of host/symbiont partnerships of freshwater symbioses (for a more comprehensive compilation see Droop [1963] and Reisser [1992])

Zoobiont	Phycobiont[a]	Obligate Symbiosis[b]	References
Amoeba:			
Acanthocystis turfacea	*C. vulgaris/sorok.* group[c] (-)	-	Kessler and Huss (1990)
Parachaos zoochlorellae	*Chlorella* sp. (-)	?	Willumsen *et al.* (1987)
Mayorella viridis	"Zoochlorella"	?	Cann (1981)
Ciliata:			
Climacostomum virens	*C. vulgaris/sorok.* group[c] (-)	-	Reisser *et al.* (1984)
Coleps hirtus	*Chlorella vulgaris* (-)	-	Esteve *et al.* (1988)
Ophrydium versatile (Bavaria, Germany)	*C. vulgaris/sorok.* group[c] (+)	+	Huss and Salzburg, unpublished
Ophrydium versatile (Massachusetts, USA)	*Graesiella* sp. (-)	?	Duval and Margulis (1995)
Paramecium bursaria	*C. vulgaris/sorok.* group[c] (+)	-	Huss *et al.* (1989a); Reisser and Widowski (1992)
Prorodon viridis	"Zoochlorella"	?	Esteban and Finlay (1996)
Stentor polymorphus	*Chlorella* sp. (-)	-	Reisser (1981)
Vorticella sp.	*Chlorella* sp. (-)	?	Graham and Graham (1978)
Porifera:			
Spongilla fluviatilis	*Chlorella sorokiniana* (+)	-	Huss *et al.* (1986)
Coelenterata:			
Hydra magnipapillata	*Symbiococcum hydrae*[d]	-	Rahat and Reich (1989)
Hydra viridis (three strains)	*C. vulgaris/sorok.* group[c] (+)	+	Huss *et al.* (1993)
Hydra viridis (one strain)	*Chlorella protothecoides* (+)	+	Huss *et al.* (1993)
Mollusca:			
Anodonta cygnea	*Chlorella* sp. (-)	?	Smith and Douglas (1987)
Rotatoria:			
Ascomorpha helvetica	*Chlorella* sp. (-)	?	Goetsch and Scheuring (1926)
Turbellaria:			
Dalyellia viridis	*Chlorella vulgaris* (-)	-	Douglas (1987)
Phaenocora typhlops	*Chlorella vulgaris* (-)	-	Eaton and Young (1975)
Typhloplana viridata	*Chlorella vulgaris* (-)	-	Douglas (1987)

[a] Parentheses indicate, whether or not the taxonomic affiliation was confirmed by molecular methods; [b] Inability of isolated algae to grow in artificial media; [c] The *Chlorella vulgaris/sorokiniana* group includes the closely related species *C. vulgaris*, *C. sorokiniana* and *C. lobophora*; [d] *Symbiococcum* was described as a new genus

In some cases, the systematic and phylogenetic position of the symbiotic algae has been determined by the application of chemotaxonomic and molecular methods which are indispensable tools for a taxonomically problematic group such as *Chlorella* (Huss *et al.*, 1989a). Chemotaxonomic characters, DNA base composition and DNA/DNA hybridisation studies showed that several symbionts of *Paramecium*, *Acanthocystis* and *Spongilla* have close affinities to the „*Chlorella vulgaris/sorokiniana* group" which includes the closely related species *C. vulgaris*, *C. sorokiniana*, and *C. lobophora* (Douglas and Huss, 1986; Huss *et al.*, 1986, 1989a; Reisser *et al.*, 1988; Kessler and Huss, 1990). Differences in cell morphology and DNA base composition indicated that the symbionts of different strains of *Hydra viridis* might belong to different taxa (Huss *et al.*, 1989b). This was later confirmed by comparative analyses of 18S ribosomal RNA gene sequences, which showed that the symbionts from three strains of *Hydra viridis* belong to the „*C. vulgaris/sorokiniana* group", whereas the symbionts from a strain isolated in Jerusalem are closely related to *C. protothecoides* (Huss *et al.*, 1993; Fig. 1). *Hydra viridis* thus gained its symbionts at least twice through independent endosymbiotic events. Although both

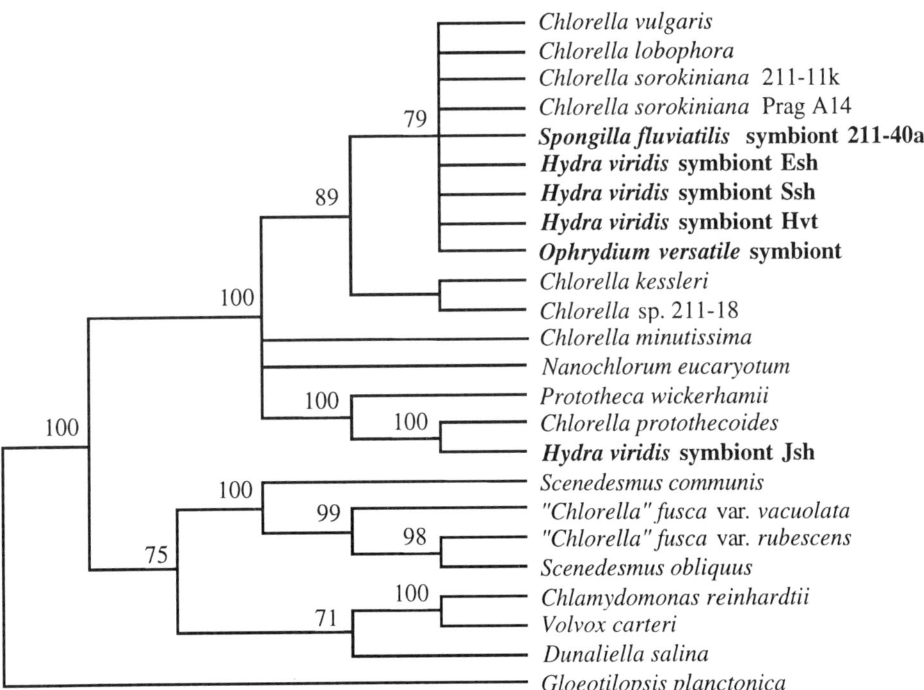

Figure 1. Maximum parsimony bootstrap analysis based on 18S rRNA gene sequences of several green algae. The cladogram shows phylogenetic relationships of native symbionts of *Hydra viridis*, *Ophrydium versatile*, and *Spongilla fluviatilis* with free-living taxa.

types of symbionts are closely related to free-living algae as indicated by a high degree of rRNA similarity, they have acquired some unknown kind of auxotrophy, as attempts to culture them *in vitro* have failed (Park *et al.*, 1967; Rahat, 1992). Recently, the symbionts of *Spongilla fluviatilis* and *Ophrydium versatile* could also be assigned to the „*C. vulgaris/sorokiniana* group" by 18S rRNA analyses (Fig. 1; unpublished). This group thus accounts for most of the freshwater endosymbionts and seems to be especially suited by some kind of preadaptation to live and coevolve in an extreme environment such as a living cell (Rahat, 1992), and to overcome the host cell's defences (Moulder, 1974, 1979). In spite of the relatively close relationship of these algae, it must be clearly emphasised, however, that none of the symbionts studied here is indistinguishable in terms of rRNA similarity from free-living strains or from each other. Considering the amount of base substitutions found in the rRNA genes, we still talk about a time scale of approximately several hundred thousand if not million years that passed since these algae descended from a common ancestor.

For continuance of a symbiotic association, it is necessary to pass symbionts from one generation to the next. The method of transmission varies widely. Some Protozoa, Porifera and Turbellaria are dependent on re-infection by phagocytosis of suitable algae. In *Paramecium* and most other Protozoa the symbionts are divided between the daughter cells at cell division. Asexually reproducing multicellular animals transmit their symbionts simply *via* buds (in Coelenterata) or gemmules (in Porifera), whereas in sexually reproducing hydra the symbionts have been shown to move from the endodermal digestive cells into the ectoderm, where they are taken up by the oocytes by phagocytosis (Campbell, 1990).

3. Host/Symbiont Specificity

Host/symbiont specificity can be tested by infecting aposymbiotic hosts with different algae. Aposymbiotic *Paramecium bursaria* and *Hydra viridis* can be generated by prolonged culturing in darkness, by photodestruction of the symbionts, or by treatment with chloramphenicol in the light (Karakashian and Karakashian, 1965; Pardy, 1976b; Rahat *et al.*, 1979). Subsequent infection experiments with several different algae showed that not only native isolated symbionts were able to establish a new symbiosis but also a number of free-living algae, mostly *Chlorella* species (Karakashian and Karakashian, 1965; Pardy and Muscatine, 1973; Muscatine *et al.*, 1975; Jolley and Smith, 1980; Rahat and Reich, 1984, 1985, 1986; Reisser, 1987a; Kessler *et al.*, 1988). After infection, three stages of compatibility were defined in the *Hydra/Chlorella* symbiosis (Rahat and Reich, 1987; Rahat *et al.*, 1989): (i) non-symbiotic algae disappeared from hydra within 1-3 days by expulsion or digestion ('transient infection'), (ii) persistent algae remained in cells of hydra for more than six days but eventually disappeared by death or dilution following host cell divisions ('prolonged persistence'), and (iii) successful 'endosymbiosis' was assumed, when reproduction of algae and transmission through buds to subsequent generations sustained the symbiotic association. In the most comprehensive of these studies (Kessler *et al.*, 1988, 1991; Rahat *et al.*, 1989), the ability of 50 strains of free-living or ex-symbiotic *Chlorella* strains to establish a stable symbiosis with the Swiss strain of *Hydra viridis* was compared with biochemical and physiological characters that are used for species delimitation in *Chlorella* (Kessler, 1982, 1984). Among these characters, only acid tolerance was correlated with symbiosis (Table 2). If we assume that the pH in the perialgal vacuoles is low, then it is convincing that only those algae can survive and propagate which can tolerate the specific conditions in this environment. This is fully compatible with the „Test Tube Hypothesis" proposed by Rahat (1989), which claims that a suitable environment must be offered for an alga to survive, be it a lake, a test tube or a host cell. Indeed, preliminary attempts to determine the pH in the perialgal vacuoles in hydra by the technique of Fluorescence Ratio Microscopy indicated a low pH of 3.5 - 4.0 few hours after infection with algae (Huss *et al.*, 1993).

TABLE 2. Symbiotic properties of *Chlorella* species, their acid tolerance, and amount of carbohydrates released to the growth medium

Species	No. of strains	Symbiosis[a]	pH limit for growth	Sugar excretion[b]
C. ellipsoidea	2	symbiotic	2.0 - 3.0	low
C. saccharophila	4	"	2.0 - 3.0	low
C. luteoviridis	1	"	3.0	low
C. kessleri	2	"	3.0	low
C. fusca var. *vacuolata* (= *S. vacuolatus*)[c]	4	"	3.0 - 3.5	intermediate
C. protothecoides	5	"	3.5 - 4.0	low
C. „paramecii"	2	"	5.5	high
C. fusca var. *fusca* (= *S. abundans*)[c]	1	persistent	4.0	low
C. mirabilis	3	"	4.0	high, intermediate
C. fusca var. *rubescens* (= *S. rubescens*)[c]	1	"	4.5	low
C. minutissima	1	"	5.5	low
C. sorokiniana	14	non-symbiotic	4.0 - 5.0	low
C. vulgaris	7	"	3.5-4.0	intermediate
C. lobophora	1	"	4.0	intermediate
C. zofingiensis	1	"	5.5	low
C. homosphaera	1	"	6.0	not determined

[a] See text for definition of symbiotic properties; [b] Maximum rate (x 10^{-10} g • ml^{-1} • h^{-1} per 10^6 cells) of excretion of maltose, glucose, or glucose-6-phosphate at optimal pH; low: 0-10, intermediate: 10-50, high: >50; data are from Kessler *et al.* (1991); [c] The varieties of *C. fusca* have been transferred to the genus *Scenedesmus* (Kessler *et al.*, 1997)

It is remarkable that free-living strains of the *C. vulgaris/sorokiniana* group show no tendency to induce symbiosis in hydra although the native symbionts are related to this group and although their acid tolerance seems to be sufficient. Obviously, cell size and pH tolerance are not the only factors that decide on a successful symbiosis. Experiments with double infections of *Hydra viridis* revealed intracellular algal competition which eventually resulted in the displacement of the supposedly less suited strain. When native zoochlorellae were present, they always remained as the sole symbionts, apparently as the best adapted (Rahat, 1985, 1992).

Alternatively, cell-to-cell recognition was claimed to be the mechanistic basis for host/symbiont specificity during the early contact phase and phagocytosis. Mainly, specific features of the algal cell surface have been discussed in this context such as antigenic determinants (Pool, 1979), net surface charge (McNeil *et al.*, 1981), or interaction of host membrane bound lectins with algal surface glycoconjugates (Meints and Pardy, 1980; Reisser *et al.*, 1982, Reisser, 1990). However, specific molecular structures that are involved in these processes have never been demonstrated, and recognition processes at an early phase of infection, if at all, have been questioned by others (Jolley and Smith, 1980; McAuley and Smith, 1982; McNeil and Smith, 1982; Rahat and Reich, 1987; Reich and Greenblatt, 1992).

4. Nutritional Interactions

Symbiotic algae release substantial amounts (40-60%) of photosynthetically fixed carbon in the form of low molecular weight carbohydrates such as maltose or glucose (Muscatine *et al.*, 1967; Cernichiari *et al.*, 1969). Excretion of maltose has been considered as an important prerequisite for the establishment of the *Paramecium*- and *Hydra/Chlorella* symbiosis (Weis, 1981; McAuley and Smith, 1982; Muscatine and McNeil, 1989), but infection experiments showed that the amount of released carbohydrates is not crucial for the infection of aposymbiotic hydra by *Chlorella* (Kessler *et al.*, 1991; Tab. 2). The released sugars are transported to the host cell and rapidly assimilated into host products (Roffman and Lenhoff, 1969; Mews, 1980; Mews and Smith, 1982; Douglas, 1987). This source of carbon and energy undoubtedly contributes to the enhanced survival of green hydra and green paramecium under conditions of limited food supply and starvation in the light (Muscatine and Lenhoff, 1965; Karakashian, 1975; Douglas and Smith, 1984). However, a predominant evolutionary advantage for the host is not obvious unless extreme and rare environmental conditions are encountered (Reisser, 1987b). For habitats where nitrogen sources are scarce, the release of maltose enables the host to divert amino acids from respiration and gluconeogenesis to growth processes (Rees and Ellard, 1989). Under normal conditions, no significant differences between the growth rate of green and aposymbiotic hydra were found (Douglas and Smith, 1983). Moreover, the algae also have nutritional demands on their host. All the nutrients required by the symbiont have to be derived from or passed through the host cell. Such movement of nutrients was shown by recovering isotopes from the algal fraction after feeding green hydra with ^{3}H- and ^{14}C-labelled food (Cook, 1972; Thorington and Margulis, 1981), demonstrating mutual exchange of metabolites between the symbiotic partners. Continuous growth of algal symbionts for at least a year was observed in green hydra kept in darkness with feeding. The growth rate of the hydra declined to a rate significantly lower than that of aposymbiotic individuals under the same conditions. These results indicate that the hydra provides all the organic and inorganic nutrients required by the symbionts and that the symbionts also may represent a cost to the host (Douglas and Smith, 1984; Smith and Douglas, 1987).

A more obvious example of host benefit from the symbiont's photosynthetic activity is the symbiosis between *Ophrydium versatile* and its *Chlorella* symbionts. This ciliate forms gelatinous colonies up to several centimetres in diameter (Winkler and Corliss, 1965). The gelatinous matrix which is highly resistant against chemical and thermical treatment, consists mainly of carbohydrates and represents a predominant proportion of colony carbon (Bücker and Hausmann, 1994). It could be shown that symbiotic photosynthesis produced sufficient carbon to support the observed growth rate of the host, the respiration and the substantial carbon content of the colony jelly (Sand-Jensen *et al.*, 1997). The formation of colonies by *Ophrydium* is thought to protect the ciliates against their normal predators (Sand-Jensen *et al.*, 1994).

5. 'Regulation' of Symbiotic Algae

Under normal conditions, the reproduction rate of the phycobionts corresponds to that of their host cell, and thus neither partner outgrows the other. Since the growth rate of host cells usually is much slower than that of the smaller algal symbionts (McAuley, 1994), a balance between the two must somehow be achieved. Mechanisms by which the size of the endosymbiont population are regulated, might be based on digestive or ecological processes (Reisser *et al.*, 1983). While digestion or expulsion of algae by the host cell probably do not play a decisive role for the regulation of the algal population size, a dynamic interaction of host and algal division rates depending primarily on ecological factors can be assumed (Reisser *et al.*, 1983). Algal growth could be limited by the amount of external nutrients such as light, host respiratory carbon dioxide, and nitrogen supply, and by the alga's ability to exploit them. Similarly, the growth rate of the host, in addition to external food organisms, depends also on internal nutrients such as oxygen, sugars, and probably amino acids excreted by the algae (Reisser, 1986). It is not clear, however, if the complex nutritional interactions resulting in a balanced growth rate of host and symbionts are part of a self-regulating system (Rahat, 1991, 1992), or if nutrient transfer to the symbionts is actively controlled by the host resulting in regulation of algal growth rate and cell division as proposed by others (cf. McAuley, 1994; Douglas and Smith, 1984).

References

Bücker, H. and Hausmann, K. (1994) *Mikrokosmos* **83**, 211-217.
Campbell, R.D. (1990) *Tissue & Cell* **22**, 137-147.
Cann, J.P. (1981) *Archiv für Protistenkunde* **124**, 353-360.
Cernichiari, E., Muscatine, L. and Smith, D.C. (1969) *Proceedings of the Royal Society of London B* **173**, 557-576.
Cook, C.B. (1972) *Biological Bulletin* **142**, 236-242.
Douglas, A.E. (1987) *British phycological Journal* **22**, 157-161.
Douglas, A.E. and Huss, V.A.R. (1986) *Archives of Microbiology* **145**, 80-84.
Douglas, A.E. and Smith, D.C. (1983) in W. Schwemmler and H.E.A. Schenk (eds.), *Endocytobiology, Endosymbiosis and Cell Biology **II***, Walter de Gruyter and Co., Berlin, pp. 631-648.
Douglas, A.E. and Smith, D.C. (1984) *Proceedings of the Royal Society of London B* **221**, 291-319.
Droop, M.R. (1963) *Symposia of the Society for General Microbiology* **13**, 171-199.
Duval, B. and Margulis, L. (1995) *Symbiosis* **18**, 181-210.
Eaton, J.W. and Young, J.O. (1975) *Archives of Hydrobiologie* **75**, 50-75.
Esteban, G.F. and Finlay, B.J. (1996) *Archiv für Protistenkunde* **147**, 181-188.
Esteve, I., Mir, J., Gaju, N., McKhann, H.I. and Margulis, L. (1988) *Symbiosis* **6**, 197-210.
Friedl, T. (1995) *J. Phycology* **31**, 632-639.
Goetsch, W. and Scheuring, L. (1926) *Z. Morph. Ökol. Tiere* **7**, 220-253.
Graham, L.E. and Graham, J.M. (1978) *J. Protozoology* **25**, 207-210.
Huss, V.A.R. and Sogin, M.L. (1990) *J. Molecular Evolution* **31**, 432-442.
Huss, V.A.R., Dörr, R., Grossmann, U. and Kessler, E. (1986) *Archives of Microbiology* **145**, 329-333.
Huss, V.A.R., Holweg, C., Seidel, B., Reich, V., Rahat, M. and Kessler, E. (1993) *Endocytobiosis and Cell Res.* **10**, 35-46.
Huss, V.A.R., Huss, G. and Kessler, E. (1989a) *Plant Systematics and Evolution* **168**, 71-82.
Huss, V.A.R., Rahat, M. and Kessler, E. (1989b) *Endocytobiosis and Cell Research* **6**, 109-112.
Jolley, E. and Smith, D.C. (1980) *Proceedings of the Royal Society of London B* **207**, 311-333.
Karakashian, M.W. (1975) *Symposia of the Society for Experimental Biology* **29**, 229-265.
Karakashian, S.J. (1970) *Annals of the New York Academy of Sciences* **175**, 474-487.
Karakashian, S.J. and Karakashian, M.W. (1965) *Evolution* **19**, 368-377.
Kessler, E. (1982) in F.E. Round and D.J. Chapman (eds.), *Progress in Phycological Research*, Elsevier Biomedical Press B.V., Amsterdam, vol. **1**, 111-135.
Kessler, E. (1984) in D.E.G. Irvine and D.M. John (eds.), *Systematics of the Green Algae*, Academic Press, London, pp. 391-407.
Kessler, E. and Huss, V.A.R. (1990) *Botanica Acta* **103**, 140-142.
Kessler, E., Huss, V.A.R. and Rahat, M. (1988) *Plant Systematics and Evolution* **160**, 241-246.
Kessler, E., Kauer, G. and Rahat, M. (1991) *Botanica Acta* **104**, 58-63.
Kessler, E., Schäfer, M., Hümmer, C., Kloboucek, A. and Huss, V.A.R. (1997) *Botanica Acta* **110**, 244-250.
McAuley, P.J. (1994) *Botanical Journal of Scotland* **47**, 97-112.
McAuley, P.J. and Smith, D.C. (1982) *Proceedings of the Royal Society of London B* **216**, 7-23.
McNeil, P.L. and Smith, D.C. (1982) *Proceedings of the Royal Society of London B* **216**, 1-6.
McNeil, P.L., Hohman, T.C. and Muscatine, L. (1981) *J. Cell Science* **52**, 243-269.
Meints, R.H. and Pardy, R.L. (1980) *J. Cell Science* **43**, 239-251.
Mews, L.K. (1980) *Proceedings of the Royal Society of London B* **209**, 377-401.
Mews, L.K. and Smith, D.C. (1982) *Proceedings of the Royal Society of London B* **216**, 397-413.
Moulder, J.W. (1974) *J. Infectious Diseases* **130**, 300-306
Moulder, J.W. (1979) in M.H. Richmond and D.C. Smith (eds.), *The Cell as a Habitat*, Cambridge University Press, Cambridge, pp. 85-96.
Muscatine, L. and Lenhoff, H.M. (1965) *Biological Bulletin* **129**, 316-328.
Muscatine, L. and McNeil, P.L. (1989) *American Zoologist* **29**, 371-386.
Muscatine, L., Cook, C.B., Pardy, R.L. and Pool, R.R. (1975) *Symposia of the Society for Experimental Biology* **29**, 175-203.
Muscatine, L., Karakashian, S.J. and Karakashian, M.W. (1967) *Comparative Biochemistry and Physiology* **20**, 1-12.
Pardy, R.L. (1976a) *J. Cell Science* **20**, 655-669.
Pardy, R.L. (1976b) *Biological Bulletin* **151**, 225-235.
Pardy, R.L. and Muscatine, L. (1973) *Biological Bulletin* **145**, 565-579.
Park, H.D., Greenblatt, C.L., Mattern, C.F.T. and Merril, C.R. (1967) *J. Experimental Zoology* **164**, 141-162.

Pool, R.R. Jr. (1979) *J. Cell Science* **35**, 367-379.
Rahat, M. (1985) *J. Cell Science* **77**, 87-92.
Rahat, M. (1989) in P. Nardon, V. Gianini-Pearson, A.M. Grenier, L. Margulis and D.C. Smith (eds.), *Endocytobiology **IV***, INRA, Paris, pp. 271-274.
Rahat, M. (1991) *OIKOS* **62**, 381-388.
Rahat, M. (1992) in W. Reisser (ed.), *Algae and Symbioses*, Biopress Limited, pp. 41-62.
Rahat, M. and Reich, V. (1984) *J. Cell Science* **65**, 265-277.
Rahat, M. and Reich, V. (1985) *J. Cell Science* **74**, 257-266.
Rahat, M. and Reich, V. (1986) *J. Cell Science* **86**, 273-286.
Rahat, M. and Reich, V. (1987) *Endocytobiosis and Cell Research* **4**, 13-23.
Rahat, M. and Reich, V. (1989) *Phycologia* **28**, 181-187.
Rahat, M., Kessler, E. and Huss, V.A.R. (1989) *Symbiosis* **7**, 117-126.
Rahat, M., Zeldes, D. and Reich, V. (1979) *Comparative Biochemistry and Physiology* **63C**, 27-30.
Rees, T.A.V. and Ellard, F.M. (1989) *Proceedings of the Royal Society of London B* **236**, 203-212.
Reich, V. and Greenblatt, C.L. (1992) *Symbiosis* **12**, 57-74.
Reisser, W. (1981) *Protoplasma* **105**, 273-284.
Reisser, W. (1986) *Progress in Protistology* **1**, 195-214.
Reisser, W. (1987a) *Annals of the New York Academy of Sciences* **503**, 316-329.
Reisser, W. (1987b) *Endocytobiology and Cell Research* **4**, 317-329.
Reisser, W. (1990) in P. Nardon, V. Gianini-Pearson, A.M. Grenier, L. Margulis and D.C. Smith (eds.), *Endocytobiology **IV***, INRA, Paris, pp. 225-228.
Reisser, W. (1992) in W. Reisser (ed.), *Algae and Symbioses*, Biopress Limited, pp. 1-19.
Reisser, W. and Widowski, M. (1992) in W. Reisser (ed.), *Algae and Symbioses*, Biopress Limited, pp. 21-40.
Reisser, W., Fischer-Defoy, D., Staudinger, J., Schilling, N. and Hausmann, K. (1984) *Protoplasma* **119**, 93-99.
Reisser, W., Meier, R. and Kurmeier, B. (1983) in W. Schwemmler and H.E.A. Schenk (eds.), *Endocytobiology, Endosymbiosis and Cell Biology **II***, Walter de Gruyter and Co., Berlin, pp. 533-543.
Reisser, W., Raduz, Z. and Wiessner, W. (1982) *Cytobios* **33**, 39-50.
Reisser, W., Vietze, S. and Widowski, M. (1988) *Symbiosis* **6**, 253-270.
Roffman, B. and Lenhoff, H.M. (1969) *Nature* **221**, 381-382.
Sand-Jensen, K., Geertz-Hansen, O., Pedersen, O. and Skovgaard, H. (1994) *Freshwater Biology* **31**, 121-130.
Sand-Jensen, K., Pedersen, O. and Geertz-Hansen, O. (1998) *Limnology & Oceanography*, in press.
Smith, D.C. (1979) *Proceedings of the Royal Society of London B* **204**, 115-130.
Smith, D.C. and Douglas, A.E. (1987) *The Biology of Symbiosis*, Edward Arnold Publishers, London.
Thorington, G. and Margulis, L. (1981) *Biological Bulletin* **160**, 175-188.
Weis, D.S. (1981) *Berichte der Deutschen Botanischen Gesellschaft* **94**, 547-555.
Willumsen, N.B.S., Siemensma, F. and Suhr-Jessen, P. (1987) *Archiv für Protistenkunde* **134**, 303-313.
Winkler, R.H. and Corliss, J.O. (1965) *Transactions of the American Microscopical Society* **84**, 127-137.

Biodata of H. Charles and P. Nardon.

Title of chapter: **Intracellular Symbiotic Bacteria within Insects.**

Dr. H. Charles is a biochemical engineer of the National Institute of Applied Sciences (INSA, Lyon, France). His PhD thesis (June 1997) was on "Molecular aspects of the principal symbiotic bacteria of *S. oryzae* (Coleoptera, Curculionidae) and study of their interactions with the host"
e-mail: <hcharles@insa.insa-lyon.fr>

Paul Nardon is a Professor in the National Institute of Applied Sciences (INSA, Lyon, France). He works on the symbiosis of the cereal weevil *Sitophilus* since 1975. This widespread insect is one of the most dangerous pests for stored and cultivated cereal in the world.
e-mail: <Lba@insa.insa-lyon.fr>

Paul Nardon

H. Charles

J. Seckbach (ed.), Enigmatic Microorganisms and Life in Extreme Environments, 651–660.

INTRACELLULAR SYMBIOTIC BACTERIA WITHIN INSECTS

H. CHARLES and P. NARDON
Laboratoire de Biologie Appliquée, INSA 406, UA-INRA 203 SDI-CNRS 5128
20, Avenue Albert Einstein, 69621 Villeurbanne Cedex FRANCE.

1. Introduction

Symbiosis is a permanent and hereditary association between two or more partners specifically distinct and, most of the time, very different from a phylogenetic point of view. A very common type of symbiosis involves bacteria, called endocytobiotes, living inside eukaryotic cells. Our investigation was limited to the endocytobiotes of insects. Most of the time they are Gram negative and non sporulating bacteria (Dash *et al.*, 1989). Such a habitat is very particular or extreme for at least three reasons: first, the bacteria, associated with their host, constitute a new biological unit, the symbiocosm, which is submitted to natural selection (Nardon and Grenier, 1993). Second, in this symbiocosm, the bacteria entirely depend on the host for their nutrition, and third, the growth of bacteria is strictly controlled by the insect (Nardon *et al.*, 1998).

In most intracellular symbiosis models, endocytobiotes are surrounded by a membrane of host origin, probably resulting from a phagocytosis event (cockroaches, aphids...). This structure, of bacteria inside vacuoles, is called a symbiosome. However in some insects, like the weevil *Sitophilus oryzae*, the bacteria lie free in the cytoplasm of the host cell, which indicates a high degree of intimacy between the two partners (Nardon and Grenier, 1989). Important problems that arise are the modalities of exchange between symbiotes and the host cell, and the mechanism of recognition. Unfortunately, very little is known about this relationship, but the possibility of transferring bacteria by microinjection from one strain to another, allows for an investigation into the specificity of the association, at least for the *Wolbachia* endocytobiotes (Chang and Wade, 1994 and laboratory results).

In insects, we must distinguish two different types of endocytobiosis (Nardon and Grenier, 1993). In "primitive symbiosis", bacteria are not present in all the insect population but, when present, they are found in numerous tissues and genital organs. However the host cells are apparently not modified. In contrast, in "integrated symbiosis", the bacteria are always present and limited to specialized cells, the bacteriocytes, and the ovaries. These bacteriocytes are generally giant and polyploid cells, transformed by the presence of bacteria (Nardon, 1988). They may be disseminated in the fat body where they divide, as in cockroaches (Brooks and Richards, 1955), or grouped to form one or several symbiotic organs, or bacteriomes, as in aphids (Ishikawa, 1989a), leafhoppers (Tiivel, 1989), or weevils (Nardon and Grenier, 1989). These traits are specific and the number, the location and the morphology of bacteriomes are perfectly controlled (Buchner, 1965; Nardon, 1988).

2. Phylogenetic position and molecular characterization

A wide variety of intracellular prokaryotes are found in association with insects; some of them are lethal for their host (*e.g.*, *Rickettsiella*) but most are engaged in symbiotic relationships. Because the endocytobiotes cannot be cultivated outside their host cells, their taxonomy has not been precisely defined. Nevertheless, recent advances in molecular biology, and especially the use of rDNA sequences for phylogenetic studies, have greatly improved our knowledge of these bacteria. At this time, insect endocytobiotes can be divided into four major groups of bacteria: (1) the flavobacteria-bacteroides group which includes the cockroach endocytobiotes; (2) the α-proteobacteria, a very homogeneous monophyletic group (2% 16S rDNA sequence divergence) corresponding to the genus *Wolbachia* (Werren, 1997); (3) the β-proteobacteria (mealybug endocytobiotes) and (4) the highly diversified group of γ-proteobacteria. Phylogenetic studies also facilitate the reconstitution of the evolutionary history of symbiosis. Hence, Bandi *et al.* (1995) showed that endocytobiotes of *Mastotermes darwiniensis*, a primitive Australian termite, are very close to the genus *Blattabacterium* (cockroach endocytobiotes). Since termites and cockoaches probably evolved from a common ancestor by the early Carboniferous period, it has been suggested that the endocytobiote integration occurred at least 300 million years ago in the primitive host insect. Concerning the α-proteobacteria group, based on sequence divergence of *ftsZ* genes, Werren *et al.* (1995) have suggested that the common ancestor of all examined *Wolbachia* lived about 50 million years ago. Finally, in the γ-proteobacteria group, the symbiotic association of *Buchnera* (aphid endocytobiotes) was estimated to be about 250 million years old, while the weevil *Sitophilus* endocytobiotes were probably integrated later (100 million years ago). These last two endocytobiotes are closely related to *Erwinia herbicola*, a free leaving bacteria associated with plant leaves, suggesting that they may be derived from an inhabitant of the plant on which host insects feed (Heddi, personal communication).

The genomic G+C content of insect endocytobiotes is generally low, regardless of their phylogenetic position (*e.g.*, *Blattabacterium*, 26%; aphid endocytobiotes, 30%; ant *Formica* endocytobiotes, 41%; leafhopper *Euscelidius* endocytobiotes, 31%). *Sitophilus* endocytobiotes are an exception with 54% C+G. It has been suggested that A+T accumulation occurred during intracellular evolution, but the molecular mechanism involved is not yet understood.

The genome of intracellular bacteria is reduced compared with that of extracellular bacteria (Herdman, 1985). This could be explained by gene deletions (or transfers to the nucleus) which may have occurred during intracellular evolution. Deleted genes may be those that became useless in the protected and stable intracellular environment. Hence, the genome size of the endocytobiotes of the weevil *Sitophilus oryzae* was estimated at 3.0Mb by pulsed field gel electrophoresis, which corresponds to about 60% of the genome size of the closest free living bacteria (Charles *et al.*, 1997).

Furthermore, endocytobiotes generally contain a single or a few rDNA operons in their genome while extracellular bacteria most often carry multiple operons (*e.g.*, seven in the Enterobacteriaceae family). *Buchnera*, as well as the primary endocytobiote of tsetse fly, possess only one copy of the rRNA operon in their genome, while *S. oryzae* endocytobiotes possess two copies of the same operon (Unterman *et al.*, 1989; Aksoy, 1995; Charles *et al.*, 1997). Such a low rDNA copy number would explain the slow growth rate of these bacteria in their intracellular environment.

3. Role of endocytobiotes

In primitive endocytobioses, particularly in insect / *Wolbachia* symbioses, the bacteria generally appear to have a weak influence on the fitness of the host (see Nardon and Grenier, 1993). Exceptions include the *Wolbachia* of *Trichogramma Bourarachae* whose presence in a laboratory strain enhances the fecundity (Girin and Bouletreau, 1995). Increased male fertility has also been reported in a strain of *Tribolium confusum* beetles following infection with *Wolbachia* (Wade and Chang, 1995). In contrast, it has been shown in *Drosophila simulans* that *Wolbachia* do not enhance productivity in infected strains but do not negatively affect the insect either (Poinsot and Merçot, 1997). However, in the same insect, as in numerous others (Werren, 1997), the presence of *Wolbachia* induces cytoplasmic incompatibility (CI) when an infected male is crossed with an uninfected female, whereas the reverse cross is normal. This incompatibility leads to high embryonic mortality. In some other strains it is bidirectional, when the male and the female harbour two different *Wolbachia* (Rousset and Solignac, 1995). Especially in the case of bidirectional CI, this incompatibility is thought to promote rapid speciation between populations which become reproductively isolated when infected by *Wolbachia* (see Werren, 1997 for extensive discussion). Another spectacular effect of *Wolbachia* is the thelytoky (production of females only) induced in *Trichogramma* (Louis *et al.*, 1993; Stouthamer *et al.*, 1993). The mechanism involved and the consequences on populations is exhaustively discussed in Werren (1997).

In specific integrated symbiosis (Nardon and Grenier, 1993), the endocytobiotes play different roles, which have been reviewed by several authors: Nardon and Grenier (1989, 1991, 1993), Douglas (1989), Ishikawa (1989a). To summarize, we can distinguish three essential distinct roles for the integrated endocytobiotes of insects.

Morphogenetic action. Generally the bacteria induce gigantism and polyploidy in host cells, except for germ cells. For instance, in the banana weevil *Metamasius*, the bacteriocytes may be more than 100 µm in diameter (Nardon *et al.*, 1985). In cockroaches, growth and division of bacteriocytes seems to be controlled by the symbiotes (Richards and Brooks, 1958). In the same way, the bacteriomes of the weevil *Sitophilus oryzae* disappear in the absence of bacteria (Nardon, 1973). In the cockroach *Blattella germanica*, extensive structural changes are also visible after partial or total elimination of symbiotes (Grigolo *et al.*, 1987).

Symbiotes as metabolite suppliers. Integrated endocytobiotes play a major role in the biology of their hosts, which are supplied with nutrients or other substances not synthesized by eukaryotic genomes. As a consequence, the presence of these bacteria greatly enhances the insect fitness through more rapid growth and increased fertility (Ishikawa, 1989a; Nardon and Grenier, 1989, 1993). Deprived of their symbiotes, aphids or cockroaches are no longer capable of reproducing. Aposymbiotic strains of *Sitophilus* can be maintained in the laboratory but with a reduced fitness. This example shows that the degree of integration of the bacteria is variable, probably as a function of the age of symbiosis, this being more recent for weevils than for aphids.

Among growth factors synthesized by endocytobiotes are vitamins supplied, for instance, to *Blattella germanica*: Pantothenic acid, riboflavin, pyridoxine and thiamine (Pant and Frankel, 1954). In *Sitophilus oryzae*, pantothenic acid, biotin and riboflavin are supplied in sufficient quantities to promote growth and reproduction. Pyridoxine and folic acid are also supplied, but at concentrations too low to enable the development of more than one generation (Wicker, 1983). The limited quantity of B vitamins in cereals may explain why aposymbiotic insects present a decreased fitness.

The supply of amino acids to the host by endocytobiotes is another good example of the role of symbiotic bacteria in adaptation to the environment. The diet of aphids is phloem sap, which is very poor from a nutritional point of view, and aphids can only survive and reproduce thanks to complementation of this diet by symbiotes (Douglas, 1996). They synthesize several essential amino acids: tryptophan (Douglas and Prosser, 1992), threonine, isoleucine, lysine (Febvay *et al.*, 1995) and methionine (Douglas, 1988a). Glutamate is the only amino acid supplied by the insect to the symbiotes (Febvay *et al.*, 1995). Concerning lipids, the role of endocytobiotes seems to be less important. Nevertheless, in aphids, it seems highly probable that sterols are synthesized by the bacteria, at least in *Myzus persicae* (Douglas, 1988b).

Interactions of symbiotes with host metabolism. This phenomenon has been principally studied in our laboratory on *Sitophilus oryzae* (see in Nardon and Grenier, 1989). Its natural food (wheat) contains too much methionine. In symbiotic weevils, the excess methionine is converted into methionine sulfoxide, whereas in aposymbiotic ones it is demethylated into the amino acid sarcosine which accumulates in the haemolymph. This second way consumes ATP. Therefore, it is clear that the symbiotic bacteria allow their

host to conserve ATP and avoid sarcosine accumulation, a function normally carried out by mitochondria.
Finally, we have demonstrated an interaction with the mitochondrial energetic metabolism. The activities of six enzymatic complexes have been compared in symbiotic and aposymbiotic insects. They are always higher in the presence of bacteria and this must be related to the supply of B vitamins, particularly pantothenic acid and riboflavin (Heddi *et al.*, 1993).

4. Intracellular living conditions of endocytobiotes

4.1. HOST CONTROL OF THE ENDOCYTOBIOTE POPULATION

The establishment of endocytobiotic relationships requires a very high host-symbiote compatibility. On the one hand, the bacteria have to avoid the defensive reaction of the insect. On the other hand, the host has to prevent bacterial invasion by controlling both the location and the density of the endocytobiotes. Among factors involved in the insect defensive system, lysosomal enzymes, hemaglutinins and antibacterial peptides have been described (Tiivel, 1993).

In amoebae symbiosis, Jeon (1995) showed that the endocytobiotes produce one protein of 96 kDa and lipopolysaccharides that could prevent symbiosomes fusing with lysosomes. As previously mentioned, the endocytobiotes of *Sitophilus oryzae* are lying free in the bacteriocyte cytosol. This supposes a full immunological compatibility between the two partners, and should hence protect the endocytobiotes from lysosomal fusion. Nevertheless, bacterial lysis occurs in the bacteriocytes of the weevil ovaries (Nardon, 1971), and the number of symbiotes is stable for a given strain, controlled by chromosomal factors not yet identified at the molecular level (Nardon *et al.*, 1998). Other factors, such the nature of the diet, can also influence the symbiote density (Nardon and Grenier, 1989). We are in the presence of what we call a microecosystem, the equilibrium of which is regulated by complex interaction factors.

4.2. STRESS PROTEIN PRODUCTION BY THE ENDOCYTOBIOTES

Stress proteins have been observed in numerous intracellular parasitic bacteria (Van der Vies and Georgopoulos, 1996). In these bacteria, the most predominant protein over-expressed in response to the intracellular environment is the chaperonin protein Hsp60 (60 kDa heat shock protein). In free living bacteria, this protein is essential for cell viability at all temperatures (Fayet *et al.*, 1989). It is known to be involved in the folding of nascent polypeptides, the assembly of oligomeric protein complexes and protein export from bacteria (Zeilstra-Ryalls *et al.*, 1991). In the case of parasite survival inside host cytosol, it has been suggested that Hsp60 could be implicated in the maintenance of essential bacterial proteins and/or the excretion of virulence factors .

The over-expression of chaperonins has also been reported in non-pathogenic bacteria, such as the endocytobiotes of tsetse flies *Glossina* (Aksoy, 1995), *Sitophilus* weevils (Charles *et al.*, 1995) and aphids (Ishikawa, 1989b). In *Sitophilus oryzae,* protein labelling experiments revealed that chaperonin represents about 40% of the total neosynthesized proteins of the endocytobiotes *in vivo*. In the aphid *Schizaphis graminum*, Baumann *et al.* (1996) showed that the amount of Hsp60 inside the endocytobiotes corresponds approximately to that found in *Escherichia coli* grown at 46°C (close to its maximal growth temperature). Finally, Morioka *et al.* (1993) reported that the Hsp60 protein of another aphid *Acyrthosiphon pisum,* is autocatalytically phosphorylated *in vitro* at elevated temperatures and can transfer radioactive phosphate from ATP to GTP. These results suggested that the Hsp60 protein could function in the symbiocosm not only as a molecular chaperone but also as an energy-coupling protein.

Very little is known about the regulation of *hsp60* gene expression in the insect endocytobiotes. In *Buchnera*, the *symSL* operon (containing the *hsp10* and *hsp60* genes) seems to be under the control of a heat shock promoter that is functional in *E. coli*. Nevertheless, the operon is not heat inducible *in vivo* and seems to be constitutively expressed thanks to an A- and T- rich region around the heat shock promoter that might facilitate transcription (Sato and Ishikawa, 1997). The common characteristic of all intracellular bacteria studied (parasite or mutualist) is that, whereas a stoechiometric ratio of 2:1 (Hsp60:Hsp10) is theoretically required in the cell, a large quantity of Hsp60 protein relative to Hsp10 was observed in the bacteria located in the host cytosol. Thus, the strict over-expression of Hsp60 may reflect physiological needs, specific for bacteria living in an intracellular environment.

From a general point of view, it is noteworthy that the comparison of stress protein expression between free living and intracellular symbiotic bacteria led several authors to assume that endocytobiosis conditions generate stress for symbiotes (Jeon, 1995). In our opinion, it is not really surprising to observe an Hsp60 protein in all intracellular bacteria. This protein is indeed essential for bacterial survival in all environmental conditions and one wonders what could represent intracellular stress for a bacterium that has been living in the host cytoplasm for a couple of hundred million years? The conservation of the Hsp60 over-expression mechanism during intracellular bacterial evolution (from the parasite to the integrated endocytobiotes and probably to the cytosolic organelle) rather reveals an essential function of the chaperonin in maintaining the equilibrium and/or the protein import/export occurring in the symbiocosm. In this sense the term symbionin, proposed by Ishikawa (1982) to qualify the chaperonin produced by the aphid endocytobiotes, seems to be particularly accurate.

4.3. INTRACELLULAR EVOLUTION OF ENDOCYTOBIOTES

Endocytobiote populations are very different from those of free living bacteria, both in terms of dynamics and evolution. They are indeed small, asexual and pass through a bottle-neck at each host generation when progeny are inoculated. They are hence clonal since no recombination can occur between lineages sequestered in different hosts and since horizontal transfers between hosts appear to be relatively rare. Moran (1996) showed that endocytobiote population structures are characterized by, first, a faster substitution rate relative to the free living organisms, second an accumulation of deleterious mutations in non neutral sites and, third, as selective constraints are reduced in the protected intracellular environment there is a greater effect of mutational bias (A-T pressure) on the genomic DNA composition (*i.e.*, on neutral and non neutral sites). Regarding this peculiar evolution, the following question comes: "why have asexual endocytobiotes not decayed to extinction?" (Hurst and McVean, 1996). Compensatory responses must be involved to explain the persistence of endocytobiotes inside their hosts. To decrease the accumulation of mutations, in the absence of recombination, the per genome mutation rate must be somehow reduced. This could be achieved by a reduction in the number of genes in the genome. Genes could be deleted (if their function in the intracellular environment becomes unnecessary) or transferred to the host nucleus. A second compensatory process may involve the Hsp60 protein that could prevent misfolding of proteins, whose tertiary or quaternary structures have been modified by an accumulation of mutations. Finally, intracellular living conditions seem to be particularly suitable for such a compensatory mechanism because (1) the host can supply the bacteria with excess enzymes or nutrients to compensate for some of the endocytobiotes failing and (2) the host can select (by elimination of the deleterious bacteriocytes) its most competitive endocytobiote populations.

As natural selection will occur both inside and outside the symbiotic association, the formation and maintenance of the symbiocosm require, first, the co-adaptation of the two partners (inside selection) and, second, the adaptation of the symbiocosm itself to the environmental conditions (outside selection). By this means, symbiosis leads to the modification of both partners and to the formation of a new biological unit (Nardon and Grenier, 1991, 1993). It is not yet clear whether intracellular bacterial populations take advantage of these associations since the integration process seems to be very costly for the bacteria. Moreover, bacterial populations might never reach equilibrium until they are totally integrated in the host cell, as has been suggested in the amitochondrial protozoa *Trichomonas vaginalis* (Horner *et al.*, 1996). Future molecular studies on these peculiar bacteria promise to elucidate this integration process, one of the most fascinating aspects of Eukaryote cell evolution. The different types of intracellular symbiosis currently observed probably represent different steps of integration of symbiotes, which finally behave as cell organelles. Such observations are in agreement with the endosymbiotic theory of the formation of the Eukaryote cell (Margulis, 1993).

5. References

Aksoy, S. (1995) *Insect Mol. Biol.*, **4**: 23-29.

Bandi, C., Sironi, M., Damiani, G., Magrassi, L., Nalepa, C.A., Laudani, U. Sacchi, L. (1995) *Proc. R. Soc. Lond.* B, **259**: 293-299.

Baumann, P., Baumann, L. and Clark, M.A. (1996) *Cur. Microb.*, **32**: 279-285.

Brooks, M.A. and Richards, A.G. (1955) *Science*, **122**: 242.

Buchner, P. (1965) *Endosymbiosis of animals with plant microorganisms*, Interscience.

Chang K.W. and Wade, M.J. (1994) *Can. J. Microbiol.*, **40**: 978-981.

Charles, H., Ishikawa, H. and Nardon, P. (1995) *C. R. Acad. Sci. Paris.*, **318**: 35-41.

Charles, H., Condemine, G., Nardon, C. and Nardon, P. (1997) *Insect Biochem. Mol. Biol.*, **27**: 345-350.

Dasch, G.A., Weiss, E. and Chang, K.P. (1989) *Bergey's manuel of systematic bacteriology*, Williams and Wilkins, Baltimore, pp. 811-833.

Douglas, A.E. (1988a) *Insect Biochem.*, **18**: 599-605.

Douglas, A.E. (1988b) *J. Insect Physiol.*, **34**: 403-408.

Douglas, A.E. (1989) *Biol. Rev.*, **64**: 409-434.

Douglas, A.E. (1996) *J. insect Physiol.*, **42**: 247-255.

Douglas, A.E. and Prosser (1992) *J. insect Physiol.*, **38**: 565-568.

Fayet, O., Ziegelhoffer, T. and Georgopoulos, C. (1989) *J. Bacteriol.*, **171**: 1379-1385.

Febvay, G., Liadouze, I., Guillaud, J. and Bonnot, G. (1995) *Arch. Insect Biochem. Physiol.*, **29**: 45-69.

Girin, C. and Bouletreau, M. (1995) *Experientia*, **51**: 398-401.

Grigolo, A., Sacchi, L., De Piceis Polver, P. Dealessi, F. and Laudani U. (1987) *Monitore Zool. ital.* **21**: 133-140.

Heddi, A., Lefebvre, F. and Nardon, P. (1993) *Insect Biochem. Mol. Biol.* **23**:403-411.

Herdman, M. (1985) *The evolution of genome size,* John Willey and Sons, New York, pp. 37-68.

Horner, D.S., Hirt, R.P., Kilvington, S., Lloyd, D. and Embley, T.M. (1996) *Proc. R. Soc. Lond. B* **263**: 1053-1059.

Hurst, L.D. and McVean, G.T. (1996) *Nature*, **381**: 650-651.

Ishikawa, H. (1982) *Insect Biochem.*, **12**(6): 613-622.

Ishikawa, H. (1989a) *Int. Rev. Cytol.* **116**: 1-45.

Ishikawa, H. (1989b) *Insect Endocytobiosis: morphology, physiology, genetics, evolution,* W. Schwemmler and G. Gassner (eds), CRC press., Washington pp. 123-143.

Jeon, K.W. (1995) *Trends Cell. Biol.*, **5**(3): 137-140.

Louis, C., Pintureau, J. and Chapelle, L. (1993) *C. R. Acad. Sci. Paris,* **316**, III: 27-33.

Margulis, L. (1993) *Symbiosis in cell evolution,* Freeman.

Moran, N. (1996) *Proc. Natl. Acad. Sci. USA*, **93**: 2873-2878.

Morioka, M., Hiromichi M. and Ishikawa H. (1993) *J. Biochem.*, **114**: 246-250.

Nardon, P. (1971) *C. R. Acad. Sci. Paris*, **272D**: 2975-2978.

Nardon, P. (1973) *C. R. Acad. Sci. Paris*, **277D**: 981-984.

Nardon, P. (1988) *Cell to cell signals in plant, animal and microbial symbiosis.* S. Scannerini S. (eds), Springer-verlag NATO ASI series H17, pp. 85-100.

Nardon, P. and Grenier, A.M. (1989) *Insect endocytobiosis: morphology, physiology, genetics, evolution.* W. Schwemmler and G. Gasner (eds), CRC Press, Washington, pp. 175-216.

Nardon, P. and Grenier, A.M. (1991) *Symbiosis as a source of evolutionary innovation: speciation and morphogenesis*, L. Margulis and R. Fester (eds), MIT press, pp.153-169.

Nardon, P. and Grenier, A.M. (1993) *Ann. Soc. Entomol. Fr. (N.S.)*, **29**(2): 113-140.

Nardon, P., Grenier, A.M. and Heddi, A. (1998) *Symbiosis* (in press).

Nardon, P., Louis, C., Nicolas, G. and Kermarrec, A. (1985) *Ann. Soc. Entomol. Fr.* **21**: 245-258.

Pant, N.C. and Fraenkel, G. (1954) *Biol. Bull.*, **107**: 430-432.

Poinsot, D. and Merçot, H. (1997) *Evolution* **51**: 180-186.

Richards, A.G. and Brooks, M.A. (1958) *Annu. Rev. Entomol.*, **3**: 37-56

Rousset, F. and Solignac, M. (1995) *Proc. Natl. Acad. Sci.* USA, **92**: 6389-6393.

Sato, S. and Ishikawa, H. (1997) *J. Bacteriol.*, **179**(7): 2300-2304.

Stouthamer, R., Breeuwer, JAJ and Luck, R.F. (1998) *Nature*, **361**: 66-68.

Tiivel, T. (1989) *Insect endocytobiosis: morphology, physiology, genetics, evolution.* W. Schwemmler and G. Gasner (eds), CRC Press, Washington, pp. 111-122.

Tiivel, T. (1993) *Endocytobiology V* S. Sato,M. Ishida, H. Ishikawa (eds), Tübingen university press, Tübingen, pp. 87-93.

Unterman, B.M., Baumann, P., Mclean, D.L. (1989) *J. Bacteriol.*, **171**: 2970-2974.

Van der Vies, S. and Georgopoulos, C. (1996) *The chaperonin,* Academic Press, pp. 137-165.

Wade, M.J. and Chang, N.W. (1995) *Nature*, **373**: 72-74.

Werren, J.H. (1997) *Annu. Rev. Entomol.*, **42**: 587-609.

Werren, J.H., Guo, L., Windsor, D.W. (1995) *Proc. R. Soc. London B*, **262**: 147-204.

Wicker, C. (1983) *Comp. Biochem. Physiol.*, **76A**: 177-182.

Zeilstra-Ryalls, J., Fayet, O. and Georgopoulos, C. (1991) *Ann. Rev. Microbiol.*, **45**: 301-325.

Biodata of **James F. White** chief author (with P.V. Reddy, C.W. Bacon and D. Cabral) of ***"The mysterious Interrelationships Between Fungi and Plants: The case of Endosymbionts".***

Dr. James F. White, Jr. is Assistant Professor of Plant Pathology at Rutgers University. He earned his Ph.D. in 1987 from the University of Texas. Dr. White conducts research on the ecology and systematics of fungal endophytes and epibionts of plants. In 1996 he was the recipient of the Alexopoulos Prize from the Mycological Society of America for his research in this field. He is author of more than 90 research articles and book chapters and coedited a book on the biotechnology of endophytes.

E-Mail: **jwhite@aesop.rutgers.edu**

J. Seckbach (ed.), Enigmatic Microorganisms and Life in Extreme Environments, 661–670.

THE MYSTERIOUS INTERRELATIONSHIPS BETWEEN FUNGI AND PLANTS: THE CASE OF ENDOSYMBIONTS

JAMES F. WHITE, JR.[1], PONAKA V. REDDY[1], CHARLES W. BACON[2] and DANIEL CABRAL[3]
[1]Department of Plant pathology, Rutgers University, New Brunswick, NJ 08903, USA
[2]United States Department of Agriculture, Richard Russell Research Center, Athens, GA 30613
[3]Departamento de Ciencias Biológicas, Facultad de Ciencias Exactas y Naturales, Universidad de Buenos Aires, 1428, Buenos Aires, Argentina

1. Background

The three major groups of terrestrial fungi (Ascomycetes, Basidiomycetes, and Zygomycetes) are believed to have evolved from a member of the aquatic group Chytridiomycetes that associated with terrestrial plants, either as a saprophyte of dead plant material or as a parasite of living plants (Pirozynski and Malloch, 1975). The earliest fossil evidence of land plants contain fungal remains that may be interpreted as chytridiomycetous and zygomycetous in nature (Kidston and Lang, 1921). These Early Devonian fungal remains have been interpreted as endosymbionts (Pirozynski and Malloch, 1975), but are probably saprophytes due to the degraded condition of host tissues in which mycelium is encountered (Kidston and Lang, 1921). It is evident that sometime after plants colonized land, symbiotic associations began to develop between plants and terrestrial fungal groups. Fossils containing roots from the Triassic were found to contain fungi (Zygomycetes) of the ecologically-important symbiotic association, known as vesicular-arbuscular mycorrhizae (Stubblefield, Taylor, and Trappe, 1987). It is unknown when or from what progenitor the Ascomycetes and Basidiomycetes evolved. It seems probable that the ancestor to these groups was zygomycetous and that their evolution was influenced by their association with land plants since they evolved as mycorrhizae, pathogens, and saprophytes of plant materials. It is within the Clavicipitaceae (Ascomycetes) that another type of fungal endosymbiont evolved. The family Clavicipitaceae contains members that are biotrophic parasites of insects, fungi, and plants. Within the groups that are parasitic on grasses, many are episymbiotic (superficial on host tissues), while some are endosymbiotic (colonize internal tissues of the host). In this chapter we will examine biological aspects of episymbiotic and endosymbiotic members of the Clavicipitaceae.

2. Evolution

The Clavicipitaceae are believed to have evolved from within the order Hypocreales

(Rehner and Samuels, 1995; Spataforo and Blackwell, 1993), an order that includes families: Clavicipitaceae, Hypocreaceae, Nectriaceae, etc.... Several features of the Clavicipitaceae distinguish it as different from other families in the Hypocreales, including: perithecia undergo *Epichloë*-type centrum development, rather than *Nectria*-type development; ascospores are filamentous and many-celled, rather than ellipsoidal and few-celled; and asci generally possess a pronounced apical thickening (White, 1993, 1997). Within the Clavicipitaceae evolutionary developments are poorly known.

3. Taxonomy

The taxonomic knowledge of the plant-infecting Clavicipitaceae is not well developed. One species, *Epichloë bertoni* is known to infect dicotyledonous plant species in the American tropics (Diehl, 1950); however, the majority of the species world-wide are graminicolous, infecting predominantly grasses. Species of genus *Epichloë* (Fr.) Tul.and its conidial state in *Neotyphodium* Glenn, Bacon & Hanlin (= *Acremonium* sect. *Albolanosa* Morgan-Jones and W. Gams) are commonly encountered in cool-season grasses of North America and Europe. Many *Neotyphodium* spp. have lost the capacity to develop the external *Epichloë* state, and produce no conspicuous fruiting structures on plants. Among these endosymbionts are the economically-important endophytes *N. coenophialum* and *N. lolii*, widespread in tall fescue (*Festuca arundinaceae*) and perennial ryegrass (*Lolium perenne*), respectively, two important species of turf and forage grasses. Genera *Balansia*, *Parepichloë*, and *Myriogenospora* infect warm-season grasses. Species of *Balansia* are most numerous in the Americas but may also be found in Asia; species of *Parepichloë* are limited to African and Asian tropics; species of *Myriogenospora* are found exclusively in the Americas.

4. Evolution of Endophytism

Two genera of the Clavicipitaceae contain endosymbionts, including *Balansia* and *Epichloë*. In genus *Balansia*, the step to endosymbiosis was made in a group of species (*B. claviceps* group) that are limited to the Americas (Diehl, 1950). Endosymbiotic *B. claviceps* and *B. obtecta* were thus derived from an episymbiont ancestor. All species of the graminicolous *Epichloë* are endosymbionts. Phylogenetic evidence suggests that episymbiotic genus *Parepichloë* and the graminicolous *Epichloë* diverged from a common anscestor that was likely also episymbiotic (White and Reddy, In press), the genus *Epichloë* acquiring the endophytic habit, and the genus *Parepichloë* retaining the ancestral episymbiotic habit. The occurrence of an episymbiotic stage in the life cycles of some *Epichloë* and *Neotyphodium* species may be a hold over from the episymbiotic habit (White, Martin, and Cabral, 1996).

5. Advantages of Endophytism

It is probable that among the benefits to the fungal symbiont resulting from a switch from episymbiosis to endosymbiosis is the following: 1) greater access to nutrients; 2) protection from dessication; 3) protection from surface-feeding insects; and 4) protection from parasitic fungi and the competition of other microbes.

5.1. GREATER ACCESS TO NUTRIENTS

In episymbiotic fungi, as exemplefied by *Myriogenospora atramentosa*, mycelium is limited to the upper surfaces of leaves. In order to produce the stroma (the fungal reproductive structure), a substantial amount of nutrients must be obtained from the host tissues. Nutrient acquisition is accomplished by the episymbiont by modifying tissues of the plant so that nutrients flow to mycelium on the surfaces of those tissies. Epidermal cells in close association to stromata are swollen (hypertrophied) and lack the waxy cuticle that normally prevents escape of water and nutrients from the leaf (White and Glenn, 1994). Because the cuticle is absent on the leaf, the episymbiont may extract nutrients from the leaf across the modified epidermal layer. In endosymbionts, the extraction of nutrients from the host is enhanced since mycelium is distributed among the internal cells of the plant tissues (Fig. 1) (White and Owens, 1992). Thus both internal and external tissues of the host may be modified to enhance flow of nutrients to the stromal mycelium (White, Bacon, and Hinton, 1997).

5.2. PROTECTION FROM DESSICATION

Episymbionts perennate as mycelium on the surfaces of plants. Meristem tillers seem to be a region of high concentration of perennating mycelium (Leuchtmann and Clay, 1988). It is reasonable to expect that the episymbiont is vulnerable to dessication if conditions become dry. This may explain why most episymbiotic species are concentrated in moist tropical climates, while endosymbionts, like *Balansia strangulans*, range into temperate regions (Diehl, 1950). In desert environments only endosymbionts, such as the asymptomatic *Neotyphodium* spp., are known to occurr. Endosymbiotic species of *Balansia*, *Epichloë*, and *Neotyphodium* may be protected from the effects of a dessicating environment since they are contained within the moist internal tissues of the plant host.

5.3. PROTECTION FROM FUNGUS-FEEDING INSECTS, PARASITISM, AND COMPETITION OF MICROBES

The surfaces of plants are scoured by insects, such as fungus mites and snails, that actively search out and consume fungal mycelium and spores. Epibiotic fungi likely fall prey to many mycophagous organisms. In addition, the air contains spores of numerous species of fungi, such as *Trichoderma* spp., that have the capacity to parasitize mycelium of other species of fungi. These propagules likely find their way to the plant surfaces where epibionts may be parasitized. Many fungi and bacteria are known to colonize the surfaces plants. These microorganisms may actively compete with episymbiontic fungi for nurients and space. Endosymbionts escape all of the potentially deletarious interactions that might occur on the surfaces of plants.

6. Formation of Stromata on Plants

In order to reproduce sexually, endosymbionts and episymbionts must produce stromata on host plants (Figs. 2, 3). The stroma generally consists of a mixture of both fungal and host tissues, interfacing in such a way as to facilitate flow of nutrients from living host tissues into the fungal component of the stroma. The fungal component of the stroma is composed of differentiated mycelia in which three tissues may be distinguished. In *Epichloë*, the fungal component includes: interstitial tissues found in spaces between the inflorescence primordium and leaves embedded within the stroma; a fungal cortex layer composed of a tight palisade of parallel hyphae that appears to function to impede loss of water and nutrients to the environment; and an outer hymenium (Fig. 3) on which spermatia (Fig. 4) are produced and following fertilization is converted to the ascomal stroma in which perithecia (Fig. 5) and asci containing ascospores (Fig. 6) differentiate.

7. Modification of Plant Tissues

Plant tissues within the stroma are generally modified, apparently to maximize the flow of nutrients into the mycelium. Mesophyll cells within stromata of *Epichloë* may be hypertrophied (White, Glenn, and Chandler, 1995). Epidermal cells trapped within the stroma show modification to the extent that it softens and fails to function as a barrier layer. Additionally, mycelium is often observen within vascular tissues within the stroma (Fig. 7). Evaporation of water has been demonstrated to occur rapidly from the mycelial surface of the stroma. Rapid evaporation of water from the stroma may be an important mechanism whereby nutrients are delivered to the mycelium of the stroma. Enhanced flow of water into the stroma to replace that lost by evaporation may facilitate the transfer sugars and other compounds in solution to the mycelium of the stroma (White and Camp, 1995).

8. Sexual Reproduction

Stromata provide the foundation on which the fungal reproductive cells are differentiated (Figs. 3-6). In the endosymbiotic *Balansia obtecta*, the stroma has been modified into a sclerotium. Stromata of *B. obtecta* germinate in the spring to form gametangia consisting of clusters of receptive hyphae and cup-shaped conidiomata bearing conidia, often classified in form genus *Ephelis*. Ephelidial conidia function as spermatia in a heterothallic mating process. Perithecia in stalked ascomatal stromata develop when spermatia of one mating type are transferred to receptive hyphae of the opposite mating type. It has been suggested that ascospores infect hosts by infection of florets (Diehl, 1950). In *Epichloë*, sexual reproduction similarly involves a heterothallic mating system. Stromata are initially covered with a layer of spermatia (Figs. 3, 4) or *Neotyphodium* conidia and receptive hyphae (White and Bultman, 1987). Spermatia are vectored between opposite mating types of the stroma by symbiotic flies, classified in genus *Phorbia*. Female flies visit several stromata, feeding on conidia and mycelium. Conidia are not digested and instead accumulate in the digestive tract of the fly. Females deposit eggs on stromata and in the process, deposit spermatia on stromata by defecating on them. After perithecia (Fig. 5) develop, ascospores (Fig. 6) are produced that are ejected from asci and are believed to also infect grasses at the florets. Plants developing from infected seeds contain the endophyte within the embryo.

9. Benefits to Hosts

Clavicipitaceous endosymbionts are so common in many grasses that it is reasonable to expect that they impart some benefit to their hosts. Numerous studies have suggested that beneficial effects on grasses may include: enhanced insect and nematode resistance, fungus disease resistance, enhanced vegetative growth, and increased drought tolerance (Clay, Hardy, and Hammond, 1985; West, Oosterhuis, and Wullschleger, 1990; Bacon and De Battista, 1990). Enhanced insect resistance has been proposed to be the result of endosymbiont produced alkaloids, such as peramine, which serve as antifeeding agents. Enhanced vegetative growth may be the result of auxins that are produced by the endosymbionts (Porter, Bacon, and Robbins, 1979). The mechanism of increased drought tolerance is believed to involve the increased ability of meristems to recover after drought conditions, however, the mechanism is not clearly understood (West, Oosterhuis, and Wullschleger, 1990). One recent study (Clarke et al., 1997) has also demonstrated endosymbiont-enhanced resistance in fine fescue grasses to the dollarspot disease, caused by fungal pathogen *Sclerotinia homeocarpa*. It has been suggested that enhanced resistance to fungal pathogens could be the result of antifungal compounds produced by endophytes (White and Cole, 1985). Koshino et al. (1989) described several fungitoxic sesquiterpenes and sterols from an endophyte (*Epichloë typhina*) of timothy grass (*Phleum pratense*) that were demonstrated to inhibit development of fungal pathogens of the grass. Whatever mechanisms account for the benefits to hosts, they translate to increased persistance and competitive capacity of endosymbiont-containing grasses in natural populations (Clay, 1988). It has been reported that pastures containing a low percentage of endosymbiont-infected plants when first established will, over a few years, drastically increase in the percentage of infected plants (Lewis and Clements, 1986; Latch, Potter, and Tyler, 1987). The selective advantage of endosymbiont-infected grasses is most pronounced when grasses are subjected to environmental stresses (Clarke et al., 1997).

10. Toxic Syndromes Caused by Endosymbionts

Due to the production of ergot alkaloids and other secondary compounds by Clavicipitaceae, host plants are often rendered toxic to mammalian herbivores. Several well-studied toxic syndromes may develop in mammals feeding on forage grasses containing endosymbiotic species of *Neotyphodium*. Cattle that consume tall fescue containing the endosymbiont *N. coenophialum* may develop a condition known as 'fescue foot'. Due to the vasoconstrictive properties of the ergot alkaloids produced by species of *Neotyphodium*, over a long period of time the flow of blood is reduced to the body extremities of animals and consequently hooves and tails may rot off. In perennial ryegrass infected by *N. lolii*, the production of lolitrems by the endosymbiont, results in

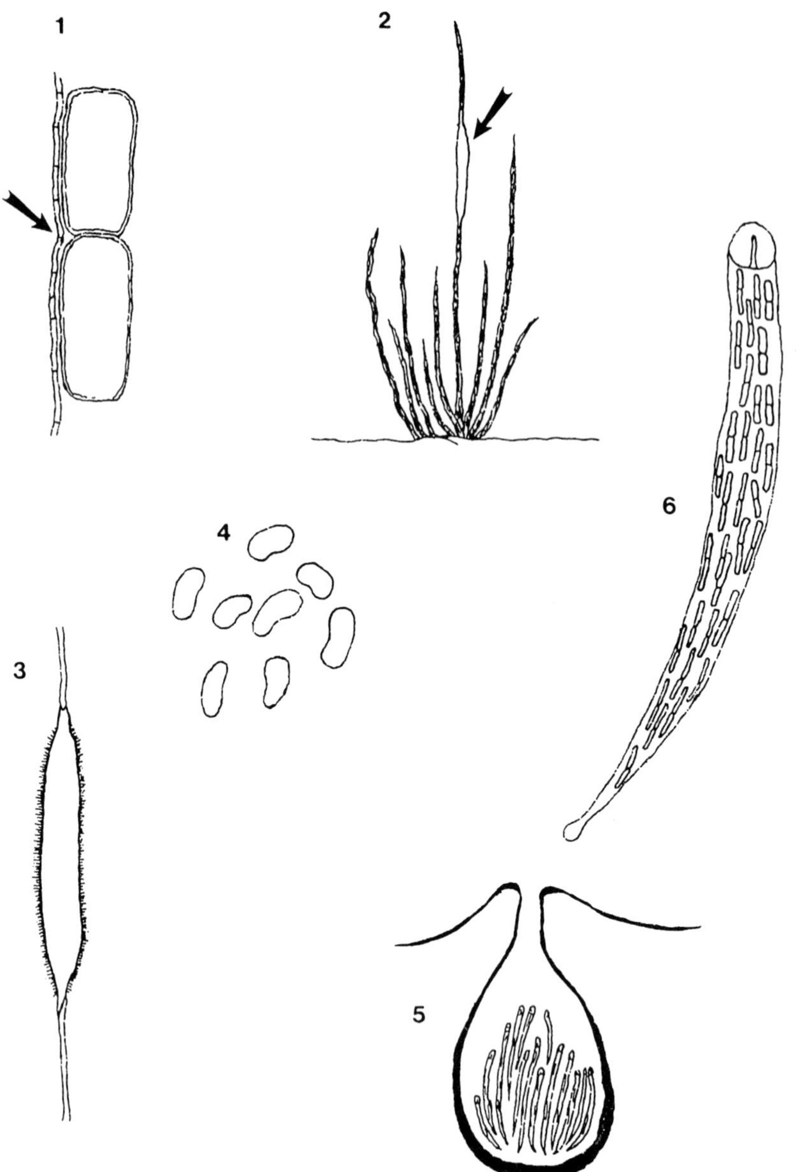

Figs. 1-6. *Epichloë baconii*. 1. Endophytic hypha (arrow) adhering to internal parenchyma cells in culm of grass, X1,000. 2. Grass plant bearing a stroma (arrow) on culm (stem). 3. Stroma with spermatial hymenium on surface, X2. 4. Spermatia, X1,400. 5. Perithecium section showing asci within, X200. 6. Ascus containing ascospores, X2,000.

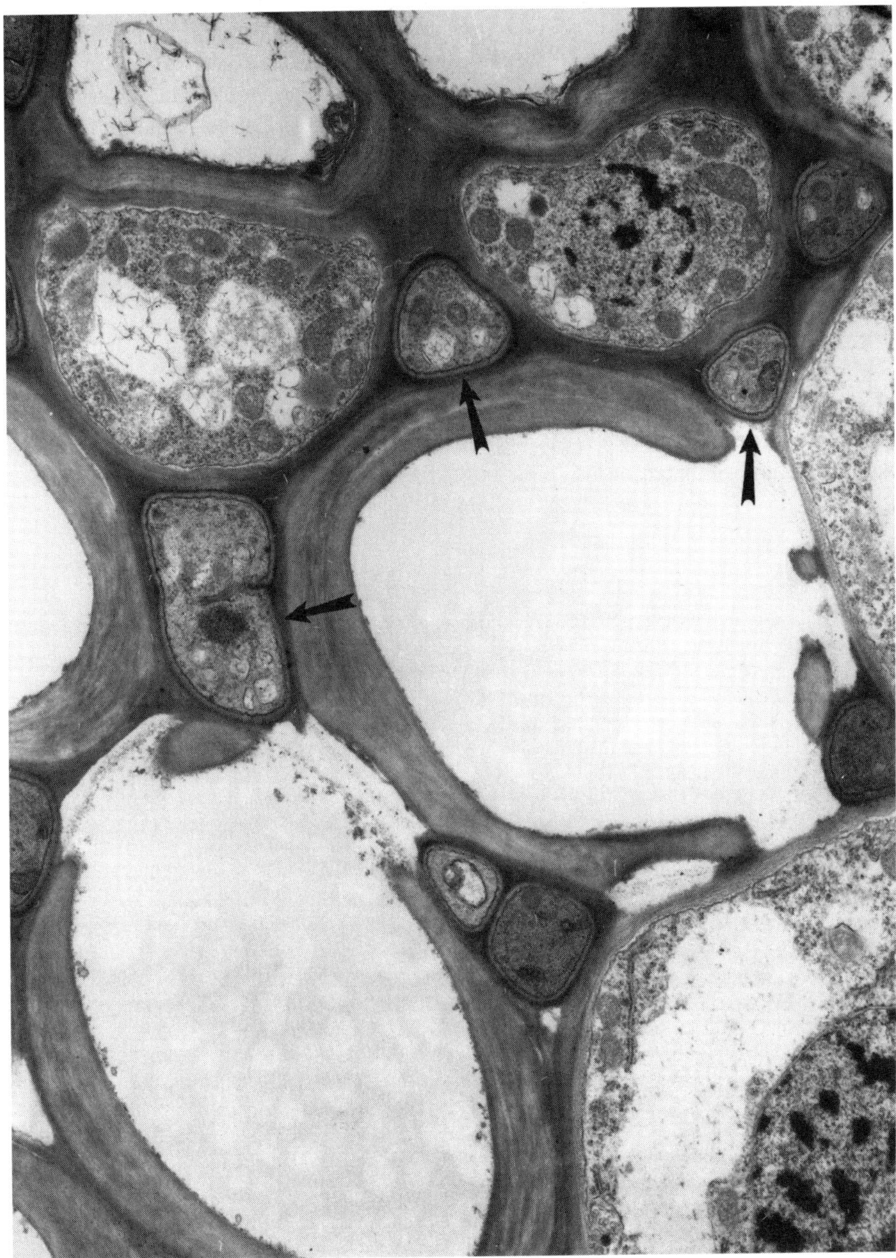

Fig. 7. Electron micrograph of *Agrostis hiemalis* showing intercellular hyphae (arrows) of *Epichloë amarillans* among developing cells of the vascular tissue, X10,000.

a condition where animals may develop uncontrollable spasms, a condition called 'ryegrass staggers'. Animals may show toxic effects until they are removed from toxic forage. Another grass, called sleepy grass (*Achnatherum robustum*), contains a *Neotyphodium* that produces lysergic acid amide (Petroski, Powell, and Clay, 1992), an alkaloid that has the effect of causing horses that consume a relatively small quantity of the grass (1% of body weight) to sleep for two to three days. In South America several grass species are infected by a *Neotyphodium* endosymbiont that causes a condition commonly known as 'tembladera'. This condition is similar to ryegrass staggers but is often fatal to poisoned animals.

11. Acknowledgements

This research was supported by PRHIDEB-CONICET, NSF DEB 96-96041, and the National Geographic Society. We are grateful to Rachna Patel (Rutgers University) for illustrations in this chapter.

12. References

Bacon, C. W., and J. De Battista (1990). Endophytic fungi of grasses. Pp. 231-256, in: Soil and Plants. eds D. K. Avora, B. Rai, K. G. Mukerji, and G. R. Knudsen. Marcel Dekker, New York.
Clarke, B. B., J. F. White, Jr., C. R. Funk, Jr., S. Sun, and D. R. Huff (1997). *Plant Disease* (In press).
Clay, K. (1988). *Ecology* 69: 10-16.
Clay, K., T. N. Hardy, and A. M. Hammond, Jr. (1985). *Oecologia* 66: 1-6.
Diehl, W. W. (1950). *Balansia* and the Balansiae in America. U.S.D.A. monograph No. 4, U.S. Government Printing Office, Washington D. C., 98 pp.
Kidston, R., and H. W. Lang (1921). *Transactions of the Royal Society of Edinburgh.* 52: 855-902.
Koshino, H. T. Yoshihara, S. Sakamura, T. Shimanuki, T. Sato, and A. Tajimi (1989). *Phytochemistry* 28: 771-772.
Latch, G. C. M., L. R. Potter, and B. F. Tyler (1987). *Annals of Applied Biology* 111: 59-64.
Leuchtmann, A., and K. Clay (1988). *Mycologia* 80: 192-199.
Lewis, G. C., and R. O. Clements (1986). *Journal of Agricultural Science, Cambridge* 107: 633-638.
Petroski, R.J., R.G. Powell, and K. Clay (1992). *Natural Toxins* 1: 84-88.
Pirozynski, K. A., and D. W. Malloch (1975). *Biosystems* 6: 153-164.
Porter, J. K., C. W. Bacon, and J. D. Robbins (1979). *Journal of Natural Products* 42: 309-314.
Rehner, S. A., and G. J. Samuels (1995). 73: S816-S823.
Spataforo, J. W., and M. Blackwell (1993). *Mycologia* 85: 912-922.
Stubblefield, S. P., T. N. Taylor, and J. M. Trappe (1987). *American Journal of Botany* 74: 1904-1911.
West, C. P., D. M. Oosterhuis, and S. D. Wullschleger (1990). *Environmental and Experimental Botany* 30: 149-156.
White, J. F., Jr., and T. L. Bultman (1987). *American Journal of Botany* 74: 1716-1721.
White, J. F., Jr., and C. Camp (1995). *Symbiosis* 18: 15-25.
White, J. F., Jr., and A. E. Glenn (1994). *American Journal of Botany* 81: 216-223.
White, J. F., Jr., A. E. Glenn, and K. F. Chandler (1993). *Mycologia* 85: 195-202.
White, J. F., Jr., T. I. Martin, and D. Cabral (1996). *Mycologia* 88: 174-178.
White, J. F., Jr., and J. R. Owens (1992). *Applied Environmental Microbiology* 58: 513-519.

VII. INDEX

INDEX

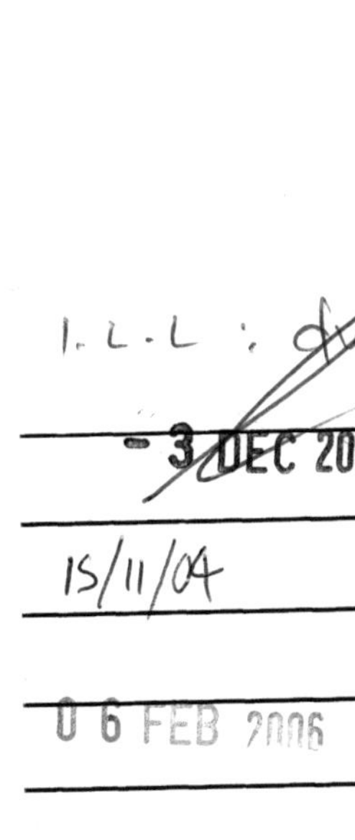
- 3 DEC 2002
15/11/04
0 6 FEB 2006